The History of
Imperial College London
1907–2007

Higher Education and Research in
Science, Technology and Medicine

Hannah Gay

Imperial College Press

ICP

Published by

Imperial College Press
57 Shelton Street
London WC2H 9HE
UK

Distributed by

World Scientific Publishing Co. Pte. Ltd.
5 Toh Tuck Link, Singapore 596224
USA office: 27 Warren Street, Suite 401-402, Hackensack, NJ 07601
UK office: 57 Shelton Street, London WC2H 9HE

Library of Congress Cataloging-in-Publication Data
Gay, Hannah
    The history of Imperial College London, 1907–2007 : higher education and research in science,
technology and medicine / Hannah Gay.
      p. cm.
  ISBN-13 978-1-86094-708-7 -- ISBN-10 1-86094-708-5
  ISBN-13 978-1-86094-709-4 (pbk) -- ISBN-10 1-86094-709-3 (pbk)
  Includes index.
  1. Imperial College of Science, Technology, and Medicine (Great Britain)

  T173.L9345 G39 2007
  607.1'1421--dc22

                            2006049718

British Library Cataloguing-in-Publication Data
A catalogue record for this book is available from the British Library.

Printed by Mainland Press Pte Ltd

Foreword

Those of us working and studying at the College today are inheritors of a proud and rich historical tradition. Our three founding Victorian colleges — the Royal College of Science, the City and Guilds College and the Royal School of Mines — federated to form Imperial College of Science and Technology in 1907. Imperial thus owes its formation, and today its continuing success, to many great institutions in their own right. In 1988 the College was renamed Imperial College of Science, Technology and Medicine, and began the formation of what is now a dynamic Faculty of Medicine. Those with whom we have merged forces in the later twentieth century are St. Mary's Hospital Medical School, the National Heart and Lung Institute, Charing Cross and Westminster Medical School, the Royal Postgraduate Medical School, Wye College and the Kennedy Institute of Rheumatology.

The Centenary is a good time for us to look back on Imperial's remarkable history and the College is fortunate that one of its former students, Hannah Gay, a distinguished historian of science, has devoted the past six years to writing its history. While carrying out the research for this fine work, she has spent many hours in the College and other archives, and interviewing past and present members of the College. For their valuable input we must sincerely thank those who have helped with the research and given their time to talk with the author.

My predecessor, Lord Oxburgh, commissioned the book and placed no conditions on its author. The result is a very engaging yet also academic book, full of interesting detail and sharp scholarly analysis. This is not an insider's tale, and is all the better for it. While critical in places, it also tells a remarkable story of success, one in which Imperial has had to overcome many difficulties to reach where it is today.

Dr Gay describes the history of the College as owing much to its 'great flux of talent' and it is in our history that we can locate the essence of our enterprising College. We can, for the first time, learn in one volume about those who built, supported and led the College, about those fascinating characters who broke the ground before us. We also learn how the College has responded to the many outside forces — political, cultural and economic — which have come its way. The book includes details of the Rectors — all determined to leave the College a better place than they found it — and of many of the academic and other staff. We are familiar with their contributions to outstanding achievement in science, engineering and medicine in the twentieth century, but here also are the stories of administrative manoeuvering and political wrangling; successful in the most part!

It is striking to learn just how closely Imperial has stuck to its 1907 charter with its clear mandate to carry out teaching and research at the highest level, to interact with industry and focus on the application of science. While the College continues to support established industries, today we have a well-established entrepreneurial spirit in our culture. Imperial has become a leader in enabling both staff and students to develop and commercialise new ideas, and many are attracted to the College for its fusion of science, technology and medicine with business and commerce. Fittingly perhaps, the Business School, home today to much of the academic drive of this activity, is our only major discipline which has not formed through amalgamation or merger — it has evolved from within.

When our century opened, Imperial was not fully part of the federal University of London and resisted joining for many years. While students received their Associateships and Diplomas from Imperial and were able to take external University degrees from the start, many at the College in its early years wanted it to become an independent degree-awarding institution. However without resources or government support, the College joined the University in 1929. The marriage was positive on the whole, but there were always moves within the College towards independence. Over the past quarter of a century, the College has been moving upwards in domestic and international rankings and in 2007

Imperial is clearly a university of international standing, ranked one of the world's 10 best universities by the *Times Higher Education Supplement.* It is therefore fitting that on our 100th birthday, 8 July 2007, the College will become an independent university and will be free to award its own Imperial degrees.

This book vividly depicts Imperial's first century and shows how well placed it is for independence. I join the author in wishing the College well as it enters its second century.

Sir Richard Sykes
RECTOR

London, November 2006

Acknowledgements

I am grateful to many people for the help I have received while working on this book. In particular I would like to thank Lord Oxburgh for thinking ahead to the centenary of Imperial College and for inviting me to write a history to mark the occasion. I have much enjoyed returning to the college where I was a student, and am grateful for having had the opportunity to learn so much about it. Since 2001 I have been a member of the Centre for the History of Science, Technology and Medicine, and would like to thank all the staff associated with the Centre for being both welcoming and supportive. David Edgerton, who headed the Centre in 2001, and who arranged for my appointment, has been especially helpful, as has his successor Andrew Warwick. Both have taken the time to read the earlier draft chapters and, in providing excellent feedback, have helped bring the book to fruition. Rob Iliffe, too, has helped in many ways, including providing useful comments on some draft chapters. I have received much support also from three departmental administrators, Ainslee Rutledge, Caroline Treacey and Robert Powell.

Former Rectors, Ron Oxburgh, Eric Ash, and Brian Flowers all discussed their time at the college with me and I am grateful for the very interesting conversations I had with them. I am grateful also to the present Rector, Richard Sykes, for taking time to talk with me about the college, and for his continuing support of the project. I have interviewed also many other members of staff, past and present. Meeting them was a pleasure and the information gained has been useful in framing the later chapters in the book. Thanks are due to the following for having given interviews and, in some cases, providing further information by e-mail: Igor Aleksander, Roy Anderson, Tony Barrett, Peter Bearman, John Beddington, David Begg, Nigel Bell, Harry Bradford, Charmian Brinson, John Burland, John Cosgrove, Jim Cunningham, Anne Dell, Richard Dickins, Dorothy Griffiths, Michael Hassell, Julia Higgins, Howard Johnson, Michael Keating, Tom Kibble, Richard Kitney, Peter Knight, Jeff Kramer, Manny Lehman, Frank Leppington, David Limebeer, David Lloyd Smith, Tidu Maini, John Monhemius, David Norburn, Diana Paterson Fox, John Pendry, John Perkins, David Phillips, Rees Rawlings, Stephen Richardson, John Smith, Rod Smith, Geoffrey Stephenson, Trevor Stuart, Chris Towler, Dorothy Wedderburn, and Gordon Williams.

I am grateful to the many others who have communicated with me via letter and e-mail. In particular I would like to thank Harold Allan, Michael Bernal, Sam Eilon, John Elgin, Jon Hancock, Richard Selley, Graham Saville and Robert Spence for their helpful contributions. Many thanks also to Robin Heath for inviting me to his home and for letting me read his father's (O. V. S. (Peter) Heath) diaries, to Alan Munro for telling me about his grandfather, Otto Beit, and to Martin Black for telling me about student and social life at the college during the 1970s, and about Masonic life. I would also like to thank David Traske for giving me a guided tour of the Wye campus; and Graham Matthews and Patricia Reader for talking with me during a visit to Silwood Park. Peter Haldane has helped with student union affairs and Rupert Neate, editor of *Felix*, has helped find photographic images. I am similarly grateful to Meilin Sancho for her help with photographs taken in the physics department.

I would especially like to acknowledge my research assistant, Yorgos Koumaridis. Not only did he find and collate the historical statistics and the data needed for the appendix, he has helped my research in numerous other ways. I could not have found a better assistant; his remarkable research skills and friendly presence have added greatly to the pleasure of the work. I would also like to thank Nigel Wheatley, Deputy Academic Registrar, and Kevin Delaney from Human Resources, who have helped Yorgos retrieve some of the college data. I have been much helped also by College Archivist, Anne Barrett, whose friendship and support throughout has been invaluable. I would like also to acknowledge the help of her assistant Catherine Harpham as well as that of special collections librarian, Hilary McEwan. Valerie Phillips, Archivist for the Royal Commission for the Exhibition of 1851, has given professional assistance and has been very supportive. She also did much copyediting for which I am very grateful. I am grateful also to Angela Darwin for her friendship and for the many intellectual exchanges we have had while working alongside each other in the archives. Also working in the archives has been David Goodgame who has helped in numerous ways. David was a new member of staff in the chemistry department when I was a student. He has recently retired, as has Bill Griffith with whom I have kept in touch over the years. Bill is a college history and history of chemistry enthusiast; it has been wonderful having someone around who actually wanted to hear about my latest archival discoveries as they happened. Both David and Bill read an earlier draft of this book and their comments led to much useful editing.

I would also like to acknowledge the help received from my colleagues, and other members of staff, at Simon Fraser University. I was able to keep my office even after decamping to London and did much writing of the book in Canada. In this connection I would especially like to thank the Dean of Arts and Social Sciences, John Pierce, as well as the former and current chairmen of the history department, Jack Little and John Craig respectively. I worked in the history department at Simon Fraser University for many years and learned much of the historian's craft there. Two of my colleagues, Michael Fellman and John Stubbs, read some draft chapters and I would like to thank them for their helpful comments. An Imperial College graduate who has worked at Simon Fraser University for many years is biologist John Webster. He told me much about the zoology and applied entomology department at Imperial, and about life at Silwood during the late 1950s and early 1960s, for which I am grateful. I would like to thank also fellow historian of science, Roy MacLeod, who read some draft chapters while visiting Vancouver and gave much useful feedback.

At Imperial College I have received help from Leszek Borysiewicz, Tony Mitcheson, and Pamela Michael. Tidu Maini has assisted both in liaison with the Imperial College Press, and more generally. I would also like to thank those at the Press who have worked on the book. They include Lance Sucharov, Katie Lydon, Kim Tan, and especially K. K. Phua (a former research student of M. A. Salam). For the index I am indebted to Judith Anderson. Neville Miles, College Photographer, scanned all the illustrations for publication. Several of his own photographs are among those reproduced in the book. I very much appreciate his help. Finally, I would like to thank my family for their support. Especial thanks go to my husband Ian and son, John, for also reading, commenting on, and copyediting, an earlier draft of the book.

Needless to say, despite all this help and despite my best efforts, there are bound to be some errors in a book of this kind. If readers do spot errors of fact, names misspelled, or anything else that needs correction, I would be grateful to hear from them. Should this book come out in electronic format, corrections can then be made.

Hannah Gay
hgay@imperial.ac.uk

Contents

List of Illustrations

I would like to thank all those who gave permission to use the illustrations listed below. Many of the illustrations are the property of Imperial College and while every attempt was made to obtain copyright permission for the others, there are a few photographs for which copyright could not be traced.

1. The Royal Horticultural Gardens (*The Royal Commission for the Exhibition of 1851*)
2. The Royal College of Science Observatory in 1893 (*Imperial College Archives, ICA*)
3. Henry Taylor Bovey (*ICA*)
4. The Marquess of Crewe (*ICA*)
5. Sir Arthur Acland (*ICA*)
6. Sir Alfred Keogh (*ICA*)
7. Professor Ernest Dalby (*ICA*)
8. Professor Hugh Callendar (*ICA*)
9. Sir Herbert Wright (*ICA*)
10. Professor Sir John Bretland Farmer (*ICA*)
11. Professor William Gowland (*ICA*)
12. Professor William Watts (*ICA*)
13. Professor Vincent Illing (*Mary Pugh, ICA*)
14. Professor William Jones (*The Phoenix, ICA*)
15. Professor Alfred Fowler (*ICA*)
16. Professor John Hinchley (*Lafayette, ICA*)

List of Acronyms

ARC	Agricultural Research Council
ARCS	Associateship of the Royal College of Science
ARSM	Associateship of the Royal School of Mines
ASTMS	Association of Scientific Technology and Managerial Staff
AUT	Association of University Teachers
BAAS	British Association for the Advancement of Science
BBC	British Broadcasting Corporation
BBSRC	Biotechnology and Biological Sciences Research Council
BP	British Petroleum
C&G	City and Guilds
CARE	Centre for Analytical Research in the Environment
CERN	Conseil Européen pour la Recherche Nucléaire (European Centre for Nuclear Research)
CGLI	City and Guilds of London Institute
CHOSTM	Centre for the History of Science, Technology and Medicine
CVCP	Committee of Vice-Chancellors and Principals
CXWMS	Charing Cross and Westminster Medical School
DIC	Diploma of Imperial College

DSIR	Department of Scientific and Industrial Research
EC	Executive Committee (of Imperial College)
EEC	European Economic Community
EPSRC	Engineering & Physical Sciences Research Council
ESA	European Space Agency
ETH	Eidgenössische Technische Hochschule (Swiss Federal Institute of Technology)
FBA	Fellow of the British Academy
FC	Finance Committee (of Imperial College)
FIC	Fellow of Imperial College
FMedSci	Fellow of the Academy of Medical Sciences
FREng	Fellow of the Royal Academy of Engineering
FRS	Fellow of the Royal Society
GB	Governing Body (of Imperial College)
HEFC	Higher Education Funding Council
HEFCE	Higher Education Funding Council for England
IBM	International Business Machines Corporation
ICA	Imperial College Archives
ICCET	Imperial College Centre for Environmental Technology
ICI	Imperial Chemical Industries
ICON	Imperial Consultants Ltd
ICPARC	Imperial College Planning and Resource Control Research Centre
ICU	Imperial College Union
ICWSA/ICWA	Imperial College Women Students' Association/later Imperial College Women's Association
IEE	Institution of Electrical Engineers
IMPACT	Imperial Activites Ltd
IMPEL	Imperial College Exploitation Ltd
LSE	London School of Economics
MIT	Massachusetts Institute of Technology
MRC	Medical Research Council

NASA	National Aeronautics and Space Administration (USA)
NERC	Natural Environment Research Council
NHLI	National Heart and Lung Institute
NPL	National Physical Laboratory
NUS	National Union of Students
OTC	Officers Training Corps
PRO	Public Records Office
RAE	Royal Aircraft Establishment
RAE	Research Assessment Exercise
RAF	Royal Air Force
RCC	Royal College of Chemistry
RCS	Royal College of Science
RNVR	Royal Naval Volunteer Reserve
RPMS	Royal Postgraduate Medical School (Hammersmith)
RSM	Royal School of Mines
SERC	Science and Engineering Research Council
SRC	Science Research Council
SSRC	Social Science Research Council
THES	Times Higher Education Supplement
UCL	University College London
UFC	Universities Funding Council
UGC	University Grants Committee
UKAEA	United Kingdom Atomic Energy Authority
UMIST	University of Manchester Institute of Science and Technology
V&A	Victoria and Albert Museum

chapter one

Introduction

This book has been written to mark the centenary of the Imperial College charter. It is intended for a wide readership, including historians of higher education, historians of science, engineering, and medicine, and historians with interests in the more general culture of twentieth-century Britain. While scholarly, the book has been written also for the interest and enjoyment of Imperial College students and alumni, and for past and present members of the staff. The 2007 centenary is a time both to look back at what has been achieved, and a time to look forward. With this in mind, I hope that the book will also be an aid to those working at Imperial, both in thinking about the college and its many disciplines, and in planning for its future. I was fortunate in being invited to write the college history by Lord Oxburgh, and doubly fortunate in that he gave no special directives and allowed me access to most of the college records and archives.[1] The Victorian records are fairly well known to scholars, but the little-used twentieth-century records, and the rich collection of personal papers, could support many more books. I hope that what I have written will encourage others to explore them further.

This book tells the story of a new kind of institution that came into being with the federation of three older colleges.[2] As will be shown, these colleges had well-formed identities already by 1907, identities that were to influence the future history of Imperial College in many ways. Why they federated, and how the resulting structure and governance of the new college contributed to its longer-term success, is a major theme of this book. Imperial College was founded by the government as an institution for advanced university-level training in science and technology, and for the promotion of research in support of industry throughout the British Empire. True to its name, the college soon built imperial links and was an outward looking institution from the start. It early laid the foundations that were to make it an international success and its connections to the larger world, to industry, and to the British state, have been major factors in its historical development. Local geography, too, has been important. The situation of the college in South Kensington, a bus ride from Westminster, close to major museums, the Royal College of Art, the Royal College of Music, the Royal Albert Hall, Chelsea and the River Thames, Hyde Park and Kensington Gardens, has been intrinsic to its development. Locality has mediated the technological spirit of the college in interesting ways.

That Imperial College was given a utilitarian mandate led to its having a distinct identity within the British university system and set it apart from other universities with more outwardly liberal aims.[3] In this, the college resembled some European and North American technological institutions which the government had in mind when creating the college. For example, the 1861 charter of the Massachusetts Institute of Technology (MIT) states that it should support, 'the advancement, development, and practical application of science in connection with arts, agriculture, manufactures and commerce'. The adoption of similar goals at Imperial spelled problems for the college in its relationship with the University of London, and for an academic staff seeking success within the wider, liberal, academic culture. How staff, with an eye to advancement in both learned societies and professional bodies, navigated their way in a college dedicated not simply to teaching and the pursuit of knowledge, but also to serving the needs of industry, is a major theme of this book. By the late twentieth century business and entrepreneurial ideals had become more widely accepted in the larger culture, something reflected in Imperial College's increased status in recent years. The college has never been wealthy and has therefore had to respond quickly and effectively to outside forces, whether cultural, intellectual, economic or political. Indeed, as will be discussed, its success relative to the far better endowed ancient universities, and to the largely better endowed civic universities, can partly be explained in terms of the way it responded to these forces.

One way of thinking about an educational institution is in terms of its alumni. Imperial College graduates have made many important contributions to science and engineering, and to public life around the world. More recently some of its graduates have contributed also to medicine. Many books could be written on their various contributions, and some have already been written. But this book focuses primarily on work carried out by people while directly associated with the college. Even then, full justice cannot be done to the many interesting men and women who have passed through the college, and to the varied and important work they carried out. This great flux of talent has been a major factor in Imperial's success. Despite being relatively poorly funded the college has always attracted good people. This was already the case in the older colleges before they federated in 1907. In part the attraction was that the colleges provided new kinds of opportunity for social advancement, something that has remained true of Imperial College to this day.

In its approach to the history of institution building, this book makes a contribution also to twentieth-century intellectual and socio-cultural history. Issues of class, gender, generation, race, ethnicity and locality intersect with the wide range of topics raised in the various chapters. The book has a large cast of characters. The many, often parallel, stories collectively tell both a human and an institutional story. Written from within the humanistic tradition, the book relates, and attempts to explain, how people came together in a new kind of venture, how they exchanged ideas and made their way in the world, and how the college to which they were attached evolved over a century. It shows how competing interests were negotiated, how social exchanges worked to generate new ideas, and how work in science and technology mediated between the natural and cultural worlds.

While writing this book I read a number of other college and university histories. Perhaps in a class of its own is the multi-volume history of the University of Oxford. In the volume covering Oxford in the twentieth century the different chapters are thematically, rather than chronologically, organized.[4] They were written by different authors, many of whom are specialists who were able to cover their topics in great depth. While a single author cannot display the same range of expertise, there is some advantage to presenting a cohesive personal account. I have followed the Oxford example in several of the following chapters and have used a thematic approach. Two other types of chapter have also been written. One type covers college governance in light of external cultural, economic, and political forces, and the other covers the history of departments and subject areas. The chapters are organized so that reading them consecutively allows for a coherent and roughly chronological history of Imperial College. However, the thematic chapters overlap with the others chronologically, and can be read as independent essays. Even the non-thematic chapters are relatively free-standing and can be read as such by those who simply wish to know more about Imperial College in a particular period. Because of this structure there may be a small amount of repetition but it has been kept to a minimum. The histories of the various medical schools that joined Imperial in the late twentieth century are not covered, but the creation of the Imperial College School of Medicine (now Faculty of Medicine) is discussed, as are some of the consequences of having a medical school at the college.

The second chapter gives a brief history of the colleges that federated to form Imperial College in 1907. While the college was founded in the Edwardian period its strong Victorian roots have influenced its history to this day. The older colleges remained strong within the federation well into the 1970s. The third chapter tells of the founding of Imperial College. In some ways this is a pivotal chapter, since how the college emerged proved important to its later development. It was created by the state, given a powerful governing body, and had a strong mandate. In 1907 Imperial College was independent of the University of London. It joined reluctantly in 1929 — and then only after much debate. The often uneasy relationship with the University is an ongoing theme in the book. It is perhaps fitting that having reached its centenary the college is to become independent once again, this time as a full degree-giving body.

Chapter four focuses on the early governance of the college and how its future was envisaged. Its initial success owes much to some very good professorial appointments, and to some active members of the governing body. The many contributions made by college staff to the war effort (1914–18) also made clear to the government that the college was a national asset, worthy of greater support. The Governing Body and the staff very much wanted the college to count in the outside world and were politically astute in moving forward. The first new subject areas to be introduced were chemical technology and aeronautics, both in response to larger political anxieties, related to Britain's industrial competitiveness and military preparedness respectively.[5] Technical optics was the next to be added, the result of concerns over Britain's optical needs at the beginning of the war. The Rector and governors also moved quickly to find the money for a student union and the provision of communal social facilities, thinking this the best way to bring together people from the three older colleges.

Chapter five is the first of the thematic chapters. It covers Imperial College during the First World War and, as with the other thematic chapters, can be read as an independent essay. Chapter seven, 'Imperial Science at Imperial College' is also thematic and shows how, in the first half of the twentieth century, the college served the needs of empire. The other thematic chapters are numbers eight, eleven and fourteen. Chapter eight covers the college during the Second World War. Chapter eleven examines corporate and social life, and is important to an understanding of the nature of the college as it evolved over the twentieth century. It is also a micro-historical and socio-cultural account of student life in London. Chapter fourteen shows how the college encouraged students to broaden their outlook, and how the curriculum began to diversify in the face of cultural and political pressures after the Second World War. It covers the history of the diversifying curriculum into the early twenty-first century. Chapters nine, twelve, and fifteen cover college governance since the Second World War and include material on the larger political and socio-cultural contexts in which the expansion and restructuring of the college occurred. Chapters six, ten, thirteen, and sixteen cover the more academic history of the college, focusing on the histories of departments — on teaching, research, and the contributions of individuals. Readers who are simply interested in the development of particular disciplines can read the relevant sections that follow chronologically within these chapters. I have looked at the departments dispassionately, focusing as much on the rough periods as on the smooth. While the overall story is one of great success, this has been achieved only by overcoming serious problems as they have arisen.

Each chapter has many, and sometimes long, endnotes. The reasons for this are threefold, the first being to include as much personal detail as possible. The many people who have worked at the college have made it what it is today and I strongly believe they should be remembered. The inclusion of much personal information in the main narrative would have disturbed the flow, and so it has been largely confined to endnotes.

Readers who seek information on a particular individual should use the index to find the appropriate endnote, in addition to any references in the main text. Second, endnotes are used in the normal scholarly way to cite reference sources where appropriate. Third, I have used endnotes for a number of scholarly asides so as to enlarge the commentary in ways that I believe will be of interest to the various readers I had in mind. While the book covers the period to 2001 in a comprehensive way, some of the chapters include commentary on the more recent period where important for reasons of clarity and continuity. The names of places and countries are given as used historically in Britain. For example, the name Ceylon is used for the period before 1972, and the name Sri Lanka for the period after.[6] There is some inconsistency in the use of people's names. In some cases I use just initials and in others I use the first name, along with surname. I have relied somewhat on my ear, and have adopted what appear to be common modes of reference in the college.

By necessity in a work of this kind I have focused more on the Rectors and professorial staff than on others. Professors are at the centres of associational groups, and earn their position by the demonstration of certain skills. But it is worth remembering that the associational group, while largely anonymous, is essential to the overall success of both the professor and the college. As mentioned, one aim of this book is to provide an understanding of the college's success. What is it that has made Imperial College the major institution it has become? Perhaps an institutional history is not ideal for this purpose, and a more anthropological look at departments, or at groups associated with successful professors would be better suited. Nonetheless, the concluding chapter adds to commentary on individual successes made in the earlier chapters by including also some abstract speculation on the overall success of the college.

Chapters two to seven cover the college history up until the Second World War. This period saw the establishment of a corporate culture, aspects of which are still recognizable today. The end of the Second World War brought many changes both within the college and in the country at large. As was the case elsewhere, people looked towards the United States for direction and were impressed by the scale of scientific and technological research being carried out in American universities. One consequence was a major increase in postgraduate and research work after the war, especially in the physical sciences, as well as in some of the engineering fields. Given their importance in 1907, it is interesting to see how little valued the biological sciences were by the late 1940s and early 1950s — before the uncovering of the structure of DNA — and how close they came to being shut down. In light of later developments this is a lesson in the importance of longer-term thinking and in how quickly old fields can be reinvigorated by new discoveries, and become of great importance.

The post-war boom saw a major expansion in the economy, and the college soon more than doubled in size. The first expansionary period is covered in chapters nine and ten. Two wartime successes, computers and nuclear technology, came to the fore. Computers, especially, were to have a major long-term influence. Indeed, during the recession that followed the boom, from the late 1960s to the early 1980s, increasingly cheap computation resulted in a major shift towards more theoretical work. It was simply cheaper than experiment, though not always rationalized that way by those claiming their theoretical approaches to be superior. Clever new uses of computers helped people through this difficult period. Computer modeling continues to be of central importance in research today and, most notably, has changed the face of engineering. The early 1970s saw also the start of major restructuring at the college, described in chapters twelve and thirteen. This restructuring is discussed in the wider political and economic context, including the student movements of the period. A further major theme of the mid to late twentieth century is the increase in interdisciplinary work. This was a response to outside funding priorities which continue to drive the trend in the founding of new interdisciplinary centres and institutes. What has further transformed the college in recent years is the addition of a School (now Faculty) of Medicine. This major expansion and some of its consequences are discussed in chapters fifteen and sixteen.[7]

Since the 1970s there has been ongoing debate in the United Kingdom over a new social contract for universities. While taxpayers are willing to support universities, they want some tangible social gain in return. The same can be said of industrial, commercial, and other investment in higher education. As yet there is no consensus of the kind that existed before the Second World War when the university sector was small. Debate over what is appropriate support for higher education, and what universities need to do in return, is ongoing. Some of the chapters covering the post-war period give an account of the social-contract debate as it has evolved to date. Keeping up with the political mood has been essential for an institution as poorly endowed as Imperial. Indeed the history of the college in the second half of the twentieth century can be seen as a series of reactions to government initiatives in higher education. As a consequence the college has been in almost permanent transition. Today, research is largely focused in the areas of medical genetics, molecular medicine, biotechnology, the environment, energy resources and conservation, information technology, nanotechnology, and new materials, but it is continually in flux. As is discussed in later chapters, the move to more interdisciplinary research has entailed some problems for traditional disciplines and for the maintenance of sensible and good undergraduate programmes.

The chapters on the earlier history of the college were easier to write than those covering the more recent period. Indeed, chapters fifteen and sixteen were the most difficult. There are several reasons for this. Greater distance allows for a clearer picture. What comes later can be used to evaluate earlier events. Further, the very helpful obituary and memoir literature is far greater for the earlier period. It is also the case that written records are usually more reliable than oral ones. For the recent period, many written records are not yet available and I have relied more heavily on interviews and direct queries in writing about it.[8] I am very grateful to all those who helped me in this way, and I very much enjoyed meeting people working at the college, as well as some who have recently retired. But the old saw about witnesses is true — they do not always agree. Memory is not always reliable. It is also uneven, sometimes vague and, at other times, enhanced. Thus, while this work has been much aided by what I have learned from talking with people it is based largely on the archival records which include the minutes of the Governing Body (now Council) and its committees, other committee minutes, annual reports, departmental and other correspondence, memoranda, and the large number of position papers generated over the years. Also of much value, and of very great interest, are the many boxes of personal papers left by people who have been associated with the college. No doubt there will be readers who disagree with some of my interpretations of this material. But the duty of the historian is to gather information as reliably as possible, tell interesting stories that give meaning to the whole, provide some interpretation of events and behaviours, and give some general analysis. The stories and analysis should be consistent with the evidence. But, as scientists know, consistency and truth are not necessarily the same. Truth is more elusive but, in being consistent with the evidence, we at least have a chance of capturing some of the truth.

One area of the college history that is not fully pursued in the following chapters is the history of the estate — though it is not entirely neglected. At the time of the Great Exhibition of 1851, the area to the southwest of the exhibition site was largely rural, much of it under cultivation by market gardeners. About eighty-seven acres of this land was purchased by the Royal Commission for the Exhibition of 1851 with profits from the exhibition. The land extended (with some interruptions) from Kensington Gore in Westminster to Harrington Road in Kensington, and from the eastern boundary of what is now the Victoria and Albert Museum to Queen's Gate in the west, with one small section extending as far as Gloucester Road.[9] The surrounding land was largely privately owned, much of it by two aristocratic families who began to speculate in building projects soon after the Exhibition closed. The Commissioners, too, speculated. Interestingly many of the houses they built were designed by architects associated with the 'domestic revival' movement. One of them, designed by Norman Shaw, is now home to the Rector of Imperial College.[10] Much of the Commission land, south of Cromwell Road and west of Queen's Gate, was gradually sold to raise money for important projects on the site. The main roads through the estate were built in 1854–6 under the supervision of Thomas Cubitt and, following Prince Albert's wish, some were originally planted with Linden trees since he thought Plane trees too commonplace. It was also Prince Albert who suggested that the area be called South Kensington.[11]

An early idea was to build the National Gallery on Commission land, but this was rejected already in 1856. Much of the land to the north of what became a major exhibition site (where the Natural History Museum now stands), and between Queen's Gate and Exhibition Road, was leased to the Royal Horticultural Society for a public garden that opened in 1861. It was an Italianate garden, with arcades and much statuary, but it never paid its way and soon fell into disrepair. By 1880 the Society relinquished its lease and the land became free for redevelopment. People who came to visit the 1862 Exhibition and the various other exhibitions mostly came on foot since the fare from central London, on the then horse-drawn bus, was an expensive six pence. This changed when the railway station at South Kensington was opened in 1868. Major construction on the Commission land began in the 1860s and, as discussed in chapter three, included a building on Exhibition Road intended as a college of naval architecture and general science. Instead it was to house the Royal School of Mines and the Royal College of Science[12] and became known as the Science Schools Building (renamed for T. H. Huxley in 1932). The South Kensington Museum (later extended and renamed the Victoria and Albert Museum), the Royal Albert Hall and the Albert Memorial were all built at roughly the same time in the late 1860s and early 1870s. The City and Guilds of London Institute built its college on Exhibition Road in the early 1880s. The Imperial Institute, which occupied much of the land later taken over by Imperial College, was constructed in the late 1880s and early 1890s. So, too, was the Royal School of Needlework on Exhibition Road.[13] A portion of this building was sublet to the City and Guilds College and, later still, the whole building (renamed for W. C. Unwin) became part of Imperial College. The Royal College of Science building, located around the corner from the Needlework School on a portion of land that the government had intended be shared with the Tate Gallery, was completed in 1906. The government had paid £70,000 for the freehold of the eight-acre site which was valued at £200,000. It changed its mind on the location of the Tate in 1899 after much lobbying by Fellows of the Royal Society, notably by Lord Kelvin and Lord Rayleigh, who wanted the entire site to be given over to science instruction. The neighbouring Post Office was rebuilt so as to house the Meteorological Office on its upper floors. Holy Trinity Church on Prince Consort Road was built in 1901–3, when the Commissioners asked the City of Westminster to take control of the road, something agreed to only on condition that the Commission first pave it with wood, which was done.

Given the later history of Imperial College it is interesting that in 1907 the Commissioners were asked to lease land to the north of the City and Guilds College for a proposed Institute of Medical Science, an idea soon abandoned. This allowed construction of the Royal School of Mines building to begin in 1909. The later history of this, and other buildings associated with the college, is discussed in other chapters. It is worth remembering, however, that in the mid-Victorian period land in much of South Kensington was rural and cheap. By investing in it, the Royal Commissioners showed great foresight. While the Commissioners worked in the public interest, at first they had few clear plans. However, they were committed to certain kinds of projects and the development of the site is important to an understanding of Victorian Britain's response to the needs of empire, industry, and advanced education in the arts and sciences.[14] The early history of the site is complex but, fortunately for the development of Imperial College, the government purchased and leased land for the expansion of the Royal College of Science and Royal School of Mines respectively; and the Commission gave the City and Guilds of London Institute a peppercorn lease for land on which to construct its college. That Imperial was later able to expand into other areas of the site has much to do with the foresight of its Rectors; notably Henry Tizard who lobbied for the takeover of the Imperial Institute (or its land base), and Patrick Linstead, who negotiated both the demolition of the Institute, and the exchange of the old Science Schools (Huxley) building for increased space in Prince's Gardens.[15]

Imperial College now sits on some of the most expensive land in the country making further expansion in the immediate neighbourhood difficult. But the purchase of Silwood Park after the Second World War, and the more recent mergers with the medical schools, and with Wye College, have led to new growth, and will allow for expansion in the future. The college is well positioned as it moves into its second century. In writing this book, and as a former student of the college, I wish it well.

...

1. Unless otherwise stated, archival sources cited in this book are held in the Imperial College Archives (ICA).

2. This is not the first history to be written of the college. A very readable short history by A. Rupert Hall, was published internally in 1982. There are a number of typescript departmental histories written by insiders (most of them ending in the 1960s) and an internal history of the City and Guilds College written for its centenary in 1985. A. Rupert Hall, *Science for Industry: A Short History of the Imperial College of Science and Technology and Its Antecedents* (London, 1982); Adrian Whitworth (ed.), *A Centenary History: A History of the City and Guilds College, 1885–1985* (London, 1985). I am indebted to the published and unpublished works.

3. Some British universities founded in the Victorian period make reference to the needs of industry in their charters; for example, the University of

Birmingham. But Birmingham, while emphasizing science and engineering, also made provision for the arts, social sciences and humanities.

4. Brian Harrison (ed.), *The History of the University of Oxford* (vol. viii); *The Twentieth Century* (Oxford, 1995). I also much enjoyed Ralf Dahrendorf's history of the London School of Economics. The LSE is smaller than Imperial and, for much of the twentieth century, its staff had many intellectual interests in common, and students were encouraged to attend lectures across the departments. This pattern allowed Dahrendorf to write what is almost a family history of the LSE. Such an approach would have been impossible for Imperial which, from the start, had very independent departments. But the two colleges have in common their modernity. Both developed new kinds of courses at the university level, and attracted new kinds of students who saw opportunities for advancement. Another successful university history is that of the University of Toronto. While different in approach to what I am attempting here, Friedland constructed his history around a number of interesting stories and episodes, organized in short chapters. My history, too, while organized in relatively long chapters, is based on an accumulation of stories arranged so as to give meaning to the whole. Ralf Dahrendorf, *LSE: A History of the London School of Economics and Political Science, 1895–1995* (Oxford, 1995); Martin L. Friedland, *The University of Toronto: A History* (Toronto, 2002).

5. Aeronautics was taught in the mechanical engineering department from 1909. The Department of Aeronautics was founded in 1920.

6. Ceylon gained independence in 1948; while the name Sri Lanka was in common use it was used officially only from 1972.

7. While medicine and medical research appear as though they will become the strongest areas of the college in this century, they came to the college only late in the twentieth century and therefore are not featured heavily in this book. The merger with St. Mary's, the subsequent mergers with other schools, and the formation of the Imperial College School of Medicine, are discussed in chapters 15 and 16. As will be seen, much medically-related work was done in the college even before the medical school mergers.

8. With electronic mail it is unclear how many written records will survive. The historian of the twenty-first century may have a more difficult task than I have had.

9. See F. H. W. Sheppard (ed.), *Survey of London* (vol. xxxviii): *The Museums Area of South Kensington and Westminster* (London, 1975). Of the original 87 acres purchased, about 18 have been taken up by roads and 17 sold for private development; 52 have been dedicated to public purposes. See also

the Reports of the Royal Commission for the Exhibition of 1851; especially the 7th Report (1889) and 8th Report (1911).

10. The house at 170 Queen's Gate, built in 1889, was originally designed for cement manufacturer, Frederick White. It included fitments, notably fire-places, taken from older houses that had been demolished on Millbank. See Hermione Hobhouse, *The Crystal Palace and the Great Exhibition: Art, Science and Productive Industry: A History of the Royal Commission for the Exhibition of 1851* (London, 2002), p. 177.

11. This was preferred over North Brompton, to avoid association with what was then considered a swampy, low-lying, unhealthy village. After Albert's death Plane trees were planted in preference to Lindens.

12. The name Royal College of Science was used from 1890 on; before that the college was known as the Normal School of Science.

13. The Royal School of Needlework had occupied a temporary building on the site since its foundation in 1874.

14. This response can be read also in the iconography of the Albert Memorial with its representation of the arts, sciences, industry and empire.

15. In some records this is called Princes Gardens, Indeed, even today the names on signs around the Gardens are inconsistent, with both versions in use. Princes Gardens is the name to be found in the *Survey of London*, *op. cit.* (9), but Prince's Gardens is used in the 1961 ordnance survey map. I use Prince's Gardens since it makes both historical and grammatical sense.

chapter two

Before Imperial: The Colleges That Federated in 1907

Introduction

Imperial College has roots in three nineteenth-century institutions, the Royal College of Chemistry, the Royal School of Mines, and the City and Guilds Central Technical College, which opened respectively in 1845, 1851 and 1885. This chapter will outline how these colleges came to be founded and, by focussing also on their cultural identities, show something of their different legacies.[1] Two of the institutions were located in London's West End: the Royal College of Chemistry (RCC), privately owned by shareholders, and the Royal School of Mines (RSM), founded by the government. Both show the infusion of the gentlemanly ideals held by members of the government, the landed gentry, high civil servants and by members of the learned societies located near by. The Central Technical College, located in South Kensington, was founded by a then newly created City organization, the City and Guilds of London Institute for the Advancement of Technical Education (CGLI). The Central displayed the ideals of the City, of guilds and of the trades, and had a different ethos from the other two colleges. However, all three institutions had to come to terms with mid-Victorian realities and did so in ways both similar and distinct. In comparing the three institutions, it will be shown how they reflected cultural proclivities important at a time when there was seen to be an increasing need for a wide variety of scientific and technical expertise. The ideal of the good scientist or engineer had to be broadly construed so as to be inclusive of the many new occupations and careers that were deemed desirable and that were then becoming available.[2]

That science education and state sponsorship began to come under public scrutiny when it did may have had something to do with Charles Babbage's difficulties in getting support for his calculating machines. He discussed these difficulties, and what he saw as the related amateurism and dilettantism in the Royal Society, in his *Reflections on the Decline of Science in England* (1830) and *On the Economy of Machinery and Manufactures* (1832). The idea of decline as it related to Britain's position relative to her industrial competitors not only fuelled the rhetoric of promoters of science and technology in the 1830s and 40s, it has done so, not always justifiably, to this day. The case of Babbage helped to nudge the largely *laissez-faire* attitudes towards science and science education in a more interventionist direction. Later in the century, there were those who held the view, publicly expressed by Alfred Russel Wallace, that scientific and technical training should be paid for by those manufacturers who profited from it.[3] Others wanted the government to do more, though many manufacturers were wary of any system that might result in the training of employees for future competitors, and in the loss of trade secrets.

Modernizers joined Babbage in attacking the dilettante membership of the Royal Society, a membership then based as much on patronage as on scientific merit. The Royal Society did, indeed, reform itself, the British Association for the Advancement of Science was founded, and the Government began to play a greater role in promoting science than earlier.[4] The educational debates, as Thomas Carlyle noted, resulted in some tension between supporters of the older metaphysical and moral sciences and those of the newer natural and physical ones. The older universities were criticized for their narrow liberal education and they, in turn, defended their turf. 'Never let us believe', wrote Edward Copleston, provost of Oriel College, Oxford, 'that the improvement of chemical arts, however much it may tend to the augmentation of national riches, can supersede the use of that intellectual laboratory, where the sages of Greece explored the hidden elements of which man consists'.[5]

Despite rhetoric of this kind, by mid-century it was clear that scientific and technical fields would have a future in higher education, though exactly what form this would take remained to be worked out. The Great Exhibition of 1851 showed the wider public what many had already grasped, that there was a connection between science, technology, education, and advances in industry. Prince Albert wanted the Kensington site that had been purchased with the Exhibition profits to be used for the permanent exhibition of scientific discoveries, and of the industrial and technical trades and arts. Others, notably Lyon Playfair and Thomas Henry Huxley, argued on behalf of technical education and for a college to be located on the site. They further argued that schools should include science subjects in their curricula, that new technical colleges be founded also elsewhere in the country, and that the older universities be reformed so as to provide higher education in scientific and technological subjects. It is in this general context that the three colleges to be discussed were founded.

The Royal College of Chemistry

The first of the three institutions to be founded was the Royal College of Chemistry. It opened a few years before the Great Exhibition, in 1845.[6] The initial promoters were two entrepreneurs, the chemist and druggist, John Lloyd Bullock, and the apothecary and medical man John Gardner. Both men had connections to Justus Liebig, the famous chemistry professor in Giessen.[7] The two saw an opportunity for economic and social advancement in the idea of a chemistry college, believing that many young men would see the utility of a training in practical chemistry and pay fees to attend. The emphasis was to be on practical training since it was recognized that chemistry was then poorly taught in Britain, largely through lectures alone.[8] Gardner and Bullock appear to have hoped that by providing practical instruction they would have the added benefit of new discoveries being made in the college laboratories. In this way they could profit from patents and licenses as well as from fees. Despite the tensions between their respective trades,[9] Gardner and Bullock saw a common cause in promoting a college. Perhaps theirs was a natural alliance between a man who had status as an apothecary and licensed medical practitioner and one whose status derived from wealth. However, to promote their enterprise they had to take on board the Liebig propaganda with its emphasis on the teaching of pure rather than applied science, that had made its way to Britain via Liebig's British students. According to Liebig pure science was to be studied both for its own sake, and because research would lead to important applications. Liebig had written much about the useful application of chemistry to agriculture, a point used to gain the support of the landed gentry and aristocracy for the new college. His *Animal Chemistry* (1840) had been translated by Lyon Playfair and was used by chemistry's boosters to demonstrate that on the Continent scientific agriculture was being taken seriously. So called 'high farming' was in fashion also in Britain — increasingly so after the repeal of the Corn Laws in 1846. It entailed a number of 'improvements' including the use of fertilizers such as imported guano, and lime, a by-product of the coal-gas industry. Edwin Chadwick, in his *Report on the Sanitary Conditions of the Labouring Population of Great Britain* (1842), suggested that sewage should be converted into manure, a popular idea with chemists at the time, but one that, in the end, proved largely impractical.[10] Gardner and Bullock promoted their college by stating the need for practical teaching devoted to the 'application of chemistry to medicine, arts and agriculture.'[11] Thomas Wakley, editor of the radical medical journal *The Lancet*, and a great Liebig promoter, persuaded a number of fellow MPs to make donations in support of the college.[12]

The idea of a practical chemistry college had to meet the approval not only of potential backers from the aristocracy and gentry but also that of academic scientists, members of learned societies, and of those, including manufacturers, who held more utilitarian ideals. It also had to attract students. In 1844 Gardner and Bullock issued a 'Proposal for Establishing a College of Chemistry for Promoting the Science and Its Application to Agriculture, Arts, Manufactures and Medicine'. This was signed by 35 men and backed by pledges from 760 others.[13] The thirty-five were well known in government, medical and landowning circles, a necessity, given the relative obscurity of the two prime movers. Gardner and Bullock's first attempt was to attach the new school to the Royal Institution, an idea supported by the Institution's two professors of chemistry, Michael Faraday and W. T. Brande. However, the managers of the Royal Institution decided against the plan, giving lack of space as their reason.[14] Eventually the shareholders purchased a ground lease and built their own school in the large grounds of a house on Hanover Square with frontage also on Oxford Street to the north.[15]

Financial support came from landowners, doctors and manufacturers, many of them druggists. Major subscribers were to have the use of a laboratory on the premises and were to be given free assistance for investigations which promised 'benefit to Great Britain, its Colonies or Dependencies'.[16] They also had the use of a library and chemical museum. Further, subscribers were to receive cheap copies of the biennial reports which, it was said, would contain much useful information. Prominent among the shareholders who took an active role in the planning and running of the college was Sir James Clark, physician to Queen Victoria. The considerable medical interest in the college was related to the fact that the Society of Apothecaries, after 1835, required some training in practical chemistry for all those wanting its licence to practice medicine. The medical community was similarly active in bringing about the University of London MD and the promotion of practical chemistry instruction at University College and Kings College. But it was chiefly land owners who were relied on to make overtures to the government and to make things happen. Two of note in this regard were Lord Ducie, owner of the Whitfield Experimental Farm near Stroud (which Liebig had visited) and William Bingham Baring (later Lord Ashburton), heir to a banking fortune and a large Hampshire estate on which experiments were encouraged. For example, Baring was one of the first to experiment with the chemical treatment of sewage for manure. Ducie lobbied Robert Peel who was also interested in scientific farming. Peel, in turn, consulted the geologist William Buckland, from whom he often sought advice on scientific matters. Buckland consulted his friends among the scientific elite and advised Peel to support a college of chemistry but not in the metropolis. It should, he stated, be sited in a provincial town, should be run by one of Liebig's assistants, and only pure scientists should teach there. However, despite Peel's interest in promoting chemical education, he believed it better to do so in existing institutions and that a college of the sort proposed should not receive government funding. So, as already noted, the college was built at private expense and in London where the two main promoters wanted it to be.[17]

By 1845 there was enough money (and a roster of important names added to a provisional council) so that a public meeting could be held to elect a permanent council and make definite plans. Prince Albert was elected President, William Baring was elected chairman and 20 vice-presidents were elected, many of whom, like Baring, were improving landowners. A. Wilheim Hofmann, one of Liebig's star pupils and former assistant, was persuaded to come to England by Prince Albert to become the first professor at the new college which opened later that year.[18]

Hofmann was paid £400 per year, plus £2 for each full-time student and *pro rata* for part timers. By its third session the college had a new building and could accommodate forty-six students working at any one time.[19] Hofmann was also given free accommodation and a private laboratory. He was not permitted to take any outside contracts without the approval of the Council, though this was readily given. As can be seen from the college register, many students came from London artisan and trades families. An early student, Thomas Hall, later taught chemistry at the City of London School. In turn, many of his students entered the college. Students came to acquire the knowledge seen to be useful in various London trades such as printing, dyeing or brewing.[20] They came also for the chemical qualifications needed for a medical license from the Apothecaries, or to acquire the knowledge needed to work as pharmacists, druggists or agriculturalists.[21] Some had aspirations to work in government science or in colonial administration, others to work as teachers. Some found academic posts of one sort or another, and some became leading chemical manufacturers.[22]

Hofmann was ambivalent. He claimed to hold pure research above all other pursuits but was also engaged in entrepreneurial chemical consultancy. Like Liebig, he was publicly critical of the British for valuing science just for its utility but he allowed students to take out patents on work carried out in the college. The ideal of pure research was absorbed by many of his students who dreamed of having what he did — namely, voice in the world of gentlemanly science; but there were other noble options.[23] Early students sent papers to Liebig's *Annalen*, but later Hofmann became a founder of the *Quarterly Journal of the Chemical Society* and encouraged ambitious students to publish there. Most students, however, attended the college for just a short time and took what they had learned back to their places of work. Hofmann negotiated well among the various interests represented by his students and college subscribers. In his 1847 annual report, he remarked on the accurate analysis of thermal waters in Bath, work on the composition of caffeine and on substances present in oil from cumin seeds. 'These investigations have been crowned by the results worthy of the great zeal and of the true love of science'. Also, 'an investigation of the neutral oil of coal gas naphtha has been undertaken by Mr. Mansfield; this research in particular affords an illustration how investigations undertaken from a purely scientific point of view may lead to results which promise to become of highly practical utility'. Hofmann and his students also helped in the national provision of pure water, the safe delivery of coal gas, and performed a variety of public analyses. In his 1851 report Hofmann noted that many young men had left the college to help their parents with exhibits for the Exhibition, implying that college students and their families were working on a project of great national importance.[24]

Imperial interests were also stressed. Gardner spoke of how the college could be of importance in the development of empire, 'an incalculable amount of mineral wealth exists in Great Britain and its Colonies, and also in India, concealed from its proprietors only for want of knowledge'. Similarly, the Earl of Clarendon, a vice-president of the college, stated that the college would help young men into useful professions 'whereby not only would individual interests be promoted, but both in England and throughout her numerous dependencies, the national character of science would be raised and the natural resource developed'.[25]

As a consequence of the broad constituency, a hybrid ideology emerged at the college bringing together the enlightenment ideal of pursuit of truth, aristocratic liberty in the form of academic freedom and time for pure research, the aristocratic ideal of shared knowledge to be met through publication, nationalist ideals, (ie. serving the nation and the empire through science), ideals related to manipulative laboratory skills which had their roots in trade traditions, and bourgeois ideals of utility and entrepreneurship. All of these, with roots in different historical periods and connections, albeit not exclusive, to different classes in society, were acceptable at a time which saw increasing numbers of people demand social recognition. Further, the fraternal community of students, teachers, and alumni was one in which young people supported each other and cooperated in promoting the ideal of the good chemist and the idea of a worthy discipline.[26]

The college copied the Royal Institution, a scientific foundation with a more clearly aristocratic value system and, soon after opening, launched a series of evening lectures in the grand manner with the Earl of Clarendon, then also President of the Board of Trade, in the chair. These evening lectures at which the college presented its face to the public further illustrate the various ideals mentioned above. For example, Gardner spoke on the educational system used at the Royal College of Chemistry emphasizing that pure knowledge was being advanced and that the education given was not just vocational but also liberal. It trained the mind such that research in any area could be undertaken and, just as at Oxford or Cambridge, the college taught men 'how to learn'.[27] But others spoke on useful matters such as calico printing, the manufacture of glass, and agricultural chemistry. A lecture by Liebig's brother-in-law, Professor Knapp, a chemical technology specialist from Brunswick (his son was later a student in the college), was entitled 'The Influence of Science Upon the Progress of National Industry'.[28] In this connection, of especial pride to college members were the later achievements of William Perkin who made and developed the first coal tar dye, mauveine, the often cited example of pure research leading to a great industry.[29]

But the Royal College of Chemistry never acquired sufficient funds to run smoothly — there were some questions concerning the financial dealings of the two founders who were early eased out of their positions of authority. But even under new management the college had financial problems. The government came to the rescue, purchased the college from its shareholders and, in 1853, amalgamated it with the Government School of Mines. The new joint institution was temporarily named Metropolitan School of Science Applied to Mining and the Arts. Within this, both of the older colleges were able to retain their names and much of their separate identities.

The Royal School of Mines

The Government School of Mines was opened in 1851 as a result of lobbying, much of it by the geologist Henry De La Beche, Director of the new Geological Survey of Great Britain.[30] The first step towards the school was the decision, made in 1837 by Lord Duncannon, Chief Commissioner of Woods and Forests, to allot some government rooms in Craig's Court, near Scotland Yard, to house the geological collection that was an offshoot of the Survey.[31] The collection, housed in the tellingly named Museum of Economic Geology, was opened to the public in 1841. The first curator of the collection, which had an analytical laboratory attached, was a Quaker, Richard Phillips FRS, who had earlier apprenticed with the druggist William Allen at the Plough Court Pharmacy in the City of London, and had become a distinguished analytical chemist. Allen's pharmacy was a training ground for talent from the East End of London, and was also a site of radical political activity.[32] When the Department of Woods and Forests took over the running of the museum, two further laboratories were added, one for chemistry and one for botany, under Phillips and Joseph Hooker, respectively. Public pressure and skilled lobbying led to the building of a fine new museum (with laboratories attached) in Jermyn Street, just off Piccadilly. This opened in 1851. In the same year De La Beche, and many others, flooded the government with memoranda urging the establishment of a mining school to be attached to the museum. De La Beche noted that the mineral wealth of the country amounted to £24 million a year — close to half that produced by the rest of Europe combined, and that British metallurgists and miners were unable to obtain the proper instruction which their European counterparts had been getting for years.[33] Despite the illogic of this argument, the time was right and the government agreed to finance mining education.

Permission for the school was given by Lord Seymour, Secretary of the Office of Woods, Forests and Works. He also gave a grant of £800 to be used for public instruction in 'the science of mining and the application of mineral products to agriculture and to the arts'.[34] The Government School of Mines and of Science as Applied to the Arts, as it was first known, opened in 1851. The school went through several name changes. In 1853 the government bought the Royal College of Chemistry from its shareholders and united it with the Government School of Mines in the newly named, Metropolitan School of Science Applied to Mining and the Arts, placing the school under the Science and Art Department of the Board of Trade. The new name reflected the wishes of those who saw the school as teaching both pure and applied science. In 1857 Roderick Murchison, who had succeeded De La Beche as Director, wanted the emphasis placed back on mining and brought the old name back. In 1861 the government and the mining industry decided that the school should modernize. A government appointed committee, chaired by Lord Granville, issued a report in 1862 recommending that the diplomas become degree equivalent. Murchison tried to get the University of London to accept his students for degrees, but was refused. Instead, in 1863, the name of the school was changed to Royal School of Mines (RSM); and the diploma course was revised and extended from two to three years. Graduates were given the title ARSM. Those who had graduated earlier were also allowed to use the letters ARSM after their names. The badge of the RSM, the crossed hammers and crown, was introduced at this time and is derived from the button design of the uniforms of the Geological Survey, a reminder of the original union of School and Survey.

All but one of the first professors were officers of the Survey. Andrew Ramsay was appointed professor of geology, Lyon Playfair (soon replacing Phillips) of practical chemistry, Edward Forbes of natural history, John Percy of metallurgy and Robert Hunt of applied mechanics; De La Beche was Director. It was a very gifted faculty and, since the professors also formed the governing council, they had much liberty. Courses were open to the public at large and certificates of atten-dance were given. Serious students who wanted an Associateship of the School had to take a series of courses lasting two (later three) years and pass examinations. In the first year only seven such students enrolled (along with twenty part-timers, mainly soldiers). Even in subsequent years full-time enrolment climbed only slowly.[35] Some of the profits from the Great Exhibition were used to fund scholarships for students, and young artisans from the East End of London were among those who won them. Students were also admitted if they had done well in the government examinations run by the Science and Art Department of the Board of Trade. Gold medallists in these exams were given free tuition. Officers in the army and navy, diplomatic or consular officers and mine managers were admitted at half-price.[36] Scholarships were donated by the Royal Family (the Duke of Cornwall Scholarships) and by members of the aristocracy. As a result of these various modes of support the student body was somewhat mixed, both with respect to family and educational background — artisans, bright school graduates, military men, mine managers, entrepreneurs, and young men who wanted to learn how to prospect for gold. One of the professors, the convivial Edward Forbes, wrote a poem about those hoping to seek their fortunes in the gold rush in New South Wales. The following is the second verse.

> Up in that street of Jermyn,
> We've been lecturing every night,
> Such Mammoniferous sermons,
> With the room at 98°
> How to get gold they demanded
> (As if we poor chaps could tell
> By whom that same is wanted)
> Yet we told them — very well.[37]

Evening classes for working men were also given to meet what Lord Duncannon called the 'great national object of educating those who are prevented by circumstance from educating themselves.'[38] The school had a lecture theatre which seated 600 and was nearly always full for these evening lectures (the fee was 6d for a set of six). In order to get in, the men (and a few women) had to give their name and address and proof of their occupation. Somehow the worker's friend, Karl Marx, was able to attend in 1863, despite his lack of workplace qualifications.[39]

The culture of the School of Mines differed from that of the Royal College of Chemistry. As there were no shareholders, the professors were much freer from the pressures of commercial or landed interests than were Hofmann and his assistants. The government did not interfere to any great extent with De La Beche or his successor, Roderick Murchison. They and the professors were supposed to be on call when advice was needed, but otherwise they could do as they saw fit. They governed their own affairs and had a collegial way of life. The professors also spent much of their social life together and formed a strong fraternal clique in the world of London science.[40] Their idealism included a notion of service to the nation that often took the form of paternalism which, as James Secord has noted, was a survival of the paternalistic system of the Geological Survey.[41] They cared for the students who came their way and tried to prepare them for work of national importance at home and in the colonies. The son of Warington Smyth, Professor of Mining and Mineralogy, remembered that his father's 'heart was always with his students, as he watched them both before and after they went out into the world. He took the greatest pleasure in telling us their doings, their hopes and triumphs in far corners of the earth'.[42] While the RSM no longer has formal existence within today's teaching structure at Imperial College, there still exists a strong sense of family, and of corporate pride among the alumni. This sense of pride is detectable throughout the twentieth century in student and alumni publications. Strong bonds were formed during student field expeditions and, after leaving the School, when working with fellow graduates, often in dangerous circumstances, in mining and metallurgical ventures around the world. The school became the major source of expertise for British efforts in geological surveying and mining and, as Roy MacLeod has put it, 'a crimson thread of kinship tied them to imperial loyalties.' Between 1851 and 1920 one-third of the 1,115 Associates worked for most of their lives outside Britain and three-quarters of them spent at least some time overseas.[43]

Student publication indicates that the Royal School of Mines was a more class conscious institution than the Royal College of Chemistry, with traces also of the military culture of which the Survey was also a product. Indeed, Murchison ran the Royal School of Mines as if it were a military operation. Like De la Beche he came to higher education and geology from a career in the army. The professors were often consulted by those holding high office. Lyon Playfair, the chemist at the Royal School of Mines was constantly in demand as was John Percy, the professor of metallurgy. Percy was an expert on iron and steel and appears to have served almost continuously on Royal Commissions during the 1860s and 70s.[44] And William Roberts-Austen, a student, and later professor, in the school, became Queen's Assay Master.[45]

While professors at the Royal College of Chemistry were occasionally consulted by the government, they earned more money, if less prestige, than their Royal School of Mine counterparts, by consulting for the water companies and for chemical industry. Professors at the Royal School of Mines, on the other hand, had the status that went with important government service. After the College and School were joined within a single institution these differences lessened. But even at the start, there was more in common between these two West End institutions than between either of them and the other institution with which they would be joined in 1907. The two discussed so far were both tied, albeit in varying degrees, to government, the governing classes, and to landed and aristocratic interests.

However, while these interests dominated the culture of the RCC and RSM it is clear that other interests were also to be included. Young graduates believed that the most noble pursuits were pure research and high government service. But most will have recognized that this was just for the few, and perhaps not for them. Because of the need for inclusion in a modern state, the ideology of serving the nation through resource development at home and in the Empire, industrial or government science, trades, entrepreneurial activities, and science school teaching, served to legitimate what most of the graduates were to do. While these types of occupation were sought also by graduates of the colleges run by the City and Guilds of London Institute, their student experiences will have been different. The old City culture and the traditions of the guilds, their monopoly on trade practice, albeit much eroded by this time, as also their power over artisanal life, led to a different kind of institution once it was decided to take part in the national effort to improve scientific and technical instruction.

City and Guilds of London Institute and the Central Technical College

In the aftermath of the 1851 Exhibition a growing awareness of foreign competition in trade and industry resulted in yet more pressure on the Government to fund higher education in the areas of science and technology. Pressure increased even further after the Paris Exhibition of 1867 when only ten out of a possible ninety awards for excellence went to Britain.[46] In an address at the conclusion of the Great Exhibition, Lyon Playfair had stated, 'raise industry to the rank of a profession' since he believed that this would encourage talent to enter technical areas. After the Paris exhibition he wrote to the chairman of the Schools Inquiry Commission (Taunton Commission), 'the one cause of this inferiority [of British exhibits] upon which there was most unanimity is that France, Prussia, Austria, Belgium and Switzerland possess good systems of industrial education for the masters and managers of factories and workshops and that England possesses none'.[47] At this time the Science and Art Department of the Board of Trade was the main agency of technical education and Playfair, who had left academic work for the civil service, was Science Secretary in the Department.

Technical education became a priority for Gladstone when he became Prime Minister. Determined to find money for technical training he looked to the City Corporation, and the Guilds of London, rather than to the learned societies of the West End, or to aristocratic largesse, for support. Wealth, accrued in part from ancient land endowments, was often lavishly displayed in the City where, in the 1860s, seventy-seven livery companies still existed. Up to the mid-eighteenth century the guilds, their masters, wardens and courts of assistants, had the power to govern trade and commerce not just in the City of London but within a large area extending even beyond the bounds of what today is Greater London. By the mid-nineteenth century, only the Goldsmiths and the Fishmongers had been able to retain any of their traditional control.[48] As a consequence, the guilds were continually re-inventing themselves.

They had largely settled on a combination of philanthropy towards people learning trades, paying for apprenticeships, funding technical schools and schools for tradespeople's children (today some of these are among the elite of Britain's schools), caring for widows and orphans, and for sick or elderly tradespeople. These were traditional guild roles but they needed to be re-thought. In so far as technical training went, a wide variety of projects were funded. For example, technical drawing classes throughout England, a system of handicraft education in elementary schools in London, a school for tailors in Golden Square and one for leather trades in Bermondsey, grants to the Society for Promoting the Employment of Women, large numbers of prizes to trades students around the country, and much more.[49] All of this co-existed with archaic tradition within the guild halls and a nineteenth-century dining club culture among the senior members.

Traditionally it had been the responsibility of the guild courts to see that apprentices were well treated and well educated.[50] After about seven years of training the apprentice submitted a 'masterpiece' which in all its parts had to be made by the apprentice alone to show that he, or in a few cases she, had mastered all aspects of the trade. This ideal of comprehensiveness lived on, even as the number of apprentices declined, and caused problems when brought face to face with the new nineteenth-century ideal of specialization. It is clear from recorded minutes of the CGLI that many a guilds' committee formed to look at what should be done about trades education in the 1860s and 1870s foundered because of the clash of these ideals.[51]

The speech given by Gladstone at Greenwich in 1875, on the occasion of the prize day for successful students in the Government Science and Arts examinations sums up well the general climate of opinion on what was then widely seen as an archaic, wasteful and largely useless guild organization. In reference to scientific and technical education, Gladstone stated,

> I should like to see a great deal of this work done by the London Companies.... What was the object for which those companies were founded? Do you suppose they were founded for the purpose of having dinners once a year, once a quarter or once a month? Do you suppose they were founded for the purpose of dealing out little sums of money to certain applicants and then having it recorded

of them how much good they had done? Nothing of the kind.... These companies were founded for developing the crafts, trades or "mysteries" as they were called. They were founded for the purpose of doing the very thing which the Government of the country, out of the taxes of the country, is now called upon to do.[52]

Gladstone, while weak on their history, knew that the guilds were very rich, though not what later became known: namely that the annual income of just the twelve major livery companies was about £800,000 in the late 1870s.[53] The income came largely from rents since all the guilds were property rich, and not just in the City of London. Many guilds had Irish estates. Gladstone's view of the guilds as archaic and wasteful was widely shared. For example, Sydney and Beatrice Webb wrote that the companies' 'magnificent social entertainments are so many social bribes to secure friends to cover their mismanagement of the great trusts in their hands'.[54]

Gladstone's speech and the threat of a Royal Commission to examine guild activities (which later materialized) forced the guilds to think seriously about what they should do. At first they did not think collectively — the traditional rivalry among the guilds was hard to overcome. For example, one of the wealthier guilds, the Clothworkers, which already helped fund the Yorkshire College of Science and the Bristol College of Science had the idea of reviving Gresham College.[55] Together with the Corporation of the City of London and the Mercers, they were custodians of the Gresham Trust. The college, founded after Thomas Gresham's death in 1575, was located in his house and with money he had left for the purpose; it was never a great success. By the nineteenth century it was nothing more than a sinecure for seven lucky professors. The Drapers Company was helping fund science classes in the Middle Class Schools in and around the City, and the Society of Arts approached the Company to take over its system of technical examinations.[56]

It was the Goldsmiths Company, which had its own college in New Cross, that suggested that all the running in different directions was unwise and that the guilds should put aside ancient rivalry and cooperate in some major new venture.[57] The wealth and strength of the individual guilds made cooperation difficult but the Goldsmiths were supported by the Lord Mayor and other City figures. The consequence was the founding of a new institution in 1878: the City and Guilds of London Institute for the Advancement of Technical Education (CGLI; incorporated in 1880). Sixteen of the guilds were major contributors and thus founders, each deciding how much it would contribute annually. There was some pride in being generous. The bureaucracy of the CGLI was cumbersome and likely reflects ancient tradition. Inclusion was intrinsic to the fraternal guild culture, hence committees proliferated and memberships were large by modern standards. Serving on the Board of Governors, in addition to the *ex officio* members, were about twenty people from the City Corporation and about three hundred liverymen.[58] Much time was spent in deciding who could be appointed to the Board and on the 17th March, 1879, it was resolved that 'HRH the Prince of Wales, Citizen and Mercer, Grocer, Fishmonger, Goldsmith, Merchant Tailor, and Clothworker be requested to accept the position of President of the Institute'.[59] This language of equality and inclusion is typical of much found in the various minute books. Three honorary secretaries and an honorary treasurer, together with the chairmen of the Council and Executive Committee, held much of the power — though subject to the discretion of the many committees.[60] A salaried Director, Philip Magnus, together with a small staff, ran the day-to-day business of the new Institute. Magnus was a graduate of University College with degrees in both science and arts and had spent several years in Berlin studying the Jewish Reform Movement. At the time of his appointment in 1878 he was a junior rabbi at the synagogue in Margaret Street, though much of his time was spent as a science lecturer in and around London.[61]

In 1878, at the invitation of John Donnelly, a sub-committee of the CGLI went to the Department of Science and Art in South Kensington for discussions. Donnelly, who headed the Department, had earlier been asked his opinion on the best way of setting up a university level technical college. In reply he sent the CGLI a letter not so much full of advice as it was with concrete estimates of what a new college would cost, and expressing opposition to a college of the type the City wanted.[62] The committee also sought the opinions of some others, including Henry Trueman Wood, Assistant Secretary of the Society of Arts, and T. H. Huxley.[63] But, as is clear from various CGLI committee and council minutes, there was resistance to paying too much attention to what outsiders wanted. Despite Donnelly's opposition it was decided to build what was then known as the Central Institution.

The first problem was in choosing a site, something that proved highly contentious. Most on the CGLI governing body wanted the Central to be in, or near, the City. But they rejected a Clothworkers suggestion that Gresham College be demolished to make room for the new college.[64] A site on the bank of the Thames near Blackfriars, favoured by some, was rejected because construction would have entailed expensive engineering works and the overall costs would have been too high.[65] Since, in the end, it proved difficult to find a suitable City site, the temptation of a peppercorn lease offered by the Royal Commission for the Exhibition of 1851, for land in South Kensington, proved irresistible. However, there was much opposition to a site so far from the City, especially since it was close to the present or planned location of various government institutions. The Normal School of Science and the Royal School of Mines had already moved to South Kensington into a building originally intended for a Naval Architecture School.[66] Also there were two museums, and the Department of Science and Art. Many in the City wanted to demarcate their work from that of the Westminster elite, and for some the move to South Kensington proved too much. The Drapers and Clothworkers had wanted their money to go more to the CGLI's Finsbury Technical College, and to regional trade schools, than to the academically oriented Central Institution. The Drapers later resigned from the Institute over this issue. For them, technical education was to be aimed at 'the class of the community most in want of it'.[67] On leaving the Institute, they joined with the Beaumont Trust, with which they already had ties, and put their money into the East London College and People's Palace (forerunners of Queen Mary College).

That the CGLI was to be given the free use of land in South Kensington was used by outsiders to pressure it to appoint a few scientific men to its committees; people who, in the eyes of the west-end elite, would give some credibility to what they saw as an uninformed collection of archaic philanthropic companies. The pressure inclined some within the guilds to pull out of any agreement with the 1851 Exhibition commissioners. One of the honorary secretaries wrote to the other, 'I think we had better now consider whether we cannot have a *distinct* wing of University College, Gower St. for technical education purposes — they like it and the City will prefer that alliance to South Kensington'. He noted that he was unwell but 'conservative reformers like you and I cannot afford to be ill and away from the scene of action'.[68] In a more constructive way, the chairman of the executive committee, Frederick Bramwell, figured out how to satisfy outside interests without giving away guild autonomy. He wrote,

> Let the guilds insert in the constitution (on grounds we can readily state) that the presidents of the Civil Engineers, the Mechanical Engineers, the Chemical Society and the Royal Society, or in turn of the Mechanical Engineers, the chairman of the council of the Society of Arts, shall be *ex-officio* governors, councillors and executive committee men and, if need be, Central Institute sub-committee men. The suggestion coming from ourselves will not hurt our respectability and these persons being all of them restricted to three year terms of office or less ... we need not fear that we should be a body corrupted.[69]

In the end, aside from the move to Kensington, all that the outside pressure achieved was the inclusion on council of four *ex officio* representatives of learned societies as Bramwell had proposed.[70] The guilds maintained control over their own institutions into the early 20th century, though wisely they took some of the advice they were given.

It was decided to build a college for two hundred students, 'to promote technical education of persons of both sexes engaged in the industries and manufacture of this country and to provide instruction in the application of those sciences and arts as are ancillary to such manufactures.'[71] This clearly foreshadows the mission statement embodied in the later Imperial College Charter. There was much pride in the new project, 'we will all share in the glory of the successful accomplishment of a grand national benefit'.[72] Alfred Waterhouse was chosen as architect and his building, no longer standing, was a model of iconography.[73] Guild insignias, stone carvings of the various trades, and crests of the major manufacturing towns of great Britain, were everywhere to be seen.[74] Unlike at the CGLI's Finsbury Technical College, where the first professors had been chosen by word of mouth, advertisements were placed and many applied. The Institute relied somewhat on outside advice in their selection.[75] William Ayrton and Henry Armstrong moved from Finsbury and were appointed professors of physics and chemistry respectively.[76] The other two original appointees were engineer, William Unwin, and mathematician, Olaus Henrici.[77] Officially it was stated that there would be five divisions or faculties at the Central: chemical technology, mechanical, civil and electrical engineering, general manufactures, architecture and building construction, and Applied Art. Instruction would be given in chemistry, engineering, mechanics, mathematics, physics, drawing, manufacturing, technology, workshop practice, modern languages and applied art. But these ambitious plans did not fully materialize.

The CGLI council resolved 'that the class of person to be taught in the Central Institution shall be those engaged (or intending to be engaged) in the manufactures of the country, whether as workmen, managers or foremen, or as principals, and also persons qualifying for teachers in technological subjects.'[78] Thus 'considerable latitude ... [was to be] allowed in the admission of different classes of students.'[79] Entry was to be by examination, and open to students aged sixteen or over. The early students came from a range of backgrounds. Many were graduates of the Finsbury Technical College and of other technical colleges and trades schools from around the country. Others were students from secondary schools which provided enough science education for their students to pass the entrance examinations, from public schools, and from the older universities.[80]

The pedagogical ideal '*la Science est une*', taken consciously from the École Centrale des Arts et Manufactures in Paris, was promoted at the Central, especially by Armstrong. All students were to learn some mathematics, chemistry, physics and mechanics, before specializing. And specialization, to begin normally in the third year, was to be research based (pure or applied), or practical (for example, the furnishing of a set of plans for the construction of a section of railway line) and not involve any formal or book-based instruction. To satisfy some of the guilds, and the recognition that 'a large artisan population exists in Chelsea and the neighbourhood', evening classes were also given. Attendance books were kept and the parents of full-time students who failed to satisfy the professors were notified. Old guild discipline was apparent in many ways.

The professors, all very gifted, all Fellows of the Royal Society, were determined to build reputations well beyond the college — something they succeeded in doing. But this ambition to be doing something more than providing advanced technical skills for young people, which was how many in the guilds saw their function, created tensions. There was continual negotiation between the professors and the CGLI committees overseeing the Central as to what the professors might be permitted. The professors both pleased and disappointed the various guild constituencies. Ayrton and Armstrong's breaking down of the distinctions between traditional academic and trades training was widely admired. Ayrton's imperialist ideals and his promotion of overseas students also found support. Taking foreign students was seen as good, taking on outside work less so. From the CGLI point of view the relatively high professorial salaries (£1000 per annum) meant that outside work was unnecessary. But the professors continually complained of needing contact with industry and slowly won the right to take on consulting work. The tight control could drive the professors to distraction on occasion, 'I have a *complete* record of the times at which I entered and left the college' wrote a frustrated Henry Armstrong.[81] But Armstrong had the full support of the guilds for another of his battles. He wanted the reorganized University of London to allow a research option (as opposed to traditional examinations) for its BSc degree. And, that examinations be weighted more towards the practical than was then usual. In 1910, and by then an employee of the new Imperial College, Armstrong attempted to get degrees for some of his students on the basis of their research (and published) work, but was unsuccessful. Also, some students whom he thought capable had failed the written examinations. Unhappy at this, he stated that the reformed university 'breaks faith with us and declines to help us in educating our students in such a manner that they may become efficient chemists able to enter works and appreciate the numerous problems demanding the constant exercise of the reflective and analytical habit of mind'.[82]

But it was Unwin, more than the other professors, who helped fashion what the City and Guilds College was to become. This was largely due to his long tenure and to the central role of engineering in the history of the City and Guilds College. Unwin was early appointed Dean of the Central, a position he retained until 1896 when a rotating deanship was introduced. In his previous position at the Royal Indian Engineering College, he had developed machinery for testing the strength of materials (such as iron girders and cement) used in India.[83] He brought this type of work to the Central, built an experimental steam engine on which successive generations of students worked, and introduced the subject of hydraulics. He published many papers on the strengths of materials (especially iron and steel), on the flow of water in pipes, and on water turbines. He was a major figure in the development of hydro-electric power, notably at Niagara Falls. All these activities drew students who enjoyed being associated with someone who provided them with so many working opportunities.

As in the Royal College of Chemistry and Royal School of Mines, one sees a hybrid ideology. Aristocrats were invited to be members of guilds and to give a certain kind of legitimacy. They, in turn, hoped to be invited; membership in a guild, especially one of the twelve great companies was much sought after by the London elite. Aristocratic presence helped the professors in that it was often aristocrats who argued for research to complement teaching. Politicians on the committees looked after national, and sometimes local, interests. Businessmen, dominant in the guilds, ensured that aristocratic idealism and archaic procedure were tempered by new ideas of utility, and that students in the college saw entering into manufacturing or commerce as noble options. The guilds also bred a sense of mystic brotherhood such that students often saw themselves as working in a common cause, helping the nation in its technical struggles with foreign competitors. Also important was the old guild ideal of craft; students were expected to acquire manual skills, not simply gain knowledge from lectures and books. Unlike the Royal College of Chemistry and the Royal School of Mines, it was not claimed that a liberal education was being provided; even though Armstrong's innovative ways of teaching of chemistry were, for example, no less 'liberal' than the ways in which chemistry was being taught in other institutions.[84] The skill of the tradesman, as Geoffrey Crossick has noted, allowed for a respectable way of life; it was an alternative to wealth or genius as a criterion for social judgement.[85] Something similar could be said of the technical skills being taught in the City and Guilds colleges.

Conclusion

The three colleges were institutions in which men, and a few women, from different classes worked together. Ideals that have been labelled for convenience as belonging to working people and the trades, as archaic guild ideals of brotherhood, aristocratic ideals of freedom of expression and freedom to pursue knowledge, enlightenment ideals of truth and progress, entrepreneurial ideals, Benthamite ideals of utility, and nationalist ideals, all combined, in different mixes, to give each of these institutions its specific character. The ideals were rooted in different traditions but remained appropriate to the Victorian age. As a consequence, within each college the exchanges were not simply scientific and technical, but also social and ideological. While I have associated the range of ideals with aristocratic, artisanal or commercial societal groups, this is a bit misleading. Ideas, after all, do not belong to any one group, whatever their visibility in different social circles. It is perhaps better to think of all the people who engaged with the three institutions as trying to expand the limits of their freedom to act and think, and that they achieved this by the collective consumption of a range of ideas and practices. The colleges were places where differences were mediated and new ideas forged. The result was the production of something new in British higher education, namely three influential colleges with interesting identities. All were successful in that students entering them, whatever their ambition or talent, could feel that they were engaged in something worthwhile.

As will be discussed in chapter three, the three colleges later met on what was largely government turf in South Kensington. Here, in 1907, the Central Institution (renamed the City and Guilds College), the Royal School of Mines and the Royal College of Science[86] joined to become Imperial College of Science and Technology. The three older colleges continued to exhibit their somewhat different identities and their historical legacies resonate even today. But the coming together of a range of constituencies in the new Imperial College reflected also the need for inclusion, the first concern of individuals in a modern political economy.

...

1. This chapter is based on two papers which contain more detail on this early history. Hannah Gay, 'East End, West End: science education, culture and class in mid-Victorian London', *Canadian Journal of History/Annales canadiennes d'histoire*, 32 (1997), 153–183; Hannah Gay, 'Association and Practice: The City and Guilds of London Institute for the Advancement of Technical Education', *Annals of Science*, 57 (2000), 369–398; (http//www.tandf.co.uk).

2. For further detail of the political, cultural and educational debates of the period see, Michael Argles, *South Kensington to Robbins: an account of English technical and scientific education since 1851* (London, 1964); D. S. L. Cardwell, *Turning Points in Western Technology: a study of technology, science*

and history, Science History Publications (New York, 1972), chapter 5; Michael Sanderson, *The Universities and British Industry, 1850–1970* (London, 1972); G. W. Roderick and M. D. Stephens, *Education and Industry in the Nineteenth Century: The English Disease* (London, 1978); Ian Inkster (ed.) *The Steam Intellect Societies: essays on culture, education and industry circa 1820–1914* (Nottingham, 1985).

3. Alfred R. Wallace, letter to editor, *Nature*, January 20th, 1870. The Putney College for Civil Engineers, founded early in the century, was an often cited example of a successful private enterprise.

4. These themes are treated in T. W. Heyck, *The Transformation of Intellectual Life in Victorian England* (London, 1982), chs. 3, 4 and 5. See also, Roy M. MacLeod, 'Whigs and Savants: reflections on the reform movement in the Royal Society, 1830–48' in Ian Inkster and Jack Morrell (eds.), *Metropolis and Province: Science and British culture, 1780–1850'* (London, 1983), ch. 3; Jack Morrell and Arnold Thackray, *Gentlemen of Science: early years of the British Association for the Advancement of Science* (Oxford, 1981). For the role of government, see Roderick and Stephens, *op. cit.* (2).

5. See Heyck, *op. cit.* (4) 64, for comments on Carlyle's essay, *Signs of the Times.* (1829); Coplestone quotation in Heyck, 68. While he wrote this in 1810, it was a sentiment that many at Oxbridge held still in the 1830s and 40s.

6. See, Gerrylynn K. Roberts, 'The Establishment of the Royal College of Chemistry: An Investigation of the social context of early-Victorian Chemistry', *Historical Studies of the Physical Sciences*, 7 (1976), 437–485; A. Rupert Hall, *Science for Industry: a short history of the Imperial College of Science and Technology and its antecedents* (London, 1982), 5; Robert Bud and Gerrylynn K. Roberts, *Science Versus Practice: Chemistry in Victorian Britain* (Manchester, 1984), 71–75.

7. Liebig was famous for his method of teaching chemistry in the laboratory which attracted students from around the world. He made important contributions to organic chemistry and to the application of chemistry to agriculture and food science. Gardner had translated Liebig's *Familiar Letters on Chemistry* and Bullock had studied with Liebig. Bullock was wealthy and owned a shop and large property in Mayfair. See also William H. Brock, *Justus von Liebig: The Chemical Gatekeeper* (Cambridge, 1997).

8. Chemistry was better taught in Scotland than in England. Thomas Graham, for example, gave practical instruction in Glasgow before moving to University College London in 1837. For chemistry teaching in this period see, William H. Brock, *The Fontana History of Chemistry* (London, 1992), chapter 11.

9. The Pharmaceutical Society of Great Britain was founded in 1841 to lobby against a proposed government bill that would have placed chemists and druggists under the jurisdiction of the Society of Apothecaries. The Pharmaceutical Society publicly supported practical training in chemistry. Professional boundaries were being fought over during the 1840s.

10. For 'high farming', see J. D. Chambers and G. E. Mingay, *The Agricultural Revolution, 1750–1880* (New York, 1966); also Bud and Roberts, *op. cit.* (6), 56.

11. Gardner and Bullock, quoted in Gerrylynn K. Roberts, 'The Royal College of Chemistry, 1845–53: A Social History of Chemistry in Early Victorian England', PhD dissertation, The Johns Hopkins University (1973), 138.

12. Gerrylynn K. Roberts, *op. cit.* (11), 184. For the radical climate in which *The Lancet* operated see, Adrian Desmond, *The Politics of Evolution: Morphology, Medicine and Reform in Radical London* (Chicago, 1989).

13. Roberts has examined the backgrounds of the promoters and donors and notes that over half were either landowners or wealthy medical practitioners and that this group provided 71% of the funds needed for the construction and running of the college. Roberts, *op. cit.* (11), 1–13 and *passim*; Roberts, *Historical Studies, op. cit.* (6), 473–483.

14. ICA; Theodore, G. Chambers, *Register of the Associates and Old Students of the Royal College of Chemistry, Royal College of Science and Royal School of Mines* (London, 1896), xlviii. (Referred to below as *Register.*)

15. There is a plaque on the building at 299 Oxford Street marking the site.

16. Quoted in Jonathan Bentley, 'History of the School of Chemistry at the Royal College of Science and Its Predecessors during the Nineteenth Century', Chemistry Part II thesis, University of Oxford, 1962, 11. It was possible to become a member of the college with an annual donation of only one guinea or a one-time donation of ten guineas, but more privileges were granted to major donors and shareholders. It would appear that little came of the service side. The professor, A. W. Hofmann, assigned his students some of the problems sent in by college donors but few donors made use of the college facilities themselves. From the Annual Reports (ICA) it would appear that Hofmann found the requests from subscribers tiresome and something he did not wish to bother with.

17. Roberts, *Historical Studies, op. cit.* (6), 440–50 and 468–9; Roberts *op. cit.* (11), 175–179 and 216.

18. Hofmann was the council's third choice. Carl Fresenius and Heinrich Will had both refused the job. See *Register, op. cit.* (14), xlix. James Clark negotiated the terms of the contract with Hofmann, ensuring an academic rather than commercial slant to the appointment — without this assurance

Hofmann would likely have declined. As it turned out Hofmann was an excellent choice. He was an exceptionally good teacher and knew how to inspire young men to reach their potential. See Hannah Gay, "Pillars of the College': Assistants at the Royal College of Chemistry, 1846–71', *Ambix*, xlvii (2000), 135–169.

19. *Register, op. cit.* (14), l. In the eight years of its existence as a private institution the college enrolled 356 students. For an account by a student of the period see, F. A. Abel, 'The History of the Royal College of Chemistry and Reminiscences of Hofmann's Professorship', *Journal of the Chemical Society*, 69 (1896), 580–596.

20. Several students came also from the major brewing town of Burton-on-Trent; *Register, op. cit.* (14).

21. It would appear that few students went into agriculture but agricultural chemistry was important to many of the college backers who wanted work to be carried out on soil chemistry and related topics.

22. In 1868, when Edward Frankland was Professor of Chemistry, he gave a report to the Select Committee on Scientific Instruction (Samuelson Committee) in which he noted that of the thirty-six students in his laboratory, twenty-five were aged between 17 and 22. Of the thirty-six, nine intended to enter chemical manufacturing, six into brewing, four into mining, three wanted to become professional chemists, one each into medicine, teaching, sugar refining, law and the church. The remainder did not know what they wanted to do. See Bentley thesis, *op. cit.* (16), 40–50.

23. Several of Hofmann's students succeeded in gaining voice in major scientific debates, and became leading research chemists; for example, William Crookes, Frederick Abel, Alexander Williamson, William Perkin, and William Odling.

24. RCC Minutes of AGMs; 1847 and 1851 Annual Reports by A. W. Hofmann.

25. Gardner and Clarendon quoted respectively in Roberts, *op. cit.* (11), 189 and 288.

26. For more on the RCC see Gay, *op. cit.* (18).

27. John Gardner, 'An Address delivered in the Royal College of Chemistry' (3 June 1846), *The Chemist*, 7 (1846), 296. The idea that a liberal education could be obtained through the study of science was promoted by other scientists at this time. For a later example see, Thomas Henry Huxley, 'A Liberal Education and Where to Find It', *Macmillans Magazine*, 17 March, 1868. In this instance, Huxley was referring to science classes being given at the South London Working Men's College, but he made the same claim for classes taught at the Royal School of Mines. It was not until about twenty

years later that the older universities began seriously to reform their curricula to accommodate science and engineering.

28. Henry E. Armstrong, 'Pre-Kensington History of the Royal College of Science and the University Problem', address delivered before the Old Students Association, September 1920, 10 (ICA Armstrong papers 22/3).

29. William Perkin (1838–1907). See Anthony S. Travis, 'Perkin's Mauve: Ancestor of the Organic Chemical Industry', *Technology and Culture* 31 (1990), 51-82.

30. For further details on the Survey, see James A. Secord, 'The Geological Survey of Great Britain as a Research School, 1839-1855', *History of Science,* xxiv (1986), 223-275.

31. At this time the Survey was still under the Ordnance Department but later it was transferred to the Department of Woods and Forests. At the 1838 meeting of the British Association for the Advancement of Science, Roderick Murchison noted the lack of a school of mines in his presidential address and called for a national inventory of mineral deposits. At the same meeting Henry De La Beche urged the government to establish a mining records office and a museum for geological and mineralogical specimens and information. In 1832 he had argued successfully for the addition of geological mapping to the Ordnance Survey. This was the origin of the Geological Survey. For further details on the founding of the School of Mines see Hall, *op. cit.* (6), *Register, op. cit.* (14), and Secord, *op. cit.* (30).

32. Allen and many other of his Quaker friends were leading slavery abolitionists. William Wilberforce met them at the Plough Court Pharmacy to plot strategy. Also, the Askesian Society (a learned society founded by East End tradesmen) had its roots in Plough Court. See Desmond Chapman-Huston and Ernest C. Cripps, *Through a City Archway: The Story of Allen and Hanburys, 1715–1954* (London, 1954).

33. The De La Beche memo is reprinted in the 1851 Prospectus for the Government School of Mines.

34. Royal School of Mines, minutes of council, 27 June 1851. Professors were allowed to keep student fees.

35. The regular lectures were attended by about fifty people. See letter from Edward Forbes to Andrew Ramsay, quoted in *Register, op. cit.* (14), xiii. In the early years the School had an average of fourteen full-time and fifty part-time students.

36. One part-time student was General Charles Alexander MacMahon, later a civil judge, in India. He was enrolled in 1879–80 while on leave. He later wrote papers for the Geological Survey of India. See obituary, *The Times,* 24 February 1904.

37. There was a major gold rush in New South Wales in the early 1850s. Poem in Andrew Ramsay papers, 5/3. This copy is amusingly illustrated by T. H. Huxley.

38. Duncannon, quoted in Bernard H. Becker, *Scientific London* (London, 1874), 184.

39. Hall, *op. cit.* (6), 4.

40. For some details of the social life, see Andrew Ramsay's Diary, (KGA/Ramsay/1/8). See also Secord, *op. cit.* (30), 241; and Hannah Gay and John W. Gay, 'Brothers in Science: Fraternal Culture in Nineteenth-Century Britain', *History of Science*, xxxv (1997), 425–53.

41. Secord, *op. cit.* (30), 237.

42. Memorial notice for Sir Warington W. Smyth, *Science Schools Journal*, Vol. III (1890), 101–4. Warington Smyth, son of a well known admiral and thus already well connected to those in government, was a consultant to mining enterprises around the world. He became Chief Inspector of Mines of the Crown and the Duchy of Cornwall.

43. Roy MacLeod, '"Instructed Men" and Mining Engineers: The Associates of the Royal School of Mines and British Imperial Science, 1851–1920', *Minerva* **32** (1994), quotation, 437; statistics, 434.

44. For example, in 1861 the Secretary of State for War appointed him to inquire into the 'Application of Iron for Defensive Purposes'. In 1867–8 he was a member of the Special Committee on the Gibraltar Shields (gun batteries), in 1871 he was appointed to the Royal Commission on Coal in the United Kingdom and, in 1875, to the Royal Commission on The Cause of Spontaneous Combustion of Coal in Ships. He was also the Superintendent of the Ventilation, Warming and Lighting of the Houses of Parliament. See W. Roberts Austen, 'Tribute to John Percy', *Proceedings of the Royal Society* (1889) xvi.

45. Sydney W. Smith, *Roberts-Austen: a record of his work* (London, 1914).

46. Adrian Whitworth (ed.), *A Centenary History: A History of the City and Guilds College, 1885–1985* (London, 1985), 1. See also, Jennifer Lang, *The City and Guilds of London Institute: Centenary 1878–1978* (York, 1978).

47. Playfair, quoted in Argles, *op. cit.* (2), 17 and 26.

48. The question of the decline of guild influence is debated. In his discussion of the institutional and social network supporting London's industrial economy in the eighteenth and early nineteenth century, Daunton challenges the established view of the guilds' decline. Martin J. Daunton, 'Industry in London: revisions and reflections', *The London Journal*, 21 (1996), 1–8.

49. See, for example, Guildhall Library, City and Guilds of London Institute (CGLI), MS 22,000 which contains minutes of meetings in which these

activities are noted. Also, MS 21,822 vol. 1 for Sub-committee D for the Industrial Employment of Women.

50. For this and other information on the guilds see George Unwin, *The Guilds and Companies of London* (London, 1963); W. Carew Hazlitt, *The Livery Companies of the City of London* (London, 1892); John Charles Thornley and George W. Hastings, *The Guilds of the City of London and Their Liverymen*, (London, date unknown, though published during reign of George V). See also Lang, *op. cit.* (46).

51. See, for example, Guildhall Library, CGLI MS 22,000; minutes of meeting convened by the Lord Mayor, July 29th, 1873. By the old customs of the City, masters were compelled to teach apprentices every branch of the trade and if it could be shown that a master had failed, the Chamberlain had the power to compel the master to do his duty. In this meeting there was much discussion on how to change the rules in light of new circumstances.

52. Gladstone, quoted in a report on his speech in *Journal of the Society of Arts*, 19 November, 1875. The word 'mystery' as used in this context is probably a corruption of 'mastery'. Anxiety over their role, was shown by the guilds even before this speech. See for example, Guildhall Library, CGLI, MS 22,000.

53. Guildhall Library, CGLI MS 21,838 contains detailed accounts of annual balances of income and expenditure. See also Royal Commission on the City of London Livery Companies (London, 1884), Volume 1, for details of the incomes of the separate guilds.

54. S. and B. Webb, quoted in Francis Sheppard, *London, 1808–1870: The Infernal Wen* (London, 1971), 21.

55. Guildhall Library, CGLI, MS 21, 834; printed memorandum from the Clothworkers, December 6th, 1876.

56. See Lang, *op. cit.* (46), 14 and Thornley and Hastings, *op. cit.* (50), 26–7.

57. Sir Frederick Bramwell FRS, Goldsmith and civil engineer, seems to have been instrumental in smoothing the path on this and future occasions. See, for example, Guildhall Library, CGLI MS 21,816, Minutes of Council, July 9th, 1884 in which Bramwell, noting the unhappiness of the Clothworkers and Drapers over various matters, urges a common front.

58. *A Short Notice of the City and Guilds of London Institute for the Advancement of Technical Education with an Account of the Proceedings at the Opening of the Central Institution by H.R.H. The Prince of Wales, K.G., President*, in CGLI, *Reports to Board of Governors, 1884–87*, vol. 1 (copy in ICA). This contains the names of governors, members of the council, and executive committee.

59. Guildhall Library, CGLI, MS 21,814 vol. 1, Minutes of the Board of Governors, 17 March 1879.

60. John Watney (Mercer), W. Phillips Sawyer (Draper) and Owen Roberts (Clothworker), secretaries; Sir Sydney Waterlow (Clothworker), treasurer. Chairman of the Council was the Lord Chancellor, the Earl of Selborne (Roundell Palmer) FRS (Mercer); Chairman of the Executive Committee was civil engineer, Sir Frederick Bramwell FRS (Goldsmith).

61. Philip Magnus (1842–1933) managed well to get both the Finsbury Technical College and the Central Technical College up and going, but ran into difficulties later with his autocratic handling of the professors. See Guildhall Library, CGLI, MS 21868/20. See also, Frank Foden, *Philip Magnus: Victorian Educational Pioneer* (London, 1970).

62. Guildhall Library, CGLI, Ms 21813: Council minutes; (Donnelly's letter of 4 August 1877 was copied into the minutes).

63. Guildhall Library, CGLI, Ms 21813. Reports from these and other consultants are on file.

64. Philip Magnus suggested that the Central be called Gresham College, and that the old name be retained, but this was rejected. See Guildhall Library CGLI, Ms 1868/20; Philip Magnus correspondence file; letter to John Watney, 5 January 1883. Gresham College was used as the administrative centre for the new Institute and remained the head office of the CGLI, albeit in new post-war premises, until 1958.

65. Guildhall Library, CGLI, Ms 21,834. This file contains reports on various possible sites.

66. For these developments see chapter 3.

67. For Drapers letter of resignation see Council minutes, 2 May 1888, *op. cit* (62).

68. Guildhall Library, CGLI, Ms 21, 864/10; John Watney letters file; letter from Owen Roberts, 21 July 1879, Clothworker's Hall (emphasis in original). Affiliation with both University College and King's College was considered.

69. Guildhall Library, CGLI, Ms 21,864/4; John Watney letter file; letter from Frederick Bramwell, 1 January 1880.

70. Council chose the presidents of the Royal Society, Chemical Society, Institution of Civil Engineers, and The Society of Arts. But the presidents served only *ex officio* on the Board of Governors and Council and were not given committee rights.

71. Guildhall Library, CGLI; MS 21,810. From *Constitution*, 1880.

72. Guildhall Library, CGLI MS 21,834, October, 1879; memo from E. C. Robins to the executive committee. Robins, an architect who specialized in colleges of science, was one of the governors and a member of the Dyers guild. He promoted the purchase of Kensington House (which he hoped to renovate) as the site for the new institution. Fortunately the purchase

was not made since, judging from contemporary descriptions, the building would have been unsuitable.

73. Alfred Waterhouse (1830–1905) who designed both the City and Guilds building and the Natural History Museum was born into a Quaker merchant family in Liverpool. He articled in Manchester and travelled widely in France and Italy before returning to work in Manchester. He came to London seeking further commissions in 1865.

74. The turret clock has been preserved and can be seen in the foyer of the Department of Mechanical Engineering at Imperial College.

75. Frederick Abel and William Perkin, both guild members and both graduates of the RCC, and T. H. Huxley who was not associated with the guilds, were much involved in the selection.

76. Henry Edward Armstrong FRS (1848–1937) had been a student of A. W. Hofmann and Edward Frankland at the Royal College of Chemistry. After two and a half years studying with Hermann Kolbe in Leipzig, he returned to London, and to both a lectureship at St. Bartholomew's Hospital and a professorship at the London Institution. He was the first professor of chemistry at Finsbury Technical College.

   William Edward Ayrton FRS (1847–1908) was educated at University College London. In 1867 he came first in the Indian Government Telegraph Service examinations. Before going to India he briefly studied electrical engineering with Sir William Thomson in Glasgow. From 1873–8 he was Professor of Natural Philosophy and Telegraphy at the Imperial College of Engineering in Tokyo before returning to London and work as a consultant and with the electrical industry. He was the first professor of physics at Finsbury Technical College.

77. Olaus Henrici FRS (1840–1911), a native of Schleswig-Holstein, began work as an apprentice and attended the polytechnic in Karlsruhe before enrolling at Heidelberg as a mathematics student. He gained a PhD in mathematical physics under G. R. Kirchhoff. In 1865, stranded in London with no money, he was given accommodation by a friend. He studied in the reading room of the British Museum while earning a living as a tutor, and slowly made his way in mathematical circles. He became an assistant at University College under T. A. Hirst whom he succeeded as professor. Henrici married the niece of A. W. Kennedy, professor of engineering at University College. Kennedy applied for the position at the Central won by Unwin. E. A. Nehan, a technician in Henrici's department, remembered him as 'very testy' and given to dressing down the staff. He also remembered Henrici wearing a round tassled smoking cap and holding a six inch cigarette holder. B/Nehan, farewell speech, 1952.

William Cawthorne Unwin FRS (1838–1933) was the son of William Jordan Unwin, Congregationalist minister and Principal of Homerton College. This college for non-conformists had City roots and was originally (1730) located in Plasterer's Hall. Unwin was educated at the City of London School and, in 1855, matriculated in chemistry at the University of London (that he did so in chemistry is likely a reflection of the excellent chemistry teacher, ex RCC student Thomas Hall). In the same year, through the non-conformist network, he was recommended to the Manchester entrepreneur and engineer, Sir William Fairbairn, who was looking for a personal assistant. Unwin learned much from Fairbairn, but took an external University of London BSc and sought an academic career. After a brief period as lecturer at the School of Naval Architecture in South Kensington he moved, in 1872, to the professorship in engineering at the Royal Indian Engineering College at Cooper's Hill near Egham in Surrey. He was appointed to the Central Institution in 1884. See E. G. Walker, *William Cawthorne Unwin* (London, 1938).

78. Guildhall Library, CGLI MS 21,819 vol.1; Minutes of Sub-Committee A (Central Institution), 29 November 1880.

79. ICA; CGLI, printed letter from Philip Magnus to the members of sub-committee A, 9 October 1883.

80. Several were Cambridge wranglers, namely students who had placed well in the University of Cambridge mathematics examinations. Students from Oxford and Cambridge came for vocational training in order to find work.

81. CGLI, Ms 21, 816 Council minutes; 25 October 1886. Report on letter from Armstrong to Watney (emphasis in original). Armstrong was annoyed that he had to account for every minute of his time and was not trusted when he more than fulfilled his contract.

82. CGLI, Ms 21,868/8; memorandum addressed to Mr. Soper, June 12, 1889. This row is discussed in *The Times*, 23 February 1910. The renamed City and Guilds (Engineering) College was by then a part of the new Imperial College. Armstrong had been the principal examiner for the London University chemistry examinations in the late 1880s and early 1890s. In the early twentieth century he won some concessions and more emphasis was placed on practical examinations for the BSc.

83. See Walker *op. cit.*(77), ch. 3. See also Herbert McLeod diary (ICA), entries for the 1870s, 80s and 90s. The diary is a good source of information on Unwin's activities at Cooper's Hill and the Central. Unwin and McLeod were close friends.

84. See W. H. Brock, *H. E. Armstrong and the Teaching of Science, 1880–1930* (Cambridge, 1973).

85. Geoffrey Crossick, *An Artisan Elite in Victorian Society: Kentish London, 1840–1880* (London, 1978), 135.

86. The Normal School was renamed the Royal College of Science (RCS) in 1890 (see chapter 3). The chemistry department (descendent of the RCC) continued to exert much influence in the RCS.

chapter three

The Founding of Imperial College

Introduction

The motivation for founding Imperial College will be discussed in this chapter, and some attempt made to understand the institution in terms of its emergence. The college was founded by the state in 1907. It was the materialization of a bureaucratic dream that, to a degree, was resisted for many years. People continued to identify with the old colleges and the formation of a new Imperial College identity was not easily achieved. Even today, if asked to identify Imperial's founding heroes, people at the college are more likely to name Prince Albert, Hofmann, Huxley, or De la Beche than the creative bureaucrats of 1907.[1] As this chapter shows, long before Imperial was founded the need to do more for technical education in Britain was gaining recognition. By the early twentieth century it had become a major political issue.

The Move to South Kensington

In 1870 a Royal Commission, chaired by the Duke of Devonshire, was appointed to enquire into 'Scientific Instruction and the Advancement of Science'. Its first report, one of five, was published in 1871 and focussed on the Royal School of Mines (RSM), the Geological Survey of Great Britain and Ireland, the Mining Record Office and the Museum of Practical Geology.[2] The RSM was seen by Commission members, including the scientists Thomas Henry Huxley, John Lubbock, George

Gabriel Stokes and the Commission's Secretary, J. Norman Lockyer, as the possible nucleus for a major national scientific and technical school. However, as the report noted, the buildings on Jermyn Street, and on Oxford Street where the associated Royal College of Chemistry (RCC) was located, were inadequate to such a vision. The Commission further claimed that there were no essential ties between the four institutions being considered in its report, and recommended that the building on Jermyn Street be given over to the Survey and to the Museum, that the Record Office be moved to the Statistical Department of the Board of Trade, and that steps be taken to re-house the RSM and the RCC, both of which had performed well over a twenty-year period and needed more space. Because of crowded conditions, many eligible students were being turned away, and the professors were demanding better facilities for their work.

One of the Commission's suggestions was that the two institutions be accommodated in the Royal School of Naval Architecture and Marine Engineering building then under construction in South Kensington. Since there were very few naval architecture students, and the projected building was rather large, this was a reasonable suggestion. But it was opposed by the Director of the RSM, Sir Roderick Murchison, who saw it as the first step down a slippery slope leading to the downgrading of mining and metallurgical work within a larger college of general science. In his view, a move to South Kensington might well serve the needs of future science teachers but not those of people wishing to become geologists or join the mining and metallurgical industries — industries he saw as vital to Britain and the Empire. But the science professors at the RSM, among them Huxley and Edward Frankland, wished to move. Their view was that for advanced research to be carried out in Britain it was first necessary to have good science instruction in secondary schools and that the expanded college should be used to train teachers. Further, the idea of having a nucleus of science and art education in South Kensington was said to have been a pet scheme of Prince Albert.[3] This claim gave some extra rhetorical force to those who supported the move.

Murchison died in 1871 and, instead of finding a new director, the government appointed Warington Smyth, Professor of Mining, as chairman of the council of professors. One year later the council advised the government that a move solely of the departments of physics, chemistry and natural history to South Kensington would be desirable, since the science professors had stated that they could no longer do their work under existing conditions. As the professor of chemistry, Edward Frankland, put it 'I have again … to state that the college has been constantly overcrowded with most urgent and diligent students who, for lack of space and the appliances of modern chemical laboratories, have received very inadequate instruction'.[4] In response to this and similar pressure the government agreed to the move and the departments of physics, chemistry and natural history (to be renamed biology by Huxley) formed the nucleus of a new institution, under the Government Department of Science and Art. The Department had been founded in 1853 with its main offices in South Kensington. It was responsible for science and art education in schools, and for instructional museums. At the time of the proposed move, the Director of Science was John Fretcheville Dykes Donnelly. A major in the Royal Engineers, he had joined the Department in 1859 after service in the Crimea. The head of the Department, Henry Cole, put Donnelly in charge of preparing the ground purchased with the profits from the Exhibition of 1851. The first task was to plan for a major art museum and some other buildings envisaged by Prince Albert. Donnelly remained on the Reserve List, retiring from the army only in 1887 with the honorary rank of Major-General. Many of the civil servants who worked with Cole, and later with Donnelly, and those working for the Board of Works, another major government presence in South Kensington were, like Donnelly, ex-military men.[5] After the First World War a special effort was made to hire more ex-servicemen at South Kensington. This legacy was important to the future history of Imperial College which, for much of the twentieth century, continued to employ both ex-military personnel and retired colonial officers in administrative positions.

The site proposed for the School of Naval Architecture has an interesting early history, illustrative of the rather haphazard planning for the newly acquired South Kensington lands.[6] It was originally intended for a building to house the art collection of the Department of Science and Art but, by 1865, the plans had changed. The naval architects were already situated in South Kensington, occupying some dilapidated buildings on the recently acquired land; their school desperately needed rehousing. In 1866, Henry Cole proposed a grandiose joint college of science and naval architecture, a scheme approved by the Treasury which allowed construction to begin slowly.[7] The building was to include large laboratories and there was consultation on their design with A. W. Hofmann, until 1865 professor at the RCC, and then with his successor Edward Frankland. Both men suggested some German models to Cole. By 1872, when the building was almost completed, the Government was faced with the Devonshire Report and the space crisis at the RSM. The result was that the new building was given to the splinter group, led by Huxley, and the westward move began.

The departments of chemistry, physics and natural history moved gradually to South Kensington, beginning in 1873. Together with new departments of mechanics and mathematics, they formed a new institution in 1881, named by Huxley the Normal School of Science. This French-inspired name was meant to imply that the college was to give prospective teachers practical training in science, but it was a name hated by the students, and by some of the professors, who demanded it be changed. The School became the far more comforting Royal College of Science (RCS) in 1890. But what of the RSM? Like Murchison, the professors in the traditional mining subjects had opposed the splitting up of the School, and were especially disturbed when geology also went to the Normal School, a move rationalized by the fact that it was a 'pure' science and that the Geological Survey was, in any case, being split off from the RSM. Further problems arose when the lease on the Jermyn Street building, used to house the metallurgical laboratory, expired. This raised the question of whether the entire School should move to South Kensington. John Percy, the professor of metallurgy, was incensed at the idea of moving, pointing out that the Office of Works had anticipated expiry of the lease years earlier and had reserved a plot of Crown land nearby for the building of a new laboratory. But his opposition was to no avail; the tide was turning in favour of South Kensington. Percy resigned his professorship in 1880 and his department joined the others in the westward migration.

The mining and metallurgical departments did not join the Normal School of Science, remaining as the Royal School of Mines. They did, however, share a Dean, T. H. Huxley, who was elected by the professors of both institutions. By 1882 they shared a joint title, Normal School of Science and Royal School of Mines. The two were often referred to as the Science Schools and, according to Huxley, were the only places in England where science education was 'trammelled neither by parsons nor litterateurs'.[8] They shared the building planned for the naval architects which, today, is the Henry Cole wing of the Victoria and Albert Museum.[9] Warington Smyth, however, continued to give his mining lectures in Jermyn Street. The professor of geology, John Wesley Judd, like his subject, was a link between the two sets of disciplines and he succeeded Huxley as Dean in 1885. William Chandler Roberts (later Roberts-Austen) succeeded Percy as Professor of Metallurgy.

By the mid 1880s the building had become overcrowded. While far too large for the naval architects, who never moved in but went instead to Greenwich, it was far too small to accommodate the various departments of the Royal School of Mines and the Normal School of Science. In South Kensington prospective students were being turned away and, for the first time, entrance examinations were set to decide student entry. The Mining Department overflowed into a number of temporary sheds on the opposite (western) side of Exhibition Road. Metallurgy, cramped in the building's basement, soon overflowed in the same direction. Geology moved to the Western Galleries of the South Kensington Museum. In 1885 the City and Guilds of London Institute opened its new Central Institution a little further north, on the west side of Exhibition Road. Chemistry, physics and mathematics were to be taught at the Central and some engineering was taught at the RSM. There was therefore some overlap in what was offered at the three colleges. The rationalization of technical education in South Kensington, as well as in the nation more generally, was seen as a problem. By the late century this problem was widely seen as critical, and the public was becoming increasingly aware of the issues.[10]

The Politics of Technical Education

Some political developments of the late nineteenth century were to change the basis by which higher technical education was funded, and to affect decisions taken on the future of the South Kensington colleges. New County Councils were created by the Local Government Act of 1888. The Technical Instruction Acts of 1889 and 1891 allowed local authorities to use the rates system to raise taxes, within limits, in support of technical instruction. The money raised could be used only for institutions approved by the Department of Science and Art — though this department only had a short time to live. Its role was taken over by the new Board of Education in 1899. The 1890 Local Taxation (Customs and

Excise Act) allowed some of the money from customs and excise duties to be used by county councils to further subsidize technical education. In 1900–01 this tax, known as 'whisky money' raised £863,847 for science and engineering education in Britain.[11]

The new role for local government was taken seriously by the London County Council (LCC) which set up its own Technical Education Board with Sydney Webb as its chairman. This Board was created, in part, as a result of lobbying by London manufacturers who claimed that there was a dearth of local technological expertise and that they were unable to hire the employees they needed. The Board issued a report in 1902 stating the need for improvements both in secondary and tertiary technical education and the LCC began to support the polytechnics, some new ones opening during this period. The Board report was widely discussed, also within University of London circles. There were some within the university, including Webb who was on the university senate, who wanted a major technical college to become part of the newly redefined teaching institution.[12] Looking toward some of the more successful European technical schools, such as the Technische Hochschule in Charlottenburg, they wished for something similar to strengthen the University of London's science and engineering faculties.[13]

The University Chancellor, Lord Rosebery, sent an open letter on this subject to the LCC in 1903, agreeing with the Technical Education Board that something needed to be done for London. He wrote, 'it is little short of a scandal that our own able and ambitious young men, eager to equip themselves with the most perfect technical training should be compelled to resort to the universities of Germany or the United States'. He further bemoaned the fact that students from the Empire chose to go to Paris or Berlin rather than London. He mentioned that a large sum of money had been offered by the firm of Wernher, Beit and Company to support a new technical college; that further funding had been offered by 'public-spirited London citizens', and that the 1851 Commission might provide a site in South Kensington.[14] He asked the LCC to support the university in creating a new institution and proposed that it provide an annual grant of £20,000 for this venture.[15] The sentiments expressed by Lord Rosebery were shared by many who were concerned that Britain was losing her competitive edge to the United States and Germany. British university education was not then geared to educating professionals in technical and applied scientific fields, though some provincial universities were making a start in that direction. That Rosebery mentioned Wernher, Beit and Co. as a possible source of funding had much to do with the politicking of Richard Burdon Haldane.[16] On one of his many visits to Germany he, too, had been inspired by the Technische Hochschule at Charlottenburg. It was, he believed, the kind of institution needed also in London and he had discussed possible funding with Beit and Wernher.[17]

While there was agreement to work with the science colleges in South Kensington, the idea then was not to unite them, but to create yet a further institution. The university already had a toehold in South Kensington, having taken over part of the Imperial Institute for offices, and an attached temporary structure for its examination hall.[18] It now saw an opportunity for further expansion. Rosebery also chaired the group of trustees overseeing the monies donated by Wernher, Beit and Co.[19] However, not all of the trustees agreed with him in wanting the new college, if there were to be one, to be attached to the university.[20] Perhaps the keenest for a new University of London college was Richard Burdon Haldane. In his autobiography Haldane notes that the House of Commons of the 1890s was 'of the old-fashioned sort and did not care much about education, especially of Continental types.'[21] Haldane, however, was an enthusiast and had been an important player in the passing of the University of London Act in 1898. He wanted to see the university go from strength to strength. 'The University of London … ought to be the chief centre of learning in the entire Empire, perhaps the chief centre of learning for the world … such as can come only in a great capital at the heart of a great country'.[22] But a new Charlottenburg type of science college in South Kensington, which he then favoured, was opposed by the new Board of Education which had already embarked on the construction of a major new building for the physics and chemistry departments of the Royal College of Science.[23] Besides, these subjects were being taught well also at the Central Institution. Robert Morant, the Board's Permanent Secretary, stated that the creation of a new college with yet further overlap would be a serious waste of money and resources.[24] London County Council, however, responded enthusiastically to Lord Rosebery's letter and wanted the scheme to go forward. The Council offered an annual grant of £20,000, subject to only a few minor conditions, in support of the scheme.

While the RCS and the Central Technical College were thriving in South Kensington, the RSM was faltering. It, too, had its lobby. In

1901 the Institution of Mining and Metallurgy formed a committee, with Hennen Jennings as chairman, to look into mining and metallurgical instruction in Britain, and other countries, with the aim of raising standards at the RSM. The Jennings Report was critical of teaching practices at the School, the cramped conditions and the lack of proper government funding, and it made a number of suggestions. These included higher entrance standards, training in mechanical and electrical engineering, and in the commercial aspects of mining, and enforced practical training during vacations. The committee wanted the government to give grants to the teaching staff to enable them to travel around the world, to view the latest in mining and metallurgical practice, and they wanted more overall funding for the institution. The report concluded that, 'recognizing the imperative necessity that the Central Mining and Metallurgical training school of the Empire should be equal to the best institutions of its kind in the world, the Institution of Mining and Metallurgy is prepared to render all the assistance in its power, in cooperation with the Board of Education, in order to attain this object.'[25] The Institution wrote also to various government officials and to Joseph Chamberlain, Secretary of State for the Colonies. It stated that, having looked at education elsewhere, 'it has been found that the majority of foreign schools devote far greater attention to actual practice and working instruction and also afford a wider and more thorough scientific training than is obtainable in any Mining School within the British Empire'. The letter also reveals especial concern with having the 'right' personnel for South African gold mines where large numbers of non-British mining engineers were then being employed.[26]

At roughly the same time as Lord Rosebery was fronting the lobby for a new technical college, Sir William Preece, who had been the chief engineer at the General Post Office and who, in 1903, was President of the Society of Arts, had been asked to chair the Bessemer Memorial Fund Committee. This Fund, sponsored by a number of major businessmen in the heavy industries, notably in iron and steel, was to raise money in memory of Henry Bessemer. The plan was to support a new mining and metallurgical school at the University of London. Haldane was involved with this too, but claimed he saw no conflict of interest between this and his working also for a more general technological college. Once again there was the prospect of university competition with yet another of the South Kensington colleges.[27]

The Founding of Imperial College

Clearly with all these proposals in the air the government needed to act. In April 1904 Lord Londonderry, President of the Board of Education, appointed a committee,

> to enquire into the present working of the Royal College of Science, including the Royal School of Mines; to consider in what manner the staff, together with the buildings and appliances now in occupation or in course of construction, may be utilized to the fullest extent for the promotion of higher scientific studies in connection with the work of existing or projected institutions for instruction of the same character in the Metropolis or elsewhere; and to report on any changes which may be desirable in order to carry out such recommendations as they may make.[28]

Sir Francis Mowatt, recently retired Permanent Secretary at the Treasury, was to chair this committee.[29] He was also one of the Wernher Beit trustees who had preferred that any new technical college be independent of the university. Also on the committee were Haldane, perhaps the keenest booster of a new college to be attached to the university, Sir Julius Wernher, and Sir William Abney, a long time South Kensington bureaucrat who had recently retired as Principal Assistant Secretary at the Board of Education.[30] Sydney Webb, in his role as chairman of the London County Council's Technical Education Board, was on the committee, as were a number of presidents of learned and professional societies.

The Board of Education received the report it wanted, namely one recommending that no new college be built in South Kensington. Inadequate though the existing space and resources were, some arrangement was to be made to bring the three colleges already in South Kensington together, and to build on what was already there. The committee was divided on whether the merged institution should have anything to do with the University of London. But before any plans for unification could go forward, agreement would have to be reached with the City and Guilds of London Institute which owned and ran the Central. As noted in chapter two, the Institute had been reluctant to build its college in South Kensington, in part because the area was a government enclave.[31] City independence was important to many in the guilds and it was prudent to expect opposition to any merger with the government-run institutions. Mowatt became ill and, while remaining on the committee, handed over the chairmanship to Haldane. Haldane, a consummate politician, had seen the direction the wind was blowing and had persuaded the various donors that the new government plans were in the national interest, and that they should redirect their grants and donations accordingly. The Board of Education, pleased with the committee's decision not to recommend the founding of a new institution, held out a carrot to the donors. It was stated that the Treasury would be asked for major new funding to support the unification of the colleges. Indeed, by November of 1905 the Treasury had offered some capital expenditure and an annual grant of £20,000. Other money, private and public, followed. Notable was a large bequest from Alfred Beit, who died in 1906. Later, in 1912, the college received another generous legacy on the death of Wernher.[32]

The final report of the Mowatt-Haldane Committee was published in 1906. It gives a good description of the state of the colleges just before the formation of Imperial College. It also gives the committee's interpretation of the intentions of Prince Albert, and how the scheme they were proposing would fulfill his vision. The Royal College of Science and the Royal School of Mines had only 283 full-time students between them in 1905–6. In addition, there were roughly as many part-time students. Many of the students had state support. About fifty of the one hundred and fifty students at the Royal College of Science were training to be teachers. Those at the Royal School of Mines wished mainly to become mining engineers or metallurgists. The City and Guilds Central Institution had about two hundred full-time students as well as many part timers. The Committee, in examining the colleges, compared conditions in South Kensington unfavourably with those at the best technical colleges elsewhere. Much attention was focussed on what were held to be superior facilities at the Massachusetts Institute of Technology, Columbia University, McGill University, and the advanced technical schools in Berlin and Zurich. But the committee saw no reason for Britain to blindly copy these institutions; rather it should take from them just what might reasonably work.

In Britain advanced technical education was a relatively new phenomenon. Many people still learned technical skills on the job and, as the committee noted, the views of parents and employers had to be respected. It was also noted that there were many well-educated young people in Britain who could take advantage of an expanded institution giving instruction at an advanced level, and that the nation would profit as a result. Further, the committee saw a great need for research as well as teaching to be carried out in the new college, claiming that 'the opportunities for research in our technological institutions are inadequate to the industrial needs of the Empire, owing not to any want of ability on the part of the professors, but to the fact that much of their time is frequently absorbed in giving of comparatively elementary instruction in Pure and Applied Science.' This had been Huxley's earlier complaint and was behind his wanting a college to train school teachers as a first priority. But the new vision was not for a teacher training college, rather it was for something closer to the North American and European models. What was needed were well equipped laboratories and workshops and that 'the staff should include, at the head of the several specialized branches of the work, men of the first rank in their profession.'[33]

Looking ahead to the new institution the committee saw the number of departments increasing indefinitely as funding allowed — something that seems to have happened. They recommended that, for the present, everything that was currently being taught at the three institutions be continued, leaving any rationalisation of existing resources until later. Further, it was recommended that the new Governing Body be allowed to act without concern for any possible future union with the University. This sticky point was seen as an entirely separate issue to be worked out at a later date.[34] The report also recommended that the government transfer its property associated with the Royal College of Science and the Royal School of Mines to the new college and that care be taken to respect the traditions, and value the prestige, of the Royal School of Mines. The site and buildings of the Central Technical College were to remain the property of the City and Guilds of London Institute.

The recommendation to delay any discussion of incorporation was opposed by the University Senate which meant that Lord Londonderry, and his successor as President of the Board of Education, Reginald McKenna, had to engage in some delicate negotiations.[35] This held up the final decision for a few months, something that led to much speculation in the columns of *The Times*, *Nature*, and other papers. A wide and substantial body of public opinion wanted something to be done for advanced technical education and the articles and correspondence shows concern over the delay. The politicians relied on Sir Francis Mowatt to be the emollient, a role he performed in brilliant fashion, smoothing the way towards a final agreement.[36] But incorporation within the University remained a sticky issue for much of the first twenty years of Imperial College's existence. As is discussed in chapter four, a *modus vivendi* was arrived at in 1926, at which point there was an agreement to statute revisions that would accommodate the needs of Imperial College. These came into effect in 1929 at which point Imperial College formally became a teaching college of the university. However, in the years that followed there were repeated moves to rescind incorporation, and to give Imperial College independent university status.

By and large all the recommendations of the Mowatt-Haldane Committee were realized, but only after lengthy negotiations, not only with the University but also with the City and Guilds of London Institute which did not wish to see the identity, or the name, of its college vanish within the larger body. A draft charter was drawn up in March 1907, and the Charter of Incorporation was officially granted on 8 July, 1907. As it turned out, both the structure and early membership of the Governing Body, as defined by the Charter, had important consequences.[37] In the early years many on the Governing Body were active in running Imperial's affairs on a day-to-day basis.[38] Since this legacy can be detected years later it is worth discussing briefly both the Governing Body's structure and its membership. There were forty members, six nominated by the Crown, four by the Board of Education, five by the University of London, five by the London County Council, five by the City and Guilds of London Institute, four by the professors at the college, two by the Royal Commission of the Exhibition of 1851, and one each by the Royal Society, the Institution of Civil Engineers, the Institution of Electrical Engineers, the Institution of Mechanical Engineers, the Institution of Mining and Metallurgy, the Institution of Mining Engineers, the Society for Chemical Industry, the Iron and Steel Institute and the Institution of Naval Architects.[39] Notably absent were representatives of the learned societies associated with the pure sciences — though scientists were appointed by the other bodies. This was a Governing Body weighted towards engineering and technology, something that had consequences for the kinds of new courses encouraged within the college, and in the prioritization of research. But none of this was clearly seen at the time. Most people were in any case agreed that the college was to focus on applied science, even within the science departments. The college received its mandate in Article II of the Charter, namely 'to give the highest specialized instruction and to provide the fullest equipment for the most advanced training and research in various branches of science especially in its application to industry'[40]. Article I of the Charter defined the name of the new college as 'The Imperial College of Science and Technology', implying that the industries mentioned in Article II were not to be just those within Britain but industries within the Empire as a whole.[41] The idea of serving imperial industry had been accepted also within the three older colleges. As we will see the early professors at Imperial took this mandate very seriously.

I will mention just the most active of the early governors, some of whom will reappear in later chapters. Among them was the Marquess of Crewe (1858–1945), nominated by the Crown and selected by the governors to be their first chairman. He remained chairman until appointed Ambassador to France in 1923.[42] Lord Rosebery was his father-in-law but this did not make him an advocate for incorporation of the college within the University, on the contrary.[43] Also active was Sir Francis Mowatt, another opponent of university incorporation, who had done so much to bring the parties together in forming the new college. Mowatt became the chairman of the college's Finance Committee and negotiated well on the college's behalf with his old colleagues at the Treasury.[44] Gerald Balfour (1853–1945), like his brother A. J. Balfour, was a politician. He had been President of the Board of Trade (1900–05), and was an active Imperial College committee member, appointed chairman of the Executive Committee in 1911. College benefactor, Sir Julius Wernher (1850–1912) was also active on this committee until his death. There was some discomfort in the college with the Wernher and Beit connection. Both had earlier been caught up in the notorious Jameson Raid of 1895, an expensive fiasco which did some damage to their reputations and their financial positions, and both had been denounced in parliament.[45]

Beit appears to have lived the life of a plutocrat. He had a large country house and a splendid mansion in Park Lane filled with art treasures. His brother, Otto Beit, who later succeeded Wernher on the governing body, was a good friend to Imperial, worked hard on many committees, and bailed the college out of a number of difficult financial situations.[46] Another governor, Sir William White (1845–1913), had studied at the Naval Architects School and continued his association with South Kensington. He had a major career in the Admiralty, including as Director of Naval Construction where he played an active role in the designing of ships for Britain's war fleet. Also appointed was Richard Glazebrook (1854–1935), then Director of the National Physical Laboratory, later to become the first professor of aeronautics at the college. Another, Francis Grant Ogilvie, began his career as a science teacher in Scotland, was appointed the first principal of Heriot-Watt College where he was also Professor of Applied Physics. Later he became Director of the Royal Scottish Museum. His fame as an administrator spread, and he was appointed to a senior position in the Department of Science and Art, later becoming Permanent Secretary at the Board of Education.[47] He was the professors' choice for first Rector and, while not chosen, was for many years an active governor.[48] Four professors were on the Governing Body: J. B. Farmer, professor of botany, W. A. Tilden, professor of chemistry, W. Gowland, professor of metallurgy and W. E. Dalby, professor of engineering. Perhaps the most active of the early governors was Sir Arthur Acland whose contributions will be discussed in other chapters.[49]

What is striking in reading the early minutes of the Governing Body, and those of the important college committees, is how much time some of the early governors spent on college affairs. From its start, Imperial College had a governing body of heavyweights, a pattern that continued through much of the twentieth century. Most of the original governors were in their fifties or sixties and most were also volunteers elsewhere, suggesting a culture in which prestige was to be gained by playing a public role in educational ventures. It was clearly worth spending some of one's time in pursuits that were not remunerative. By the later twentieth century the functions performed by these status-seeking volunteers were being performed by paid professionals, and a paid secretariat. Even from the perspective of the early twenty-first century, with our heightened class, feminist, and post-colonial sensitivities, these men appear worthy. Clearly something of the old value system still remains.

The college had been created by the state at a time when anxiety over Britain's industrial competitiveness was high, and when there was widespread demand that more be done for technical education in the country. It drew students from a wider cross section of the public than did other universities — though few came from very poor families. As mentioned, the City and Guilds College was seen by its founders as a place to which young men who, in earlier times, might have been guild apprentices, could come to be trained. The Royal School of Mines which had drawn boys from the public schools when it offered geology and was connected to the Geological Survey, had some difficulty in continuing to attract them to its mining and metallurgy courses in the new college. Nonetheless, the RSM had proportionately more students with a public school education than the other two colleges, many of them from families with mining interests. Women were admitted to Imperial, as they had been to all of the colleges before they federated, but few enrolled. Despite this, Imperial in its early years was an interesting site of class mixing and social mobility. Ambitious students, often only a few years after graduation, were able to gain access to important government, military, imperial, and industrial circles, something that for most would have been impossible without connections made at the college.

As we shall see, the early professors took some cues from the successful men on the governing body. They, too, gained much personal prestige by rendering direct government (including military), industrial and colonial service, much of it without remuneration. But, as the century progressed, the academic staff looked more to the learned and professional societies for confirmation of their worth. By mid-century these societies, by offering prestigious awards and fellowships, became the major arbiters of what counted as good in both science and technology. This trend was, perhaps, a consequence of technical expertise becoming more widely distributed, not just in Britain but worldwide. The role of professor had continually to be redefined and, over time, older kinds of voluntarism declined, though committee work of various types increased. True to its roots, Imperial College continued to value research for its utility, whether industrial, military, medical or other. In this the college was ahead of its time. The gentlemanly ethos of the older universities and of the learned societies, in which disinterested scholarship was valued above all else, has slowly given way to an academic culture which is now more generally attuned to the needs of government, manufacturing and commerce. In this, universities are reflective of a larger cultural trend in which new business values have become normative in society. In retrospect we can see that Imperial College was a trendsetter in this regard.

Since its foundation in 1907, a pragmatic and utilitarian approach has been normative.

...

1. This has much to do with the revisionist history of Patrick Linstead who wrote a number of historical articles in which he emphasized the role of Prince Albert as founder. He also overstated the role of Lord Haldane in the early twentieth century. For Linstead on Prince Albert see, for example, a lecture he gave on the centenary of Albert's death, 'The Prince Consort and the Founding of Imperial College', *Nature* vol. 193 (13 January 1962), 107–13. The Prince Consort is described as 'the Founder of the Imperial College itself'. Clearly this was the view of a biased organic chemist who identified the Royal College of Chemistry with Imperial.

2. Royal Commission on Technical Instruction, vol. 1 (London, 1871). For a summary of the Commission Report see letter from Commission members to *Nature*, 30 March 1871, 421. See also, Michael Argles, *South Kensington to Robbins: An Account of English Technical Education since 1851* (London: 1964), ch. 4.

3. Prince Albert had wanted museums and exhibits of the arts and sciences on the Royal Commission site; he had not been an advocate of colleges, on the contrary. T. H. Huxley and some of his colleagues at the RSM were the principal backers of an educational presence.

4. Frankland quoted in Margaret Reeks, *Register of the Associates and Old Students of the Royal School of Mines and History of the Royal School of Mines* (London, 1920), 111.

5. From the start Prince Albert wanted efficient people in charge at South Kensington. He eschewed aristocratic amateurism in favour of people like Cole who had left Christ's Hospital at age fourteen and made his way in the world (Cole had been a major force behind the creation of the Penny Post and the Public Record Office). Cole, the son of an army officer, had many soldiers among his staff. He had been especially impressed by the work of the Royal Engineers in finding efficient solutions to the technical problems associated with putting on the Great Exhibition. 'Officers of the Engineers were the type of men he liked to have around him'. See F. H. W. Sheppard (ed.), *Survey of London* (vol. xxxviii): *The Museums Area of South Kensington and Westminster* (London, 1973), 71–8. Quotation, Anthony Burton, *Vision and Accident: the story of the Victoria and Albert Museum* (London: V&A, 1999), 83.

6. The construction history of the schools and museums in South Kensington is complex. In the 1860s there were a number of national collections housed

in temporary buildings in the South Kensington area. These collections were the forerunners of both the Victoria and Albert Museum and the Science Museum collections. Among them was one of ornamental art that Henry Cole had helped accumulate when in charge of the Board of Trade School of Design at Marlborough House. The School (forerunner of the Royal College of Art) and its collection moved to South Kensington in 1857, into the first major home of what was then called the South Kensington Museum. This building, widely viewed as ugly and industrial, soon acquired the derisive name the 'Brompton Boilers'. By the 1860s the museum acquired a new facade, and a quadrangle in a style roughly reminiscent of the Italian Renaissance. The quadrangle, unlike the Boilers (which were later moved to house the Bethnal Green Museum), still stands as part of the Victoria and Albert Museum. This complex was home both to the museum and art school, and its later extension (originally intended for the naval architects' school) became home to the Royal School of Mines and Normal School of Science. The Cromwell Road extension to the museum, designed by Aston Webb, opened later in 1909. By the 1870s the various national collections had been augmented by articles left by exhibitors at the 1851 and other international exhibitions, and from a number of other sources. Notable were some fine British paintings (now in the Tate Gallery) moved from the overflowing National Gallery (forerunner to the present gallery in Trafalgar Square). The Eastern and Western Galleries of the Museum were constructed in 1871–3, the Eastern Galleries were on the West side of Exhibition Road (originally named Prince's Gate, this road was built through the centre of the site acquired after the 1851 Exhibition). The Western Galleries were close, and parallel, to Queen's Gate. Both galleries extended from just south of the Imperial Institute Road (now Imperial College Road) to Prince Consort Road and coexisted with a number of storage sheds. Two cross galleries also existed, one from roughly where the present entrance to the Science Museum is located and the other just south of Prince Consort Road. The former held Frank Buckland's museum of fish culture for a number of years. The original intention was that the Galleries running north and south would house international exhibitions. But, when these ventures lost money, the galleries were used to house the museum's Indian Collection, the growing national collection of science materials, and some of the overflow from the Royal College of Science. Francis Galton's Anthropological Laboratory was, for a while, located in the Western Galleries. Remains of the galleries can still be seen at the western (Imperial College Road) entrance to Imperial College. A more permanent fixture on the site, the Natural History Museum, a division of the British Museum, was built in 1873–80. The National Training School of Music was opened in 1876 in the house later occupied by the Royal

College of Organists. It was renamed the Royal College of Music in 1892 when it moved into its present building designed by Alfred Waterhouse. The first part of a permanent Science Museum was completed only in 1928. For more on the South Kensington site see Sheppard, *op. cit.* (5), especially ch. 4; Burton, *op. cit.* (5); 'Reports of the Commissioners for the Exhibition of 1851', (7th Report, 1889); Hermione Hobhouse, *The Crystal Palace and the Great Exhibition of Art, Science and Productive Industry: A History of the Royal Commission for the Exhibition of 1851* (London, 2002). See also Sophie Forgan and Graeme Gooday, "A fungoid assemblage of buildings': Diversity and Adversity in the Development of College Architecture and Scientific Education in Nineteenth-Century South Kensington', *History of Universities* vol. XIII (1994), 153–92. It was H. G. Wells who aptly described the local buildings of his student years as a 'fungoid assemblage'.

7. Cole wanted a college of science to complement the (later Royal) College of Art which had already been constructed.

8. Leonard Huxley, *Life and Letters of Thomas Henry Huxley*, vol. iii (London, 1903), 233.

9. The principal architect was Captain Francis Fowke RE, though his plans had to be modified by others after his death in 1865.

10. See Peter Alter, *The Reluctant Patron: Science and the State in Britain, 1850–1920* (Oxford, Hamburg and London, 1987), chapter 2.

11. Argles, *op. cit.* (2), 35.

12. The University of London Act of 1898 redefined the university as a teaching as well as an examining body.

13. It was Webb who coined the term a 'London Charlottenburg'.

14. The Commissioners were themselves thinking about how to provide more for science on the site. See Hobhouse, *op. cit.* (6), chapter 6.

15. Rosebery's letter appeared in *The Times*, 29 June 1903. Archibald Philip Primrose Rosebery, Earl of Rosebery (1847–1929) was a major Liberal politician, Foreign Secretary in Gladstone's government and, after Gladstone's retirement, briefly Prime Minister. He split from the party under Lloyd George. Rosebery was the first Chancellor of the reorganized University of London. According to Beatrice Webb, it was her husband, Sydney, who drafted Rosebery's letter. See Barbara Drake and Margaret I. Cole (eds.), Beatrice Webb, *Our Partnership* (London, 1952), 6.

16. Richard Burdon Haldane, later Lord Haldane of Cloan (1856–1928). Haldane was a lawyer and philosopher who became Liberal MP for East Lothian in 1885. He was a political acolyte of Lord Rosebery and a friend of Beatrice and Sydney Webb. Haldane was a great promoter of university education, not just in London. Within the Liberal Party he was a

major imperialist. He became Secretary of State for War under Campbell Bannerman and helped reshape the Army and War Office after the debacle of the South African War. 'National Efficiency' was a watchword of the time to which he fully subscribed. He headed the War Department also in Asquith's government, but had to resign in 1915 when accused (wrongly) of pro-German sympathies. He became Lord Chancellor in the first Labour coalition government of 1924.

17. See Richard Burdon Haldane, *An Autobiography* (London, 1931), pp. 91–2 and 142–4; Dudley Sommer, *Haldane of Cloan: his life and times, 1856–1928*, (London, 1960); Eric Ashby and Mary Anderson, *Portrait of Haldane at Work on Education* (London, 1974), chapter 3. In my view the authors of this latter book see Haldane's role in the creation of Imperial College as more central than the evidence warrants. Haldane himself tended to think of the college as his creation and, through his own memoir and correspondence, has persuaded many others to believe the same. Haldane played an important, but not pivotal, role in the formation of Imperial College. As will be shown here and in chapter four, in many ways he was seen as a hindrance by those who had different views on the type of institution needed.

18. The large 'temporary' building had been erected as a changing room for Queen Victoria and her retinue when she came to open the new Institute. It came to be known as the Great Hall (this name was later taken over for the large hall in the Sherfield Building). For more on the Imperial Institute see chapter seven.

19. The trustees included people who were important in the early history of Imperial College and later volunteered much of their time to helping the college in its early years. Among them were, Julius Wernher, the Duke of Devonshire, Arthur Balfour, Sir Francis Mowatt and Richard Burdon Haldane.

20. A. Rupert Hall, *Science for Industry; a short history of Imperial College of Science and Technology and its antecedents* (London, 1982), ch. 2.

21. Haldane, *op. cit* (17), 88.

22. Quoted in Sommer, *op. cit.* (17), from a speech given by Haldane to the Old Student Association of the Royal College of Science.

23. This building on Imperial Institute Road (now Imperial College Road) was designed by Aston Webb. The government paid the Royal Commissioners for the Exhibition of 1851 £70,000 for the site which had been valued at £200,000. The site was also to include the Science Museum Library and a building at the north eastern corner for a post office (ground floor) and the Meteorological Office (upper floors). See 'Reports of the Commissioners of the Exhibition of 1851' (8th report, 1911).

24. Sir Robert Laurie Morant (1862–1920) was the chief architect of the important 1902 Education Act. He became Permanent Secretary of the Board of Education in 1903 and remained in the post until 1911. See also Hall, *op. cit.* (20).

25. ABC/8/1; Institution of Mining and Metallurgy, 'Technical Instruction Proposals' addressed to the Council of the Royal College of Science and Royal School of Mines, 9 June 1902.

26. B\ Le Neve Foster; C. McDermid (Secretary of Institution of Mining and Metallurgy) to Chamberlain, 14 November 1902.

27. B/ Le Neve Foster; see Preece to C. Le Neve Foster, 10 February 1903.

28. ABC/7/2; minutes and 1906 report of the Board of Education Committee chaired by Mowatt/Haldane. For discussion of a preliminary report of this committee see 'Imperial College of Applied Science', *The Times*, 3 July 1905.

29. Sir Francis Mowatt (1837–1919) was Permanent Secretary at the Treasury, 1894–1903.

30. Sir William De Wivesley Abney FRS (1843–1920) served in the Royal Engineers and joined the Science and Art Department in 1877. He later became a senior civil servant at the Board of Education. He carried out research in spectroscopy (produced one of the first infra-red spectra), colour vision, and photography, and had a large laboratory in the South Kensington Museum.

31. See also Hannah Gay, 'Association and Practice: The City and Guilds of London Institute for the Advancement of Technical Education', *Annals of Science*, 57 (2000), 369–398.

32. Sir Julius Charles Wernher (1850–1912) was born in Darmstadt and came to London after the Franco-Prussian War to work as a diamond buyer. In 1876 he visited Kimberley, bought shares in a diamond mine, and then land that later became very valuable. His partnership with Alfred Beit began in 1879 and their interests extended to the gold fields of the Witwatersrand. They both gave generously to the college and to other educational institutions. Wernher, a member of the Haldane committee, donated £25,000 to Imperial College and left also a bequest valued at £150,000. Beit left his De Beers shares and £50,000 to Imperial College. In the 1930s when De Beers were in financial difficulties and paid no dividends, the college income dropped by about £5000 a year, causing some difficulty. See G. Seymour Fort, *Alfred Beit: a study of the man and his work* (London, 1932); Jamie Camplin, *The Rise of the Technocrats* (London, 1978), Thomas Pakenham, *The Boer War* (London, 1979), and Raleigh Trevelyan, *Grand Dukes and Diamonds: The Wernhers of Luton Hoo* (London, 1991).

33. *Final Report of the Departmental Committee on the Royal College of Science* (Cmd 2872, 1906).

34. A minority addendum to the *Report* stated that a Royal Commission should be appointed after five years to consider the relation of the new institution to the University. A second addendum, by a yet smaller minority of two, recommended immediate incorporation within the University.

35. Reginald McKenna (1863–1933), Liberal politician, was President of the Board of Education at the founding of Imperial College in 1907. He later held other cabinet posts including Chancellor of the Exchequer (1915–16). After retiring from politics he became Chairman of the Midland Bank.

36. Government gave its consent to the new college in 1905 but the parties, notably the CGLI, had difficulty in coming to a final agreement.

37. For a report of the first meeting of the Governing Body, the election of the chairman and founding of committees, see *The Times*, 13 July 1907. See also GB minutes, 1907. The interests represented on the governing body remained roughly the same until very late in the twentieth century. The Governing Body was unusual among university institutions in the large number of governors nominated by external bodies. As Lord Oxburgh put it 'a full Council meeting was like a gathering of the United Nations' (personal communication).

38. It is interesting, for example, that the advanced prospectus for the year 1908–9 lists the names of the Rector, College Secretary and members of the governing body but does not list any of the professors. And, very much later, in the prospectuses of the 1940s, governing body members and their accomplishments are stressed over those of the professors, though professors were listed by that time.

39. That there should be five representatives from the University of London was contested but finally agreed on. Changes were made to the college charter in 1916 so as to allow representatives from countries of the empire to sit on the Governing Body.

40. Charter of Incorporation, 8 July, 1907. The University of Birmingham had a similar clause in its first charter. Joseph Chamberlain had wanted that university to specialize in science applicable to industry.

41. That the new college was to serve imperial interests was implied from the start, but its name emerged only slowly. In early discussions the proposed college is referred to in many ways including, the 'New' or 'Royal' college, or simply the 'College of Technology'. The Board of Education report of 1906 refers to the 'Imperial' college even though the title of the report names it the 'Royal College of Science'. The latter was clearly impolitic given that the C&G and RSM would never have agreed to the name. In the end, the

name Imperial College of Science and Technology was chosen by Morant and his bureaucrats at the Board of Education.

42. Robert Offley Ashburton Crewe-Milnes, Marquess of Crewe (1858–1945), was educated at Harrow and Trinity College, Cambridge, had been a viceroy of Ireland and was to become Colonial Secretary and Leader of the House of Lords (1908–11) at the time when the South African Union was formed. He then became Secretary of State for India and, in 1916, was for a short period President of the Board of Education. In 1917 he became chairman of the London County Council. He was a keen supporter of the new Imperial College. See James Pope-Hennessy, *Lord Crewe: The Likeness of a Liberal* (London, 1963).

43. See his speech on this subject, given at the annual dinner of the Institution of Mining and Metallurgy; *The Times*, 13 March 1914, p. 8. He wanted Imperial to become a separate technological university.

44. This was no easy task as government support was not always reliable and there was much debate over how much the government, and how much the London County Council, should be supporting the college in its early days. See GB correspondence with Board of Education 1907–12.

45. Jean Van der Poel, *The Jameson Raid* (Oxford, 1951). Beatrice Webb, who wrote disparagingly of imperialist capitalism, snobbishly noted, 'What a significant fact was the appearance of the South African millionaire gold-diggers, who dominated London "society" as well as the City …. I must admit that even the Webbs accepted their gracious hospitality in return for their benefactions to the London University, the London School of Economics and the Imperial College of Science!' (Webb, *op. cit.* (15)). Some people at the college had similar attitudes. But not former Normal School student, H. G. Wells, who painted an unflattering and poorly disguised portrait of the Webbs in his *The New Machiavellians* (London, 1911).

46. Sir Otto John Beit (1865–1930) was born in Germany and, like his brother, had a number of business interests in Southern Africa. He founded the Beit Memorial Trust for Medical Research, and established the Beit Fellowships at Imperial College in memory of his brother Alfred. He was a member of the Governing Body, 1912–30, and was created a baronet in 1924.

47. Sir Francis Grant Ogilvie (1858–1930) was also Director of the Science Museum (1911–1920), and Principal Assistant Secretary, Department of Scientific and Industrial Research (1920–1922).

48. B/ Gerald Balfour; correspondence: A. William Tilden to Balfour, 20 December 1907. letter stating that Ogilvie was approved by the professors for Rector.

49. Arthur Herbert Dyke Acland (1847–1926), baronet and Liberal politician, had a fine record of supporting educational reform. He had been President of the Privy Council Committee on Education, and a member of the cabinet in Gladstone's last government. He was responsible for the Science and Art Department, organized the schools inspectorate, and introduced compulsory playtime into the school day, He negotiated the Regent's Park site, and was instrumental in the founding of Bedford College. Acland retired from active politics in 1895 but continued to devote himself to educational causes. During the First World War he was Deputy (Acting) Rector of Imperial College.

chapter four

Governance and Innovation, 1907–43

Governance and Early Growth

After coming together in 1907 to form Imperial College, the three older colleges retained much of their identity: their names, their staff, and their diplomas.[1] The work of forming an identity for Imperial College began only slowly and the early Rectors and Governing Body had to think creatively about how to proceed. The assets of the new college, not large, were valued at £986,664 and its income from ongoing grants and fees was approximately £65,000.[2] It possessed just three buildings, one shared by the Royal College of Science (RCS) and the Royal School of Mines (RSM) on the east side of Exhibition Road (the Science Schools building),[3] the City and Guilds (C&G) College on the west side of Exhibition Road, and the new building for physics and chemistry on Imperial Institute Road (the RCS building). The latter housed also the college administration, the Science Museum Library, a book and stationery stall run by Lamley and Company[4], a refectory, and a few other services. The Science Schools building housed the biological sciences, the RCS mathematics department, and the mining and metallurgy departments of the RSM. The geology department, then in the RCS, occupied the Western Galleries of the museums complex.

By 1908 the Governing Body was ready to appoint a principal officer for the college and, turning to Scottish and European models, decided on the title Rector.[5] Henry Taylor Bovey, a civil engineer and Dean of the Faculty of Engineering at McGill University, was chosen.[6]

He resigned two years later because of ill health and was succeeded by Sir Alfred Keogh.[7] Bovey's appointment led to a lapse in the joint deanship of the Royal College of Science and the Royal School of Mines, but the deanship of the City and Guilds College continued. Alexander Gow was appointed the first College Secretary and, by May 1908, a salary scale was agreed for both academic and support staff.[8] The academic scale began at £150 for demonstrators, £300 for lecturers, £400 for assistant professors (later readers) and £800 for professors. Lab boys were paid 10 shillings per week in their first year, rising to 15 shillings in their third year, and typists one shilling per hour. The top of the professorial scale was £1000 and six of the continuing professors were paid that amount.[9] After the First World War, there was a major inflation and a new salary scale had to be introduced. Junior staff members saw their salaries increase by about one-third, though not in real terms. The original scale proved unsustainable and professors appointed later were paid less (in real terms) than those in office in 1907. During the economic depression of the late 1920s and early 1930s, senior staff had to take pay cuts.

Just two women were on the professional staff at the start; Martha Whiteley was appointed demonstrator in chemistry, and Margaret Reeks, daughter of an earlier registrar of the Royal School of Mines, was appointed Draughtsman.[10] Reeks soon asked for her title to be changed and the governors agreed to her being called Technical Artist. Her role was to give some technical drawing classes to students, prepare lecture illustrations on large calico cotton sheets which were displayed in the front of classrooms, and to help with illustrations for staff and college publications. It is a minor but interesting point that with only two women on the teaching staff, and only about three women students per year, the governing body was soon dealing with a request for more women's lavatories in all three buildings, suggesting that there was a growing number of women research and support staff hidden from the records. The minutes also reveal that before the First World War a disproportionate number of the women students were given fee remissions and special grants to attend courses. This was because of a widely held perception, which may well have been true, that the women were poorer than the men, and because of a desire to see more women students at the college.[11] The number of women students increased slightly during and after the First World War, to roughly 12 per year by 1930 (most of them in chemistry and botany), but then declined.[12] In 1938 only 11 women were among the 1094 full-time students. The fact that the college as a whole still attracted only few women drew criticism from the University of London and was a much discussed issue during the 1920s.[13] In 1940 the total student number dropped to 672 due to military service and to the fact that the college took no first year students that year. By 1943 the college had resumed all of its first year courses and student numbers reached 1026, including 29 women.[14]

Until after the Second World War most undergraduate courses at Imperial College were based on a four-year progression. Students with little background in their chosen field took the first year courses, those better prepared entered the second year programme. Intermediate examinations were given at the end of the first year and associateship examinations at the end of the third year.[15] Students could use the fourth year for advanced (post-associateship) work and earn the newly created Diploma of Imperial College (DIC).[16] The first diplomas were presented in 1912 when the largest number (eight) went to students who had completed an advanced course in railway engineering under W. E. Dalby, Professor of Mechanical Engineering and Dean of the C&G College.[17]

In 1909, discussion over the role that the City and Guilds of London Institute (CGLI) would play in the running of its engineering college came to a head. A body, known as the Delegacy, was set up to oversee the interests of the C&G College and there was much debate among the various guilds as to representation on this body. The first Delegacy had nineteen members of which eight were appointed by the Imperial College governing body, eight by the CGLI and three by the Goldsmiths Company.[18] The quorum, however, was only six. It was agreed that the Delegacy would be responsible for academic decisions having to do with the engineering programme, would make all official announcements regarding the C&G College, appoint its academic staff, and collect student fees. The general supervision of work, and of student discipline, would be the responsibility of the Rector. This division of responsibilities was a recipe for future problems. Already in 1911, Keogh complained to the Delegacy that the C&G professors treated him as their servant. He insisted on further control over them and over the courses offered.[19] The Delegacy later relinquished much of its power to the Rector but lasted, albeit in diminished form, until 1976.[20] In addition to a typical Board of Studies, the CGLI insisted that there be a separate Engineering Board to advise the Rector on academic matters. This was in part because guild members saw their college in a different light from the other two colleges. They wanted an emphasis, though not exclusive, on undergraduate education, and that help be given to the type of young person who might earlier have been a guild apprentice. The Engineering Board was to ensure that this vision not sacrificed to postgraduate education, that the courses met the requirements of the professional bodies, and that the professors kept up ties with prospective employers.[21]

The early governors, and not just those who served also on the Delegacy, had a very hands-on approach to the new college. The fact that the first Rector, Henry Bovey, was a sick man made their work more onerous, but even after Sir Alfred Keogh was appointed the governors continued to carry out much of the administrative work. In 1911 Keogh sought a major reorganization in governance procedures. He asked for two small executive committees to be formed in order that decision-making become more streamlined. The Governing Body agreed, and an executive committee and a finance committee were elected. Many of the most active governors served also on these committees. For example, Gerald Balfour, chairman of the executive committee, and Francis Mowatt, chairman of the finance committee, continued to put in many hours of work. The governors had further obligations when, in 1914, Keogh was given leave for war work. One problem they faced was that fees were not bringing in as much money as had been projected, a problem exacerbated once the war began. Since the college was not attracting as many students as it had hoped to some of its courses, the professors pressured the governing body for more publicity. In 1913, Otto Beit agreed to work on this; he also financed three fellowships of £150 per year for research students.[22] In 1914 it was decided to allow the children of staff members to pay only half fees, and more women were encouraged to apply. During the war the first scholarships intended solely for women were donated by *The Common Cause*, voice of the National Union of Women's Suffrage Societies.

Discussions on expansion and future land acquisition began immediately after federation, both with the government and with the 1851 Exhibition Commissioners. Slowly more land was released which allowed for the planning of new buildings. Since the governors wanted to build social as well as academic bridges between the colleges, it was early decided to build a student union.[23] In addition, the governors were actively seeking funds to build a hostel. At the end of the war they turned down an offer to lease Crosby Hall in Chelsea. The London County Council offered the college a five-hundred year lease, at a low rate of £141 per year.[24] A large sum of money would have been needed to convert the building for hostel use and the governors preferred to think in terms of building a hostel closer to the college. It was, perhaps, a missed opportunity. More serious was the failure, in 1919–20, to purchase the old lodge site and land at the north-west end of Exhibition Road, land now occupied by the Royal Geographical Society, and by a block of flats. The college made an offer to buy the lodge and land but then had second thoughts, offering as an excuse the existence of an old covenant. According to this, access to the land could only be had from Kensington Gore and, by implication, not Exhibition Road. The college refused to renew its offer even after the government stated that the covenant would be declared void by act of parliament. Sir Arthur Acland thought his fellow governors were making a serious mistake in fearing to invest for the future and stated that later generations might well be critical of this lost opportunity.[25]

In addition to the Union, two other major buildings had their start at the beginning of the war and were ready for occupancy some time after the war ended; one was an extension to the C&G College, largely funded by the Goldsmiths' Company.[26] An extension was urgently needed since the number of students working in the C&G Waterhouse building had exceeded its capacity already by the 1890s. The other new building was for the RSM, made possible by the Wernher and Beit benefactions, the Bessemer Memorial Fund, and by government funding.[27] The Advisory Board of the RSM, anxious over the loss of the School's older identity, was concerned with the iconography of its new building and asked that the names of Prince Albert, De La Beche, Murchison, Lyell, Percy, Bessemer and Siemens be inscribed on its front.[28] Later, when William Watts, the first professor of geology at Imperial, retired (albeit from the RCS), he requested that the names of John Wesley Judd and Andrew Ramsay be added, and they were. The front was really a frontage, the chance for architect Aston Webb to show his artistic skills. Behind the facade was an assortment of poorly designed laboratories. As geology professor Percy Boswell later put it, 'I lived to curse the memory of Aston Webb, whose incompetence had resulted in a well lit internal corridor, contrasted with working rooms nearly all of which required artificial lighting and services like gas, water and drainage safely buried under concrete floors'. Webb had 'sacrificed everything to his elevation'.[29] Behind the new building was the Bessemer Memorial Laboratory in which semi-industrial scale metallurgical processes could be carried out. This laboratory was used both for the instruction of students and by industries wishing to test new processes. A similar laboratory was opened in the chemistry department where students were taught how to prepare semi-industrial scale quantities of organic chemicals. The Whiffen Laboratory, named for William Whiffen, an old RCS student who donated the money for its construction, was opened by A. J. Balfour. In his speech Balfour is reported to have said, 'instead of performing their experiments in test tubes and -er- that kind of machinery the students will be able to carry out their tests in -er er- retorts and the like mechanisms'.[30] These two laboratories met the then fashionable ideal of giving students, as closely as possible, an industrial experience while at the college.

John Farmer the professor of botany, was soon able to persuade the governors that a new botany building was needed. He wanted space for major new sections in biochemistry and bacteriology, and for an existing section devoted to the study of the physical properties of woods and fibres under Percy Groom.[31] That Farmer took the entire department in the direction of applied botany was seen as somewhat revolutionary. Few academic botanists were then carrying out applied work and, for most, the herbarium was still their central focus. Farmer's work, discussed further in other chapters, was of interest to a number of major industries, notably rubber and cotton, and it was to these that the college turned, after the war, for the funding of its botany building. The Royal College of Science had a history in rubber research under chemistry professor W. A. Tilden.[32] Several of his students worked in Ceylon as rubber chemists. Herbert Wright, a botany student who had worked also with Tilden, and who carried out a study of the rubber tree in Ceylon, later made a large fortune by investing in rubber and became a major figure in the industry.[33] Wright and Farmer were able to persuade the Rubber Growers Association to help finance a new building. Wright was elected to the Governing Body and, a loyal old student, supported the college in financial and other ways. Later he helped finance yet another building for botany, the Plant Technology Building, on the west side of the Beit quadrangle, which was officially opened in 1923. This five-storey building housed biochemistry, bacteriology and plant pathology, as well as the student union snack bar. With all the new space the college could grow.

Imperial during the 1914–18 war is the subject of a separate chapter, but while Keogh was away on war-related work the college was left in the hands of the governing body. One of its members, Sir Arthur Acland, took charge of day-to-day affairs in what he later termed 'our shrunken condition'.[34] This was a difficult time for universities throughout Britain; all were being asked to make cuts by the Treasury.[35] But Imperial was able to build in certain areas such as technical optics and meteorology which were seen as central to the war effort. Further, friends of Henry George Plimmer FRS endowed a chair of plant pathology at the college in 1915 which Plimmer occupied until his death in 1918.[36] The chair then lapsed until 1928 when it was funded from government sources and William Brown, a former research student, became the professor.

The college remained open to students throughout the war and plans for post-war reconstruction were much discussed. Acland was keen that the government come to terms with what he believed was its failure to take science seriously. He claimed that this had led to wartime problems and mentioned the then often cited example of the sale of lard to Germany despite warnings from scientists that it would be used in the manufacture of explosives. Governmental ignorance of science, Acland believed, would lead to yet further problems in peacetime. In a memorandum to the Privy Council's Advisory Committee on Science he wrote,

> not only are our highest Ministers of State ignorant of science, but the same defect runs through almost all the public departments of the Civil Service. It is nearly universal in the House of Commons. … This grave defect in our national organization is no new thing. … Our success now in the very difficult time of reorganization after the war depends largely on the possession of our leaders and administrators of the scientific method and the scientific habit of mind.

This, he claimed, could only come about by a change in the education given to the class from which high officials are recruited.[37]

While Acland displayed a technocratic bias, the government recognized that there was some truth in criticisms such as his. It began planning educational reform already during the war.[38] Universities were hit hard by the war; fee income had dropped, and inflation ravaged both endowments and grants from local authorities. The government agreed to help out, but only if the universities accepted a standing committee as an intermediary. The result was the creation of the University Grants Committee (UGC) which came into being after the war to dispense government funds.[39] University independence was protected in that the funding was by block grant and not earmarked. The UGC reported to the Treasury until 1964 after which it (and its successors) reported to the Department for Education and Science (and its successors). Increased government funding allowed some college plans to materialize.[40] Otto Beit offered money for the much desired hostel in anticipation of matching funding from the UGC, but had to rescue the project with yet more money when insufficient funding was forthcoming. The hostel, with accommodation for fifty students, opened in 1926 but places were in such demand that plans to expand it began almost immediately. In 1927 Beit promised a further £20,000 provided it was matched by other donors, or by the government. Again he had to give more than promised.[41] By 1930 hostel space had doubled and the union building had been expanded. The administration moved from the RCS building into the new union expansion. Furnishings for the hostel were chosen by Professor Brereton Baker and his wife, Muriel. Baker, Chairman of the Hostel Committee, retired soon after the new hostel space opened and donated the clock which is still in the Beit Quadrangle.

Admission to the hostel was contested when some Indian students complained, with justification, of being excluded because of race. This precipitated much discussion of the position of 'oriental' students. Lord Buckmaster, Chairman of the Governing Body, expressed the wish that wealthy Indians would endow a hostel of their own, arguing that since money for the hostel had been given by an Englishman, the English had prior rights to the spaces. But not all the governors agreed with this point of view. In the end the Governing Body voted to allow some Indians admission, provided they were British nationals.[42] The hostel, and the quadrangle where it is located, were named for Beit after his death in 1930.[43] Tennis courts were built in the quadrangle.[44] Squash and Fives courts were built against the wall of the south museum gallery.

College finances continued to be a problem. There was even the suggestion of inefficiency, prompting the government to appoint a committee in 1922, chaired by Lord Meston, to look into the financial management of the college and its future needs. The committee's 1923 report was mildly critical of the college's financial administration. Further, in the committee's view, there were too many research projects being funded with strings attached. Meston recommended decreases in government funding to departments with heavy industrial involvement. However, he recommended that, overall, the college should receive considerably more funding.[45] Accountability in research was, and remains, important. It was an especially tricky problem for a college such as Imperial trying to get off the ground without an endowment, and with insufficient funding for what it wished to do — something Meston, Oxford educated and a former senior civil servant in India, did not fully appreciate. Nor, it would seem, did he appreciate the terms of the college charter. While universities should be places on which the wider public can depend for disinterested and reliable information, this does not mean that they not work with industry on its problems. It makes sense for much industrial research to be carried out within universities; but the problem, then as now, was how to have industry pay for such research while not influencing its results. In return for financial support some limited secrecy provisions are, perhaps, justified. But ethical problems related to industrial support were not given serious thought in the 1920s, and Meston was perhaps naively critical of what he saw. Today such problems, and not just at Imperial, have become more acute.

The Founding of the Chemical Technology Department

The Governing Body early set up a sub-committee to examine what new areas of science or technology should be added to the existing college departments. The chair of this committee was Sir Arthur Rücker FRS, Principal of the University of London, and a former professor of physics in the RCS.[46] The committee made a number of recommendations in its report of 1908, foremost being that a department of chemical technology be added. The report stated that American and German superiority 'in certain branches of [the chemical] industry is due to the scientific organization of their factories' and proposed that a four year course be designed to prepare students for positions 'on the scientific staff of chemical works and factories producing paper, fibres, textiles, sugar, starch, foodstuffs, glass, ceramic wares, coke, gas or oil.'[47] Long before the founding of Imperial College, the professor of chemistry at the Central Institution, Henry Armstrong, had arranged for courses in applied chemistry to be given in his department, and had wanted the City and Guilds of London Institute (CGLI) to open a chemical engineering department at its college. This never happened. Armstrong's department was slowly phased out after 1910 and he moved to the RCS, working there for a few years before retiring. Even after his retirement Armstrong remained an important figure in college affairs and his desire to see applied technical chemistry become the first area of expansion was realized.

T. E. Thorpe, the professor of chemistry, supported the Rücker committee's report. Some lectures in chemical technology were given in the chemistry department during the 1909–10 session by W. A. Bone, Professor of Fuel and Gas Industries at the University of Leeds.[48] Bone travelled to London once a week until 1912 when he was appointed Professor of Fuel and Refractory Materials on a permanent basis.[49] The Rücker committee's recommendation for a four year associateship course was not followed, but Bone headed a new sub-department of chemistry and gave an advanced (post-associateship) DIC course. When Aston Webb was asked to design a building for chemical technology Bone, who did not like the design of the Royal School of Mines, and who especially disliked the idea that the little money available be spent on exterior decoration, insisted on, and was given, a very plain box-like building with basic services delivered according to his own specifications. The building, refitted and still functional today, was constructed behind the Royal College of Music situated on Prince Consort Road, and some land adjacent to it was held in reserve for a planned extension.

It was, perhaps, a close thing that Bone came to the college since he cancelled a planned trip on the *Titanic* only at the last minute. An energetic professor, he wrote many letters to the college secretary which reveal much about working conditions at the time. By 1914, after bitterly complaining about the many hundreds of letters he had to write himself, he finally had access to a typist. He was lucky. In 1921 Hugh Callendar, professor of physics, complained to the Rector, 'I wish I had a secretary. I get so sick of typing and my machine is out of order'.[50] Secretaries, one per department, only became the norm by the mid 1920s. Bone also complained about having to share a phone line with the botany department and with some others, and that the demands on the line were so great he could never make calls when he wanted. This complaint is born out by a story told of the professor of botany, John Farmer; namely that on one occasion he threw his phone into a waste-paper basket, frustrated at having to wait too long for a trunk (long distance) phone call to be put through.[51] Even as late as 1929 the chemistry department had only one phone, situated in the messenger's box in the main entry hall.[52] Before the days of plentiful telephones, communication within the college was often achieved by yelling. Alan Hill, who joined his father on the technical staff of the geology department, remembered especially Professor W. R. Jones going to the back stairs and yelling for assistance still in the 1930s and 40s.[53]

Bone's main interests were in fuels and in combustion processes at high pressures.[54] Of his vocation he stated,

> born and bred as I was on Teesside, amid the roar and smoke of iron smelting, steel furnaces, rolling mills, engineering works and shipyards, where thousands of tons of coal were daily swallowed up and the night sky reddened with the blaze thereof, from childhood upwards I have ever been a fire worshipper.

In 1927 Bone invited a group of colleagues to his home to celebrate the publication of a book he had written together with his colleague and former student, D. T. A. Townend. Like many others this group mistakenly confused fire-worship with Zoroaster and named itself the *Zoroastrians*. They met on a regular basis to discuss research.[55] Bone was among a generation of Imperial professors who took seriously the college charter, believing that his role was to further science as applied to industry. But, when he wanted to give some public lectures on 'Coal and Fuel Economy', and to invite some prominent governors such as Lord Crewe and Lord Haldane to chair the sessions, he was discouraged by Keogh who wrote, 'the college [has] never gone in for anything of this kind and I feel sure the feeling generally would be that our efforts should lie in the direction of turning out new knowledge rather than dealing with subjects that are of an economic character.'[56] There was still debate on how the charter should be interpreted and the exchange between Bone and Keogh is illustrative of the tensions existing in the new institution over what its identity might be.

Given his field, Bone was naturally interested in refractory materials and an early appointee in this connection was G. I. Finch.[57] Finch was also an experimental photographer and he converted part of the tunnel under Prince Consort Road (much of the tunnel was used as a firing range by the College Rifle Association) into a dark room. Another early appointee in chemical technology was G. W. Himus who ran the coal and fuel laboratory in the RCS basement.[58] Himus remembered that in 1913 there was very little apparatus for the students and that they had to provide their own drawing equipment. Once a year he would relive his earlier power station days (he had run the main power station in Shanghai), don a boiler suit, and demonstrate the use of a large gas generator that had been donated by Robert Mond. This event entailed the burning off of methane gas, and the flame was visible in the neighbourhood for three days.

Thorpe also wanted some lectures in chemical engineering to be given to fourth year chemists and, in 1910, J. W. Hinchley was appointed to lecture on the design of chemical plant.[59] In 1917 there was concern on hearing that University College was planning to raise money for a William Ramsay memorial laboratory in chemical engineering, despite the fact that the University had an agreement with Imperial not to duplicate this field. Bone insisted on expanding quickly and Hinchley was appointed permanently to a professorship in a newly independent chemical technology department, though still in the Royal College of Science.[60] As Bone wrote to Gow, 'Mr. Hinchley has proved himself … and his outside consulting practice as a chemical engineer has already established for him a good connection with chemical factories'.[61] The future culture of the department of chemical engineering and chemical technology was much determined by these two men and their often uneasy relationship.[62] Bone clearly saw the kind of scientific research that he was pursuing as superior to the practical engineering of chemical plant taught by Hinchley. Hinchley, on the other hand, thought that fuel research should be a sub-section of chemical engineering. D. M. Newitt later wrote that when he joined the department in 1921 he was astonished at the frankness with which the two criticized each other, but that 'it was a stimulating and exciting place with both Bone and Hinchley inspiring respect and loyalty among their students'.[63] Bone was also in open warfare with rivals in his own field; among the most notorious of his quarrels was one with his old student R. V. Wheeler, Professor of Fuel Technology at the University of Sheffield. But, despite his acerbic manner, Bone built an excellent department and left a major legacy.

Hinchley encouraged his students to found a Chemical Engineering Society at the college and its first president, H. A. Humphrey, later designed the Billingham works for ICI.[64] When D. M. Newitt later applied for the new Courtaulds Chair of Chemical Engineering, he noted that 'Hinchley's teaching was of a pioneering and imaginative character' and that 'the inherent soundness of his methods' was shown by their continued use at Imperial.[65] Another area which was developed in the new department was blast furnace research. This was supported by the British Iron and Steel Federation and, at its request, research in this field continued long after Bone's retirement.[66]

While the story of this department is one in which an applied science gradually welcomed engineering into its midst, there remained problems of conflicting vision. As will be discussed elsewhere, these problems were later exacerbated when an undergraduate course was introduced at the end of the Second World War, and for which the chemical engineers did most of the teaching.[67] On Bone's retirement in 1936, A. C. G. Egerton succeeded him as head of department. Unlike Bone, Egerton was a decentralizer and managed to ease some tensions by giving people more autonomy than they had earlier experienced.[68] But it was a difficult transition. As Bone wrote to Egerton, 'the fact that I am to retire … does not imply in any way that I am relinquishing either my scientific work or industrial connections. On the contrary.'[69] Bone was a reluctant retiree but, since he died shortly afterwards, further disputes over space and facilities were averted.[70] In 1939 Tizard moved the department into the C&G and, in 1942, it was renamed Chemical Engineering and Applied Chemistry (Chemical Engineering in 1954, and Chemical Engineering and Chemical Technology in 1959).

The Founding of the Aeronautics Department, 1920

Professor Dalby introduced some lectures on aeronautics in the mechanical engineering department after having been inspired by a display of aeroplanes during a visit to Paris in 1909. With the coming of the war the possibility of a diploma course in aeronautics was discussed.[71] Recognizing that aeroplanes would play a role in the hostilities, the Air Board decided it wanted a professorship of aeronautics at the University of London, though this came about only after the war. In 1917 the government set up a Civil Aerial Transport Committee to consider all aspects of air transport, including the education of specialists. Richard Glazebrook, Director of the National Physical Laboratory and a member of the Imperial College governing body, was on this committee.[72] It recommended that just two higher educational institutions be funded to teach aeronautics, and that one of them be in London (the other was at Cambridge). The Air Board found someone willing to donate £25,000 to the University of London, and the prospect of a new chair was held out as a carrot to Imperial College in the hope that it would join the university. But the governors were unprepared to relinquish any of Imperial's independence and did not think £25,000 was enough money to endow a new department. Dalby stated that at least £50,000 would be needed and Glazebrook added that the aeronautics laboratory at the National Physical Laboratory cost about £15,000 per year to run and that Imperial's would cost no less.[73] The Governing Body wanted the donor, later revealed to be Basil Zaharoff, to give more, or for the Air Board to find further donors. The governors also insisted that they be the ones to select any holders of the chair, were it to come to Imperial College.[74] Negotiations took a long time and the professorship was delayed until after the war as a result. Imperial made a few concessions, the main one being that the Zaharoff chair in aeronautics would be a University of London chair, not an Imperial College one.[75] More money was forthcoming, including an annual operating grant of £8,500 from the government. After the war the Air Ministry gave the college four aeroplanes, though only one had an engine. The Ministry also wanted to donate an aircraft engine testing house that had been built on Office of Works land in South Kensington, but too many of the college's neighbours had complained about the noise for that kind of work to continue in peacetime. However, the Air Ministry maintained a laboratory in South Kensington until 1931.[76]

At first the department was placed in the Royal College of Science; it was later moved to the City and Guilds college by Henry Tizard.[77] Space was found in a building in Lowther Gardens owned by the 1851 Commissioners, and Richard Glazebrook was appointed professor. Many of the early lecturers were visitors who brought their expertise to the college. Glazebrook's appointment was seen as an interim measure and he retired after only three years to be succeeded by Leonard Bairstow, a former RCS student and professor of aerodynamics in the department.[78] Bairstow's work on the design of wind tunnels, and of aircraft, contributed much to aircraft stability. Two of his female assistants, B. M. Cave-Browne-Cave and E. D. Lang, carried out many of the calculations for his work.[79] Lang remained a research assistant in the department from 1925 to 1946 when she became an assistant lecturer in mathematics, retiring in 1953.

While under Bairstow, the department moved from Lowther Gardens into the Science Schools building on Exhibition Road, and a wind tunnel came into use there in 1925. The department was adversely affected when the Air Ministry decided to move its South Kensington laboratory to Farnborough in 1931, and apparatus that had been used in teaching was no longer available. But the department soon recovered and offered advanced courses in aerodynamics, aeroplane structures, meteorology and air navigation. Many of the students were Royal Air Force (RAF) officers and part of the department's work was carried out at the Royal Aircraft Establishment (RAE) at Farnborough. The RAF accepted that the associateship awarded by the department was equivalent to a degree for the purpose of commissions. Many pioneers of the British aviation industry were ex-students of the Imperial and Cambridge departments. In 1943 one of the Cambridge graduates came to Imperial as Zaharoff Professor. Arnold Hall, under thirty years of age at the time, was a specialist in jet propulsion on which he had worked with Frank Whittle while at the RAE.[80]

Aircraft which, in the 1920s, could reach just over 16,000 ft, were used for meteorological studies at the college. The Air Ministry had taken over the Meteorological Office during the war and supported a new School of Meteorology at Imperial. The subject received a boost due to the rise in civil aviation. By the late 1930s, about ten research students were being admitted annually, and many qualified applicants were turned away.[81] The meteorologists had their own space in the Meteorological Office building, at the southern corner of Imperial Institute Road and Exhibition Road. The building also housed the Office's fine library and had a full range of observational instruments on its roof.[82] The first professor of meteorology, Sir William Napier Shaw, came from the Meteorological Office and was assisted by Elaine Austin and by David Brunt, a future professor — both seconded to Imperial College.[83] In 1924 Shaw was succeeded by Sir Gilbert Walker, an eminent meteorologist who had headed the Observatory Services in India.[84] The School then sought transfer to the physics department, but no space was available and it remained part of aeronautics until 1932. In that year aeronautics moved to the C&G and meteorology stayed in the RCS as a subdepartment of physics.

The Relationship between Imperial College and the University of London, 1907–29

In 1907 Imperial College was not part of the University of London.[85] Many hoped it never would be. The City and Guilds of London Institute (CGLI) was a major voice against Imperial College joining the University. Indeed members of the Delegacy were able to persuade the Governing Body to pass a motion that 'the autonomy of Imperial College should be maintained and that its relations to all the universities of the Empire should be identical'.[86] This motion was prompted by the fact that, in 1910, the government had set up a Royal Commission on University Education in London, with Richard Burdon Haldane, who favoured the union of Imperial College with the university, as its chairman. The Royal Commission had more than Imperial to consider; some other colleges had uneasy relationships with the university. The main problem for Imperial was that it was underfunded and did not have the means to go its own way without further government support. In the early twentieth century the college looked to the governments of the colonies, and to industry, in an effort to find funds, and hoped to persuade the government to grant it the right to award its own degrees. Increasingly, degrees were something that employers in technical industries demanded, and that students sought.[87] In this connection, also in 1910, Richard Glazebrook put a resolution before the governing body which was approved: 'it is desirable to establish in London a Technological University in which degrees may be obtained without further examination'. The implication was that Imperial College would become that university and that its associateship diplomas would become degrees.[88] The college also had the support of several provincial universities in seeking independent university status.[89]

The Governing Body agreed that Imperial College would make representation to the Royal Commission and Keogh, the new Rector, carried out its wishes. He was asked to emphasize that the college 'has been charged with carrying out a very special work involving great national responsibilities;' and that the college charter specifically gave the Governing Body 'the responsibility of providing for technical education in relation to imperial and national industries and to provide for them on the highest scale'.[90] Further, the college needed to establish connections not just in London but with 'universities in the provinces and in the colonies'. Its students should come from around the Empire and 'it should in fact be an imperial college'. Since the University of London had no mandate to meet these industrial and imperial needs, 'it is undesirable to attempt amalgamation'. Indeed 'it is of vital importance that the Imperial College should be kept from incorporation … [we need] to retain our independence and the possibility of working out our own salvation'.[91] The new college needed also educational flexibility 'free from those restrictions which … are always imposed when the curriculum has to be adapted to the theoretical preconceptions of those with little understanding of the needs of our students, or of the aims of the college'. The governors did not want the Royal Commission to place over the college 'another authority whose aims are not directed towards the improvement of science and technology'.[92] 564 students signed a petition against amalgamation with the university.[93] The views of the governors and students were presented to the Royal Commission in February 1911, but the Commission did not go along with them. It did, however, recommend that the associateship examinations be accepted by the university as being degree equivalent. It also recommended that faculties, rather than colleges, be the ruling parties within the university and that Imperial might become a new faculty of technology. This radical suggestion impacted the sensitive issue of identity not just of Imperial, but of all the London colleges. It was so unpopular that the Commission's report was largely ignored and Imperial's situation remained unchanged.[94]

On the matter of examinations there were on-going discussions with the University of London. Imperial's first year examinations were accepted by the university as equivalent to its intermediate examinations. According to Professor Callendar, 'it is well recognized … that the standard of the first year's work in chemistry and physics is beyond that required at the London Intermediate Examination and is much higher than our sister colleges attempt to reach as the result of one year's work'.[95] As far as the college was concerned, the final associateship examinations should have been accepted also, something the university adamantly refused to do. That Imperial College students had to prepare for, and then take, a second set of examinations in order to gain an external degree was a major bone of contention. Professors in the three associated colleges were insistent on not giving up the associateship diplomas, and in not changing their teaching patterns in order to prepare students for the existing university degree examinations.[96] They were also unwilling to give up the freedom of admitting students as they chose. Many of the professors disliked the university matriculation require- ments which included having to pass examinations in subjects such as Latin, which gifted students without conventional schooling might well fail.[97] Indeed, in this period some very successful science and engineering students, some of the Whitworth Scholars for example, had very little conventional schooling, and had been admitted on the basis of demon- strated practical skills or success in the technical exams set by the Board of Education or the CGLI — 'some of the most brilliant students', according to Alfred Keogh.[98] Further, Imperial could demonstrate that those among its students who *did* choose to take the additional degree examinations performed better, on average, than did students from other colleges. At first only a minority chose to take degrees but, by 1929, when Imperial finally joined the university, about seventy-five percent of students were sitting both sets of examinations.[99]

The First World War postponed any resolution of these problems but they flared up again in 1919 when members of the Governing Body visited A. J. Balfour and H. A. L. Fisher, the President of the Board of Education, to present their views. Past and present students signed yet another petition, and many Imperial professors signed a letter to *The Times*, stating that Imperial College should be given the right to award its own degrees.[100] Correspondence in *The Times* in this period reflected also the views of those opposed to Imperial's independence, much of it coming from the university. Keogh responded, insisting again that Imperial be given the right to award degrees, something he believed was amply justified by the quality of education and training given at the college.[101]

The problems between the college and the university continued to fester until a temporary agreement was reached in 1926.[102] Keogh's successor, Thomas Holland, believed that arguments for Imperial's inde- pendence would go nowhere and wanted to come to terms with the university.[103] In 1923 he asked for a conference on some proposals he had drawn up and which, three years later, formed the basis for an agreement which formally came into effect in 1929 when revised statutes, accepted by Imperial College, were legally in place.[104] By the mid-1920s the staff at Imperial were worn down by the dispute. As John Farmer put it, 'it was clear that the London chaos could not go on indefinitely'.[105] Like Holland, he wanted matters to be settled, but on condition that departments kept control over their courses and that the Imperial charter not be subverted. By and large these conditions were met. Imperial maintained its right to award associateships to people who did not meet the university matriculation requirements.[106]

But for all of this to happen L. N. G. Filon, the Vice Chancellor, had to be wooed. One of his complaints was that Imperial had never been part of the Officers Training Corps (OTC).[107] Holland promised that if Imperial's associateship examinations were accepted as degree equivalent he would start an infantry company at Imperial. He kept his word and asked Major (later Colonel) P. C. Bull and G. C. Lowry to organize a company. This was not a popular move and there was much opposition to the OTC. *The Phoenix* ran a number of negative articles and H. G. Wells, true to form, joined the debate. He wrote, 'I cannot imagine what a good RCS student can do for his country or himself by joining the OTC. His duty is to give all his time to his scientific work, the most important work in the world'. And again, 'I must cling to my clear cut persuasion that this OTC stuff is a waste of time, even for men desirous of military efficiency, that it simply accustoms men to certain rather ridiculous antics and to mechanical subservience to incompetent leading, and the less of it the RCS student has, the better for him and for his country and for mankind'.[108]

The dispute with the university is interesting and illustrates tension between those holding different educational ideals. The university senate and council, seeking respectability for the university in its newly formed teaching capacity, favoured a traditional view of what it means to be an educated person. The senate, largely composed of people with backgrounds in the classics and humanities, was uneasy with technical subjects entering on an equal basis with the arts, pure sciences, medicine and law.[109] Many on the senate held the view that Imperial College students lacked a broad general knowledge because of the narrow entry requirements. Some technical subjects including mining, it was believed, belonged in trade schools, not universities. But, by 1926, faced with the overall success of Imperial, the University agreed to change its matriculation and degree requirements, and accordingly its statutes, so as to accommodate the kinds of courses offered at the college. The University approved the college syllabuses, including those in new degree fields such as mining, and agreed that a single set of final examinations could be used both for the associateship and the university degree. As a result, the college continued to set its own examinations but submitted them to the scrutiny of external examiners chosen by the university.[110] There were some face-saving conditions for the university; for example, associateships were formally to be given to those who had passed what were now labelled university examinations. In gaining the right to set its own examinations, Imperial set a pattern later followed by the other large colleges in the university. The college could also examine for the MSc and PhD degrees and keep its DIC as well as its associateships.[111] Matriculation requirements were adjusted gradually; only by the mid-1950s was Imperial in full compliance.

In becoming a college within the university, Imperial had to allow the university court to have the final decision on new courses of study, and on new professorships. This power was largely nominal but was exercised when serious funding issues were at stake. As to appointment decisions, these continued to be made by the college and the Delegacy, though for important positions university-wide committees were appointed. The college had to allow any government funding to be channelled through the university court, a decision that was reversed in 1964 when the college gained direct access to the University Grants Committee. To smooth the process of integration, the university was generous during the early years of the new agreement and helped fund many new positions at the college. The university also began to recognize Imperial professors as university professors, a process that took several years. Some of Imperial's assistant professors, however, were unwilling to change their title to that of reader, and refused university recognition.[112] Despite the agreement, Imperial's relation with the university has never been easy. As will be discussed in other chapters, the question of separate status, and of the creation of a technological university, has arisen many times since 1929.

From the Late 1920s to the Second World War: The Tizard Years

In 1929 Holland was succeeded as Rector by Henry Tizard.[113] Four years later Tizard joined the Aeronautical Research Committee, beginning the work that would eventually draw him away from Imperial. While at the college Tizard thought seriously about its future expansion. In 1933 he began negotiations to take over the building occupied by the Royal School of Needlework which was situated to the south of the Waterhouse C&G building on Exhibition Road.[114] As a result, the Needlework

School was given a ten-year tenancy of reduced space in its old building, the remainder of the building being taken over by the college. Tizard gave some of the new space to the aeronautics department which he moved to the C&G, and the rest was used for the expansion of the other engineering departments. For example, with the 1933 appointment of A. J. S. Pippard to the chair of civil engineering, the Goldsmiths and Clothworkers Companies gave generously to help the department open a new structures laboratory.

Lord Buckmaster retired as chairman of the Governing Body in 1934 and there was then much discussion as to whether the college had reached a sufficient stage of maturity to elect a scientist to replace him, something Tizard wished for. However, Lord Crewe, Imperial's first chairman, was asked to make enquiries among the governors, and to look rather for a suitable man of affairs. Lord Linlithgow was elected shortly afterwards. Two years later he was appointed Viceroy of India at which point Lord Rayleigh, recently retired from the physics department, became the first scientist to chair the governing body. Rayleigh and Tizard had a good working relationship. Tizard was outspoken and warm, though with a somewhat caustic wit. He was a man of vision with many plans afoot, and was helped in forwarding these by Rayleigh and by G. C. Lowry who was appointed to replace Gow as College Secretary in 1934.[115] Given the economic difficulties of the mid-1930s, Imperial was fortunate in having this resourceful trio in control. One of Lowry's first tasks was to negotiate the sale of the athletics grounds to the Wembley Council — a forced sale, since the council needed the land for other purposes. In 1936 he negotiated the purchase of new grounds at Sipson Lane, Harlington, intended both for athletics and for a future biological field station. The existing field station near Slough was considered too small, but the college's request to close it down and move it to Harlington was opposed by Ministry of Agriculture personnel.[116] They were not prepared to support the college's request for funds from the Treasury for this purpose. Tizard, who noted that the college was not responsible to the Ministry of Agriculture, found support for the move within the Royal Society and the Ministry moderated its position. The reason for the Ministry's opposition is complex. It had supported research at the Slough field station since its opening but on the condition that, were the field station to close, the proceeds of any sale would come to it. The college held that were the field station to be moved this condition did not apply, and that the sale of land at Slough should not be interpreted as the closing of the field station. The Ministry disagreed and Slough was not then sold. Despite this, the Harlington land was purchased. It was the closest available large space that the college could afford. Thirty-two acres were intended for sporting facilities and fifty-six acres for the hoped for field station. But the matter of the field station and its location remained unresolved until after the Second World War.

To help with planning at the South Kensington site, Tizard made a careful study of land usage in the immediate college area and commissioned a three-dimensional wooden model to help him envisage future possibilities. There were ongoing negotiations with HM Office of Works on the allocation of land in South Kensington, but the government was hearing also from supporters of advanced art education in London. Indeed, it had agreed to provide a new building for the Royal College of Art (RCA). A site that had earlier been purchased in Cornwall Gardens was sold, having been deemed too small. Instead, the government was considering three other possibilities all within the so-called island site.[117] This was unacceptable to Tizard who viewed the island site as essential to the future expansion of Imperial. He was, however, willing to give up the RCS building on Imperial Institute Road in exchange for the consolidation of Imperial's hold over the island site. After consultation with the Board of Studies and other interested parties, Tizard wrote to the Office of Works stating Imperial's conditions. They were: that the island site be properly surveyed as to drainage and suitability for further construction, that three houses on Queen's Gate be purchased for the college so as to allow entry from there, that the site already occupied by Imperial College in the Western galleries be given to the Chemical Technology Department for a new building, that new buildings be funded for those departments still in the Huxley building, and for the physics and chemistry departments should the RCS building have to be given up. Space was to be given also for new squash and tennis courts, and for the college boilers. Further, if at any future date the Imperial Institute were to be vacated, Imperial would have the first option to take it up, or to demolish it for future development.[118] Had the war not intervened, these proposals might well have been accepted by the government. As it turned out, they formed the basis for new negotiations after the war.[119]

As an interim measure, between the sale of the Wembley grounds and the development of sports facilities at Harlington, Tizard took over some more space in the Western galleries, vacated when the War Museum left in 1937. He wanted to build a gymnasium and, while this never happened, the space was used for various indoor sports. A college boat house in Putney was completed in 1938 and became a major sporting venue. It was opened by Viscount Falmouth who had donated a large sum towards its construction, and who was to become Chairman of the Governing Body in 1947.[120] Tizard was a great supporter of the Boat Club and his witty speeches given at club dinners were long remembered. On one occasion, after several members of a crew that had represented the college at Henley failed their examinations, he jokingly referred to 'eight men with but a single thought — if that.'[121]

In addition to being forward-minded, Tizard had a good eye for how money might be saved or made. Anticipating more recent moves in this direction, Tizard insisted that professors charge full overheads and labour costs, including their own, for any outside work undertaken.[122] And he asked Professor Lander, Dalby's successor as professor of mechanical engineering, to look into the college's energy needs and how they might best be met. At the time these were being provided by the Kensington and Knightsbridge Electric Lighting Company which had its office on Prince Consort Road on the east side of the hostel. But, on Lander's recommendation, this company was replaced by the Gas, Light and Coke Company which bore the cost of installing both new coke-fired, and electrode, boilers.[123] There was some acrimony as a result of dropping the services of a college neighbour, but the new boilers were more economical and, in addition, had some pedagogical use. They gave Lander and his students the opportunity of testing their relative efficiency, and statistics to demonstrate this were kept for a number of years. Another move to increase efficiency was in the separation of the union and refectory services. Tizard asked Mr. and Mrs Peacock to run the college refectory services which, under the union, had been losing money for some years.[124] This was a good move, and the Peacocks ran things well until the end of the Second World War. At that point, Colonel Bull, a lecturer in the chemistry department, took over the reorganization of the refectory, hired new staff and soon had them serving about 900 people each day. But even this did not meet the demand and the wartime canteen, set up in the RCS building, continued to serve meals for several years after the war had ended. A solution was found only when a large refectory was opened in 1950 in the newly named Ayrton Hall, located at the southern end of the old Royal School of Needlework building (renamed for Unwin). Despite all the growth since 1907, the total value of the college's buildings at the beginning of the Second World War was estimated at only £1,250,000 and their contents at £350,000.[125]

In 1939, Kingsley Wood of the Air Ministry asked the college to release Tizard for war work This was no surprise since Tizard had already informed the Governing Body of his impending departure. The Air Ministry offered £2000 per year for a replacement and the popular professor of chemistry, J. C. Philip, recently retired, agreed to return to the college as Deputy Rector.[126] At the start of the war Tizard sent a memorandum to all departments stating that, given that so much of the college revenue derived from government sources, it was to the advantage of the college that departments undertake work for the government. And, while all the costs had to be shown, charges were to be kept as low as possible.[127]

...

1. Governing Body (GB) minutes, 12 July 1907.
2. Fees were charged per short lecture or laboratory course until after the Second World War.
3. This was renamed the Huxley building in 1932 and, today, is the Henry Cole Wing of the Victoria and Albert Museum.
4. Lamley and Company sold college textbooks from its South Kensington shop after the closure of the RCS stall, continuing to do so into the 1970s.
5. The Rector's salary was not to exceed £2000 per year. GB minutes, 10 January 1908.
6. Henry Taylor Bovey FRS (1850–1912) was born in Devonshire and educated at Queen's College, Cambridge. A civil engineer who specialised in bridge design and hydraulics engineering, he worked for the Merseyside Docks and Harbour Works before moving to Montreal in 1877 as Professor of Civil Engineering and Applied Mechanics at McGill University. He was a founder member and President of the Canadian Society of Civil Engineers and was President of the Royal Society of Canada.
7. Sir Alfred Keogh (1857–1936) was born in Ireland and was educated at Queen's College, Galway. He studied medicine and entered the army, becoming surgeon-major in 1892. During the South African War he ran a general hospital and, in 1902, was appointed Deputy Director General of the Army Medical Services, Colonel and Surgeon General in 1904,

and Director General in 1905. He was a brilliant administrator who made major reforms to the Royal Army Medical Corps. He helped found the Royal Army Medical College on Millbank, introducing postgraduate training in military medicine in 1907. (This college opened in 1903 on the site of the old Millbank Penitentiary. Chelsea College of Art and Design took over the premises in 2001.) Keogh also founded an army school of hygiene and brought teaching hospitals more generally into the Territorial Army organization. He was knighted in 1906 and was Rector of Imperial College 1910–22.

8. Alexander Gow (1869–1955), was educated at Borough Road College on a Queen's scholarship. He then became a science teacher at County High School, Isleworth, before going to Cambridge where he took a degree. Gow had been Principal of the Blackburn Technical School before moving to Imperial where he was paid an annual salary of £400.

9. H. L. Callendar, S. H. Cox, J. B. Farmer, W. Gowland, W. A. Tilden, and W. Watts. The governors also decided that the college working hours would be 9.30 am–4.30 pm in winter, 10.00 am–5.00 pm in summer and would include a half day on Saturday (GB minutes, 13 February 1908).

10. Both women had been on the staff before 1907 and both were paid £150 per annum. Reeks joined the RSM in 1893. She published a history of the RSM in 1920. Whiteley had been a science teacher at Wimbledon High School before coming to the RCS, where she later carried out research for the University of London DSc. She was appointed a demonstrator under Tilden in 1904. She became an assistant professor in 1920, the first woman to reach that rank. She retired (as a reader) in 1934 and was paid a special retirement allowance to bring her retirement income above £300. This was not a unique case as the college often helped retirees who were hard up through no fault of their own. See GB minutes, 8 June 1934. A. A. Eldridge, 'Martha Anne Whiteley', *Proceedings of the Chemical Society*, June 1957, 182–3. Some of Whiteley's work is discussed in chapter 5.

11. See, for example, GB minutes, 12 November 1909.

12. While the numbers were low, chemistry had more women research students than any other department. See 16th Annual Report for year ending 31 July 1923.

13. See Executive Committee (EC) Minutes for the year 1924, for example.

14. See Annual Reports for 1937, 1938, 1940 and 1943. 1026 was the total number of students. This included part-timers. Full-time students numbered only 780. The number of women students began to rise slowly after the war, but fell again in the 1950s before beginning a slow rise from

1958 onwards. The proportion of women, however, increased significantly only from the 1970s.

15. This practice changed shortly after the Second World War. At that point the three year course became the norm. All students entered into what had been the second year course, a level by then standard for first year entrants. Final exams were taken at the end of the second year, and during their third year students had some latitude in choosing lecture courses and short research projects.

16. The diploma's acronym (DIC) was at first disputed by the Institute of Chemistry. GB minutes, 14 June 1912: letter from Raphael Meldola, President of the Institute of Chemistry, reprinted.

17. William Ernest Dalby FRS (1862–1936), known as Ernest, was born in London and entered the Stratford Works of the Great Eastern Railway at the age of fourteen. He attended evening classes and, at the age of twenty, won a scholarship to the Science Schools, followed by a Whitworth Scholarship one year later. After gaining his associateship, he took a job with the Crewe Works of the London North-Western Railway and rose to a senior position in the company. In 1890 he decided on a change of course and accepted an invitation to join Professor Ewing's teaching staff at Cambridge. He spent six years at Cambridge and was awarded an Hon. MA. In 1896 he was appointed to the chair of mechanical engineering at Finsbury Technical College. He succeeded William Unwin in the University of London chair of civil and mechanical engineering at the Central Technical College (later C&G) in 1904. He was a specialist in the balancing of engines, a subject for which he had an international reputation. See C. E. Inglis, 'William Ernest Dalby', *Obituary Notices of Fellows of the Royal Society*, vol. 2 (December, 1936).

18. In 1939 three representatives from the Clothworkers Company were added. Goldsmiths and Clothworkers were privileged because of their record of generous donation.

19. B/Acland; correspondence file: Keogh to Chairman of Delegacy, 7 January 1919. It wasn't just the City and Guilds professors who resented the Rector's powers. In 1911 John Perry wrote to Keogh stating 'I have never in sixteen years had to send a student out of my lecture room and never had the slightest want of discipline'. But, for the first time he had to interrupt a lecture because students were whispering which he was sure would not have happened if he had control over discipline. But, 'if my only power of punishment is to ask you to see a student, I am afraid that I shall never … use this power … a professor who depends upon his Rector or Principal doing what he ought to be able to do himself is usually a very

poor disciplinarian.' Perry to Keogh, 7 November 1911, quoted in 'History of the Mathematics Department' (typed manuscript in ICA, pp. 23–4).

20. The CGLI agreed to the indefinite suspension of the Delegacy in 1976 with certain provisions relating to nominees for the Imperial College Governing Body, and that any revisions to the charter be agreed by the CGLI, and that both the Clothworkers and Goldsmiths Companies be represented in any future charter discussions.

21. GB minutes, 29 November 1909.

22. Beit fellowships, created in memory of Alfred Beit, continue to this day. More recently the Beit Trust, together with Imperial College and the Foreign and Commonwealth Office (Chevening Fund) have provided scholarships for students from the old Rhodesias (Zambia, Zimbabwe) and Malawi; Beit donations have also helped the Imperial College Trust, a charity dedicated to the benefit and welfare of students, and to the support of extra-mural activities.

23. For more on the union and student activity see chapter 11. The union was built on Prince Consort Road, then still paved in wood.

24. Crosby Hall is a medieval building which was removed from the City and rebuilt on the Chelsea Embankment in 1910. For discussion see GB minutes, 28 June 1918.

25. GB minutes, 12 December 1919 and 27 May 1921.

26. The Goldsmiths donated first £50,000 and later a further £37,000 towards this building. Because of delays with the departure of military personnel who occupied the building during and after the war, it was not formally opened until 1926. E. A. Nehan, a technician in the mathematics department from 1902–52, stated that before the Goldsmiths extension was built there was an 'oasis of green' with a horse driven cab-rank at the curb of Exhibition Road. During construction of the extension Nehan witnessed a horse and cab standing in the rank disappear as the ground gave way. Horse and driver landed unharmed in the C&G boiler house. B/Nehan, farewell speech, 1952.

27. The Union, Goldsmiths' extension, and the RSM were all designed by the architect Aston Webb who had earlier designed the RCS building.

28. GB minutes, 12 November 1912.

29. B/Boswell; typescript memoir, pp. 63–4. Webb was an expert in facades and had designed the new frontage for Buckingham Palace. What Boswell failed to mention was that before the Second World War the glass roof of the RSM was often whitewashed in summer to prevent the building from overheating. In the war a bomb smashed the glass roof.

30. This seems less scientifically literate than one would expect from Balfour, but is reported by A. A. Eldridge in 'Some Personal Recollections' (B/ Eldridge). Arthur Alfred Eldridge (1889–1970) was educated at the RCS and later joined the chemistry staff, becoming a senior lecturer in the department. He was active in college life.

31. Percy Groom FRS (1865–1931) was educated at Mason's College and Trinity College, Cambridge. He worked briefly at the University of Bonn before his appointment as Professor of Botany at the Imperial University of Whampoa, China, in 1889. There he became interested in the economic botany of timber trees. He returned to Britain in 1892 and, after a brief period in Oxford, helped to set up the sub-department of plant physiology under Sir Isaac Bayley Balfour in Edinburgh. He resumed his interest in economic botany on being appointed head of the department of botany at the Royal Indian Engineering College at Cooper's Hill near Egham. When this college closed in 1901, Groom moved to become head of the botany department at the Northern Polytechnic Institute. He was appointed assistant professor at Imperial in 1908 and retired in 1930. His line of work was then discontinued, in part because it was being carried on at the Forest Products Research Station at Princes Risborough. Later, work on timber and wood preservation was reintroduced into the department by J. F. Levy, who joined the staff in 1946, one year after graduating from Imperial. For Farmer see chapters 6 and 7.

32. William Augustus Tilden FRS (1842–1926) served a five-year apprenticeship with Barnsbury pharmacist Alfred Allchin, and took an external BSc at the University of London. He taught at Clifton College where he began research on the terpenes and made important discoveries which anticipated both the cracking of petroleum and the synthesis of artificial rubber. In 1880 he was appointed the first professor of chemistry at Mason's College, Birmingham; he succeeded T. E. Thorpe at the RCS in 1894. He retired in 1909 when Thorpe returned to the professorship.

33. Rubber was a major industry in Ceylon before becoming one in Malaya. The Mycological Laboratory was named after Wright. Farmer was a persuasive man and two other members of the Governing Body gave generously towards the botany building. They, too, had laboratories named after them, namely the Robert Kaye Gray Plant Pathology Laboratory, and the Francis Ogilvie Physiology Laboratory (later renamed for Vernon H. Blackman).

34. GB minutes, 11 December 1914. In 1914, 300 students and 59 staff members left to join the forces. Roughly 600 students were still in attendance. The largest enrolment during the war was in the Royal College of Science and its largest department was chemistry with about 65 associate-

ship students. By comparison, physics had about 15, and botany 14. The second largest department at Imperial was mechanical engineering with about 55 students. For further enrolment details see appendix. Quotation, GB minutes, 22 June 1917.

35. Letter from the Treasury sent to all universities dated 29 July 1915. See EC minutes, 8 October 1915. Gow replied to the Treasury pointing out the many cuts that had already been made. As discussed in chapter 5, it was unclear who should be paying the salaries of those engaged in war work.

36. Henry George Plimmer FRS (1856–1918) had worked under Pasteur before becoming a lecturer in pathology at St. Mary's. He was Director of the Cancer Laboratories at the Lister Institute (1902–15) before being appointed Professor of Comparative Pathology at Imperial. Plimmer was an eminent microscopist and comparative pathologist.

37. B/Acland; memo to Privy Council Committee, 1916. This type of view, anticipating the more overt technocracy of a later period, is discussed in Anna-K Mayer, 'Reluctant Technocrats: Science Promotion in the Neglect-of-Science Debate of 1916–18', *History of Science* vol. 43 (2005), 139–59. Some people spoke out against the expansion of science education. Seeing science as some kind of Germanic vice, they did not wish to risk 'fostering the very spirit against which the war was being fought' and claimed that 'elite education would not be remodelled into a support system for a society masterminded by expertise'. Quotations, pp. 153 and 157.

38. *Scheme of Organization and Development of Scientific and Industrial Research*, (Board of Education, 1914–16, Cmd. 8005); *Report of the Committee of the Privy Council for Scientific and Industrial Research, 1915–16*, (Cmd. 8336). The Board of Education Committee which produced the first of these two reports was chaired by H. A. L. Fisher (1865–1940), President of the Board of Education. The report recommended the downgrading of classical education in schools, especially Greek, and the relative upgrading of scientific instruction both in schools and universities. Fisher who had read Greats at Oxford was a major educational reformer. Before the war he was Vice Chancellor of Sheffield University and, after a career in government, returned to Oxford as Warden of New College. See H. A. L. Fisher, *Our Universities* (London, 1927) in which he praises the University of London for introducing new technical chairs such as those in applied chemistry, concrete, and glass technology. After the war, Fisher helped enable a scholarship scheme for postgraduate student exchanges between Imperial College and American universities which began in 1919. About twelve students a year travelled to the USA; MIT was the favoured destination.

It was also Fisher who set up the commission under Herbert Asquith to recommend conditions under which Oxford and Cambridge might receive grants from the government. The resulting modernization, with its huge expansion of the sciences, meant that the two older universities became serious competitors to Imperial for government funding after the war. By the 1960s Oxford had the largest physics and chemistry undergraduate schools in the country.

39. Its predecessor was the Committee on Grants to University Colleges. One of the first things the UGC did was to establish the income and expenditures of all universities with the exception of Oxford and Cambridge. According to the UGC Return of 1920–21, the wealthiest universities were Birmingham, Sheffield and St. Andrews. The plan was to give proportionately larger grants to the poorer universities, but the economic depression of 1921–2 meant that this did not fully materialize. After the war much support for university scientific and medical research came also from the Rockefeller Foundation

40. H. A. L. Fisher to Crewe, House of Commons, 16 April 1919 (GB minutes, 9 May 1919). Letter informing Crewe that the college grant was to be increased from £30,000 to £52,000, plus a non-recurring grant of £28,000 for 1919–20 (the total UK universities grant that year was £1m). In addition, the Ministry of Reconstruction gave capital funding to the college.

41. Otto Beit's total donations to the college were £77,000.

42. GB minutes, 18 November 1932. There was a student Oriental Society and its membership was largely Indian. The Society won the support of the Indian High Commissioner in its protest. John Jones, the Registrar, sympathized with Indian students having difficulty finding accommodation, and invited a few of them to become lodgers in his home. He opened his office to students during the lunch hour and acted as an unofficial counsellor. Very formal in manner, he retained the Victorian habit of working in grey top hat and grey morning coat, turning to dark suits only at the end of his career. He retired in 1927 but continued to invite Indian students to lodge with him.

43. This was not the original intention, which had been to name the hostel after T. H. Huxley. In 1932 it was decided instead to name the original Science Schools building on Exhibition Road after Huxley. Huxley was also remembered in a series of lectures named after him but these were later discontinued when supporters of other famous professors wanted memorial lectures also in their names. In 1934 Julian Huxley approached the college on behalf of the recently widowed Mrs. Leonard Huxley who

wished to sell a large volume of Huxley papers to some public institution. She was not well off and asked £4000 for the papers that were then valued by the British Museum at £2500–£3000. Julian Huxley tried to make the offer more attractive by agreeing to include some Huxley memorabilia, such as his academic gown. The college agreed to try and raise money through private donation but was unwilling to purchase the papers with college funds. In the end only £1250 was raised and Mrs. Huxley accepted that amount. The papers came to the college and formed a rich nucleus around which the college could build its archive. Emeritus Professor William Watts, who had built an excellent geological library in his own department, was chairman of the Records Committee and began the task of consolidating and cataloguing the college records. A muniments room was founded on the fourth floor of the Beit building. The Watts Library in Geology at this time contained close to seven thousand bound volumes, periodicals and many maps including 1770 mounted maps of the Geological Survey of Great Britain and Ireland. After Otto Beit's death the Wernher and Beit connection with the college continued when Harold Wernher, son of Julius, succeeded Beit on the Governing Body.

44. Tennis had long been a favourite sport at the college/s but students had to go to Battersea Park to play before the courts were built. The Lawn Tennis Club was formed in 1910.

45. Meston Report (copy in ICA). It recommended that the standing grant be increased from £15,000 to £68,000. In exchange the college was to balance its budget without depletion of its capital, something it had not been doing. Meston stated that the current salary scale was fair and should not be cut.

46. It was Arthur Rücker who had earlier dealt with attempts by three different railway companies to construct a railway under Exhibition Road. He argued that precision measurements at the RCS and C&G would be impossible were a railway to be built. In 1893, after about five years of the physics and engineering professors presenting arguments against the proposals to various bodies, Rücker and J. Norman Lockyer appeared before a Select Committee of the House of Commons to testify against the passing of the Clapham Junction and Paddington Railway Bill. Lord Kelvin, then President of the Royal Society, also gave evidence. The Select Committee was unanimous in supporting the cause of science. But, a year later, another proposal was put forward by the railway companies for an electric railway; this time with assurances that the return current would be completely insulated. So, Rücker had a further fight on his hands and once again was successful. He departed for the principalship of the University

of London in 1901. An earlier proposal from the District Railway Co. had more lasting results. This company first proposed a pneumatic railway, and then a foot subway, to be constructed from its South Kensington Station to the Albert Hall. Permission was given for the latter in 1885 and a (still existing) subway was partially completed, allowing covered access to the south-east entrance to the Royal Horticultural Society Gardens and to the Inventions Exhibition. But the 1888 construction of Imperial Institute Road caused a break in the covered passage. Plans to continue the subway to the Royal Albert Hall were never realized; though, while the arcades still stood, covered access to the Hall was possible.

47. See M. de Reuck, 'History of the Department of Chemical Engineering and Chemical Technology', 6. (Typescript in ICA.) Chemical technology was then seen as a major German success story and many believed that Britain needed to catch up.

48. In 1911 Thorpe issued a report on how the chemistry department should be organized and suggested two divisions 'pure' and 'applied'. The pure division was to have three professors in inorganic and general, analytical, and organic chemistry. The applied division was to be built around the work that visiting Professor Bone was organizing but was to include also instruction in the construction of chemical works and factory management. He also wanted a chemical museum with exhibits of both historical interest and of chemical products of current interest. KC/9/1

49. William Arthur Bone FRS (1871–1938) was a native of Teesside, a lover of cricket and of Wagner. A Methodist, he was educated at Quaker schools before studying chemistry at Owen's College. He was a research student under H. B. Dixon at the University of Manchester and then, under Victor Meyer, at Heidelberg. In a letter (26 August 1927) congratulating Bone on the publication of his *Flame and Combustion of Gases* (co-authored with Donald T. A. Townend; London, 1927), Dixon wrote that when, fifty years earlier, he repeated Bunsen's experiments he little expected the English School in flame spectroscopy to develop as it had, and that Bone was now the foremost exponent of the school. Bone and Townend's book, dedicated to Dixon, became a standard work in its field. So, too, did a second work that Bone co-authored with Townend and D. M. Newitt, *Gaseous Combustion at High Pressures* (London, 1929). Bone donated all his book royalties to the National Children's Home and Orphanage. Newitt and Townend, both research students under Bone, submitted a joint thesis for their DIC, possibly the first joint thesis at Imperial. For Dixon letter see B /Bone; scrap book containing reviews of Bone's publications. See also

G. I. Finch and A. C. G. Egerton, 'William Arthur Bone, 1871–1938', *Obituary Notices of the Royal Society of London*, vol. 2 (1939), 587–611.

50. Quoted in Kathleen W. Greenshields and W. D. Wright, 'Historical Introduction' to a booklet published on the occasion of the opening of the new physics building, 1960. Bone and Callendar were not alone in having to write their own letters, most professors had to do so. Staff had letter presses on their desks to make copies. Another point of contention, about which Bone often wrote to Gow, related to the 'noise' from the neighbouring Royal College of Music (RCM). He stated that the organ and orchestral music drifting out of windows, that the RCM refused to close, distracted his students. See, for example, KCT/1–4 Bone to Gow, 29 June 1920.

51. See KCT/1–4; Bone to Gow, 13 December 1917. J. W. Hinchley, professor of chemical engineering, was not given a phone extension because the line was overloaded.

52. F. W. L. Croker memoir. Typed manuscript from a tape made in 1980 by Francis Walter Leslie (Les) Croker who came to the chemistry department as a laboratory boy in 1926. In 1949 he was made Laboratory Superintendent for the physical and inorganic chemical laboratories and later for the entire department. The tape was made for Bernard Atkinson who was working on a departmental history at the time. Croker states that the first secretary was hired only in 1932 when Miss Hornsby came to work for Professor Briscoe. Croker also states that it was he who typed up Derek Barton's doctoral thesis. For Barton, see chapters 10 and 13.

53. See Mary Pugh, *The History of the Geology Department, Imperial College, 1958–88* (typescript, 1995), 64.

54. Bone carried out his high pressure work in a room hung around with rope curtains as a security measure. Chemical reactions were carried out in small chambers encased in steel. The chambers' construction resembled that of naval guns and D. M. Newitt, when a student of Bone, worked with a naval gun expert on improving their design. Safety measures were discussed in a letter to the Rector when Bone wanted to increase pressures from 500 to 1000 atmospheres. His research was funded by both ICI and Nobel. KCT/1–4, Bone to Tizard, 20 June 1931.

55. Quotation in Finch and Egerton, *op. cit.* (49), from a lecture given to the Society of Chemical Industry. Members of the *Zoroastrians* included Bone's Imperial colleagues, Armstrong, Baker, Finch, Newitt, Fraser and Townend and three people later associated with the college: A. C. G. Egerton then still at Oxford, Eric Rideal at Cambridge, and Henry Tizard then at the DSIR. For book see note 49.

56. KCT/1–4, Keogh to Bone, 17 September 1918.

57. George Ingle Finch FRS (1888–1970), born in Australia and educated at the *Eidgenossche Technische Hochschule* in Zurich, was appointed Demonstrator in Fuel and Refractory Materials in 1913. Before coming to England he had managed a *Badische Anilin* (BASF) works in Germany. He became professor of applied physical chemistry in 1936 and by then had a research group working in the area of electron diffraction. Finch is perhaps best known in college history as a mountaineer, member of the 1922 Everest expedition (his dispatches were published in *The Times*), and leader of student mountaineering expeditions. Finch pioneered the use of oxygen breathing apparatus at high altitudes. An improved version of his apparatus was used by Sir Edmund Hilary and Tenzing Norgay on their successful ascent of Mount Everest in 1953. Finch resigned from Imperial in 1952 and went to India as Director of the National Chemical Laboratory. See M. Blackman, 'George Ingle Finch', *Biographical Memoirs of Fellows of the Royal Society*, vol. 18 (1972), 223–239.

58. Geoffrey Wilfred Himus (1894–1964) was an afficionado of Lewis Carroll and his students would include allusions to Carroll in examination papers, hoping to gain extra marks. A graduate of the RCS, Himus retired in 1956. For further details see, M. de Reuck, *op. cit.* (47), 19 and 83–5.

59. John William Hinchley (1871–1931) was educated at Lincoln Grammar School, leaving to become an apprentice engineer at the Reliance Iron Works. He came to the Science Schools on a National Scholarship in 1893, was an excellent student and very active in union affairs. Later Hinchley headed the Siamese Metallurgical Services (part of the new Siamese Mint) in Bangkok. On returning to England he worked as a consultant and part-time lecturer at the Battersea Polytechnic before being appointed at Imperial. He was a founder of the Institute of Chemical Engineering in 1922. Hinchley left his excellent library to the department which was much enlarged after the Second World War by donations from A. C. G. Egerton. It was named the Egerton-Hinchley Library in 1960. See also, Edith Hinchley, *John William Hinchley: chemical engineer* (privately printed by Lamley and Co., 1935). The author was Hinchley's widow. She claims that Hinchley did very well at his grammar school but that his family was very poor and could not afford his matriculation examination fees. They saved for the fee three times but when the money had to be surrendered for a third time, Hinchley gave up any idea of going to university and became an apprentice. He won the National Scholarship while on a vacation course in London. (From 1886–91 the matriculation fee at the University of London was £2. While this is about £170 in today's terms,

the real difference between then and now is that late Victorian society was very much poorer overall, and poor families had almost no discretionary income.)

60. The department began with two professors; less ambitious than the original projection in which it was stated that the department would consist of 5 sub-departments each with a professor: fuel and heat, chemical processes, work and factory apparatus, drawing and design of chemical plant, and handicrafts, statistics, economy and library. See de Reuck, *op. cit.* (47).

61. KCT/1–4; Bone to Gow, 18 May, 1917. The emphasis on works experience was pushed further in 1917 when Bone arranged with the Skinningrove Iron Co. in Yorkshire for students to work there in the summer vacations. After the First World War this company took three new graduates each year and gave them three years management training, taking in all parts of the works.

62. The men were exact contemporaries, and also exact contemporaries of the Albert Hall.

63. Newitt, quoted in de Reuck, *op. cit.* (47), 90.

64. Humphrey was also renowned for having organized a spying mission to *Badische Anilin* (BASF) in 1919 in order to secure details of the German ammonia plant at Oppau. Since Hinchley made his students go on fact-finding missions to chemical plants, Humphrey was likely well trained. He was later made a Fellow of Imperial College. For more on Humphrey's role at ICI see W. J. Reader, *Imperial Chemical Industries: a history* (London, 1970).

65. ULP/Chemical Engineering; statement by D. M. Newitt, 1945.

66. This Federation provided funding for many of Newitt's research students carrying out work on blast furnaces, and also for some research students in the chemistry department after the Second World War.

67. To see how such problems were worked out at Imperial and other institutions see Clive Cohen, 'From Science to Engineering Science: chemical engineering in Britain, Germany and the United States', MSc dissertation, Imperial College London, 1994. The college had been thinking of introducing an undergraduate programme in chemical engineering earlier, but when Hinchley died in 1931 his chair lapsed and the idea was not taken up until after the war. In 1946 the chair was newly funded by Courtaulds.

68. Sir Alfred Charles Glyn Egerton (1886–1959) was educated at Eton and Christ's College, Oxford before entering University College London as an undergraduate chemistry student under William Ramsay. After some post-graduate work in France and Germany he became an Instructor at the Royal Military Academy at Woolwich before returning to Germany for further

work with H. W. Nernst. During the First World War Egerton held a senior position in the Ministry of Munitions and was a major figure in the setting up of explosives factories. After the war he returned to Oxford, and a readership. He was appointed at Imperial in 1936. In 1938 he became Secretary of the Royal Society and was much engaged there during the Second World War. He was also a member of the War Cabinet Advisory Committee and was sent to Washington in 1942 as temporary director of the British Commonwealth Central Scientific Office. Egerton was knighted in 1943. See D. M. Newitt 'Alfred Charles Glyn Egerton', *Biographical Memoirs of Fellows of the Royal Society*, vol. 6 (1960), 39–64. For more on Egerton see chapter 10.

69. KCTI 1–4, Chemical Engineering and Chemical Technology correspondence file; Bone to Egerton, 8 April 1936.

70. On Egerton's appointment, Finch was promoted to professor and Newitt to assistant professor; E. A. Guggenheim had come in 1936 as lecturer in chemical thermodynamics but left for a chair at Reading in 1945. A. G. Gaydon came from physics to join Egerton in working on the spectroscopy of flames. Egerton also initiated work in low temperature chemistry and gas liquefaction.

71. The role of aeronautics during the war is discussed in chapter 5.

72. Sir Richard Tetley Glazebrook (1854–1935) was educated at Liverpool College and Trinity College, Cambridge. Fifth wrangler in 1876, he carried out research under James Clerk Maxwell at the Cavendish, and was appointed Demonstrator, later Assistant Director, by Lord Rayleigh. He was briefly Principal of University College Liverpool until, in 1899, he became Director of the new National Physical Laboratory at Bushey House. He was Zaharoff Professor of Aeronautics at Imperial, 1920–23.

73. The NPL had an aeronautics section because Haldane, when Secretary of State for War, had wanted it set up to support flying in the services. Glazebrook, the director of the NPL, took an interest in Dalby's early courses and invited some of his students to work at the NPL. Some were supported by the Women's Aerial League (the women's arm of the Aerial League of the British Empire) which gave three scholarships to the college. One of the early recipients was Herman Shaw, later Director of the Science Museum. See also chapter 5.

74. GB minutes, 14 December 1917. Zaharoff gave money to other universities also; for example he endowed the Marshall Foch Chair of French Literature at the University of Oxford.

75. This was not a precedent since there had been University of London chairs in the RCS and C&G colleges before federation. W. E. Dalby, for example, held a University of London chair in engineering. However, once the university became a teaching institution it was not keen to see its chairs go to colleges that had not federated. Zaharoff was a bit of a stickler on the terms of his donation and at one point wanted the college to return his money. But the government intervened and he changed his mind. After the death of Cambridge engineer, Bertram Hopkinson, who had been expected to take the chair, Zaharoff asked that Glazebrook be the first professor. While Glazebrook did not have specialized knowledge in aeronautics he was rightly seen as a highly capable administrator who would appoint good staff.

76. See EC 28 February and 23 May 1919. The site of the testing house was dedicated for future expansion of the Science Museum.

77. Given that Dalby had introduced aeronautics as a subject in the City and Guilds the placement of the new department in the RCS seems odd. Lowry, in his memoir, speculates that this was because Keogh had trouble with the independence of the C&G college, especially with Dalby who was not accepting of his authority. B/ G. C. Lowry, 'Pearls before Swine', 8 (ICA typed memoir).

78. Leonard Bairstow FRS (1880–1963) was born in Halifax and won a scholarship to the Royal College of Science in 1898. One of his fellow students was Henry Wimperis who later held important positions in the Air Ministry and was an influential figure in the early development of radar. Wimperis served on the Governing Body of Imperial College and declared Bairstow to have been 'the most brilliant student that had been produced by the College' over several decades. Bairstow moved to the NPL after completing his associateship and some further advanced work. There he began work in aerodynamics and over the course of his career made a significant contribution to improving both the control and stability of aircraft during flight. During the First World War he worked for the Air Board on the design of aircraft and worked on his book *Applied Aerodynamics* (London, 1920). A second much enlarged edition appeared in 1939. He was appointed to a professorship at Imperial in 1920 and became Zaharoff Chair and head of department three years later. See G. Temple, 'Leonard Bairstow', *Biographical Memoirs of Fellows of the Royal Society*, vol. 11 (1965), 23–40.

79. Before electronic computers became commonplace many women were hired as calculators. Bairstow had a number of women assistants, the two mentioned being the best known.

80. Whittle worked at Power Jets, a company founded to exploit his inventions, also located in Farnborough. Arnold Hall came with superlative letters of recommendation from his Cambridge professors, B. Melvill Jones, and G. I. Taylor. For more on Hall see chapter 10.

81. D. Brunt, 'The Position of Meteorology at Imperial College'; memo submitted by the college to the university in May 1938. Appendix A to P. A. Sheppard, 'A History of the Department of Meteorology, 1920–52' (typescript in ICA).

82. Sheppard, *op. cit.* (81),1. The Meteorological Office occupied the floors above the Post Office. The space now belongs to the Science Museum.

83. Sir William Napier Shaw FRS (1854–1945) was born in Birmingham, the son of a manufacturing goldsmith. Educated at King Edward VI School in Birmingham and at Cambridge, he stayed at Cambridge for some years after graduating, as fellow and tutor at Emmanuel College. He was appointed Demonstrator (along with Glazebrook) at the Cavendish when Lord Rayleigh was appointed Professor. Shaw became Secretary of the Meteorological Office in 1900, and Director, 1905–20. Austin, a major collaborator, returned to the Meteorological Office when Shaw retired from Imperial. William Napier Shaw (with the assistance of Elaine Austin), *Manual of Meteorology*, 4 vols. (Cambridge, 1926–31).

84. Sir Gilbert Thomas Walker FRS (1868–1958), a senior wrangler, held a fellowship at Trinity College, Cambridge before joining the Indian Meteorological Office in 1903. At Trinity he tutored Robert Strutt, later also a professor at Imperial College. Walker had many interests including one in 'primitive' hunting devices. In this connection he had made a study of the boomerang and earned the nickname 'Boomerang Walker'. He approached the new (to him) field of meteorology in statistical fashion, collecting and analysing masses of data. By 1920 he had arrived at a global view of the seasonal variation in climate and described his famous discovery of the Southern Oscillation as follows: 'there is a swaying of press[ure] on a big scale backwards and forwards between the Pacific Ocean and the Indian Ocean'. He linked this and similar oscillations with increases and decreases in rainfall, though never able to fully reach his goal of predicting when monsoon rains would fail. The Pacific atmospheric circulation, linked to sea surface temperatures, is now named the Walker Circulation. See, Brian Fagan, *Floods, Famines and Emperors: El Niño and the fate of civilizations*, (New York: Basic Books, 1999), pp. 19–22, quotation, 20. See also G. I. Taylor, 'Gilbert Thomas Walker', *Biographical Memoirs of Fellows of the Royal Society*, vol. 8 (1962), 167–74.

85. The College and the University were, however, neighbours since the University had its offices in the Imperial Institute building until moving to Senate House in 1936. Imperial College had been designated a School of the University which, in the older usage, simply meant that it was a place where students who had matriculated at the university could receive an approved course of study. The Royal College of Science and the City and Guilds College, but not the Royal School of Mines, had earlier been approved Schools of the University. Being an approved school did not imply any formal association, or that the university had any control over the college's affairs.

86. GB minutes, 11 November 1910.

87. For example, the Indian Forest Service insisted on its senior employees having degrees. Earlier the ARCS was accepted.

88. GB minutes, 9 December 1910.

89. B/Acland; correspondence; Acland to Crewe, 19 February 1913.

90. GB minutes, 9 December 1910

91. Memorandum by Rector for presentation to the Royal Commission, submitted to GB in July 1910.

92. GB minutes, 9 December 1910. See also Rector's Memorandum.

93. This was the first of several student petitions. See note 100 below.

94. See ABD/ 5–11 for Royal Commission on the University of London, Imperial's submission and correspondence. Also, EC minutes, 22 March 1918 for further student proposals for an independent technological university.

95. Callendar, quoted in W. D. Wright, 'History of the Physics Department', 17. (This typescript manuscript exists in several versions with different page numbers. The one cited here is in the possession of the physics department.)

96. Hugh Callendar, the professor of physics was an exception; courses in his department prepared students also for the University of London examinations. A. A. Eldridge recalled that when he graduated in 1910, examinations were still written with quill pens, specially provided. (Eldridge, 'Some Personal Reflections', Diamond Jubilee Lecture, Imperial College Chemical Society, 21 February 1956).

97. Unlike the three Imperial colleges, University College London and King's College London had traditions of entry examinations in literary and classical fields, even for science students.

98. Rector's memorandum, July 1910. W. E. Dalby, Dean of the City and Guilds College, was a living reminder. He left school at age fourteen for a seven year apprenticeship with the Great Eastern Railway, had taken

examinations at the Science Schools winning an entry scholarship and later the Whitworth scholarship. Other examples of eminent people failing to matriculate are Arthur Fage FRS, a mathematics student who later made a major career in the aeronautics section of the NPL, and O. V. S. Heath FRS, who failed the university entrance examination in mathematics, but was accepted at Imperial after taking a trivial *viva voce* exam in front of Andrew Forsyth. Forsyth clearly thought biologists did not need mathematics. Interestingly, Heath later used mathematics in his work and was able to learn what was needed. See O. V. S. (Peter) Heath diary, 3 October 1920 and *passim*. Diary privately owned by Robin Heath.

99. In 1923 eleven City and Guilds students passed the degree examinations, but failed the diploma examinations. See Marjorie de Reuck and Edwin Glaister, 'A History of the Mechanical Engineering Department (typed manuscript, 1960), 48–50.

100. See *The Times*, 30 January 1919, p. 3; 17 and 31 December 1919, pp. 18 and 6. One year later the students petitioned the Prime Minister, David Lloyd George.

101. *The Times*, 8 January 1920, 6.

102. The final push towards unification on the university side came after a report by delegates of the University of London who visited the college in 1924. One of the visitors, Sir Alfred Ewing FRS, was strongly in favour of the university recognizing Imperial's associateship diplomas for its degrees. He also stated that the college should be able to keep its four diplomas. He found the programme at Imperial well balanced though, like Meston (see above), he pointed out the dangers of too close a relationship with industry and noted that undergraduate training at Imperial leaned too much in the direction of utility to future employers at the expense of students being given fundamental scientific knowledge. See Annual Report, year ending 31 July 1924.

103. Sir Thomas Henry Holland FRS (1868–1947) was born in Cornwall and educated at the Normal School of Science. He received his associateship in geology in 1888 and became a research student with Professor Judd. He worked briefly at Owen's College before joining the Indian service in 1890. In 1908 he became Director of the Indian Geological Survey. In 1910 he returned to England to become Professor of Geology at Manchester but, in 1916, was seconded back to India for war work where he became Minister of Munitions. He returned to Manchester in 1918 and, in 1922, succeeded Keogh as Rector at Imperial. He was active in many professional societies, was President of the Institution of Mining and Metallurgy, Chairman of the Empire Council of Mining and Metallurgy, President of the Institution

of Petroleum Technologists, and Chairman of Royal Society of Arts. Holland left Imperial in 1929 to become Principal and Vice Chancellor of Edinburgh University. He had just been reappointed for a second term as Rector when he decided to leave; perhaps he took this decision because he was not given the reasonable salary increase that he asked for. Tizard, his successor, received a salary of £2500 plus a £500 entertainment allowance. Holland had received £2000 and no entertainment allowance.

104. In 1924 the Board of Education set up what became known as the Hilton Young Committee to review the federal structure of the university. Holland made strong representation to the committee on the burden of the double examination system. Some of the committee's recommendations were incorporated in the revised statutes of the University of London Act of 1929.

105. Board of Studies minutes, 21 June 1923

106. Gradually, after the Second World War, all students had to meet newly revised university matriculation standards. The RSM was the last to fully comply, doing so only by 1953. Compliance meant the end of the intermediate (first) year and, following the lead of the other two colleges, the RSM gave only three year courses from that date. Later in the century four year degrees were reintroduced.

107. This was seen as especially offensive since the University of London OTC drilled in South Kensington where the university had its administrative offices.

108. See *The Phoenix*, vol. XI (1925), 41. Second quotation from letter to the editor, *The Phoenix* vol. XI (1926), 154. This intervention should be understood also in the context of a national debate on the OTC movement taking place at the time. Bertrand Russell was another opponent, but General Smuts found the OTC ideals not unlike those of the new League of Nations. After the war, militarism was a touchy subject at Imperial as elsewhere. But Bull stated that for Britain to maintain its position as a world power a defensive organization was essential (*ibid.*). For the formation of the OTC see Lowry memoir (*op. cit.* (77)), pp. 13–15. Some students had for years joined the University OTC independently, and there had been some activity earlier at Imperial when metallurgist, William Merrett, trained a corps of engineers recruited from the RSM, RCS and C&G.

Phillip Charles Bull (1889–1960) was educated at Harrow School and the Royal College of Science. After serving in the war he joined the chemistry department in 1919. He also ran a chemical consulting company in the Fulham Road together with some other staff members and ex-students. Independently wealthy, he lived in rather grand style on Queen's Gate. He was an excellent photographer and, as a close friend of Wallace Heaton, always had the latest equipment.

109. Some colleges within the university did offer engineering degrees, but their students had to pass matriculation examinations in some arts subjects and had to take some arts or humanities instruction while at the university.

110. Before the agreement, while Imperial's entrants did better, on average, than those from other colleges, about eighteen students failed the degree examinations each year. After 1926, the number of failures declined dramatically to about one or two per year. The new examination system consisted of two parts. Students entering the first year courses still took the intermediate examination at the end of their first year. Part One was taken at the end of their second year and Part Two at the end of the third year. A fourth year was then available for more specialist work without examination. Those who entered the college exempt from the intermediate examinations could take Part One at the end of their first year (the second year course programme) and Part Two at the end of their second year. This latter pattern became the norm in 1947, a three year degree programme became standard, and the old first year programme was abandoned. Only after the Second World War did regulations come in that required all degree students to spend at least three years in study. Prior to this, students could take both their associateship and degree examinations in their second year and then leave. The new regulation meant that there was a much greater demand for the specialized courses given in the third (fourth for those still entering without higher school certificates) year.

111. The PhD degree was introduced at the University of London after the First World War. The government encouraged universities, many of which were reluctant, to introduce the degree so as to keep research students in Britain rather than have them seek the degree elsewhere, notably in Germany and the USA. The PhD soon replaced the DSc in popularity.

112. There had been a similar problem earlier when Instructors at the RSM were reluctant to exchange that title for Lecturer in 1907. Professors and assistant professors not recognized by the university were never addressed by their Imperial titles in university correspondence but always as Dr., Mr. Mrs. or Miss.

113. Sir Henry Thomas Tizard FRS (1885–1959) was born into a naval family and grew up near the dockyards at Chatham. His father had been the navigator on the Challenger expedition. Tizard had expected to enter the Navy but was excluded after a negative eye examination. He won a scholarship to Westminster School and from there moved to Oxford where he

studied both mathematics and chemistry. On the advice of his chemistry tutor, N. V. Sidgwick, he spent a year in Berlin with H. W. Nernst after taking his degree. It was in Nernst's laboratory that he first met Frederick Lindemann whose later career was to intertwine with his own. In 1914 he joined the Royal Flying Corps and was put in charge of a team testing the performance of service aircraft. In 1916 he became a qualified pilot. A major influence at this time was the Cambridge engineer, Bertram Hopkinson FRS, who rationalized research in military aeronautics and moved this work to two neighbouring locations in Suffolk, Orfordness and Martlesham Heath. Tizard was put in charge at Martlesham, supervising research on structures and aerodynamics. Later he moved to London with Hopkinson and when, in 1918, Hopkinson was killed while piloting his own aircraft, Tizard was appointed to succeed him as overall Controller of Research and Experiment. Tizard was a victim of the 1918 influenza epidemic which ended his military career. He was demobilized and returned to Oxford, to a fellowship at Oriel, and to chemistry. His work on liquid fuels and spark ignition found many industrial backers. He was promoted to a university readership in 1920 but did not stay long. In the same year he accepted an offer from Sir Frank Heath to become Assistant Secretary at the new Department of Scientific and Industrial Research (Sir Francis Grant Ogilvie, member of the Governing Body, was Principal Assistant Secretary at this time). Tizard succeeded Heath as Permanent Secretary in 1927 (Tizard's successor as Secretary in 1929 was Sir Frank Smith, FRS, a former RCS physics student). In 1929 he was appointed Rector of Imperial College. Heath was appointed to the governing body in 1931 and was influential in College affairs. For more on Tizard see chapters 6 and 8. See also R. W. Clark, *Tizard* (London, 1965).

114. The Royal School of Needlework building, which opened in 1901, was the project of Prince and Princess Christian of Schleswig-Holstein (she was the third daughter of Queen Victoria). An earlier neighbour of the Central Technical College, and supported also by the CGLI, was the small Art Woodcarving School which had space for twelve students a year who were given free tuition. It opened in 1884 but did not survive long.

115. Geoffrey C. Lowry (1894–1974) was College Secretary 1934-1958. He was educated at Westminster School and Magdalene College, Cambridge, where he received a first-class degree in political economy. After military service during the First World War he worked first for the Ministry of Labour and then as Superintendent of a Working Men's College. A former Captain of Boats at Cambridge, he was a major supporter of rowing at Imperial.

116. For the acquisition and sale of the field station at Hurworth, Slough, see chapter 7.

117. The island site is the land between Imperial College (formerly Institute) Road and Prince Consort Road, bounded by Exhibition Road in the east and Queen's Gate on the west. Lowry gives a good description of the site in Tizard's period. While the buildings on its perimeter were largely permanent, the interior was filled with temporary structures of various kinds. At the centre was the Great Hall of the university linked to the Imperial Institute. This was the most grandiose of the temporary buildings having been originally constructed as a changing and rest room for Queen Victoria when she opened the Imperial Institute. See B/ Lowry, *op. cit.* (77).

118. See GB minutes for October 1937 with letter from Patrick Duff, HM Office of Works. Rector's reply, 22 November 1937.

119. Tizard's plan for the site included fewer and smaller buildings than were erected during the Jubilee expansion. There was to be a central boulevard allowing a clear view of the Albert Hall from the southern end of the site. There were also plans for the construction of a rather grand residence for the Rector.

120. Evelyn Hugh John Boscawen, eighth Viscount Falmouth (1887–1962) was educated at Eton and Trinity College, Cambridge. He served in the army during the First World War, was an accomplished engineer, and sat on many boards, including the advisory councils of the DSIR and Fuel Research Board. During the Second World War he headed the Fire Division of the Ministry of Home Security, for which several Imperial College people worked. He had major land and china clay interests in Cornwall and was active in the affairs of the London County Council. He was appointed to Imperial's governing body in 1932, becoming chairman in 1947.

121. B/Tizard, typed obituary notice (anon.)

122. Professor Fortescue, head of the department of electrical engineering, took objection to charging for his own time, but Tizard insisted he do so. Tizard had problems trying to sort out Professor Munro's many contracts and thought Munro had taken on too much. But he did insist on proper charges and laid down the rules. See, for example, correspondence in 1930 with regard to work being done by Munro for Horlicks on insect infestation in their malted milk stocks; KZ/9/4/1.

123. This company had works on Horseferry Road, a site often visited by students. The chemical technology department received a research grant from the company. While Sir David Milne Watson, the president of the

company, was happy with the contract he was annoyed when the college installed electrical heating for the Union building. The most recent upgrade in the college heating system was completed in 2000 with a new heat and power plant built in partnership with London Electricity. This latest venture was supposed to save the college up to £900,000 a year (Annual Report, 1999–2000).

124. J. G. (Jimmy) Peacock joined the army on leaving school; he served in Africa and India and received some training in military engineering. After about four years of service he came to the RCS and RSM as assistant to his father who was Clerk of Works (in 1907 Clerk of Works at Imperial). Peacock succeeded his father in 1908 and worked with Aston Webb on the construction of the Bone Fuel building, the Beit Hostel, the boat house in Putney, the laboratories at Hurworth, Slough, the Whiffen Laboratory, and on a number of other construction projects. During the war he held a commission with the Royal Flying Corps, continuing to serve until 1920. In 1925, when the Union was in financial difficulty, the Rector, Thomas Holland, asked him to become Manager. He and his wife brought stability after some years of chaos. Peacock was a well-known college figure, a great supporter of sporting events and other student activities.

125. GB minutes (vol. 36) 1942. This valuation was in anticipation of war damage and insurance claims.

126. See B\Tizard\1. Philip agreed to a salary of £1000. The college used some of the extra money to increase the salary (and responsibilities) of G. C. Lowry, the College Secretary. However he, too, left to serve in the war but returned to the college on the death of Philip in 1941. H. J. T. Ellingham was appointed Deputy Secretary to assist him.

127. GB minutes, circular letter from Tizard, 20 November 1939.

chapter five

Imperial College during the First World War

Governance and College Life

Imperial College was still young when hostilities broke out in 1914. The war seriously disrupted college life and it was feared by some that with the loss of so many to the services, and to war duties of various kinds, the college would have to close and that it might never reopen, at least not as a unity of the three older colleges. That Imperial survived owes much to the determination of the Governing Body, and to the fact that the college staff proved useful to the war effort in many ways. War opened people's minds, especially of those working in government, to the utility of scientific and technical expertise.[1] Scientists and engineers came increasingly to be seen as useful to the country's economic and general well being. However, according to J. D. Bernal, the utility of science in war was not recognized overnight. He recalls a leading scientist who, on offering to organize a meteorological service for the army, was informed that, 'the British soldier fought in all weathers'.[2] Like many others, the army generals did not fully understand what science had to offer. While the collaboration of scientists and other civilians in the war effort went far beyond anything that had happened earlier, people at Imperial bemoaned the fact that students were still being used as cannon fodder when they could have been used in far more constructive ways. That this was not more widely understood is puzzling given new technical weaponry such as tanks, zeppelins, aeroplanes, submarines, big guns, high explosives, motorized transport, and poison gas.

Only after much agitation, and after the fall of the Asquith government in May 1915, did Lloyd George's coalition government establish a number of new boards and departments, including the Admiralty Board of Invention and Research and the Ministry of Munitions, to deal with technical matters.[3] An Air Inventions Committee was formed in 1917. As we shall see, people from Imperial worked for these, and for other government departments during the war.[4]

One of the first people to leave the college for war related work was the Rector, Sir Alfred Keogh.[5] In September 1914 he volunteered and was appointed Chief Commissioner of the Red Cross Society, beginning work in France. Initially he was given leave for just a few months, but Lord Kitchener requested that he return to the War Office in London and resume his former post as Director General of the Army Medical Services.[6] This meant that Keogh's leave from Imperial became indeterminate. He had a very good press which reflected well on the college, and he was seen as making a great contribution to the overall health of the army. He introduced many new hygiene regimens and organized the setting up of military hospitals. His activities were keenly followed in the student journal, *The Phoenix*. But Keogh's departure was a major blow. He had made a good start at unifying the three colleges but much remained to be done. The war also meant fewer students, and the major reduction in college income was a serious problem for the new institution.

The Governing Body appointed one of its own, Arthur Acland, a former minister of education (in Gladstone's final government), to act as Deputy Rector.[7] He deserves much of the credit for keeping the college running in difficult times. Already in 1914 about 300 students and 60 staff members had left to join the armed forces, the former a significant number given that the students remaining numbered only about 600. In 1915 a total of 932 were registered at the college but, of these, 470 were in the armed services and 66 were engaged in other war work.[8] Proportionately more army volunteers came from the Royal School of Mines than from the other two colleges, and they largely joined the infantry (a RSM tradition). Many were killed in the retreat from Mons. In addition to the sharp drop in undergraduate student and staff numbers, many research assistants, fearing conscription as privates, left to work in munitions factories.[9] The conscription of young men was very divisive when introduced in 1916.[10] Further, the military authorities invited volunteers of increasing age as the war progressed, volunteers in the 40–50 age group by 1917. This proved too much for the College Secretary, Alexander Gow, who wrote to Acland that he could not see how the college could continue. Gow advised Acland to refuse staff the permission to volunteer and wondered 'to what extent the Governors are justified in accepting fees of students for courses of impaired efficiency'.[11] The Governing Body passed a resolution to refuse staff permission to volunteer, but later rescinded it on receiving a plea from H. A. L. Fisher, President of the Board of Education.[12]

The difference of opinion between those at the college and the Oxford educated Fisher is suggestive of a divide in attitude — perhaps describable as one between the middle- and the high-brow.[13] Those at Imperial did not see the war so much in terms of service, duty, sacrifice or glory, but rather as something that was perhaps unnecessary, and that was in any case being conducted inefficiently and irrationally.[14] It was held that if better attention were paid to technical matters and the 'scientific method' the country would be in better shape. Many at Imperial believed that their skills, and the college facilities, were not always being used in the most productive ways. Hew Strachan has discussed the ways in which the war was envisaged in 1914 and claims that for many it was a war of ideas.[15] Certainly this was the view of former Normal School of Science student, H. G. Wells. In his essay *The War to end the Wars*, Wells discussed the 1914 war in ideological terms as a war of liberal democracy against Prussian militarism and autocracy. Victory, he believed, would allow what he saw as progressive and democratic ideas to spread around the world.[16] Wells, a frequent visitor to his old college, wrote a number of articles for the student journal, *The Phoenix*, promoting the utility of science and technology in both war and peace. He urged Imperial students not to volunteer for military service believing that their being well trained in science and engineering would serve the country better. While Fisher will have shared some of Wells's liberal views, his patriotism was differently constructed.

The antipathetic positions of Gow and Fisher illustrate that views held at the college, while undoubtedly patriotic, did not mesh entirely with the needs of the country as seen from the perspective of the governing elite.[17] This was a major point of tension and the college was not yet sufficiently mature to easily withstand the strain. Acland wrote numerous letters throughout the war to officials at the Board of Education and the War Office asking that students and staff be exempted from military service, but with limited success. While many Imperial students did volunteer, the college authorities differed from those at Oxford where the utmost was done to help in mobilizing young men, including scientists.[18] Even after Lord Derby brought in a scheme for exempting students who were making a contribution to the war effort, most Imperial students were found not to qualify, causing much annoyance, even anger, at the college.[19] At best, some could briefly postpone military service.

Despite college ambivalence towards military service, many of the academic staff joined students in volunteering for the armed forces. Not all were able to use their technical expertise to the full as army volunteers, but some did. For example, applied mathematician Herbert Klugh joined the RNVR and spent the war engaged in mine sweeping at which he became expert.[20] He kept to this work, helping to clear Britain's coastal waters, even after the war had ended. Stephen Dixon, the professor of civil engineering, became a personal assistant to Sir Henry Fowler at the newly formed Ministry of Munitions, but continued some part-time work at the college.[21] In 1917, even though past middle age, Dixon enlisted and received a commission as a transport officer and was sent to France. After ten days he was transferred to the Royal Engineers and spent much of the rest of the war building railways behind the British lines, something at which he was an expert.[22] There were few students left in his department. The second year course should have enrolled thirty-six, but twenty-six had left to join the armed forces. Three of the department's four postgraduate students left to work in an explosives factory. Since a similar situation existed also in mechanical engineering, the two departments resumed the combined engineering programme (offered earlier) for the duration of the war.

Acland asked Keogh to write letters to the press detailing the work of the college, stressing its mandate and emphasizing Imperial's central importance to the Empire, and that it should be left to do its work even in wartime.[23] Articles were sent to the press detailing the importance of the college's work to the war effort and Britain's industrial progress.[24] Imperial scientists were not alone in bombarding the press with complaints about the failure to properly mobilize scientific talent. Some press coverage was given to a 'Memorandum from Men of Science' making much the same point and lamenting the authorities' failure to understand that science and technology had an important role to play in the war. In particular it stated that highly trained personnel should not be used for normal army service.[25] By the end of the war the college had lost about one-third of its teaching staff and more than half of its registered students to military service. But, despite this, most courses were offered throughout the war, albeit in diminished form.[26]

Further problems were created when the government commandeered space within the college to house various war related operations. As with other universities, Imperial became something of a military camp.[27] For example, the War Office pay office took over the top two floors of the new C&G's Goldsmiths' extension, as well as rooms in the RSM. The new engineering structures laboratory in the extension was taken over by the Admiralty Invention and Research Board. It was not used for research, but was handed over to the Irish and Scots Guards for the housing of recruits. The Board also took over the C&G's hydraulics laboratory and the sumps were said to have been used as baths by the new recruits.[28] Some soldiers were billeted in the RSM and, according to one witness, in the morning 'one had to step over sleeping soldiers lying in the corridors'. The tops of museum cases were used as dining tables and the soldiers discarded all sorts of scraps into specimen drawers. Mice followed and ate not just the scraps but also some of the specimens.[29] The Admiralty also took over part of the physics workshop which was used for experimental work on bomb-sighting. The Machine Gun Section of the Army Pay Corps took over the RSM examination hall, and the Air Board housed its drawing office and printing press in the Science Schools building, much to the inconvenience of the mathematics and zoology departments.[30] The Royal Flying Corps drawing office was not small; it employed roughly twenty men and thirty women and occupied several rooms in the building.[31] The college felt an obligation to house also various volunteer groups such as the Women's Emergency Corps, and a group of French scientists working on explosives for the French government in London. About fifty refugee students, mainly from France, Belgium and Russia, were taken in and given financial assistance in the form of reduced fees and help with accommodation. Help was also given to empire students who were invalided out of the armed forces and unable to return home. Some refugee professors were also accommodated; for example, Professor Hubert, a mining engineering professor at the University of Liège, came with some of his research students.[32]

All this activity made running the day to day operations of the college difficult. Added to this were problems with the police when rules were broken, usually having to do with issues such as the failure to black-out windows, or follow fire prevention codes.[33] Night patrols to watch for fire were instituted.[34] Some students joined a unit of the Officers Training Corps attached to the Royal Engineers.[35] Not yet willing to be associated with the University of London, most students shunned the units there.[36] Some of the staff and students joined citizens volunteer units. Geologist Percy Boswell described his activities as the member of a group, calling itself the Artists Rifles, which drilled and practised skirmishing in Richmond Park.[37] Most of the women students volunteered for the Land Army, spending their vacations on farms to which they were assigned.[38] Lord Crewe, chairman of the governing body, ordered all college personnel to carry identification. Official badges were issued by the Admiralty and War office to those engaged in war work for the government.[39]

The Founding of the Department of Technical Optics

The war brought a major new subject area to Imperial, namely technical optics. By the early twentieth century Britain had lost its lead in the manufacture of optical instruments. Many of the small craft firms that had been leaders in their field during the late eighteenth and early nineteenth century were gone. Further, even the manufacture of optical-grade glass had declined. For example, only one firm, Chance Brothers of Birmingham, manufactured glass suitable for telescopes and range finders. Even before the war these problems were seen as serious. Many people, including Silvanus Thompson, Principal and Professor of Physics at Finsbury Technical College, were trying to help modernize the optical industry.[40] Thompson had visited the Carl Zeiss works in Jena and he promoted new approaches to instrument making both at his college and among the London trades. He was very critical of how optics was being taught in universities, 'they call it optics but it is really mathematical gymnastics applied to the optical problems of one hundred years ago'.[41] In 1905, at his urging, the London County Council (LCC) set up an Institute of Technical Optics and the Northampton Polytechnic in Clerkenwell began offering courses. But the start of the war turned the shortage of good optical instruments, and the fact that people in the armed forces were not trained in their proper use, into a national crisis.[42]

Imperial College saw an opportunity and, in 1915, petitioned the Board of Education to fund a course in technical optics.[43] In 1916, the Board agreed to increase its grant to the college for this purpose. A committee, which included college personnel and representatives from the government and the optical trades, was set up to consider how to proceed. It was agreed that courses for the optical trades would continue at Northampton Polytechnic but that Imperial would become a centre for research and advanced teaching. In 1917 the Department of Technical Optics was founded and ran a large summer school to attract possible students. This was attended by a motley crowd including Cambridge wranglers and trades apprentices.[44] It must, however, have been a success, since sixty-six students registered when the full course began in 1918, and a further fourteen took the correspondence course set up at the same time.[45] The fact that within a very short period this new department gained three professorships is testimony to its perceived importance. The first to be appointed was the head of department, F. J. Cheshire, Director of the LCC Institute of Technical Optics and honorary head of the optics department at the National Physical Laboratory.[46] Cheshire was also Scientific Director of the Optical Department at the Ministry of Munitions where he became Deputy-General. He understood well the military requirements for optical instruments. But even before his arrival, Imperial was not entirely lacking in optical expertise. Spectroscopy was a major research area in the physics department, a field pioneered by J. Norman Lockyer and continued by his student Alfred Fowler, by then a professor.[47] Fowler made a number of specially commissioned optical instruments during the war and one of his research students, L. C. Martin, became a lecturer under Cheshire.[48] Fowler also made monthly diagrams of moonlight illumination which were much appreciated. A captain working at the Admiralty wrote to him, 'our monthly distribution is nearly 125 copies … their fame has gone beyond [the British latitudes for which they were designed] and we have frequent requests for similar productions'. And, from the Meteorological Office, then situated in South Kensington, he received several letters stating that the War Office wanted diagrams for all future months since they were so useful in issuing instructions.[49]

In the geology department Percy Boswell was carrying out research on glass sands. Boswell became an associate of the new department of technical optics and taught some classes there. He also began visiting

glass manufacturers, advising them on which sands to use in manufacturing optical grade glass.[50] In this connection he worked under the
Ministry of Munitions which employed a number of outside specialists to visit glass and steel works to see whether they were working to
maximum capacity. Up to 1914 manufacturers in these areas were very
secretive about their methods, but a consequence of mandatory wartime
inspections was that expert advice became more appreciated, and more
eagerly sought, after the war.[51] A. E. Conrady, lens designer to the
leading British microscope firm of the day, W. Watson and Sons, was
appointed Professor of Optical Design — though he continued to work
also for Watson's on submarine periscopes and gun sights. In addition
to possessing great technical ingenuity, Conrady turned out to be a very
good teacher. He remained at Imperial until his retirement in 1931.[52] A
third professor, A. F. C. Pollard, was appointed in 1919 to give instruction in the mechanical design of optical equipment.[53] Several companies
donated equipment to the new department which was located on the
top floor of the old Science Schools building. Though the department
expanded yet further by hiring some junior staff after 1918, it reached its
peak in student enrolment already in 1920. After the war the field was
no longer seen as a priority and student numbers declined rapidly. The
undergraduate course in technical optics was discontinued, though the
two-year postgraduate course, run by Pollard, was retained and research
in the field continued. Pollard offered also a second postgraduate physics
course in the design of optical instruments. Cheshire remained as head
of department until 1925, but after Conrady's retirement in 1931 the
department collapsed into a section of the physics department under the
nominal headship of A. O. Rankine. In fact, Rankine's chief interest was
in acoustics and geophysics. Martin, by then an assistant professor, taught
optics for the physics associateship and BSc course.

The Chemists' War

Perhaps of all the departments at Imperial it was the chemistry department that contributed most to the war effort though, as this chapter
illustrates, the old adage that the First World War was a chemists' war
is not entirely correct.[54] However, the use of poison gas was seen as an
almost singular horror and came to play a central role in the collective
memory of this war. Chemists were put to work both on the offensive use

of gases, and in finding methods of defence. But they were engaged also in a range of other activities. As William Pope, Professor of Chemistry at Cambridge and former student of Henry Armstrong at the C&G, put it, 'practically every chemical technologist and every academic chemist in the country' was engaged in war related work.[55]

Chemistry professor, James Philip, was appointed Chairman of the Royal Society Committee on Synthetic Drugs and coordinated the preparation of pharmaceuticals in colleges and universities throughout the country.[56] His own department was his model. Jocelyn Thorpe, professor of organic chemistry, had offered his services at the start of the war and his research students were immediately set to work figuring out German patents, especially for anaesthetics.[57] Within a few months Thorpe had some major production lines in operation, manufacturing drugs for use mainly by the troops.[58] In April 1915, Thorpe was asked by Keogh to go to Brighton Pier where Herbert Brereton Baker, the professor of inorganic chemistry, was carrying out some work for the Admiralty.[59] Thorpe was to inform Baker that intelligence had arrived indicating that the Germans were planning to use poison gas against the troops in France. It was rumoured that either arsine or arsenic trichloride would be used.[60] According to his diary, Baker immediately informed Keogh that, in his view, those gases would not be effective since they would not spread well, and would probably decompose rapidly once the shells containing them had exploded. Nonetheless he agreed to develop a respirator that would protect against them. One week later Baker was urgently summoned to meet Keogh at the War Office. By then gas warfare had begun and Baker was asked to go to the front with Keogh and J. S. Haldane to identify which gases had in fact been used.[61] Passports for Baker and Haldane were rapidly issued but, since there was no time to develop photographs, their thumb prints were used instead. In his diary Baker described the complicated manoeuvres needed to get out of Dover Harbour which had been netted and boomed to catch enemy submarines. Once across the channel they were driven to army headquarters in St. Omer, and then to the front.[62] On arrival Baker recruited one of his own members of staff, B. M. Jones, who literally had to be pulled from the trenches, to help set up a laboratory behind the line.[63] The next day Jones appeared for work in a kilt, uniform of the London Scottish regiment in which he had enlisted as a private.

The Director of the Central Laboratory, chosen by Keogh, was William Watson, assistant professor of physics at Imperial.[64] The main purpose of the laboratory was gas defence and Watson, together with his staff, ran many risks during the course of their work. They tested an evolving range of gas masks and other protective devices for the widening range of gases that they feared would be used.[65] After Baker's arrival, he and Jones were driven by army ambulance to a dressing station near the front and later confirmed that the gas used at Ypres had been chlorine. This was suspected from the evidence given by the survivors, and was confirmed by the analysis of a chemical deposit left on buttons taken from the uniform of a Canadian soldier. The soldier had barely survived the gas attack and was, according to Baker, 'a hero if ever there was one'.[66] Baker thought that phosgene could have been used in addition to chlorine. On returning to London, he worked on possible absorbents for that gas. A consequence was that helmets containing sodium phenate were issued to soldiers considered to be in danger.[67]

In his diary, Baker gives an interesting account of a respirator test. The headquarters for this work was the Army Medical College at Millbank. It was there that final tests of equipment, and decisions on what to issue to the troops, were made. On one occasion Haldane was sceptical of a helmet developed by members of the Newfoundland Medical Corps, and favoured by Baker. Keogh ordered a test. Baker wore the Newfoundland helmet which contained hypo-soaked cotton waste, while Haldane wore the gas mask that he favoured, one developed at the Central Laboratory in France. The two men were exposed to chlorine in a gas chamber. According to Baker, Haldane 'retired after four minutes and had not stopped coughing when I came out twenty minutes later'. But Haldane 'was intent on crabbing [the Newfoundland helmet]' and criticized its bulk, saying 'no one could move in it'. Baker then put on the helmet and ran up and down a passage. 'I must confess I was a bit pumped but I made as light of it as possible'.[68]

From this and other passages in his diary one can infer that there was some animosity between Baker and Haldane. Baker repeat-edly describes Haldane as remote and impractical. On one occasion he proudly notes that Kitchener told him that Haldane was 'too philo-sophical' and that 'you are the sort of man I want'.[69] Baker worked on respirators throughout the war and tested them at Imperial College. Encouraged by Lord Kitchener, he also investigated new chemicals to put in shells and hand grenades. These were tested by dropping them from aircraft at a Royal Flying Corps station outside London. Further, animals such as sheep and pigs, some wearing gas masks, were placed in trenches and then had shells containing various chemicals lobbed at them.[70] Baker contributed to the war effort also in many other ways. One of his major achievements was a device for ventilating carbon monoxide from pill boxes after the firing of guns, allowing for much greater efficiency in the use of artillery weapons. Like many others who worked for the Ministry of Munitions, Baker was rightly annoyed by the lack of proper recognition of this, and his team's many other achievements, once the war was over.[71]

Thorpe also worked on poison gases, but in production rather than detection or defence. In a memoir, Lady Thorpe recalls her husband returning from a meeting with Lord Kitchener and saying, 'I am going to dislike myself, I am going to be ashamed of being British'.[72] He turned down a Lt. Colonelship offered him by Lloyd George, rightly believing that it would have left him with even less control over what he did, but he agreed to carry out research in the production and properties of poison gases. He went to France to determine the behaviour of gases in trenches, and then had some experimental trenches constructed at Imperial.[73] He put Martha Whiteley in charge of trench warfare research in South Kensington,[74] and moved his main operation to Porton where the production and testing of a range of chemicals was carried out.[75] Experimental gas attacks were conducted on Salisbury Plain and J. N. Sugden, one of Baker's assistants, was responsible for collecting and analysing air samples from various sites.[76] Sugden also spent part of the war working on water purification, and on the delivery of safe water to the troops.[77] He and others built mobile purification units on trucks, and on barges, and delivered them to the fronts.

Thorpe reported that spies came to Porton disguised in various ways, one as a Russian general. He himself was arrested, and briefly imprisoned as a suspected spy when, without proper identification, he went to a research station at Dungeness, a place where trench gas research was also being carried out.[78] Students, too, often complained of being arrested as spies. This was especially the case for civil engineers doing surveying work, and for geologists who carried around ordnance survey maps. Thorpe received a number of honours for his work, was knighted at the end of the war, and was appointed also Officier de la Légion d'Honneur. After the war he built a lively research department and many people came to study with him. Whiteley, who had been synthesizing barbiturates before being put to work first on Thorpe's phenacetin production line, and then in charge of the trench warfare unit, was labelled in the Press as 'the woman who makes the Germans weep' because of her work with tear gases.[79] She had a small staff working with her, including G. A. R. Kon who moved to the chair of organic chemistry at the Chester Beatty Research Institute after the war.

A. A. Eldridge recalled working on many of the departmental war projects: the drug factory, the water purification units, and the trench warfare unit.[80] He also worked on the production of new antiseptics and helped Boswell with his research on glass sands. But mostly he helped Baker with research into respirators, measuring the rate of penetration of chlorine, phosgene and other gases through various fabrics. Muriel Baker suggested the use of ladies stockings stuffed with cotton waste which led to some useful developments.[81] The first respirators tested at Imperial were flannelette bags with celluloid windows. It was Professor Philip who developed the first box mask with a wood charcoal filter.[82] One former student remembered that he and others were put to work 'running backwards and forwards along the corridor with our heads shrouded in blue flannelette bags saturated in various chemical absorbents'.[83] He also remembered having to evacuate, seal, and then pack glass globes which were sent to the Front for the collection of gas samples. The Ministry of Munitions equipped one of the department's laboratories for the analysis of spent German gas shells, and it was there that the properties of mustard gas were first studied by, among others, Frances Micklethwaite.[84] Eldridge remembers her and her co-workers 'popping into the corridor every now and again to see how the experimental blisters on their arms were getting on'.[85] Christopher Ingold was a research student under Thorpe at the time, working also on war gases, but was soon sent to liaise with people in Glasgow where much gas work was also being done. One student remembers Ingold gassing himself with phosgene while still at Imperial, and that he was rescued by Hilda Usherwood whom he later married.[86]

There were other chemical factories in operation aside from the one producing drugs. H. V. A. Briscoe, who later returned to the college as professor of inorganic chemistry, had been a research student under T. E. Thorpe, and during the war was a junior staff member. He designed a factory for the manufacture of thorium nitrate and a number of rare earth compounds. He ran a pilot operation at Imperial while supervising industrial production elsewhere.[87] Much of Briscoe's work was secret and it is still unclear exactly what he did.[88] He was remembered later as having been a very competent experimentalist, with a wonderful set of lecture demonstrations.[89]

William Bone, the head of the sub-department of chemical technology, was a pacifist. While helping in the war effort he drew the line at poison gas work. Bone worked on fuel problems related to the war, and was Chairman of the British Association Committee on Fuel Economy. He trained some students for senior work in munitions factories and, more than most professors, tried to have his students exempted from military service. To help in this he made them sign forms stating that they were working on matters of national importance and that '[I] place my services unreservedly at your disposal'.[90] Because Bone saved some students from the armed services by insisting on the importance of their work, he could expect (and received) long hours of work in exchange. His students and paid assistants worked on Saturdays, and gave up all vacation time for the duration of the war.[91] A course on explosives was started in the department by G. I. Finch who had been appointed in 1913. But one year later he joined the army, as did five of his students. Four others left to join the Admiralty cordite factory at Holton Heath, near Poole in Dorset. Bone appointed two women lecturers because of staff shortages.

Other Scientific and Technical Contributions

One area to receive a boost because of the war was aeronautics. Interest in this field existed even earlier at the college. In 1909 Ernest Dalby, Dean of the C&G, went to Juvisy near Paris to attend an exhibition celebrating Blériot's cross-channel flight. There he saw a number of aircraft, some in flight demonstrations. The Governing Body encouraged work in this area and, already in 1909, was putting out feelers to see where funding might come from. Much came from the government, but the very first outside financial support came from the Women's Aerial League[92] which, in 1910, donated a £50 per annum scholarship for three years. The first recipient, F. H. Bramwell a post-associateship student at the college, carried out his work at the National Physical Laboratory (NPL). He was further subsidized by the governing body which gave him £100 for his research. By the start of the war, Bramwell was second in command at what had become a serious aeronautics department at the NPL. He then took a commission in the Royal Naval Air Service and worked for the Admiralty on the performance and stability of seaplanes, and some early bombers. Curiously he did not see much of a future for aeroplanes after the war and joined the staff of Imperial Chemical Industries where he later became a senior executive. As is discussed in chapter four, an aeronautics department was set up at Imperial only after the war. But by 1914 Imperial's advanced aeronautics diploma (DIC) course in mechanical engineering had already trained a number of people who became engaged in war-related activity. Dalby was urged by senior civil servants and military officers to train yet more.[93] Many of the early graduates worked at the Royal Aircraft Factory (later Royal Aircraft Establishment) at Farnborough, an institution with which the aeronautics department continued to have close ties once the war was over.[94] The Factory had a small wind tunnel and the control and stability of aeroplanes was a subject of intense study.[95] Aeroplanes played a surprisingly important role in the war and, by dropping bombs, caused a higher number of civilian casualties than expected.[96] This, and the fact that airships and aeroplanes could not easily be kept from British airspace, became a serious concern. Little could be done about it at the time though members of the electrical engineering department gave it their best. They adapted storm detectors to forecast the direction of attack by zeppelins. J. T. Irwin cut a huge parabolic acoustical reflector (about thirty feet in diameter) in the cliffs at Dover which, with a Tucker hot-wire microphone (invented during the war) at its focus, was used, sometimes successfully, to detect oncoming aircraft.[97] Acoustical work of this kind was important also in the detection of enemy submarines and gun batteries. As will be seen in chapter eight, the aircraft detection problem was later taken up by Henry Tizard, a future Rector of Imperial, when he became engaged with the development of radar before and during the Second World War.

The head of the department of metallurgy, Harold Carpenter, was on the Royal Society War Committee and was a senior advisor at the Admiralty. He arranged for the Bessemer Laboratory to be placed at the service of the war effort, and materials were manufactured there for the Ministry of Munitions. For example much work was done on aluminium alloys, and fifteen tons of an arsenical compound, and a similar quantity of tungsten ore, were treated there in 1916. The latter was for munitions being manufactured by the High Speed Steel Alloys Co. in Sheffield.[98] W. R. Jones, a former geology student who later became professor of mining geology, returned to the college from Malaya in 1915 and advised this company on tungsten deposits. He knew how to locate tungsten ores and, not long after his return, was sent to Burma as director of the company's mining venture there. His principal role was to ensure the shipment of ores to Britain. Similarly, Vincent Illing, who later became professor of oil technology, was sent to Trinidad to locate oil fields and ensure the shipment of potash and some other minerals used in paints. Laboratories in Illing's oil technology section of the geology department were put at the disposal of the government which installed an oil testing and refining laboratory for the Air Ministry. His colleague, geologist Morley Davis, together with a number of research students, spent the war at the Royal Geographical Society preparing a one-millionth map of Europe for the War Office, as well as many contour maps depicting railways and roads. The Corrosion Research Committee of the Institute of Metals had earlier introduced work at the college that led to the metallurgy department becoming a leader in corrosion studies. With the arrival of war many metallurgists turned their attention to finding ways of minimizing corrosion in naval vessels, then a serious problem. Three such research groups formed. After the war one group transferred to the Admiralty Engineering Laboratory, and in 1928 another transferred to the National Chemical Laboratory; but one group remained and, sponsored by the Non-Ferrous Metals Research Association, continued working at Imperial for many years.[99]

The physics department carried out some major research into X-rays during the war, while at the same time manufacturing X-ray units for the army medical services. W. J. Colebrook, superintendent of the physics department workshop, supervised their manufacture as well as super-vising work in the Air Ministry workshop.[100] R. J. Strutt volunteered to do X-ray work in the field and took some of the Imperial units to field hospitals in France where he taught people how to use them.[101] One unit, paid for by donations from staff and students, was sent as a gift to Keogh in recognition of his services. The head of the physics department, H. L. Callendar, led a research group working on fuels and engines with the aim of increasing the speed of fighter aircraft.

The war brought home the need for certain kinds of research also in botany, and the Royal Society formed a Natural Products Committee of which Professor Farmer was chair. He was also chairman of the Food Investigation Board's Cold Storage Section, and research in this area was carried out at the college largely by women research students and assistants.[102] Botanists working under V. H. Blackman attempted to increase food supplies in ways which, in retrospect, seem strange. For example, they conducted large scale field experiments testing the effect of high frequency electrical discharges on the growth of wheat, hoping to increase yields.[103] The government heavily supported this research and experiments were extended to other crops, though with little success. The professor of plant pathology, Henry Plimmer, taught courses in tropical hygiene to military personnel as well as to students, and was a member of the Trench Fever Committee.[104] Biochemists under Professor Schryver worked on adhesives for aircraft, another well-funded area of research. Professor Groom became a member of the Air Board Inspection Department. He worked with the forestry service attempting to find timbers suited to the manufacture of aircraft.[105] On one occasion he reported on the 'progressive deterioration' in the quality of timber used by the Germans for their aeroplane frames and propellers.[106]

Entomologists under Professor Lefroy worked on insecticides to control the housefly, a project sponsored by the London Zoo. Lefroy was looking for something to replace the toxic arsenical insecticides then in use. A fly room was set up and two women assistants, Miss Lodge and Miss Jackson were left to do the work when Lefroy left for active war service. With the rank of Lieutenant Colonel, the commission refused by Jocelyn Thorpe, Lefroy worked on the control of insects for the army medical service. He returned to civilian life in 1917 and undertook the examination of insect pests in Australian grain stores to help meet concerns over the serious losses in stocks waiting for shipment to Britain.[107]

In addition to training people in aeronautics, Dalby and his staff in mechanical engineering worked for the Admiralty Board of Invention on alternatives to the use of steam propulsion in warships. Gas turbines and the internal combustion engine were being considered, but neither technology materialized at that time. They worked also on diesel engines for submarines, and ran a munitions manufacturing plant in the department's workshops. The plant operated sixty hours per week; staff and about twelve students worked there and, as in chemical engineering, they had no vacation time during the war. Over nine thousand items: gauges, fuses, machine parts etc. were manufactured. Much of what was made went to the Royal Aircraft Factory, and the operation made a small profit for the college.[108] One of the governors, Cyril Jackson, thought this type of work was a waste of high talent and that the manufacture of munitions should have been carried out in a factory. However, Acland believed the college had a duty to comply with government requests for things needed at short notice. The professor of electrical engineering, Thomas Mather, was a member of the London Munitions Committee responsible for coordinating this type of manufacturing work at various London colleges.[109] Within his own department, in addition to the acoustical aircraft detection work mentioned above, research students worked on carbons for searchlights, attempting to produce more focussed beams. One wartime casualty was the department's aerial, used to receive radio signals from distant places. Strung between the C&G building and the Colcutt Tower, it was ordered down at the start of the war.

All this extra work had to be paid for and became a bone of contention within the college. Staff who were actually employed by government ministries, or those who had joined the armed services, were no longer paid by the college. Those who received pay from the government while working at the college had any difference in pay made up by the college. For patriotic reasons the Governing Body did not wish to ask much of the government by way of remuneration for college employees. But they also feared losing staff by not paying them bonuses for the extra work. Many among the college staff, including technicians, typists, and tradespeople, wanted increases in their wages for the overtime work they were being asked to do. Professor Callendar reported unrest among low-paid workers in the physics department. They were willing to serve their country by carrying out overtime work, but not for nothing. The government brought in a war bonus pay scheme in 1917 at which point the Governing Body was able to extract a bit more money from the Admiralty. Technicians received between 5s and 10s per week pay increases.[110] More generally, bills for work were sent to the appropriate authorities and small amounts of overtime pay were awarded to college staff. However, the government bonus scheme discriminated against women who were to be given bonuses worth about two-thirds of what men received for the same work. This caused a problem at Imperial and, after some agitation, the

Governing Body agreed to uphold Article IX of the Charter in which it is stated that 'neither sex nor opinions on any religious subject should qualify or disqualify any person … for any emolument in or in connection with the Imperial College'. Imperial was, and remains, a very male dominated institution, and over the years there is little doubt that there has been sexual discrimination in matters of advancement, and in other ways. But in 1917 it was agreed that overt pay discrimination could not be allowed, and that the college had to find ways to pay the women working in the munitions factory, and in other war-related work, bonuses equal at least to those given to single men.[111] At the end of the war, some people were given one-time monetary awards for their wartime service. Martha Whiteley, for example, was awarded an extra £100 in addition to her £250 salary. The college had a smaller salary bill during the war but it also had a much smaller income and was spending much of it on war-related projects.

Coda

We have seen that by the end of the war, a majority of students and a large number of the staff had joined the armed services. Many of the students were reluctant conscripts but the staff who joined up were mostly, like the Rector, willing volunteers. College administrators made various attempts to stem the flow, not only for the sake of the college, but also in the belief that it was in the country's best interests to use its technically trained citizens in other ways. In this they were at odds with many among the national and governing elites but, by the next war, had largely proven their point. This chapter has illustrated how quickly technical expertise was redirected during the war, and has shown something of the many and varied contributions made by college personnel. These contributions illustrate both great ingenuity and the wide range of expertise that existed at the college in this period. At Imperial it was not just a chemists' war. But the chemists were very important and, at the Central Laboratory in France, in South Kensington, and at Porton, they (together with physicist William Watson) played a central role in gas research. That they did so is related to the fact that the Rector, Sir Alfred Keogh, had the overall responsibility both for protecting the troops, and for the treatment of gas victims. He and Kitchener looked to Imperial for expertise in this field. The chemists' experience in this war, and their work with poison gases, led also to interesting new chemical research in the postwar period.

Keogh returned to the college in 1918 shortly before the armistice. He had been away for three and a half years and his return was a major event. It was marked by two large receptions: one, in the evening, attended by senior college staff, and by four Secretaries of State for War, including David Lloyd George, the Prime Minister. The following day there was a reception given by students and staff in the Union. Keogh was popular with students not only because of his war work and great reputation, but because he was a details man, remembered people's names, and took note of their problems. No doubt these skills contributed to his being a highly gifted administrator. Armistice Day brought great joy to the college. Professor Farmer, however, was so engrossed in his botany lecture that he failed to stop at eleven a.m. despite the sound of the Guilds' clock, and the jubilation and noise outside.[112] Eventually his students rebelled and stormed out of the room. The staff and students at the RSM were all gathered on the roof of their building and, when the clock struck, they unfurled flags and sounded maroons. Below, a major street party began. The college declared an official two-day holiday but it took much longer for normal work to resume.

About 2200 students and 103 staff had served in the armed forces during the war. 309 died and 356 were wounded.[113] Many of the staff who came to the college after the war had fine war records, whether in government service or in the military. For example, Keogh's successor, Sir Thomas Holland, appointed in 1924, had been Minister of Munitions in India. Alfred Egerton, who succeeded Bone in 1936, had been a civil servant at the Board of Invention, sorting through the many suggestions that poured in, before becoming Director of the Department of Explosives at the Ministry of Supply. He helped design and erect the National Explosives Factories and worked on the synthesis of ammonia. Hyman Levy, who joined the mathematics department in 1920, had been an instructor in the Royal Flying Corps and, in that capacity, met many Imperial people during the war.

After the war people's minds turned both to rebuilding and to remembering. The alumni societies of the three colleges tried to give assistance to those who needed it, helping both with small sums of money and in finding work.[114] A college memorial committee was appointed, chaired by Professor Philip. It was decided to raise money for a sports field, something the college did not yet have. Sports were not then a major Imperial activity; sporting events and teams were mainly associated with the three old colleges. Before the war sports were played in Hyde Park, Kensington Gardens, and at local sports grounds such as the Chelsea Football Club's ground at Stamford Bridge. Enough money (about £10,000) was raised to purchase some land in Wembley.[115] The commonly held sports field was both symbol and mark of a new start at bringing the three old colleges together. But there was no single memorial tablet, each college erected its own. Despite this, the experiences of people both on and off the battlefield had been shared, and those who had been through the war returned to the college determined to build it anew. Imperial College, which many thought would not survive the war, was given a new lease on life.

...

1. Already in 1916 the government appointed a committee of enquiry, chaired by J. J. Thomson, to look into science in the educational system of Great Britain. Francis Ogilvie, who played a major role in the founding of Imperial College, and who was a member of the Governing Body, was on this committee.

2. J. D. Bernal, *The Social Function of Science* (Cambridge, MA, 1967), 171.

3. Asquith resigned after the Dardanelles disaster. For the new Board and Ministry see, *History of the Ministry of Munitions* (HMSO, London, 1922); C. Wrigley, "The Ministry of Munitions: an innovatory department', in K. Burk (ed.), *War and the State: The Transformation of British Government, 1914–18* (London, 1982); M. Pattison, 'Scientists, Inventors and the Military in Britain, 1915–19: The Munitions Inventions Department', *Social Studies of Science*, 13 (1983), 521–68. See also Roy MacLeod and E. K. Andrews, 'Scientific Advice on the War at Sea, 1915–17: The Board of Invention and Research', *Journal of Contemporary History*, 6 (2) (1971), 3–40. As this latter paper points out, the first months of the war demonstrated Britain's unpreparedness in the production and supply of munitions.

4. For further information on science and the war see, D. S. L. Cardwell, 'Science in World War One', *Proceedings of the Royal Society of London* A342 (1975), 447–56; G. Hartcup, *The War of Invention: Scientific Developments, 1914–18* (London, 1988); D. E. H. Edgerton, 'Science and War' in R. C. Olby, G. N. Cantor, J. R. R. Christie and M. J. Hodge, *Companion to the History of Modern Science* (London, 1990); Keith Vernon, 'Science and

Technology' in Stephen Constantine, Maurice W. Kirby and Mary B. Rose (eds.), *The First World War in British History* (London, 1995), ch. 4.

5. For more on Keogh, see chapter 4.

6. Kitchener (Secretary of State for War) and Keogh were friends. Both were born in Ireland and both had served in the South African war.

7. For Acland see chapter 3.

8. GB minutes; 10 December 1914 and 14 May 1915.

9. B/Acland; see Gow to Acland, 27 February 1917.

10. The stalemate on the Western Front was punctuated by a series of Allied attempts to break through. By the summer of 1915 the Western Front army was much enlarged and growing. This led to the need for increased recruitment and, in the end, conscription.

11. B/Acland; notes by Gow, 1917. See also, GB minutes, 25 May 1917. Acland also wrote to the Secretary of State for War on more than one occasion asking for Keogh's release, eventually receiving assurance that he would be released early in 1918. See GB minutes, 8 February 1918.

12. GB minutes; resolution February 1917, rescinded May 1917. H. A. L. Fisher, Vice Chancellor of the University of Sheffield, had been brought into the government at the beginning of the war. For Fisher see also chapter 4.

13. I attach no value judgement to the terms high- and middle-brow, and use them loosely and descriptively. They are to be preferred over 'upper-' and 'middle-class' since high and middle-brow attitudes cut across class lines.

14. As has been noted elsewhere proportionately more RSM students had been educated at public schools than had the students in the other two colleges. This, perhaps, is the reason for the higher number of RSM volunteers at the start of the war. The student journal *Phoenix* represented college ambivalence well. Traditional patriotism was on display alongside the criticism.

15. Hew Strachan, *The First World War* (vol. 1) (Oxford, 2001), ch. 12.

16. Later Wells envisaged a technocratic dictatorship, the 'Air Dictatorship', made possible by new technical weaponry. Scientists and engineers, he believed, had a responsibility to ensure that this did not happen and that their inventions were used for progressive purposes.

17. This was the case despite the fact that Fisher had done much to modernize educational curricula in schools and had pressed for greater emphasis on scientific subjects. After the war he promoted reform in the same direction at Oxford.

18. See J. M. Winter, 'Oxford in the First World War', in Brian Harrison (ed.), *The History of the University of Oxford; vol. viii, the twentieth century* (Clarendon, Oxford, 1995), 8. At the University of Glasgow all students

and staff not engaged in essential medical duties were encouraged to join the army. See Michael Moss, J. Forbes Munro and Richard H. Trainor, *University, City and State: The University of Glasgow since 1870* (Edinburgh, 2000), 132–5.

19. Prime Minister Asquith had wanted to avoid conscription and Derby was asked to come up with a scheme to avoid it. But this proved unworkable and conscription was introduced in 1916. Derby then had the task of prioritizing essential civilian work and ordering conscription accordingly.

20. For a brief account of his activities see *The Phoenix* vol. 1 of new series (1915), 57.

21. The first Minister of Munitions was David Lloyd George which may account for much early activity in this ministry. Naming this government department a 'Ministry' was a first in British politics. The term was taken from the French.

22. See his reminiscences in *The Phoenix* vol. 42 (1933), 20. See also Joyce Brown (ed.), *A Hundred Years of Civil Engineering in South Kensington* (London, 1985), 43–6. On retirement, Dixon returned to live in France.

23. B/Acland; see Acland to Gow, 18 January 1916.

24. See, for example, an article written on the occasion of a visit to Imperial by Members of Parliament. 'Science in Harness: the work of Imperial College, an ally of industry', *The Times*, 17 May 1916, 2.

25. See B/Acland; letters and 'Memorandum from Men of Science'. This memorandum, signed by 36 eminent scientists including Imperial professors J. C. Philip. W. A. Tilden, T. E. Thorpe, and H. E. Armstrong (all chemists, and all but Philip then retired) was one of many that appeared around this time, not just at Imperial. A further memorandum was initiated at Imperial by three of the engineering professors: Dalby, Mather and Dixon. See also, letters to the editor on 'Neglect of Science', *The Times*, 2, 10 and 21 February 1916.

26. Herbert Dingle (appointed professor of natural philosophy in 1938), recalled that he was just one of two students attending Professor Strutt's third year course in 1916, but that Strutt lectured as though there were a class of 100.

27. The Ministry of Munitions focussed its takeover of laboratories at Imperial, Cambridge, Birmingham, St. Andrew's and at the City and Guilds' Finsbury Technical College.

28. B. G. Manton 'History of the Civil Engineering Department, 1884–1956' (typescript in ICA), 25.

29. P. G. Boswell quoted in David Williams *et. al.* 'History of the Geology Department' (typescript in ICA), 31–2.

30. GB minutes, 12 May 1916, 10 May 1918; EC minutes, 12 November 1915 and 12 October 1917.

31. The Air Board was asked to cover the expense of installing lavatories for the women.

32. GB minutes; 27 November 1914.

33. For example, the police closed the physics workshop down on one occasion for failing to black out. B/Acland; Gow to Acland, 1 February 1916.

34. There was concern over possible zeppelin raids and that, since the college was harbouring military personnel, it would not be covered under the Hague Convention, and was a possible target. GB minutes, 25 June 1915.

35. This was under William H. Merrett (1871–1938), an assistant professor of mining. Merrett was a classics scholar who later came to the RSM to study metallurgy. He became a junior assistant to Roberts Austen at the Royal Mint before returning to the RSM as an instructor in 1901. He worked at the college for forty-three years and was a good friend to students, helping many find work. Boswell describes Merrett in his memoir (B\Boswell) as 'by far the kindest and most unselfish man I ever met'. Merrett had a distinguished war record and trained this OTC unit until 1939 when it was disbanded. The OTC had been established in 1910 during the Haldane army reforms. A University of London OTC unit was formed at Imperial only after the college joined the University in 1929.

36. At that time the University of London offices were located in South Kensington and its OTC units met and drilled there.

37. B/ Boswell memoir, 94–5.

38. See, for example, report in *The Phoenix*, vol. 4 new series (1917), 10.

39. GB minutes, 9 July 1915.

40. See Hannah Gay and Anne Barrett, 'Should the cobbler stick to his last? Silvanus Phillips Thompson and the making of a scientific career', *British Journal for the History of Science*, vol. 35 (2002), 151–86. Thompson had earlier been a student at the RSM. For technical optics at Imperial College see Gilbert E. Satherthwaite, 'Optics at Imperial College: the early years', (Imperial College, printed pamphlet, 1997). See also L. C. Martin, 'The Work of the Technical Optics Section at the Imperial College', *The Scientific Journal of the Royal College of Science*, vol. xx (1949), 88–91.

41. Quoted in Satherthwaite, *op. cit.* (40).

42. Roy MacLeod and Kay MacLeod, 'War and economic development: government and the optical industry in Britain, 1914–18', in J. M. Winter (ed.), *War and Economic Development: essays in memory of David Joslin* (Cambridge, 1975). The Northampton Polytechnic later became

the Northampton College of Advanced Technology and is now City University.

43. See EC minutes, 5 February 1915; GB minutes, 12 February 1915, and 13 October 1916.

44. Wranglers were the holders of first class degrees in the mathematical tripos at Cambridge. See Andrew Warwick, *Masters of Theory: Cambridge and the rise of Mathematical Physics* (Chicago, 2003).

45. Martin, *op. cit.* (40). See also W. D. Wright (ed.) 'History of the Physics Department' (typescript in ICA, 1960), 20–23.

46. Frederick John Cheshire (1860–1939) was a Yorkshireman who was educated at Birkbeck College and at the RSM. He was a civil servant in charge of optics at the Patent Office, and a lecturer at Birkbeck College, before his distinguished wartime work.

47. In 1914, at the outbreak of the war, Fowler and one of his students, William Edward Curtis (later FRS and a professor at King's College London), were in Riga on their way to Kiev to observe a solar eclipse. They had to abandon their equipment and had a difficult journey back to England. The equipment survived the war and was later returned. For more on Fowler see chapters 6 and 8.

48. Louis Claude Martin (1891–1981) was given a professorship in 1944 and retired in 1951.

49. See KP AA/7–11, Fowler correspondence. H. P. Douglas (Admiralty) to Fowler 30 August 1917; H. G. Lyons (Met. Office) to Fowler, March 28, May 27, April 24, 1916. Many more requests were made, including for moonshine maps for France.

50. Boswell came to know Pilkington's at St. Helen's well in this period and the company began manufacturing optical grade glass after the war. See Boswell memoir (B/Boswell). Before the war much of Britain's optical grade glass had been imported from France and Belgium. The Ministry of Munitions published a document by Boswell, 'British Resources of Sand suitable for Glass-making' (1917). Two thousand copies were printed. The ministry later published further work by Boswell on the same subject. For more on Boswell see chapter 6.

51. In this connection, Boswell mentions Pilkington's at St. Helen's, Vickers at Barrow and Armstrong-Whitworths on Tyneside. B/Boswell memoir.

52. Alexander E. Conrady (1866–1944) was educated at the University of Bonn. He left Germany for work in Canada and the USA before settling in England where he set up a small factory in Yorkshire making dynamos and lamps. On the side, he studied the design and manufacture of telescope and microscope objectives. In 1902 he became a full-time lens designer at

W. Watson and Son. He was the author of a standard two-volume text, *Applied Optics and Optical Design* vol. 1 (London, 1929) and vol. 2 edited and completed by Rudolph Kingslake (London, 1958). In his memoir, Boswell describes Conrady as 'fairly tall, slim, very neat, and [with] a well kept tapering black beard … a suave, polished continental' who, every summer, took his family on a boating holiday on the Thames. His three daughters were students at Imperial. Kingslake was married to Conrady's daughter, Hilda, also a specialist in optics.

53. Alan Faraday Campbell Pollard (1877–1948) was educated at University College London, and served as a captain in the Royal Flying Corps during the war. He had pre-war experience in the optical industry and was appointed Professor of Optical Engineering and Instrument Design in 1919. He retired in 1944 but volunteered to remain at the college as manager of the physics workshop, then geared up for yet another war effort.

54. The idea of the 'chemists' war' has been challenged by Kevles. See Daniel Kevles, *The Physicists: the history of a scientific community in modern America* (New York, 1979). He claims that the submarine was the most important technical invention of the war and thus the war belonged to physicists and engineers as much as to chemists. But this view fails to take into account the perception of horror, and the association of chemists with what, at the time, was viewed as the worst of modern weaponry.

55. W. J. Pope, 'The national importance of chemistry', in A. C. Seward (ed.), *Science and the Nation* (Cambridge, 1917). Pope played a major role in gas warfare. The government, aware of chemistry's importance, helped merge various companies to form ICI during the war. In addition British Drugs Ltd. and British Dyes Ltd. were founded. See also Roy MacLeod, 'The Chemists Go to War: The Mobilization of Civilian Chemists and the British War Effort, 1914–18', *Annals of Science* 50 (1993), 455–81.

56. The Royal Society directed much of its efforts towards the war. Three Imperial professors were on the Royal Society War Committee: H. Brereton Baker, R. J. Strutt and H. C. H. Carpenter.

57. Thorpe was a member of the Advisory Council of the Privy Council's Committee for Scientific and Industrial Research, forerunner of the DSIR. For Thorpe see also chapter 6.

58. At first Thorpe and his colleagues manufactured anaesthetics and pain-killers for the Admiralty, including 40 lbs of phenacetin and 1½ lb of beta-eucaine. A little later they diversified and prepared 60 lbs of hexa-methylenetetramine, a chemical used to make explosives. Thorpe's work formed a blueprint for work sponsored elsewhere by the Royal Society

Committee on Synthetic Drugs. Large scale production of mustard gas and of acetone took place under former RCS student, W. H. Perkin, at the new Dyson Perrins Laboratory at Oxford. Perkin also supervised the formation of British Dyes Ltd., founded in 1915.

59. H. B. Baker diary 1915–16 (PRO WO 142/281). Baker does not give the date but it must have been around 22 April 1915. The diary has few page numbers and few dates. It is episodic but reveals much about gas defence at the time. I am grateful to Tom David for telling me of the diary's existence, and for letting me read a chapter from his doctoral dissertation (in progress) on the Central Laboratory in France.

60. Later, in 1916, Baker asked some of his chemistry staff to attempt the liquefaction of arsine in steel tubes. This work was carried out in an alley behind the chemistry department. On one occasion Baker found a tube lying on the ground 'for every passerby to go and smell ... if it had exploded it might have done considerable damage. We took it up on the roof and put it in a bucket of water. Bubbles came out showing it was leaking'. Baker and his staff also examined enemy steel shells but never found one containing arsine. See Baker diary, *op. cit.* (59).

61. John Scott Haldane (1860–1936) was a leading authority on the physiology of respiration and had carried out work on deep sea divers. Haldane was the brother of R. B. Haldane and the father of J. B. S. Haldane (who joined him in some of the war work). For a comprehensive account of the use of poison gases in the First World War see L. F. Haber, *The Poisoned Cloud: Chemical Warfare in the First World War* (Oxford, 1986). Haber mentions that Haldane was outspoken on the incompetence of senior army officers during the war. Nonetheless he sees Haldane as more important to gas defence work than Baker. Having read some of the available documents on chemical warfare deposited in the National Archives, I suspect Haber is incorrect on this point. It may well be as a result of Haldane's outspokenness that both Kitchener and Keogh favoured Baker (see also note 63 for Baker's connection to Harold Hartley). Baker was awarded the CBE for his war work in 1917.

62. After having seen the problems caused by the initial gas attack, Haldane returned to London and wrote a report which Asquith read to the House of Commons. In his diary Baker noted that Asquith had mentioned that he (Baker) was continuing to investigate the problem. Baker stayed at the front for several more days before returning to London and reporting to Keogh on what needed to be done in respirator work.

63. Brian Mouat Jones FRS (1882–1953) was educated at Dulwich College and Balliol College, Oxford. He had taught chemistry at Government College, Lahore, before coming to Imperial as an assistant professor. Like many other members of staff he volunteered as a private, but Baker arranged for him to get a commission so that he did not have to return to his billet at 8.00 pm every evening. After the war Jones was responsible for the dismantling of the Central Laboratory and its removal to Porton Down. He then became Professor of Chemistry at Aberystwyth and, shortly after, was appointed Principal of the Manchester College of Technology (1921–38). He was Vice Chancellor of the University of Leeds (1938–48) and, during the Second World War, returned briefly to direct the laboratory at Porton.

It is interesting that Baker, who also studied at Balliol, had taught chemistry at Dulwich College when Jones was a pupil there. Sir Harold Hartley, too, had been a student of Baker's at Dulwich and had then gone on to Balliol where he had been Jones's tutor. In 1918 Hartley was made Controller of the Chemical Warfare Department. Before and after this appointment, Hartley worked periodically at the Central Laboratory in France. He had much communication with Baker also in London. Clearly networks were important. Working at the Central Laboratory was another of Baker's Dulwich pupils, C. E. Brock, who, according to Baker, had been 'an idle boy in the Science V'. Brock was put to work on offensive weapons in a field laboratory. He experimented with various gas bombs. None came to much though, according to Baker, some sheep lost their lives. Pigs, however, were more resilient (see also note 76). Throughout the war Jones wrote to Baker about new developments at the front, both about possible new gases in enemy shells, and on the use of shells filled with the SK lachrymator developed at Imperial College. (These shells were manu- factured and filled in a factory at Calais.) In his letters to Jones, Baker adopted a paternal tone, continually reminding him not to over expose himself to gases in the field and laboratory. See Baker diary, *op. cit.* (59) and correspondence file PRO 142/Cl 28/15.

64. The Central Laboratory was in a Lycée, located near General Headquarters, that had been commandeered by the army. Watson worked closely with his Imperial colleagues, Baker and Jones. There is some interesting wartime correspondence between them in PRO (WO 142/Cl 28). William Watson FRS (1868–1919) was educated at the RCS and was appointed demon- strator in physics in 1890, assistant professor in 1897, and professor in 1915. Students gave him the nickname 'Mercury Bill'. Before the war, Watson had engaged in several research projects, including a study of the

'deterioration' of the earth's magnetic field (he also assisted the magnetic survey of the British Isles), development of the petrol motor, and colour vision. He was the author of two widely used textbooks, notably *Practical Physics* (London, 1906), based on his laboratory teaching at the RCS. During the war Watson was given the rank of Lt. Colonel and later was awarded the CMG; but, sadly, exposure to poison gases led to his early death in 1919. See Wright, *op. cit.* (45). See also H. L. Callendar, 'William Watson, 1868–1919', *Proceedings of the Royal Society* A 97 (1920), i–iii.

65. For example, in 1916 there was concern that chloropicrin (trichloronitromethane) was going to be used by Germany and research was carried out on a new type of respirator.

66. Baker diary, *op. cit* (59). The French had used chlorine gas earlier, in 1914. The first major use was by the Germans against French and Algerian, and then Canadian, troops at Ypres in April 1915. The gas cloud at Ypres killed about 5000 men and injured many more. Given the heavy losses it is fortunate that a truly reliable delivery system for poison gases was never developed.

67. These helmets required a tube valve to keep expired carbon dioxide away from the phenate. Goggles were attached to protect also from tear gas. Later hexamine replaced the sodium phenate.

68. Quotation from Baker diary, *op. cit* (59). At this stage (late 1915) there was much fear that chlorine gas would be dropped on London from Zeppelins. Sir William Ramsay wrote a pessimistic report on this but Baker, who had carried out some tests dropping chlorine bombs from aircraft over the marshes at Shoeburyness, argued that much of the gas was destroyed on impact and did not travel far. Much of Britain's chlorine was manufactured by the Castner Kellner Company. Sir George Beilby, a director of the company, was an important consultant on gas warfare at the War Office and he agreed with Baker.

69. There are several jabs at Haldane in Baker's diary. For example, noting that Haldane read Herbert Spencer's *Moral Philosophy* in bed as an aid to falling asleep, he wrote 'it takes all sorts to make a world'. On their departure for France he noted that Haldane was ridiculously upset at having lost eighteen shillings that he had placed in a paper bag in his pocket (money later found). Baker put this over concern with lost money down to Haldane's 'Scottish heritage'.

70. Tests involving larger quantities of gas were carried out at Porton Down.

71. Who was and who was not properly recognized appears to have been somewhat haphazard. Much of this had to do with turmoil at the Ministry of Munitions at the end of the war when Winston Churchill took over the

portfolio. He was unpopular within the ministry which soon fell apart, with negative consequences for many who should have been on the various honours lists. Looked at from this distance more scientists at Imperial should have received honours. However, Baker, Whiteley, Thorpe and Philip were recognized.

72. Memoir by Lady Thorpe. Typed copy in ICA; B/Thorpe.

73. The trenches were dug outside the chemistry department and a small temporary building was constructed alongside them for gas analysis etc.

74. This consisted of examining the properties of gases in the trenches, and in finding ways to ventilate the trenches; also chemical ways for neutralizing some of the dangerous gases. Hertha Ayrton, widow of C&G physics professor W. E. Ayrton, invented a fan for ventilating trenches which had limited use.

75. One of these was ethyl iodoacetate, the lachrymator codenamed SK (for South Kensington where it was first made). See J. F. Thorpe, 'History of SK', (typescript, 10 February 1919) PRO/MUN5/385/1650/7. On one occasion Baker was summoned to the War Office to give a demonstration of chemical warfare preparations to the King and the Prince of Wales. Baker noted in his diary that when the Prince of Wales appeared rather bored with chlorine tanks and respirators, he gave him a first-hand demonstration of the SK lachrymator. The Prince of Wales had asked to be exposed to the gas and was given just enough to make him weep.

76. At Porton they also did some work with gases in trenches. One rather sad, but interesting, discovery was that dogs died in these trenches but pigs would bury their snouts in the soil and usually survived.

77. This work, supervised by Baker, was carried out with Colonel Horrocks of the Army Medical Service. The smaller mobile plants were sent by lorry and ferry to France. Larger ones were taken to their locations, including Mesopotamia (Iraq) by barges. (See brief description in Baker Diary). On 16 October 1916, Lt. Sugden, was in 'high glee'; the barge unit was 'working like a charm', and the water was 'beautifully clear'. The plants contained sedimentation, alum injection, and filtration units, and could also chlorinate and dechlorinate water supplies. This technology had been pioneered much earlier at the Royal College of Chemistry under Hofmann and Frankland, and such work continued at the Royal College of Science, and later at Imperial. Sugden, who oversaw the plants being set up in Iraq in 1917, was later promoted to Captain in the Royal Engineers. He joined the academic staff in 1919 and taught the first year laboratory course, becoming something of a legendary figure in the department. See

H. J. T. Ellingham, 'James Netherwood Sugden, 1894–1944', *Journal of the Chemical Society*, (January 1945). Sugden was killed in 1944 by a flying bomb that landed near the Natural History Museum.

78. B\Thorpe; Lady Thorpe's memoir.

79. A. A. Eldridge, 'Martha Annie Whiteley, 1866–1956', *The Record*, (September 1956), 13–4. Whiteley received the OBE for her war work.

80. A. A. Eldridge, Diamond Jubilee Lecture of the Imperial College Chemical Society, 21 February 1956.

81. Baker diary, *op. cit.* (59). From the diary one has the impression that Muriel Baker helped her husband by working in a laboratory at their home. She was able to find ways of making more concentrated solutions of the various chemicals and had a number of suggestions for better kinds of absorbents. The early phenate respirator that the Bakers developed was given to General Russell who ordered its mass manufacture.

82. Philip was awarded the OBE for his war work. Animal charcoal had been used earlier but Philip demonstrated that activated wood charcoal had better absorption properties. Animal charcoal, made by the carbonization of bones, contains only about 10% carbon.

83. Chemistry Department correspondence; O. L. Brady to Eldridge, 26 March 1958.

84. This work was very important since mustard gas is neither visible nor, as is the case with chlorine, are its effects immediately noticeable. Detection of this gas, and signalling soldiers to put on their gas masks before it was too late, was a major concern. Various hooters and rattles were used to sound the alarm. When Micklethwaite arrived at the college as a lecturer in 1914 she worked first in Thorpe's drug 'factory'. She also worked closely with the chemist Sir Gilbert Morgan.

85. Eldridge, *op. cit.* (80).

86. Chemistry department correspondence; F. Dickens to Eldridge, 21 March 1958; and Ingold to Eldridge, 25 March 1958. Usherwood was a research student under Martha Whiteley. As a research student Ingold was awarded a Beit fellowship both because of his brilliance and his poverty. Baker stated 'he is an exceptional man and I believe he will have a distinguished future'. (Board of Study Minutes, July 7, 1915.) Ingold became a lecturer in the department after the war and was later professor at Leeds and then at University College London. He published over 700 papers in organic chemistry and was widely seen as a genius in that field.

87. I am not sure, but think these compounds, many of which are luminescent, could have been used for the manufacture of mantles for various lighting devices used by the troops.

88. This is true also of his work in the Second World War which is discussed in chapter 8.

89. Mrs Hay, who worked as a tea lady in the department for thirty-one years, remembered that later, when Briscoe was professor of inorganic chemistry, she took him his tea, toast and *Daily Mirror* each morning, and then fed the rats that he used for gas experiments. B/Eldridge; Mrs. O. Hay's reminiscences.

90. B/Bone; form in correspondence file.

91. Before the First World War the average working week was 56 hours. After the war it came down to 48 hours.

92. Auxiliary to the Aerial League of the British Empire.

93. For example, see GB minutes, 30 June 1916.

94. For more on the early development of aircraft for war see D. E. H. Edgerton, *England and the Aeroplane: An Essay on a Militant and Technological Nation* (London, 1991).

95. The scientific personnel at the RAE were drawn largely from Imperial and Cambridge. Among those from Cambridge were George Thomson, later head of the physics department at Imperial, and Bennett Melvill Jones who, after taking the mechanical sciences tripos, came to Imperial for the one-year DIC course. He later became Mond Professor of Aeronautics at Cambridge. The Mond chair was founded at roughly the same time as the Zaharoff Chair. Henry Tizard, was an exception and joined the RAE from Oxford. Like Melvill Jones, he was much influenced by Cambridge engineer, Bertram Hopkinson. Hopkinson would probably have become the first Zaharoff professor had he not been killed in an air accident in 1918.

96. Official statistics list 1415 people as having been killed by bombs delivered from the air (from aeroplanes and airships), and 3367 as having been wounded. See *The Times*, 2 February 1916 for an account of a zeppelin raid when over 200 bombs were dropped.

97. William Sansome Tucker (1877–1955) was a student at the RCS and gained an ARCS (and an external London BSc) in physics in 1903. He returned to the college as a part-time research student under Hugh Callendar in 1909. During the war he worked at the college on various projects and was awarded a DSc in 1915. His microphone was used also in the detection of gun positions. See W. S. Tucker 'A selective hot-wire microphone' *Philosophical Transactions of the Royal Society* A 221 (1921) 389. See also, D. Zimmerman, 'Tucker's Acoustical Mirrors: Aircraft Detection before Radar', *War and Society* 15 (1997) 73–99; Roy MacLeod, 'Sight and

Sound on the Western Front: Surveyors, Scientists and the "Battlefield Laboratory", 1915–1918', *War and Society* 18 (2000) 23.

98. B/Acland; see Gow to Acland, 17 January 1916; GB minutes, 12 May 1916.

99. M. S. Fisher, 'History of the Metallurgy Department, 1851–1963' (typescript in ICA).

100. Colebrook had joined the physics workshop when it had been created in the RCS by C. V. Boys. He retired in 1932.

101. Strutt (together with Imperial's H. B. Baker and H. C. H. Carpenter) was a member of a panel of scientists advising the Admiralty Board of Invention. A. J. Balfour, then First Lord of the Admiralty, was responsible for the Board. He named Admiral Sir John Fisher as its chairman, and while Fisher stated that 'the war is going to be won by inventions', Strutt recorded his disappointment that the Admiralty did not take the scientific panel's advice more seriously. One of Strutt's inventions to be dismissed was an electrically heated undergarment for airmen. The Admiralty, understandably, was more supportive of Carpenter's work on preventing corrosion in naval vessels. See wartime papers in B/Rayleigh. Quotation, MacLeod and Andrews, *op. cit.* (3), 9.

102. One of Farmer's assistants was Hilda Mary Judd, daughter of Professor Judd, geologist and former Dean of the RCS. Hilda Judd had been a student in the RCS (1901–4).

103. B/Acland; see internal correspondence and that between V. H. Blackman and H. Frank Heath of the Board of Education, 1916–17. By and large food shortages were less of a problem than in the Second World War. Food rationing was introduced in stages only from January 1918. Potatoes and bread were never rationed.

104. Plimmer's death in 1918 was probably due to tetanus, or to some other infection related to his work on 'trench fever'. His widow donated the stained glass window later installed in the Queen's Tower.

105. Groom identified what he thought would be the best airplane timber, namely the Sitka Spruce. This tree was mainly harvested on the Queen Charlotte Islands, British Columbia. While Groom made enquiries of the lumber mills in Masset and Ocean Falls, they were not equipped to mill the timber appropriately. As one of his correspondents put it, 'Pacific Mills Ltd. of Ocean Falls cuts a large quantity of Sitka Spruce logs of excellent quality which all goes into pulp'. Groom recognized that the government would have to send advisors to the mills which would need to re-tool were production to become a reality. B/ Groom. Loren L. Brown to Groom (no date).

106. Anon., 'History of the Botany Department, 1851–1960' (typescript in ICA), 32.

107. R. G. Davies, 'History of the Entomology Department', (1995, typescript in ICA) 6.

108. GB minutes, annual report, July 1915.

109. Before the war Mather rarely lectured because of a speech impediment. But with the loss of so many of his staff he began to do so, reportedly with great success.

110. B/Acland; Gow to Acland, 11 September 1917.

111. See FC minutes, 10 November 1916 and 25 January 1918, EC minutes 7 June 1918, and GW/15 'War Bonus Review, 1917–18'. Higher bonuses were given to married men.

112. The kind of solemnity that accompanied future Armistice Day ceremonies, with the two minutes of silence rigorously observed, was only gradually acquired. By the 1960s this, too, had largely gone though, interestingly, it has made a comeback in recent years.

113. GB annual report, August 31, 1919. The ratio of dead to wounded was extraordinarily high. During the war it was customary for universities and colleges to publish lists of casualties in the newspapers just before Christmas each year. Imperial College was no exception. The national average for war deaths among serving university students and staff was 12%. For Oxford and Cambridge it was 20%, reflecting the greater number of volunteers early in the war. For Imperial the number was roughly 14%. See J. M. Winter, *The Great War and the British People* (London, 1986).

114. There is a file of letters (AB5) giving details of the kind of help sought by, and for, ex-servicemen after the war. One letter is from Frances Parker, Lord Kitchener's sister, who helped run the K. K (Kitchener of Khartoum) Empire Association, and was 'keen on our men getting on after the war'. She wrote on behalf of a man who had served for five years in France and who wanted to study entomology but had no money.

115. This land had to be sold in 1935 when Wembley Council needed it for other purposes. In 1936 new grounds at Sipson Lane, Harlington, intended both for athletics and for a future field station, were purchased.

chapter six

Continuity within the Three Old Colleges, 1907–45

Introduction

While Imperial College early added new departments in chemical technology, technical optics, and aeronautics, other areas of the college continued to develop along lines already set in the late-Victorian period. In the early twentieth century the old college structures were still strong but, by the end of the Second World War, people began to identify more with their departments, and increasingly also with Imperial College. This chapter records some of the academic developments within the departments of the three old colleges during the first half of the twentieth century. The departmental histories of this period are very much tied to the interests of the departmental heads who had almost complete control over the type of work carried out.

The Royal College of Science

Biological Sciences

Before the botany department acquired its new buildings, space for the biology departments in the old building on Exhibition Road was very tight. In 1907 botany, under John Bretland Farmer, had only two rooms and much of its teaching was carried out in other places, notably at the Chelsea Physic Garden. Farmer was a highly energetic professor, and by the time he retired in 1937 botany had two buildings, four professors, two assistant professors and six lecturers and demonstrators.[1] In 1911 he brought V. H. Blackman to the college as professor of plant physiology and pathology.[2] After the First World War, Blackman headed a research institute in plant physiology, supported by the Ministry of Agriculture and later by the Agricultural Research Council (ARC). That this institute came to Imperial was a consequence both of Farmer and Blackman's politicking and of the realization, after wartime food shortages, that Britain needed to carry out more agricultural research.[3] The institute remained an integral part of the botany department for many years. Some of its work was carried out at Rothamsted where students worked on physiological problems related mainly to fruit production. In connection to the new interest in food science, S. B. Schryver was appointed assistant professor of biochemistry in 1913.[4] Schryver, a specialist in plant proteins, was promoted to a professorship in 1919 and remained in the botany department until his death in 1929. He was succeeded by his former student, A. C. Chibnall, who received funding from ICI and began a comprehensive research programme on the biochemistry of forage crops.[5] Chibnall recollected working on the nitrogenous constituents of green leaves at the Chelsea Physic Garden. He claimed that when he began his research there was no equipment, except for an old balance with a three-foot beam that had been left at the Physic Garden by Francis Darwin in the 1890s, and that he had to make all his own apparatus. Chibnall stated that despite the equipment problems, 'Schryver gave me a good grounding in the protein chemistry of the day, for he was a stimulating talker and well versed in the pre-war literature'.[6] Successively, Schryver and Chibnall headed an internationally renowned group of plant protein chemists and, among other things, pioneered the analysis of amino acids. Chibnall resigned in 1943 and moved to Cambridge. The biochemistry chair then lapsed until 1961 and the appointment of Ernst Chain.

Botany got off to a good start despite its having relatively few students. Its graduates filled important positions around the world, and some closer to home.[7] In this early period botany brought in more research funding *per capita* than any other department. As is discussed in chapter seven, Farmer received support from a number of imperial industries. For example, both the Rubber Growers Association and the Empire Cotton Growing Corporation supported research at the college, the latter until 1958. Much work was carried out also for government bodies such as the Food Investigation Board, the Forest Products Research Board and grants came from the Agricultural Research Council. When botany's second building for plant technology was ready in 1923 it was opened by the Duke of Devonshire, Secretary of State for the Colonies — appropriately, given all the empire-related work being carried out.

From 1929, the year that Imperial College formally joined the university, quinquennial inspection visits were made. The report issued after the 1935 inspection noted the excellence of the botany department. It recommended that the ARC Research Institute in Plant Physiology be expanded, and that the university help with more funding to enable botanists to use the college field station. Up to then the station had been used almost exclusively by zoologists and applied entomologists.[8] The botanists worked also with people in other departments; for example there was collaboration with engineers on work relating to the refrigeration of foodstuffs, lighting in greenhouses, and the effects of electrical discharges on the growth of cereal plants.

Farmer was succeeded as head of department by V. H. Blackman who continued also to head the research institute in plant physiology. Blackman had several women students working on fruit production, among them Helen Archbold (later Helen Porter FRS, and Imperial's first woman professor). On Blackman's retirement in 1943, F. G. Gregory headed the research institute and the *de facto* separation of the biology department was formalized.[9] Botany, and zoology and applied entomology, became two separate departments and, after the departure of Chibnall to Cambridge, biochemistry was reduced to just a small section of the botany department. During the Second World War the botany department buildings in South Kensington were occupied by military personnel. Once again courses were given at the Chelsea Physic Garden, and at the Slough field station. The maintenance of plant stock was difficult, but William Brown, professor of plant pathology, coped well with this problem and saved much.[10]

Zoology had a shakier start than botany, despite the fact that the discipline had been founded at the RCS by T. H. Huxley. After the 1905 death of zoology professor, George Howes, the RCS did not immediately appoint a successor and was considering closing down the course. Instead, Professor Dendy from King's College was asked to give lectures at the college. He did so until 1909 when the Governing Body decided the course would continue. Adam Sedgwick was then appointed to the chair.[11] In this period the other zoology departments in London drew most of their students from the medical schools. Imperial was an

exception and had few students until Sedgwick decided to train entomologists for work overseas. As Sedgwick wrote to the Rector, 'we shall
be expected to have a supply of properly trained men to take up posts
in connection with agriculture [in Britain and the Empire].' He rapidly
expanded the department by appointing H. M. Lefroy, C. C. Dobell,
E. W. MacBride, H. M. Fox, L. Doncaster and E. Hindle, and managed
to lay a good foundation for zoology, especially applied entomology, before
his premature death in 1913.[12] At that point the governors decided to
re-form the biology department with sections for botany, zoology, and
applied entomology, and the Rector appointed Farmer as head. An associateship course in applied entomology was approved in 1919.[13] Biology
students of this period appear to have been well integrated in their department. This was in part because members of staff, notably Farmer and
MacBride, arranged regular weekend field trips to places near London,
as well as to more distant places during the summer vacations.[14]

Harold Maxwell Lefroy was appointed to the chair of applied entomology in 1912 and helped relieve the space problem by doing much
teaching at his home in West London.[15] There was much travelling
back and forth. According to his obituarist, Lefroy was 'an ardent and
dangerous motorist. It was an experience to be driven by him ... as
fast as the car could go, through London streets in company with a
biscuit box full of noxious living insects, a few glass bottles of poisons
and a cylinder of some lethal gas'.[16] Lefroy was given permission to
take on a number of outside posts, such as Entomologist to the Royal
Horticultural Society, which allowed students to carry out field work also
at the Society's Wisley gardens.[17] E. W. MacBride, who had been an
assistant professor since coming to the college from McGill University in
1909, was promoted to the chair vacated by Sedgwick's death in 1913.[18]
MacBride, a specialist in marine invertebrates, gave lectures on heredity
which were attended by large numbers of part-time students, many of
whom came from colleges within the University of London. MacBride
remained a Lamarckian all his life and, while his anti-Darwinian stance
was still respectable in 1909, it became decreasingly so as the years
progressed.[19] One of his students, H. G. Cannon, stated 'when the
Lamarckian atmosphere got too exciting, I could always run across the
road to cool down in the old brick building at the back of the [Natural
History] Museum'.[20] MacBride inherited the rooftop greenhouse from
the botanists and used it to house animals (invertebrates) at constant

temperatures. He had much difficulty with the Board of Works because its electricians, who still worked for the V&A, had control over the circuits in the building and he had to call them even for blown fuses. Sometimes the government turned off the electricity without warning, leading to the loss of specimens.[21] MacBride was a keen eugenicist and gave lectures at the college for members of the Eugenics Society.[22] He tried to boost student numbers in zoology by arranging a wide range of lecture courses for outside groups, especially during the summer vacation. Teachers, and people on furlough from jobs overseas with the Colonial Office, or working for other government departments, were among those who enrolled.[23] One of the lecturers appointed during MacBride's period as head of department was Lancelot Hogben. Perhaps better remembered today for his left-wing politics and popular science books than for his zoology, Hogben held the departmental licence to perform experiments under the Vivisection Act.[24]

After Lefroy's death in 1925 Frank Balfour Browne was appointed as head of the sub-department of entomology.[25] With MacBride's retirement looming, a college committee was struck to consider, once again, whether courses in zoology and applied entomology should continue. With hindsight one can see that matters were taken out of its hands when J. W. Munro was appointed assistant professor.[26] Munro, ambitious and energetic, was not prepared to see entomology disappear. There was tension between Balfour Browne and Munro, especially when Munro sought industrial support for his research on a scale that Balfour Browne found distasteful. Both men wrote to the Rector, Thomas Holland, who appears to have taken Munro's side in the dispute. In a letter to Professor Farmer, Holland wrote that if things were as Munro described, Balfour Browne 'must submit or get out. The new entomological work is strictly in accordance with the declared policy of the College as laid down in the Charter.' In other words applied industrial work was to be encouraged. Balfour Browne resigned in 1929.[27] He was a wealthy man and needed neither the aggravation nor the salary. Munro was promoted as his successor and, when MacBride retired in 1934, became head of a newly created department of zoology and applied entomology.

Munro, needing someone to identify and study the pests that he was attempting to control, had brought O. W. Richards to the college in 1927.[28] In 1934 Richards, together with zoologist H. R. Hewer, began teaching a course in evolutionary theory and biogeography. This replaced

MacBride's somewhat idiosyncratic course, though Richards, while not a Lamarckian, was no orthodox Darwinian.[29] Munro wanted to bring a chair in parasitology to the department, and discussed it already in 1936. But this plan only came to fruition in 1954 when B. G. Peters was appointed to a new chair one year after Munro's retirement. At the end of the Second World War things again looked bleak for zoology, but once more Munro showed his mettle. He persuaded Southwell to turn over the plant technology building, which had been occupied by the military during the war, to zoology and applied entomology.

Physics

By 1907 physics and chemistry were installed in their new premises in the building they were to occupy until the college expansion of the 1960s. H. L. Callendar was head of the physics department and had spent much time getting the space that had been allocated to physics properly fitted.[30] Callendar's fields of study included thermodynamical problems related to steam at high pressures, optical pyrometry, and radioactivity. He also pioneered the development of platinum thermometry.[31] These broad interests allowed connections to be made to people working in the chemical technology, mechanical engineering and aeronautics departments. For example, together with Ernest Dalby, he and his research group worked on problems having to do with detonation in internal combustion engines. Callendar was the first to give a satisfactory explanation of engine 'knock'. This work, carried out in a laboratory which had been set up at the college by the Air Ministry, contributed to increasing the speed of fighter aircraft during the First World War. The laboratory remained at Imperial until 1934 when it moved to Farnborough.

In 1908 a new chair in physics was created to which R. J. Strutt was appointed.[32] He worked mainly on the radioactivity of minerals in relation to the earth's internal heat, and on the related accumulation of helium in minerals. The First World War disrupted the routine of the physics department such that it needed consolidation and reorganization in the years after 1918. Strutt, who had turned his attention to the production of medical X-ray units during the war, resigned the professorship in 1921. Having succeeded to the barony, he decided to follow in his father's footsteps and resume his research privately.[33] At about the same time, the colour vision laboratory was closed after the death of Sir William Abney.

This laboratory, housed in the South Kensington Museum, had been the site of the Board of Trade sight tests for the mercantile marine. Its personnel and equipment were merged with the technical optics department. The older, pre-Imperial, tradition of astrophysics continued under Alfred Fowler.[34] He had succeeded Norman Lockyer in the chair, and continued to work at the Norman Lockyer Observatory. The observatory, located on land now occupied by the Science Museum, was not well equipped, but in the 1920s students could still study the night sky in South Kensington.[35] Of Fowler's other research space, Herbert Dingle recalled,

> Mr. Fowler's laboratory was a table in a frequently-used lecture theatre, in the corner of which a dark room had somehow to be improvised. For apparatus he had one small spectrograph of a type we should now consider suitable for a promising child's stocking at Christmas and, for encouragement, the collected indifference of most of those who might have been concerned. In such circumstances was begun that remarkable series of researches which lie at the basis of modern physical astronomy — the identification of the bands in the spectra of red stars; the detection of magnesium hydride and other compounds in sunspots; the solution of the problem of comet-tail spectra; the laboratory production of "cosmic hydrogen" and many contributions to the experimental foundation of spectroscopy which have made our astrophysics department a goal of pilgrimage from all continents of the earth.[36]

Callendar died in 1930 and George Paget Thomson was appointed to succeed him.[37] In 1937 Thomson was awarded the Nobel Prize for physics, for having earlier demonstrated the wave properties of electrons (and thus wave-particle duality) by the study of electron diffraction in crystals. Work stemming from this was brought to the college and research began on the surface properties of solids using new electron diffraction approaches. After the discovery of the neutron in 1932, and Enrico Fermi's work in Rome, Thomson became interested in neutron generation, and in developing techniques for measuring slow neutron flux.

When the Air Ministry Laboratory moved to Farnborough in 1934, a new School of Applied Geophysics moved into the vacated space. Bridging physics with geology, work in this area was carried out under

A. O. Rankine, and was funded by the Department of Scientific and Industrial Research (DSIR).[38] Mining ventures in Britain and the Empire then largely relied on German and American expertise in geophysics. Under the agreement with the DSIR the college was to be responsible for giving instructional courses designed to train geophysicists to meet the demands of prospecting companies. Rankine hired J. McGarva Bruckshaw, whom he had met when working on the Imperial Geophysical Experimental Survey in Australia, as his research assistant. The School gave two postgraduate courses, one for geologists and the other for physicists. Geophysics remained attached to the physics department until 1955 when it moved to geology under Bruckshaw.

In 1924 Sir David Brunt became professor of meteorology.[39] In 1932, when the Meteorological Office closed its South Kensington office, the meteorology department joined the physics department, but it was located separately in the old Huxley building. Postgraduate courses were given and the majority of students came from abroad. To encourage more Britons, Brunt wanted to create an institute of meteorology to carry out contract work for the Meteorological Office, and train its staff. The Rector, Henry Tizard, supported this idea but it was rejected by the university senate as having no academic merit. Meteorology became an independent department once again in 1939 (to be closed down and returned to physics as a small section in the 1970s).

Chemistry

W. A. Tilden, the first head of chemistry at Imperial, retired in 1909. He was succeeded by T. E. Thorpe who had headed the RCS department earlier.[40] Thorpe, who retired in 1912, had interests mainly in inorganic chemistry and during his tenure the department's traditional strength in organic chemistry was temporarily eclipsed. Thorpe was an authoritarian and, like many of the professors who began their careers in the mid-Victorian period, he toured the teaching laboratories each day wearing grey silk hat and morning coat. By the early twentieth century this was seen as a somewhat intimidating and increasingly archaic practice.[41] Thorpe was succeeded by H. Brereton Baker and, in 1914, the department took on its tripartite structure with sections in physical, inorganic and organic chemistry. Baker headed the inorganic chemistry section. Like his predecessor he was seen as very Victorian, but kinder and less authoritarian.[42] He was known as an expert experimentalist and much of his work was on the effects of traces of moisture on chemical reactions. He introduced the concept of intense drying — he dried things for years — and understandably received the epithets 'dry Baker' and 'his imperial dryness'. Baker, like many of the early professors, had a gifted woman assistant. Margaret Carlton carried out much of the research in his laboratory and was acknowledged when Baker won the Davy Medal from the Royal Society in 1924. She became a lecturer in the department only in 1946, retiring in 1960. This limited career progress was typical for women of her generation.[43] While several of the women working at the college in this period were acknowledged as gifted scientists, they were not seen as serious candidates for professional advancement.

J. C. Philip, who earlier had been appointed to boost physical chemistry in the department, was given a chair in 1914.[44] At the same time, one of Thorpe's old students and namesake, Jocelyn Thorpe, was appointed professor of organic chemistry.[45] During his tenure organic chemistry recovered some of the glory that it had lost since the retirement of Tilden. Thorpe attracted some outstanding research students and young lecturers to the college, including E. H. Farmer, C. Ingold, A. W. Johnson, G. A. R. Kon and R. P. Linstead. When Baker retired in 1932, Philip and Thorpe became joint heads of department, and H. V. A. Briscoe was appointed professor of inorganic chemistry.[46] Under Briscoe, high-vacuum work, begun earlier by H. J. Eméleus, became a major research technology. Analytical chemistry was another early strength and the main analytical laboratory was modernized before the Second World War.[47]

The interwar period saw a dramatic rise in the importance of chemistry in the industrial and commercial world. The 1920s saw a number of mergers due to rising production costs, with large companies such as ICI and Unilever setting the stage for the rise of the British chemical industry. Recognition of the utility of chemistry led to the manufacture of a wide range of new consumer goods, and the resulting industrial activity was reflected in expanding chemistry departments across the country. Chemistry experienced its best decades of the century from the 1930s to the 1970s.[48] In 1938 Jocelyn Thorpe was succeeded by another fine organic chemist, Ian Heilbron.[49] Before coming to the college, Heilbron, typical of prospective professors with bargaining power, insisted on improvements. He refused to come to Imperial unless the semi-industrial Whiffen laboratory was converted into a modern organic chemistry research laboratory with space for 24 research students. In a letter to Tizard he wrote that the department's laboratories were not equipped for serious research and that they failed to 'provide even the bare minimum of services and essentials. The research students are crowded into a number of unsuitable rooms with restricted bench space and bad lighting … the conditions are altogether deplorable'.[50] This was an exaggeration, but the Whiffen laboratory was renovated and, while physically moved since Heilbron's time, it has remained a staging ground for major careers in organic chemistry. Another early field of interest was agricultural chemistry under assistant professor A. G. Pollard who taught mainly at Slough.

Mathematics

Mathematics was taught at two of the old colleges after 1907. A mechanics section in the C&G College had a laboratory which was the centre of mathematical instruction for engineers. Herbert Klugh, who had taught in the old C&G department, continued to teach in the laboratory after the foundation of Imperial. Strictly speaking it was an outpost of the mathematics department, now concentrated in the Royal College of Science, but was not seen that way by the engineers.[51] Andrew Forsyth, a former Cambridge professor, was appointed Professor of Mathematics in 1913.[52] He was chosen because the Rector believed that pure mathematics should be taught in the RCS in order to raise Imperial's standing in the mathematical world. But not everyone agreed. It was argued that the RCS had no history of pure mathematics teaching and that it was unnecessary in a college of science and technology. Applied mathematics had been taught in the RCS by the much loved John Perry who had given instruction also in practical mechanics and electricity.[53] Paradoxically it was in the C&G College that *both* pure and applied mathematics were taught, by Olaus Henrici until his retirement in 1908. Henrici disapproved of the Cambridge approach to mathematics teaching and promoted a largely continental European style. The appointment of Forsyth was unpopular, in part because of Henrici's influence, in part because Perry was forced into an unwished-for early retirement, and in part because the Rector tried to move all mathematics teaching to the RCS mathematics and mechanics department (mechanics was not dropped from the title until

1951). W. E. Dalby, Dean of the C&G College, was adamantly opposed to this move, and opposition by engineers lasted well into the 1950s. At first Dalby refused to have C&G students taught mathematics in the RCS. In 1925 he reluctantly agreed that first-year students be taught there. The mechanics laboratory which the engineers continued to see as *their* mathematics department, remained in the C&G College until Imperial acquired the Royal School of Needlework building in 1934. Rearrangements then led to the further consolidation of mathematics and mechanics in the old Science Schools building (named for Huxley at this point), though some of the mathematics teaching staff still belonged to the C&G.[54]

Shortly after joining Imperial, Forsyth arranged for a chair to be given to another ex-Cambridge man, his friend Alfred North Whitehead.[55] Whitehead, who worked at the college for ten years, had switched his main interest from mathematics to philosophy already before his arrival at the college. He assumed the departmental headship in 1923 but left one year later to take up a chair in philosophy at Harvard University. Forsyth and Whitehead were both very accomplished men and, while at Imperial, continued to publish in their respective fields. But they did little to encourage the growth of pure mathematics or original research among students. There were few associateship students in the mathematics and mechanics department which largely provided service teaching for the other disciplines. On Whitehead's departure the Governing Body decided once again to appoint an applied mathematician as head. The Rector chose Sydney Chapman, a specialist in the kinetic theory of gases and an expert on the physics of the upper atmosphere.[56] Chapman's complex calculations were facilitated by a Hollerith machine and by several women computing assistants. Surprisingly, given his interests, he appears to have had no major collaborators in other departments. But he modernized the curriculum and introduced students to some Einsteinian theory. Hyman Levy who joined the department in 1923 succeeded him as head in 1946.[57] The C&G staff members, still rebellious, claimed that they worked in the Department of Mechanics and Mathematics.

City and Guilds College

When the City and Guilds' Central Technical College joined Imperial its building, designed for 200, was bursting at the seams with close to 300 students. The Goldsmiths extension was opened after the First World War and it, too, soon filled. Already in the early twentieth century Imperial College was the largest engineering college in the country. By comparison, just twenty-eight people graduated with the mechanical sciences tripos at Cambridge in 1905, though the numbers grew slowly after the First World War. The engineering sciences department at Oxford was even smaller with only about ten graduates a year in the interwar period.[58] In 1907 the C&G College had a department that combined civil and mechanical engineering, another that combined physics and electrical engineering, and departments of chemistry and mathematics. W. E. (Ernest) Dalby was Dean of the C&G College and Professor of Engineering. W. H. Ayrton was Professor of Physics, Olaus Henrici was Professor of Mathematics, and H. E. Armstrong was Professor of Chemistry.[59] By 1912 only Dalby remained. Following recommendations made by the Wolfe Barry Committee, physics (but not electrical engineering), mathematics and chemistry were moved to the RCS and, in 1913, Dalby's department was split in two when separate civil and mechanical engineering departments were created.[60]

Mechanical Engineering

Stephen Dixon was appointed to head the civil engineering department and Dalby remained Dean of the C&G and head of mechanical engineering. Unhappy at the split, Dalby did not cooperate with Dixon and went so far as to physically seal off his department from civil engineering. During the war Dalby stepped up the training of students in aeronautics and moved the department towards work on internal combustion engines, boiler design, and the properties of liquid fuels. Mechanical engineers took in a range of outside work much of it having to do with the railways and, for the Admiralty, on steam ship boiler design. This was at a time when the Admiralty was wondering how to replace the steam engines in many of its ships. The type of work encouraged by Dalby allowed for cooperation with Callendar in physics, with fuel researchers in chemical technology, and with V. C. Illing who ran the course on oil technology in the geology department. Fuel research was something that the Rector, Sir Alfred Keogh, promoted as an area for bringing people together, and for attracting funds from industry. Clearly it fitted a time when liquid fuels and the internal combustion engine were becoming increasingly important. In 1913 the Institution of Petroleum Technologists (later the Institute of Petroleum) was founded, and the DSIR founded its Fuel Research Station. Since Dalby had already introduced work in aeronautics, the stage was set for future research also in aircraft engines and fuels. Also successful was his DIC course in advanced railway engineering.

In 1931 Dalby was succeeded by C. H. Lander who had been Director of the DSIR Fuel Research Station.[61] Lander was an inveterate smoker and allowed smoking in the department, something Dalby had banned for many years. According to Hugh Ford, who was an undergraduate in this period, Lander was a very poor lecturer and when he made mistakes the class would stamp on the floor. At the end of each lecture Lander would beat a hasty retreat to his assistant waiting in the wings with a lighted cigarette at the ready.[62] But Lander was in other respects highly capable and he re-equipped the laboratories with a grant from the Mercers Company. Two of his former colleagues, Owen Saunders and Margaret Fishenden, joined him at Imperial and, together, they set up a broad-based research programme in heat transfer. Saunders was appointed to a new Clothworkers Readership in applied mathematical physics.[63] Fishenden was given just an honorary lectureship, and was appointed to a readership only in 1947, after Saunders succeeded Lander as head of department.[64] Research in heat transfer gave mechanical engineering at Imperial an excellent reputation for many years. Saunders, encouraged to work on jet engines by Tizard, studied gas flow at supersonic speed through channels and pipes, and collaborated with people in aeronautics.

At the time that Lander was appointed head of department, Edward Frank Dalby Witchell (1880–1956) was promoted to a professorship. It appears that many students of this era associated mechanical engineering with Witchell. He had been a student in the department and continued to be a major presence, almost its life and soul, until his retirement in 1946. Though not a major research scientist, Witchell was a good lecturer in thermodynamics, took a keen interest in student activities, was treasurer of the student union for many years, and provided witty musical entertainment at departmental social events.

Electrical Engineering

In 1907 William Ayrton was professor of physics at the C&G, but he died in 1908. Physics teaching, as mentioned, moved to the RCS and Ayrton's department became focussed solely on electrical engineering. Ayrton had, for years, worked in the field of electricity and was succeeded by his colleague Thomas Mather who, like him, was a good designer of electrical measuring instruments.[65] The department had two assistant professors, one of whom, G. W. Howe, was a notable radio specialist.[66] Already in 1913 he successfully received signals from Glace Bay, Nova Scotia, by using an aerial suspended between the Waterhouse building and the Colcutt Tower. He left for a professorship in Edinburgh in 1922. S. Parker Smith, the other assistant professor, was responsible for teaching heavy electrical engineering and worked on the design of heavy equipment. The division of the department into sections for light and heavy electrical engineering, anticipated by the activities of these two assistant professors, was later formalized in the department's course offerings.

Even before the founding of Imperial, the electrical engineering department had taken over some space in the basement of the Royal Needlework School and, by 1904, had built a fine laboratory there; equipped with heavy generating machinery, it was known as the dynamo room. The department also occupied the basement of the Waterhouse building, where space was devoted to electrical measuring instruments, work on telephony and telegraphy, and to lecture theatres. The 1914–18 war saw much innovation in communication technology and many students came to the college as a result. Enrolments shot up to over fifty a year in the early 1920s. The department also had many occasional students from other Imperial departments and was a centre for interesting interdisciplinary exchanges, a pattern that continued through the century. In their final year the heavy engineering students were put in charge of the department's generating plant — each student had a two-week stint, learning how to stoke boilers, and care for the steam engines, generators and motors.

When Mather retired in 1922 he was succeeded by C. L. Fortescue who had been professor of electrical engineering at the Royal Naval College in Greenwich.[67] While not carrying out much research, Fortescue modernized the department, was a very good teacher who gave lecture demonstrations in the old Faraday manner, and took an interest in student affairs. He brought in many outside specialists to give lectures to advanced students. But the economic recession of the late 1920s made it difficult to keep this up without some sacrifice. Fortescue decided to dismantle the generating plant, sell the equipment, and in its place set up a new high voltage laboratory with equipment for the testing of insulating materials. This work was supported by the British Electrical and Allied Industries Research Association. Instead of running the generating plant, third year students specializing in heavy electrical work had to take a short course in heat engines and thermodynamics in the mechanical engineering department. By 1928 there were four sections in the department: power, electrical machine design, electrical measurement, and telephony and telegraphy. But the C&G building had not been designed with all this electrical work in mind. When the building was surveyed in 1933 it was found that the electrical engineers had caused serious structural damage to the building with all their wiring.[68]

The telephony and telegraphy option was so popular that Fortescue wanted to form a separate sub-department, and give students specializing in this field more instruction in mathematics. Ideally, he believed, there should be two separate departments: power (heavy electrical machinery), and light electrical engineering, including communication technologies. Indeed, this latter area was to become a major focus in the department. The creation of two departments had been suggested much earlier in the Wolfe Barry report of 1907, but the suggestion was not taken up. Rather, a compromise was reached to allow greater specialization in the final year. Fortescue, himself, was interested in radio and asked permission to give intensive short courses for people wishing to join the manufacturing industry or become radio operators. Permission was granted in 1939 and Fortescue trained many people who carried out war work in wireless communication. He retired in 1946 and was succeeded by Willis Jackson.

Civil Engineering

The civil engineering department continued the Unwin tradition of work in surveying, materials and the testing of structures, hydraulics and turbine design. These were the major fields in the associateship course. The new professor, Stephen Dixon, had to equip two new large structures laboratories and, a few years later, also the Hawksley Hydraulics Laboratory located in the basement of the new Goldsmiths extension.[69] For this he had much help from both Dalby and Unwin. In 1920 a Board of Trade committee, commenting on the lack of information on river flow in Britain, encouraged Dixon to begin an annual project for third year students. For many years they worked on a hydraulic survey of the Severn River, gauging it at Bewdley, a project that became one of Dixon's main enthusiasms (Unwin had earlier measured water flow in the Thames). First year civil engineers learned surveying in Hyde Park and Kensington Gardens, and puzzled onlookers with their antics. When asked what they were doing they enjoyed causing distress with the stock response that they were surveying for a new railway line. After the Second World War much of the surveying work moved to Silwood Park and, during their summer vacation, students carried out surveying further afield, often in the New Forest. Typically, second year students carried out a six-mile railway survey in the Wyn Forest close to where the third year students were gauging the Severn. The field courses were much enjoyed and helped build camaraderie among students and staff. In the college Dixon undertook the testing of various water turbines. In addition, he acquired a wire drawing plant and tested the strength of wire ropes, carrying out research also on their magnetic properties for the Safety in Mines Research Board. In 1928 the new Maybury Chair in Highway Engineering came to the department, financed by the Paviors Company and the Institution of Municipal and County Engineers. Since funding for the chair and the DIC course was insecure it was unclear whether the position would last. However the appointee, R. G. H. Clements, was soon testing all kinds of road surfacing materials at the college and the chair was funded until his retirement in 1946.[70]

One of Dixon's students, Sir Harold Harding, an accomplished engineer and future member of the Governing Body, remembered Dixon (nicknamed Dikko) as 'a man at the top of his profession … who did not seem unduly obsessed with mathematical matters, so there was still hope for us'.[71] Dixon retired in 1933 and was succeeded by A. J. S. Pippard for whom 'mathematical matters' were very important.[72] Unlike the mechanical engineering department, civil engineering did not fully modernize at Imperial until after 1945. Pippard began to change things on his arrival, but financial constraints prevented him from doing more. He found the department very poorly equipped, 'there was not a single experiment suitable for instruction in the behaviour of structures and I realized I had to start from scratch to build up even the semblance of a decent structures laboratory'.[73] Like Heilbron, Pippard had refused to take the professorship unless Tizard agreed to modernize the laboratory. Money for this was forthcoming from the Clothworkers, allowing Pippard to set up also a small models structures laboratory in which students learned modern analytical techniques.[74] Pippard disapproved of much of the applied work carried out by Dixon. For example, his having used the laboratory testing machine to test the strengths of roofs of underground mines for the Mines Research Board, work Pippard thought more suited to a government laboratory. Pippard's headship marks a turn from the heavily industrially-based work of the early period towards an emphasis on mathematical modelling and engineering science.

Since the senior engineering professors, Bairstow, Lander, Fortescue and Witchell all retired at the end of the war, Pippard was soon thrown into a senior role and became Dean of the C&G. An energetic man, he built a fine department while serving also on many committees external to the university. He early appointed C. M. White to the readership in hydraulics and together they reorganized undergraduate teaching in the department after the war. When not engaged in administrative duties, Pippard carried out research in engineering structures together with his assistant Letitia Chitty. They had met while working together at the Admiralty Air Department during the war. According to Pippard, Chitty was the methodical calculator while he was the intuitionist thinker.[75] In 1934 Pippard appointed S. R. Sparkes, one of his former students at Bristol, as demonstrator in structures.[76] Pippard was a close friend of the Professor of Geology, H. H. Read, and, like Dixon, was interested in the application of geological ideas to civil engineering. In this connection he, together with Professor Clements, brought the field of soil mechanics to the department after the war. During the 1940s reinforced concrete was increasingly used as a construction material and the Cement Makers offered the college a chair in concrete technology to which A. L. L. Baker was appointed in 1945.[77]

Geology and the Royal School of Mines

Geology

William Watts, who held the chair of geology from 1906 to 1930 was a great stickler for detail. When he retired he gave the Rector a list of seventy possible successors with long notes on each![78] Unlike the early professors in mining and metallurgy, Watts stayed long enough to have a major impact on the new college's affairs. He was a much loved professor and, on his retirement, the college raised money both for a Watts medal, and for a portrait by William Rothenstein to be hung in the RSM. Earlier, when Professor Judd took the department to the Western Galleries of the South Kensington Museum, he had divided the space there in two; one part for the geology course leading to the ARCS, and the other for applied geology courses for RSM students. From 1910 lectures were given also in engineering geology by Herbert Lapworth, assistant city engineer in Birmingham, who had worked on the construction of the water delivery system from the Rhyader reservoirs in Wales to Birmingham. His lectures were aimed also at civil engineers. At about the same time Arthur Wade began giving lectures on petroleum geology.[79] And, given the then general interest in petroleum and internal combustion engines, Wade persuaded Keogh to set up a committee with a view to establishing a new ARSM in his field.[80] Keogh was keen to support this, in part for reasons already mentioned, and in part because he was seeking ways to bring the three colleges together. He believed that encouraging work that related to fuels in several different disciplines was one way of doing so. In 1913 V. C. Illing came from Cambridge as a research student and was soon appointed demonstrator in the new petroleum geology course.[81] In 1923 the course title was changed to oil technology. Illing was appointed Professor of Oil Technology in 1937, continuing to build what became a world-renowned school in his field.

After the First World War, Thomas Holland had wanted to move the entire geology mix into the RSM but was opposed by Watts. As a result geology stayed with the RCS, though physically moving, along with its sub-departments, into the RSM building. Watts, C. G. Cullis, and the professor of mining, William Frecheville, drew up the curriculum for a new ARSM course in mining geology.[82] Geology thus had three sub-departments, two offered courses leading to the ARSM, while

its core course (the least popular) led to the ARCS. Watts, in making
the argument that his department remain in the RCS, postponed some
problems for later. He, himself, carried out 'pure' geological research and,
as his Royal Society obituarist noted, had made the Charnwood Forest
a place of pilgrimage for students of stratigraphy worldwide. He was an
excellent teacher and many of his students had major careers in universities, with geological surveys, or in the mining and petroleum industries.[83]
One of his appointees in 'pure' geology was John William Evans, who
came to the college aged fifty-seven, as lecturer in crystallography and
petrology. Evans had had a colourful career overseas, mainly in Brazil,
Bolivia and India, before becoming Governor of the Imperial Mineral
Resources Bureau. He was multilingual, highly knowledgeable and, if
the many tales told of him are true, highly eccentric.[84] Another notable
figure of this early period was Arthur Holmes, a pioneer in the radiometric dating of rocks who established the geological time scale from a
determination of lead/uranium ratios. On receiving the Wollaston Medal
in 1956 he stated, 'when I was a student, 100 million years was still
regarded as an extravagant estimate of the age of the earth; today 4,500
million years is a minimum estimate'.[85]

When Watts retired, one of his old students, Percy Boswell,
succeeded him.[86] That the department needed to rethink its role was
brought home by a 1935 report resulting from the first University of
London visitation.[87] Its 'pure' research was no longer seen as impressive and much of its equipment was labelled 'outdated'. For example,
the microscopes being used by geology students were over fifty years
old. However, the oil technology course under Illing was seen as highly
successful and further applied work was encouraged. In 1938 Boswell
retired early because of ill health and was succeeded by yet another
of Watts's old students, H. H. Read. Read, too, had been Herdman
Professor at Leeds before coming to Imperial and according to Boswell,
was 'always boast[ing] of his plebeian contacts and love of the soil', though
he came from Kentish yeoman stock and a family who looked askance at
his descent into geology from the 'great calling of farming'.[88] Read was
also remembered as having been a vigorous editor of *The Phoenix* in his
student days, and his writing skills were often remarked upon. Read was
a major mineralogist and petrologist, and pioneer in a new era of structural geology made possible by the introduction of physical techniques
such as radiometric age determination. He restored the reputation of

'pure' geology at the college and trained many fine geologists. In 1941 W. R. Jones, who had earlier lectured on ore microscopy at the college, succeeded Cullis as professor of mining geology.[89]

Mining

In 1907 the professor of mining was Samuel Herbert Cox who had been appointed in 1906 after the early death of Sir Clement Le Neve Foster.[90] Even before the formation of Imperial, Cox and William Gowland, the professor of metallurgy, had given much thought to revising the mining and metallurgy courses which had fallen behind the times.[91] Plans for equipping the new building were already underway. Cox also wanted his students to have regular access to a mine in order to learn underground mine surveying and other technical skills. Tywarnhaile mine in Cornwall, near Redruth and the Camborne School of Mining, was purchased in 1909.[92] Before the 1914-18 war Imperial was one of the few institutions allowed to grant certificates to mine surveyors, and mine surveying continued to be an important part of training in the mining department for many years. Students also visited and worked in mines during their vacations. Many went to the Rio Tinto copper mines in Spain with which the department had connections.[93] Cox retired in 1913 and was briefly succeeded by William Frecheville, a graduate of the RSM who had worked at mining sites around the world.[94] S. J. Truscott, who had been educated at the RSM, was appointed professor of mining in 1919 and was the longest serving of the early mining professors, retiring only in 1935.[95] In the same year L. H. Cooke was appointed to a new professorship in mine surveying.[96] Truscott wrote to Keogh saying he wanted to reform the mining course and pay more attention to coal and iron mining, but this only occurred after his retirement with the appointment of his successor, J. A. S. Ritson.[97] Despite this appointment the main focus of the department remained hard rock metalliferous mining. Indeed, while Ritson's appointment was seen as a timely move towards coal by some, Ritson himself was seen as needing more experience in metalliferous mining and was sent to visit gold and copper mines in South Africa before being allowed to take up the professorship.[98] Emphasis in this department was very much on practical experience for students as well as for professors. According to a 1932 report, graduating students were expected to have completed a minimum of 720 hours practical work.[99] While much of this was carried on in the laboratory, students were expected to work also at Tywarnhaile, and to visit and work in operational mines during their vacations.

Metallurgy

In 1907 the professor of metallurgy was William Gowland.[100] At first metallurgy was offered only as an advanced level DIC course and many students came from the Ordnance College at Woolwich to take it. Professorial and student exchanges between the RSM, RCS and Woolwich were common, having begun earlier when all three colleges were under the Board of Education.[101] Metallurgy became a full four-year course only after Gowland was succeeded by William Carlyle in 1911.[102] The department then assumed equal status with mining to which it had earlier been subservient. The metallurgists moved from their cramped quarters in the basement of the Science Schools building to new quarters in the Bessemer Laboratory and were given further space in the new RCS building. Relatively, the space for metallurgy was commodious given that in the early years there were only about twenty associateship students per year. This number shrank to about four during the war years, rising back to twenty after 1918, and stabilizing at about thirty-five by the 1930s. Carlyle stayed for only two years and, in 1913, Harold Carpenter was appointed to succeed him.[103] Carpenter had been professor of metallurgy at Manchester but, compared to Carlyle, his experience was seen as too narrowly academic. Before being allowed to take up his new position he was sent on a one-year trip, at the college's expense, to visit some major metallurgical works worldwide, and to learn something about extraction metallurgy. Carpenter's own research was in physical metallurgy. He and a close woman research collaborator, C. F. Elam, worked on crystal growth in metals and, together, they won many honours.[104]

Students in Carpenter's department were given some theoretical training, some information on mining economics and tropical hygiene, but the pedagogical emphasis, as in mining, was on the practical. There was much laboratory work including, at first, 450 hours of assaying. There were regular visits to metallurgical works and, in the fourth year, instruction was given in the new Bessemer Laboratory. There, on a semi-industrial scale, students roasted metal ores in a reverberatory furnace and smelted the products in a blast furnace.[105] Just before the Second

World War the department evacuated to Swansea where its work was integrated with the department there. As a consequence new ideas were later brought back to the college. Under Carpenter's successor, C. W. Dannatt, the Bessemer Laboratory was completely refurbished, with a grant from the Nuffield Foundation, and the department reoriented itself for work in extraction metallurgy.[106]

Conclusion

How departments were set up in 1907 influenced their future development. It very much mattered what kind of person was appointed to head a department. The botany department was fortunate in having J. B. Farmer. Highly capable, he also had the connections to build a major research department from a very small base. Already in the late nineteenth century, he had anticipated the Imperial College charter, and had turned his own research in a direction that helped serve imperial industries such as coffee, tea, cotton and rubber. The strong foundations that he laid in plant physiology and plant pathology stood the department well for much of the twentieth century. Similarly, Adam Sedgwick, though at Imperial for only a few years before his premature death, made the important decision to build a department specializing in applied entomology. He appointed some outstanding people, specialists who laid the foundation for the college's major strength in this area. This foundation was later consolidated by J. W. Munro.

The chemistry department was already very strong in 1907 because of a pattern laid down by A. W. Hofmann and E. Frankland at the Royal College of Chemistry (RCC). The older tradition was absorbed by chemists both at the RCS and C&G and the departmental merger went well. The early professors of chemistry were all major figures who built on, and enriched, the RCC tradition. Physics and mathematics had more complicated starts since their early history in the RCS was shakier than that of chemistry, and because the merging of the departments from the C&G led to philosophical disputes, especially in mathematics, over the kind of department needed. The physics department strengthened after the appointment of G. P. Thomson, but physics and mathematics truly came into their own only after the Second World War. Despite this, both departments early had very gifted professors on their staffs — A. Forsyth, A. N. Whitehead, H. L. Callendar, R. Strutt, and A. Fowler were all outstanding. Also strong in the early period was meteorology.

Callendar's work illustrates that interdisciplinary work was part of Imperial's ethos from the start, and that it was not something invented in the later twentieth century. Indeed, Keogh encouraged much interdisciplinary work, in part because he saw new opportunities such as those connected to internal combustion, motorized transport, and petroleum studies, but also because he was actively seeking ways to unify the three colleges. The founding professors at the Central were all excellent, and, when the college joined Imperial, it was already a very strong engineering college. But emphasis both in research and pedagogy was very much on the applied and the practical. While the later turn to engineering science under Lander and Pippard was important, it is worth noting that many of the earlier graduates had very successful careers.

Since Dalby remained head of mechanical engineering, and was somewhat resentful of the split, work on structures and hydraulics, introduced earlier by Unwin, did not at first move smoothly to civil engineering. They were seriously pursued only after Pippard took over as head of department. Further, because of Dalby's long tenure, there was more continuity in mechanical than in civil engineering. During the First World War Dalby set the ground for later work in the area of fuels, thermodynamics and heat transfer. When the C&G physics department split up, what remained behind was electrical engineering. Mather, staying true to the spirit of Ayrton, changed little in this department. Fortescue modernized the undergraduate curriculum and introduced work in the area of communication, but the department had to wait until after the Second World War for a new start under Willis Jackson.

Geology had a difficult transition at the founding of Imperial, something that can be traced back to its earlier separation from the Geological Survey. Professor Watts had to think of ways to meet the ethos of Imperial College and did so by introducing a number of applied courses. But he was unwilling to think of his field as anything other than a 'pure' science, and so insisted on the department remaining in the RCS. As is discussed elsewhere, this had repercussions throughout the twentieth century. There is some irony in that the department's later problems were exacerbated by the 1938 appointment of one of the best 'pure' geologists of the twentieth century. H. H. Read built a major research school at the college. But, as will be discussed in later chapters, the identity of the department became increasingly unclear due to the growing strength of applied areas introduced both early in the century by Watts, and later by others.

Mining and metallurgy both had a difficult start but gradually modernized. Mining specialists found common ground with people working in the applied geological fields, and metallurgy moved away from the ultra-practical. Students who had once spent many hundreds of hours doing assays began to learn about physical metallurgy and extraction metallurgy, as these fields began to acquire a solid theoretical base.

While there were few women students, there were a surprising number of women research assistants working at the college in this period. Many women were hired as calculators notably in the mathematics and engineering departments. Others helped carry out the major research programmes of the professors, including in the laboratory sciences. Especially notable in this period are Elaine Austin (see chapter four), Margaret Carlton, Constance Elam, Letitia Chitty and Maud Norris (later the wife of Professor O. W. Richards). Norris worked first with Professor Lefroy and was a major lepidopterist (see chapter seven). What is interesting is that all five were recognized as highly capable research workers, and as very good scientists. Their professors often did their best to secure good incomes and working conditions for them. The cases of Martha Whiteley, first appointed a demonstrator in chemistry, and Margaret Fishenden appointed honorary lecturer in mechanical engineering, are different. Whiteley was well integrated in her department, taught undergraduates, had research students of her own and became the first woman assistant professor (reader). Fishenden had problems of integration within mechanical engineering though she was recognized as an excellent technologist and gave undergraduate lectures. This was unusual since, in this period, most men thought that women were unsuited for the teaching of male undergraduates. Further, it did not enter the minds of their male colleagues that women should, or could, enter a normal academic career path. That they could be equals in that sense, run research sections of their own, become administrators, or even head a department, was then largely unthinkable. Most of the academic women staff employed at the college remained research assistants. Whiteley and Fishenden were among the very few exceptions. Until looking into these cases, I had not recognized the distinction made between research and other abilities. Like many others I had assumed that, until later in the century, women were largely seen as unsuited to science *tout court*.

...

1. John Bretland Farmer FRS (1863–1944) was educated at Magdalen College, Oxford and came to the Royal College of Science in 1892 as assistant professor under T. H. Huxley. On Huxley's retirement in 1895, Farmer became Professor of Botany. Huxley had been Professor of Biology but was replaced by professors of botany and zoology (T. G. B. Howes, 1853–1905). Farmer and Howes became joint heads of the RCS department of biology. For the new buildings see chapter 4; for more on Farmer's work see chapter 7.

2. Vernon Herbert Blackman FRS (1872–1967) was educated at the City of London School, St. Bartholomew's Hospital Medical School, and St. John's College, Cambridge. Before coming to Imperial he was Professor of Botany at Leeds. He helped develop the Fruit Research Station at East Malling and some of his students worked there. One of his students, Eric Ashby, remembered Blackman as 'puffing his outsize pipe as he joined us for afternoon tea in the laboratory'. (Typescript of Ashby's Blackman Memorial Lecture, 21 May 1968, in ICA).

3. Already in 1912, Parliament had voted to fund the Development Commission to promote research in agriculture. Imperial College was selected to work on plant physiology and grants for research and salaries came from the Ministry of Agriculture and Fisheries.

4. Samuel Barnett Schryver FRS (1869-1929) was educated at University College London and in Leipzig, where he gained a PhD.

5. Albert Charles Chibnall FRS (1894–1988) was educated at St. Paul's School and Clare College, Cambridge but the First World War and military service interrupted his studies. After the war, the University of London allowed ex-servicemen to register for a PhD with only Part 1 of the Cambridge tripos, so Chibnall came to Imperial where he took an external PhD under Schryver. Known as 'Chibs', he won the Huxley medal and, on graduation, worked briefly at the Yale and Connecticut Agricultural Station, and then at University College London. He returned to Imperial College in 1929 after the death of Schryver, when he was appointed assistant professor. Chibnall carried out work on the fatty and nitrogenous substances in leaves and was appointed to a chair in biochemistry in 1937. In 1943 he succeeded former Normal School of Science student, Sir Frederick Gowland Hopkins, in the Sir William Dunn chair of biochemistry at Cambridge. For more on Chibnall see his memoir, 'The

Road to Cambridge', *Annual Review of Biochemistry*, vol. 35 Part 1 (1966), 1–22. In it he noted that, 'compared to what I was to encounter later at Cambridge, the keynote of the administration at the Imperial College was simplicity', (11). He appreciated that professors were very much left to their own devices. Two of Chibnall's Cambridge research students, F. Sanger and R. R. Porter, later became Nobel laureates.

6. Chibnall, *op. cit.* (5), 4.

7. For example T. G. Hill became professor of botany and plant physiology at University College London.

8. GB minutes, 20 May 1936.

9. Frederick Guggenheim Gregory FRS (1893–1961) was remembered as a 'vivid personality', a born rebel with a 'bubbling flow of thoughts' on any number of subjects, an intellectual, and fine musician. He entered his scientific projects with great fervour and passion which impressed generations of students and generated much loyalty, but also some resentment. He was born Fritz Guggenheim, to a German immigrant family living in London, and entered Imperial in 1912 as a chemistry student. Swayed by J. B. Farmer's lectures in botany, he switched subjects and appears to have carried off most of the major prizes on graduating. Gregory joined Blackman's research team and was sent to Cheshunt to work on the physiology of greenhouse crops. It was there, in 1916, that he changed his name by deed poll, a consequence of being taunted for his German name during the war. He applied his botanical and mathematical skills to the analysis of the effects of a range of variables on plant growth. In 1929 he was appointed assistant professor and began work on plant stomata, work developed further by his student O. V. S. Heath, later professor at Reading, and by his assistant, H. K. Porter, later professor at Imperial. Blackman retired from the Directorship of the Institute in 1943 and Gregory was appointed Acting Director. He resented this, but in 1947 he succeeded to the chair and full directorship. See Helen K. Porter and F. J. Richards, 'Frederick Guggenheim Gregory', *Biographical Memoirs of Fellows of the Royal Society*, vol. 9 (1963), 131–53. See also memoir by O. V. S. (Peter) Heath (handwritten, copy in ICA).

10. William Brown FRS (1888–1975) was born in Dumfriesshire and educated at the University of Edinburgh where he studied both mathematics and botany, winning university medals in both disciplines. At Edinburgh he was a contemporary of two other Imperial professors, J. W. Munro and H. Levy. Brown came to Imperial as a research student under V. H. Blackman in 1912 and, in 1916, was appointed a research assistant at the Institute of Plant Physiology. Brown spent his entire career at Imperial

where his major work was in fungal physiology and in fungal plant parasitism. He succeeded V. H. Blackman in the chair in 1937. Brown grew tens of thousands of lettuce plants and infected many with fungal parasites. Those not infected he sold to the benefit of the department's overseas travel fund.

11. Adam Sedgwick FRS (1854–1913), great nephew of Darwin's geology professor of the same name, was educated at King's College London where he began studies towards an MD. He moved to Trinity College, Cambridge before completing his MD and took a degree before beginning research under Michael Foster and F. Maitland Balfour at the School of Physiology. He was appointed Professor of Zoology at Cambridge in 1907 and was a specialist in invertebrate morphology. See J. Stanley Gardiner's obituary of Sedgwick, *The Zoologist*, March 1913. See also chapter 7.

12. Letter to Rector (Henry Bovey), dated April (no day), 1909; (KZ9/1.2). Sedgwick's health was never good since he suffered from *spina bifida*. He died of tuberculosis. Fox, Doncaster, and Dobell all became Fellows of the Royal Society and were specialists in animal physiology, genetics, and parasitology respectively. Edward Hindle, also later FRS, soon left Imperial for the chair of biology at the medical school in Cairo. Later he founded the School of Virology at Mill Hill. Fox joined Hindle in Cairo in 1919 and was to become Professor of Zoology at Bedford College. For Lefroy and MacBride see below.

13. Keogh wrote a letter to *The Times* in 1919 (letter drafted by Lefroy) stressing the importance of entomology for the economy and for public health. He encouraged parents to let their children study in this field. He received a flurry of letters from interested parents as a consequence. One correspondent donated his large Lepidoptera collection to the college. See correspondence of November 1919 in KZ/9/2/1.

14. O. V. S. (Peter) Heath describes some of these weekend field trips in his diary. For example, in the Spring of 1923 they visited Box Hill collecting mosses, had a guided tour of the Zoo with Professor MacBride, a tour of Kew Gardens with Professor Groom, an outing to Rothamsted, an entomology outing to Caterham together with Bedford College students, supervised by Professor Lefroy, a trip to Carters Seeds, a mycology outing to Oxshott, and two geology outings to Caterham. Professor Farmer arranged longer summer field trips to Snowdonia as well as annual botanical outings to his home in Gerrards Cross. The Heath diary is privately owned by his son, Robin Heath.

15. Harold Maxwell Lefroy (1877–1925) was educated at Marlborough School and King's College, Cambridge, He worked as an entomologist in the

department of agriculture in the West Indies and, in 1901, was appointed Imperial Entomologist in India. He built a large entomology department at Pusa. He had taught periodically in South Kensington before 1912, the year he was appointed Professor of Entomology. For more on Lefroy, see Helen Lefroy and Laurence Fleming's typed biographical manuscript in the ICA. For Lefroy's work on pest control see chapter 7.

16. Obituary, *The Times,* 15 October 1925, 16.

17. The Royal Horticultural Society (RHS) had an early connection to the colleges in South Kensington. When the Normal School opened, the Society's Gardens were still in South Kensington. William Thiselton Dyer, who was to become Director of the Royal Botanical Gardens at Kew, worked at the gardens and taught botany at the Normal School.

18. Ernest William MacBride FRS (1866–1940) was born in Belfast and educated at Queen's College, Belfast and at Cambridge. After graduation he spent a year at the marine zoological station at Naples, returning to Cambridge as a research student and demonstrator under Sedgwick. A popular president of the Cambridge Union in 1891, his debating skills were in evidence all his life, and he was remembered for his abundant flow of conversation. He was appointed Strathcona Professor of Zoology at McGill in 1897 where he worked until his appointment at Imperial.

19. See Peter J. Bowler, 'E. W. MacBride's Lamarckian Eugenics and Its Implications for the Social Construction of Scientific Knowledge', *Annals of Science*, 41 (1984), 245–260.

20. Quoted in J. E. Smith, 'Herbert Graham Cannon, 1897-1963', *Biographical Memoirs of Fellows of the Royal Society*, vol. 9 (1963), 55-68. Cannon was later Byer Professor of Zoology at Manchester.

21. KZ/9/2/; see MacBride to Gow, 7 January 1930.

22. Leonard Darwin, President of The Eugenics Education Society, asked MacBride to give a year's course to train members to become independent investigators and field workers in eugenics. A set of twelve lectures was agreed on and the arrangement continued for several years. Attendees were charged a laboratory fee of £1. In 1913, fourteen students took the course, nine of them were women; see MacBride to Rector, 20 November 1912, and Gow to MacBride, 14 July 1913. KZ/9/2/1

23. This brought in much needed money to the college. Lefroy did the same, though less enthusiastically. The Colonial Office paid £1.1s a week for its men to take courses in insect control.

24. Lancelot Hogben FRS (1895–1975) had a strict Methodist upbringing as the son of a fundamentalist preacher. He was educated at Trinity College, Cambridge where he became an active Fabian and left religion behind.

He was appointed lecturer in zoology at Birkbeck College in 1917, having just spent three years building bungalows for war victims. He came to Imperial in 1919 where he found a soulmate in Hyman Levy who taught him mathematics. In the footsteps of H. G. Wells, both men believed that social good could be brought about by a scientific way of thinking. They shared this belief with their more famous contemporary, J. D. Bernal, after whom it became labelled 'Bernalism'. Hogben left Imperial in 1925 and worked in a number of locations before taking a professorship in Birmingham where he specialized in medical statistics. His best known work is probably *Mathematics for the Million* (London, 1936). See G. P. Wells, 'L. T. Hogben, *Biographical Memoirs of Fellows of the Royal Society.* vol. 24, (1978), 181–221. See also Gary Werskey, *The Visible College: A Collective Biography of British Scientists and Socialists of the 1930s* (London, 1988).

25. William Alexander Francis (Frank) Balfour Browne (1874–1967), was educated at St. Paul's School and Magdalen College, Oxford. He taught biology at Queen's College, Belfast, and at Cambridge University before coming to Imperial. He had formed an interest in water beetles already as a child and this remained his principal research focus. Three monographs written by him on the subject were published by the Ray Society. Balfour Browne was a keen bicyclist and while at Oxford his machine was maintained by William Morris, the future Lord Nuffield.

26. James Watson Munro (1888–1968) was educated at Dundee High School and the University of Edinburgh where he was a contemporary of William Brown and Hyman Levy, also future professors at Imperial. On graduating he went to Tharandt, near Dresden, to work with K. L. Escherich, a leading forest entomologist, where he gained a doctorate. Munro won a scholarship to do further research work at Imperial College under Lefroy, but war interrupted his studies. He served in France before being asked to join a Cambridge research team working on the scabies mite. After the war he joined the Forestry Commission and proposed a programme in forest entomology. He returned to Imperial in 1926. Munro's work, including at the two college field stations, is discussed further in chapters 7, 8 and 10.

27. See 1928 correspondence between Balfour Browne and Munro and correspondence between them and Holland; also Holland to Farmer, 17 September 1928 (KZ/9/5). It is clear there was much acrimony. But, as Munro pointed out, he had brought a lot of grant money to the department and had taken much of the responsibility for running the course since Balfour Browne was only a part-time professor. Holland appears to have been something of a realist. He saw how money might come to the college and accordingly nursed Munro's ambitions.

28. Owain Westmacott Richards FRS (1901–84), son of the Medical Officer for Croydon, was educated at Hereford Cathedral School and Brasenose College, Oxford where his tutor was Julian Huxley. He was a research student under E. B. Poulton and, like him, a great specimen collector. He became a specialist in systematic entomology and insect ecology. Systematics at Imperial was facilitated by a close association with the Natural History Museum and thus access to the national collections. One of Richards' collaborators was Nadia Waloff with whom he worked on grasshopper and broom insect populations. Richards became a reader in 1937 and, in 1953, succeeded Munro in the chair, and as head of department. From some contemporary accounts it would seem that Richards was a very different head from the energetic and outgoing Munro. He has been described as very smart and aware of his authority, yet shy, aloof, and somewhat ethereal. See Richard Southwood, 'Owain Westmacott Richards', *Biographical Memoirs of Fellows of the Royal Society*, vol. 33 (1987), 539–71.

29. See O. W. Richards and G. C. Robson, *The Variation of Animals in Nature* (London, 1936). Hewer was a major figure in the department. He ran the undergraduate programme and led many field trips to the New Forest, to Oronsay where seals were studied, and to other locations.

30. Hugh Longbourne Callendar FRS (1863–1930) was educated at Marlborough School and Trinity College, Cambridge. He was Professor of Physics at Royal Holloway College (1888–93) before moving to McGill University to set up the new physics department there. He returned in 1898 as Quain Professor of Physics at University College London. He succeeded Arthur Rücker as professor of physics in the RCS in 1901, but was later frustrated with lack of money for equipment for the new building. See letters from Callendar to Gow, 1908–9. A specialist in thermodynamics, his major published work was *The Properties of Steam and Thermodynamic Theory of Turbines* (1920). He was remembered also as one of the college's many early motoring enthusiasts, and as a kindly man who helped people financially. See obituary in *Proceedings of the Royal Society*; series A, vol. 134 (2 January 1932), xviii–xxvi; and L. H. Callendar 'H. L. Callendar — Instrument Engineer', *The Chartered Mechanical Engineer*, February 1966.

31. W. Siemens constructed the first platinum resistance thermometer.

32. Robert (Robin) John Strutt FRS (1875–1947), later fourth Baron Rayleigh, was educated at Eton and Trinity College, Cambridge. He was a long-term member of Governing Body, and its chairman, 1936–47. While at Cambridge his mathematics tutor was Gilbert Walker, later Professor of Meteorology at Imperial.

33. The third Baron Rayleigh had equipped two large laboratories at the family estate, Terling Place in Essex.

34. Alfred Fowler FRS (1868–1940) was a Yorkshireman who came to the Normal School of Science in 1882, with a scholarship from the Keighley Trade and Grammar School. He was possibly the youngest student ever admitted to the college. Though specializing in mechanics, Fowler was drawn to J. Norman Lockyer and, on becoming Lockyer's assistant, reinvented himself as an astrophysicist. Together they carried out major work in solar, stellar and cometary spectroscopy. When Lockyer left the college, Fowler was appointed assistant professor and, in 1915, professor of astrophysics. He travelled widely, observing a range of solar eclipses, a pattern begun by Lockyer who earlier had organized many eclipse expeditions. Fowler was the first Secretary of the International Astronomical Union in 1919, and Royal Society Yarrow Professor (1923–34).

35. There was more than one tower in the observatory. In one there was a transit instrument on loan from Greenwich, in another a twenty-inch reflector for radiation experiments, on loan from the Permanent Eclipse Committee. In 1910 a five inch equatorial telescope was installed for training students. According to D. M. Levy, a student in the Royal School of Mines, 1900–05, staff would be in attendance at night and students could make observations under expert guidance. Observational work continued for some years after the formation of Imperial College. See B/ D. M. Levy, 'Reflections of a "Minesman" of some sixty years ago' (Memoir, 1955). Levy's son J. F. Levy was a professor of botany at the college. During the day, students would be taught how to use theodolites and sextants — also on the roof of the RCS building. Peter Heath recorded that on one occasion they were supposed to be studying the sun but spent more time observing the pigeons. And that the roof was so hot that it hurt to stand on it; the students retreated to the lavatories below and put their heads under cold water. O. V. S. (Peter) Heath diary, 23 May 1921. Diary privately held by Robin Heath.

36. Herbert Dingle, quoted in Kathleen W. Greenshields and W. D. Wright, (historical introduction to booklet published for opening of new physics building, 1960). Herbert Dingle (1890–1978) was educated at Imperial where he was a student of Professor Fowler. He joined the staff after gaining a DIC. When Dingle was appointed to the chair in 1938, it was renamed the chair of natural philosophy and the chair of astrophysics was discontinued. However, work in spectroscopy continued under two younger physicists, W. R. S. Garton and A. G. Gaydon. Dingle acted as head of department during the war when George Thomson was working for the

government. He left for a chair in history and philosophy of science at University College London in 1946. Dingle was known as something of a scientific heretic for his criticism of Einstein's Special Theory of Relativity. His turn to philosophy probably had something to do with the manner in which the scientific community ignored or dismissed his ideas. See his *Science at the Crossroads* (London, 1972).

37. George Paget Thomson FRS (1892–1975), son of physicist J. J. Thomson, had been a professor in Aberdeen before coming to Imperial. At Aberdeen he began the work that would win him the Nobel Prize for physics in 1937 (with C. J. Davisson). Imperial's first Nobel Laureate, Thomson built up a school of electron diffraction, applied to the study of surfaces. In this he was assisted, to a degree, by G. I. Finch, Professor of Applied Physical Chemistry in the Department of Chemical Technology. But Finch had his own rival group in electron diffraction and the two had a number of intellectual disputes. Thomson later moved to other work in nuclear physics. See also chapters 8 and 9. P. B. Moon, 'George Paget Thomson', *Biographical Memoirs of Fellows of the Royal Society*, vol. 23 (1977), 543–7.

38. Alexander Oliver Rankine FRS (1881–1956) was educated at University College London where he joined the staff on graduation. Rankine was given a professorship at Imperial in 1921 after having carried out submarine detection work during the war. His work at Imperial on the Eotvos gravity meter led him to construct magnetic analogues for use in mineral prospecting. See also chapter 7.

39. Sir David Brunt FRS (1886–1965) was educated at the University College of Wales (Aberystwyth) and Trinity College, Cambridge. He joined the Meteorological Office after work with the Army Meteorological Services. He came to Imperial as an assistant to Sir William Napier Shaw and carried out research in the area of cloud physics, and on the motion of the air. He won many awards, was on many boards and committees and, on retirement, became Chairman of the Electricity Supply Research Council. See O. G. Sutton, 'David Brunt', *Biographical Memoirs of Fellows of the Royal Society*, vol. 11 (1965), 41–52. See also chapter 8.

40. Sir Thomas Edward Thorpe FRS (1845–1925), known as Thomas when young and Edward later, was educated at Owen's College under Henry Roscoe, and at Heidelberg under Robert Bunsen. Appointed Professor of Chemistry at the Andersonian College in 1871, and at the Yorkshire College of Science (Leeds) in 1874, he succeeded Edward Frankland at the RCS in 1885. In 1894 he left to become Government Chemist and head the new government chemical laboratory (later National Chemical Laboratory), returning to Imperial in 1909 when his successor W. A. Tilden retired.

Thorpe was a man of wide scientific interests. His *Dictionary of Applied Chemistry* which went through several editions was a standard work for many years. The fourth edition was revised and enlarged after his death by two of his former students, Martha Whiteley and Jocelyn Thorpe, both on the department staff. During the Second World War, Whiteley removed all the *Dictionary* papers to Cambridge for safekeeping. B/Ellingham, box 3; see Whiteley to Ellingham, 29 October 1940.

41. Martin Onslow Forster FRS (1872–1945), assistant professor of organic chemistry, was similarly intimidating. G. W. Himus, recollected his 'terrifying appearance … in morning coat, striped trousers, patent leather boots, jet-black hair and moustache with not a single whisker out of place … an appalling manifestation of assistant professorial magnificence'. But top sartorial honours in this early period went to chemistry lecturer G. A. R. Kon, later professor at the Chester Beatty Research Institute. He was described in the *Evening Standard* as the 'best dressed scientist in London'. Standard wear for students at this time was grey flannel slacks, white shirt and grey pullover. But to enter the library a jacket was needed. Quotations in E. R. Roberts (ed.), 'History of the Chemistry Department', (typescript, 1963), 81–2 and 93. Forster had earlier been a chemistry student at the C&G under H. E. Armstrong. He resigned from Imperial under protest when matters for organic chemistry, which had been in decline under T. E. Thorpe, did not immediately improve under H. B. Baker. He later moved to Delhi as Director of the Indian Institute of Science.

42. Herbert Brereton Baker FRS (1862–1935) was educated at Balliol College, Oxford and then spent twenty years as a schoolmaster at Dulwich College, carrying out research in his spare time. One of his students at Dulwich was Sir Harold Hartley who claimed that Baker encouraged his interests in the history of chemistry. (Aside from his many academic and industrial achievements, Hartley published work in the history of chemistry and for many years was editor of the historical *Notes and Records of the Royal Society*.) Baker's research at Dulwich won him election to the Royal Society and appointment to the Lees readership at Oxford where he stayed briefly before moving to the chair at Imperial. Baker and his wife, Muriel, were great supporters of student residences and gave many hours of their time towards making these a reality. G. C. Lowry, in his memoir, remembered Baker, Farmer, Watts and Dalby as being the most influential professors in college life during the 1920s (B\Lowry). See also chapters 4 and 5.

43. Martha Whiteley, a lecturer (later reader) in the chemistry department in this period was an exception. See chapters 4 and 5.

44. James Charles Philip FRS (1873–1941), the son of a minister of the Church of Scotland, was educated at the University of Aberdeen and was later a research student under H. W. Nernst in Göttingen, and then under H. E. Armstrong at the C&G. During his tenure at Imperial the number of papers published in physical chemistry increased from about one a year to about twelve. He studied aqueous and non-aqueous solutions by classical phase equilibria and electrochemical methods. He introduced electrochemistry and surface chemistry, both later departmental strengths. He was a much loved professor who took great interest in student affairs, and his vocal talents were much in demand at college functions. He was Acting Rector during the Second World War, 1939–41, and died in office.

45. Sir Jocelyn Field Thorpe FRS (1872–1939), had been a student both at King's College London and the RCS during the 1890s and had a PhD from Heidelberg. At Imperial he succeeded Martin Forster, a wealthy man who had paid for much of his own equipment. When Thorpe arrived, Professor Baker asked the Rector to provide some extra funding for him since he could not be expected to follow in Forster's footsteps with regard to supplies. Thorpe was a major college personality around whom stories gathered, such as that in his student days he excited the neighbourhood by throwing large lumps (about 1 lb.) of sodium into nearby park ponds on Guy Fawkes Night. Thorpe's work is discussed further in chapter 5.

46. Henry Vincent Aird Briscoe (1888–1961), (known as Vincent) was educated at the City of London School and the RCS. He was an active contributor to *The Phoenix*. He held a lectureship at Imperial, and professorships at Sir John Cass Technical Institute and Armstrong College (Newcastle), before returning to the chair of inorganic chemistry. He worked on a range of problems, notably on ones related to heavy water. His work at Imperial led to accurate methods of determining deuterium in small samples of water or gas. His obituarist described Briscoe as having been a keen motorist whose knowledge of London streets 'would not have disgraced a taxi driver'. See *The Times*, 25 September 1961. See also chapters 5 and 8.

47. Earlier analytical chemistry was taught by H. Chapman Jones who served under all the professors from Hofmann to Baker.

48. Chemistry has been revived more recently by pharmaceutical activity in the post genomic age, and by research in new materials.

49. Sir Isidor (Ian) Morris Heilbron FRS (1886–1959) was educated at the Royal Technical College in Glasgow where he was later appointed professor; he came to Imperial in 1938 from a professorship in Manchester. He worked on a variety of projects, perhaps the best known being his contribution to the synthesis of Vitamin A, manufactured by Hofmann La Roche from

1946. Heilbron's lectures were remembered as highly entertaining and he was especially adept at playing on the rivalry between British and German organic chemists. He was knighted for his war work in 1946. After he left Imperial College Heilbron helped found the Brewing Industry Research Foundation and became its first Director. See also chapters 8 and 10.

50. KC 10/2; Heilbron to Tizard, 19 February 1938. When A. J. (John) Greenaway visited the department in 1936 he recognized some of the old fittings that had been brought over from Oxford Street when the Royal College of Chemistry moved to South Kensington in 1873. Greenaway (brother of artist Kate Greenaway) had been a student of A. W. Hofmann and joined the staff of the Royal College of Chemistry. He moved to the Normal School where he remained on the staff until 1881. See *Record of the Royal College of Science Association* (June 1936), 27.

51. There was some justification in this view since the City and Guilds could still appoint mathematicians provided that they were committed to teaching also in the RCS.

52. Andrew Russell Forsyth FRS (1858–1942) was educated at Trinity College, Cambridge and was senior wrangler in 1881. In 1882 he became Professor of Mathematics at the new University of Liverpool, the city in which he had grown up. Later he returned to Cambridge as a lecturer before succeeding Arthur Cayley as Sadleian Professor of Pure Mathematics. Forsyth published a six-volume work, *Theory of Functions* (1893–1917), and led the fashion for function theory in Britain. Despite his own success he wanted to abolish the order of merit system in the mathematical tripos. In 1910 he resigned his chair at Cambridge; his resignation was related to his affair with the wife of physicist C. V. Boys. They later married but disapproval of this liaison led the couple to move to Calcutta for three years before Forsyth's appointment at Imperial. Forsyth was also an unusually gifted linguist. E. A. Nehan, a technician in the department, found Forsyth a far more sympathetic boss than Henrici whom he described as 'a very testy type'. B/Nehan, farewell speech, 1952.

Sir Charles Vernon Boys FRS (1855–1944) was educated in physics at the Royal School of Mines and was appointed assistant professor in the physics department after it became part of the Normal School (later RCS). He also held the post of Metropolitan Gas Referee. Boys was a prolific inventor of instruments and new techniques in applied physics. One of his tricks was the making of quartz fibre by shooting arrows laden with molten quartz across the laboratory. He was among the first Fellows of Imperial College, appointed in 1932.

53. The Executive Committee set up a committee to look specifically at mathematical education which, after much debate, recommended the expansion of pure mathematics and the termination of Perry's contract. For more on Perry, See Hannah Gay, 'John Perry', *Dictionary of Nineteenth-Century British Scientists* (Thoemmes Continuum, 2004), 1584–5.

54. See GB Annual Report, 1925; also Report of Mathematics Joint Committee.

55. Alfred North Whitehead FRS (1861–1947) was educated at Trinity College, Cambridge and was fourth wrangler in 1883. He is best known for having published, with Bertrand Russell, *Principia Mathematica*, 3 vols. (Cambridge, 1910–13). He rejected Einstein's General Theory because of philosophical objections to the idea of a non-homogeneous geometry of space time. He set out his own theory in *Principle of Relativity with Application to Physical Science* (Cambridge, 1922).

56. Sydney Chapman FRS (1888–1970) was educated at the Royal Technical Institute, Salford, and at Manchester and Cambridge Universities. On graduation he worked at the Royal Greenwich Observatory under Frank Dyson where he developed his interest in geomagnetism. After the First World War he became Professor of Mathematics at Manchester, moving to Imperial in 1924. He broadened the scope of the mathematics department with the help of W. G. Bickley, W. H. McCrea, W. G. Penney and H. Levy. The ongoing work of unifying the department was left to Levy when Chapman became Sedleian Professor of Natural Philosophy at the University of Oxford in 1946. Chapman's atmospheric researches at Imperial can be followed in his (with J. Bartels) *Geomagnetism*, 2 vols. (Oxford, 1940). He ended his career as Director of the Geophysical Institute in Alaska. See T. G. Cowling, 'Sydney Chapman, 1888–1970', *Biographical Memoirs of Fellows of the Royal Society*, vol. 17, 53–89.

57. Hyman Levy (1889-1975) was educated at Edinburgh University and had begun doctoral studies at Göttingen when war broke out and he had to escape. During the war he worked at the National Physical Laboratory on aeronautics and became a flight instructor at the Royal Aircraft Factory. He joined Imperial in 1920. Levy was active in the Labour Party during the 1920s and 30s, and persuaded the Party to set up an advisory committee on science of which he became chairman (1924-30). Like many left-wing scientists of his generation he strongly believed that science should be used to improve the human condition. After a meeting with N. I. Bukharin in 1931, Levy became a member of the Communist Party and was one of its better publicists. While critical of the Party after the Hungarian uprising of 1956 he did not resign, but was expelled two years later for having

criticized the USSR for its persecution of Jews. See John Stewart, 'Levy, Hyman (1889–1975), mathematician and socialist activist' (*ODNB*, 2004); also Werskey, *op cit.* (24).

58. J. R. Tanner (ed.), *The Historical Register of the University of Cambridge* (Cambridge, 1917), 972–80. See also T. J. N. Hilken, *Engineering at Cambridge University, 1783–1965*, (Cambridge, 1967); J. B. Morrell, 'The Non-Medical Sciences, 1914–39' in Brian Harrison (ed.), *The History of the University of Oxford, vol. VIII, The Twentieth Century* (Oxford, 1994), 139–63.

59. For the early professors see chapter 2. Armstrong was an excellent teacher and many of his students had stellar careers. For example, M. Forster, N. Haworth, F. Kipping, A. Lapworth, W. H. Perkin, J. C. Philip, W. Pope, and R. Robinson. Lapworth, Kipping and Perkin married three sisters (née Holland). It could be argued that the C&G chemistry department, which was closed down, was superior to the one at the RCS in 1907.

60. The committee had been appointed by the Governing Body in 1907 to oversee the transition of the engineering departments and recommend future development. Its membership, under chairman John Wolfe Barry, included representatives from the City and Guilds of London Institute (CGLI), William Unwin, Sir Alexander Kennedy, professor of engineering at University College London, and D. S. Capper, professor of engineering at King's College. Sir John Wolfe Barry (1836–1918), son of architect Sir Charles Barry, was a major civil engineer, whose work included Tower Bridge and numerous docks; among them were the Surrey Commercial, Natal Harbour, and the Barry docks in New South Wales. He was a long-time member of the CGLI and served on the Governing Body of Imperial College.

61. Cecil Howard Lander (1881–1949) left school for an apprenticeship in the engineering department of the Manchester Ship Canal. He then entered Manchester University where, after gaining a BSc, he became a research student and worked on heat flow in engines. During the First World War he joined the RNVR and contributed to the development of paravanes, and electric firing and detection gear for anti-submarine work. He was the inventor and patentee of the Q Mark 5 paravane for which he received a government award of £1000 in 1922. Lander became Director of the DSIR Fuel Research Station in 1923. He served on the Royal Commission on the Coal Industry in 1925. Lander was an excellent violinist and had considered a career in music.

62. Hugh Ford, *op. cit.* (63), 382. Stamping on the floor during lectures was customary and could signify either approval or disapproval.

63. Sir Owen Alfred Saunders FRS (1904–93) was educated at Birkbeck College London (he was among the final intake of full-time day students), and Trinity College, Cambridge where he held the senior scholarship. Rutherford offered to take him on as a research student and is reported to have been rather taken aback when Saunders refused, saying that he did not wish to pursue atomic research. The true reason may have been a shortage of money. Instead Saunders joined the DSIR Fuel Research Station in East Greenwich where he worked on a project relating to the performance of industrial furnaces. The project had been suggested jointly by Henry Tizard (then Director of the DSIR) and C. H. Lander. Tizard became Rector of Imperial in 1929 and, when Dalby retired, invited Lander to become professor of mechanical engineering. Tizard continued to mentor Saunders, encouraging also his later work related to jet engines. Saunders and Fishenden were co-authors of *The Calculation for Heat Transmission for Engineers* (1932) which became a classic. Saunders was also a gifted magician and member of the Magic Circle. He sometimes gave magic shows at the College. See Hugh Ford, 'Sir Owen Saunders', *Biographical Memoirs of Fellows of the Royal Society*, vol. 41 (1995), 379–394.

64. Margaret Fishenden (1889–1977) was educated at the University of Manchester where she gained a BSc in physics (1909) and a DSc (1919). She was a lecturer in meteorology at Manchester (1910–15) and then was appointed Director of the Domestic Heating Research Laboratory for the Manchester City Corporation. She moved to the Fuel Research Station of the DSIR in 1923. It is not clear why Fishenden accepted just an honorary lectureship at Imperial (1932–45) since, from the start, she set up and ran the heat transfer and thermodynamics laboratory jointly with Saunders, and was an important research scientist in her own right. In 1945 she was appointed senior lecturer and, in 1947, Reader in Applied Heat. As with Margaret Carlton, she was taken seriously as a research scientist but was not at first seen as a serious contender for academic advancement. After she retired in 1954 Fishenden continued lecturing for a few years. Without further evidence one can only speculate that this was to make ends meet. She was a single mother and earlier discrimination had not given her the opportunity to build a good pension. When she finally ended lecturing she cut herself off completely from engineering, and from the department. Fishenden was also an accomplished bridge player.

65. Thomas Mather FRS (1856–1937) was born near Preston and was apprenticed to a firm of engineers in the town. He won a Whitworth scholarship and entered Owen's College at the age of twenty-two. A further scholarship brought him to the Royal College of Science and, on graduating, he

became William Ayrton's assistant at Finsbury Technical College. He transferred to the Central Technical College with Ayrton in 1884. On Ayrton's death he was appointed acting head of department, made permanent in 1910 with his appointment as Professor of Electrical Engineering. For the kind of teaching programme Ayrton set up at Finsbury and later at the Central see, Graeme Gooday, 'Teaching Telegraphy and Electrotechnics in the Physics Laboratory: William Ayrton and the creation of an Academic Space for Electrical Engineering 1873–84,' *History of Technology*, 13, (1991), 73–114.

66. The electric arc had long been a subject of study in the department under Ayrton. W. Duddell and Hertha Ayrton were both known for their work in this field. Howe continued this tradition and applied the arc to wireless technology.

67. Cecil Lewis Fortescue (1881–1949) was educated at Cambridge and at the Royal Naval College Greenwich where he was professor from 1911. Fortescue was remembered as a charming man and as being very kind to students, but as being always in the minority when it came to debates over academic policy.

68. KEE 9; Dendy Watney to Lowry, 30 March 1933.

69. Stephen Mitchel Dixon (1866–1940) was educated at Trinity College, Dublin and worked for various railway companies before turning to academic work in Canada where he became professor of civil engineering at the University of New Brunswick in Fredericton, and then at Dalhousie University. In 1905 he returned to Britain and the chair of civil engineering in Birmingham before coming to Imperial in 1913. He has been described as lively and very witty. His students appear to have loved him and many had major careers, including a later professor at Imperial, Sir Colin Buchanan.

Charles Hawksley donated £4000 to the civil engineering department in memory of his engineer father, Thomas Hawksley. The money was used in the construction of a modern hydraulics laboratory, fully completed only after the war. In the 1930s the college raised some money by allowing film companies on the premises. Professor Fortescue complained about this saying that a film he had seen, made by Gaumont, trivialized the hydraulics laboratory by showing it in a sequence between a farcical golf match and a scene on Margate beach.

70. Raymond George Hubert Clements (1880–1953) was educated at Heriot-Watt College and first worked for the Scottish Ordnance Survey. He then moved to work as a highway and town planning consultant. During and after the First World War he was responsible for some 3000 bridges admin-

istered by the Army of the Rhine. His chair was named for Sir Henry Maybury, Director General of Roads in the new Ministry of Transport.

71. Quoted in Joyce Brown (ed.), *A Hundred Years of Civil Engineering in South Kensington: the origins of the Department of Civil Engineering of Imperial College 1884–1984* (Imperial College, London, 1985), 57.

72. Alfred John Sutton Pippard FRS (1891–1969) was known as Sutton. He was educated at Yeovil School and the Merchant Venturers College, Bristol. He won an industrial bursary from the Royal Commission for the Exhibition of 1851 and apprenticed as a civil engineer before joining the Pontypridd and Rhondda Valley Joint Water Board. While working there he wrote an MSc thesis on masonry dams. At the beginning of the First World War he was on a list of civil engineers that came to the attention of H. C. Watts, a fellow student at Bristol, who was then working in the technical section of the Admiralty Air Department. Watts told his boss that Pippard was the most brilliant student of his year whereupon Pippard was invited to join the group and began working on the stresses of aircraft parts; (see H. C. Watts, *Journal of the Aeronautical Society*, vol. 70 (1966), 68). Pippard built a reputation in this field and co-authored a standard text (A. J. S. Pippard and J. L. Pritchard, *Aeroplane Structures* (London, 1919)). He worked also on airship structures, while holding chairs first at Cardiff and then Bristol, and was working on the R101 airship when it crashed in France in 1930. While the crash was not due to structural failure, it marked an end to this work. Airship work had attracted many young engineers and was a major stimulus to structural engineering more generally. Several other Imperial College figures were involved, notably Richard Southwell, a consultant to the R101 design team. Southwell had a major theoretical dispute with another team member, John Fleetwood Baker (later Lord Baker). As a result Baker left the Royal Aircraft Works at Cardington and joined Pippard in Bristol where he took a PhD. Pippard and Baker co-authored *The Analysis of Engineering Structures* (London, 1936) which became a standard work. In 1943 Baker was appointed to the chair at Cambridge. Pippard had been a visiting lecturer in aeronautics at Imperial in 1919–22, and was appointed to the chair of civil engineering in 1933. Pippard left a memoir (typescript copy in the ICA) in which he described his own work as 'opportunistic and not of lasting value'. But he was undoubtedly a major figure in his field. See also A. W. Skempton, 'Alfred John Sutton Pippard', *Biographical Memoirs of Fellows of the Royal Society*, vol. 16 (1970), 463–478.

73. B\ Pippard; typed memoir, chapter 9.

74. Clothworkers donation of £20,000 for the 50th jubilee of the C&G in 1935.

75. See B\Pippard, typed memoir; section titled 'reminiscences'. Letitia Chitty (1897–1982) studied mathematics at Newnham College, though her course was interrupted by the war. After the war she returned to Cambridge, gaining first class honours in the mechanical sciences tripos — the first woman to do so. Pippard's view of their working relationship was prejudiced, despite his admiration for Chitty. She was not just a 'methodical calculator' but a good theoretician in her own right. She should have been encouraged to publish more on her own, and should, perhaps, have been given a professorship. Together, Chitty and Pippard worked on many projects, notably on bridge girders, suspension cables, the voussoir arch, and on the structural problems of dams. Chitty was the first woman to be awarded the Telford Gold Medal and continued her work on dams under A. W. Skempton (see chapter ten). For Chitty's wartime memories see, *Journal of the Royal Astronomical Society*, vol. 70 (1966), 67–8. Obituary, *The Times*, 8 October 1982.

76. Stanley Robert Sparkes (1910–1976) was educated at Bristol University and at Imperial College under Pippard. Sparkes worked for the Teesside engineering firm Dorman Long before coming to Imperial. (This firm employed many of Imperial's civil engineering graduates in the early twentieth century, and many of them worked on bridge construction around the world.) Sparkes, a specialist in steel structures, stayed at Imperial for over forty years. He was appointed Reader in 1947, and Professor of Engineering Structures in 1958. Sparkes was a major educator and keen Rugby enthusiast. He had played with the Wasps as a young man and for many years was president of the college rugby club. His later career was hampered by increasing blindness.

77. Concrete technology and soil mechanics are discussed in chapter 10.

78. William Whitehead Watts FRS (1860–1947) was born in Shropshire. He won a scholarship to Cambridge and, after graduation, spent some time as an extension lecturer at Cambridge before joining the Geological Survey of Ireland. He returned to academic life as an assistant professor at Mason's College in Birmingham before succeeding J. W. Judd at the Royal College of Science in 1906. His best known work, begun in 1896 for the Geological Survey and published posthumously, was *Geology of the Ancient Rocks of Charnwood Forest, Leicestershire* (London, 1947). Watts made headlines when he challenged Sir William Ramsay's claim that Britain's coal reserves would be exhausted in 175 years. (Ramsay made the claim in his presidential address to the British Association, at Portsmouth

in 1911.) Watts made many contributions to Imperial College life, donated his library to the department, and helped found the college archives. He was an early devotee of the motor car, and owned a Morris-Oxford which he reputedly spent as much time under as in. Students enjoyed driving with him on field trips. Watts donated a student prize to Imperial, named after John Wesley Judd. He brought George Sweeting to the college as his clerical assistant in 1908. Sweeting became departmental librarian in 1932, a position he held for thirty years. Sweeting was much loved by the students and took especial care of those from overseas. See B/Lowry and B/Boswell for typed memoirs in ICA; and P. G. H. Boswell, 'William Whitehead Watts' *Obituary Notices of Fellows of the Royal Society*, vol. 6 (1948), 263–76.

79. By then John Cadman (later Lord Cadman) had introduced his pioneering course in petroleum mining at the University of Birmingham.

80. See G. D. Hobson, 'The History of the Oil Technology Course and Its Offshoots' in M. Ala *et al.* (eds.), *Seventy-Five Years of Progress in Oil Field Science and Technology* (Rotterdam, 1990).

81. Vincent Charles Illing FRS (1890–1969) was educated at Cambridge and Imperial College. He came to Imperial as a research student under Watts to work on Cambrian faunas. But Illing became more interested in Cullis's lectures on mining geology, and those given by Arthur Wade on petroleum geology. He became a demonstrator, then lecturer in 1915, and began a survey of British oil shales believing the Kimmeridge oil shale to be commercially viable. Only later did he learn of the vast deposits elsewhere in the world. Also in 1915 he first visited Trinidad to study its oil geology, the start of both his and the college's close ties with the Trinidadian oil industry. Illing later ran a consulting company for the petroleum industry from Alfred House, opposite South Kensington underground station, and was an important figure in the tapping of oil fields in many parts of the world. In 1947 he negotiated the compensation ($110 million US) for Shell-Mex when the Mexican oil industry was nationalized. See N. L. Falcon, 'Vincent Charles Illing', *Biographical Memoirs of Fellows of the Royal Society*, vol. 16 (1970), 365–84. See also chapters 5 and 7.

82. Charles Gilbert Cullis (1871–1941) was educated at King Edward VI Grammar School in Birmingham and at Mason's College where he was a student of Lapworth, Poynting and Tilden. In 1890 he won a scholarship to the RSM and was Murchison medallist in 1891. He joined the RSM staff before the formation of Imperial College, and became professor of economic mineralogy in 1914. His title was changed to professor of mining

geology in 1930. Cullis was a popular teacher who attended many student functions.

83. Among those who made major academic careers were the Williams twins, Howel, later a professor at Berkeley, and David, later a professor at Imperial; also Oliver Bulman and James Stubblefield who submitted a joint PhD thesis to the University of London. Bulman, who was later Woodwardian Professor of Geology at Cambridge, claimed that in retrospect he was glad to have failed to gain an entrance scholarship to Cambridge and that he had the chance to study under Watts and E. W. MacBride at Imperial, and under D. M. S. Watson, the professor of vertebrate paleontology at University College London (and father of Imperial geology professor Janet Watson). See James Stubblefield, 'Oliver Meredith Boone Bulman, 1902–74', *Biographical Memoirs of Fellows of the Royal Society*, vol. 21 (1975), 175–95.

84. See David Williams, 'History of the Geology Department' (typed manuscript, 1963) for more on the department, including anecdotes about Evans.

85. Arthur Holmes FRS (1890–1965) was educated in physics at Imperial College where he was a research student under R. J. Strutt before joining the geology department in 1912. He later held a chair at Durham before being appointed Regius Professor of Geology at Edinburgh. He was made a Fellow of Imperial College in 1959. Quotation in Williams, *op. cit.* (84), 100.

86. Percy George Hamnall Boswell FRS (1886–1960) was a school teacher in Ipswich for several years. He studied the geology of East Anglia largely on his own, but with the encouragement of William Watts to whom he had written for advice. The two became lifelong friends. Largely self-taught, Boswell took an external University of London BSc in 1911, and worked as a geologist despite serious problems with his eyesight which hindered him throughout his life. At the age of twenty-six he enrolled as a research student at Imperial College having won a Remanet Studentship. He gained his DIC in 1914 and DSc in 1916. He was a demonstrator in the geology department, and carried out research on sands and sedimentary rocks, until appointed Herdman Professor of Geology at the University of Leeds in 1917. He succeeded Watts at Imperial in 1930. In 1934 he was given leave to join the Leakey Expedition to East Africa. Boswell left a typed memoir which is in the ICA. See also chapters 5 and 11.

87. Regular five year 'visitations' were a statutory requirement (Statute 114, University of London Act (1929)). The 1935 visitors were Ernest Rutherford, former Normal School of Science student, Sir Frederick

Gowland Hopkins, and Frank Horton. Aside from their criticisms, they noted that the college provided facilities 'on a scale that few universities or colleges can approach'. For details of their report see GB minutes, 20 March 1936. Rutherford had earlier (1917–25) been a member of the Governing Body, as representative of the Government of New Zealand.

88. Quotation in B\Boswell (typed memoir, p. 60). Herbert Harold Read FRS (1889–1970) was educated at the RCS and gained the ARCS in geology in 1911. In 1914 he joined the Geological Survey. During the First World War he served at Gallipoli, in Egypt, and at the Battle of the Somme when he was invalided out. After the war he joined the Geological Survey in Scotland. He was appointed to the Herdman chair of geology at Liverpool in 1931. See also, J. Sutton, 'Herbert Harold Read, 1889-1970', *Biographical Memoirs of Fellows of the Royal Society*, vol. 16 (1970), 479–93.

89. William Richard Jones (1880–1970), was educated at the RSM and then worked as a government geologist in the Malay States before becoming Managing Director of the High Speed Alloys Mining Company in Burma during the First World War. His research was mainly on the geology of tin fields, but he also gained a reputation for helping Welsh miners suffering from silicosis. Until the 1920s compensation was given only to those who had been working with rocks containing more than 50% free silica, but Jones showed that the disease was associated with a hydrated fibrous silicate of potassium and aluminium. This material, known as sericite, did not qualify under the compensation law. Jones successfully fought for a change in the law and was thenceforth dubbed 'Sericite Jones'. He also designed a miner's mask which supplied clean air from a compressed supply. On a visit to China, Jones had made a study of the materials used in the manufacture of porcelain and his study of kaolin deposits informed his later work on the china clay industry in Britain. In 1948 he was awarded the CBE for helping to rehabilitate the industry in Cornwall. He appears to have been much in demand by the Labour government after the war and there is some interesting correspondence in the ICA between Jones and various government ministers including Stafford Cripps, Hugh Dalton, and Emmanuel Shinwell. The letters cover a range of matters relating to British mineral industries other than coal and oil. Jones appears to have been a favourite of Henry Tizard and, from the tone of Tizard's letters, they must have been close friends. The friendship likely blossomed during the war when Jones was in charge of the RSM maintenance party. Tizard spent many nights with this group in the RSM. Both men were also active in the Masonic lodge at Imperial. Tizard often sought Jones's advice, even after returning to Oxford as Master of Magdalen College. Jones's responses to technical

questions are models of clarity and display much detailed knowledge. See B/Jones.

90. Samuel Herbert Cox (1852–1920), known as Herbert, had been a student of Andrew Ramsay at the RSM. He later became Inspector of Mines in New Zealand and was an expert in mine surveying.

91. See chapter 3 for a discussion of problems with the RSM courses.

92. GB 5 February 1909 and 24 May 1912. The mine was no longer working and was purchased for a small sum. In 1912, land surrounding the mine was purchased for surveying courses. When buildings on the mine site were not in use by students they were rented out to others. For example, local religious services were conducted in the lecture room.

93. At this time Spain produced about 10% of the world's copper and the mines were largely controlled by British interests.

94. William Frecheville (1854–1940) was educated at a technical school in Berlin and at the RSM. In 1912, after extensive mining experience in South Africa, India and the United States, he was appointed professor of mining.

95. Samuel John Truscott (1870–1950) was educated at the RSM and had experience in mining in many parts of the world, notably in South Africa, before his appointment as professor of mining.

96. Lewis Henry Cooke (1870–1929) was educated at the RSM. See obituary in *The Times*, 30 August 1929, 14.

97. Truscott to Keogh 17 June 1919. The move to coal was prompted by expansion of the domestic coal industry at this time (90% of Britain's energy came from coal until after the Second World War). From later correspondence it would appear that even in 1935 there was ambivalence over the move to coal mining. J. A. S. Ritson (1887–1957) was a Durham man with experience of the coal fields there. He had been an Inspector of Mines before becoming professor of mining at Leeds from where he came to Imperial. Supporters of Ritson were keen to point out his impressive war record. He had served in the Durham Light Infantry, commanded a battalion, and won the DSO. It would appear that this helped win him the appointment. (ULP/Mining).

98. See chapter 7.

99. Report of the Committee to Review the Courses in Mining Engineering at the RSM; 1932. (GB Minutes, 1932.)

100. William Gowland (1842–1922), had worked for the Broughton Copper Company in Manchester before being seconded to the Imperial Mint and Imperial Arsenal, Japan. After twenty years in Japan he returned to England and joined the RSM as professor in 1902.

101. The metallurgist, Hilary Bauermann, professor at Woolwich, was well known at the RSM where he often gave lectures. A former RSM student, Bauermann had a reputation as an adventurer and had worked in many parts of the world. He left a generous bequest to Imperial intended, in part, to allow graduating students to follow in his footsteps, travel worldwide, and view mines, mineral deposits, and metallurgical practices.

102. William Arthur Carlyle was educated at McGill University in mining and metallurgy before working in mines in Colorado, and in British Columbia where he became Head of the Department of Mines. He returned to McGill where he taught for a few years before becoming General Manager of the Rio Tinto Copper Mines in Spain, a factor in the long association of this company with Imperial. Many students had summer vacation work experience, and were to find their first jobs, with this company. Carlyle was very popular and his experience highly valued. That he left Imperial so soon (for consultancy work) was widely regretted. However, he continued to take students to Rio Tinto in Spain during the summer vacation.

103. Sir Henry Cort Harold Carpenter FRS (1875–1940) was educated at St. Pauls and Merton College, Oxford. After a first class degree in chemistry he became a demonstrator at Owen's College and in 1901, having worked with Sir William Roberts Austen at the RSM, was appointed head of the department of chemistry and metallurgy at the National Physical Laboratory. In 1906 he was appointed to the chair of metallurgy at Manchester. As his name suggests, he was a descendent of Henry Cort of puddling fame. He was also the grandson of the scientist W. B. Carpenter and the nephew of Estlin Carpenter, Principal of Manchester College, Oxford. After the early death of his father, his uncle Estlin played a major role in his upbringing. See C. A. Edwards, 'Henry Cort Harold Carpenter', *Obituary Notices of Fellows of the Royal Society,* vol. 3 (December 1941), 611–25.

104. Constance Fligg Elam (later Tipper) (1894–1995) came to Imperial in 1911 from Cambridge where she had passed the engineering sciences tripos examinations. A research student under Carpenter, she worked on a method of strain annealing which brought them fame. She stayed at Imperial until 1928 before moving to Cambridge as a Reader in the engineering laboratory. In 1920 Carpenter wanted the college to raise Elam's salary from £250 to £300. Her work is 'so good that she is fully entitled to the advance I have recommended'. But the request was refused. In 1923 she was supported by the DSIR at a salary of £300. Carpenter to Gow, 3 November 1920 (K MET 9–10.)

105. William Hume Rothery, one of Carpenter's students, and later the first professor of metallurgy at Oxford, went through this training. When, in 1944, Richard Southwell, not yet fully loyal to his new college, consulted him as to whether a friend of his should accept a position in the metallurgy department, Hume Rothery replied, 'if the RSM metallurgy is today as it was when I was there, I should advise … against accepting a post. It was a first-class school of technology … but the teaching staff were so divorced from research work that the latter suffered.' This prompted Southwell to attempt to modernize metallurgy at the college and, after the war, he began asking around for advice.

106. Cecil William Dannatt (1892–1961) was a student at the RSM, graduating in 1914. After military service he returned as a research student. On graduation, he worked in the metallurgical industry for a few years before joining the staff in 1937, as assistant professor of extraction metallurgy. Carpenter died in 1943 during the department's evacuation to Swansea. Dannatt became acting head of department and was appointed permanent head when the department returned to London in 1945. The Institution of Mining and Metallurgy, wanting a production metallurgist, not another physical metallurgist, to be in charge of the department, were happy with his appointment. Dannatt took a leading role in establishing the Nuffield Research Group in Extraction Metallurgy at the RSM, discussed in chapter 10. Dannatt, who retired in 1957, was a serious competitive tennis player. See obituary, *The Times*, 15 April 1961.

chapter seven

Imperial Science at Imperial College

Land of Hope and Glory, Mother of the Free,
How shall we extol thee, who are born of thee?
Wider still, and wider, shall thy bounds be set;
God, who made thee mighty, make thee mightier yet![1]

Introduction

The sentiments expressed in 'Land of Hope and Glory', while not universally held, were typical of the period in which Imperial College was founded.[2] While the bounds of Empire extended considerably as a result of the First World War, expansionist views appear to have been less persuasive by the 1920s. Commenting on past attitudes towards imperialism, historian John MacKenzie, writing in the 1990s, stated 'the eras of triumphal justification, moral outrage, and apologetic sensitivity are passing away'.[3] I hope he is right and, with a post-colonial experience stretching back to the end of the Second World War, that it is possible to think anew about imperial matters. How to enter the discourses of earlier generations and engage with their prejudices is always problematic; but this chapter will illustrate something of the attitudes and interests of people working at Imperial College in the first half of the twentieth century. Like others engaged on imperial projects, those at the college were infused with a sense of patriotism. Some wished to advance the interests of Britain, and those of colonial settlers of British descent. Others, themselves from overseas, were concerned to carry newly acquired expertise back to their own countries. It is perhaps banal to suggest that the legacy of scientific and technological projects such as theirs, affecting millions of people around the world, has been mixed.[4]

The morality of empire will not be discussed here, but an interesting question is whether science needs some kind of imperialism in order to progress. Today the whole world has become the object of scientific investigation, and the results need to be shared. An aspect of this is the building of networks with chains of reference working their way around the world. This cannot happen without some people assuming leadership roles, that is without at least a modicum of cultural imperialism. As will be shown, H. G. Wells, a former student at the Normal School of Science, understood this. But the intertwining of science and imperialism is problematic. Today many in the post-colonial world see imperialism as 'bad', and science as, on the whole, 'good'. Post-colonial countries have not abandoned science, even while recognizing the associated problems of rapid modernization. While today we debate the legacy of imperial science, those who earlier worked on scientific and technical projects in the setting of empire rarely questioned what they were doing. Our views may well be infused with a post-colonial sensitivity, but this should not lead us to conclude that scientists and engineers working in the early twentieth century were naive. Rather, they saw problems in many parts of the world, believed that they had the means to address some of them, and that in doing so they could improve the lives of others while advancing the interests of their own country.

Western science and technology are major cultural forces. As Daniel Headrick put it, they were among the, 'tools of empire'.[5] But how were these tools used? Lord Lugard's 'dual mandate' that both the productivity of the British Empire and the lot of its colonial subjects be improved, was taken seriously at the college.[6] So, too, the view of Lewis Harcourt, Colonial Secretary from 1910–15, that economic and scientific development were synonymous, and that Britain would enhance also the 'civilized' world by bringing economic development to others.[7] Views such as these were reflected in the 1907 Imperial College Charter of Incorporation. According to Article II, the college was to give 'the highest specialized instruction, and to provide the fullest equipment for the most advanced training and research in various branches of science especially in its application to industry'. But, the name of the college, defined in Article I, was intended to imply that the industries in question were to be those of the empire, not simply those within Britain.[8] People in government understood this and came to the college for help. For example, in 1920, on receiving complaints from imperial industries relating to the shortage of well trained personnel, Lord Milner, the Colonial Secretary, wrote to the Rector, Sir Alfred Keogh, emphasizing the need for the training of expert workers. He was assured that 'the full resources of the college would be placed at his disposal to further the ends in view which are coincident with the work and policy of the college'. In a further letter Keogh wrote,

> as the Imperial College has been developed with a special view to meeting Empire requirements it has been particularly active in matters to which you drew attention ... London is indeed the Mecca of the Colonials and this college has taken an outstanding position in providing and in training men for scientific service overseas.[9]

Empire and science were encapsulated in the college motto, chosen by the Governing Body in 1908, '*scientia imperii decus et tutamen*'. Line 262 from Book V of Virgil's *Aeneid*, it was translated, not strictly literally, as '[scientific] knowledge is both the ornament and safeguard of the empire'.[10] That the college coat of arms incorporates this Latin motto, alongside the Royal Coat of Arms, was taken seriously by a generation of staff and students better versed in the classics and heraldry than are their counterparts today. Keogh had a rather grand vision, 'Imperial College is not a building it is an idea'. He envisaged Imperial College professors giving instruction not simply in London but around the Empire. He also stated that Imperial students should be able to travel elsewhere for the latest instruction; for example, to Liverpool and to Oliver Lodge's old laboratory, for instruction in wireless telegraphy. There should be 'cooperation between Imperial College and every centre of technical education in the Empire'.[11]

In 1907 no one on the college's governing body specifically represented the empire. This changed during the First World War when, in 1916, seven new members were added to be chosen by the Secretary of State for India, the governments of the Dominions of Canada and New Zealand, the Commonwealth of Australia, the Union of South Africa, and the Colony of Newfoundland.[12] In 1951 the charter was further

Illus. 1–2; top, the Royal Horticultural Society Gardens (1861–80), the future site of many Imperial College and museum buildings. The Royal Albert Hall, which opened in 1871, was built to the north of the conservatory shown at the top of the picture. Bottom, the Royal College of Science Observatory (later named for Norman Lockyer) in 1893, with the Science Schools building on the opposite side of Exhibition Road. The students were learning how to use sextants and theodolites.

Illus. 3–6; top left, Henry Taylor Bovey, the first Rector of Imperial College; top right, the Marquess of Crewe, the first Chairman of the Governing Body; bottom left, Sir Arthur Acland, member of the Governing Body and Deputy Rector during the First World War; bottom right, Sir Alfred Keogh, the second Rector, in Surgeon General's uniform.

Illus. 7–10; top left, W. E. (Ernest) Dalby, the first Professor of Mechanical Engineering and Dean of the C&G College; top right, Hugh Callendar, the first Professor of Physics; bottom left, Sir Herbert Wright, former student of botany, major donor, and member of the Governing Body; bottom right, Sir John Bretland Farmer, the first Professor of Botany.

Illus. 11–14; top left, William Gowland, the first Professor of Metallurgy; top right, William Watts, the first Professor of Geology; bottom left, Vincent Illing, Professor of Oil Technology; bottom right, William Jones, Professor of Mining Geology.

Illus. 15–17; top left, Alfred Fowler, Royal Society Yarrow Professor of Physics, with his telescope; top right, John Hinchley, the first Professor of Chemical Engineering; bottom, Alexander Gow, the first College Secretary.

Illus. 18; cartoon recollecting Armistice Day (D. R. H. Moore, *The Phoenix,* 1921). Push-Ball matches were a popular student activity of the period and took place all over London.

Illus. 19–20; top, a 1927 meeting of the Zoroastrians at the St. Alban's home of William Bone, Professor of Chemical Technology. Front row from left, D. T. A. Townend, H. B. Dixon, H. B. Baker, H. E. Armstrong, A. Smithels, H. James Yates. Middle row, from left, D. M. Newitt, W. E. Stocking, W. A. Bone, R. Mond, G. I. Finch, D. A. Winter. Back row from left, F. R. Weston, R. P. Fraser, H. Hartley, D. S. Jerdan. (Members not present for picture, included A. C. G. Egerton, E. Rideal and H. T. Tizard.)

Bottom, the Bone building in 1915 before the extension and upper floors were added, with the drawing office hut at front left, gas generator front centre and 3,000 cu. ft. gas holder to the right. The gas generator plant was a gift of Robert Mond.

Illus. 21–23; top, buses on Prince Consort Road with students and supporters on their way to a RSM v C&G rugby match in 1912. Note the CTC flag showing that the old name, Central Technical College, was still in use. Bottom, the Engineers Cup, RSM v C&G, in 1923 with (right) enthusiastic onlookers.

Illus. 24–27; top, Sports Day at Duke of York's regimental headquarters in Chelsea, 1934; left, the 100 yards dash; right, the 120 yards hurdles.
Sports Day at Motspur Park, 1937; centre, Ted Coulson, master coach, encouraging the C&G tug-of-war team; bottom, high jump.

Illus. 28; Cartoon by Grey Moon depicting members of the chemistry department (*The Phoenix*, 1925).

Illus. 29–31; top, Grey Moon cartoons; left, Miss Holme, the President of the Women's Association (*The Phoenix*, 1926). Note the change from 'Ladies' to 'Women's' on the door, a comment on the fact that women students had asked to be called 'women' not 'ladies'. Right, Jimmy Peacock kicking out the old when he took over the catering services (*The Phoenix*,1925). Bottom, cover illustration (*The Phoenix,*1924).

Should all the Phase Rule be forgot
And never brought to mind
I ken the ballads o' this Scot
Will bring back Auld Lang Syne.

Illus. 32; cartoon of James Philip, Professor of Physical Chemistry, by Grey Moon (*The Phoenix*, 1925). Philip was Deputy Rector during the Second World War until his death in 1941, and was known for entertaining staff and students with songs at college events and at his home.

Illus. 33–34; chemistry department staff, 1933-4, with Henry Tizard; top from left, H. J. Emeleus, M. Carlton, P. C. Bull, R. P. Linstead, G. A. R. Kon, H. V. A. Briscoe, J. C. Philip, H. T. Tizard; bottom from left, J. F. Thorpe, M. A. Whiteley, E. H. Farmer, A. King.

Illus. 35−37; top left, chemist, Martha Whiteley, the first woman to become an assistant professor (reader). Top right, mathematicians, from left, Professor Hyman Levy and Professor Alfred North Whitehead. Bottom from left, entomologists, Professor James Munro, Professor Frank Balfour Browne, and Professor Owain W. Richards, 1927 (the year in which Richards was appointed a lecturer).

Illus. 38; Department of Botany 1937, front row from left, H. W. Buston, L. E. Hawker, A. S. Horne, B. D. Bolas, F. Howarth, F. Y. Henderson, R. J. Tabor, W. Brown, V. H. Blackman, F. G. Gregory, A. C. Chibnall, S. G. Paine, C. A. Pratt, G. E. Blackman, S. E. Jacobs.

Illus. 39; Sir Alfred Egerton (standing on right) and J. H. Burgoyne in the combustion laboratory, chemical engineering and applied chemistry department, with students working in the background (date unknown, but late 1940s).

Illus. 40–43; top left, Sir Henry Tizard; top right, Sir Richard Southwell; bottom left, the clock tower on the old City and Guilds Waterhouse building with the Imperial Institute to the left, 1960; bottom right, the Imperial College Home Guard "C" Company, on guard duty at Buckingham Palace, 1944.

Illus. 44–46; top, the Goldsmith's Extension with the City and Guilds building in the background, from a painting by R. T. Cowern (1962). Bottom left, Sir Leonard Bairstow, Zaharoff Professor of Aeronautics; bottom right, Professor Sydney Chapman, head of mathematics (photograph taken later in his career).

amended to reflect the new Commonwealth status of India, Pakistan and Ceylon. Each government was invited to select one governor, as was the Secretary of State for the Colonies.[13] The Governing Body was also given the right, as it saw fit, to invite further nominees from any other self-governing colony or territory of the Commonwealth. Those governors appointed in 1916 were asked to promote the college and encourage students from their countries to come to London to study. Various scholarship schemes were instituted to enable this, for example the Dominion Research Fellowships which, the Rector claimed, would result in the 'vivifying and strengthening of the "Imperial" link which already exists in the representation of the seven [*sic*] Dominions on the Governing Body'.[14]

The 1916 charter amendments were a reflection of the fact that the government, and the early staff, took seriously the name Imperial College and had begun to think what it meant in practice to be imperial. The charter helped give direction to work carried out, and was instrumental in the formation of a college identity. Imperial staff and students saw the college as being distinct from other institutions of higher education and, during the Edwardian period, its name gave it both weight and dignity. By the end of the century the name had become somewhat anachronistic, being more associated with excellence in science, medicine and engineering than with empire.[15] But, among the early college staff, many eagerly sought projects that showed their support of the empire. Adam Sedgwick, the professor of zoology, came to the college in 1909 specifically because it was an 'imperial' college. He had been Professor of Zoology at Cambridge, and while in India, on a health-related leave of absence, had an epiphany. He decided to take the Imperial College chair because he thought it 'a duty to his science and to the Empire' that he train young people for imperial service.[16] The colonies were then seen as frontiers both for employment and for discovery. Lectures were given in tropical hygiene and students were encouraged to seek work overseas.[17] These early policies helped establish an ethos at the college which has left an imperial trace.

In 1907 South Kensington already had a major imperial presence, namely the Imperial Institute, a close neighbour of the new Imperial College. The idea for the Institute, a project dear to the heart of the Prince of Wales, later Edward VII, was to house a permanent exhibition of Empire under one roof, 'to which the populace would flock to wonder at the benefits colonial rule afforded them'. The Institute would exhibit, 'the vast area, the varied resources, and the marvellous growth during her Majesty's reign, of the British Empire'.[18] The emphasis was on the industrial resources of the Empire, and the best of its natural and manufactured products were to be on display. The architectural competition for the building attracted sixty-six entrants and the winner was T. E. Colcutt who described his design as a 'free rendering of the Renaissance style'.[19] Most students of my generation (the early 1960s) viewed the building as a monstrosity; with hindsight, I think we were wrong.[20] When the building opened in 1893 over 25,000 people were present. At the ceremony Queen Victoria completed an electric circuit to the Queen's Tower and fifty changes were rung on the Alexandra peal of ten bells.[21]

The Institute had a scientific bent from the start. Its first Director was Sir Frederick Abel, one of Hofmann's first students at the Royal College of Chemistry. Before coming to South Kensington he had been Chief Ordnance Chemist at Woolwich.[22] A scientific and technical department at the Institute was founded with the financial help of both the Goldsmiths' Company and the Royal Commission for the Exhibition of 1851. Two large laboratories were constructed on the top floor of the building, used mainly for the testing of natural product samples, and for assaying minerals sent by colonial geological surveys. By 1905 the Institute was receiving over 200 requests for analytical and related scientific work each year.[23] The maintenance of the Institute was funded by such work, by private donations, mostly from the colonies, and by exhibition entry fees. But money was always a problem and, by the time Imperial College was founded, the government had taken over the running of the Institute which earlier had been close to financial collapse. Space was rented out, much of it to the University of London for its administration offices, and for an examination hall.

During the First World War the scientific role of the Institute regained some importance because of renewed interest in the resources of empire. Research was carried out on materials such as rubber, oilseeds, cotton, natural product drugs, and minerals. A huge inventory of colonial samples was acquired, used both for educational and commercial purposes.[24] Dyes were a problem at the start of the war since Britain had relied mainly on German imports. Natural products from the empire were explored in the hope of finding substitutes; for example, a dye for khaki uniforms.[25] None of this work was very successful and, in 1917, Lord Haldane, chairman of the Dominions Royal Commission, recommended closure of the Institute. For this he came under attack in the press and was suspected, not for the first time, of having hidden German sympathies.[26] After the war, the Institute was caught up in a new move to propagandize the benefits of empire, something which the dominions and colonies were more willing to support financially than they were science.[27] In this connection one of the bodies then formed, the Empire Marketing Board, was to have an influence also at Imperial College.[28] At the Institute, it funded the construction of a cinema which opened in 1927. This became the Institute's greatest public attraction, showing films on topics such as British Columbia salmon packers, the Canadian Pacific Railway, tea cultivation in Ceylon and cotton cultivation in Africa. That cotton was being increasingly cultivated in Africa had much to do with Sir Alfred Jones, a Liverpool shipping magnate who, in 1902, brought together a group of people with interests in the cotton trade. He wanted to promote the growing of cotton within the empire and the British Cotton Growing Association was formed as a development agency. Later, one of its offshoots, the Empire Cotton Growing Corporation, played an important role in funding research at Imperial College.

These quasi-governmental moves were paralleled by more official ones. During the war the government recognized the need for more formal ways of promoting scientific research. The Department of Scientific and Industrial Research (DSIR), the Agricultural Research Council (ARC), the Medical Research Council (MRC) and the Colonial Research Council (CRC) were among the results. The last of these was formed in 1919 and, along with the others, albeit less munificently, was an important source of funding at Imperial.[29] In 1922, it helped in the founding of the Imperial College of Tropical Agriculture in Trinidad, with which several Imperial College scientists had an association, and where some graduates found work.[30]

Science and Empire

In what follows the focus will be on scientific and technical work carried out in the metropolis, though many among the early Imperial staff carried out work elsewhere in the Empire. Robert Stafford has written of the huge volume of geological data flowing into Britain during the nineteenth century from colonies and dominions around the world.[31] The flow continued into the twentieth century and Imperial College geologists were active, both in its maintenance, and in analysing the data that arrived. But the data flow was not simply geological; college botanists, zoologists, engineers, miners, metallurgists, chemists and even physicists received information, and all contributed to imperial science and engineering. As will be shown, while data flowed in, young graduates flowed out. Scientific and technical knowledge moved in many directions. But first, let us briefly turn back to a time before the three colleges federated, namely the period of the South African War.

In February 1900 students at the Royal College of Science (RCS) and Royal School of Mines (RSM) led a spontaneous demonstration on hearing of the Relief of Ladysmith. Many students were away serving with the Generals Sir George White and Sir Redvers Buller, but those at home poured out of their classrooms, the miners with their emblematic picks and spades, to be joined in procession by students from the Central Technical College, Royal College of Art, and Royal College of Music. The art students had earlier made a huge triumphant lion out of plaster which was used to lead a parade of about 1000 students who marched first to the Albert Memorial, and then to the Colonial Secretary's house in Prince's Gardens. They roared for 'Jos', then an iconic figure with his orchid and eyeglass, but he was not at home. Cheers were given and patriotic songs were sung until the students were addressed by the minister's son, Austen Chamberlain. After hearing from him they marched to the Knightsbridge Barracks where they sang Rule Britannia, and then to Baden Powell's house at Hyde Park Corner where they sang 'For he's a jolly good fellow'; then on to Buckingham Palace, the War Office, and the Cadogan Place home of Sir George White (defender of Ladysmith), for more of the same.[32] In 1908, a memorial to those from the RSM who lost their lives in the South African War was erected at the head of the staircase in the new Royal School of Mines building. Professor Judd was the moving force behind this, and he persuaded about 300 students and members of staff to subscribe.

While the demonstration was in part a spontaneous release of tension, the form that it took illustrates a type of imperial enthusiasm that lasted until the sobering events of the First World War.[33] After 1914 imperialist ideals were still highly regarded at the college, but were manifested differently. Pacifist articles began to appear in student newspapers and journals; but there was still much admiration for students who left for work in the Empire, and for professors whose research supported imperial industries. It was widely held that the application of scientific and technical expertise would help Britain bring peace, order, and progress to the world. This is how H. G. Wells, himself a former student at the Normal School of Science (later Royal College of Science), saw things. He wrote,

> I have lived through a lot of Imperialism. ... In the days which culminated in the Boer War I was a strong imperialist ... I am now an anti-imperialist, but my case is that it is imperialism which has changed and not I. ...Then as now my ends were cosmopolitan. In the days before the Boer War the Empire was ... a great free trading system. [And], our Imperial diffusion gave us enormous advantages for scientific and educational work. ... The essential task ... for Empire was to think, teach, intercommunicate and unify ... through the systematic perfection and realization of a liberal ideology, that would unite first the Empire and at last the world in a common world aim.[34]

Wells believed this aim had been corrupted by Joseph Chamberlain, and by those who supported his tariff reforms. Wells was a firm believer in free trade and was antipathetic towards the nationalism which he believed lived alongside economic protectionism. But this did not stop him from promoting his vision of Empire.[35]

At the start of the First World War Wells wrote articles in the college journals, and gave speeches advising students not to volunteer. He claimed that their scientific work was of greater importance to the nation than any military contribution they might make.[36] After the war he addressed a meeting of Imperial College students at the Central Hall, Westminster. The title of his address was 'What is Education?'. He began with a telltale jab at the university. Imperial College, he stated, can turn out men 'whose breadth of knowledge compares favourably with that of graduates of various universities ... we are not barbarians'. But even Imperial students had a responsibility to learn something of history so as to be responsible 'citizens of our great Empire'. Only by understanding its past,

> can we hope to work for [the empire's] endurance. [Oxbridge dons]
> may stand a better chance of entering the 'talking shop' than we
> do, but what would they have to talk about were it not for the
> men who go out into the far corners of the Empire to provide
> both subjects of discussion and the money to carry on. We require
> a better education than a mere talker does ... let us be 'Imperial'
> in all things.

He urged students to serve the empire in technical ways but, to be worthy servants, they needed more than could be found 'in the works of Perry or in the works of our other illustrious professors'. They needed also to read of the lives and work of those who had served the Empire before them.[37] Wells's speech was well received but it was not until after the Second World War that formal humanities education came to Imperial. His celebration of what later came to be known as the 'technocratic', was music to the ears of many at Imperial and, until his death, he was a frequent guest and invited speaker at major college events.[38]

Serving in the 'far corners of the Empire', was a pattern already well established in the three colleges that came together to form Imperial College in 1907. The Governing Body saw it as its duty that this pattern be maintained. In 1911 Arthur Acland, chairman of the Education Committee, asked all the college staff to report on work they were doing in relation to the industries of the Empire so that he could write a memorandum to be used both in raising public awareness of their work, and in seeking further research funds.[39] That Acland should have been concerned about this is understandable also in the context of Imperial College's dispute with the University, and its need to find a distinct role.[40] Every department reported that it was carrying out Empire related work. When a Parliamentary Committee visited the college during the war, Acland, then Deputy (Acting) Rector, told the MPs that the Governing Body was appalled at Britain's neglect of subjects such as economic botany, and stated that the 'vegetable industries rely almost entirely on German science' and 'what was practically an employment agency for the scientific work of our Empire existed in the University of Berlin'. The governing body, he claimed, had made the most strenuous efforts to develop science at the college with the Empire in mind.

> Nothing but a sincere conviction that the most vital interests of the Empire were at stake has kept alive the enthusiasm alike of the Governors, the Professors and the students (for even the latter thoroughly understand) … London being the heart of the Empire, and this Institution having been made the largest and most fully equipped scientific College of the Empire, it is truly an Imperial College of Science.[41]

A year later, when members of the British Imperial Council of Commerce visited the college they received much the same message from the Rector, Sir Alfred Keogh.[42]

Biological Sciences

Acland was much influenced by John Bretland Farmer, Professor of Botany, and especially supported Farmer's many schemes. Farmer had come to the Royal College of Science in 1892 as assistant professor under T. H. Huxley. Shortly after his arrival he went on a fact-finding tour to India and Ceylon where he noted the spread of plantation agriculture. Recognizing the potential problems of monoculture (coffee had already fallen victim to fungal disease in Ceylon), he came back to London and designed an applied botany course in which students would learn about plant diseases. In 1910, Farmer asked the governing body to consider opening a department in plant physiology,

> to see the College meet a pressing demand for the adequate training of men who will go out to various parts of the Empire as competent advisors on the many matters of immense financial and economic importance which are continually arising, especially in relation to agriculture as pursued under tropical conditions.

He mentioned that Herbert Wright had offered £1000 to start a fund to help establish such a department.[43] Wright had been a botany student in the Royal College of Science and had studied also some chemistry under W. A. Tilden. Tilden was engaged in research on the chemistry of rubber and several of his students became rubber chemists in Ceylon and later Malaya.[44] Wright went to Ceylon after graduating and carried out research on the rubber tree. He later made a large fortune by investing in rubber and was prepared to help finance applied botany at the college. Further, he was active in the Rubber Growers Association, a cooperative which supported research and was willing to do so also at Imperial.[45] Wright was a serious college booster, especially after being appointed to the governing body in 1918, and was in favour of the college becoming an independent technical university. He wrote to Acland,

> throughout our Empire there is a call for a much closer relationship between science and industry. … If the British Empire is to fully utilize its present natural resources at home and overseas, if we want our technologists to place our country first in the world for increased production of all commodities, we must ourselves raise the status of scientific industrial research and give our students the fullest and highest recognition.

He also mentioned that City men needed a push since they didn't understand what was needed to build a first class technical university.[46]

After Acland had collected information from the staff on their various imperial projects, he sent a document to the Board of Education. While the document makes no mention of a new department of plant physiology, it did, among other things, ask for funding for two new botanical chairs, one in plant pathology (visiting lecturers had been teaching this subject at the college) and one in the technology of woods and fibres. Plant pathology, he wrote, 'is intended to meet the necessities of the Vegetable Industries abroad and at home'. It should relate to 'rubber, cocoa, tea, cotton and various other industries all of economic importance'.[47] The other chair was to focus on the structure and quality of timbers grown in Britain and the Empire and was intended for Percy Groom who was already teaching at the college.[48] Groom had become interested in the economic botany of timber trees when professor of botany at the Imperial College of Whampoa in China. This interest had been furthered during his professorship at the Royal Indian Engineering College at Cooper's Hill, near Egham in Surrey. He appears to have tested woods for governments throughout the Empire — all were interested in what might be the best uses for their timber trees.[49] The government provided funding for the two chairs, and V. H. Blackman came to the college as professor of plant physiology and pathology. After the war Blackman headed a new research institute in plant physiology attached to the botany department and supported by the Ministry of Agriculture.

Wright and Farmer were able to persuade the Rubber Grower's Association to help finance two new buildings for botany. That large sums of money were available was a consequence of the boom in rubber related to the production of tyres for the new automobile industry.[50] In this connection the Association later funded research on the resurfacing of highways, with a view to making them better suited to automobile traffic. This work was carried out in the civil engineering department under Professor Clements.[51] The first botany building, later named for Farmer, was completed shortly after the First World War and the second, known as the Plant Technology Building, opened a few years later, in 1923. This building housed plant pathology, bacteriology and biochemistry until the Second World War.[52] With the new buildings the department began to think seriously about how to expand and began to attract other sponsors. For example, in 1923 the Secretary of the Empire Cotton Growing Corporation wrote to the Rector stating that the Corporation was thinking how best to increase cotton production within the Empire and had come to the conclusion that more basic science on the cotton plant was needed. It was willing to support research in 'plant genetics, plant physiology, entomology, mycology, plant pathology and soil physics … with a view to forming such a group of specialists at home to whom cotton agriculturalists abroad might look for assistance'. At first the Corporation offered just £1000 in research scholarship money, but was willing to offer more if the college responded favourably.[53] The Corporation supported research in plant physiology and plant pathology until 1958.

Students were encouraged to enter applied botanical fields and, in this connection, articles began to appear in the student paper, *The Phoenix*. In 1926 an article entitled, 'Openings and Prospects in Economic Botany', told students that there was much demand for experts in the field, that initial salaries of £400 were usual and, 'for an active man, keen on outdoor pursuits, fond of games and sport, the life is congenial and no more dangerous to health than life at home'. Young botanists would be doing useful work 'combatting diseases' and 'raising improved varieties of plants', and would have the opportunity of 'studying the varieties of vegetable and animal life with which the tropics abound'. And, wherever they decided to go, they were bound to find someone from Imperial College.[54] Farmer remained head of the department until his retirement in 1928 and was Director of the Biological Laboratories from 1911. Having taken the decision to train students in mycology, plant pathology and physiology, he ensured for them something close to a monopoly on appointments in botanical and agricultural departments in the dominions and colonies. He was knighted in 1926 and, on his retirement two years later, testimonials came from all over the world.

The letter from the Empire Cotton Growing Corporation mentions also a willingness to support entomology, another early strength at Imperial. This subject had been taught even before 1907, on an occasional basis, by Harold Maxwell Lefroy. After Professor Sedgwick decided to make applied entomology the main focus of the zoology department, Lefroy was appointed permanently as Professor of Entomology (in 1911) and an associateship course in applied entomology was introduced. In his inaugural lecture Lefroy stressed the importance of applied entomology to India and the colonies, and gave statistics on the destructive cost of insect pests. He had much experience in the area, having worked as an entomologist in the Imperial Department of Agriculture in Barbados, where he helped curb moth borer infestations in sugar cane. From 1901, he was Imperial Entomologist in India and built a large entomological department at the Agricultural Research Institute in Pusa.[55] Lefroy was a major consultant on insect problems around the world, including to the Indian silk industry. Gradually he began to turn research at the college in two directions: towards pests in stored foods (especially grains) and towards sucking insects that attacked orchards. In 1917 the Royal Commission on Wheat Supplies successfully asked for his release for a year to work on grain stores in Australia.[56] Work on hemipterous sucking insects was carried out at his home, and in an old apple orchard nearby, where students reared and studied a range of insects. One of his research students was James Munro who was to become a major force in applied entomology at the college.[57] Lefroy wrote several letters to government bodies stressing the importance of his field of work to the empire, and promoting Imperial College as the main scientific centre for applied entomology in England, a plausible, though debatable, claim.[58] But Lefroy was unhappy with his treatment at the college and complained about restrictions on his consulting work. In a letter to the Rector, threatening to resign, he pointed out that he provided the main insect collections for student work, he paid for the Saturday morning spraying excursions, and he provided his own library for student use.[59] Despite his concerns, Lefroy remained at the college and, by 1920, his students were to be found working around the world in places such as India, Trinidad, Nyasaland, Rhodesia, Gold Coast, Malay States, Uganda and Tanganyika.[60]

Lefroy died in 1925, the result of poisoning by one of his fumigants.[61] His successor was a more traditional entomologist, Frank Balfour Browne, who specialized in water beetles. Balfour Browne appears to have made a distinction between those students who, in his view, were fit for work in the empire and those who were not. In a letter to a schoolmaster he wrote, 'I don't think I made myself clear with regard to seeing students before accepting them. It is only those who expect me later to find them jobs that I must see before accepting, as the type of man is very important for colonial entomological jobs, and the Imperial College has got to make a reputation for supplying men as good as those from Cambridge. I will accept anyone with *good character* for the ordinary course, or for research'.[62] Balfour Browne was an interim professor, appointed while the Governing Body considered what to do about the zoology courses. While applied entomology had drawn students under Lefroy, the general zoology course was not well subscribed. Balfour Browne worked only four days a week and James Munro, Lefroy's former research student, was appointed in 1926 as an assistant professor to help maintain the applied entomology associateship course. Munro's most recent job had been with the forestry department at Oxford, where he had continued his work on timber beetles begun earlier under Lefroy.[63] In temperament he appears to have been very different from the rather casual Balfour Browne. Highly energetic, he made sure that the applied entomology course would continue. During the war, he had worked briefly at Cambridge on a team trying to figure out how to combat the scabies mite but, on returning to Imperial, was determined to take the college in a direction where there would be no overlap with work being done at Cambridge which, aside from the London School of Hygiene and Tropical Medicine, was the only other major academic centre in applied entomology in Britain.[64] He decided to continue what Lefroy had begun, namely work on pests that infested stored foods such as grains. But for this he needed more experts to join in the research. Aside from O. W. Richards, who was appointed to a lectureship in 1927, Munro was not able to appoint others until after he succeeded to the professorship. In the meantime he looked around for possible work and, following in Lefroy's footsteps, began to take on consulting work in the London docks.

In 1924 Lefroy had been asked by Messrs Weber, Smith and Hoare, wharfingers, to help with an infestation of cacao moths (*Ephestia elutalla*) in a shipment of cacao beans from West Africa. By the summer of 1926 the problem had escalated and, as Munro put it,

> the warehouses presented an extraordinary sight. The cacao bags and the walls were covered with silken webbing — spun by wandering caterpillars — which could be peeled off in pieces the size of a tablecloth — moths emerging from the cacao rose in clouds so dense as sometimes to obscure the roof.

And, he stated, the masses of caterpillars on the floor made it difficult to walk without slipping.[65] Looking for a possible research sponsor, Munro took Stephen (later Sir Stephen) Tallents, Secretary of the Empire Marketing Board, to view the infestation. Tallents immediately looked into ways to help finance a research group, and into finding a suitable location for fumigation experiments, and for breeding insects which needed to be securely isolated. The result of his and Munro's efforts was the acquisition of a field station at Hurworth near Slough and the opening of a pest control laboratory. One of the first to work there on cacao moth problems was Munro's student Miss Noyes. Finding the funds to purchase the site was not straightforward, but Munro was able to persuade the Empire Marketing Board to provide most of the money.[66] He also sought the help of Guy (later Sir Guy) Marshall, Director of the Imperial Bureau of Entomology, a body with which Lefroy had had a close connection, and this brought further sponsorship. Marshall told Munro that there were serious problems with the storage of Australian dried fruits, and with tobacco storage at the Imperial Tobacco Company's Rhodesian operations. Marshall suggested research funding could come from those quarters, and indeed it did. The Empire Marketing Board also acted as an informal employment agency, writing to the college on many occasions seeking employees for its various business sponsors.[67]

Even before the field station opened, Munro had begun work experimenting with different fumigants at the docks. Balfour Browne, who had a different view on what proper academic research entailed, was not pleased with Munro's close ties to commercial concerns. But Munro had the backing of the Rector, Thomas Holland, and the work continued. The tension between Munro and Balfour Browne led to Balfour Browne's resignation and, in 1930, Munro became the professor of entomology. On the retirement of the head of the zoology department, E. W. MacBride, in 1934, Munro became the head of a combined zoology and applied entomology department. By this time the Depression had led to the inability of the Empire Marketing Board to keep up its annual payments, but the new Rector, Henry Tizard, appealed successfully to the DSIR for additional funds.[68]

On his appointment to the chair, Munro found the staff he needed. He brought in W. S. Thomson to help O. W. Richards in work on the taxonomic and physiological aspects of the pests he was trying to control. He appointed G. V. B. Herford, a recent research student at the college, to be in charge of insect stocks, and M. J. Norris was given the not always easy task of rearing caterpillars in large numbers. She was to become a major expert on the reproduction of *Lepidoptera*.[69] In the mid 1930s Munro appointed G. S. Fraenkel, a German Jewish refugee, who helped lay the ground for insect physiology at the college before leaving, in 1948, for a chair at the University of Illinois (Urbana). In his fumigation work, Munro collaborated with A. B. P. Page, a member of the chemistry department who later moved to applied entomology. Two other chemists joined the team: C. Potter and O. F. Lubatti. Lubatti had worked for the Colonial Medical Services in Hong Kong and had taken the advanced DIC chemistry course in foods and drugs. Potter developed a pyrethrum spray that was very effective in some of the London warehouses. Together the entomologists and chemists worked on cotton pests in Africa,[70] and on the storage of grains, dried fruits (Australia), copra (Malaya, Gold Coast and New Zealand), tobacco (Australia and Southern Rhodesia), Kola nuts (Sierra Leone), spices, especially nutmeg (Grenada), and potatoes (Cyprus). Another project was the fumigation of secondhand clothing and woollen blankets being sent to the colonies. The entomologists cooperated also with those in the agricultural chemistry and foods and drugs sections of the chemistry department. Advanced diplomas (DIC) were given in these fields and students from around the world came to take the courses.

By the Second World War Imperial College was widely recognized as an important centre in applied entomology.[71] But during the war much of Munro's work was redirected when the Hurworth field station was taken over by the ARC. He turned his attention to bed bug and mosquito control.[72] As discussed in chapter ten, after the war the college needed to find a new field station. Munro played a central role in the acquisition of Silwood Park. There, work continued in the manner that he and Lefroy had pioneered. A 1974 report indicates that Imperial College staff were then still working on pest control, though nematology and some other research areas had been added to applied entomology. In that year, thirty-two overseas students were working at the renamed Commonwealth Overseas Spraying Machinery Centre, and others were working on a range of problems including leaf-cutting insects on Trinidadian cocoa plants, cotton pests in Malawi, rice borers in Malaysia, and a number of human parasitic diseases such as *Leishmaniasis* in Sri Lanka and *Trypanosomiasis* in several parts of Africa.[73] The tradition continues to this day.

Geology and the Royal School of Mines

Well before becoming part of the new Imperial College, those at the Royal School of Mines had worked on imperial projects.[74] As early as 1868 Huxley had stated that those holding the RSM associateship were 'instructed men' acting as 'centres for the diffusion of science throughout this country and the colonies'.[75] While geology was later situated in the Royal College of Science, the staff continued to train people for geological surveying work around the world. The first two heads of the geology department at Imperial, William Watts and Percy Boswell, were 'pure' geologists with interests mainly in the geology of Britain, but many of their students left for work in geological surveys overseas.[76] Others worked in the mining and metallurgical industries of empire. Watts also took in work from the overseas surveys, and members of his department examined and assayed specimens. Staff and research students were paid for this work, supposedly done in their spare time.[77] Watts saw also the need for applied geological work and, together with Professor Frecheville, head of the mining department, set up courses in petroleum geology (later oil technology), mining geology, and engineering geology. Vincent Charles Illing who came from Cambridge in 1913 to study with Watts, later built the oil technology course into one with an international reputation, arguably the best of its kind. Illing was born in the Punjab, and had grown up there and on Malta. During the 1914–18 war he began what was to become a long association with the island of Trinidad, both for him and for his students. He was sent there to study the geology of the Naparima region with a view to oil exploration and the development of an industry.[78] While many Imperial students later found work in the Trinidadian oil industry, Illing also trained Trinidadians for survey work, mineral separation, and chemical analysis.[79] He became a major consultant in the oil industry worldwide, and his students profited from this. Later in life, Illing turned his attention to North Sea gas and other British projects.[80] Work in oil and gas fields worldwide continues at Imperial to this day, engaging people from a number of different departments.[81]

Charles Gilbert Cullis, Professor of Mineralogy and Mining Geology, was largely responsible for the mining geology course begun in 1918. Many of his students left for work as mineral prospectors in the empire. He also took on a number of empire related contracts; for example, he directed a mineral survey on the island of Cyprus.[82] His course was later taken over by W. R. Jones, a former student at the college who returned as Professor of Mining Geology in 1941.[83] On graduating, Jones had joined the Geological Survey of Malaya, becoming an expert on tin ores. He returned to Britain in 1915, only to be sent off again, this time to Burma where he was asked to explore tungsten deposits needed for steels used in the manufacture of high-speed machine tools and in munitions. At the college he further developed the field of ore microscopy and set his students to work on a variety of ores for a number of mining companies.[84]

After the First World War the total numbers of undergraduate students in the different branches of geology settled to roughly eight in pure geology, twenty-eight in oil technology, and eighteen in mining geology. Thomas Holland, who had succeeded Keogh as Rector, was himself a geologist and encouraged the applied work. Holland had been a student at the RSM before leaving for work in India where he made a major career. He became Director of the Geological Survey of India and a high civil servant in the Indian government.[85] He successfully redirected the Survey so that it better served Indian industrial interests, and his experience in this connection was valued at the college. In London, Holland maintained his imperial connections, among other things as chairman of the Empire Council for Mining and Metallurgy. The generation of geology students who graduated during his term as Rector included many who similarly left to seek work in the empire. Eight became professors of geology at empire universities, forty became directors of geological surveys, and others became employees in empire universities, geological surveys and mining companies.[86] Among them was Robert Shackleton, a Beit Fellow in 1932–4, who had a career trajectory that was typical of those who became academic geologists at Imperial in mid-century. Shackleton left the college to become a geologist in Fiji. At the start of the Second World War he returned to work briefly at the college before being seconded to the Government Mining and Geological Department in Kenya, where he carried out studies on the Mingori gold belt. After the war he returned briefly to a chair at Imperial.[87] His wife, Peigi Wallace, also joined the staff as a lecturer in paleontology.

Mining and metallurgy students were as likely as the geologists to find work in imperial settings. Something of the spirit of the early period can be seen in a speech given by Professor Truscott at the annual dinner of the Royal School of Mines in 1921. Looking back he noted how he and his colleagues realized the importance of certain industries and how exciting this was for staff and students alike, 'the tremendous factors of Africa's gold, Canada's nickel, Straits tin and Rhodesia's chromite in the mineral wealth of the world. The Broken Hill deposit in Australia was the largest lead-zinc deposit in the world …'.[88] The professors of mining and metallurgy were expected to have had experience in mining overseas. The first two to be appointed after 1907, William Frecheville in mining and William Carlyle in metallurgy, both met this criterion. Frecheville had worked at a number of sites including at gold and cobalt mines in the Transvaal and as a mines manager in the gold fields at Mysore. Before his appointment in 1912, he was already a major consulting engineer. Carlyle, a graduate of McGill University, and later professor of metallurgy there, had extensive experience managing mines in British Columbia and, during the late 1890s, had been in charge of the Department of Mines in the province. Later he gained extensive knowledge of the Rio Tinto copper mines in Spain where he was Manager. Future appointees were expected to have similar experience, something that conflicted with the fact that these disciplines were becoming increasingly academic. In this connection there were problems also with students who favoured professors with experience in the mining and metallurgical industries over those with purely academic achievements. Such professors, it was widely held, had good connections and could prepare students better for the kinds of work available. Carlyle's work with the Rio Tinto mines helped in it becoming a major employer of Imperial students. Frecheville's experience in the South African gold mines, similarly helped many find employment there.[89]

As discussed in chapter six, Harold Carpenter, appointed to succeed Carlyle in 1911, was seen as too narrowly academic. So, too, was Professor Truscott's successor, J. A. S. Ritson, appointed in 1935. Before being allowed to take up their new positions they were both sent, at the college's expense, to visit respectively metallurgical and mining works in the empire. After an extended visit to gold and copper mines in South Africa Ritson reported,

> I met nearly every available consulting mining engineer on the Rand and in the Rhodesias; I met the majority of mine managers … in fact I had a wonderful trip, which I hope will bear such fruit that the Royal School of Mines Mining Department will prosper during my tenure as Chair of Mining. The contacts I made in Africa will, I feel certain, stand me in very good stead in the future and should be of great assistance in helping me to place students.[90]

This latter point was important. The reputation of the mining and metallurgy departments was such that they were often the first point of call when companies looked for employees. Mining companies also donated money to the college and supported student bursaries. For example, in 1930, the British South Africa Company, the Rhodesian Anglo-American Company and the Bewana M'Kubara Company, wanting suitably trained British mining geologists, metallurgists and mining engineers, offered both research scholarships, and entry scholarships from selected public schools.[91] Clearly not only the training but also the family background of future employees was seen to be important.

Geophysics was a new field introduced at Imperial in 1921 with support from the DSIR. A. O. Rankine, an expert in acoustical and seismic methods for locating oil deposits, who had visited oil fields around the world, was appointed to head the section, then attached to the physics department. As mentioned in chapter six, under the agreement with the DSIR, the college was to be responsible for giving instructional courses designed to train geophysicists to meet the demands of empire prospecting companies which, up to then, were having to rely on American and German expertise in this area. Rankine left in 1937 to take a senior appointment with the Anglo-Iranian Oil Company.[92] This company, forerunner to British Petroleum, had close connections to the college, funded many research projects, and gave employment to many graduates, not just in geophysics. *The Record of the Royal College of Science Association* (1937), marked Rankine's departure with a rhyme:

> If Rankine prefers travel
> >   To academic toil,
> No one of us will cavil,
> >   At the fact that he's struck oil.

Engineering

Arthur Acland's 1911 memorandum to the Board of Education stated many other needs in addition to those of the botany department discussed above. One area that he considered important was applied chemistry. The professor of chemistry, T. E. Thorpe, had long promoted applied chemistry and held the view that this should include technical work on rubber, sugar, plant alkaloids and the fermentation industries. Thorpe had appointed William Bone to give lectures on fuels and Bone later became the first professor of chemical technology. Bone had a grand vision and discussed his postwar plans with the Rector already during the war. He wanted to build a department on a 'scale commensurate with the responsibility of the College to the Empire'. This, he stated, entailed a department with four sections, one devoted to fuel technology and chemistry,[93] and to the study of refractory materials, a second to chemical engineering, a third to electrochemistry, and a fourth to the carbohydrates, fats, oils and rubber. Not all these plans came to fruition but over three hundred students from the empire/commonwealth took courses in the chemical technology (later chemical engineering and chemical technology) department between its founding in 1912 and 1960. Most came from India and Pakistan and many later held important positions in their home countries. For example, L. A. Bhatt specialized in fuel technology and later founded the chemical technology department at the University of Bombay. He became a senior civil servant and oversaw the rapid increase in chemical manufacturing in India during the Second World War. Bhatt was also a major consultant and, in 1950–2, was

President of the Indian Chemical Manufacturers Association. Another graduate, S. K. Sircar, joined the Bararee Coke Works at Kusunda, rising to become its manager. This company began to produce a range of high quality by-products of coke, and Sircar became a major consultant in this field. Later, Alfred Egerton, George Finch, and Dudley Newitt trained a new generation of young Indian technologists and engineers. Egerton also chaired a committee reviewing the working of the Indian Institute of Science in Bangalore. In a memoir of Newitt, Peter Dankwerts wrote, 'when I was at Imperial College one might have felt that the Raj had never been deposed. There were a great many diligent Indian students who could be found working there even on Christmas Day.' He also noted that whenever Newitt visited India his former students flocked to him.[94] After the Second World War, Imperial College formed a relationship with the Delhi College of Engineering and Technology; senior staff from Imperial visited Delhi to give advice on courses and many of the Delhi staff came to Imperial for training. George Finch, an Australian by birth, worked in several areas of chemical technology and had many Indian students engaged in the study of oils and fats.[95] He resigned from Imperial College in 1952 and went to India as Director of the National Chemical Laboratory.

The other engineering departments were also engaged in a range of imperial projects. Electrical engineers, specializing in the heavy side of their discipline, helped in generation and transmission problems worldwide, and others worked on wireless and telegraphy projects. Professor Fortescue had many Indian connections and was instrumental in forging ties between Imperial and a number of Indian educational institutions, notably the Indian Institute of Science in Bangalore.[96] Electrical engineers continued to work in Commonwealth countries into the later twentieth century. Willis Jackson, the head of the department in the 1960s, encouraged his students to gain such work experience. In 1968 he noted that students in his department had tackled subjects such as the use of TV for educational purposes in India, methods of generating and distributing electricity in Sierra Leone, and that he intended to send one group of students to Zambia to study electrification in rural areas, and another to Ghana to study telecommunication systems. His aim was to extend college work in this direction over the longer term.[97]

The Civil engineers, too, had many imperial connections. Bridge building and railway construction around the world were of continuing interest to the college staff and the tradition of working in the empire was well entrenched. William Unwin, who had been professor of civil and mechanical engineering before the formation of Imperial College, had worked on major bridges in India and elsewhere in the Empire, and on hydro-electricity generation projects in the Himalayas and at Niagara Falls. His knowledge of hydraulics was used also by the Australian government when pumping over five million gallons of water a day, over three hundred miles, to the Coolgardie goldfields. Unwin's successor, W. E. Dalby, and the first professor of mechanical engineering at Imperial, was a railway specialist. Records show that many students from the Dalby period took up railway appointments both at home and overseas, especially in India and East Africa.[98] But others worked on dams, hydrology and irrigation and, later, civil engineering students took the new science of soil mechanics around the world.[99]

The first professor of civil engineering, Stephen Dixon, had been a professor in Canada and had been involved with a number of Canadian railway and bridge building projects. Given that, like Dalby, he was a railway specialist, it is not surprising that many students continued to enter bridge construction and railway work in the 1930s and 40s. For example, Dixon's student, J. F. Pain, was awarded the Manby Premium for a paper on the Sydney Harbour Bridge, built by Dorman Long. The chief designer and engineer for the bridge was a former City and Guilds student, Sir Ralph Freeman.[100] In 1924, shortly after graduating, Pain went to Sydney. From there he wrote to some of his old student friends that the first thing people asked him on arrival in the city was what he thought of their beautiful harbour, and that he didn't dare tell them that he was 'out here with a view to disfiguring all this natural beauty with a large bridge'. Almost thirty years later Pain and fellow ex-student K. E. Hyatt were back in Sydney. Hyatt reported that this time they were being encouraged to go and see the 'beautiful' harbour bridge 'with which the local inhabitants seem rather taken up and which, on inspection, was vaguely familiar to brother Pain and self'. This correspondence is part of a larger set written by members of the XXI Club, founded in 1921 by C&G students who had fought in the First World War and had come to Imperial in 1918 and 1919. On graduation, knowing that they would be working far from home, they decided to keep in touch by means of a correspondence club. The club is a microcosm of engineering life in the twentieth century and included its share of men working in the empire. In addition to the bridge builders, who worked not only in Australia but around the world, building bridges across the Nile, Limpopo, St. Lawrence, and many other major rivers and harbours, other members worked on a variety of overseas projects from railways to power generation, from fortifications to telegraphy and telephony. To celebrate the club's fortieth anniversary one member, J. C. G. Vowler, composed a poem:[101]

> For four decades we've been about the world,
> And every continent has known our name,
> From land to land our flag has been unfurl'd,
> And distant shores have echoed back our fame.
>
> On Empire's fields our lot was fairly cast,
> We've laboured under many a foreign sky,
> ....

From Empire to Commonwealth

It is, perhaps, not surprising that when Patrick Linstead was Rector of Imperial and, during the expansionary period of the college in the early 1960s, considered changing its name, he received letters urging him not to do so. One came from Richard Foot, an ex-student and retired Brigadier living in New South Wales. He wrote,

> I hope the word 'Imperial' will somehow be incorporated in any new title. The word 'Empire' and its associated adjectives have become almost dirty words in these days. Yet there was once a British Empire, and 'Imperial College' is one of the few verbal relics of it that remain. The college was founded in the heyday of that Empire and may still be a monument of it, and many of its first students and graduates, between 1907 and 1918, gave their lives in the defence of the idea of empire as we knew it then.[102]

This letter would have had resonance for members of the XXI Club, and for others of that generation. Many were then just retiring from senior positions in government, academia, the engineering professions, and industry.

The end of Empire led to a major shift in Britain's position in the world, something that took time for people to accept. Britain had to adjust to new realities, and Imperial College had to adjust also. Already in 1949 Sir Henry Tizard, Imperial's former Rector, then Chief Scientific Advisor to the Ministry of Defence, warned Britons 'we are a great nation, but if we continue to behave like a great power we shall soon cease to be a great nation'.[103] After the Suez debacle of 1956 many people came round to Tizard's point of view. At Imperial College, Commonwealth ties replaced those to Empire. For example, the head of the physics department, P. M. S. Blackett, was much engaged in helping what were then labelled 'developing' countries. In a speech given at the University of Leeds, Blackett played on Dean Acheson's quip 'Britain has lost an empire but has not yet found its role', and stated 'Britain wisely let an empire go and then unwisely forgot about it'.[104] Blackett's point was that Britain should play a major role in the development of her ex-colonies and that universities still needed to train students for work overseas, and to train overseas students for work in their own countries. He believed that universities and research institutes in many Commonwealth countries needed to be built up, so that in the longer run ex-colonies could become truly independent. This was a typical view at Imperial. Blackett, himself, had a major interest in Indian science and technology but, more critical than most, he believed that India had been conserved in its 'backwardness' under British rule. He spent much of his later life working with the Indian government in the development of major technical projects.[105] Blackett also held the view that scientists whose research work had dried up often still had much to offer, and that they should be encouraged to work in the ex-colonies, either in higher education or in the running of technical enterprises. He was able to find work for many people doing just that. His colleague Abdus Salam, another Nobel Laureate and theoretical physicist of world stature, had similar views. He devoted much time to the promotion of technical projects in his native Pakistan and opened a summer college there where the country's needs were discussed, and plans for projects laid. He also played a role in setting up Pakistan's research councils, modelling them on those in Britain. Both Blackett and Salam invited scientists from 'developing' countries and many, especially from India and Pakistan, came to the physics department to study or carry out research.[106] As with Blackett in India, Salam was consulted on the setting up of Pakistan's atomic energy industry's infrastructure. With hindsight we see this differently, but in the 1950s and 60s nuclear fission was believed one of the best ways to produce the energy needed for rapid industrialization, and as a necessary precondition for development and modernization.

The above examples could be multiplied many times, but space does not allow for more on imperial science at Imperial College. It is worth mentioning, however, that there has been an imperial legacy also in the pattern of administrative appointments. For example, J. M. Corin, Imperial's Financial Secretary from 1949 and later College Secretary, was employed in the Indian Civil Service from 1930–46. His successor as Financial Secretary, A. E. Savage, had twenty-two years experience with the Colonial Service in Nyasaland. M. J. (Mickey) Davies, who was appointed College Secretary in 1962, was born and educated in South Africa. A Rhodes Scholar, he later joined the colonial service and worked first as Secretary to the Governor of Tanganyika, and then became also District Commissioner in the Arusha District. He retired from the colonial service in 1961 when Tanganyika became independent. A later College Secretary, John Smith, came to Imperial in the 1970s after a career in the colonial, then foreign, service. The end of empire released many competent people from their jobs at a time when universities were expanding. Those entering non-academic positions at Imperial contributed much to the college ethos, and their stories inspired many students to seek work overseas.

Imperial College, today, is a world, rather than an imperial or Commonwealth institution. It has strong ties worldwide and its students and professors are drawn from many countries. Its graduates, too, can be found almost anywhere. But this is a pattern based on a century-long history. Well into the 1980s college calendars and prospectuses continued to stress Commonwealth ties. Only by the mid 1990s did Europe trump the Commonwealth in calendar space allotted. A strong imperial trace is still detectable but, as Rupert Hall noted, 'the College never had an adequate income, nor an adequate endowment for research and development, to enable it to approximate to the idea of an institution vast enough and diverse enough to inspire and guide the scientific and technical developments of the British Empire.'[107] The early ambition of Keogh, Acland, and others, that the college be 'a Mecca', *the* imperial college, to which all would flock for expertise, or for instruction, was an overweening one — especially given the resources then available.

Nonetheless, much was achieved and the college truly lived up to its name. While the early staff members could not foresee how short the life span of the Empire would be, in following the paths they did, they helped fashion the future.

...

1. These words, by Arthur C. Benson (1862–1925), were written in 1902, in close collaboration with Edward Elgar who incorporated the famous tune from Pomp and Circumstance March No. 1 into the 'Coronation Ode' for Edward VII in 1902. The Ode was performed also at the coronation of George V in 1911.

2. Ronald Hyam, in his 'The British Empire in the Edwardian Period', in Judith M. Brown and William Roger Louis (eds.), *The Oxford History of the British Empire: The Twentieth Century* (Oxford, 1999), claims that the sentiments were not widely shared. He notes there was a pessimism about the future of the empire and much defeatist talk in periodicals and newspapers. Student journalism during the Edwardian period at Imperial College, and at its antecedents, gives a different impression. While the South African War was not without its critics, the Empire was seen in a largely positive light — perhaps because so many students hoped for employment overseas. Their imperialism appears to have been envisaged in cultural and economic, rather than political or military, terms. It entailed the spread of British ideas, education, commerce, and the bringing of material improvement.

3. John M. MacKenzie (ed.), *Imperialism and the Natural World* (Manchester, 1990), 2. See also John M. MacKenzie, *Propaganda and Empire: The Manipulation of British Public Opinion, 1880–1960* (Manchester, 1984).

4. Science and empire is a large topic on which much has been written. For a range of essays in a single volume see, Roy MacLeod (ed.), 'Nature and Empire: Science and the Colonial Enterprise, *Osiris*, vol. 15 (2000). See also, Richard Drayton, *Nature's Government: Science, Imperial Britain and the 'Improvement' of the World* (New Haven and London, 2000); P. Palladino and M. Worboys, 'Science and Imperialism', *Isis*, vol. 84 (1993), 91–102; Richard Drayton, 'Science, Medicine and the British Empire' in Robin W. Winks (ed.), *The Oxford History of the British Empire*, vol. V (Oxford, 1999), 264–75; Roy Macleod, 'Passages in British Imperial Science: From Empire to Commonwealth', *Journal of World History*, vol. 4 (1993), 2–29.

5. Daniel R. Headrick, *The Tools of Empire: Technology and European Imperialism in the Nineteenth Century* (Oxford, 1981). See also Headrick, *The Tentacles of Progress: Technology Transfer in the Age of Imperialism, 1850–1940* (Oxford, 1988).

6. Frederick John Dealtry Lugard (1858–1945) was a soldier and administrator of empire. A major figure in the administration of several African colonies, he was also responsible for setting up Hong Kong University in 1911.

7. Michael Worboys, 'The Imperial Institute: the state and the development of the natural resources of the Colonial Empire, 1887–1923', in MacKenzie (ed.), *Imperialism and the Natural World, op. cit.* (3), 169.

8. For the role of the Board of Education in the naming of Imperial College, see chapter 3.

9. EC minutes, 28 January 1921. GB minutes, 11 November 1921. Lord Crewe, chairman of the Governing Body, wrote about the college work to Winston Churchill when he succeeded Milner as Colonial Secretary in 1921. He noted that if Imperial was to fulfill its mandate, then government funding had to increase. Churchill replied that he would appeal to the Treasury against any cuts to the universities. EC minutes, 10 February 1922. This correspondence should be read in the context of a then recent report by the Committee on Research in the Colonies, chaired by Lord Chalmers.

10. The words *decus et tutamen*, found on the rim of the pound coin, were first used on crown coins in 1662. The words were a safeguard against coin clippers in the days when the coinage was made of precious metals.

11. Keogh, quoted in C. A. M. Smith, 'The Imperial College of Science and Technology, South Kensington S. W.', *Engineering* (1912) (series of articles on the new college), p. 4.

12. See revised charter (7 July 1916).

13. By 1951 Newfoundland was no longer a colony but a province of Canada. The British Commonwealth of Nations was founded earlier, in 1931, by the Statute of Westminster.

14. These scholarships were instituted in 1925 with funds from the 'Dominions' and from donors in Britain. Two scholars came from each of Canada, Australia, India, New Zealand, Australia and South Africa. One woman was among the first group in 1925, botanist M. J. Collins from Australia. See GB minutes, June 1923, lvii.

15. Elsewhere 'Imperial' was being dropped from names. For example, at Oxford and Cambridge the chairs of Imperial History were renamed Commonwealth History.

16. J. Stanley Gardiner, obituary of Sedgwick, *The Zoologist*, March 1913. See also chapter 6.

17. Usually people from the Royal Army Medical Corps came to give lectures on tropical hygiene. For example, Lieut-Col. C. Melville gave courses beginning in 1911. EC Minutes, 29 September 1911.

18. The Institute was intended as a memorial of the Queen's Jubilee to represent the 'Arts, Manufactures and Commerce of the Queen's Colonial and Indian Empire'. See 'Reports of the Commissioners for the Exhibition of 1851' (7th Report, 1889), 26. For quotation see MacKenzie, *Propaganda and Empire, op. cit.* (3) 122–3. An early twentieth-century venture with the empire in mind was the British Science Guild. See, Roy MacLeod, 'Science for Imperial Efficiency and Social Change: Reflections on the British Science Guild, 1905–1936', *Public Understanding of Science*, vol. 3 (1994), 155–93.

19. Colcutt, quoted in MacKenzie, *Propaganda and Empire, op. cit.* (3) 125.

20. By the early 1960s the building had acquired layers of soot and dirt and looked very black, nothing like the restored Queen's Tower, the only surviving part of Colcutt's large edifice, looks today.

21. The peal of bells in the Queen's Tower was named for Alexandra, Princess of Wales, by their Australian donor, Elizabeth Millar, (each individual bell being named for a member of the Royal family). For details of the opening see MacKenzie, *Propaganda and Empire, op. cit.* (3), chapter 5. The bells are rung today on Royal anniversaries such as the Queen's birthday.

22. His successor at the Institute was another chemist, Wyndham Dunstan, who had earlier worked at St. Thomas's Hospital.

23. Worboys, *op. cit.* (7), 172. Later, in 1923, Professor Farmer complained to the College Secretary, Alexander Gow, that the Institute was trying to offload this work on to the college. He wrote, 'the whole of our department is organized on a basis foreign to the ordinary *ad hoc* sort of enquiry' addressed to the Institute. Excess botanical work of this sort, he stated, should be directed to Kew — but he was willing to help where he could. KB/ Botany Department correspondence; Farmer to Gow 8 March 1923. When the Imperial Institute closed in 1958, its work on natural products was taken over by the Tropical Products Institute which also took over the Tropical Stored Products Centre, descendent of Professor Munro's pioneering work at Imperial College's first field station at Hurworth, Slough.

24. MacKenzie, *Propaganda and Empire, op. cit.* (3), chapter 5.

25. Both the dye and drug industries had given way, somewhat, to German industry before the war. The government helped create both British Dyes and British Drugs during the war and khaki and other dyes were, by the end of the war, produced largely synthetically.

26. Suspicion had been voiced also earlier when he was Secretary of State for War. These attacks were unfounded.

27. See Worboys, *op. cit.* (7).

28. In 1926 Prime Minister Stanley Baldwin found it impossible to promote imperial preference through tariffs on food imports. The Empire Marketing Board was an idea of the Colonial Office to help solve this problem. The Board was a collective, supported by several colonies and dominions, to further their commercial interests with respect to agricultural products. £250,000 of the Board's income (about 25%) was set aside for research. The leading figure in promoting these interests was nutrition expert, Walter Elliot MP FRS, who became Minister of Agriculture in 1932. By then the Depression had put an end to the wealthy times and the Board was closed down in 1933. In its place the Imperial Bureau of Agriculture was founded, principally for information exchange. It supported research but on a more limited scale. Elliot, who later became Minister of Health, had earlier received funding from the Empire Marketing Board to study nutrition among Scottish schoolchildren. 1500 were given milk at school, anticipating later school milk programmes. Later the Commonwealth Bureau of Agriculture moved to Silwood, and became a college neighbour.

29. The DSIR also funded research on colonial problems.

30. The Imperial Department of Agriculture was set up first in Barbados and moved to Trinidad in 1922 when the Imperial College of Tropical Agriculture was founded. Similar institutions had been founded earlier in other parts of the Empire. John Farmer, Imperial College's professor of botany, was closely involved in the founding of the college in Trinidad. In 1924 the Assistant Director of Agriculture in Trinidad, Imperial College graduate W. G. Freeman, held a dinner party at St. Augustine at which 12 old Imperial students were present. Two were botany professors at the new college, others were chemists working for the Empire Cotton Growers Association and the British Empire Sugar Research Association, and entomologists working under Freeman. They sent a greetings telegram to Farmer. See *The Phoenix* (March 1924), 61.

31. Robert A. Stafford, 'Annexing the Landscapes of the Past: British Imperial Geology in the Nineteenth Century' in MacKenzie, *Imperialism and the Natural World, op. cit*, (3). See also Suzanne Zeller, *Inventing Canada: Early Victorian science and the idea of a transcontinental nation* (Toronto, 1987) and Peter Anker, *Imperial Ecology: environmental order in the British Empire, 1895–1945* (Cambridge, MA, 2001).

32. *RCS and RSM Journal,* (March 1900). Also B/Levy; see memoir of D. M. Levy, a mining student at the time.

33. It is interesting that when the Rector, Sir Alfred Keogh, was seconded to head the Royal Army Medical Corps during the war, the student paper saw this as *imperial* service, 'our Rector has returned to the service of Empire'. *The Phoenix*, new series vol. 1 (1915), 2.

34. H. G. Wells, *Imperialism and the Open Conspiracy* (London, 1929), 7–9 and 11.

35. Joseph Chamberlain was Colonial Secretary 1895–1903. He was no enthusiast for the Disraelian idea of Empire but was keen on promoting an Empire of British settlement, and British emigration to places such as Canada and Australia. In support of this vision, he wanted tariffs and protectionist policies. Chamberlain was a former Liberal but broke with Gladstone over Irish Home Rule. For the position of 'Liberal Imperialists', including Lord Haldane, on issues of empire see, H. C. G. Matthew, *The Liberal Imperialists: The ideas and politics of a post-Gladstonian élite* (Oxford, 1973), chapter 5.

36. In this he was supporting the college administration and many senior professors. Wells's son, George Philip Wells FRS, was also a student at the college. He studied zoology under E. W. MacBride but moved to Cambridge for his degree.

37. H. G. Wells, speech, delivered in 1921, partly transcribed in *The Phoenix* VI (1921), 98. John Perry was Professor of Mathematics and Physics in the Royal College of Science until 1912. Wells failed to acknowledge that Oxford, at least, educated many colonial governors and high government officials. Up until the Second World War Oxford also trained many students from the Empire (roughly 16% of its students) but mainly in arts subjects. Unlike at Imperial the number of Commonwealth students at Oxford dropped dramatically after the Second World War. New governments preferred local training for their civil servants but continued to favour British training in science and engineering. Imperial maintained its roughly 20% of overseas students (as did the London School of Economics). For Oxford, see Daniel L. Greenstein, 'The Junior Members, 1900–1990' in Brian Harrison (ed.), *The History of the University of Oxford* (vol. viii); *The Twentieth Century*, (Oxford, 1995), chapter 3.

38. Along with G. B. Shaw, another early visitor to the college, Wells was seen as a progressive post-Victorian voice.

39. B/Acland 71/1; 'Memorandum on the Development of Research Work in the Imperial College in relation to the industries of Empire'.

40. See chapter 4.

41. B/Acland; written remarks for presentation by Acland to visiting MPs. For a report of this visit see *Nature* vol. 97 (25 May 1916), 264. Acland also

stressed the need for changes in the public schools and that classics should give way more to science education.

42. B/Keogh; statement on visit, 7 June 1916.

43. Organization Committee minutes, 26 May 1910.

44. Rubber was a major industry in Ceylon before the plantation industry grew in the Malay Straits region. By 1940 13.2% of plantation acreage in South East Asia was budded rubber. See G. Stafford Whitby, 'Looking Back over Fifty Years of Rubber Science', *Proceedings of the Institution of Rubber Industry*, vol. 7 (1960), 155–75. In the ICA there is a typescript memoir (1958) of the early days of rubber chemistry by Stafford Whitby who was then a professor at the Institute of Rubber Research, University of Ohio at Akron. He entered Imperial College in 1908 and, on graduating, left to work with rubber companies in Malaya, Sumatra and Java. In his memoir he mentions that several other Imperial College students were working there also. Another major rubber chemist was Ernest Harold Farmer FRS, a research student under Jocelyn Thorpe, who joined the staff of the chemistry department in 1924. In 1932 he left for work with the Rubber Growers Association and, in 1938, became Senior Chemist at the newly formed Research Association of the British Rubber Producers, becoming Assistant Director in 1941. See Frank L. Warren, 'Ernest Harold Farmer, 1890–1952', *Journal of the Chemical Society* (May, 1954), 1654–59. Also, G. Gee, 'Ernest Harold Farmer, 1890–1952', *Obituary Notices of Fellows of the Royal Society*, vol. 8 (November 1952), 159–69. See also Lucille H. Brockway, *Science and Colonial Expansion: The Role of the British Royal Botanic Gardens* (London, 1979), chapter 7.

45. The Rubber Growers Association was formed by British owners of estates in Ceylon and Malaya. The British Rubber Producers Association, with which Wright, and later E. H. Farmer, were closely involved, was set up to encourage research. On the board of this Association were Norman Haworth and Eric Rideal who promoted rubber research at Manchester and Cambridge respectively. Haworth, a Manchester graduate, had studied also some chemistry at the C&G under Armstrong, and under Tilden at the RCS where he later became a demonstrator.

46. B/Acland; Wright to Acland, 5 December 1918.

47. In 1920, the botany department won a diploma at the Fifth International Rubber Exhibition for their exhibit on diseases of the rubber plant.

48. B/Acland; 1911 document.

49. B/Groom. Box 5/1 for example. Groom left several boxes of papers related to his work on timbers.

50. The Rubber Growers Association appealed to its members for donations towards the new building. The response was international with money coming in from plantation organizations also outside the British Empire. Over 270 rubber companies made donations and some donations came also from other industries such as tea, copra and coffee. Together they contributed about £30,000 towards the first botany building. The Mycological Laboratory in the new building was named for Wright who made a further donation. Farmer was persuasive and two other members of the governing body who also gave generously towards the botany building had laboratories named after them, namely the Robert Kaye Gray Plant Pathology Laboratory, and the Francis Ogilvie Physiology Laboratory (later renamed for Vernon H. Blackman). The Rubber Growers also supported latex research in the chemistry department. See GB minutes, 10 December 1920. and EC minutes, 15 February 1935.

51. EC minutes 1 November 1935. One of Clements's research students, Rajinder Nath Dogra, had been an undergraduate in the civil engineering department (1929–31). As Chief Engineer in Chandighar he was responsible for setting out the modern city and road system. He became the first Director of the Indian Institute of Technology in Delhi and was made a FIC in 1970.

52. After the Second World War the zoologists and entomologists took over the space. One of the first graduates to have worked in the new building was Raymond Henry Staughton who gained his associateship (and an external BSc from the University of London) in plant pathology in 1924. His first job was as a mycologist with Ceylon Rubber Research. Later he worked on cotton plant diseases in East Africa. He became Professor of Horticulture at Reading University, but ended his career as Principal of University College, Ghana.

53. Botany department correspondence file; L. G. Kilby to Rector, 22 May 1923.

54. Article by RJT in *The Phoenix*, vol. 11 (1926), 102–3. In 1929 John Farmer wrote a similar article noting that over the previous eight years the Colonial Office had 425 job openings having to do with applied botanical work, and with salaries in the £500–£1100 range, plus pensions. There were also a few senior posts open each year. He noted that only 3% of his ex-students had to give up work in tropical countries for health reasons. See *The Phoenix*, vol. 13 (1929), 134–7. In 1928 *The Phoenix* promoted engineering work in Nigeria (vol. 13, no. 4).

55. For more on Lefroy's work in Barbados and at the Agricultural Research Institute in Pusa see Headrick (1988) *op. cit.* (5), 216–18.

56. Another request two years later was refused and led to tension. Many at the college thought that Lefroy spent too much time away on consulting work.

57. One piece of work conducted by Munro under Lefroy's supervision was an investigation into the life history of the Death Watch Beetle that was then endangering the roof of Westminster Hall. Lefroy had been asked to do this work by the government and received much public recognition for clearing the roof timbers of the beetle. After this success, Lefroy opened a small production plant in Hatton Garden where one of his assistants, Elizabeth Eades, supervised the production of 'woodworm fluid' and many bottles were sold. This was the start of the firm Rentokill Ltd.

58. See examples in KZ9/3/; see also Zoology Department correspondence; letters from MacBride to Gow about activities within the department of zoology.

59. Zoology Department correspondence; Lefroy to Keogh, 16 June 1919.

60. GB Minutes see, for example, annual report 1920–1.

61. Undergraduate O. V. S. (Peter) Heath kept a diary and noted (21 May 1924), 'Lefroy has gassed himself again and won't be able to lecture any more this term'. Lefroy died in October of the following year, a few days after poisoning himself yet again. It is unclear whether this was a suicide or an accidental poisoning. On graduation, Heath worked in Trinidad and South Africa, before returning to Imperial College for doctoral studies under F. G. Gregory. He joined the staff as a lecturer in botany before leaving for the chair of horticulture at Reading in 1957. Heath's career was not unlike that of R. H. Staughton's earlier. See note (52) above. Heath's diary is in the possession of his son Robin Heath.

62. B/ Balfour Browne; Balfour Browne to Roebuck, 11 September 1926. Emphasis in original.

63. There was a small forestry department at Oxford. In 1924 the Imperial Forestry Institute was established there, well supported by the Colonial Office.

64. The major research areas at Cambridge were medical parasitology and domestic crop pests.

65. Munro, quoted in R. G. Davies, 'A short History of Entomology at Imperial College' (1995 typescript, in ICA) 18–21.

66. According to Gough, the EMB provided £10,625 to purchase the land and an annual operating grant of £3574. See H. C. Gough, 'Professor J. W. Munro: the founder of the Imperial College Field Stations and of research on stored products pests in the United Kingdom', *Antenna: Bulletin of the Royal Entomological Society of London* vol. 14, January 1990, 10–15.

67. For example, letter from A. Ryan of the EMB to H. Tizard, 2 June 1930, wanting to hire a botany graduate to work in helping to set up an Indian Tea Association laboratory in plant physiology at Tocklai (see KB\Botany Department correspondence file).

68. The Empire Marketing Board closed in 1933. One of the DSIR inspectors, Sir Frederick Gowland Hopkins, a former student at the RCS, was impressed by Munro's insect breeding programme at Hurworth. The field station also received grants from the Carnegie Foundation, and the University Grants Committee (Gowland Hopkins was on that inspection team also).

69. Maude Norris and Owain Richards later married. For more on Richards see chapters 6, 8 and 10.

70. Munro helped with a major bollworm infestation in Tanganyika in the 1940s; (see entomology departmental correspondence; Munro to Rector 22 July 1952). In this letter Munro wrote, 'if the entomologists have increased the populations of tropical countries by checking insect borne diseases (and the College and College trained entomologists were in large responsible for this) then it is now up to them to increase the food supply'. See also, Annual report of GB, 1937.

71. During the 1930s the five permanent members of the department published several books and over 130 papers. Except in medical entomology, the domain of the London School of Hygiene and Tropical Medicine, Imperial College led the field in applied entomology in London, arguably in the country. But Cambridge, and some government laboratories, had more specialists in the pests of crop plants. After the Second World War, Imperial moved also into medical entomology when Peter Boreham, a specialist in tropical diseases, came to work at the college. In 1977 he was among the first to warn of the dangers of new insect borne diseases such as malaria, ebola, lassa and marburg viruses, coming to Britain as the result of increasing travel. See ICA, newspaper clippings file for Boreham's Pharmaceutical Society Award lecture, 1977.

72. Bed bugs had been studied earlier. In 1934 Munro gave a BBC radio talk on bed bugs, fleas and other domestic pests and received an avalanche of mail from across the country. See Davies, *op. cit.* (65), 21. See also chapter 8 for Hurworth and for Munro's work during the Second World War.

73. ICA; Report entitled, 'Imperial College at Silwood Park: Department of Zoology and Applied Entomology; Contributions to Developing Countries, 1971–3' (1974).

74. The founding of the RSM coincided with discovery of gold in California and Australia, and much English capital flowed into gold ventures there

and in South Africa, as well as into copper mining in Chile. The City of London became the hub of mining enterprise throughout the world and remains an important financial centre for mining and oil exploration today. Gold, silver and copper were the main focus of the RSM in its early days and continued to be important after the founding of Imperial College. Oil was another early focus. For the pre-Imperial period see Roy MacLeod, "Instructed Men' and Mining Engineers: The Associates of the Royal School of Mines and British Imperial Science, 1851-1920', *Minerva* vol. 32 (1994), 422–39; James A. Secord, 'The Geological Survey of Great Britain as a Research School, 1839–55', *History of Science*, xxiv (1986), 223–75; Robert A. Stafford, *Scientist of Empire: Roderick Murchison and the Imperial Theme in Nineteenth-Century British Geology* (Cambridge, 1989). For some personal histories see Margaret Reeks, *Register of the Associates and Old Students of the Royal School of Mines and History of the Royal School of Mines* (London, 1920).

75. Huxley was addressing a government inquiry and is quoted in Reeks, *op. cit.* (74), 107.

76. For example, N. R. Junner, an 1851 Science Research Scholar from the University of Melbourne, who came to study with Boswell, later became Director of the Geological Survey of Sierra Leone and then of the Gold Coast.

77. See, for example, EC minutes 23 March 1917. On this occasion specimens came from the Gold Coast. Sometimes the colonial surveys took over laboratories in the department. For example, the Nigerian and Gold Coast Surveys in 1921 (see EC minutes, 28 January 1921).

78. The interest extended to the mainland of South America and Illing became an expert also on Venezuelan oil geology.

79. N. L. Falcon, 'Vincent Charles Illing, 1890–1969', *Biographical Memoirs of Fellows of the Royal Society*, vol. 16 (1970), 365–84. See especially, 373–4.

80. According to his obituarist N. L. Falcon, Chairman of the Gas Council, Illing was an optimist on North Sea gas well before most people. Illing ran a consultancy business alongside his college work and many students found work on his consulting projects.

81. A mid-twentieth-century example is William Daniel Gill, a specialist in the geology of the Himalayas and of Pakistan, who discovered a number of oil fields during the Second World War. He was appointed to the chair of oil technology in 1961. See chapter 10.

82. Annual Report of GB, 1929.

83. See also chapter 6.

84. Others at the College were concerned with mines dust. H. V. A. Briscoe, professor of chemistry in the 1930s, was honoured with a gold medal from the Consolidated Goldfields of South Africa for his research on the properties of industrial and mines dusts. His assistants, two women and one man, were awarded premiums of forty guineas. EC minutes, 29 April 1938.

85. During the First World War Holland was Minister of Munitions in India.

86. David Williams, 'History of the Department of Geology' (1958, typescript manuscript in ICA); see statistics in appendix. There are some memoirs written by early students who worked on imperial projects. For example, one by E. J. Wayland, who entered the college in 1907, and went to Uganda in 1919 to set up the African Seismological Observatory (B/ Wayland).

87. Shackleton found the department with its different specialities too diverse and soon moved to the Herdman chair in Liverpool. A non-Imperial graduate of this generation who had a similar career trajectory was G. R. (Rex) Davis, a graduate of Rhodes University who came to Imperial as Professor of Mining Geology in 1967 after a varied industrial career. He had been a mining geologist in Zambian copper mines, chief geologist at the Kilembe copper/cobalt mine in Uganda, and had worked for the Canadian firm, Falconbridge, establishing an exploration company in Southern Africa.

88. SDA/ Report on RSM dinner, 27 May 1921.

89. These connections lasted. For example, R. N. Pryor left to work at the Rio Tinto mines after graduating in 1948. He became Chief Mining Engineer before leaving to become a mining consultant and, in 1968, returned to Imperial as Professor of Mining where he taught mineral production management. See obituary by Marston Fleming, *Nature* vol. 282 (1979), 890.

90. GB minutes, 22 November 1935. Letter reprinted.

91. Annual Report of GB, 1930.

92. Formerly the Anglo-Persian Oil Company. Oil was discovered in Iran in 1908 and the British government acquired a major interest in the then new company. It became British Petroleum (BP) in 1954. BP was privatized and sold for £6 billion in 1990. It is now the third largest oil company in the world.

93. Bone wanted attention to be paid also to fuels used in other parts of the world. His colleague J. W. Hinchley was working on the dewatering of peat and Bone thought attention should be given also to wood distillation, and lignites.

94. Peter Danckwerts, 'Famous Men Remembered: D. M. Newitt', *The Chemical Engineer* (October 1984), 66–7. It was William Bone who made the first major effort to bring Indian students to the college after the First World War and persuaded the Rector to seek more funding from the government to support them. Newitt had served with a Sikh regiment in Mesopotamia and received the MC. He is reputed to have established a record for the largest fish to have been caught in the Tigris. In his article Danckwerts also recounts that Newitt was on an international committee to link the calorie to the Joule. Every time the committee members achieved another significant decimal place they celebrated with a dinner in a European capital. This will have suited Newitt who was something of a gourmet and wine connoisseur. See also EC minutes, 22 November 1918.

95. For some examples, see Annual Report of the GB, 1924.

96. One of Imperial's early chemistry professors, Sir Martin Forster, had left to become Director of this Institute, dying in office in 1923.

97. B/Jackson; letter from Jackson to Lord Murray of Newhaven, chairman of the UGC, 2 May 1968.

98. For example, W. P. Burford, who became Deputy Chief Engineer of the Great Indian Peninsula Railway, and A. E. Hamp, Chief Engineer, Kenya and Uganda Railways and Harbours. The railway tradition continued with some later students; for example, C. E. Salberg who became Chief Engineer of the Assam-Bengal Railway.

99. For example, A. M. R. Montague, Secretary, Central Board of Irrigation in India; C. G. Hawes, a pioneer of soil mechanics in India, H. W. D. Taylor, engineer on Lloyd Barrage, Sukkur, and Sir Alfred Chatterton and Sir Thomas Ward, both major irrigation engineers in India. Imperial College engineers and geologists worked on many major dam projects such as the Volta Dam in West Africa and the Mangla Dam in Pakistan. Tema Harbour in Ghana was another major project involving engineers from Imperial.

100. Freeman taught at the college during the First World War. J. F. Pain led a group of about forty engineers working for Freeman's company, Freeman Fox and Partners, retained by Dorman Long to design the bridge. Among the forty were several Imperial graduates, including civil engineer Gilbert Roberts (later FRS), and his wife, Elizabeth Nada Hoira, a mathematics graduate. Another was K. E. Hyatt. Pain, who later became a Director of Dorman Long, was reputedly a strict disciplinarian. At the time, the bridge across Sydney Harbour included the largest steel arch in the world.

101. XXI Club correspondence; October 1962 bulletin.

102. B/Foot; Brigadier Richard Foot (Rtd.) OBE MC, to Linstead, 7 October 1964. Foot was a graduate of the RSM.

103. Tizard quoted in Peter Clarke, *Hope and Glory: Britain 1900–1990* (London, 1997), 233. The context of Tizard's remarks was Britain's entry into the nuclear arms race.

104. P. M. S. Blackett, 'The Universities and the Developing Countries'; convocation lecture, University of Leeds, 1954. Blackett made many such speeches. For example, 'Science, Technology and World Advancement' was given at the University of St. Andrews on 3 November 1961 (James Irvine Memorial Lecture) and his Hinchley Memorial Lecture given to the Institution of Electrical Engineers, 15 February 1961, 'Science, Technology and Developing Countries'.

105. See Robert S. Anderson, 'Patrick Blackett in India: military consultant and scientific intervenor, 1947–72', Part One, *Notes and Records of the Royal Society* 53:2 (May 1999), 253–73, and Part Two, 53:3 (September 1999), 345–60.

106. In 1995 Abdus Salam gave his Ettore Majorana Erice Science for Peace Prize of £30,000 to the college to provide help for overseas research students.

107. A. Rupert Hall, *Science for Industry: a short history of the Imperial College of Science and Technology and its antecedents* (London, 1982), 59.

chapter eight

Imperial College during the Second World War

A few years ago it would never have occurred to me — or I think to any officer in any fighting service — that what the RAF soon came to call a 'Boffin', a gentleman in grey flannel bags, whose occupation in life had previously been something markedly unmilitary such as biology or physiology would be able to teach us a great deal about our business. Yet so it was.[1]

— Air Chief Marshall Sir John Slessor

Preparing for War

Imperial College staff were associated with some of the big scientific stories of the war — the atomic bomb, radar, operational research, penicillin, DDT, and the development of jet propulsion — as well as with more routine problems related to civil defence, food production and storage, fire protection, munitions and intelligence.[2] Some of this work will be discussed below but, first, some general points will be made about the college in wartime. Despite Chamberlain's optimism, the Munich crisis of 1938 prompted the staff to prepare seriously for war.[3] The College Secretary, Geoffrey Lowry, recalls making war preparations and that the Rector, Henry Tizard, set up a War Emergency Committee which included some senior professors, administrators, and technical staff. Plans were drawn up for the safe storage of records and dangerous chemicals, for bomb shelters, and for maintenance parties to supervise activities in the various college buildings. Also considered was how to secure the perimeter of the college, including blocking access by underground ducts and tunnels that linked some college buildings to others in the area. The Board of Education advised all major colleges in London to draw up evacuation plans, and Lowry negotiated a possible college evacuation with former Rector, Sir Thomas Holland, who was then Vice-Chancellor and Principal of the University of Edinburgh.[4] It was agreed that, in the event of war, the chemistry, physics and engineering departments would move to Edinburgh, and that Imperial staff would use the university facilities, while teaching their own students. In the end these plans did not materialize and most of the college departments stayed in London. Exceptions were the metallurgy department which moved to the University of Wales at Swansea, where it stayed until 1943, and some of the mining courses which were given at the Camborne School of Mining in Cornwall.

Imperial was the only major college in London to remain largely in place. Other colleges were evacuated, including the Royal College of Music, causing one Imperial student to note that he missed the 'multifarious sounds of warring orchestras' when walking down Prince Consort Road.[5] As in the First World War, Imperial College provided a haven for a number of refugees, both students and professors. This began immediately after Hitler came to power in 1933, but only slowly.[6] In 1933 Tizard wrote to Professor Fowler, 'there has been no general resolution of the Governing Body about displaced German Jews, and at present I am dealing with each case at it arises.' Fowler had petitioned Tizard both for a general resolution, and on behalf of a particular woman research student, Charlotte Kellner, who was the first refugee student to be admitted. Near the end of the war she was appointed to a lectureship. Fees were waived for her and for most of those who came later.[7] Sydney Chapman, head of the mathematics department, was another who worked hard on behalf of refugees. Chapman, a pacifist who changed his views after Hitler came to power, was publicly anti-fascist.[8] Anti-fascist political activity of the 1930s was a complicating factor in the college's response to Jewish refugees. The Governing Body was antipathetic to the college taking a political stance and wished to continue promoting student exchanges with German universities. Outspoken professors such as Chapman and fellow mathematician Hyman Levy may well have been an embarrassment in the mid-1930s, though later they were much admired for their integrity. Once the war began, the government began clamping down on enemy aliens. Some refugees were denied access to parts of the college which were declared restricted zones and others were asked to leave the college.[9] By the start of the war Tizard was more firmly behind efforts to bring refugee scientists to Imperial. However, he rationalized his support in terms of utility to the war effort rather than on humanitarian grounds.[10]

Early in 1939 mobilization orders were distributed to college personnel, but they were soon rescinded and followed by new orders. Les Croker, then a technician in the chemistry department, remembered that in the initial confusion over the department's possible evacuation, hazardous chemicals, solvents, and acids were poured down the drain, and that later there was a frantic effort to replace the scarce materials that had been wasted in such dubious fashion.[11] The chemical technology department managed things better by securing its chemicals and inflammable liquids outdoors, in galvanized iron dustbins, and its gas cylinders in an outside shed.[12] Croker also recalled Lowry sending instructions that all windows be taped, then deciding that there would be less blast damage if windows were left open, then countermandering this order when he thought that open windows would let in poison gases.[13] Harry Jones remembered the mathematics department being given rather strange instructions. The old Huxley building had many windows that bore no rational relationship to the floor plan. The tops of the windows were at knee height and the bottoms just below the ceilings of the floor below. Since this building had no air raid shelter below ground, the instructions were to sit on top of the desks at the sound of an air raid siren. In this way, it was claimed, people would be largely out of the path of any flying glass.[14]

The original evacuation plans caused some serious problems.[15] Believing that space would be available, the Governing Body gave much of it away to the government for war-related activities. For example, the military took over the hostel, the two botany buildings, and space in some of the other buildings. The result was not unlike that in the First World War, namely the college was heavily occupied by outside people engaged in the war effort. To help make room, first year students were not admitted in 1939, and undergraduate botany students once again took classes at the Chelsea Physic Garden. Some botanists moved to the college field station at Slough, others to the agricultural research station at Rothamsted. Since Professor Munro had been instrumental in setting up the field station at Slough, the zoologists and entomologists were already well established there. With more botanists moving in, there was some tension between Munro and the head of botany, William Brown, over the use of land and greenhouse space. Tension increased when pressure was placed on the college to give up its land at Harlington, site of the college sports grounds, and of some biological fieldwork. The Middlesex War Agricultural Committee wanted the land for growing vegetables.[16] Since the college had just invested heavily in the construction of playing fields and a sports pavilion, it was reluctant to hand any of the land over. In an attempt to avoid doing so, one of the agricultural chemists, A. W. Marsden, was given the title, Head of the Grasslands Research Association, and was instructed to make it look as though some important science was being done at Harlington.[17] This ruse was only partially successful; about seven acres were given over to the Middlesex committee for growing vegetables, and some further land was given to the Harlington Council for air raid protection trenches. The college was saved from having to give its sports fields over to agriculture when, in 1940, the War Office requisitioned the sports pavilion and stationed a mobile unit of the Scots Guards there. This was because the Harlington property was one of the largest flat areas without buildings near London, and was viewed as a possible landing site for an airborne parachute invasion. The playing fields were requisitioned for the recreation of troops stationed in the pavilion and nearby. At South Kensington, the college squash courts became a temporary prison when members of the Japanese embassy were held there in police custody while awaiting repatriation.[18]

Daily Life at the College

On 1 September 1939, when war was imminent, the maintenance parties were called in. Some members were given rooms in the hostel, others slept in the various college buildings and cooked their own meals. Their job was to patrol the buildings both night and day, and do their best to ensure the safety of staff and students.[19] Several bomb shelters were built, shored up by sandbags, and students and staff retreated to them when they heard the air raid siren. One shelter was built in the basement under the structures laboratory in the City and Guilds (C&G) building. The senior administrators had a bunker under Prince Consort Road. The shelters were liberally supplied with emergency food rations,[20] with buckets of water, fire extinguishers and stirrup pumps in case of fire, and with shovels, picks, crowbars, and hurricane lamps in case there were bomb victims needing to be rescued. After the quiet months of the 'phoney war', the maintenance parties had to be especially vigilant during nighttime bombing raids, watching for fires and damage to buildings. An interesting feature of these parties was that normal hierarchies were disturbed. For example, at the Royal School of Mines (RSM), the professors had to take orders from Bill Guiver, the head cleaner, who was appointed fire captain.[21] Since the maintenance parties lived on the site they needed some recreational activities. These, too, led to new associations. For example, a billiards table was purchased and placed in the RSM building. Sammy Sylvester, another cleaner in the building, gave snooker lessons to the academic staff. He was an exceptional player and Tizard, who enjoyed billiards and snooker, often played with him. Another good snooker player and member of the maintenance party was George Sweeting, the geology department's librarian. Tizard often spent nights at the college, even after he had taken leave as Rector, and is remembered as sitting up half the night chatting with members of staff. All this activity provided the opportunity for true camaraderie across class and professional lines; and, given that there were also women in the maintenance parties, across gender lines as well.[22]

Major bombing of London began in 1940–41, but the college did not suffer serious damage. The RSM was hit and Professor Ritson's rooms were demolished, but not much else. At the time, Ritson was seconded to the Ministry of Mines as Director of Production. A half-ton high-explosive bomb fell through the roof of the C&G College, and through several floors of the building, before coming to rest in the boiler room. It did not explode but broke some water pipes, causing some flooding. F. J. Cutt, a maintenance engineer and plant superintendent, and John Watt a member of the workshop staff, bravely deactivated the bomb.[23] A hut at the Slough field station, used by the agricultural chemists, was hit, as was the boat house in Putney. The boat house had been taken over by a government demolition squad which A. E. Coulson described as 'a pretty tough lot … and none too clean'.[24] Considerable glass damage occurred in South Kensington once the 'doodle-bug' raids began in 1943–4, and Professor Egerton was one of those badly cut by flying glass.[25] A Home Guard unit, which included college students and staff, was stationed in the old Huxley building under the command of Major W. H. Bevan and, at the end of the war, a grateful college put his name forward for a state honour.[26] The unit, which was attached to the King's Royal Rifles, went regularly to Bisley to learn shooting with a variety of guns, and to practice throwing hand grenades. They also carried out guard duty at Buckingham Palace, and for General de Gaulle.

Students and staff were expected to bring gas masks to work and regular inspections were held. It would seem that not everyone took gas attacks seriously since, on occasion, up to one hundred people were without their supposedly mandatory masks.[27] Staff and students were instructed to recycle (salvaging as it was then called) paper, and the college was admonished on one occasion for having had waste paper among its refuse, something then illegal.[28] The number of students at the college declined after the government introduced conscription for eighteen year olds in 1942, and stabilized at about 1000.[29] Despite there being fewer students, the loss of staff who had volunteered for military service, or who had been seconded to do war-related work, was a serious problem. Aspects of this will be discussed below but, by 1943, the problem was acute. Richard Southwell, then Rector, wrote many letters to various government authorities making a case for staff retention.[30] Even the problem of the reservation of women for certain occupations was a worry. On this matter Southwell wrote to Lord Hankey in the Privy Council Office that the college's 'inability to employ even a few more women makes a wholly disproportionate difference to our capacity to carry on'. Here he had in mind not so much teaching staff or research assistants, but secretaries, catering, and cleaning staff. And, as to urgent war work, he wrote, 'we on our side contend, of course, that our work is direct war work and that the Government has asked us to do it; but despite their evident goodwill the Ministry of Labour does not seem able to raise us in their priority list'.[31]

In many ways things went on as usual despite the war. The Governing Body, this time under the chairmanship of Lord Rayleigh, once again played an important role in keeping things going. Student union activities were curtailed but not completely. There were freshers dinners each year with the customary pep talks; and dances were organized, usually by the Imperial College Women's Association. On one occasion, at a dance to raise money for the Red Cross Prisoners of War fund, a cabaret was put on by US troops.[32] Some sporting activity continued, despite the occupation of the Harlington grounds. Surprisingly, foreign students continued to arrive at the college during the war; for example, several from China came to the chemical engineering and aeronautics departments for postgraduate work.

While the utility of science in warfare was not taken very seriously before the First World War, by the late 1930s the idea that scientists could contribute to military success was taking hold; though, as Southwell's letters imply, not quickly enough for him. Before the war academic scientists and engineers throughout the country had been identified, in order that they could be called on should they be needed. University vice-chancellors were asked by the government to register all their scientific staff and, by 1939, a card index of about 80,000 names had been compiled.[33] Imperial College identified the skills of its own staff, with some curious results. For example, a woman who worked in the chemistry department was described as having a 'good physique' and as someone who was good at 'searching and filing literature'. Some heads of departments included the wishes of their staff. One biochemist was listed as having an objection to poison gas work, but as willing to work on food supplies. Alfred Egerton, head of the chemical technology department, wrote of his colleague George Finch that he was an excellent linguist, had a good knowledge of munitions, and had been carrying out research for the Admiralty.[34] Hyman Levy identified himself as a fluent speaker of German and several of its dialects, expert in mathematics as it related to aerodynamics, and on statistics as they related to standardization of production. H. V. A. (Vincent) Briscoe, the professor of inorganic chemistry, noted his own skill in the operation of chemical plant (something he had done during the First World War), and that he had a knowledge of sympathetic inks and secret writing. And William Penney, not knowing what the future held for him, put himself forward as suited to any job involving ballistics and strengths of structures.[35] During the war the staff were called on in various ways, though whether because of the vice-chancellors' list is questionable.[36]

Radar, Operational Research, and Nuclear Science

As in the First World War, the Rector of Imperial College played a very important role in the war effort. Henry Tizard's most significant work was in connection with radar.[37] The problem of air defence was already well recognized by the end of the First World War. After Hitler came to power in 1933 it was seen to be critical. In 1934 Frederick Lindemann, then Dr. Lee's Professor of Experimental Philosophy at Oxford, wrote to *The Times* critical of the RAF's reliance on bomber aircraft. In his view, this implied a first strike policy, something he believed to be wrong.[38] Henry Wimperis, Director of Scientific Research in the Air Ministry agreed. Wimperis had studied aeronautics at Imperial earlier in the century and had been concerned with air defence for many years. It was he who suggested that there be an air defence sub-committee to the Committee of Imperial Defence, and that it be composed of scientists. The resulting Committee for the Scientific Survey of Air Defence had as its mandate to consider the various options for aerial defence and recommend action. The new committee included Wimperis, A. V. Hill, a Nobel laureate in physiology, meteorologist (and former Imperial College student) A. P. Rowe, P. M. S. Blackett, then at Manchester, and Henry Tizard who was appointed chairman.[39] As a scientist, Tizard was not in the same league as Hill and Blackett, but his managerial skills were well recognized and, on this committee, they proved invaluable.[40]

When, in 1934, Wimperis suggested founding the committee he already knew of Robert Watson-Watt's work at the National Physical Laboratory's radio research station at Slough, on what would later become radar.[41] Tizard's success as chairman of the committee was that he was able to take Watson-Watt's ideas, despite facing a range of opposition, and help make them a reality. Opposition to Tizard came from many quarters, most famously from Lindemann (later Lord Cherwell), because it was thought dangerous to put almost all the aerial defence eggs in a single radar basket. Not unreasonably, Lindemann wanted Tizard's committee to back also some other options. Among his suggestions was the detection of night time bombers with infrared radiation, and the parachuting of bombs to explode in front of incoming fighter aircraft.[42] In retrospect the second of these, and some of the suggestions coming from other sources, appear bizarre. For example, one proposing that 'death-rays' be aimed at incoming bomber pilots.[43] The story of radar, and its competitors, has been told many times[44] and will not be repeated here, but it is worth pointing out that Tizard was not the only member of Imperial College to engage in radar-related work. Many were involved both at the college and at the various government research stations. For example, Professor Fortescue and some of his staff in the electrical engineering department began work in 1940 on the development and manufacture of new types of phase-shifting transformers which were an essential part of the radar equipment then being used by the Admiralty.[45] Fortescue together with William Penney, then still in the mathematics department, also ran radio courses for students, many of whom became operators during the war. Others from Imperial worked at the Telecommunications Research Establishment at Malvern and were engaged in developing radar for offensive purposes such as guiding bomber aircraft to their targets.

Tizard, who had been thinking seriously about radar for five years before the war, found the pressure of his committee work increasingly burdensome once the war was underway. In addition to the air defence committee, he was a member of the Advisory Council on Scientific Research and Technical Development, and on a number of its committees.[46] He therefore took leave from Imperial. After Churchill became prime minister and turned to his friend Lindemann for technical advice, Tizard's influence in government circles waned. He resigned from the Air Ministry and was appointed to head a British scientific mission to the United States, setting off for America just as the Battle of Britain began, and radar was beginning to show its mettle.[47] Later Tizard clashed yet further with Lindemann and Churchill when he opposed the strategic bombing campaign of 1942, though not strategic bombing *per se*. He left government service in 1943, resigned from Imperial, and retreated to Oxford when his old college, Magdalen, elected him President.[48] After the war, and with a new government in office, he returned to government service and, as chairman of both the Defence Research Policy Committee in the new Ministry of Defence, and the Advisory Council on Science Policy, became one of the most influential scientists in the country. His loss to Imperial was deeply felt since he was a dynamic Rector and had many plans afoot before the war began. James Philip, recently retired from the chemistry department, and a much loved college figure with a strong sense of duty, stepped in as Deputy Rector when Tizard took his leave of absence. Unfortunately, Philip died suddenly in 1941. Lowry, who had been away on military service, was then released back to the college, and together with chemist, H. J. T. (Tom) Ellingham, who had acted as College Secretary in his absence, ran the college on a day to day basis until Richard Southwell was appointed Rector in 1942.[49]

Blackett, who was to come to Imperial in 1953, was aligned with Tizard both on the matter of radar, and on strategic bombing. Because of this he, too, lost influence in matters of air defence after Churchill came to power. But Blackett found other ways to help the war effort, and made a major contribution to the field of operational research.[50] The story of Blackett's role in operational research is not really an Imperial story since Blackett was then still at Manchester.[51] In the war setting, operational research meant finding ways of making the best use of military equipment and personnel by means of the careful analysis of data. Or, as Charles Goodeve put it, using 'quantitative common sense'.[52] But military chiefs were doubtful that anything very useful would come from a team of boffins.[53]

While Blackett was later to oppose the acquisition of nuclear weapons by Britain, he was involved in wartime discussions leading to the making of the atomic bomb.[54] In a 1939 letter to *Nature*, Otto Frisch had outlined his and Lise Meitner's understanding of some earlier experiments in nuclear fission. In the same year a letter from Frederic Joliot focussed on the excess neutrons produced during fission. The letters led to increasing speculation that the construction of a bomb of unprecedented magnitude was possible.[55] One of the speculators was the head of Imperial's physics department, G. P. Thomson. In 1939 most British scientists were sceptical that an atomic bomb could be constructed; even those who thought it theoretically possible doubted its practicality. Thomson was not a sceptic and informed the government that something should be done immediately to secure uranium supplies. Since radium had earlier been extracted from large quantities of uranium-rich ores found in the Belgian Congo, he thought that the spent ores were likely to be stored somewhere in Belgium and should be located. Tizard was a sceptic, but he supported the idea of the quick and preemptive purchase of uranium ores. As it happened the Americans were ahead of them on this, and much of the ore formerly owned by the Belgian Union Minière was already in the United States. For his own work, Thomson managed to secure one tonne of uranium ore from stocks intended for British potteries.[56]

Inspired by Joliot's letter, Thomson, returned to work in his laboratory at Imperial with a team of scientists including P. B. Moon, J. L. Michiels, C. E. Wynn-Williams and M. Blackman.[57] Morris Blackman has described some of this work in his memoir of Thomson. Blackman's group was given the task of studying the multiplication of slow neutrons in a 'pile' using natural uranium oxide and paraffin as moderator, an experiment that Blackman later remarked would 'send present day safety officers into hysterics'.[58] Thomson and his co-workers came to the conclusion that a chain reaction would not be possible using simple uranium ore, but that it would be were the ore to be enriched with U235. Frisch, by then working in Birmingham with Rudolf Peierls, showed that a chain reaction would be possible also were metallic uranium to be used. He and Peierls calculated what the critical mass of uranium needed to start a chain reaction would be — much smaller than had previously been imagined. They shared this information with Tizard who, in turn, shared it with Thomson and a few other scientists. As a result of all the work showing that a bomb was possible, Thomson formed what was known as the Maud committee.[59] The committee, which he chaired, met at the Royal Society. Its members included Blackett, Lindemann, P. B. Moon and a number of other scientists, notably John Cockcroft and Mark Oliphant.[60] They set to work thinking about an atomic bomb and how it could be made. By 1941 most of the committee members agreed that a bomb could be built, but disagreed on the timetable needed. On this point Blackett was more conservative than the others and was firmly of the opinion that Britain would be unable to build a bomb in a period of less than five years. Tizard, when consulted, agreed with this minority view. There was much activity among the Maud Committee scientists to get the government to take the new ideas seriously. Lindemann reported to Churchill and received the now famous reply, 'Although personally I am quite content with existing explosives, I feel we must not stand in the path of improvement'.[61] Amazingly, despite Churchill ordering a stepping up of activity, leading to the involvement of some British industries, and of many more scientists and engineers, secrecy was largely preserved until the first bomb was dropped on Hiroshima.

A new 'Directorate of Tube Alloys', the clever code name for the British bomb effort, was instituted by the DSIR, behind the back of the Maud committee, and negotiations began with the Americans. Collaboration was slow at first but, by 1942, British scientists began to migrate across the Atlantic. Tizard's earlier mission to the United States (British Technical Mission) during which many scientific exchanges occurred, had been very successful and had opened doors for a later mission headed by James Chadwick.[62] Thomson, too, had made connections with nuclear scientists on his visit to the United States in 1941 but, in 1942, he disengaged from the atomic bomb project after a brief period as head of the British Scientific Office in Montreal. He remained Scientific Advisor to the Air Ministry, and continued his neutron research at Imperial. Roger Makins, the future Lord Sherfield, future chairman of the United Kingdom Atomic Energy Authority, and future chairman of Imperial's Governing Body, was seconded from the British embassy in Washington to help Chadwick's mission. Much of the British effort was centred in Canada but, after the Quebec Agreement of August 1943, close cooperation with the Americans began. After the war Thomson, who had been working on thermonuclear reactions at the college (he was beginning to think of the H-bomb), considered it wrong to continue doing secret work at a university, and transferred his research team to Aldermaston.[63] In 1952 he left Imperial to become Master of Corpus Christi College, Cambridge.

That British nuclear work was being conducted in the United States and Canada meant that many of Britain's brightest scientists were recruited to work there. Two among the Imperial College contingent were William Penney and Geoffrey Wilkinson. Penney had been a mathematics student at the Royal College of Science, and had doctorates from both Imperial and Cambridge. In 1936 he was appointed assistant professor of mathematics at Imperial, but was seconded part-time to the Ministry of Home Security when the war began, and to the Admiralty in 1940. For these bodies Penney worked on a variety of projects, including working with others from the college on fire retardant foams. But it was his theoretical work on the effects of explosions that brought him to the attention of Tube Alloys. He had also impressed military people with his ability to explain technical matters, such as what would happen to a Mulberry Harbour when hit by blast waves coming from explosions at different locations.[64] Penney joined the Manhattan project late in the war — in 1944. At first, Lord Falmouth, chairman of the college Executive Committee, refused a request from Sir Edward Appleton of the DSIR to release him. Penney was one of the few people left teaching mathematics at the college and it was unclear to the governing body how useful he could be to the war effort.[65] Falmouth replied to Appleton, 'we could suggest the names of several other people who, we think, would be equally suitable for the work you have in mind.' But Penney was released, became very popular with the Americans, and won the respect of General Groves, head of the Manhattan Project. In his memoir Groves stated 'throughout the life of the [Manhattan] project vital decisions were resolved only after discussion with the men I thought were able to offer the soundest advice'. He then listed five people whom he regularly consulted, Penney was among them.[66] Penney's role was to calculate the effects of blast and shock waves. Apparently he carried all kinds of mathematical shortcuts in his head which enabled him to do the work more quickly than anyone else. After Hiroshima he examined the effects of the blast, and from them calculated the yield of the explosion, a calculation that has never been seriously challenged.[67] He became something of a legendary boffin because of his astounding memory, and because he rarely wrote anything down. After the war the authorities insisted he stay in government service until his knowledge was passed on, at least to the Americans. Turning down a chair at Oxford, Penney worked for the Armaments Research Establishment at Fort Halstead, and then at Aldermaston where he became the 'father' of Britain's own atomic bomb. He later followed Makins as Chairman of the United Kingdom Atomic Energy Authority and returned to Imperial as Rector in 1967.

Wilkinson was a student when the war began. After graduating in 1941 he immediately joined the Tube Alloys project at Montreal and at Chalk River. He worked together with John Cockcroft — both had earlier attended the same school in Todmorden, Yorkshire — as well as with the spies Allan Nunn and Bruno Pontecorvo.[68] Wilkinson's role was the identification of fission products, work that helped enlarge his already good knowledge of basic descriptive chemistry. Through this work he met Glen Seaborg, with whom he worked at the Livermore Laboratory in California for a brief period after the war. Wilkinson had wanted to return to Britain immediately after the war ended, but his old professor, H. V. A. Briscoe, advised him to first get out of nuclear chemistry where he would forever be under the thumb of physicists. Wilkinson took the advice, accepted an assistant professorship at Harvard, and began working on organo-metallic compounds. In 1955, on Briscoe's retirement, he was appointed professor of inorganic chemistry at Imperial. He continued with the work begun at Harvard, and was awarded the Nobel Prize for chemistry in 1975 (jointly with E. O. Fischer).

The Aeronautics Network

It is interesting that scientists at Imperial were drawn into both radar and atomic bomb work because of a network which had its roots in aeronautics and the first course in that field offered by Professor Dalby at Imperial in 1909. As discussed in chapters four, five and six, aeronautics brought Imperial people into close contact with those working at the National Physical Laboratory, with members of the Royal Flying Corps (later RAF), and with people at the Royal Aircraft Establishment at Farnborough. Bonds that were formed during the First World War remained strong, and were reinforced with the threat of war in the 1930s. Young men who had performed well during the First World War were in positions of authority by the start of the next war and, by then, a complex network of personal connections had been formed. Among those associated with the network were former student Henry Wimperis at the Air Ministry, Tizard, Thomson, Blackett, and future Rectors Richard Southwell and Roderic Hill.[69] All were connected also to others in important government, military and technical positions, and they were all immediately drawn into the war effort.

A new generation of aeronautical experts entered the mix during the Second World War. For example, A. A. Hall who came to Imperial as professor of aeronautics during the war, though continuing part-time work at the RAE on the arming of aircraft. In 1950, he wrote a memoir in which he discussed both the technical developments of the war and the associated personnel.[70] He noted that by the end of the war about one-third of the national effort was devoted to military aviation. He claimed that the Spitfire and the Lancaster bomber were the most effective aircraft of their type and noted also the many new inventions coming forward in the aviation industry.[71] Most notable were jet propulsion, and turbine-propellor driven transport aircraft. In his view all these developments owed much to people trained at Imperial College and at Cambridge. Up until the 1950s, these two institutions were alone in offering specialist university degrees for those entering the aviation industry. One of the earliest Imperial graduates to move into aviation was Frederick Handley Page, an industry pioneer. He retained a close connection to the college throughout his life and was a long-serving member of the Governing Body. Already during the First World War, he employed several graduates. Later he employed many more. Another notable former student, Harold Roxbee Cox, had a major career in both government and the aviation industry.[72] He provided work for several Imperial graduates at Power Jets Ltd. during and after the Second World War.

C. H. Lander, the head of mechanical engineering, brought Owen Saunders to the college from the DSIR Fuel Research Station where they had both worked before the war. Saunders, appointed to a readership, was very much under the influence of Tizard who encouraged him to work on the performance of aircraft piston engines at high altitudes. Saunders pioneered the use of liquid oxygen injection into the air intake of fighter aircraft, and was the first to make it actually work in practice. His work on Spitfire engines helped increased their mileage. The field of heat transfer, which Lander brought to Imperial, was given a major boost after gas turbines became a reality. Soon Saunders was working on gas turbine engines for aircraft and his work also brought him into close association with Frank Whittle and Harold Roxbee Cox at Power Jets Ltd.[73] Saunders also helped develop gas turbines for ships, as well as pioneering rocket motors using hydrogen peroxide fuel. He was seconded to the Directorate of Turbine Engineering in the Ministry of Aircraft Production in 1942 and worked for the Ministry until the end of the war. Saunders was a member of a number of top secret committees including, with Tizard, of the Big Ben Committee which focussed on V2 rockets and guided anti-aircraft projectiles. Related to this, Tizard and Saunders were major figures in the development of the Woomera Rocket Range in Australia. Saunders served also on the Admiralty's chemical advisory panel on boilers and furnaces, and was chairman of the Admiralty panel on marine propulsion. Lander chaired a committee looking into combustion chambers of jet propulsion gas turbines for a novel wartime use, namely the dispersal of fog over airfields — a project known as FIDO.[74] Much of the work directed by Lander and Saunders was designated secret, and entry to many areas of the mechanical engineering department was restricted.

There was a close connection also between aeronautics and meteorology.[75] Additional facilities came to meteorology at the start of the war since its importance for navigation and other purposes was well understood. P. A. Sheppard came from the Meteorological Office at this time and was appointed Reader. But, shortly after, both he and the professor, Sir David Brunt, were seconded back to the Air Ministry in order to run a training school at the Meteorological Office. Brunt was appointed Principal Technical Officer at the Ministry and, from 1940 onwards, appears to have carried out a number of secret missions, mostly across the Atlantic.[76] The crash training program at the Meteorological Office meant that there was a glut of trained meteorologists after the war, and the department at Imperial fared less well as a result.

A. J. S. Pippard, head of the civil engineering department, was yet another person with aeronautical roots. Pippard had engaged in airship and aeroplane structures work during the First World War and, shortly after, became a lecturer in aeronautics at Imperial. He then moved to professorships in Wales and Bristol before returning to Imperial as Professor of Civil Engineering in 1933. Unlike in the First World War, Pippard felt his skills were wasted in this war. He sat on a number of committees, including the Structural Precautions Committee of the Ministry of Home Security, concerned mainly with civil defence. Well before the start of the war this committee considered what might be the number of civilian casualties inflicted from the air. Pippard was shocked by the official estimates, numbers actually reached only by allied bombing of German cities. At the start of the war, Pippard was seconded to work at the Ministry of Home Security's Civil Defence Research Unit at the Forest Products Research Station at Princes Risborough, but complained that not enough work was sent his way. In his memoir he mentions the boredom, and his annoyance at having been taken from his work at Imperial. After a year away he was allowed to commute to London to teach his courses. He believed that civil engineers were undervalued and that their skills were not being put to proper use during the war. In this connection he recorded that a near riot occurred when C. P. Snow visited Imperial. Snow, then head of scientific personnel for the Ministry of Labour, came to the college to 'direct' young graduates into suitable national service. He told young civil engineers that they were not wanted in the Royal Engineers, and that each county needed only one civil engineer on its staff. He upset many by saying that they should join the army in non-engineering units.[77]

Pippard thought that something exciting might at last happen when Barnes Wallis showed up at the college and told him of a scheme to wreck the Möhne dam. He wanted Pippard's advice on how a large piece of masonry might be destroyed by means of an explosive applied at the water face. Pippard agreed to conduct some experiments at Princes Risborough. But, while experiments were carried out there, and at a Welsh reservoir, Pippard was excluded from them — something he much regretted.[78] One rather harebrained scheme in which he played a marginal role was code named Habbakuk. Like the story of the dam busters this, too, has been told many times.[79] The project, in which Richard Southwell played a greater role than Pippard, was the brainchild of inventor Geoffrey Pyke. The idea was to build a floating runway for aircraft needing repairs or refuelling. It involved separating a large flat block of ice from the arctic shelf. One such was towed to Corner Brook, Newfoundland. There the ice was strengthened using Pyke's invention which involved the micro-reinforcement of ice with wood pulp. Large wooden rafts were to be used as moulds and the rafts were then to be fused together. The resulting giant structure was to be towed to wherever it was needed. Pippard reported unfavourably on the project which was eventually dropped.[80]

Other Imperial College civil engineers perhaps felt more fulfilled in their war work. S. R. Sparkes worked on the defence of vital industrial plant, and of dams in India; C. M. White who, after the war, was promoted to the new chair of fluid mechanics and hydraulic engineering, worked on the design of pneumatic breakwaters and Mulberry harbours; and Professor Clements worked for the Ministry of Home Security where he was Deputy Chief Engineer, responsible for the construction of deep bomb shelters, and for camouflaging reservoirs. He retired from Imperial at the end of the war.

Chemistry and Chemical Technology

As in the First World War, the chemists and chemical engineers were in great demand for both the normal and clandestine war effort. D. M. Newitt, who was to become professor of chemical engineering after the war, was officially appointed Director of Research at the Inter-Services Research Bureau.[81] In fact this was a cover; he was Director of Research for the Special Operations Executive (SOE) from 1941–5, responsible for the production of a range of weapons (including some that were bacteriological), poisons, hidden radio devices, methods of forging documents, camouflage and disguise, etc.[82] This work, strictly clandestine, was carried out at a number of secret locations, including at the Natural History Museum which housed a camouflage workshop run by an expert in film costumes and disguises.[83] 'Continental' personal effects, such as toothpaste and cigarettes, to be used by people smuggled into occupied Europe, were designed there. There was a small 'continental' clothing factory around the corner on Queen's Gate where clothes were not only made, but also 'aged'. Other production units were sited elsewhere.

Newitt, himself, worked on a range of sabotage devices. For example, on sophisticated time fuses for explosives, and the fitting of timed bombs on corgi motorcycles and other such delivery vehicles. His group also made contrivances for derailing trains, and contributed to the design of midget submarines to be armed with explosive devices. One of the submarines is said to have been of use in destroying Norway's heavy water plant and another is said to have played a role in damaging the *Tirpitz*.[84] It is not clear how much this kind of work really contributed to the overall success of the war; but, according to Peter Dankwerts, Newitt had a schoolboyish love of guns and explosions and much enjoyed his wartime job. When working on a device for disabling time bombs by the questionable means of firing bullets at them from the correct angle, Newitt caused major damage to the Park Lane house where this work was being carried out.[85]

Newitt's colleague, G. I. Finch, a scientific advisor to the Ministry of Home Security, also worked on incendiary bombs, though not for clandestine use. Rather he, together with William Penney, sought methods of protection against such bombs. Finch also developed a flash bomb used by the RAF for night photography. Another colleague, R. P. Fraser, who had earlier developed a high-speed camera for the study of flames, headed a small research group at Imperial working on new weapons. Interested in fuel flowing at supersonic speeds through nozzles, his work had application in rocketry, flame throwing, and gun development. Fraser's mini rockets, weapons such as the 'crocodile' and the 'wasp', were tested in the department's wartime 'flying-bomb unit' and at a drained reservoir near Staines. The college unit was housed in a Nissan hut erected on the Rector's lawn, and named the Jet Research Laboratory.[86] Fraser recovered a doodle-bug (V1 rocket) from the Staines reservoir and soon figured out how its propulsion system worked — setting alight a poplar tree on the Rector's lawn in the process.[87]

Chemists H. V. A. Briscoe and H. J. Eméleus, too, began the war by working on incendiary weapons and on methods of sabotage.[88] W. C. Gilpin, a student who worked under Briscoe, remembers making pyrotechnic material for decoy fires and trying out chemical mixes at a film studio which had a fire-making machine. He noted that on one occasion they created an explosion and blaze so great, 'that any aircrew seeing it might well be excused for thinking they had hit the Royal Arsenal'. But the authorities were not satisfied and wanted the flames to be more multicoloured, so various metal salts were added to the mix.[89] As in the first war, Briscoe was engaged in some secret work, though little is known about it. I suspect, but cannot prove, that he was working alongside Newitt in some of the SOE technical work. He is said to have secreted equipment into the Royal College of Science building that picked up radio communications from various South Kensington embassies; and he added a fictitious woman to his staff, with instructions that if anyone came looking for her the War Office was to be notified immediately. Les Croker remembers Briscoe removing seals from official documents with dental floss so that they could be replaced without detection.[90] D. H. R. Barton began working with Briscoe after he gained his PhD under Professor Heilbron in 1942. Barton is said to have devised a method of sending secret messages by having them invisibly tattooed on the skins of East Asians, something he demonstrated on a Chinese woman volunteer. He also worked for Briscoe on the separation of hydrogen isotopes as halides. Barton was later seconded to Albright and Wilson in Birmingham where this work continued.[91]

More mundane work was also carried out. In the chemical technology department much of it related to domestic heating and ventilation, and was sponsored by the Ministry of Works anxious to conserve coal supplies. Interestingly, as in the First World War, it was the head of this department who was most insistent on keeping his staff and students from military service. Alfred Egerton held the view that trained people should be used in their technical capacities, that research teams were especially valuable in times of war, and that they should not be dispersed. Egerton, who later wrote about his department's activities during the war, was able to retain some major research groups, including his own.[92] His own research group focussed heavily on the liquefaction and storage of methane and succeeded in making a working carburettor for methane. It was used by a Midland bus service during the war. In pursuing this line of work, Egerton founded a low temperature technology group at Imperial that became renowned.

The professor of organic chemistry, Ian Heilbron, was a small lively man said to have had a great capacity for projecting his personality. He spent much of the war as an advisor in the Cabinet Office and left the day-to-day running of his section to the assistant professor of organic chemistry, E. R. H. Jones.[93] Each night, however, Heilbron returned to issue instructions and to sleep in the department. Jones ran the organic chemistry teaching laboratory, also designated as the place where poison gases would be identified were they to be used by Germany — they never were, at least not in warfare. Research students were put to work examining various gases that could be used, and in seeking possible protective agents. Because the work was secret, they had to wait until after the war before publishing their theses.[94] Jones and Eméleus were both appointed Senior Gas Advisors for London, and Jones ran courses at Imperial for prospective gas examination officers, instructing about fifty people a week who came for the week-long courses. Practice drills were performed outside, and on the department roof; but only after Jones had assured the College Secretary that the drills would cause no 'risk or inconvenience' to people working nearby.[95] Another organic chemist, Leonard Owen, later a professor in the department, synthesized the anti-arsenical drug, BALINTRAV — the start of his work on organo-sulphur chemistry which forever filled the building with its characteristic odours.[96] Owen also worked on nitrogen mustard gases which fortunately were not used in war, but later proved useful in chemotherapy. Some others from the chemistry department worked at Porton Down making the arsenicals for which Owen was seeking the antidotes.[97]

In addition to his work in the Cabinet Office, Heilbron was Scientific Advisor to the Department of Scientific Research at the Ministry of Supply and, later in the war, Scientific Advisor to the Ministry of Production. He was on many committees and, like Jocelyn Thorpe in the first war, directed work at Imperial on the synthesis of drugs that previously had been imported from Germany. His research team worked also on several important chemical problems of the period: the industrial synthesis of vitamin B1, and of vitamin A; the purification of penicillin (a very difficult problem);[98] and, as will be discussed below, the production of DDT. The vitamin A work was given a major boost by the work of his student R. A. Raphael. As E. R. H. Jones remembered it,

> Ralph carried out the only [experiment] ... in my whole experience the result of which I can still visualise. We put away a colourless solution one evening and the next morning it was bright yellow. But it was the discovery of a spectacular isomerization which made possible the synthesis of vitamin A and the carotenoids by Otto Isler in Basel.[99]

Biological Sciences: Food Supplies and DDT

James Munro, the professor of applied entomology, worked with Heilbron on DDT late in the war, but first he helped in conserving scarce food supplies and other organic commodities. In 1938 Tizard invited Munro to lunch with Sir William Beveridge, then seconded to Whitehall from his position as Master of University College, Oxford. Beveridge told Munro that he had been asked by the government to look ahead to the securing of food supplies in case of war, and that he was especially concerned with grain and flour supplies. He wanted Munro's help.[100] Munro then approached the DSIR suggesting that an intrusive examination of food stores was necessary, since many businesses hid their food pest problems. He appointed a member of his staff, G. V. B. Herford, to conduct a grain survey, and he kept lobbying the government until it finally came round to his point of view on inspections. As he put it, 'after two years of scepticism regarding the danger of insect infestation, during which time the Ministry of Food has disregarded advice given by me…[it] is now faced with a very serious problem'.[101] Munro had his own problems. In 1940 the DSIR took over the college field station at Slough.[102] Further, the DSIR appointed Herford as Director of what then became *its* pest infestation laboratory. Munro resisted the takeover of his research empire, but to no avail. Many of the college's zoologists and entomologists together with Albert Page, a chemical pesticide specialist, moved to work with the government scientists at Slough for the duration of the war. This was possible because the zoology and applied entomology department had very few undergraduate students at the time (biology students were not, as a rule, protected from conscription) and the staff were not needed in South Kensington. The Ministry of Food also took over the Imperial Hotel at Colwyn Bay in Devon, and some Imperial staff carried out research on insecticide spraying there.[103] Zoologist Humphrey Hewer was appointed Chief Rodent Officer for the Ministry of Food.[104]

Munro was somewhat disgusted by the turn of events (with some justification since he probably knew more about food infestations than anyone else). For the rest of the war he devoted much time to mosquito control. In this he had the support of Professor Egerton who was advising the cabinet, and was well aware of problems having to do with malaria affecting the troops. Further, Heilbron was pressuring the government to support research into a substitute for the natural product insecticide pyrethrum, the spray form of which had been developed by an earlier generation of Imperial College scientists. Japanese and Macedonian supplies had been cut off, and those from Kenya were proving insufficient. Heilbron put his team to work analysing a Geigy product called Cesarol, and figured out that just one of the two isomers of its principal ingredient was an active insecticide.[105] Arrangements were made with Geigy for the compound, later known as DDT, to be made under licence in Britain. Teams were then sent to tropical countries to test its efficacy. Munro had a team working in British Guyana, where sugar cane workers were seriously affected by disease-carrying mosquitos. The spraying of DDT in mosquito infested areas controlled malaria and yellow fever to a degree never before experienced. While this new line of research led to tension between Munro's team and the DSIR, Munro was able to get funding from the Colonial Office and the Agricultural Research Council (ARC), funding which continued after the war ended.

Botanists, too, were much concerned with the food supply. In this connection G. E. Blackman was supported by the ARC to work on herbicides and weed control.[106] Blackman also cooperated with people at the Royal Horticultural Society Gardens at Wisley on the growing of certain plants for insecticide production. After the war, some of the selective weedkillers developed by the botanists had a major effect on plant husbandry worldwide.[107] Like Munro, Blackman lobbied the government and the ARC, making the point that biological knowledge was not being applied to the war effort as it should. For example, he pressured the government to cultivate soybeans for protein production. Some research on their cultivation was carried out at Imperial, and the government seriously considered growing them on a large scale. But, since many food shipments were still getting through, the idea was shelved. Blackman was Secretary of the Biological War Committee, a committee sponsored by the British Ecological Society, the Society of Experimental Biology and the Association of Applied Biologists. Several other members of the Imperial College staff were members, including O. W. Richards and Professors Chibnall and Munro. The committee met regularly and had as its mandate the sending of useful suggestions to government departments. For example, methods of destroying migrant starlings, of combatting warble fly which attacked cow hides, of extracting medicinal agents from native plants, and a method for identifying good weed seeds for poultry feed. They also suggested which varieties of rose hips should be collected for their vitamin C content. Syrups made from these were issued to parents who were supposed to give their children a daily dose.[108] There were problems for the botanists in retaining experimental plant stock that had been built up before the war. The head of the department, William Brown, managed to move over 5000 specimens preserved in alcohol to Slough, and to conserve living specimens that needed to be regularly propagated.

Munitions

As in the First World War, the mechanical engineering department's workshop was turned into a munitions factory. After Dunkirk, Lander was convinced that the war situation was so urgent that the workshop could no longer be retained for student instruction. A member of the Governing Body, he persuaded his fellow governors to convert it for the manufacture of munitions. It was decided to ally with munitions firms rather than contracting directly from the government, a decision made in consultation with the Ministry of Supply.[109] The workshops of other C&G and RSM departments joined with mechanical engineering to form a small specialized factory at the college. Placed under the direction of Professor Witchell, with the various workshop superintendents acting as managers, its first order was for close to four thousand fuse keys for anti-aircraft weaponry. In addition, gun components, gun carriages, mortar parts, etc. were made. About forty people worked there, including three Belgian refugees and eleven women. One of them, Miss James, was a 'woman of advanced years' who wanted to 'do her bit' for the war effort.[110] In the workshops students, staff, and some government employees worked together, for up to 72 hours a week. The students were paid 9½d per hour and, over the course of the war, the college received £78,136. 11s. 1d. for this work but made little profit. There was also a physics workshop factory under the direction of Professor Pollard. Pollard, recently retired, volunteered to stay and supervise the manufacture of munitions. In 1943 eight men and three women working under his direction were making moulded plastic pinions to replace metal in the clock mechanisms of time fuses, de-icing equipment for aeroplanes, and electrical generators for Hammersmith Hospital. They also made some precision optical devices for which Pollard's optical skills were important.

The Governing Body believed it had a duty to give bonuses to members of staff with increased workloads due to the war, especially to those working in the munitions workshops. A remuneration scheme that the government had in place for the civil service was considered but, as in the First World War, it was rejected because it discriminated against women who were given lower bonuses than men. The Governing Body at first decided to seek a legal opinion on the clause in the charter against sexual discrimination but, in the end, appear to have left things to the Rector's discretion; there is no evidence of any discrimination in war bonuses.

Southwell Comes to Imperial

R. V. Southwell was appointed to succeed Tizard in 1943.[111] Southwell had created a small but good engineering science department at Oxford — with staff who had largely moved there with him from Cambridge. On arriving at Imperial Southwell, said to have been 'amiable, friendly, clubbable and approachable', began to think of post-war reconstruction.[112] Already by 1943 he set up a committee to consider how to celebrate what he saw as the college's 'centenary' in October 1945. He also wanted to use the occasion of what was actually the centenary of the Royal College of Chemistry to launch a college appeal.[113] Not everyone supported the planning of a celebration before the war had ended; the pessimists wanted to wait for the 1951 anniversary of the Royal School of Mines. But Southwell went ahead and received a promise from King George VI that he and Queen Elizabeth would attend the celebration. Even before the appeal was launched some new funding came to the college. For example, in 1943 the Cement Makers Federation funded a chair in civil engineering for an initial period of ten years. But Southwell's main aim was to raise money to improve the corporate life of the college and, indeed, that is where much of the appeal money later went.

Clearly Southwell had entered a very different environment from the one he left in Oxford. There he had been able to experience not only life as the fellow of a college, he had also headed a small and congenial engineering sciences department. To help 'improve' social life at Imperial, Southwell moved into the warden's flat in the hostel after having it enlarged and modernized — something possible even in wartime. From there he introduced a number of changes to the corporate life of the college which are discussed further in chapter eleven. In thinking about the academic side he was influenced by a 1944 visit to the Massachusetts Institute of Technology (MIT). There he signed an agreement with Karl Compton, President of MIT, to develop a strong relationship between the two institutions after the war, including the exchange of postgraduate students and staff.[114] Southwell noted that the MIT administration was made up largely of seconded professors, a pattern virtually unknown in Britain at the time. He admired the fact that MIT's huge endowment allowed it to act with little outside interference. But he preferred the British methods of higher education and came back believing that Britain could take a lead in technical education after the war, despite financial hardship.

Just as in the First World War, administrators at the college were frustrated by civil servants who, they believed, lacked an understanding of the importance of college work. That the training of young scientists and technologists would contribute not only to the war effort, but also to the postwar recovery was not, it was claimed, fully appreciated. Planning for the recovery moved into high gear in 1944 when the Ministry of Education appointed a special committee to look into the nation's postwar needs in higher technological education.[115] Despite the committee's recommendation that there be some prioritization of science and technology education, the tug of war over human resources was fierce.[116] Southwell claimed that Ministry of Labour bureaucrats did not understand the national need for technically trained personnel. He tried to persuade them to exclude young scientists from being conscripted after the war, even though colleges and universities did not have the capacity to receive all the students who were expected to return from military service.[117] Professor Heilbron wrote a private letter on the same subject to Sir Stafford Cripps, asking him to ensure that the 'needs of our industries for graduates in science and technology two years hence have been properly surveyed in the light of events since May 1945'. Sending qualified young men into the army, he wrote, would result in a 'ridiculously inefficient use of an institution' devoted to 'supporting British industry at this critical time'.[118]

Southwell thought seriously of closing down the biological sciences at the end of the war, indicative of how little he understood their importance, or their history at Imperial College. He shared the more general 1940s belief in the overriding importance of the physical sciences. His reasoning for closure was varied. For a start very few students came to the biology departments during the war, and none had entered in 1944. This meant that the departments were not entitled to any deferments from military service for students wishing to enter in 1945. To be fair, Southwell did complain about this, but nonetheless thought that the biological sciences were no longer viable. A. C. Chibnall, the professor of biochemistry, who for much of the war had been absent carrying out his vegetable protein research at Dartington Hall in Devon, had left for a chair at Cambridge. Professors Brown and Munro were due to retire in the early 1950s, and Southwell was disinclined to give limited college space to their departments. A major rubber research institute, planned for the top floor of the West Beit building, had fallen through; and the DSIR, having taken over the Slough field station, was unwilling to give it up. Southwell thought that the pre-war plan to move the college field station to Harlington should be dropped. An engineer, perhaps he believed that Imperial should be principally an engineering college. Fortunately Professor Munro was made of sterner stuff. He insisted that there was a zoology and applied entomology department, and that space would have to be found for it in South Kensington, even if not in its old location in the Huxley building which, during the war, had been occupied by the army, the meteorology department, and the statistical section of mathematics.[119] Munro also insisted that the college not give up its Slough field station without a fight, and that if the government wished to hold on to it, Imperial should be given the funds to buy more land. His insistence paid off. His department was given the West Beit building which had been occupied by biochemistry and some areas of applied botany before the war. And, as is discussed in chapter ten, the government eventually agreed to pay compensation for the Slough field station.

By the end of the war land for the new Royal College of Art building had been found outside the island site, and the prospect of Tizard's 1937 plan being realized became more likely. Things looked hopeful and, reporting on the year 1945–6, Southwell stated, 'the college is not likely ever to have a more eventful year'.[120] His long-planned celebration, held in October 1945, was a great success. As he had optimistically predicted, the war had recently ended and people were in celebratory mood. The King and Queen arrived at the Albert Hall, for the ceremony which was designed to boost the image of the college. Accompanied by uniformed

Imperial College Senior Training Corps Cadets, the royal couple moved in procession around the main corridor before mounting the staircase. At the top they stood under spotlights while those in the Hall below sang the National Anthem. The promenade area was filled with students and the seats were occupied by members of staff in academic dress, visiting dignitaries, their wives and husbands. The King glowingly referred to the contributions the college had made to the war effort, and speeches were made outlining future plans and remembering past glories. In the evening there was a dance in the Union, attended by the King and Queen. There was an open house for two days and, according to Southwell, 'we could have filled the college for a week'.[121]

The overall success of the various events gave Imperial a major boost in confidence and increased public awareness of its work. Major donations came in, including one for a chair in chemical engineering from Courtaulds Ltd. Unilever donated funds to help with improving the college's corporate life and, when the appeal closed in December 1945, its £200,000 target had been reached. Not all was plain sailing however; there were major problems with students and staff being demobilized, large numbers of resignations and retirements, and the other readjustments necessary after the war. With MIT in mind, Southwell reinstated the Deanships in the RCS and RSM and persuaded the professors that the Deans should be elected for short periods of office, as had been the case in the C&G since Dalby's retirement. By 1947 the student numbers had rebounded to 1326 and included 61 women.[122] That year saw also the return of some foreign students and students from the empire.

The earlier celebrations on VE Day had been boisterous and somewhat reminiscent of Armistice Day at the college in 1918. Just as at the end of the First World War, old college identities mattered. Three separate war memorials were unveiled in the months that followed. The casualty lists were proportionally much shorter than in the previous war, and the mood was to look forward and forget. As elsewhere in society, the atrocities of this war were not fully acknowledged until later in the century. Imperial looked back on its contribution to the war effort with some pride. The war had helped to identify its strengths, especially those in chemistry, chemical technology, applied entomology, physics, and mechanical engineering. The work carried out, and the wide range of expertise shown, established the college, more definitively than before, as Britain's leading educational institution in science and engineering.

But peacetime brought new challenges and the college had to reposition itself in light of changing economic realities. Southwell retired after six years as Rector and was knighted on his retirement. In 1948 there was a new Rector, Sir Roderic Hill, thinking seriously both about expansion along the lines first proposed by Tizard and, like Southwell, about ways to expand the social and cultural horizons of Imperial's students. These plans, and the Jubilee expansion, are covered elsewhere.

...

1. Quoted in Harold Wilson, 'Blackett Memorial Lecture', given at Imperial College, 3 December 1975.

2. For a general account of scientists and engineers in this war see Guy Hartcup, *The Challenge of War: Britain's Scientific and Engineering Contributions to World War Two* (New York, 1970). For a comprehensive account of military technology used in the war see M. M. Postan, D. Hay and J. D. Scott, *Design and Development of Weapons: Studies in Government and Industrial Organization* (London, 1964). Few Imperial College scientists were engaged in intelligence work, but one or two did work at Bletchley Park.

3. People at the college were aware of the political events in Germany already in the early 1930s. Many of them had connections to scientists and scientific institutions there.

4. For the evacuation plans and later cancellation see documents in GXK.

5. *The Phoenix* vol. 48, March 1940, 2.

6. Imperial's response was relatively slow and somewhat independent of national efforts. For example, a more determined effort was made by Lionel Robbins who set up the Academic Freedom Fund at the LSE for Jewish refugees in 1933 and actively sought funding. In the same year a national body, the Academic Assistance Council, with Lord Rutherford as its president, drew support from several universities, and from some colleges of the University of London, but not initially from Imperial. The Council helped academic refugees find work and was largely funded by British academics and by British Jews. By 1934 the University of London, as a whole, had found places for 67 refugee academics. Imperial, acting independently, accommodated several refugees, not all with financial support. They included Professor Plusansky from Warsaw who came with some of his civil engineering students. A. J. S. Pippard took an interest in Polish refugees, helping them set up a separate Polish College in South Kensington. This later merged with Battersea Polytechnic. F. A. Paneth, a

major German scientist, came to work in the chemistry department with a grant from Imperial Chemical Industries, but later moved to a chair at Durham. Several refugees found room in the Chemical Technology Department, notably Professor Timmermans, a distinguished physical chemist from Brussels, whose work on the liquefaction of hydrocarbons fitted well with work being done in the department. He arrived with several of his research students. Günther Nagelschmidt joined the geology department and Otto Klemperer (cousin of the conductor) came to work in the physics department. One of the few who obtained a permanent position at Imperial was Gottfried S. Fraenkel, a German Jewish refugee, who helped lay the ground for insect physiology at the college before leaving, in 1948, for a chair at the University of Illinois (Urbana). He came to Imperial as the result of persistent petitioning by the head of zoology and applied entomology, J. W. Munro. Munro fought successfully also on behalf of Nadia Waloff who was appointed to a lectureship. Her situation was different in that she had been resident in Britain for a number of years, but as a citizen of the USSR was viewed as an alien. The chemistry and chemical engineering departments appear to have taken in the largest number of student refugees (perhaps because both heads of department, H. V. A. Briscoe and A. C. G. Egerton, respectively, were actively concerned with the plight of Jewish refugees). Among the students was Franz Sondheimer FRS, a German Jewish refugee who came top of his chemistry class and stayed to work for a PhD under Heilbron. Sondheimer had a major career in organic chemistry which took him from Imperial, via Syntex in Mexico, and the Weizmann Institute, to a Royal Society chair at Cambridge. Another student refugee chemist, Heinz Peter Koch, was given a scholarship and, in 1940, became editor of the student magazine, *The Phoenix*. For a contemporary account of Jewish scientist refugees, see E. Rutherford, 'The Wandering Scholars: Exiles in British Sanctuary', *The Times*, 3 May 1934. See also. M. G. Ash and A. Söllner (eds.), *Forced Migration and Scientific Change: emigré German-speaking scientists and scholars after 1933* (Cambridge, 1996). For a more detailed account of what happened at Imperial College, see Charmian Brinson, 'Science in Exile: Imperial College and the Refugees from Nazism — a case study' (*Leo Baeck Yearbook*, 2006).

7. Charlotte (Lotte) Kellner, disowned by her Nazi husband, came to work with Alfred Fowler. She and Rudolf Peierls had been student friends in Berlin and she stayed with Peierls in Cambridge when she first came to Britain. From the available records it would appear that Fowler was the first professor to pressure Tizard to give Jews asylum. (See Physics and

Technical Optics correspondence file; Fowler-Tizard correspondence, 1933.) Later the head of the physics department, G. P. Thomson, also petitioned Tizard on behalf of Kellner, and on behalf of other refugee students. When Thomson became more heavily engaged in government war work, Herbert Dingle was appointed acting head of department. Dingle continued to petition, helping also to bring in some financial support for refugees.

8. Chapman actively helped a number of refugees, including providing some temporary accommodation in his own home. After the war, Chapman, H. Levy and G. P. Thomson were discovered to have been on Hitler's 'black list', the *Sonderfahndungsliste GB*, as among those to be arrested, and presumably executed, after Germany's planned invasion of Britain, See T. G. Cowling, 'Sydney Chapman, 1888–1970', *Biographical Memoirs of Fellows of the Royal Society*, vol. 17 (1971), 53–89.

9. One person who was asked to leave in 1940 was Nicholas Kemmer, a Beit Scientific Research Fellow (1936–8), and demonstrator in the mathematics department (1938–40). Interestingly he went on to have a major career in Britain. At Cambridge he was the PhD supervisor of two of Imperial's most important postwar theoretical physicists, P. T. Matthews and M. A. Salam. Later he became Tait Professor of Natural Philosophy at Edinburgh. Charlotte Kellner also had to leave Imperial in 1940, only temporarily, until she acquired British citizenship the following year. James Philip, the Deputy (Acting) Rector wrote letters on behalf of German refugee staff and students and was largely successful in getting government rulings overturned. As a result, removal from the college was in most cases only temporary.

10. See Brinson, *op. cit.* (6).

11. Francis Walter Leslie (Les) Croker, taped memoir (1980); transcribed in Bernard Atkinson, 'Department of Chemistry: Recent History, 1960–89' (electronic disc in ICA). Croker was hired as a Lab Boy in 1926. He became Laboratory Superintendent for inorganic and physical chemistry in 1949, and for the entire chemistry department in 1956. He was much involved in the training of technicians in London and was made an Honorary Associate of Imperial in 1978.

12. Chemical Technology and Chemical Engineering correspondence file; G. Himus to A. C. G. Egerton, 31 August 1939.

13. Croker memoir, *op. cit.* (11).

14. See Sir Nevill Mott, 'Harry Jones', *Biographical Memoirs of Fellows of the Royal Society*, vol. 33 (1987), 333–4.

15. The decisions on evacuation were taken by the college in consultation with the Board of Education. There were three letters from the Board between

the 10 and 16 September 1938, advising first move, then stay, then move. A letter in October advised staying, which is what happened despite further correspondence on the subject.

16. This demand came in 1940. The agricultural chemists were already growing some vegetables at Harlington, as well as at the Slough field station, and made a profit of about £45 per year doing so throughout the 1940s. Professor Munro worked hard to get improved rates of pay for labourers who worked on the vegetable plots and on other projects at Slough. The Rector agreed to a pay raise of one penny to 1s 3d an hour in 1941. Munro also won some sick leave for farm labourers working for the college. See B\Munro; correspondence file, letters to Ellingham 1940–41.

17. Croker memoir, *op. cit.* (11). Croker helped Marsden in the work which was intended to help improve the quality of sheep pastures known to be short on some important nutritional elements. Marsden and Croker spread chemicals around the various plots, harvested the grasses, and then tested them to see whether some of the needed elements had been taken up.

18. B/ Lowry; G. Lowry memoir, 'Pearls Before Swine' (typed manuscript).

19. D. H. Hey FRS, who came to the chemistry department with Sir Ian Heilbron, remembered that he was asked to organize teams of Imperial students to act as fire watchers also for some other buildings in London. For example, at Burlington House where the Royal Society and Chemical Society were both located. Geoffrey Wilkinson, who came to Imperial as a student in 1939 recalls that his first task was to fill sandbags in Imperial Institute (now Imperial College) Road. Wilkinson has left a handwritten memoir (B/Wilkinson) which includes details of the college during wartime.

20. The list of emergency food rations for the RCS building maintenance party is interesting. The food was to be used only in emergency and included coffee essence (but no tea), large stocks of flour, cooking fat and margarine, tins of condensed milk, soups, baked beans, peas, herrings, salmon and fruit salad, 48 lbs of Ryvita, 12 tins of cream crackers, 18 tins of biscuits, marmite, jars of meat paste, 84 boxes of cheddar cheese, marmalade, jam, piccalilli, and HP and OK sauces.

21. The head of the RSM maintenance party was Professor W. R. Jones, one of Tizard's closest friends at Imperial.

22. The adventures of the various maintenance parties have been recorded with pride in a range of correspondence, and in several memoirs. See especially Lowry, *op. cit.* (18).

23. The government had a 'War Damage' payment scheme and the college received money for repairs to buildings. F. J. Cutt was made an Honorary

Associate of Imperial College in 1979. In his citation the College Orator, Brian Spalding, stated that in the late 1920s, when Thomas Holland resigned as Rector, Cutt applied for the job. At the time he was a fifteen-shilling-a-week lab boy in mechanical engineering. Somehow, he was summoned for an interview and, when asked about his qualifications, stated that he sang in the church choir, knew how to pump the organ and sometimes read the lesson. Recognized by Professor Dalby as one of his lab boys, Cutt was asked whether he could also do a professor's job. He is reputed to have replied that he 'could push bits of paper about pretty well'. The prank did Cutt no harm. He was soon given a pay raise, perhaps for initiative, and remained at the college for the next fifty years.

24. XXI Club correspondence; letter from A. E. (Ted) Coulson, 27 September 1939. Coulson was a porter and maintenance man in the C&G. He was a well-known college figure who took a great interest in student affairs and coached the C&G tug-of-war team to many successes. During the war he was a member of the C&G maintenance party. He wrote regularly to XXI Club members telling them of college affairs.

25. A doodle-bug (V1 rocket) fell on the neighbouring 66/68/70 Exhibition Road and twenty people were killed. The results were described in a letter from Major Bevan to Ellingham (no date; see B/Ellingham, box 3). Some college personnel were unlucky with bombs falling elsewhere. For example, James Netherwood Sugden, a senior lecturer in the chemistry department, was killed when a bomb fell outside the Natural History Museum. Mrs Hinchley (Professor Hinchley's widow) was killed when her home was hit.

26. For the Home Guard see B/Ellingham, box 3. H. J. T. Ellingham was a chemist and played a major role in the RCS maintenance party and also in the Home Guard. He left the college soon after the war to become Secretary of the Society for Chemical Industry, later returning to the college administration to help with the post-war expansion. See also Croker memoir *op. cit.* (11).

27. Chemical Technology and Chemical Engineering correspondence file. See, for example, Lowry to Egerton, 2 March 1942, on the gas mask inspections on 24 and 27 February.

28. Chemical Technology and Chemical Engineering correspondence file; H. Ardern (City of Westminster Highways Department) to Lowry, 27 April 1942.

29. In 1939 conscription applied only to men aged 20 and 21. Others, including eighteen-year olds, could volunteer. As in the First War, Imperial students were encouraged not to volunteer but to continue with their studies.

30. Of the approximately 110 members of the teaching staff about one-third worked over half-time for the government (civilian or military branches) for the duration of the war. Others worked for less time. Of the junior staff, several joined the armed service. But there was a serious shortage also of non-academic staff many of whom joined the armed services or moved to other work. Women, too, moved to other war-related work. In 1942 national service for women from age 20 was introduced making it difficult for the College to hire young women for the range of jobs they usually filled. The problem was not just the loss of clerical and academic staff in South Kensington. Outdoor labourers at the field station were extremely hard to come by, as were low-level support staff in London. The Rector received many letters from department heads asking for help. See files GXA-N for some of the details.

31. GXF-GXH; Rector's file; Southwell to Hankey, 17 May 1943. This complaint echoes those made by college staff during the First World War.

32. See 'Social Life at Imperial College', *The Phoenix*, vol. 51 (1943). For more on student life during the war see chapter 11.

33. Making such a list had been suggested to the government by some Fellows of the Royal Society. See Guy Hartcup, *The Effect of Science on the Second World War* (London, 2000), 7. In 1914 there were far fewer scientists at universities than in 1939. It has been estimated that in 1902 there were still only about 300 academic positions for scientists in British universities. See Peter Alter, *The Reluctant Patron: Science and the State in Britain, 1850–1920*, (tr. Angela Davis: New York, 1987), 225. But, as David Edgerton has pointed out, by the Second World War, it was the many engineers and scientists in Britain's military and state research establishments that were the backbone of the technical war effort. See his, *Warfare State: Britain, 1920–1970* (Cambridge, 2006).

34. The Chemical Technology department was renamed during the war and, for a short period, was known as Chemical Engineering and Applied Chemistry. Egerton retained his title as Professor of Chemical Technology while remaining head of department.

35. See GXA-D for information on the skills inventory, September, 1938.

36. It is difficult to pin down exactly what the government paid for the services of Imperial staff seconded to the war effort. The college was expected to pay any difference in salary. There is a record in the case of George Paget Thomson, head of the physics department, whose work will be discussed below. From 1939–42 he was paid £900 per year by the government. Since his salary at Imperial was then £1800, the college paid the difference. See GXE.

37. Tizard was also engaged with a number of other problems including chemical warfare. See B/Tizard; papers include some war time papers relating to the British Technical Mission and to his work for the Ministry of Supply. The bulk of Tizard's wartime papers are at the Imperial War Museum.

38. *The Times*, 8 August 1934, 11.

39. Tizard was already serving on the Advisory Committee on Aeronautics.

40. See chapter 4 for Tizard's earlier career. He was appointed Rector of Imperial College in 1929.

41. R. A. Watson-Watt had suggested that radio location of aircraft might be possible and had begun to work on the problem. Radar is an acronym for Radio Detection and Ranging, coined in 1940 by people in the United States Navy. Before that, the system developed in Britain was known as RDF (Radio Direction Finding). Radar was not a single invention; rather it was something that developed with input from many hundreds of people. The all-important cavity magnetron generator (of microwaves that could be focussed and projected over long distances) was developed (refined, not invented) at Birmingham University and at British Thomson Houston in Rugby — under government contract in both places. Tizard's genius was in bringing the many people involved together.

42. Work on infra-red detection was conducted by R. V. Jones but the project was terminated in 1938. See R. V. Jones, *Most Secret War* (London, 1978). Lindemann was supportive of radar but wanted also alternatives to be developed.

43. One Imperial student, on graduating from the mechanical engineering department, went to work at the Royal Aircraft Establishment at Farnborough in 1941 and recalls 'a unique blend of idleness and incompetence', and being put to work on air defence schemes he thought were 'nutty'. See B\Oldroyd; P. W. J. Oldroyd, typed manuscript (1979).

44. For the history of radar and Tizard's role, see, for example, David Zimmerman, *Britain's Shield: Radar and the defeat of the Luftwaffe*, (Stroud, 2001); Guy Hartcup, *op. cit.* (33); Tom Schachtman, *Terrors and Marvels: how science and technology changed the character and outcome of World War II*, (London, 2002); for an appraisal of the various Tizard-Lindemann disputes by a contemporary see, C. P. Snow, *Science and Government* (Cambridge, MA, 1961). Snow, especially in his account of the strategic bombing debate, is now rightly seen to have been heavily biased against Lindemann. For another account favouring Tizard see, P. M. S. Blackett, 'Tizard and the Science of War', *Nature*, vol. 185 (1960), 647–53. Blackett also stressed

Tizard's contribution to operational research. See also Adrian Fort, *Prof: the Life of Frederick Lindemann*, (London, 2003).

45. Anon, 'Electrical Engineering at Imperial College' (typescript history of the department, ca. 1965). Thirty of these transformers were built and delivered to HM Signals School in Portsmouth by December 1940.

46. The Advisory Council was chaired by Lord Cadman. Imperial College chemistry professor, I. M. Heilbron, was also a member. The Council was under the Ministry of Supply and was responsible for advising on research establishments under the Ministry. These included research departments at Woolwich (munitions, metallurgy, ballistics and signals), Porton Down and Sutton Oak (chemical defence), Fort Halstead and Aberporth (rocketry and projectiles), Christchurch (air defence, notably radar) and Biggin Hill (air defence, notably search lights and sound location). Some of Tizard's papers (then top secret) relating to his work on the Council are in the ICA (B/Tizard).

47. In the USA, Tizard, John Cockcroft and some other Britons met with the National Defence Research Committee chaired by Vannevar Bush. The British were keen to have the use of the American Norden bomb sight. In exchange, some of Britain's scientific secrets were shared with the Americans, including radar work. Tizard took a model of the cavity magnetron with him. It was taken to the Radiation Laboratory (Rad Lab) at MIT where scientists made a number of further radar breakthroughs, though not until after the Battle of Britain was won. By 1939 a chain of radar stations had been set up along the South East coast of Britain. In 1940 Fighter Command was able to relieve aircraft from reconnaissance missions and use them for the interception of incoming bombers. See also David Zimmerman, *Top Secret Exchange: The Tizard Mission and the Scientific War* (McGill Queen's Press, 1996).

48. Tizard was the first scientist to become the head of an Oxford college.

49. Despite his age, Lowry volunteered for military service and became an instructor at a RASC officers school in Bournemouth. After Philip's death the Governing Body asked the army for Lowry's release which was granted. Ellingham, too, was very stretched during the war. He worked for the Home Guard, and later for the Ministry of Supply advising on materials to substitute for those no longer available.

50. It is said that the term 'operational research' was coined by A. P. Rowe, former college student and fellow member of the Committee for the Scientific Survey of Air Defence.

51. Blackett's war work is discussed in Bernard Lovell, 'Patrick Maynard Stuart Blackett, Baron Blackett of Chelsea, 1897–1974', *Biographical*

Memoirs of Fellows of the Royal Society, vol. 21 (1975), 1–115. See also, Mary Jo Nye, *Blackett: Physics, War, and Politics in the Twentieth Century* (Cambridge, MA and London, 2004).

52. For more on operational research during this period, see Maurice W. Kirby, *Operational Research in War and Peace: The British Experience from the 1930s to 1970* (London, 2003), chapters 3–5 (quotation, p. 3). Operational research was undoubtedly a novel contribution to the war effort but it is possible that its role has been overstated and somewhat romanticized by scientists such as Charles Goodeve, Henry Tizard and J. D. Bernal, who were engaged with it, and by commentators such as C. P. Snow. See Edgerton, *op. cit.* (33), 205–6.

53. Boffin, a naval term of abuse, was borrowed by the RAF and given its modern meaning during the war, perhaps to put scientists in their place. When some of Blackett's calculations suggested that concentrating aircraft in certain areas would cause enemy submarines to dive so frequently that their air and battery supplies would become exhausted, he was opposed by those who saw no reason to divert several bomber squadrons for that purpose, even though U-boats were taking a heavy toll at this time. British shipping losses reached 600,000 tons per month in 1942. Air Chief Marshall Sir Arthur Harris supposedly exclaimed, 'are we fighting this war with weapons or slide rules?' To which Churchill is said to have replied, 'that's a good idea, let's try slide rules for a change'. Quoted in a letter to the editor, *Daily Telegraph* (5 February 1965) from Sir Charles Goodeve FRS. Goodeve wanted to emphasize how important scientists had been to the war effort. He was a Canadian who came to study at University College London where he later held a chair in physical chemistry. He joined the RNVR during the war, founded the Royal Navy Scientific Service, and became Deputy Controller of Research and Development for the Admiralty. He was renowned for his technical expertise in weapons development. After the war he became Director of the British Iron and Steel Research Association, and founded the Operational Research Club and the associated *Operational Research Quarterly*. Goodeve served also on the Governing Body of Imperial College.

54. After the war Blackett's political views were out of favour, including within the Attlee government. He publicly claimed that the United States had dropped the atomic bombs on Japan for political, not military, reasons (in order to keep the Russians from invading) and wanted Britain to stay out of the nuclear arms race. This latter view was not shared by most of the scientists who had worked on the atomic bomb, Penney and John Cockcroft for example. The 1948 publication of Blackett's *The Military*

and Political Consequences of Atomic Energy, led to yet further antagonism. Blackett succeeded G. P. Thomson as head of the physics department at Imperial in 1953 and, with the election of the Harold Wilson government in 1964, he once again found political favour and re-entered the public arena. His *Studies of War: nuclear and conventional* (London, 1962) was favourably received. By that time people such as Herman Kahn of the Rand Foundation were using Game Theory to support various 'first strike' scenarios. This frightened many people and contributed to the growth of the Peace Movement. Blackett joined other scientists discussing science and peace related issues at the Pugwash conferences (the first was held in Pugwash, Nova Scotia, 1957).

55. For the history of Britain's role in the making of the atomic bomb see Margaret Gowing, *Britain and Atomic Energy, 1939–45*, (London, 1964).

56. In light of this it is interesting that Thomson had earlier rejected a proposal made by Leo Szilard in 1934 that he come to the college to start work on fission. The proposal may well have seemed fantastic to Thomson at the time. His own work began only in 1939. For Szilard's suggestion, see Brinson, *op. cit.* (6).

57. Philip Burton Moon (1907–94), an ex-student of Thomson's, held a lectureship at Imperial from 1931, and was appointed to a chair in Birmingham in 1938. He returned to work at Imperial during the war. Charles Eryl Wynn-Williams had been an assistant to Rutherford at Cambridge. He came to Imperial in 1935 as an assistant lecturer and retired as a Reader in 1970. According to R. V. Jones, Wynn-Williams was the father of digital computing since, when at Cambridge, and as a pioneer in the design of circuitry for electric counting, he invented the 'scale of two' counter. Wynn-Williams spent much of the war at Bletchley Park working on 'colossus'. He also designed the physics department's telephone exchange. See R. V. Jones, op. cit (42).

Moses Blackman FRS (1908–83), known as Morris, was born in Cape Town and educated at Rhodes University College, the University of Göttingen (DPhil, 1933), Imperial College (PhD, 1935), and Trinity College Cambridge (PhD, 1937). When Hitler came to power in 1933, Blackman left Germany and came to Imperial as a Beit Fellow to work with Sydney Chapman and George Thomson on electron diffraction. He joined the staff in 1937 and was promoted to a professorship in 1970. He has left an account of the work carried out under Thomson in 1940–1. In it he claims that Thomson was responsible for organizing the experiments to measure secondary neutron densities resulting from fission, but did not take an active part in the laboratory. Blackman was the theoretician of the group. From 1942 Blackman worked also for the Ministry of Home Security on the properties of firefighting foams. These foams occupied several of the Imperial College staff during the war, including mathematicians Sidney Chapman and William Penney, and chemical technologist, G. I. Finch. They worked in an office of the Ministry on Horseferry Road, Westminster. M. Blackman, 'Sir George Thomson at Imperial College', *Icon*, February 1976 (copy in ICA). See also D. W. Pashley, 'Moses Blackman'', *Biographical Memoirs of the Fellows of the Royal Society*, 33 (1987), 48–64.

58. See Blackman, *op. cit.* (57) and P. B. Moon, 'George Paget Thomson', *Biographical Memoirs of Fellows of the Royal Society*, vol. 23 (1977), 543–7. The pile never went critical but their confidence was, perhaps, premature. In his memoir, Blackman also notes that Thomson was a 'real head of department', meaning that he took a close interest in all that was being done and expected the staff to give reports every day. With his extensive knowledge of mathematics and physics Thomson was able to oversee a wide range of work.

59. Named after a person mentioned in a telegram from Nils Bohr, though not in connection to nuclear physics.

60. Several on the committee had earlier been students of Rutherford at Cambridge.

61. Quoted in Gowing, *op. cit.* (55), 106.

62. Another important scientific ambassador of this period was Alfred Egerton.

63. One of his team, R. Latham, remained at Imperial to work on plasmas. On the removal to Aldermaston see Rector's correspondence; Thomson to Linstead, 24 February 1958. This letter contains also a summary of Thomson's wartime work on thermonuclear reactions carried out at Imperial.

64. See Lord Sherfield, 'William George Penney, 1909–1991', *Biographical Memoirs of Fellows of the Royal Society*, vol. 39 (1994), 282–302. Mulberry harbours were floating harbours, some of which later assisted the Allied landings on D-Day. Imperial College civil engineers C. M. White and A. L. L. Baker were much involved in their design.

65. See GXE; Falmouth to Appleton, 8 May 1944. Earlier in 1944 Appleton had asked for the release of chemist H. J. Eméleus. This was refused but the refusal was overridden by the Treasury. Earlier Falmouth wrote to Lowry, 'we have the right to decide what is meant by extreme necessity otherwise there is no protection whatever for the college'. This was in response to what Falmouth saw as the last straw, namely a request by the

War Office to permanently second Sydney Chapman. (GXE; Falmouth to Lowry, 25 April 1943). Viscount Falmouth joined the Governing Body in 1932 and was chairman of the executive committee throughout the war. He was an accomplished engineer and a senior civil servant in the Ministry of Home Security where he, too, worked with Imperial personnel on measures to counter incendiary bombs. Lord Falmouth was also a London County Council alderman. He volunteered much of his time to college work during the war and was very anxious about the secondment of staff. He became chairman of the Governing Body in 1947. For Falmouth see also chapter 4.

66. See Sherfield, *op. cit.* (64).

67. Cambridge physicist, G. I. Taylor, one of the major theoreticians on the Manhattan project, had predicted the energy of the atomic bomb explosion. In 1942 Harry Jones, then a reader in the mathematics department, accompanied Taylor on a mission to the United States to visit research centres concerned with the detonation of explosives. Like a number of others from Imperial, Jones was a member of the Physics of Explosions Committee of the Ministry of Supply. He carried out some research on explosives at Fort Halstead, together with William Penney.

68. For more on Wilkinson see chapter 10.

69. Hill worked with Thomson and Southwell on problems relating to aircraft structures during the First World War.

70. Sir Arnold (later Lord) Hall went on to have a major career at the interface of the aviation industry and government. Typescript memoir in ICA.

71. This industry was an interesting amalgam of state and private ownership. As David Edgerton has pointed out, in the interwar and war period, the Air Ministry was the largest funder of research and development in Britain. *op. cit.* (33), 118.

72. Harold Roxbee Cox (1902–97), created Baron Kings Norton in 1964, became a research student at Imperial after having taken an external BSc at the University of London in 1922. After gaining a PhD and DIC in 1924, he worked on the Airship R101, joined the Royal Aircraft Establishment at Farnborough, and returned to Imperial College as a lecturer in 1932. The war saw his rise in government, from head of the air defence department at the RAE, to Director of Special Projects at the Ministry of Aircraft Production. He also became Chairman and Managing Director of Power Jets Ltd. which was set up privately at Farnborough in 1936 to exploit Frank Whittle's ideas on jet propulsion, and later nationalized. After the war Roxbee Cox became Chief Scientist to the Ministry of Fuel and Power. Later in life he was Chancellor of Cranfield University. He

was the author of numerous papers on aircraft structures, gas turbines, and civil aviation.

Cranfield was set up as a postgraduate college for aeronautical science after the war, a decision taken by Stafford Cripps who had been Minister for Aircraft Production. It was located at the RAF station at Cranfield in Bedfordshire, where a cadet college already existed. Its first Principal was Imperial College graduate Ernest Frederick Relf FRS (1888–1970). Like many of his generation Relf had earlier moved to aeronautics from physics via the mechanical engineering department. He worked first under Bairstow before moving to the NPL where he rose to become Superintendent of the Aeronautics Department.

73. Whittle had attempted to interest others in his jet propulsion ideas many years earlier. One person who rejected them in 1929 was Alan Arnold Griffith (1893–1963) who was in charge of the Air Ministry Laboratory, located at Imperial College until it was moved to Farnborough in 1931. Later, in 1936, Griffith changed his mind and issued a favourable report on Whittle's ideas. Power Jets contracted British Thomson Houston to build a jet propulsion unit, including also an internal combustion turbine. Tizard, then Chairman of the Aeronautical Research Committee, insisted that a series of official tests on jet propulsion be carried out. See Postan, Hay and Scott, *op. cit.* (2), 178–87.

74. FIDO was an acronym for Fog Investigation and Dispersal. A. O. Rankine was much involved in this project and tested the system in a wind tunnel constructed in the Empress Hall, Earls Court. The system was then tested at various airfields by an ex-student, A. C. Hartley (C&G 1908–10). A number of Polish refugee students were also involved. The system worked quite well but was too expensive, consuming about 100,000 gallons of petrol per hour per airstrip. After the war Lander became Dean of the Military College of Science at Shrivenham.

75. Meteorology began as a section of aeronautics, then became an independent department. Later it became a section of the physics department.

76. Brunt's total salary was paid by the government for almost the entire duration of the war. See GXE; Rector's correspondence file.

77. B/ Pippard; autobiographical memoir (typed). Professor E. H. Brown remembered that Snow was similarly dismissive at Nottingham when he came to interview students there. Snow told Brown that civil engineers 'like mud: there is mud on airfields so we will send you to the RAE at Farnborough'. FIC citation for Brown, 1999.

78. B/Pippard, see Pippard memoir; also, J. E. Morpurgo, *Barnes Wallis: A Biography* (London: 1972).

79. For example, see W. S. Churchill, *Second World War*, vol. V (London, 1953); Simon Bray, 'The Habbakuk Project', *Icon*, (October, 1979), 11; Dorothy M. C. Hodgkin, 'John Desmond Bernal, 1901–71', *Biographical Memoirs of Fellows of the Royal Society*, vol. 26 (1980), 56–8.

80. Pyke was apparently inspired by Inuit ice sleds reinforced with mosses. The ice/wood pulp material, called pykrete, was resilient. Revolver bullets bounced off it, and it was very slow to melt.

81. Dudley Maurice Newitt FRS (1894–1980) studied for an external BSc from the University of London while working at Nobel's Ardent factory. After distinguished military service in the First World War he came to study with William Bone at Imperial College. During the Second World War he was Reader in high pressure technology. He was a highly gifted, and much honoured, scientist, around whom many stories gathered. When he retired in 1961 he was given a major farewell dinner. For his post-war work see chapter 10. See also chapters 4 and 7; and A. R. Ubbelohde, 'Dudley Maurice Newitt', *Biographical Memoirs of Fellows of the Royal Society*, vol. 27 (1981), 365–78.

82. See W. J. M. Mackenzie, *The Secret History of SOE: The Special Operations Executive 1940–45* (London, 2000), 233 and 722–4. The SOE was a relatively small operation, employing about 4000 people in total.

83. The museum was closed during the war and used as a training centre in disguise and camouflage; and for training SOE operatives who were to be sent into the field.

84. The *Tirpitz* was damaged in 1943 while in a Norwegian fjord by charges released from a small submarine, but was sunk in 1944 by heavy bombing. Mini-submarines such as the X-type, Welman and Welfreighter, were tested at the Staines Reservoir. The latter two types were originally conceived for operations against the Japanese. Also produced was something called the 'Sleeping Beauty', a motorized submersible canoe. See Mackenzie, *op. cit.* (82).

85. See Peter Danckwerts, 'Famous Men Remembered', *The Chemical Engineer*, (October 1984), 66–7.

86. The Rector's Lawn occupied the space on which the Roderic Hill building was constructed in the 1950s.

87. Nigel Greatorex, 'Imperial College at War', *Icon*, (October, 1975).

88. Emeléus was among those who later went to work for 'Tube Alloys' at Oak Ridge Tennessee, on the electromagnetic separation of uranium isotopes. After the war he became a reader, later professor of chemistry, at Cambridge. See Norman N. Greenwood, 'Harry Julius Emeléus', *Biographical Memoirs of Fellows of the Royal Society*, vol. 42 (1996), 123–

50. H. V. A. Briscoe, joint head of the chemistry department, was upset by Emeléus's departure. He had no idea of the work being carried out in the USA and complained in a letter to the head of the chemistry department in Birmingham, that the United States had many inorganic chemists who could do whatever Emeléus was assigned to do. GXE; Rector's correspondence file; Briscoe to W. N. Haworth, 29 Jun 1944. One of Emeléus's students was G. R. Martin who came to Imperial in 1938 and who, later, was a professor at the University of Kent. Martin worked for Emeléus on chemical warfare agents and was sent to ICI Billingham to pursue this work after graduation. He has written an interesting memoir of Fritz Paneth which includes details of wartime work at the college (copy in ICA).

89. B/Eldridge; A. A. Eldridge correspondence file. W. C. Gilpin to Eldridge, 1 April 1958. In 1958 Gilpin was working for the Magnesite Co. in Hartlepool.

90. For hidden woman, and dental floss, see Croker memoir, *op. cit.* (11).

91. For tattooing see Barton's FIC citation, 1980. Briscoe was later of great help to Barton at the start of his academic career. For more on Barton see chapters 10 and 13.

92. Report included in M. de Reuck, 'History of the Department of Chemical Engineering and Chemical Technology, 1912–39' (typescript, 1960).

93. E. R. H. (later Sir Ewart, but known also as Tim) Jones FRS (1911–2002), was educated at the University College of North Wales (Bangor). He came to Imperial from Manchester with Ian Heilbron and was appointed lecturer, then assistant professor (not all staff used the alternative title, reader, at this time). A gifted organic chemist he returned to Manchester, and a chair in chemistry, in 1947. He was later appointed Waynfleet Professor of Chemistry at Oxford. Jones played a major role in the making of science policy after the war.

94. B/Eldridge; A. A. Eldridge, 'Some Personal Recollections'.

95. GX\A-D; Jones to Lowry, 27 September 1939.

96. British Anti-Lewisite (BAL) was developed at Oxford early in the war for the treatment of skin lesions caused by contact with the chemical warfare agent lewisite. BALINTRAV (delivered intravenously) was a further development in this line. Leonard Owen FRS (1914–99) later had a laboratory in the department named for him. See obituary, *The Times*, 10 November 1999.

97. B/ Eldridge; G. R. Martin to Eldridge, 11 May 1958. Martin, later a professor at Durham and then Kent, worked at Imperial during the war and made several trips to Porton. He records making 50 lb of arsine on

his first trip, under a tarpaulin shelter in the middle of Salisbury Plain. He and others also made a range of fluorides, on a larger scale than at Imperial, and tested their toxicity on rats. For Martin, see also note 88 above.

98. A. H. Cook, a reader in the department, worked closely with Heilbron and was the person most engaged in the penicillin work. Ernst Chain, later awarded the Nobel Prize (with Florey and Fleming) for his work on penicillin, came to Imperial as Professor of Biochemistry in 1960. Chain's contribution was in the successful purification of penicillin, work carried out while working at Oxford. The product was first used successfully at the Radcliffe Infirmary. It took the work of hundreds of chemists, both in Britain and the United States, before production on a significant scale was achieved.

99. Heilbron was renowned for his work on vitamin A and was much helped by Jones and Raphael. Synthesis was finally achieved in Switzerland, in the laboratories of Hofmann La Roche in 1946. Interestingly, the chemistry of vitamin A is closely related to that of rubber; so, in pursuing this work, Heilbron was continuing an Imperial tradition. For quotation see Leslie Crombie, 'Ralph Alexander Raphael, 1921–98', *Biographical Memoirs of Fellows of the Royal Society*, vol. 46 (2000), 465–81. Raphael later held chairs at Glasgow and Cambridge.

100. As discussed in chapters 6 and 7, Munro was a specialist in combatting pests in stored grain supplies. The Ministry of Agriculture had a number of other campaigns to raise food production such as 'ploughing up' and 'digging for victory'. The number of allotments increased from 815,000 in 1939 to 1,400,000 in 1945. See Alan F. Wilt, *Food for War: Agriculture and Rearmament in Britain before the Second World War* (Oxford, 2004), 189. Food rationing was introduced in 1940 when Lord Woolton took over the Ministry of Food. Beveridge headed the Food Rationing Committee.

101. Applied Entomology correspondence file; Munro to Ellingham, 16 September 1941.

102. To make matters worse, the zoology and entomology space in the old Huxley Building was taken over by an expanded meteorology department and an expanded statistics section of mathematics. After the war a large part of one of the botany buildings was given to zoology and applied entomology. But, the Ministry of Agriculture, which took over the DSIR laboratory, refused to move out of the Hurford (Slough) Field Station after the war. See also chapter 10.

103. The Ministry of Food moved many of its offices to Colwyn Bay.

104. Humphrey Hewer (1903–74) came to study at Imperial in 1920, joining the academic staff under E. W. MacBride in 1926. At that time he specialized in colour changes in fish and salamanders. He was also active in the British Association for the Advancement of Science. During the war he became a reader, and late in his career, professor. After he retired in 1965 he was appointed Chairman of the Ministry of Agriculture and Food's Infestation Control Laboratory's Advisory Committee and, in 1967, chaired the Ministry's Advisory Committee on Farm Animal Welfare. Hewer was a major college figure and the college gave a dinner in his honour in 1970.

105. Together with A. H. Cook, Heilbron developed a new form of chromatography for this work. For more on this, see E. R. H. Jones, 'Professor Ian Heilbron', *The Scientific Journal of the Royal College of Science*, vol. 19 (1949). Cesarol, later known as DDT (dichlorodiphenyl trichloroethane), was introduced as an insecticide by Swiss chemist P. H. Müller in 1939. The compound had been prepared 65 years earlier by O. Zeidler but he had been unaware of its insecticidal properties. Muller was awarded the Nobel Prize in 1948 for Physiology or Medicine, for making a break with traditional arsenical insecticides and introducing a chemical which had a major positive effect on health worldwide. Later some of its more negative environmental properties were recognized and it was widely banned. In recognizing the isomeric properties of the compound, and in developing a separation technique, Heilbron contributed to a more effective agent.

106. Geoffrey Emett Blackman FRS (1903–80), was the son of V. H. Blackman (who retired in 1937). He resigned in 1945 to take the Sibthorpian Chair of Rural Economy at Oxford. See J. L. Harley, 'G. E. Blackman', *Biographical Memoirs of Fellows of the Royal Society*, vol. 27 (1981), 45–82. Blackman was assisted in his wartime work by his student, A. J. Rutter, a future head of the botany department.

107. See 'Review of the Department of Botany' (1951); parts of this are reprinted in *The Record*, January 1952.

108. B/Blackman; The 1941–2 minutes for the Biological War Committee are in the G. E. Blackman file.

109. One of the principal firms in this connection was Messrs. Benham and Sons, a firm of heating engineers that had contracted with the government to supply various munitions.

110. Quotation from 'Report on Munitions Manufacture, 1940–45' (September, 1945); report presented to the Governing Body by Professor E. F. Witchell; see GB minutes, 21 June 1946. For a student's memoir of this work see,

B\Oldroyd; P. W. J. Oldroyd, 'Work at the City and Guilds College during the Summer Vacation of 1940' (typed manuscript 1979).

111. Sir Richard Vynne Southwell FRS (1888–1970), educated at Trinity College Cambridge, had served in all three services during the First World War (RASC, RNVR and RAF) before moving to become Superintendent of the Aerodynamics Department at the National Physical Laboratory. There he carried out research on the structures of rigid airships, moving to Cambridge as a lecturer and fellow of Trinity College in 1925. In 1929 he was appointed Professor of Engineering Science at Oxford and a fellow of Magdalen College. Southwell's work was largely theoretical. His relaxation theory was (and remains) of considerable importance in the mechanical theory of structures. See Derman Christopherson, 'Richard Vynne Southwell', *Biographical Memoirs of Fellows of the Royal Society*, vol. 18 (1972), 549–65.

112. In the early 1930s Oxford's engineering department had 5 academic staff and only 12 graduates a year. Cambridge had a staff of 27 and about 100 graduates per year. See Jack Morrell, *Science at Oxford, 1914–39: Transforming an Arts University*, (Oxford, 1997), 98–102; quotation, 98. Three of Southwell's Oxford students later became professors at Imperial, Willis Jackson, D. G. Christopherson and H. B. Squire.

113. Southwell believed hostilities would be over by October 1945. There was much optimism about the progress of the war by the time he came to the college. In part this was because of the surrender of the German army at Stalingrad in January 1943, in part because of US involvement in the war, and in part a recognition that offensive bombing had become much more effective due to developments in the radar guidance of aircraft to targets. A land invasion of France could, at last, be envisaged. Nationally, too, postwar planning took off in the latter part of 1943. See chapter 11 for more details on the 1945 celebration and Southwell's attempts to improve social life at the college.

114. See Southwell, letter to editor, *The Times*, 14 January 1944, 5. Compton had visited Imperial the previous summer. There had been student exchanges with MIT also after the First World War.

115. Special Committee on Higher Technological Education, appointed April 1944, submitted its report to the Minister, Richard Law, in 1945; (HM Stationery Office, 1945). Southwell was on the committee.

116. See also chapter 9.

117. See, for example, Southwell's letter to the *Times*, 27 March 1946. After the war the matter of technical education, as a national priority, was debated in the House of Commons (Hansard, 22 March 1946). The debate shows

a clear dichotomy between those who saw education as an arm of the state and who wished to see more money for technical education, and those who saw education as serving the individual, and were against privileging science. For more on this see chapter 9.

118. Chemistry department correspondence file; Heilbron to Cripps, 15 September 1945. Cripps had earlier studied chemistry at University College London. Alfred Egerton, who had been a fellow student at UCL, was married to Cripps's sister and supported Heilbron and the Rector on this issue.

119. Munro had the support of the Colonial Office which funded much of his research.

120. 'Rector's Summary for 1945; GB Minutes vol. 39.

121. Southwell; Rector's Report 1945–6. Theodora Cooper (née Bamber) has posted a memoir on the Imperial College website (Alumni News/Alumni Memoirs, 24 April 2006) in which she notes the 'tremendous times' in 1945. She also mentions that she is in the photograph taken of the King and Queen in the Royal Albert Hall (see illustration 46). She is to the left of Queen Elizabeth. On Bamber's right is fellow student, Joan Bridges.

122. The government had agreed that if the universities allowed large numbers of ex-servicemen to enroll, then much new funding would be forthcoming. But numbers increased also because the 1944 Education Act allowed all children to stay at school until university age. Those gaining university places were increasingly funded by county councils making attendance possible.

chapter nine

Expansion: Post-War to Robbins, 1945–67 (Part One)

Introduction

After the war there was a willingness on the part of government to spend more on universities and to see them expand. Before 1939 the Treasury spent only about £2m per year on universities. By 1946 this had risen to about £5m and by 1950 to about £55m.[1] By the late 1940s the view that much of this new money should go towards science and technology was gaining support. This was in part a response to the rise of the USA and USSR as superpowers, and the recognition that some sort of technological race was afoot. As Patrick Linstead put it in 1957, 'a highly technical war gave place to a highly competitive peace, in which the standard of living and the stability of the currency could only be maintained by greater technological efficiency'.[2] In 1953 Imperial College was selected by the government for major expansion and, by 1963, it had almost doubled in size. 1963 was also the year that the Robbins Committee published its report recommending that access to universities be greatly increased. Included among its recommendations was yet further expansion of Imperial College. This chapter includes discussion of the political background to the major post-war expansion at the college. It also focusses on some of the major innovations at the college during this period. Perhaps the most important of these were the reintroduction of biochemistry, the introduction of nuclear science and engineering, and of computing and control. Also to be discussed in this chapter is the Robbins Report, and its impact on the college.

According to sociologist, A. H. Halsey, the end of the war brought a 'convulsion of social conscience' and debate over what kind of society Britain should become.[3] Central to the discussion was education, including what the nation's priorities should be with respect to higher education. Even before the war had ended, many within university circles argued that since arts and humanities students had received no exemption from military service during the war, it was their turn to attend university once the war was over; and that arts faculties, neglected for many years, should be given priority in funding.[4] Others argued that the national need for scientists and engineers should override such considerations. This, under-standably, was the view of Imperial's Rector, Richard Southwell. For a start he was very much against prospective science and engineering students being conscripted into the army after the war and sent many letters to people in government urging exemption. He also pushed hard for science and technology to be given preference over the arts and, in a letter to *The Times*, stated that the country could not afford to make 'fairness the sole criterion on which matters be judged at a time when national recovery is said to demand more scientists'.[5] As mentioned in chapter eight, Southwell had visited the Massachusetts Institute of Technology in 1944 to gain some ideas on how to plan for the future. Perhaps the main lesson of this visit was that Imperial should focus much more on postgraduate courses, and on research. A college committee was struck to make recommendations to the University of London and the University Grants Committee (UGC). In his submission Southwell piled on the rhetoric. He stressed the need for 'a continual flow of young technologists [into industry which, no less than the army] will always have need of subalterns [for whom] it must look mainly to the technical departments of universities'. For this to happen, he argued, there had to be 'larger buildings, more attractive stipends, larger staffs ... and [these] are not mere "claims that must wait their turn"'. Failure could bring about 'collapse of our national economy'.[6] This kind of lobbying was successful. The government accepted the 1946 Barlow Report, and its claim that far too few university trained scientists and engineers were being produced in Britain.[7]

Many people, not just those at Imperial, were critical of the ways in which young scientific talent had been used during the war, and critical also that things were not being quickly rectified once the war was over. As journalist Ritchie Calder put it 'we snatched our children from the classrooms and turned them into radar specialists' with the result, he claimed, that many young people were only semi-skilled and were having difficulty finding work. This was a major overstatement, but he was expressing a widely held concern over a lack of skilled workers. He approvingly noted that Sir Henry Dale, the retiring President of the Royal Society, was encouraging senior scientists to return quickly to their colleges and universities from their wartime positions in order to educate the next generation of youth. Calder was critical of J. D. Bernal, still chairing committees at the Ministry of Works, and P. M. S. Blackett still with the Committees of Scientific Manpower and Atomic Energy, when they should have been back at Birkbeck and Manchester encouraging young people in research.[8] This type of criticism will have resonated with Southwell who was having great difficulty getting his staff back. The release of people from government service involved the clearing of major bureaucratic hurdles. In addition there were many vacancies to fill; Southwell had especial problems with the engineering chairs. Professors Fortescue and Witchell had recently retired and Professor Lander had resigned. In the country many chairs of engineering were unfilled, and the field of possible replacements was not strong. Southwell was also concerned about aeronautics, and possible competition from the newly opened RAF postgraduate college at Cranfield. Many of Imperial's pre-war aeronautics students had been RAF and Naval Air engineering officers. To add to his problems, Imperial had been without a biochemistry professor since A. C. Chibnall moved to Cambridge in 1943. The biochemistry situation was acute; only one biochemist (H. W. Buston) remained on the Imperial staff at the end of the war. Since, in addition, no chemistry students had taken the conversion course for several years, biochemistry was allowed to lapse. As Southwell put it, 'some day ... a man of "professional timber" will appear ... and then will be the time to reconstitute the chair'.[9]

The reports of two government committees were especially important in framing higher education in Britain immediately after the war. Both favoured giving priority to science and engineering. A Ministry of Education committee, reporting in 1945, stated 'the position of Great Britain as a leading industrial nation is being endangered by a failure to secure the fullest possible application of science to industry and ... this failure is partly due to deficiencies in education'. The report is interesting in being also one of the first to suggest both a need for management studies in Britain, and that more colleges be given university status.[10] A second committee, chaired by Sir Alan Barlow, had been asked to report on the 'use and development of our scientific manpower and resources during the next ten years' and to address the immediate problem of demobilization and the training of technical personnel. In its 'Report on Scientific Manpower', published in 1946, the committee made the interesting observation that while only 2% of the population attended university, about 5% of the population had an IQ equal to the top half of those graduating. 'If university education were open to all on the basis of measured intelligence alone, about 80% would come from those children who started their education in the public elementary schools and only 20% from those whose education had been in independent schools'. The report also led to renewed speculation that Imperial College might become part of a separate university since, looking to South Kensington, it stated, 'the association of groups of colleges … with new foundations in complementary faculties and residential facilities [could] form a complete university institution'.[11] In the immediate post-war period Imperial students joined the debates on the college's future, its pedagogy and its cultural identity. The idea of a new university, soon abandoned as an immediate proposition, was popular while it lasted. Students criticized the existing curriculum and complained of having to sit through too many poorly delivered lectures. If there was to be no new university in South Kensington, students wanted increased integration with other University of London faculties so that they could attend classes also in disciplines other than science and engineering.

In 1945 Herbert Morrison gave a speech to the BAAS in which he declared that science must have a social purpose and that it must find its aims and aspirations from inside a general national economic plan.[12] This was not a universally welcome message. *The Times* commentator stated that there should be room also for 'carefully chosen men to pursue with freedom that sublime curiosity into nature's workings which prompted Newton and Faraday, and to inspire the young with some better ideal than bigger and better refrigerators'.[13] This was a piece of gentlemanly rhetoric, albeit with a grain of truth. The ethos of Imperial College fitted better the new government's utilitarian ideals and, within a few years, the college was promised funds for a major expansion. Southwell and his successor, Roderic Hill, had successfully lobbied the UGC and the government for this.[14] While the government had supported growth in science and technology education immediately after the war it decided to give it a further boost with a major expansion in one or two institutions of higher education. But, until announced in the House of Commons on 29 January 1953, it was not known which college or university would be chosen.[15] Imperial was asked to expand from 1650 to 3000 students by 1962 and the Chancellor of the Exchequer invited the college to submit its plans to the UGC, for construction to begin in 1957.[16] There had been some discussion of moving Imperial out of London but the metropolitan ties were strong and it was decided to expand around the existing site. It would appear that Imperial's closest competition came from a proposal for a completely new technical university, favoured by those in the government who wanted to avoid what they saw as bureaucratic inefficiency in existing universities.[17] Indeed, one of the arguments made in Imperial's favour was that its three-college structure allowed for more individual initiative than was possible in more centralized institutions.[18] After the announcement, Hill stated that he saw the development as the most important since the founding of Imperial.

> As we all know, Imperial College is a federation of three colleges, the histories of which run well back into the nineteenth century. Speaking personally, I feel entitled to say at this early stage, that I shall make the attempt to base the new and larger life of Imperial College just as firmly on the spirit and traditions of the three colleges as ever.

And, echoing Churchill, he stated, 'we now find ourselves … at the end of the beginning'.[19]

Governance and Innovation

Hill had studied fine arts and had planned to become an architect before war intervened in 1914 and sent him in a different direction. As Rector, he thought deeply about the nature of technical education, and of the need for scientists and engineers to make their work and aims intelligible to others. In this he was attuned to what was a major post-war discussion on the rapid pace of technical change, and on society's capacity to deal with it. The general mood was captured a little later by C. P. Snow in his lecture on the 'Two Cultures', not that he was original.[20] In pointing to the gap between those educated in the arts and the sciences, and to the problems it posed, he was simply resurrecting an old issue discussed already before the war by H. G. Wells and others. The debate over how scientists should be educated to be responsible citizens was especially lively in student publications of the 1950s. Hill engaged with the students, encouraging them to feel that they 'were part of the show', that they could have influential careers, and that their work was of consequence.[21] He wanted them to learn to be socially relaxed so that in their working lives they could be at ease within the corridors of power.[22] But how to educate Imperial students with such goals in mind? Hill is perhaps best remembered for importing some older, liberal, university ideals into the college and introducing Imperial students to a wider cultural world. He initiated both a series of lunchtime lectures on non-technical subjects, and the Touchstone weekends at Silwood at which students could discuss a range of topics, many of current interest, together with experts.[23] He also paid attention to the college library, and improved Imperial's relationship with its South Kensington neighbours, especially with the colleges of art and music. With some difficulty, he managed the return of the Harlington land that had been expropriated during the war, allowing the resumption of normal sports activity. In the interim, he presented a cup for a Hyde Park relay race to be competed for by the three constituent colleges.[24] He also improved the college health services and founded the Holland Club, a dining and social club for non-academic staff named for the third Rector. Hill also began to think about the college administration and appointed Imperial's first Pro-Rector, H. H. Read, professor of geology, in 1951.[25]

Even before the major government announcement of 1953 there was some expansion in South Kensington. The Royal School of Needlework lease for part of its old building (renamed the Unwin Building) expired at the end of the war. Much of the vacated space was used for a refectory known as Ayrton Hall. This relieved pressure on the crowded refectories in the Union building, as did removal of the Union Library to the Unwin building, and the bulk of the administration offices to 178-9 Queen's Gate. Hill also planned a new building for aeronautics and chemical engineering. After a major career in the air force, he was naturally much concerned with aeronautics and, in a 1949 note to the Governing Body, mentioned how the war had showed Britain's inadequacy of expertise in aviation: 'we were found wanting'. He stated that the plans for the new building were 'modest in the extreme' and that the United States was installing more and better wind tunnels than anything being proposed for Imperial. He noted that in Professor Hall, 'we have one of the most distinguished, if not the most distinguished of the younger men in this field', but that to keep him 'we will need to provide adequate facilities'.[26] As it turned out, the new building was not enough to keep Hall; he had resigned by the time it was opened. The building, named for Roderic Hill, was opened by the Queen Mother, three years after Hill's death. The opening was the major event of the fiftieth anniversary celebration of the College Charter in 1957. At a Mansion House dinner to mark the Silver Jubilee, Lord Hailsham, Minister of Education, stated what Imperial people had long wanted to hear, namely that he firmly believed that technology and engineering should be on the 'university side of the higher education fence'.[27]

Hill died suddenly in 1954.[28] He was succeeded by R. P. Linstead who became the first Rector to be appointed from the college staff.[29] Linstead was a major figure at the college. He had made his mark already as a student, as a young lecturer, and again on being appointed Professor of Organic Chemistry in 1949.[30] He was a member of the Governing Body and had taken a prominent part in planning the college expansion under Hill, especially in thinking about academic growth points. He continued what Hill had begun, namely planning the expansion, rationalizing and modernizing the committee structure, and devising new channels for formal communication. In his 1955 inaugural address, Linstead stated that he wanted the college to train people for modern industries and cited nuclear engineering and automated control engineering as examples. He also wanted to train more science teachers fearing that after the post-war 'baby bulge' the flow of good students might dry up. He wanted one-third of postgraduate students to come from overseas. He also wanted to continue Hill's project of broadening the curriculum, and that students be 'more exposed to what London has to offer'.[31] Linstead was much helped with his planning by the College Registrar, E. F. Cutcliffe, who had an encyclopedic knowledge of both college affairs and government regulations.[32]

The expansion plans were to increase space for the RCS by 39%, for the RSM by 87%, for the C&G by 100% and to include space also for about 125 interdisciplinary research students across the three colleges. Further, there were to be new residences, refectories, and increased social space. Construction began with the addition of new storeys to the Union building, the RSM, and the Goldsmith's extension of the C&G, and of new sections to the chemistry and chemical engineering buildings. But the major focus was on new buildings for the island site, including a good central library.[33] More than anything Linstead wanted demolition of the Imperial Institute buildings which he saw as a 'sad jumble'.[34] The Warburg Institute was the last remnant of the University of London's earlier occupancy, but a number of other institutions, such as the Institut Français, were tucked away in various corners. Linstead wanted the government to make good on an earlier agreement with Henry Tizard that Imperial should have the use of the Imperial Institute or its land base. Linstead wanted the land but had to fight for demolition both within and outside the college. He insisted that planning for new buildings on the site go ahead, believing that with plans in place resistance would vanish. But resistance did not quickly diminish. The Royal Fine Art Commission put up a good defence of the Imperial Institute building in a 1955 report to the Treasury and, led by John Betjeman (later Poet Laureate), mounted a publicity campaign to save the building. In 1955 the government was still undecided and the college had to tread carefully in case the conclusion was reached that things were simply too difficult, and that expansion of the college at another site or worse, the expansion of a different institution, might be preferable.[35] Interestingly the proposed demolition of the C&G Waterhouse and Unwin buildings did not meet opposition.[36] The college architects, Norman and Dawbarn, were asked to draw up an alternative set of plans, should permission for demolition of the Imperial Institute not be given.[37] But these were not needed once a final settlement was reached in June 1956 — though Betjeman and others continued their protest until the Institute building finally came down. Compromises were made; Imperial agreed to keep the Colcutt (Queen's) Tower and not to build as densely on the island site as was originally planned.[38] Of the tower a student wrote with some prescience, 'as time passes, no doubt we shall come to value the incongruous phallic symbol of bygone Victoriana that Mr. Betjeman has forced into our lap'.[39]

Linstead also made a timely and persuasive application to the UGC for the acquisition of land in Prince's Gardens, and for funds to develop the area as a site for student residences.[40] In the end, with the Institute demolished (except for the central tower), land for the college in South

Kensington jumped from 9.3 to 28 acres. During the reconstruction there was much chaos and dislocation but Linstead appears to have kept things going by reminding people of the prospect of 'fine new buildings' which 'will perform without fuss or exaggeration, but we hope not without elegance, a complicated and necessary task of our time'.[41] During his tenure he saw a doubling in the number of full-time students. He was the first to introduce pastoral care at the college by appointing special tutors to help with student's personal problems — one tutor for each of the constituent colleges and one for women students. The academic staff also doubled to about 600 and the average number of professors per department rose from 2 to 4.5. One-third of the professors were Fellows of the Royal Society.

Nuclear Science and Engineering

While the building plans were still in flux, Linstead had to think also of the academic side. Here his priorities were clear: nuclear engineering and science, biochemistry, computing and control, and the history of science and technology. Linstead was especially keen on a nuclear future for the college. He invited Sir John Cockcroft to be the special visitor at the 1956 Commemoration Day and Cockcroft gave a speech extolling the glorious future of atomic energy. 'Calder Hall is only a beginning' he told the graduating students and that, by the time most of them were forty years old, the energy supply of the United Kingdom would be almost completely nuclear. Linstead appears to have shared this view of things and more than once repeated Cockcroft's 1958 claim that in four years time nuclear energy would be as cheap as that derived from coal. He was determined that Imperial be at the forefront of nuclear research, and of the training of personnel for the new industry. At the time it appeared that there would be a huge demand for graduates in this field.[42] In wanting Imperial to be a leader in this area he had allies both in Owen Saunders, the head of the department of mechanical engineering, and Patrick Blackett, the head of physics. Blackett had been persuaded to come to Imperial by Hill when the expansion was announced. He was a promoter of nuclear science also at the national and international level, urging an undecided government to join in the founding of the European Centre for Nuclear Research (CERN). In the 1950s Blackett was promoting research into novel military uses of nuclear science and gave several speeches warning of Soviet progress in military technology.[43] Further support for nuclear science and technology work at the college came from Sir Roger Makins (later Lord Sherfield), Chairman of the UKAEA, and a member of the Governing Body.[44]

In 1956 J. M. Kay was appointed to a chair of nuclear power in the mechanical engineering department.[45] Kay came to Imperial intending to work on the design problems of nuclear power generation, and to encourage interdisciplinary nuclear research with other departments, especially with chemical engineering. Kay developed a one-year MSc course which began in 1957, with twenty-four students chosen from a large field of applicants, and intended to meet the needs of engineering graduates wishing to enter the field of nuclear power station design.[46] Kay had close contact with the Central Electricity Generating Board and with the UKAEA, both of which offered help in developing the course. A further nuclear technology course, with a range of options, began in the renamed Department of Chemical Engineering and Chemical Technology in 1959, under G. R. Hall, Reader in Nuclear Technology.[47] This course was intended to give general training in the chemical, chemical engineering, and metallurgical aspects of nuclear technology. The metallurgical aspects were also of interest to J. G. Ball, a pioneer in plutonium metallurgy. Ball, appointed Professor of Physical Metallurgy in 1956, had previously worked at the UKAEA plutonium laboratory at Harwell which he had helped design. The courses in both departments were backed by the departments of physics and mathematics which offered training in the theoretical aspects of nuclear science. Russian classes were also offered with an eye to student exchanges and visits (where possible) to nuclear installations in the USSR.[48]

In the late 1950s there was much discussion as to whether universities should have nuclear reactors for research purposes. Aside from safety, one argument against such a move was that the kind of simple reactors one might build for universities were not likely to be anything like the ones that nuclear engineers would encounter in the larger world. Kay desperately wanted a research reactor for Imperial. He submitted plans which various government committees simply sat on and, in 1960, he wrote to *The Times* stating that the delay made him 'frustrated beyond words'.[49] But other universities were jumping on the bandwagon. Thus deciding where, if anywhere, academic nuclear reactors were to be sited was a major problem for the government. In the end it looked likely that one would be built for the University of London but whether it would be Kay's design, or that of its Queen Mary College rival, was not clear. As an editorial in *Nuclear Engineering* rather awkwardly put it, 'if the negotiations over the University of London reactor were to be used as the plot of a play, film or even comic opera it is doubtful that the critics could forbear the use of a few sarcastic comments ... [on] the tortuousness of [the] theme and ... overall lack of realism.'[50] The DSIR committee looking into the designs preferred the one proposed by Queen Mary College, but politics intervened and Imperial College was given the go-ahead to construct its 100kW Consort Reactor. Adding to all the confusion, at the point this decision was made Kay resigned and moved to the nuclear industry. P. J. Grant, appointed reader in nuclear power in 1959, had to pick up the pieces.[51] The reactor was built at Silwood Park by the General Electric Company and came into operation in 1965. It was used for research in nuclear science and technology, reactor engineering, neutron physics, and solid state physics. In 1971 the addition of radiochemical and physical laboratories broadened the areas of research.

Biochemistry

Linstead began discussions for the reintroduction of biochemistry in 1955. In this, too, he was aided by Blackett who chaired a committee seeking candidates for the head of a new department. Its first choice was N. W. Pirie and, when he declined, an informal approach was made to Ernst Chain.[52] Chain's demands were for a large 'national institute' and, since nothing on such a scale was envisaged, Linstead and Blackett approached some other people to see whether they would be interested in the chair, but with no success. In 1957 Linstead visited Chain in Rome for further discussions, and Blackett visited him the following year. Chain's conditions were still exacting. He wanted a major new laboratory and the provision of accommodation nearby.[53] He also wanted a fermentation plant of the type that existed at his institute in Rome. But this was an expensive installation and one that many people considered more suited to an industrial than a university setting. Chain was appointed in 1961 and, despite opposition to the plant, he and Linstead campaigned for funding. They persuaded the Wolfson Foundation to donate £350,000 towards equipping new laboratories, including a fermentation plant. Further funding came from the Fleming Memorial Fund and the SRC.[54]

In addition, the MRC agreed to set up a Metabolic Reactions Research Unit in the department. Linstead planned the new building as the first stage in the redevelopment of the RCS site. It was completed early in 1965 and formally opened by the Queen Mother in November of that year.[55]

Chain's interests were in the metabolism of both micro-organisms and higher animals. His wife, Anne Beloff-Chain, who had similar interests, was appointed Reader and worked in the MRC Unit. The department placed much emphasis on the development of new methods of preparing microbial metabolites on a semi-industrial scale, and in developing the instrumentation for their analysis. Teaching was to be only at the postgraduate level, but there were problems with the courses that Chain proposed. The first was a two-year MSc course in general biochemistry for which Chain thought recognition would be a mere formality. But the SRC claimed that the course was simply a disguised undergraduate course and totally unsuited for students transferring to biochemistry from other disciplines. Chain then proposed three one-year courses and the SRC, with much cajoling from Linstead, agreed to approve two of them, one in biochemical methods and the other in microbiological science; the third in animal biochemistry was turned down. Approval for the two was given reluctantly by the Council which, with some justification, thought that the fermentation plant should be the platform for courses in the department; namely, they wanted courses in fermentation chemistry or biochemical engineering.[56] But for Chain the fermentation plant was simply a tool in a larger biochemical project. Perhaps the major lesson he had drawn from his earlier work was that rapid metabolic change is possible. People close to death could be rescued almost instantaneously with penicillin. Much of Chain's later work was directed at isolating biochemicals that produce rapid change in living organisms. Younger members of his team such as Steven Rose and Harry Bradford, later a professor in the department, were put to work perfusing different organs with biochemicals and carefully recording the consequences. Any natural product of interest could then be synthesized on a larger scale in the pilot plant. This work had much support from major drug companies, Beechams even providing Chain with a chauffeured car always at his disposal.

Chain, a friend of Harold Wilson, saw himself as working in the 'white heat' of a technological revolution.[57] But the early level of funding could not last and, within a few years, he was frustrated by lack of money. The sudden death of Linstead, who had given him much support, was a major blow and he fared less well with Linstead's successor William Penney. The department had developed on both a microbiological and a physiological biochemical basis, consistent with Chain's view that biochemists should work at the level of the organism. He was dismissive of the chemical approach to cellular biology, an unfashionable attitude at a time when molecular biology was on the rise. As a result, when Chain retired in 1973, the new Rector, Brian Flowers, was faced with the difficult problem of what to do next: build on what existed or change direction? This will be discussed elsewhere, but in retrospect it appears that Linstead's judgement failed in this case.[58] He was very keen to get a new department of biochemistry going, and for it to forge links with existing departments. This is understandable. But he had difficulty finding a suitable professor and should perhaps have delayed. But, swayed by the idea of catching a Nobel laureate, he went ahead without looking carefully at where the discipline was heading. Chain was a brilliant scientist and full of enthusiasm for the various projects under his direction. Perhaps it was his time in Rome which led to his talking also much with his hands. He will have been persuasive but, in giving in to his demands, Linstead did not think clearly about the discipline, or of the financial and academic consequences of having a highly capitalized special-purpose unit built for a professor with only twelve years remaining before his retirement.

Computing and Control

Both nuclear science and engineering, and Chain's approach to biochemistry, entailed very costly installations. So, too, Linstead's third major priority, computing science. But in this field things moved more slowly than he hoped and did not really gather speed until after his death in 1968. However, as a consequence of work in this field during the war, there was much interest in computers at the college already in the 1950s.[59] For example, some people in the mathematics department designed and built a digital relay computer, the Imperial College Computing Engine (ICCE 1), a project envisaged by Tony Brooker and developed by Keith Tocher and Sydney Michaelson. One of their research students was M. M. Lehman, a former undergraduate in the department, who designed an arithmetic unit for what was intended to be an electronic version of ICCE 1 (ICCE 2 was never built). While this work had the active support of Hyman Levy when he was head of department, it did not have that of his successor, Harry Jones. Jones took the view that research that included computer hardware did not belong in a mathematics department and should be left to computer manufacturers such as Ferranti. Further, he believed that the chances of acquiring a good commercial machine would be poor if the department already had its own, even if slow and inefficient. He therefore closed the work down. With hindsight this may look to have been a prescient decision but, at the time, the development of hardware, software, and mathematical logic were inextricably linked. The fact that the departmental focus was on machine architecture should not have been held against the group. Other universities were supporting this type of research and, as R. J. Cunningham has suggested, a better solution would have been to transfer the group to electrical engineering.[60] In any event the college lost a highly gifted group of people who left and went their separate ways, Lehmann later returning to a chair in computer science.[61]

There was interest in computing also in the electrical engineering department, though more in analog than digital computers in this period.[62] John Westcott, a student of Willis Jackson in the late 1940s, spent a year (1947–8) at MIT where he came under the influence of Norbert Wiener and began work in cybernetics.[63] On his return he continued working in this field, spending much of his time at the General Post Office Research Station at Dollis Hill where similar research was being carried out. When Arnold Tustin became head of department in 1955 he planned for a major control section in the new electrical engineering building. But this took time, and Tustin was extremely frustrated by delays to the building due to Betjeman's success in saving the Colcutt (Queen's) Tower.[64]

Linstead gave Westcott the further task of planning computing facilities for the entire college. In 1960, IBM approached the college with an offer of a 709 computer, the largest thermionic valve machine they had then built. Its size and power consumption made it a questionable asset and Blackett advised Linstead to refuse the offer. Two years later, after Westcott had done some consulting for IBM, the company made a new offer, this time of a 7090 computer (a solid state machine) and offered help also with the running costs. The college was interested but needed University approval which entailed lengthy negotiations. When the problems seemed close to resolution, Westcott recognized that the college needed more computer expertise. He approached a number of people around the country to see whether they would come to Imperial, but only succeeded in getting them chairs at their own institutions. Computer expertise was so thin on the ground that universities were unwilling to let their experts leave and readily gave in to their demands. In the end Imperial managed to attract Stanley Gill who was appointed to a chair in computing science in 1964.[65] He was also appointed Director of a new College Computer Unit, situated in the electrical engineering department.[66] In 1966 he was joined by W. S. Elliot who was appointed to a second chair in computing, and given the responsibility of developing a new MSc course.[67] Earlier, Gill, Elliot and Lehman had worked together at Ferranti on the Pegasus and Mercury machines.

That the college had made a relatively early start in computing science influenced the Flowers Committee on Academic Computing. In 1964 this committee had been asked to look into the provision of university computing nationally. The Committee recommended that a South East Regional Computing Centre be located at Imperial College. But the University of London intervened, agreeing to the type of computer recommended by the Committee, but not to its being located at Imperial. Therefore the CD 6600 went to the University and, as with the Ferranti Mercury and Atlas machines, it was located near Senate House. It is possible that the Malet Street site had been chosen earlier in part because Professor Jones had ended the work of the computer builders in his own department. Had he been more supportive the university might well have built its computing centre at Imperial. Despite the setback, Westcott received funding from the SRC to purchase a high-speed link to the Atlas machine. As is discussed in chapter thirteen, the Computer Unit evolved into the Department of Computing and Control in 1970.

History of Science and Technology

Linstead's fourth priority was more modest and was in line with Hill's legacy of broadening the courses available to Imperial students. Linstead wanted a professor in the history of science or technology to set up a small postgraduate centre, and to provide optional lectures for undergraduates. The University of London agreed to a chair in 1961. Even earlier, some lectures in the field had been given at the college. For example, G. J. Whitrow gave lectures in the history and philosophy of mathematics; and A. W. Skempton gave lectures on aspects of civil engineering history.[68] During a visit to the United States, Linstead had asked Derek J. de S. Price whether he would be interested in returning to Britain to take up the new chair, but Price declined. The eminent historian of science, Charles Singer, was then consulted on other possible candidates, and on how the department should be structured. Skempton, who was very supportive of Linstead's venture in creating Imperial's first humanities chair, visited Singer to discuss matters. Singer was then renting (from Daphne du Maurier) a large house overlooking St. Austell's Bay in Cornwall. There was plenty of space for his vast library and it was reputedly there that the decision to offer the chair to A. R. Hall was made. Hall and his wife M. B. Hall, appointed a senior lecturer, came to Imperial in 1962.[69] The Halls instituted two MSc courses and trained some good PhD students; but they failed to plan for their own succession and, by 1980, the department was in decline.[70]

The Robbins Report

Linstead worked extremely hard in seeing through the expansion, and in running the college during its period of rapid growth. But further expansion was to follow. The initial expansion has been called the Jubilee Expansion since major construction began in 1957, the silver jubilee year of the college charter. But no sooner was the new construction near completion than the Macmillan government set up the Committee of Enquiry into Higher Education, with Lionel Robbins as its chairman.[71] Linstead was a member of the committee. The Enquiry had its roots in the Education Act of 1944 which made secondary education a realistic possibility for all.[72] The result of this Act was that local authorities set up a range of schools, including many new selective grammar schools which had as part of their mandate the education of prospective university entrants. By the late 1950s the new system came under criticism as its built-in biases became apparent.[73] The 'eleven plus' examination which decided entry into the grammar schools was shown to favour, albeit not exclusively, children from middle-class families and was not a true measure of aptitude. Further, the grammar schools existed alongside a privileged independent sector. And, as Peter Clarke has put it, 'England's selective school system [was] married to an even more selective university system'.[74] There was much pressure on the government to set up a commission to look into the future of the university system. Lord Simon, for example, raised the issue many times in the House of Lords.[75] By 1961, when the Robbins Committee was set up, only about 4% of the population reached university, and only about 8% received any form of higher education — percentages widely seen as far too low.[76] But expansion of the universities was well under way before Robbins, and the Treasury was already having difficulty dealing with the existing bureaucratic machinery.[77] The problem for the Committee was to find solutions both to the ethical questions of access and the bureaucratic problems of delivery. Its terms of reference were,

> to review the pattern of full-time higher education in Great Britain, and in the light of national needs and resources, to advise Her Majesty's Government on what principles its long-term development should be based. In particular to advise, in the light of those principles, whether there should be any changes in that pattern, whether any new types of institution are desirable, and whether any modifications should be made in the present arrangements for planning and coordinating the development of the various types of institution.[78]

The committee did not work in a vacuum. The American state university system, well on its way to serving a high percentage of the population, was at the back of many people's minds. Further, it would appear that almost everyone in Britain had an opinion on higher education, and the public debate was lively. Some, like Kingsley Amis, believed that 'more will mean worse', but that was a minority view. Even the young Tories of the Bow Group in their 'Campaign for Education' favourably quoted Geoffrey Crowther in a 1963 booklet, 'all to sixteen, half to eighteen and one-fifth to twenty-one'.[79] The Labour Party submission to Robbins proposed that universities be open to all qualified students and that the number of universities be increased to seventy-five. Another committee, the Scientific Manpower Committee chaired by Solly Zuckerman, had reported to the Ministry of Labour in 1962. It noted that universities were 'bursting at the seams', and had already recommended expansion of the system.[80] One year later Robbins recommended a more than doubling of university places by 1980.[81] The thinking behind this was indeed the same as that of the Labour Party; namely that there be places for all young people who were qualified and wished to attend university. Doubling the number of places was seen as a start. That more is better was a simple article of faith for many people. Further, that only a fraction of the nation's intellectual potential had been tapped was widely accepted, and no doubt also true. Robbins gave no explicitly economic justification for his recommendation but appears to have held the view that justice demands there be places for all those who could benefit from higher education. Blackett was somewhat critical of Robbins. He supported greater access, but did not agree that there be sixty universities, or that university status be given some local colleges. Indeed, he thought there were already too many universities (new universities of the Sussex generation were on stream) and that quality would be better maintained by the expansion of those already in existence.[82]

Echoing the earlier views of Herbert Morrison, Robbins stated that attention should be given to forecasting the needs of the country and opening up places accordingly. Higher education was to be made more systematic and statistics which had largely been ignored, if they existed at all, were now to be routinely collected and centralized. 1963 was also the year in which Harold Wilson gave his famous speech to the Labour Party Conference claiming that 'a Britain that will be forged in the white heat of [the scientific] revolution will have no place for restrictive practices and outdated measures on either side of industry'.[83] For Wilson, the principal role for new state-funded science was to be in the service of industry.[84] He wanted universities to produce well qualified graduates prepared to join industry and that private industry take on a greater role in funding research and development. This was not exactly how many scientists understood the new 'scientific revolution', hoping rather for more state-sponsored 'pure' scientific research in both universities and state research laboratories. Such a vision did not materialize in the decade that followed.

Given the political climate of the early 1960s, it is not surprising that Robbins envisaged a major expansion in science and technology in the university system.[85] In this connection, and of especial interest to Imperial, was the recommendation that there be five special institutions for advanced scientific and technical education along the lines of MIT or Delft. The five were to include Imperial College. Robbins had asked that all academic institutions write to the committee outlining their methods of internal governance and how these might be improved. But it was London University, not Imperial, that had been asked to respond. Imperial had to send its views to the University. Before doing so Linstead consulted his senior colleagues as well as members of a committee that he had set up in 1962 to plan for the 1970s.[86] To help forward the discussion he used his insider knowledge of the Robbins Committee and wrote an internal memo entitled, 'Relationship of the Imperial College to the University of London and the Possible Formation of Technological Universities'.[87] In this he noted that during the period 1951–3, when the issue had been much discussed (not for the first or last time), it was decided that Imperial should remain within the University of London. Linstead listed the advantages and disadvantages of Imperial's relationship with the University and basically saw the existing problems as constitutional in nature; namely that Imperial College's 'importance in the national educational scene is not reflected in the practices and procedures of the university'. Current arrangements within the university, he stated, needed reform. As we have seen in other chapters, this was a perennial Imperial cry, but one not without justification in this instance. What Linstead wanted was for Imperial to gain direct access to the UGC, and that its Rector be a member of the Committee of Vice-Chancellors and Principals (CVCP). Given that new universities with far smaller budgets than Imperial had access, these demands were reasonable. Linstead stated that if his two objectives could be achieved, the question of Imperial becoming an independent technical university could be set aside. But, for bargaining purposes, Imperial's views were to be kept secret.

Linstead wrote to the Principal of the University of London stating first that Imperial had drawn up some tentative plans for both short and longer term expansion in light of the Robbins recommendation that it become a Special Institution for Scientific and Technological Education and Research.[88] He envisaged Imperial's student numbers increasing to 4700 by 1973. But for this to occur, the Jubilee expansion would have to be completed, further sums be granted for new capital works, adequate recurrent grants be in place, and the maintenance grants for postgraduate students be increased.[89] Linstead saw any future expansion as a natural extension of the Jubilee scheme. The college wished to grow further, with major increases in the intake of students in mathematics, chemistry and physics. But it also wished to diversify. In his letter, Linstead mentioned the new history of science and technology department and expressed a wish for the addition of departments in management studies and architecture. Further, the college was thinking about 'clusters' such as earth sciences, applied biology, materials science and technology, and social sciences. After outlining future plans, Linstead expressed the college's concerns with governance and with changes he believed were needed to the University statutes. In this he was persistent and among his major achievements were statute revision, the gaining of direct access to the UGC, and a seat on the CVCP.

Linstead rightly saw the Robbins Report as a major signpost to what lay ahead for higher education in Britain and, as a Committee member, was firmly supportive. In 1963, one day after his report was published, Lord Robbins was Imperial's Special Visitor at Commemoration Day. In his speech Robbins did not discuss the Report but expressed his concern that 'the conduct of those who rule over us is sufficiently informed by technological and economic knowledge'. He wanted to see improvement in the communication of new research and findings. In this he echoed the earlier concerns of Hill, concerns which were widely shared and had a major influence in the development of non-technological studies at Imperial (see chapter fourteen). But, as will be discussed elsewhere, problems in the world economy during the 1970s, a decline in the attraction of a university education, and demographic change, brought Robbins' expansionary vision temporarily into question.

The 'Feudal' System in South Kensington

In 1959 Linstead visited a number of American universities and wrote some notes on his experiences. He thought the way in which academic appointments were made in the United States was 'ridiculously democratic' and could not lead to excellence. He was especially shocked by the democratic procedure in professorial appointments at the University of Chicago where all department members had a vote.[90] This exemplifies Linstead's fundamental conservatism. Imperial College was, in many respects, an innovative institution. It had brought new disciplines and new kinds of people into the academy, had pioneered the acceptance of new kinds of accomplishment for matriculation, and had had its mining course accepted as a degree course. But the break from a paternalistic tradition, both with respect to departmental governance and towards students, proved more difficult. At Imperial very few had a say in academic appointments. The fact that the Rector had almost complete control over the appointment of heads of department compensated for a lack of power at the centre. Since department heads owed their position to the Rector they, like barons in a feudal system owing allegiance to the king, were respectful of the Rector's position and carried out his instructions which, by and large, were minimal.

Great care was taken to appoint dynamic heads who could carry their departments forward with little help from the centre. This was necessary in a college that had few resources and no serious endowment. For many years the system worked well. The head was expected to choose research areas, oversee the curriculum, and handpick his staff — though after 1929 the appointment of professors and readers required the approval of the University. While expected to be autocrats, department heads had also to be paternalists. Being a good head meant promoting something akin both to family and fraternal culture at the departmental level. Within the department, each member of staff, and all the students, had their place. People, of course, were very conscious of where they stood in the hierarchy but, in well-governed departments, being low in the pecking order was compensated by the recognition that one was part of a worthwhile enterprise, that one belonged. The system was propped up by various membership rituals, and by well-orchestrated social events. It was a system not without its problems, even before the war. By the late 1950s it was beginning to fall apart.

I am not sure when the term 'baron' first came into use to describe heads of departments. But its use was widespread in the 1950s and 60s and, by the end of that period, had become a term of abuse. Before the war the college and its departments were small. All the professors in the college knew each other; indeed most of the academic staff knew each other. By the 1960s departments had grown considerably and it was no longer possible for a department head to know everyone in his fiefdom. Further, given the appointment of so many young academic staff and technicians, old procedures came increasingly under fire. Many of the younger staff had visited the United States, and some had carried out postdoctoral research there. They had witnessed the informal and democratic climate at American universities and, unlike Linstead, they largely approved. The system at Imperial College which gave the head so much, albeit not total, control over personnel and working conditions, was resented. Those who had finally reached the pinnacle of a department were not always sympathetic to the demands of younger staff. They did not need to hold department meetings or set up committees. They could seek advice from their favourites, or not at all. And they had almost sole control over promotion. The Rectors, bound up with the major expansion, were no longer in touch with junior staff or with students and, without good lines of communication, were dependent on the word of department heads. Further, since money was flowing in to departments as never before, heads had more discretionary power in the 1950s and 60s than at any time before or since. Those that acted best in these new circumstances appear to have chosen one of two options — or somehow combined the two. One was to retain a sense of family or brotherhood in the face of historical circumstances working against this. Perhaps it was easier to retain some of the older norms in disciplines such as botany or geology where fieldwork involving groups of staff and students promoted bonding and a sense of group identity. The other option was to agitate for the promotion of staff members, something that Professors Blackett and Jones were able to carry off better than anyone else. Some heads of department, failing to recognize that their position would be enhanced, not challenged, by promoting their staff and creating new professorships, paid a price for not doing so. Further, as numbers began to climb, staff and postgraduate students began to identify more with their sections than with their departments. Professorial power increased relative to that of department heads. By the end of the 1970s baronial power had largely ended. The movement away from permanent headships, not unique to Imperial, was one of the many historical consequences of Robbins, of greater access to universities, of the political movements of the late 1960s, and of a changing economy.

...

1. UGC, 'Report on University Development, 1947–52' (Cmnd 8875).
2. *The Phoenix* (Summer, 1957); excerpt of speech by Linstead on occasion of the silver jubilee of Imperial College. Sputnik, in 1957, was both a symbolic and very concrete reminder that the race was on. But the technological race was associated with the economy more generally.
3. A. H. Halsey, *The Decline of Donnish Dominion: The British Academic Professions in the Twentieth Century* (Oxford, 1992), chapter 2; quotation, 45.

4. Before the war science and engineering students represented about 25% of the total student population in the UK. The numbers were low; about 9000 in science and about 4500 in engineering.

5. *The Times*, 27 March 1946. Other staff joined in the letter writing campaign. For example, in an earlier letter to *The Times* (14 January 1946), H. V. A. Briscoe, H. Levy and G. I. Finch noted that highly qualified students were unable to come to the college and that 'there seems urgent need to overhaul regulations that needlessly waste the intellectual wealth of the people'.

6. Rector's papers; draft submission by Southwell, 1945.

7. For Barlow Report see below.

8. Peter Ritchie Calder, 'Science and the State', *The New Statesman and Nation* (8 Dec 1945), 354. This criticism was made despite Calder, Blackett, and Bernal all belonging to the informal group Tots and Quots, founded by Solly Zuckerman in 1931 and revived at the start of the war. This group published (anonymously) *Science in War* (London, 1940) advocating a scientific approach to the war. Later Ritchie Calder joined Blackett and Bernal as part of an informal group of scientists and engineers which met under the auspices of the Labour Party, often at the Reform Club. Their aim was to develop a scientific and technical policy for the country. Others attending included C. P. Snow, J. Bronowski, H. Florey, J. Huxley, and Imperial's D. M. Newitt. Harold Wilson attended as the representative of the party leader, Hugh Gaitskell. Later, Wilson delivered the Blackett Memorial Lecture at Imperial, on the occasion of the extension of the physics department and its being renamed the Blackett Laboratory (3 December 1975). Wilson noted that Blackett was appointed to the National Research Development Corporation (NRDC) and played an active role there until he moved to work for the new Ministry of Technology in 1964. Blackett was very much behind the creation of this Ministry and was a major advisor to the Ministers Frank Cousins and Tony Benn. The NRDC, founded in 1948, was Stafford Cripps's idea. He wanted to keep British inventions in Britain and not have them developed and manufactured elsewhere.

9. See Rector's papers; Southwell's memorandum to chairs committee 'The Future Organization of Biochemistry', 24 March 1944. Interestingly Southwell was against bringing in someone with expertise in animal physiology/chemistry, 'hardly an appropriate field of work for Imperial College', but this is what later occurred with the 1961 appointment of Ernst Chain.

10. Ministry of Education, Special Committee Report (chaired by Lord Eustace Percy, Rector of the Newcastle Division of the University of Durham) (HMSO, 1945). Southwell was a member of this committee. Since the government had little control over what subjects were taught in universities,

it moved to expand technical college education by introducing the Colleges of Advanced Technology in the postwar period.

11. The Scientific Manpower Report was commissioned by Herbert Morrison. Blackett was a member of the Barlow Committee and had earlier written to Morrison stating that the report by Vannevar Bush on science education in the US, namely, *Science — the endless frontier* (NSF Washington, 1945) could provide a valuable lead for Britain. See also Jean Bocock, Lewis Baston, Peter Scott and David Smith, 'American Influence on British Higher Education: Science, Technology and the Problem of University Expansion, 1945–63', *Minerva*, 41 (2003), 327–46; also discussion in *The Phoenix* (June 1947), 5–7.

12. Morrison was Home Secretary in Churchill's wartime cabinet and became Deputy Prime Minister in the Attlee government.

13. *The Times*, 10 December 1945, 45.

14. Sir Roderic Hill (1894–1954) succeeded Southwell as Rector in 1948 after serving for thirty-two years in the RAF. He was the son of M. J. M. Hill, professor of mathematics at University College London and a Vice Chancellor of the University. Roderic Hill, too, later became Vice Chancellor. Hill was a student at the Slade School of Fine Art before joining the army in 1914, and the Royal Flying Corps in 1916. After the war he worked at the Royal Aircraft Establishment in Farnborough and at the Staff College, Andover. In 1930–2 he was Chief Instructor of the University of Oxford Air Squadron and Richard Southwell, who was a good friend, helped him to a fellowship at Brasenose College. From Oxford, Hill moved to become Deputy Director of Repair and Maintenance at the Air Ministry and, in 1936–8, was Air Commodore, commanding forces in Palestine and Jordan. In 1938 he was promoted to Air Vice Marshal and appointed Director of Technical Development at the Air Ministry. There he worked under Air Vice Marshal Tedder, succeeding Tedder as Director General in 1940. Hill thus came into contact with all those working on radar during the Second World War. His main job, however, was the armament of Spitfire aircraft. He was also sent across the Atlantic to inspect aircraft there and ensure their readiness for combat in Europe. In 1942 he was appointed to head the RAF Staff College at Gerrards Cross and, after thinking himself overlooked, was promoted to Air Marshal in 1943 and placed in charge of a fighting group preparing for the invasion of Europe. In 1944–5 he was promoted to Commander-in-Chief, Fighter Command. After the war he stayed at the Air Ministry coordinating training throughout the RAF. He joined the Imperial College Governing Body in 1945 (also on the Governing Body were some of his former colleagues at the Air Ministry: Ben Lockspeiser, Henry Wimperis

and Ronald McKinnon Wood). The succession of Imperial College Rectors, Tizard, Southwell, Hill (all three were friends) is rooted in the history of aeronautics, the RAF and the RAE. See also Prudence Hill, *To Know the Sky: The Life of Air Chief Marshal Sir Roderic Hill* (London, 1962).

15. In 1966 P. M. S. Blackett was quoted as saying, 'it was much maligned old Cherwell who did it.' In other words, it was Lord Cherwell who persuaded the Conservative government to choose Imperial. See *Observer*, 2 October 1966, colour magazine, 29. Records show, however, that Cherwell at first supported the founding of a completely new institution along the lines of MIT. See PRO; UGC 7/905.

16. In 1956 the UGC had £3.8m in its capital programme. By 1963 this had grown to £30m. This period of expansion was during Keith (later Lord) Murray's tenure as chairman of the UGC. According to Carswell, Murray was 'in manner large, benevolent, persuasive, in action almost inexhaustible, he was a convinced and consistent expansionist'. It was during his tenure that the government agreed to the construction of seven new universities, and for the university colleges at Newcastle and Dundee to gain independence. As a consequence, the old quinquennial system of university funding soon began to show signs of collapse. For example, after the settlement of 1957, grant negotiation was reopened in exchange for universities taking in more students than had been planned for. Murray also persuaded the government to make a major increase in academic salaries (about 28.5% between 1957 and 1960) and that universities should collectively set up the Central Council for Admission. See John Carswell, *Government and the Universities in Britain: Programme and Performance, 1960–80* (Cambridge, 1985), chapter 2. For Murray's views on the expansion of Imperial College see the report of his Commemoration Day Address of 1962, *Nature*, vol. 196 (24 November 1962), 710–14.

17. These issues were discussed in many places. See, for example, the discussion within the CVCP (PRO: UGC 7/864 and 7/905). In the end the UGC was against the building of a new technological university, hence the agreement to a major expansion of Imperial College and to smaller expansions elsewhere.

18. *Hansard*, vol. 510, No. 42, 29 January 1953. By this time a Conservative government was in office but it accepted the advice of the UGC and the Labour government plans for higher education. There had been seven years of reports and discussion leading to the announcement. In addition to the Percy and Barlow committees mentioned above, there were the Tizard Advisory Council on Scientific Policy (which recommended increasing postgraduate courses and research), the Weeks National Advisory Council

on Education for Industry and Commerce, and the Hankey Committee on Technical Personnel. The latter had advised against too hurried an expansion of academic facilities. In addition the reports of many professional bodies and learned societies were taken into account. Hill was very much against any idea of Imperial becoming an independent university. He had strong personal ties to the University of London and became Vice Chancellor in 1953.

19. Hill, quoted in *The Phoenix* LX (Spring, 1953), 6–7.

20. C. P. Snow, *The Two Cultures and the Scientific Revolution*, (Rede lecture, Cambridge, 1959). This lecture deplored the cultural gap between 'literary intellectuals' and scientists. As David Edgerton has pointed out, this was an odd comparison given that Snow's lecture focussed on the question of why few scientists reached the administrative levels of the civil service. See David Edgerton, *Warfare State: Britain 1920–1970*, (Cambridge 2006), especially chapter 5. For more on the 'two cultures' debate see chapter 14 below.

21. Prudence Hill, *op. cit.* (14), 248.

22. This opens up the huge area of science in the corridors of power discussed by Edgerton, *op. cit.* (20). Edgerton reminds us of the existence of large and well-funded state and military research centres, and of the powerful positions held by many scientists in the state bureaucracy. However, the idea that scientists had a difficult time in reaching the top, and that the country was exclusively run by an elite educated in the arts and classics was a well entrenched view, hard to dislodge even today.

23. This aspect of Hill's legacy is discussed in chapter 14. Curiously, given Hill's desire to see Silwood used for cultural activities, Professor Munro records having to fight with Hill to preserve the hall and library as social space in the main building, and prevent them from being converted into laboratories. B/Munro, Munro memoir (typed manuscript).

24. Today the Hyde Park Relay is an inter-university event.

25. Read had the vague duty of 'providing some assistance to the Rector at the top level of the college'. The Governing Body accepted clearer job descriptions for this and other administrative positions in 1968.

26. Rector's correspondence; Hill to GB (no date) 1949. See Edgerton, *op. cit.* (20), for details of the size and efficiency of Britain's interwar and wartime aircraft industry, and on Britain's aviation expertise. In both cases, contrary to Hill's claim, Britain compared well with its competitors.

27. See 50th Annual Report (1956–7); report of dinner, 27 May 1957. Hailsham was invited back to give the Commemoration Day Address in 1961. For a report on this see 'Research in the Universities', *Nature*, vol. 192 (2 December 1961), 787–91. Hailsham praised Imperial for finding the balance

between 'usefulness and popularity', implying that utility *per se* was still unpopular in academic circles. He also mentioned that three-quarters of the government's science budget went into defence-related research. He stated that universities should try to encourage more industrial support for science and technology since industry was 'not motivated by war or the fear of it'. In this he anticipated the position of the later Wilson government.

28. According to his daughter (see Prudence Hill, *op. cit.* (14)). Hill suffered from high blood pressure for many years, and was in very poor health during the last eighteen months of his life. But to most people at Imperial his death came as a shock.

29. Sir Reginald Patrick Linstead FRS (1902–66), known as Patrick, was educated at the City of London School and at Imperial College. He was a research student under Jocelyn Thorpe and, on gaining his PhD, worked briefly for the Anglo-Persian Oil Company before returning to join the staff at Imperial in 1929. In 1938 he was appointed to the Firth Chair of organic chemistry at the University of Sheffield, leaving two years later for a chair at Harvard University. During the war he was involved with the liaison of Anglo-American work, returning to London in 1942 as Deputy Director of Scientific Research at the Ministry of Supply. After the war he was appointed Director of the Chemical Research Laboratory at Teddington. He succeeded Ian Heilbron as professor of organic chemistry at Imperial in 1949. Linstead had considerable administrative skills and was a good committee man. He sat also on many boards and committees outside the college. For example, he was a governor of the London School of Economics and a Trustee of the National Gallery. According to Ellingham, Linstead was 'quiet, modest and perhaps rather shy … scrupulously fair … urbane, thoughtful and kind, [with] a happy disposition and natural charm'. See B/Linstead, H. J. T. Ellingham, 'Sir Patrick Linstead: biographical notes'. See also D. H. R. Barton, H. N. Rydon and J. A. Elvidge, 'Reginald Patrick Linstead, 1902–66', *Biographical Memoirs of Fellows of the Royal Society*, vol. 14 (1968), 309–337.

30. J. S. Anderson FRS remembers Linstead as 'a lecturer in his first year, with a pink-cheeked nervousness that was hard to recall in his later years'. Quoted in B. G. Hyde and P. Day, 'John Stuart Anderson 1908–1990', *Biographical Memoirs of Fellows of the Royal Society*, vol. 38 (1992), 4.

31. GB Minutes, 49th Annual Report (1955–6) includes Linstead's inaugural lecture.

32. Eric F. Cutcliffe (1904–78) was Assistant Registrar, 1937–40, and Registrar, 1940–66. He was known in the college for his physical fitness and his love of ballroom dancing. He especially enjoyed the annual summer ball at Silwood.

33. Before the major expansion most people relied heavily on departmental libraries and on the Science Museum Library. At first the Central Library was principally an engineering library but, by the late 1960s, and after major purchases also in the sciences, most disciplines were well represented.

34. Inaugural lecture, *op. cit.* (31).

35. An interesting debate took place in the House of Lords with a range of opinions on the demolition represented. Viscount Falmouth spoke on behalf of Imperial stating that too much time had already been lost and demolition should be allowed to proceed. See report in *The Times*, 14 March 1956, 4. In the Commons a major college ally of this period was Austen Albu MP. An Imperial College graduate in mechanical engineering, he was active on the Parliamentary Science Committee and was a member of the college's governing body. Albu was later appointed Minister of State for Economic Affairs in the Wilson government. This minor job was a disappointment for a man who had expected to become Minister for Science and Technology in a Gaitskell government. But Albu and Wilson were not on good terms and Albu's political career did not advance.

36. When the Waterhouse building was demolished a small parchment and a sealed jar were found under one of the entrance hall columns. The parchment recorded the laying of the foundation stone in 1881 by the Prince of Wales (the future Edward VII); the jar contained some items of the time: sets of Troy and Avoirdupois weights, a set of 1881 coins and newspapers of the day. All are in the Imperial College archives.

37. These plans involved the complete demolition of the RCS building among other things.

38. *Hansard*, House of Commons, 21 June 1956. The announcement by H. Brooke, Secretary to the Treasury, also confirmed Imperial's possession of the North, East and South sides of Prince's Gardens. As to the island site, a new lease of 999 years from 1956 was arranged with the Royal Commission for the Exhibition of 1851 for all the old and new property to be occupied by Imperial. The cost was £5000 a year, rising to a maximum of £25,000 over the following ten years. (See Eleventh Report of the Commissioners, Cmd. 1502, 1961). Some expense, and local civil engineering skill, were required to keep the tower standing after the rest of the building was removed. After finally accepting the demolition, the Royal Fine Arts Commission, wanting to preserve the possibility of restoring the axis between the Natural History Museum and the Royal Albert Hall, did not want Imperial covering the central area of the island site with tall buildings. Some compromises on the original plans were made, but the idea of a central axis, included in Tizard's earlier plans, was lost. For plans before the decision to keep the Colcutt

Tower was made, see *Nature*, vol. 177 (21 April 1956), 723–6. The article 'Imperial College of Science and Technology: Proposed Expansion Scheme', makes much of the need for Britain to respond to the threat posed by the advanced state of science and technology education in the Soviet Union. For article on revised scheme with the Colcutt Tower in place see 'Imperial College of Science and Technology: Revised Expansion Scheme', *Nature*, vol. 178 (8 June, 1957), 123–24. The final plans were revised in only minor ways from those shown in this article.

39. *The Phoenix* LXXI (Autumn, 1956), 13–14.

40. Prince's Gardens was acquired in 1956 and added 6 acres to the college estate. The first new hall of residence was Weeks Hall on the north side, largely paid for by Vickers and named for Lord Ronald Weeks, Chairman of Vickers Nuclear Engineering Ltd. During the war, Weeks had been Director General of Army Equipment. Shortly after, new halls of residence (Falmouth, Keogh, Selkirk and Tizard), refectories and a bar were built on the South side, later collectively known as Southside.

41. B/Linstead; Linstead quoted in H. J. T. Ellingham 'Sir Patrick Linstead: biographical notes'. The UGC had exacting space standards. It allowed 150 sq. ft. for a lecturer's room, 225 sq. ft. for a reader, 300 sq. ft. for a professor, and 3.5 sq. ft. per student in lecture theatres, etc.

42. In the 1950s Britain was planning for twelve nuclear power stations and also had a major research and production programme in military applications such as nuclear-powered submarines, atomic and hydrogen bombs and their delivery.

43. See report of a speech given at Trinity College, Cambridge on the nuclear arms race; *The Times*, 3 March 1956. Blackett believing that a nuclear war could only be won with, for him, an unthinkable first strike policy, placed himself in the somewhat paradoxical position of wanting a nuclear solution to a nuclear problem.

44. Roger Mellor Makins (1st Baron Sherfield) FRS (1904–1996) was the son of Brigadier Sir Ernest Makins MP. The Makins family were college neighbours and for many years lived on Queen's Gate in a house that stood on a site now occupied by the mathematics and computing departments. Makins was educated at Winchester and Christ Church, Oxford where he read history and formed a friendship with Frederick Lindemann. On graduation he won a fellowship to All Souls and was soon called to the bar. But his career was made in Whitehall. During the Second World War he was a minister in the British Embassy in Washington and, through Lindemann, became involved with Tube Alloys and the British contribution to the atomic bomb. Makins returned to the United States as ambassador in 1952. He was

Joint Permanent Secretary at the Treasury, 1956–60 and, in 1960, moved to become chairman of the UKAEA. He appointed William Penney as his Deputy Chairman. On his retirement, Penney succeeded him as chairman. Sherfield became Chairman of the Governing Body in 1962, and later he invited Penney to become Rector of Imperial College. Sherfield was a very clubbable man and reportedly a very good dancer. Brian Flowers saw him as a 'towering figure, with a razor sharp mind, robust good health and prodigious vigour; but he radiated calm confidence, exquisite manners and good humour. Never domineering he was always dominant; on the Governing Body many were strong-minded and experienced men and women, but all accepted his leadership without question.' For quotation and further information, see Lord Selbourne, 'Roger Mellor Makins (1st Baron Sherfield), 1904–1996', *Biographical Memoirs of Fellows of the Royal Society*, vol. 44 (1998), 267–78.

45. John Menzies Kay (1920–) was educated at Sherborne School before becoming an engineering apprentice at the Crewe Locomotive Works. He left his apprenticeship in 1939 for Trinity Hall, Cambridge where he took the mechanical sciences tripos in 1941. After a further graduate apprenticeship at Rolls Royce he returned to Cambridge and took a PhD in fluid mechanics under B. Melvill Jones, graduating in 1949. He was a lecturer at Cambridge before leaving in 1952 to become Chief Development Engineer at the UKAEA Industrial Group at Risley Lancs. where he was involved in the design and construction of nuclear reactors.

46. Later, when employment prospects in this field looked bleak, Penney initiated a joint course with Queen Mary College, thus consolidating weakened resources. Sir Stanley Brown, Chairman of the Central Electricity Generating Board wrote to Penney 'there is a huge reservoir of both young and old nuclear experts in the UKAEA who may need redeployment … both we and our consortia will inevitably be more concerned with these people than with new entrants'. See KN; Brown to Penney, 5 December 1968.

47. Geoffrey Ronald Hall (1928–2001) was educated at the University of Manchester and then worked in the chemistry division at Harwell rising to Principal Scientific Officer. In 1956 he was seconded to work on nuclear power in India where he designed and set up the Trombay Laboratory and briefly headed the radiochemistry division. He was appointed Reader in Nuclear Technology at Imperial in 1958 and was promoted to a professorship in 1963. He left Imperial to become Director of the Brighton Polytechnic in 1970.

48. There was some hesitation over this given the events of 1956, and the fact that the college was housing 17 Hungarian refugee students at 12 Prince's Gardens. Dennis Gabor petitioned Linstead to help the Hungarian refugees, which he did both by fund raising and by providing accommodation.

49. *The Times*, 19 October 1960.

50. 'Reactor Merry-Go-Round', *Nuclear Engineering*, (August, 1961), 2.

51. Peter John Grant (1926–99) was promoted to Professor of Nuclear Power in 1966. He was educated at Cambridge and had been Chief Physicist at the atomic energy division of the General Electric Company before coming to Imperial in 1959. He had to take over from Kay as head of the nuclear power section in 1961 and to oversee construction of the reactor, designed jointly by Kay's nuclear power group in mechanical engineering and people at the General Electric Company. Grant had been involved in this while an employee of GEC, and was responsible for the shielding design and, as it turned out, for much of the commissioning of the reactor. The cost was approximately £200,000. Grant later became involved in University of London affairs and was Deputy Vice Chancellor, 1990–2.

52. Sir Ernst Boris Chain FRS (1906-79) was born in Berlin and studied chemistry at the Friedrich Wilhelm University. He came to England the day Hitler came to power, became a student at University College London and then, recommended by J. B. S. Haldane, moved to work with Sir Frederick Gowland Hopkins at the Sir William Dunn School of Biochemistry at Cambridge. After two years he moved to the Sir William Dunn School of Pathology at Oxford to work with Howard Florey. There he carried out the work on penicillin that led to his being awarded the 1945 Nobel Prize for Medicine (with Alexander Fleming and Florey). Chain played a major role in developing the fermentation and fractionation processes needed to purify penicillin which, when upscaled in the United States, led to the successful manufacture of the drug. Chain was annoyed that the MRC refused to fund this research, furious when, after his success, Oxford refused to consider patent protection, and critical of what he saw as an amateur response to his work more generally. He was especially critical that no one was prepared to fund the setting up of a large-scale fermentation plant in Britain. These frustrations took him away from Britain and, from 1948–60, he directed the International Research Centre for Chemical Microbiology at the Istituto Superiore di Sanità in Rome where a large fermentation plant was built. Gradually British drug manufacturers, such as Beecham, set up similar fermentation plants and Britain, once again, became a centre for antibiotic research. Chain joined the faculty at Imperial in 1961, was knighted in 1969, and retired in 1973. He was an excellent pianist and had considered making

a musical career. See Ronald W. Clark, *The Life of Ernst Chain: penicillin and beyond* (London, 1985); obituary, *The Times*, 16 November 1979.

53. Chain asked the College to rent a house in the neighbourhood with 'seven bedrooms, three bathrooms, three reception rooms (one large and suitable for music), dining room, study and large kitchen'. The college could not afford this kind of largesse. Later, after discussions with the UGC and the Treasury, it was decided to include a penthouse flat in the new biochemistry building. Chain was to be charged rent, to be deducted from his salary. Linstead took steps to ensure this benefit was for Chain only during his tenure, and not for any of his successors (his immediate successors did live in the flat but it was later converted for laboratory use). The Wolfson Foundation agreed to include the flat in its building, and agreed also that after Chain retired the college could do as it pleased with the space. See Lloyd Davies, memorandum on the history of the biochemistry department flat, 14 December 1972.

54. The total cost of the building was £1.5m.

55. Chain's papers have been deposited in the Wellcome Library and contain much information on the founding of the department and design of the biochemistry building.

56. See ULP/Biochemistry correspondence files. The microbiological course did develop into one on fermentation science and technology.

57. Wilson actually used the term 'scientific revolution' but that is not how it was (or is) remembered. See Edgerton, *Warfare State*, *op. cit.* (20), 230.

58. Perhaps this is unfair. Had Chain produced another wonder drug one might well think differently. He thought he had found such a drug in the antiviral chemical, statalon, which has remarkable antiviral properties but proved too toxic for human use.

59. For a good account of the more general history of computer science see Michael Mahoney, 'Computer Science: the search for a Mathematical Theory', in John Krige and Dominique Pestre (eds.), *Science in the Twentieth Century* (Amsterdam, 1997), chapter 31.

60. See R. J. Cunningham, 'Computing' in Adrian Whitworth (ed.), *A Centenary History: a history of the City and Guilds College, 1885–1985* (London, 1985).

61. Michaelson, who was the top Imperial mathematics graduate in 1945, later held the chair of computing science at Edinburgh. Brooker later held the chair at Essex and Tocher later headed the operational research section at United Steel, one of the foremost centres in the field. ICCE1 worked reliably, but very slowly, until 1957.

62. At this time it was not yet clear that digital computers would prove superior. Differential equations, for example, could be solved more quickly using analogue machines.

63. John Hugh Westcott FRS FREng (1920–) was apprenticed with British Thomson Houston Ltd. in Rugby and took an external University of London degree in electrical engineering. During the war he was seconded to the Radar Research and Development Establishment at Malvern where he worked on the development of electrical circuits for radar equipment for coastal and field artillery. After the war he was sent to Germany to assess destruction of industrial units in the Ruhr and to investigate what research had been done on radar. He came to Imperial in 1946 on a Senior Studentship from the Royal Commission for the Exhibition of 1851. He joined the Imperial staff as lecturer in 1950 and was appointed Professor of Control Engineering in 1961. Westcott applied his knowledge of control engineering also to economic modelling. In this connection he worked for the government and was much involved with the Club of Rome. He and three friends founded Feedback Ltd., a successful company that made automation kits, control units, and test equipment for use in university teaching as well as in industry. In an article in the *Sunday Times* (22 September 1968) Westcott was described as 'a tall thin man who could look austere if he stopped smiling long enough'.

64. Tustin, who had been appointed as a control specialist, worked also on electric traction. Earlier he had worked on machines for trolley buses and trams at Metropolitan Vickers. This company later amalgamated with others to form Associated Electrical Industries.

65. Stanley Gill (1926–75) was educated at Cambridge. For several years he worked at the National Physical Laboratory and then at Cambridge carrying out research in punched card computing and computer programming. From 1953–64 he worked for Ferranti, in charge of the Atlas programming team. He was Professor of Computing at Manchester from 1962–4 (the first chair of computing in the United Kingdom). From 1966–70 he was Advisor on Computers to the Ministry of Technology. He has left an interesting collection of papers located both in the ICA and in the Science Museum.

66. The college could use only one-third of the time on the new computer. IBM used one-third and associated institutions the other one-third. 34 other universities and 25 schools and colleges of the University of London had access to the machine — a measure of the scarcity and expense of early computing facilities.

67. William Sydney Elliot (1917–2000) was educated at Cambridge where he took the physical sciences tripos. His PhD research at the Cavendish was

interrupted by the war when he went to the Air Defence Research and Development Establishment at Christchurch, Hants. to work on radar. In 1947 he joined Elliot Brothers (no relation) where he led a team designing the 401 computer (now in the Science Museum) and another team on the Nicholas computer used in computing the trajectories of early guided missiles. He joined Ferranti in 1953 and was responsible for the design and development of the Pegasus machine. In 1956 he moved to IBM and helped to set up their laboratory near Winchester. He began converting American machines for the UK market. But this work was not to his liking and he returned to Cambridge working on a joint project with Ferranti on the Titan Computer. At Cambridge he began work on computer-aided design, a topic he wanted to develop at Imperial. But his Imperial career was not a happy one and he agreed to reduce his responsibilities and return to Cambridge part-time.

68. Gerald James Whitrow (1912–2000) was educated at Christ's Hospital and Christ Church, Oxford where, after graduating, he was senior scholar. While at Oxford, he was much influenced by the cosmologist E. A. Milne. He joined Imperial as a lecturer in 1945 and received a conferred chair in applied mathematics in 1972. Whitrow had an interest in the history and philosophy of timekeeping and among his books on the subject are, *The Natural Philosophy of Time* (1960) and *The Nature of Time* (1975). Whitrow is remembered also for his very loud voice which could be heard 'from miles away'. For Skempton see chapter 10. Both Whitrow and Skempton were occasional attendees at the Sunday afternoon salons held at the Drayton Gardens home of the French historian of technology, Jean Gimpel.

69. Alfred Rupert Hall (1920–) was born in Stoke-on-Trent and educated at Cambridge. He was a professor of the history of science at Indiana University before coming to Imperial. Both Halls are specialists in the period of the Scientific Revolution. See also transcript of interview conducted by A. K. Mayer with Rupert Hall and Marie Boas Hall, 21 January 1998, (ICA). As with the Chains, the Halls were given a college flat (at 179 Queen's Gate — see KH/2/2). Nancy Skempton (Skempton's wife) rebound some of Hall's old books and restored others, including a copy of one of Robert Boyle's works in 5 folio volumes. Judith Niechcial, *A Particle of Clay: the biography of Alec Skempton, civil engineer* (Whittles, Caithness, 2002), chapter 5.

70. The department was revived as the Centre for the History of Science and Technology during the Rectorship of Eric Ash, and with the appointment of D. E. H. Edgerton as head of department. See chapter 14.

71. Lionel Charles (Lord) Robbins (1898–1984) held the chair of economics at the London School of Economics for many years. Before the war he had

defended traditional monetarism against Keynes, but during the war he ran the semi-socialist war economy as Director of the Economics Section of the Offices of the War Cabinet. A major public figure, he was on the boards of many public bodies. According to Carswell, Robbins was a large impressive man with a mane of silvery hair who had a 'friendly, comforting confidence, and disagreement was tolerated but made no impression ... he intended from the first that his report should mark a great advance' (*op. cit.* (16) 28).

72. Before the war 80% of the population left school at 14 and entered the labour force. The new Act raised the school leaving age to 15. A new foundation was also laid for the financial support of students at university. In 1950 less than 2% of the 18–21 age cohort attended university. Professional training for engineers was still largely by apprenticeship/pupilage on the job.

73. Many of the grammar schools were excellent and helped students from poor backgrounds to achieve much. This was especially the case with some of the older grammar schools that existed even before the war and were brought into the new system. Two examples of eminent Imperial professors who came from poor backgrounds and received an excellent education in pre-war grammar schools are Geoffrey Wilkinson and Willis Jackson (both discussed elsewhere). They were among the fortunate; many children from poor families did not receive an education commensurate with their abilities. While, in principle, one can sympathize with Anthony Crosland's wish (when Secretary of State for Education) 'to destroy every fucking grammar school in the country', in practice such sympathy would be misplaced since nothing better emerged. The new comprehensive system was very uneven and in many areas of the country educational opportunities declined. (More grammar schools were closed during Heath's Conservative government, when Margaret Thatcher was Minister of Education, than earlier.) Crosland had little sympathy for the university elite but was concerned with other areas of higher education and is perhaps most responsible for the binary divide that shaped policy into the 1990s. 'On the one hand we have ... the autonomous sector, represented by the universities ... [and] colleges of advanced technology. On the other hand we have the public sector, represented by the leading technical colleges and the colleges of education' (Crosland, Woolwich speech, 27 April 1965). The rise of the polytechnics and, later, their licence to grant degrees, stem from Crosland's policies.

74. Peter Clarke, *Hope and Glory: Britain 1900–1990* (London, 1997), 287.

75. Lord Ernest Simon, Manchester engineer, city politician, and former chairman of the University of Manchester Council.

76. Between 1918 and 1945 only one of Britain's Prime Ministers (Baldwin) had a university education; since 1945 only three (Churchill, Callaghan and Major) have not. That 4% of the age cohort were attending university by 1961 (2% in 1950) was a consequence of the Education Act.

77. The 'new' universities of the 1960s (Sussex, East Anglia, Stirling etc.) were already under construction when the Robbins Committee was formed.

78. The Robbins Report: *Report of the Committee on Higher Education* (Cmnd 2154, 1963). For more on the role of the state in higher education before Robbins see R. O. Berdahl, *British Universities and the State* (Cambridge, 1959). For an excellent account of the Robbins Committee by a government insider, and on what followed from the Report, see Carswell, *op. cit.* (16). See also Brian Salter and Ted Tapper, *The State and Higher Education* (Ilford, 1994) and Richard Layard, John King and Claus Moser, *The Impact of Robbins* (Harmondsworth, 1969). That there was only one member of the Robbins Committee from Oxbridge (Helen Gardner) was not surprising given that one of its purposes was to bring greater equality to the British university system. In fact the Report was hard on Oxford and Cambridge, and both universities soon put in place committees to report on their own situations.

79. Kingsley Amis, 'Lone Voices: views of the "fifties"', *Encounter*, vol. 15 (1960), 8–9 (quotation in increasing font-size capitals in original). Ralph Smith and Robert Eddison, *Talent for Tomorrow* (Bow Group pamphlet, London, 1963). Crowther had been Chairman of the Central Advisory Council for Education and, in that capacity issued the Crowther Report (1959) which recommended raising the school leaving age to sixteen. The quotation was taken from the report. Crowther was the editor of *The Economist* and earlier had been a member of Tots and Quots (see note 8). He was also a major proponent of the Open University.

80. One year before Zuckerman's Report, the Advisory Council on Scientific Policy had published, *The Long-Term Demand for Scientific Manpower* (Cmnd. 1490, 1961). It foresaw a levelling out in demand for scientific and technical manpower in the 1970s. But Zuckerman's committee, which included Willis Jackson from Imperial, had to rethink this prediction in light of the 1961 census and new estimates from employers. They predicted shortages in mathematically trained people, and in electrical and mechanical engineers. They also saw a trend in scientists moving into managerial work. For a discussion of Zuckerman's Report, *Scientific and Technological Manpower in Great Britain*, (1962), see *The Times*, 17 October 1963.

81. The principal recommendations were for an expansion of university places from 216,000 to 560,000 by 1980–1 with intermediate targets of 328,000 by

1967–8 and 392,000 by 1973–4. It suggested 60 institutions be of university status, under an enlarged UGC reporting along with the various research councils to a new Minister of Arts and Science. Ten Colleges of Advanced Technology and some teacher training colleges were recommended to get university status or to become faculties in neighbouring universities. The intention (not achieved) was to bring the total number of universities in Britain to 56 by 1968, the year in which the actions of university students was to dominate the news worldwide. The recurrent costs were estimated to be £743m per year by 1980–1, and that a capital expenditure of £1420m over 17 years would be needed.

82. A new bureaucracy, the Universities Central Council on Admissions was created in the 1960s to bring some standardization to the admissions process. By 1989 there were 52 universities (an increase of 22 from pre-Robbins) and 82 colleges and polytechnics under the new funding councils.

83. Harold Wilson, 'Labour and the Scientific Revolution', (speech given at Scarborough Conference, 1963). For a contemporary view see also, Norman J. Vig, *Science and Technology in British Politics* (Oxford: Pergamon, 1968).

84. Including the defence industry and what Wilson thought was needed in the context of the Cold War.

85. While the number of science and technology places did increase, the output of science and technology students from secondary schools could not keep apace the expansion. The end result was that increases in the number of arts and social sciences students outstripped those in science and engineering. This had much to do with universities needing to show that they had achieved the growth targets set.

86. This was called the 1970 Committee. Membership was restricted to those under fifty years of age and excluded heads of department with the exception of Hugh Ford who chaired the committee. A number of younger heavyweights sat on this committee, including D. H. R. Barton, C. C. Butler, B. G. Neal, and J. H. Westcott. Their recommendations were largely instituted in the late 1960s and early 1970s, as will be discussed in chapter 10.

87. See GCB1/1-6. See also B/Linstead, box 4, for much on Robbins Committee.

88. According to Carswell (*op. cit.*, note 16 chapter 6), Linstead was the force behind this decision. The five institutions chosen were Imperial College, the University of Strathclyde, the Manchester College of Science and Technology; one College of Advanced Technology yet to be specified, and one completely new foundation. They soon took on the acronym, coined by Linstead, the five SISTERS. But only the three named institutions

were funded and not to the degree Robbins had envisaged. For letter see B/Linstead/box 4.

89. Linstead to Principal of University of London, 21 November 1963. At this time Imperial was educating about 1 in 13 of Britain's scientists and 1 in 9 of its engineers.

90. B/Linstead; notes on a visit to Harvard, Princeton, MIT, Chicago, California Institute of Technology, Yale, Columbia, and Cornell; Spring, 1959. He did not care for a system in which department members voted on the appointment of new staff, and on department heads or, worse, chairmen.

chapter ten

Expansion: Post-War to Robbins, 1945–67 (Part Two)

Continuity and Growth

As discussed in chapter nine, new fields such as nuclear power and biochemistry came to Imperial after the war, but there was growth also in existing areas. This chapter describes some of the principal developments within the various departments in the postwar period. Southwell had especial difficulty filling chairs in engineering but he early made one internal promotion and one new appointment that were to be of major significance in the postwar history of the college. Owen Saunders was promoted to succeed C. H. Lander and became head of the department of mechanical engineering in 1947. Willis Jackson, whom Southwell had known at Oxford, was appointed to succeed C. L. Fortescue as head of the department of electrical engineering in 1946.[1] As the interesting correspondence between them indicates, Southwell had some difficulty in persuading Jackson to come to Imperial.[2] But he was successful and, in the 1950s and 60s, Jackson, Saunders, and the head of the physics department, P. M. S. Blackett, were perhaps the most powerful academic voices in the college. Their views transcended their own departments which, in some important respects, became models for the rest.

Electrical Engineering

Jackson found conditions in his new department very poor and thought that much of the equipment belonged in the Science Museum. However, as his obituarists put it, within a very short period 'he created the most active and forward-looking electrical engineering department in Britain'.[3] Imperial adopted a three-year undergraduate course after the war and Jackson was instrumental in designing the new course for electrical engineers.[4] This included a common first-year course for students in aeronautics, electrical, civil and mechanical engineering. In 1948–9 Jackson arranged for some electrical engineering students to devote one day a week to study at the London School of Economics and, as is discussed in chapter fourteen, was a great promoter of a broader education for engineers. He also introduced specialist courses for postgraduates and encouraged research which, during the war, had been close to non-existent. By the time he left in 1953 (he was to return in 1961) he had persuaded the DSIR, and many of his industrial contacts, to support research groups in electrical materials (Jackson himself worked on the properties of dielectrics), microwaves and ultrasonics, electron physics, plasmas, information theory, and control.[5] In his pedagogy, and in promoting research, Jackson was concerned to bridge what he saw as a gap between universities and industry. During the eight years he was away from Imperial he developed a range of educational projects at Associated Electrical Industries that were to inform his later pedagogical philosophy. During his absence the head of department was Arnold Tustin who, too, had earlier worked under Fleming at Metropolitan Vickers.[6] In 1959 Tustin, wishing to spend the few years before his retirement back in the laboratory, asked to be relieved of the headship. Linstead then invited Jackson to return, but had to compete with Cambridge where Jackson had also been offered a chair. Linstead was successful and Jackson returned in 1961. Tustin moved to head the heavy electrical engineering section, by then in the new 'high block' building that he had planned and nursed to completion.[7] In the new building Tustin and B. J. Cory introduced a one-year MSc course in electric traction, designed to meet the needs of British Railways after its decision to electrify main lines and suburban services.[8] The electric traction work was later taken on by Eric Laithwaite who, on succeeding Tustin in the chair of heavy electrical engineering in 1963, brought his novel ideas on the linear induction motor to Imperial.[9]

Colin Cherry came to Imperial with Jackson in 1947 and fostered research in the technical and theoretical aspects of telecommunication, including digital signal processing, coding theory, routing strategies, and global communication.[10] Cherry was one of the first to promote the use of TV in the classroom and developed closed-circuit TV to include also some other institutions. Linstead was cautious about these initiatives and did not allow them to develop far. Cherry, like Westcott, had come under the influence of Norbert Wiener when spending a sabbatical year at MIT. While a very private person, he was a gifted communicator and he popularized Wiener's ideas in Britain. He became known also for his writing on the psychological and sociological aspects of communication, and for his BBC radio lectures on information theory. He attracted students from around the world.[11] Communicating with the man himself was, however, not always easy. He was renowned in the college for 'failing to receive' administrative notices.[12] In 1958 he became Henry Mark Pease Professor of Telecommunications, a chair funded by Standard Telephones. Together with his colleague Dennis Gabor, another major contributor to the field, he organized several international symposia on information theory.

Gabor was a genius. He was clearly something of an enigma to his students. As Eric Ash, his first research student, and a future Rector of Imperial, put it,

> When Gabor moved to Imperial College in 1948, he presented a remarkable and, at first, unfathomable phenomenon to his research students. He had little comprehension of the difficulty encountered by ordinary mortals in attempting to gain an appreciation of his aims in a particular research endeavour, or to follow a leaping thought line without the help of stepping stones. His lecture courses on electromagnetic theory, statistical physics, the plasma state, electron optics seemed memorable but hardly capable of assimilation — at least until the realization dawned that they were not so much lectures as master classes. It was necessary to 'know' the subject before attending these occasions; but then the experience was enormously worthwhile. ... There was no doubt amongst his students that Gabor could and should receive the Nobel Prize for something — we debated just what it might be.[13]

Ash and his fellow students were correct in predicting a Nobel Prize for Gabor. It was awarded in 1971 for the invention of holography. Appointed Mullard Reader in Electron Physics in 1948, Gabor was given a chair in 1958.[14] He was much more than just an electrical engineer; he was a physicist, an outstanding inventor, a polymath. He was also an elitist. Gabor believed the Robbins philosophy to be totally misguided. Increased university access, he wrote, was 'a terrible notion of our democracy ... the biggest idealistic blunder of the century'.[15] He wanted a rigorous education for the elite and publicly expressed his hatred of the permissive society. Gabor wrote on many topics and, like Jackson, turned increasingly to engineering pedagogy. He was a founder member of the Club of Rome and, like many others at the time, given to futurology. While believing that computers would never totally replace human minds, they would, he claimed, cause serious problems. He envisaged a nightmare future of a leisured world with only a few people working, 'a world for which we are socially and psychologically unprepared'.[16]

After Jackson returned to the college in 1961 he introduced a further MSc course in electrical engineering applied to medicine. This course, first suggested by Tustin, was designed to train people both for a new manufacturing industry, and in the use of new electronic medical instruments in hospitals. The course was run by Bruce Sayers who had earlier worked with Cherry on binaural audition.[17] A research group was set up, focused mainly on instruments for monitoring cardiac function, and the interpretation of neuro-physiological signals. The group soon received worldwide attention. Jackson also returned to the question of modifying the undergraduate programme, but gave this task to one of his new staff members J. C. Anderson, a specialist in materials, appointed Reader in 1961.[18] The day-to-day running of the department was also left very much in Anderson's hands as Jackson turned more to events on the national stage. He became a major spokesperson on broadening the education of engineers, joined many government committees, including taking the chairmanship of the TV Advisory Committee in 1962.[19] Typical of his speeches in the 1960s was one given at the BAAS meeting in 1967. Young people were turning away from science and engineering in the later 1960s, and Jackson attributed this to wider issues such as environmental degradation, nuclear weapons and the problems of nuclear waste. He early talked of the responsibility of scientists and wanted science and engineering students to be exposed to the social sciences so that they could think constructively about the social consequences of emergent technologies. Jackson was a workaholic and was at the centre of science and engineering politics in his day. He collapsed at his desk in 1970 and died a few hours later. In introducing many short postgraduate courses, and in considering the broader education of his students, he anticipated well the future direction of the college.

Mechanical Engineering

Owen Saunders had come to Imperial before the war with Professor Lander and Margaret Fishenden and, as discussed in chapter six, they immediately began to modernize the department. This was especially true of the thermodynamics section in which heat transfer, the field Lander brought to Imperial, was central. But the department lost many of its staff during the war and it fell to Saunders to rebuild. He introduced a range of new activities in addition to nuclear power, discussed in chapter nine. Included were fluid mechanics, lubrication, plasticity and production engineering. Further, Saunders sat on the Aeronautical Research Council which supported departmental research on rocket motors and the use of hydrogen peroxide fuel. Much work at this time was focussed on gas turbines, supported by the Ministry of Fuel and Power and by the British Shipbuilders Association. Peter Grootenhuis was appointed in 1946 to work in this area.[20] However, by the 1960s Saunders was in something of a quandary since his prediction of rapid development in gas turbine technology failed to materialize and A. G. Smith, who had developed a postgraduate course in this field, had moved to a chair at Nottingham. It was not clear which way to move. In the end Saunders decided to continue promoting fluid mechanics and to wait and see whether it would be steam or gas turbine technology that would eventually capture the market. By the time Penney became Rector, the nuclear power section built up earlier with the backing of Linstead was in difficulty and Professor Kay had resigned from the chair of nuclear power. Similar problems existed at Queen Mary College and it was decided to consolidate nuclear work. Walter Murgatroyd moved from Queen Mary College to a chair in nuclear power in 1967.[21] Linstead's rosy predictions for nuclear power never materialized and this section diminished in importance.

While several of Saunders' post-war appointees were to influence the future of the department, none was to do so more than Hugh Ford. Ford had been one of Saunders' research students before the war, and in the interim had gained much experience in industry.[22] He returned to the college in 1948 as Reader in Applied Mechanics. Three years later he was given a professorship. Together, Saunders, Ford, and lecturer Russell Hoyle, planned the post-war expansion of their department into the new building. This was especially difficult since the mechanical engineering building was constructed in stages on the site of the old Waterhouse building in which the department was housed. Already one of the largest departments in the college, student intake increased fourfold once the new accommodation was in place. Ford built up many connections to industry and, as will be discussed in chapter fourteen, was concerned that his students learn also about the social and managerial aspects of engineering. Saunders made many other appointments in the late 1940s and 50s. Among them were Sam Eilon, another of his former PhD students, who joined the staff in 1954 to help to set up a Production Engineering Section, and Brian Spalding who earlier had established a combustion group at Cambridge under Sir William Hawthorne, someone with whom Saunders had close connections.[23] J. M. Alexander was appointed to a readership in plasticity.[24]

Saunders cooperated with Jackson on the first-year engineering course and, in this connection, appointed Derman Christopherson to a chair in applied science with special reference to engineering.[25] Christopherson arrived in 1955 but stayed only five years. He was succeeded by Bernard Neal who, like Christopherson, took an interest in the undergraduate curriculum.[26] In all, Saunders managed to build an impressive department which, given the depredation of the war, had almost a new birth. In 1964 he resigned the headship on becoming Pro-Rector. When Linstead died suddenly in 1966, he became Acting Rector until the appointment of William Penney in 1967.[27] He was succeeded as head of department by Hugh Ford.

Civil Engineering

After the war A. J. S. (Sutton) Pippard built what became the leading postgraduate civil engineering department in the country. On his appointment Pippard, together with Stanley Sparkes, set up a new structures laboratory funded by the Clothworkers. After the war this was modernized; its centrepiece was a model of the Verwoerd Dam on the Orange River, which Sparkes claimed was the most sophisticated model for dam testing at the time. In 1953 Linstead put Sparkes in charge of the college Planning Office, and later appointed him Director of Building Works for the Jubilee expansion. But work in structures carried on, having been strengthened in 1948 with the appointment of J. C. Chapman. Chapman worked in the area of composite structures, notably of steel and concrete and, after the war, was engaged in a major project relating to ships' hulls for the British Shipbuilders Research Association.[28] The new department building, largely planned by Sparkes and Chapman, allowed their own and other fields to expand yet further in the 1960s. Much of the apparatus for the structures laboratory in the new building was built internally by departmental technicians, notable among them J. E. (Jack) Neale who had been with the department since 1948. J. I. Munro, a practising engineer, was appointed to a lectureship in 1957 and, in 1963, on a visit to the University of California at Berkeley, was converted to the systems approach by George Dantzig and Lofti Zadeh. On his return he gave the first course in computer programming in the department and, in 1974, founded the new systems and mechanics section.[29]

In addition to Pippard just one other professor remained in the department at the end of the war, namely A. L. L. Baker, appointed to a new chair of concrete technology in 1943.[30] Baker was a tall man and there was barely enough headroom for him to work comfortably in his basement laboratory. But he worked there for over twenty years before moving into the new building which was fitted with one of the best concrete laboratories in Europe. He enjoyed large engineering projects, and thinking also about the social and environmental problems they posed. It is not surprising therefore that his postgraduate course on the theory and design of concrete structures drew students from around the world. He fostered strong ties to industry and many of his students came to the college on industrial bursaries. They were to play a major role in the not always (aesthetically) welcome spread of concrete technology worldwide in the second half of the twentieth century.

In 1948, C. M. White was appointed to the newly created chair in fluid mechanics and hydraulics engineering.[31] He introduced a very successful one year DIC course in hydro power with support from the English Electric Company. Structures and fluid mechanics were the main areas of research in the department but Pippard soon began to add new fields. One was public health engineering, supported by a grant from the Rockefeller Foundation in 1948. A joint DIC course with the London School of Hygiene and Tropical Medicine was first given in 1950 and attracted eight students.[32] It was run by F. E. Bruce, a former student who had studied public health at Harvard and was a consultant to the World Health Organization.[33] In 1955 Pippard introduced a further DIC course in engineering hydrology under the direction of P. O. Wolf, then a lecturer in the hydraulics section.[34] The course was timely since there was much public anxiety over the rate of increase in the world's population, the adequacy of the world's water resources, and over recent major flooding in Britain.[35] The course was interdisciplinary, with input from botany, geology, geophysics, meteorology, and mathematics. It later became an MSc course under the direction of J. R. D. Francis who had joined the staff in 1946 and succeeded White as professor of hydraulics in 1966.[36] In 1951 Hugh Dalton, then Minister of Local Government and Planning, invited Pippard to become chairman of a committee which was set up to consider pollution in the tidal reaches of the Thames. The committee took nine years to report and was much criticized for this in the Press. But Pippard claimed that time was needed for pollution research (carried out largely at the government Water Pollution Research Laboratory). In the end the committee's report was of major significance. It basically stated that the Thames was a filthy sewer and forced the government to take action. As a result the water quality of the Thames gradually improved.

Professor Clements, a specialist in the design of roads, had wanted to introduce the new field of soil mechanics to the department before the war. In this he had the support of Pippard, but plans were delayed until the war ended. In 1945, the year he retired, Clements invited Karl Terzaghi, the pioneer/founder of the new field of soil mechanics to give a series of lectures at the college. But, an Austrian citizen, Terzaghi was not then given a visa to enter Britain.[37] Also in 1945, Pippard invited A. W. Skempton, a former student, to come to the college on a part-time basis and introduce this new field.[38] In the same year Pippard appointed A. W. Bishop to an assistant lectureship.[39] Soon Skempton became a full-time senior lecturer and, by 1950, he and Bishop had developed a postgraduate course in soil mechanics.[40] Skempton built a research group which soon gained international recognition. He was rewarded with a chair in soil mechanics in 1955 and succeeded Pippard as head of department in 1957. While Skempton appears to have overshadowed Bishop, their careers were interdependent. According to Skempton their 'views on soil mechanics were very similar; we both valued field work and, if anything, he was more intent on the practical engineering applications.

But in some respects our talents were complementary. I certainly knew far more about geology while he was far the better designer of apparatus and a remarkably skilled experimentalist'.[41]

The civil engineering department has long had close relations with large engineering firms such as Binnie and Partners, John Mowlem, Ove Arup, and Wimpey. In the case of Binnie, Geoffrey Binnie first approached Pippard when his family's company was involved with construction of the Dokan dam in Iraq, built mainly to aid the irrigation of land in the region of Kirkuk. Pippard put together a team of theorists which included Letitia Chitty with whom he continued his close working relationship. The team used Southwell's relaxation methods in their work and the result is widely considered to have been one of the most complex pieces of three-dimensional stress analysis undertaken by civil engineers before the use of electronic computers.[42] Later, again through Binnie, Skempton was asked to lead a soil mechanics team on the very challenging Mangla Dam project, on the river Jhelum in Pakistan. This earth dam, constructed in a seismic area, was to provide one-third of the power requirements of Northern Pakistan by the 1970s. Begun in the late 1950s, the project took close to twenty years to complete, employed about 2000 engineers and over 30,000 workers.[43] N. N. Ambraseys, a seismic specialist, and future professor, was a member of the team. He and Skempton later worked with Binnie on the Mornos dam and reservoir that supplies Athens with drinking water. Another dam specialist was E. H. Brown who was consulted on numerous projects including the Aswan Dam power house.[44]

Professor Clements had earlier fostered interest in transport problems and, in 1962, when long-time reader B. G. Manton retired, Skempton decided to expand in this area.[45] In 1963, C. D. Buchanan was appointed to a chair in transport made possible by a grant from the Nuffield Foundation.[46] A few years later Barbara Castle, Minister of Transport, set up a committee on urban transport planning under the chairmanship of Lady Sharp.[47] Sharp's report recommended the establishment of a major teaching centre in the field of transportation, and that it should offer a range of two-year postgraduate courses. For reasons that are not entirely clear, Imperial did not see itself in this role and responded to the Minister by stating that several centres should provide such courses, and that it could be one of them. Buchanan was disappointed. He wanted to see his section expand, and to take in more postgraduates on short courses.

But there were space problems and Skempton was unenthusiastic.[48] Buchanan resigned the chair in 1972 and left to become Director of a new School for Advanced Urban Studies at the University of Bristol. While at Imperial he tried to have the area around the college declared an 'environmental area' and stated that with the 'worst effects of cross-filtering traffic eliminated' it has the potential for 'creating a wonderful sector of London'.[49]

Chemical Engineering and Chemical Technology

Sir Alfred Egerton was appointed to the chair of chemical technology, and as head of department, in 1936. In 1937 the University of London approved an undergraduate course in chemical engineering.[50] Students entering the undergraduate course were trained first in basic science and engineering by staff in other departments and by staff in the chemical engineering department only in their third year. This led to some problems of association until the southern extension to the Roderic Hill building was completed in 1966. The new space allowed the first and second year students to join their department which then began giving its own engineering and chemistry courses to the students. When the undergraduate course opened in 1937 it attracted only four students but after the war it grew significantly, in tandem with the expansion of Britain's chemical industry. This expansion was reflected also in the growth of chemical engineering departments nationally. In 1950 there were only seven such university departments and together they graduated about 120 students each year. By 1970 there were twenty-three departments graduating about 750 students a year.[51] D. M. Newitt, appointed to the Courtaulds chair of chemical engineering, took charge of the course after the war. He developed it in ways that, according to him, still owed much to the earlier teaching of J. W. Hinchley. Newitt was joined by J. M. Coulson who returned to the college from managing a wartime explosives factory, and by J. F. Richardson, another former student. Coulson and Richardson wrote a successful course textbook before leaving the department, Coulson for a chair at Newcastle and Richardson for a period in industry.[52] In addition to the academic staff, Newitt was much assisted by J. S. Oakley a technician who came to work in the department after working on bomb disposal during the war. His interest in explosives and other interesting devices was shared by Newitt and they worked well together. During the Cold War it was Oakley who brought a number of Ministry of Defence contracts to the department, for things that might more properly have belonged in the mechanical engineering department, or perhaps in a James Bond film. They were jokingly referred to as 'hi-tech pocket tools and cutting keys for a certain sort of burglar' and a device 'for removing without trace seals on diplomatic bags'.[53] Oakley became Superintendent of the Laboratories and Workshops in the department in 1968.

Newitt promoted the chemical engineering course in various ways. As he put it,

> no industry has made such spectacular advances or has a more assured future than that which we, as chemical engineers, are most closely associated. Wide ranges of non-ferrous alloys and plastics, improved artificial fibres and synthetic rubbers, new pharmaceutical products, insecticides and fertilizers, and a new chemistry based on olefines and acetylenes are some of the results of the war-time exertions which await development.[54]

The department had moved to the C&G College in 1937 and so had become, officially, an engineering department. This move and the new undergraduate course marked the beginning of the relative rise in importance of chemical engineering with respect to chemical technology. But the balance between these two divisions remained in flux and continued as a source of tension for some years. In the immediate postwar period Egerton built up a major research group in low temperature work and continued to promote work introduced by Bone in the fields of flame spectroscopy, combustion and fuels. But much of Egerton's attention was elsewhere, outside the college. As Chairman of the Scientific Advisory Council of the Ministry of Fuel and Power from 1948–53, he was involved in the long-term planning of the newly nationalized coal, gas, and electricity industries. When he retired in 1952, Newitt became head of department and A. R. P. Ubbelohde, a former Egerton student, was appointed to a chair in chemical thermodynamics.[55] The relationship between Newitt and Ubbelohde was strained and when, on Newitt's retirement in 1963, Ubbelohde became head of department, the question of the identity of the department came very much to the fore. This is not surprising given that the period was one of transition. The old guard was being replaced by a new generation. Newitt was a brilliant man and remained active in research for a further ten years, but his apparatus was seen as obsolete. High-pressure work was continued by J. S. Rowlinson, appointed not as an engineer, but as professor of chemical technology in 1961.[56] When Ubbelohde became head of department he attempted to persuade Shell to fund Rowlinson in his desire to study reaction mechanisms at pressures of 50,000 atmospheres (as compared to Newitt's 15,000 atmospheres) but was unsuccessful.

P. V. Danckwerts was appointed to a chair in chemical engineering science in 1956.[57] At the time the college was looking for a reader to replace Coulson, and Danckwerts was seeking a return to academic life from a senior position with the UKAEA's Industrial Group. The college was keen to appoint him but a readership would not do. Linstead persuaded the University of London to allow a third chair in the department. Danckwerts was expert in a range of fields, from chemical kinetics, to flow problems in chemical industry, to the disposal of nuclear waste. The title of the new chair is interesting in that it suggests an attempt to bridge the divide between the scientists and engineers. While Danckwerts stayed for only three years, he did point the department in a direction that was to prove highly fecund in the future. His interest in flow and process engineering led him to give a series of lectures on the subject. And, at his suggestion, the department began to modernize its undergraduate course; to include both more analytical and more engineering content. When, in 1959, he moved to the Shell Chair of Chemical Engineering at Cambridge he was succeeded by K. G. Denbigh, who had been Professor of Chemical Technology at Edinburgh.[58]

The problems between Ubbelohde and Newitt had almost led to the splitting of the department, something Denbigh was against. Indeed, he had refused to come to Imperial were that to happen. However, to safeguard his position he insisted on a separate budget line for chemical engineering; he also insisted on a name change for the department. From the time of his appointment it was to be known as the Department of Chemical Engineering and Chemical Technology. Denbigh, like Danckwerts, was interested in modernizing the undergraduate course. Despite his own standing as a thermodynamicist, Denbigh's pedagogical views differed from those of Ubbelohde who tended to favour basic science teaching, and theory over practice. Denbigh was able to show that Imperial students were finding it more difficult to gain job placements in the late 1950s and early 1960s and blamed this in part on the lack of proper engineering training in the undergraduate programme.[59] On Newitt's retirement, Denbigh became the Courtaulds Professor but resigned the chair in 1966 after difficulties with Ubbelohde. Ubbelohde was a rather imperious head. He was supported by his secretary, Georgina Greene who, with his approval, took over much of the departmental administration. She wielded power over the staff and was understandably resented. Most resented was that she controlled access to Ubbelohde. Greene and Ubbelohde sat in adjoining offices with a half-silvered mirror between them. Several people remember the pantomime that was played out in front of the mirror for the sake of Ubbelohde on the other side. Because of this experience, the department hesitated before placing power in the hands of any professional administrator after Ubbelohde retired. This meant more work for the academic staff; but eventually, realizing that Greene and Ubbelohde were *sui generis*, and that the problems associated with them were unlikely to be repeated, they recognized that having a departmental administrator was, after all, a good idea.

It is probably fair to state that by 1960 the department had lost its reputation as the leading chemical engineering department in the country.[60] The members of staff were known more for their considerable achievements in physical chemistry and chemical physics than for chemical engineering. In addition to the people already mentioned, A. G. Gaydon, appointed to a chair of molecular spectroscopy in 1962, was further proof of this.[61] But Danckwerts' pointer towards process engineering was to be given a major boost with the appointment of Roger Sargent as professor of chemical engineering science in 1962. Sargent, a former student in the department, had spent time in France with an air liquefaction company before returning as senior lecturer in 1958. His interest in process engineering had very practical consequences both for industry and for the future history of the department.[62] To aid his work he pushed for more instruction in computing, and a course was introduced in 1964. While other research in the department during the 1960s was still heavily focussed on aspects of kinetic theory, thermodynamics and spectroscopy, the application of these fields to the practical needs of industry gradually increased. But, as Ubbelohde put it in a 1967 statement to the SRC, while departmental research was often inspired by industrial problems, 'our primary research purpose is the obtaining of new knowledge, not trouble shooting for industry'.[63] Tellingly, however, the southern extension of the department, completed in 1966, included space for a pilot plant laboratory.

Aeronautics

Arnold Hall was appointed head of the aeronautics department in 1944 in succession to Leonard Bairstow.[64] He was twenty-nine years old at the time and the college had high hopes for him. His departure in 1951 to become Director of the RAE was a blow, but Hall made much progress towards building the department before he left. In 1949 he introduced a new undergraduate aeronautics course, though students spent much of their first two years studying mechanical engineering and only special-ized in their third year.[65] This may have seemed sound pedagogically but it had a psychological flaw. Young aeronautics enthusiasts were often put off their subject by having to wait too long before learning much about it. Hall managed to find funding for a new aerodynamics laboratory which, with obligatory wind tunnels, opened in 1951.[66] He also wanted a good aerostructures laboratory, but had to make do with existing facilities until the completion of the Roderic Hill building in 1957. Hall claimed that aeronautical structures were unlike others and that the instruction given in civil engineering, for example, would not serve his students well.[67] Funding for a readership in aeronautical structures was forthcoming and Hall promoted one of his senior lecturers, J. H. Argyris, to the position in 1950. Argyris had come to Imperial in 1949 and soon made a name for himself in the new field of 'aero-elasticity'.

When Hall resigned, the Rector, Roderic Hill, took advice on how to proceed from Sir Frederick Handley Page. Handley Page's attitude was not altogether academic; he wanted to see more nuts and bolts training and encouraged the construction of new wind tunnels for this purpose.[68] He also gave Hill advice on whom to appoint. Hill decided on H. B. Squire but had to compete with the Cambridge department to appoint him.[69] Both departments had been founded at the end of the First World War but, by the early 1950s, Imperial's department was larger. In 1951–2 it had a total of 69 undergraduate and 29 postgraduate students.[70] Squire was a recognized expert in aerodynamics and his research covered a wide range of topics from jet propulsion to aircraft noise to seaplane aerody-namics, to pioneering theoretical work on the heating effects of high speed airflow. He had wanted to introduce a third stream of aeronautical engineering into the department but was unsuccessful. It remained bipartite with its aerodynamics and aeronautical structures divisions.

Argyris was given a chair in 1955 and proved a difficult, albeit gifted, colleague.[71] In 1957 he negotiated a formal division of the department to coincide with the move into the new building. His structures division was to constitute a sub-department with rights to 45% of the departmental staff. Further, Argyris was to have control over all staff appointments in his sub-department — though Squire would have a veto. Squire appears to have been someone who retired from conflict and he allowed power to slip from his hands. Argyris wrote what perhaps is the all-time record number of demanding letters to college Rectors during his years at Imperial. For example, early recognizing the potential of computers for use in dealing with the geometrical complexity of aircraft structures, he detailed his computing requirements in 1957. They were way beyond anything the college could then provide. As Argyris put it, 'I know that it is not possible to order a Ferranti Pegasus from one day to another but I feel that some urgent effort should be directed at the realization of such a plan'.[72] One year later he offered his resignation on being appointed Director of the Aeronautics Department at the Institute for Statics and Dynamics, University of Stuttgart. Very good computing facilities were offered him there. With hindsight it can be said that Linstead should have accepted the resignation. But he wanted to keep Argyris and asked him to consider staying at Imperial on a part-time basis. In the event a very complicated part-time arrangement was worked out, which included the use of a college flat at 177 Queen's Gate. Aside from Argyris, the only people to clearly profit from the new arrangements were some postgraduate students who learned Argyris's methods and had access to the Stuttgart computer for their calculations. William Penney, Linstead's successor, wanted to end the arrangement but was unable to do so; his successor Brian Flowers found a way. Argyris and the college came to an amicable parting.[73]

With Argyris being absent for much of the time the departmental administrative duties fell heavily on Squire and his successor P. R. Owen.[74] Squire had seen through the move to the new building but took his own life in 1961. It fell to Owen to build the department in its new premises. During the interregnum between the death of Squire and Owen's appointment in 1963, Argyris persuaded Linstead to set up a rotational headship entailing a complex division of power between the two professors. Owen agreed to this before taking up his chair, signing a document to that effect in 1962. Owen was to be the first head for a period of three years beginning in January 1963. In fact the rotation did not occur and Owen remained in place, though Argyris's position in the department was never entirely clear.

Much of what happened during the 1960s was related to the fate of the British aircraft industry. At the end of the war the industry had to meet new challenges in both military and civil aviation. In the important sphere of civil aviation, the Americans took the lead. Britain did develop turbine-propellor and jet propelled aircraft but its lead was transitory and the demand for personnel by the aircraft industry declined dramatically until the coming of Airbus many years later.[75] Owen saw the decline as an opportunity to broaden the department's outlook. He made fluid dynamics a central discipline and encouraged its application to a wide range of industrial and environmental problems. For example, he was interested in the aerodynamics of tall buildings, in flow related to power-plant heat exchangers,[76] the hydrodynamics of off-shore structures, and air flow around fast moving land and water-based vehicles. In this connection the department had earlier helped Donald Campbell in developing the *Bluebird* which broke seven water speed records between 1955 and 1964.[77] Further work on racing car aerodynamics led to work with the automobile industry more generally. Owen was able to expand his staff and open new laboratories. One was a hypersonics laboratory under J. K. Harvey who had worked with hypersonic wind tunnels at Princeton. Two of the younger men Owen brought to the department in the 1960s, to work on the range of new problems he envisaged, P. W. Bearman and J. M. R. Graham, were later to become professors and heads of the department.

One of Owen's innovative moves was the setting up of the Physiological Flow Unit in 1966. The roots of this lie in Farnborough where the Institute for Aviation Medicine was neighbour to the RAE. Scientists interested in cardiovascular and respiratory problems related to flight saw common ground with those at the RAE interested in fluid mechanics. Both Owen and M. J. Lighthill, who held a Royal Society research professorship at Imperial from 1964, had worked at the RAE and had come into contact with the physiological work being carried out at the Institute.[78] Lighthill's mandate at Imperial was to encourage interdepartmental activity, something he succeeded in doing. At the time he was working also with Colin Caro, a specialist in cardiovascular and respiratory physiology at St. Thomas's Hospital. Their collaboration proved fruitful, but both men believed the work of applying fluid mechanical ideas to blood circulation would best be forwarded in a permanent inter-disciplinary unit. Lighthill put this idea to Linstead who, along with Owen, agreed to provide support. The unit, administratively part of the aeronautics department, was given premises at 11 Prince's Gardens, but moved to the Roderic Hill building when further space for laboratories was needed.[79] Caro became Director of the Unit which began with only two permanent staff members, though a number of visiting academics and several research students were attached. Gradually the SRC and the MRC came to support work being done in the Unit, mainly on cardio-vascular problems such as arteriosclerosis. Instruments capable of being inserted into blood vessels and measuring blood flow were developed. Courses were given and many of those who attended were from medical fields. The Hayward Foundation helped fund a new laboratory which opened in 1976, but the Unit's financial situation was tenuous.[80]

Physics, Mathematics, and Meteorology

George Thomson continued as head of the physics department until his retirement in 1953. After the war, he headed a research group in thermo-nuclear processes which, as he later wrote to the Rector, began working in that field in 1945. 'I was interested in what could be done with forms of nuclear energy other than fission, and began to think about the possi-bilities of a hydrogen bomb. However, I quickly tired of this and turned to the possibility of producing controlled energy by fusion'.[81] By the 1950s, he had moved this group to Aldermaston, recognizing that secret work did not belong in the university. Another nuclear physics research group began work under the direction of Samuel Devons who came to the college in 1950.[82] Even before his arrival, work with particle accelerators had been carried out. One early model, in use before the war, generated both proton and deuteron beams. It was poorly shielded, and was situated in open laboratory creating 'noise, vibration and sparks!'. It was said that while the accelerator was running, 'sparks could be drawn from the hot water heater pipes in the basement spectroscopy lab' below.[83] However, after the war, radiation hazards were finally recognized and a new, more powerful (2 MeV), Van de Graaff generator was installed with better protection. B. W. Wheatley was appointed Imperial's first radiation protection officer in 1949, later moving to CERN to become radiation protection officer there. Devons inherited the second machine and improved its performance such that it provided beams for a large group working on nuclear spectroscopy.

The largest research group in the immediate post-war period was in technical optics, a field with a long history at the college. By the 1950s the group was under the direction of W. D. Wright.[84] W. T. Welford, a good designer of optical systems, had joined the group in 1947 and, in 1960, the section was strengthened with the appointment of C. G. Wynne, one of the most distinguished lens designers of his generation.[85] Another major figure in this period was Harold Hopkins, a pioneer in fibre optics and inventor of an illuminated endoscope.[86] A. G. Gaydon and W. R. S. Garton continued to be active in spectroscopy during the 1950s and were joined by Lady Anne Thorne in 1955.[87] Thomson had earlier set up a group under D. H. Perkins to work on cosmic rays. High altitude balloons coated with various photographic emulsions were used in this work, some of which was carried out on the *Jungfrau* in the Swiss Alps.[88] From the late 1940s the plan for geophysics, then under J. McGarva Bruckshaw, was that it be moved to Geology as soon as space had been added to the RSM building. This occurred in 1955. Smaller departmental clusters were engaged in acoustics, heat, and electron diffraction, and a number of people were working on their own projects.

The arrival of P. M. S. Blackett as head of department in 1953 augured major change.[89] Blackett was persuaded to come to Imperial by Hill who told him that the planned expansion would give him the freedom to build the department as he wished. While many were surprised that Blackett left Manchester, it appears that he wanted to be closer to the centres of power in London and that he and his wife enjoyed London life.[90] Before taking up the chair Blackett wrote to Hill stating that, in his view, the department was engaged in too many fields of research and that he would like 'by concentration of effort … to produce fewer and stronger groups, some of which, it is to be hoped, will be first class'. He wrote that he wanted to bring C. C. Butler with him from Manchester and that Butler would build a group in cosmic ray research. He also wanted a group working on paleo-magnetism, his own more recent area of interest, and intended bringing J. A. Clegg from Manchester to help him build this field.[91] In 1952 Blackett was hesitant about supporting nuclear physics because he thought serious work in the area would be too expensive. But he expressed the wish to continue support for those areas of research that he thought were good at Imperial, namely technical optics, spectroscopy, electron diffraction, and acoustics.[92]

Blackett decided to close down research in so-called low energy physics, whereupon Devons resigned taking the Van de Graaff generator with him. Instead, Blackett supported what became known as high-energy nuclear physics and asked Butler to set up a group in this field. Butler later recorded his first impressions of the RCS building: 'although I knew the Science Museum Library, the physics department on its west side was unknown to me. First impressions were not very encouraging! There were long wide corridors, vast teaching laboratories and the largest men's toilet I have ever seen. The research space was scattered about the building and it was very difficult to assess its extent'.[93] Butler brought his cloud chamber group with him from Manchester and was appointed to a readership (he became a professor in 1957).[94] He hired Derek Miller, a technician who worked with him on the development of the bubble chamber.[95] Together they, and others, built a small liquid hydrogen bubble chamber which began operating in 1958, the first of its kind in Western Europe. Imperial then joined with a number of other universities in a consortium, led by Butler, to build the larger British National Hydrogen Bubble Chamber. Butler's group designed the optics for that project. Like his professor, Miller was soon drawn into the centre of particle physics research, work that took them often to Hamburg and to CERN in Geneva. The construction of bubble chambers, and the scanning of particle tracks on films, was very expensive. Scanning required new computer technology some of which was developed at Imperial by a later group leader and head of department, Ian Butterworth.

Harry Elliot was another who came with Blackett from Manchester. It was he, not Butler, who set up research primarily in cosmic rays. Later his section moved towards astrophysics more generally.[96] Elliot was soon promoted and, by 1960, was a professor. He had a licence to carry out cosmic ray work in a disused part of Holborn Underground Station (an Act of Parliament was required for this) but, in 1964, just as he was getting some good results, he was asked to leave.[97] Elliot moved his magnetometer and other apparatus to the observatory at Pic du Midi. He also helped design some British instruments for the first international scientific space satellite, Ariel 1. His own group built a cosmic ray detector for the satellite which was launched by NASA in 1962. Work in the field of satellite instrumentation has remained a major strength of the physics department.

To complement the work of Butler and Elliot, Blackett wanted some theoreticians in his department and, to the dismay of the mathematics department, persuaded two of its stars, M. A. Salam and P. T. Matthews to join him.[98] Bringing them to the department brought also great prestige. The two were already well on their way in building a leading centre for research in theoretical physics. Brilliant and dynamic, Salam was a font of ideas and was well complemented by Matthews who had a keen sense for which of his many ideas might actually work. The two attracted many students (35 postgraduates in 1959) and their weekly seminars attracted famous visitors, including Murray Gell-Mann and Steven Weinberg. T. W. B. Kibble, a future head of department, joined Salam's group in 1959. Blackett was concerned that Salam might return to Pakistan and wrote to Linstead, 'I would not be surprised if he ended up as Prime Minister!'.[99] He wanted Linstead to be as generous as possible to Salam's group and, in that connection, was able to acquire a professorship for Matthews. That the group was able to expand depended also on the move into the new physics building.

The loss of Salam and Matthews was hard for the mathematics department. Before the war Chapman had modernized the syllabus and introduced new material in mathematical physics and wanted to expand in that area.[100] His successor, Hyman Levy, continued in the same direction. The department had a good reputation for numerical analysis and, before the electronic computer age, calculations were done on hand-cranked machines largely operated by women. Levy was succeeded as head by Harry Jones who took over in 1955 at the start of the expansionary period at the college.[101] Some growth had been possible immediately after the war, when the zoology department left the old Huxley building, but Jones soon began drawing up plans for the new Huxley building on Queen's Gate — though it was his successor who moved in. The mathematics department had finally severed its link to mechanical engineering and had changed its name from the department of mathematics and mechanics to the department of mathematics in 1951. It still provided much service teaching for other departments, but began attracting more students of its own — about 20 per year by the early 1950s and about 40 per year by the end of the decade. In addition, there was some funding for new areas in applied mathematical research.

As discussed in chapter nine, Jones closed down the computer builders. He early made a distinction between computer construction and computer use, and encouraged only the latter. Jones's own work in the theory of metals made the department a world centre in that field. Fluid mechanics, notably gas dynamics, received a major boost, as did the areas of probability and statistics, in part a consequence of air defence work. Links to production engineering were forged by the statisticians led by George Barnard.[102] The Rector and the University wanted the department to expand in the direction of pure mathematics, something Jones resisted. In a letter to Linstead in 1956 he stressed 'the danger that a school of pure mathematics in this college might actually deflect some of our best men from the useful fields of applied science which we, in the national interest, are trying to develop.'[103] But Linstead prevailed and W. K. Hayman was appointed to a new chair of pure mathematics in the same year.[104] Four years later another pure mathematician, K. F. Roth, joined the staff.[105] Much of this expansion was possible as a result of the department acquiring space in Prince's Gardens in 1964.

The other occupants of the old Huxley building were meteorologists. Blackett made a raid there too. He invited B. J. Mason to move and set up a cloud physics group in his department.[106] At the end of the war the head of meteorology, a small postgraduate department, was still Sir David Brunt. Brunt, who was Secretary of the Royal Society, appears to have spent more time there than at the college. Further, the Meteorological Office's decision not to use Imperial as its training centre (something that had been considered) meant that the department shrank in importance. It is understandable that Mason made the move. Brunt retired in 1952 and his successor, P. A. Sheppard, was unable to sustain the department over the longer run.[107] It collapsed into a section of physics on his retirement in 1974. But some of the meteorology staff joined the mathematics department. Notable among them was R. S. Scorer, Professor of Theoretical Mechanics, who built up research in fluid mechanics.[108] Jones perhaps learned something from Blackett. By the time he retired he had built a major department with eleven professors.[109]

Blackett was undoubtedly an empire builder and, while favouring 'pure' research, was aware of the importance of Imperial's links to industry. In building his department, he kept an eye both on the new transistor industry and on commercial applications of work in electron physics. In this connection he reinstated the chair in instrument technology that had been allowed to lapse and appointed J. D. McGee in 1954. McGee soon had a major research group working on the detection and intensification of faint optical and X-ray images and in developing instruments for use in astronomical observatories and space satellites.[110] He claimed that when he arrived at the RCS he had to throw out a lot of old Wimshurst machines from the space allotted to him. Perhaps they dated back to the 1870s when Frederick Guthrie had a large collection. Blackett also supported the work of one of Thomson's young appointees, Bryan Coles, who, after Bruckshaw's geophysics group had moved to geology, was given space to develop research in solid-state physics.[111] Coles learned much from Harry Jones, and developed a course for metallurgy students. After the retirement of Thomson, Morris Blackman's team in electron diffraction merged with Coles's group. Solid-state physics took off in the 1950s and Coles was at the forefront of the field. He attracted many students, including some from the United States. N. F. Mott joined his group as a senior research fellow on retiring from the Cavendish.[112] Coles spent his entire career at Imperial and, as will be seen elsewhere, played an important role also in college administration.

Another area that Blackett and Linstead thought worth promoting was biophysics. In the 1950s many physicists were turning to work on biological problems. As Pnina Abir-Am has put it, after the war biophysics was 'a potential scientific and cultural redeemer of the physicists' conscience and tainted skills'.[113] Imperial wanted to catch the new biophysics wave, though not all the biologists were impressed. As botanist P. H. Gregory put it in a memorandum to Linstead, 'it would be naive to expect the walls of Jericho to fall as soon as a team of physicists attack biological problems'.[114] But Linstead, keen to support Blackett, approached a number of foundations for funding, and asked a number of eminent scientists whether they would be interested in starting a new group. He had hoped to attract J. C. Kendrew from Cambridge but, as Lawrence Bragg advised him, that would have meant creating more than a section. He would have needed a new department.[115] J. H. Randall, professor of physics at King's College was also contacted since he had moved into biophysics and reputedly wanted to give up running his department. But Linstead was unable to promise space for more than 25 doctoral students, fewer than Randall wanted. Biophysics had to wait.

As mentioned, Blackett's own work at Imperial was in geomagnetism, a field he had begun to explore in Manchester. His larger interest was in the theory of continental drift and whether geomagnetic data could be used in testing the theory.[116] Blackett personally led much of this work at Imperial, at least until the election of Harold Wilson's government in 1964 and his own election as President of the Royal Society in 1965. From then on his main focus was on science and industrial politics in his role as scientific advisor to Tony Benn at the Ministry of Technology. The day-to-day leadership of his geomagnetism group was taken by J. A. Clegg. The team collected rock samples from all over the world to be tested using both Blackett's and J. McGarva Bruckshaw's magnetometers. The Imperial group also helped to set up a geophysics laboratory at the Tata Institute in Bombay, and magnetic data from many Indian rock samples was added to the rest.[117] Gradually enough data was collected to support Alfred Wegener's 1915 theory of continental drift. Since the theory was receiving also much support from elsewhere, Blackett's good judgment in picking research topics had placed him yet again at the centre of a new and exciting area of scientific research. While not in the same category as the particle physics and cosmic ray research that was a spin-off from his Cavendish work, his work in this new field was nonetheless a major achievement.

Construction on the new physics building began in 1957. According to Butler, Blackett was not interested in planning for it. He had wanted chemistry to move and for physics to expand into the space vacated in the RCS building.[118] Linstead must have had similar feelings since, when he unexpectedly became Rector, he decided that it was physics, not chemistry, that would move. Physics was given a large bomb site with frontage on Prince Consort Road for construction of its new building. Blackett delegated much of the planning to W. Thompson, the departmental superintendent, and to two of his academic staff, Bryan Coles and N. R. Barford; but he kept an eye on the overall design realizing that the building would be identified with his name. Blackett insisted that the library and space for undergraduates be on the lower floors and that the bulk of research be conducted higher up. He consulted Professor Skeaping of the Royal College of Art on a decorative lintel to be placed over the main entrance. Skeaping created a suitably iconic design on a slab of Irish black limestone. The building was officially opened by Sir John Cockcroft on 20 October 1960.[119] Moving in, along with the staff, were about 300 undergraduates and about 100 postgraduate students — numbers that were soon to grow.

Blackett had great presence, was tall and very good looking. Not only did his appearance suggest natural authority and leadership, he had the authority that came from outstanding service in both world wars, from his having been a student at the Cavendish during its golden years, and as a Nobel Laureate. As a leader he had both autocratic and democratic instincts. The former can be seen both in his dictating which fields (usually well chosen) his young staff would engage in, and in not holding department meetings.[120] Blackett believed in giving his staff independence only up to a point. Perhaps he also gave them the advice encapsulated in the following poem.

> If you can organize your lab
> nor lose the power to think
> and deal with academic chores
> nor drive yourself to drink.
> If you can make a test of all your findings
> and risk it on one test of yes or no
> and lose, then
> you'll stay a prof my son.[121]

His democratic instincts can be seen in his insistence that there be a single common room in the new building open to all the staff, including secretaries, technicians, and postgraduates. It was a place where all the research groups could come together and exchange ideas over tea or coffee. In a radical departure, he listed the physics staff in alphabetical order, not according to rank, in the college calendar.[122] Further, Blackett was willing to delegate power and he led the way in setting up the multi-professorial department.[123] He had seven professors in place before moving into the new building, at a time when the physics department at University College London had only one. By the time he resigned the headship in 1963 (two years before his retirement) and was succeeded by Butler, the department had ten professors. Multiplying professors was a recipe for success soon copied by other departments.

Chemistry

The chemistry department had inherited a constitution that allowed for a form of governance that differed from the other departments in the college. Since 1914 the professor of organic chemistry had headed one section, and the professor of inorganic chemistry had headed the inorganic and physical chemistry sections. There was no overall head of department. This arrangement did not meet the approval of Roderic Hill who, when he arrived at the college, put a plan before the Governing Body for an overall head, and that the head should be the senior of the two professors, H. V. A. Briscoe. This was a major blunder, though Hill's choice can be understood. Briscoe went a long way back. He had links to the Edwardian college having come to the department as a student in 1906, and had been professor of inorganic chemistry since 1932. He had also served the college well during the war.[124] However, the professor of organic chemistry, Sir Ian Heilbron, who also had a fine record of wartime service in science, was a far more distinguished chemist than Briscoe. When he heard of the plan, he resigned. He was deeply offended and wrote to Lilian Penson, Vice Chancellor of the University, complaining that 'it is stated that the Professor of Organic Chemistry will work under the direction of the head of the department. This in effect makes him scarcely more than a Reader or Senior Lecturer'.[125] The next most senior organic chemist, Reader A. H. Cook, resigned in protest and left with Heilbron. Together they set up a major research laboratory for the brewing industry.[126] There is little doubt that the wider chemistry community sided with Heilbron since the college had great difficulty in replacing him. No organic chemist of repute applied for what many saw as the premier organic chemistry chair in the country.[127] Hill was acutely embarrassed, but the matter was resolved by some of the governors. They persuaded R. P. Linstead to return to his old college as professor. Linstead, who had established a good reputation in the 1930s with his discovery of phthalocyanine and allied compounds, was appointed in 1949. On his return he reinitiated work on biologically important molecules, and his research on porphoryn and chlorophyll type compounds was supported by the Rockefeller Foundation.[128]

Linstead recognized that Hill's plan would not do. In light of the fact that a new professor in physical chemistry was soon to be appointed, and that Briscoe was to retire in 1954, he devised a new departmental constitution to take effect in that year.[129] Linstead's constitution was tripartite, but met Hill's requirement that there be an overall head. The three sections, inorganic, organic and physical chemistry would each have a professor and separate budgetary provisions. One of the professors was to be head of department with responsibility only for broad policy, and for coordinating the work of the three sections in the undergraduate course. Linstead became the first head, but Hill died in 1954 and so Linstead served only one term under his new constitution before becoming Rector. This left the chemistry department in a difficult position. R. M. Barrer had been appointed to the chair in physical chemistry only in 1953, and a successor to Briscoe had yet to be found.[130] Further, after the Heilbron debacle, it was not clear that a major organic chemist could be attracted to the chair. There was little option for Linstead but to offer Barrer the headship, which he accepted. Linstead then arranged the appointment of a young man from his own research group, E. A. R. Braude, to the chair in organic chemistry.[131] Braude had worked with some good organic chemists at Imperial, namely Heilbron, E. R. H. Jones and Linstead, but had not developed his own line of research. Only recently promoted to a readership, his appointment to the chair was questionable. Shortly after, G. Wilkinson was appointed to succeed Briscoe in the chair of inorganic chemistry.[132]

The record is scanty but it would appear that Braude and Barrer did not get on, and the burden of the organic chair was too great. Braude took his own life one year after his appointment. This tragedy opened the organic chemistry chair yet again. Once again the department had to look to one of its own and D. H. R. Barton was appointed.[133] The department now had the three professors in place who would lead it to a period of great success. Barrer was a remote head of department but the divisions functioned well as independent units until the discipline of chemistry needed to be reconfigured in the 1970s. Already by 1951 Briscoe had overseen the reconstruction of the cavernous old analytical laboratory. The addition of an extra floor allowed the construction of two new laboratories and, together with other renovations, three modern teaching laboratories, named for eminent chemists earlier associated with the department (Crookes, Frankland and Philip), awaited the new professors. Briscoe also arranged for the agricultural chemistry laboratories located in huts at Harlington to be re-equipped after the war. Work in that field continued until the late 1950s when it was allowed to lapse.

Wilkinson and Barton were both highly competitive; they built major research teams and were to become Nobel Laureates.[134] Barrer, more retiring, also built a major research team and was nominated for the Nobel Prize. Wilkinson was a leader in the renaissance of inorganic chemistry after the war and trained over 70 PhD students during his career. At Imperial he furthered his work on the transition metals and made major contributions to the fields of coordination chemistry and catalysis.[135] He was a man in a hurry. As his obituarists put it, 'the spirit of his research group was more like that of an urgent gold-rush in the West than the scholarly and disciplined calm expected in academia.'[136] His work was aided by new nuclear magnetic resonance (nmr) and other spectroscopic techniques, and he early appointed staff to work in these areas. They included D. F. Evans and L. Pratt who came from Oxford to develop work in nmr.[137] According to Wilkinson the department was soon able to acquire an nmr machine since Lord Todd, who sat on the SRC, was fed up with all the big money going to physics and was delighted to be able to support a major chemistry project for a change.[138] Two other Oxford graduates, D. M. L. Goodgame and M. Goodgame, appointed lecturers in 1961 and 1962 respectively, were soon joined by W. P. Griffith. The three brought a further range of spectroscopic and inorganic chemical expertise to the section.[139] Wilkinson also appointed a theorist, John Avery, later a professor in Copenhagen, who, according to Bernard Atkinson the departmental historian, 'was greatly appreciated but little understood'.[140] Wilkinson carried his staff with him and promoted their interests. He brought a North American informal style to the department and was called Geoff even by his students — unusual at that time. His pioneering work led also to his writing (with F. A. Cotton) a highly influential textbook, *Advanced Inorganic Chemistry* (1962). He finished revisions for the sixth edition of this book in the week before he died.

While there were several small research groups in the physical chemistry section, the main thrust was in surface science. Barrer had a large team working on the absorptive properties of zeolites but surface science was important in the department even before he was appointed. F. C. Tompkins, a reader (later professor), led a research group studying the chemisorption of gases on thin metal films.[141] Work related to Barrer's was carried out also by L. V. C. Rees, a future professor. The section also had a long history of research in electrochemistry which, in the 1950s and early 60s, was directed by G. J. Hills (later Sir Graham Hills, Vice Chancellor of the University of Strathclyde). Joining him in 1960 was a future professor in the department, M. Spiro.[142]

Barton's organic chemistry section was the largest of the three. One of his colleagues, L. N. Owen, was given a chair in 1961.[143] Like Wilkinson, Barton made use of new physical methods in his study of complex organic compounds. Perhaps with an eye on the pharmaceutical industry, he turned increasingly to the development of new methods of synthesis and to the preparation of complex compounds under strict stereochemical control. This work was much admired and Linstead had difficulty keeping him at the college. Hearing Barton's complaints about poor facilities, and of the many job offers he had received from the United States, Linstead may well have regretted allowing physics to construct its new building before chemistry.[144] Barton was a workaholic and, perhaps not realizing that other people require more sleep than he did, thought a seventy-hour work week a reasonable norm for his research students. For many years, until he gave up smoking, students were warned of his approach by the odour of his cigars. As with Wilkinson, Barton attracted many students who went on to fill important positions in the chemistry world.[145]

Biological Sciences

Even though the biological sciences were not his priority, Southwell was especially critical of the fact that, after the war, there was no deferment quota for young men wishing to enrol in botany and zoology when the college had the capacity to educate twenty a year in each field, and when demand for biologists by the Ministry of Agriculture and the Colonial Office was high.[146] Getting nowhere with government officials on this matter he seriously considered dropping the biological disciplines; but J. W. Munro, Imperial's energetic professor of applied entomology, ensured that his department was not closed down simply because of a temporary lack of students. He was even able to persuade Southwell to hand the plant technology building over to the zoologists. Munro was also successful in pressuring the government to pay for the Hurworth field station that had been taken over by the Ministry of Agriculture during the war.[147] Southwell then allowed Munro to search for a new site, provided he not use any of the college's precious petrol ration in doing so. Together with one of his students, A. E. H. Higgins (later a lecturer at the college), Munro cycled around the Thames valley and found the Silwood estate near Ascot. Despite opposition from the botanists, who wanted an alternate site, the Governing Body sided with Munro and instructed the college solicitors to purchase Silwood.[148] The Army Pioneer Corps, in occupation at the time, was, according to Munro, 'the roughest of all the service', and hard to dislodge. But Munro, again resourceful, arranged to 'bump into' Henry Tizard at the Athenaeum where he explained the problem.[149] Within two weeks the Army had agreed to leave, allowing a major expansion in entomological work after the war.

The entomologists were especially anxious to get started at Silwood.[150] Work on DDT during the war led to a huge expansion of the insecticide industry, and support for the work of Munro and his team grew.[151] Ashurst Lodge was purchased a short time after Silwood and A. Page and O. F. Lubatti, the fumigant chemists in the entomology section, moved there. They were soon joined by B. G. Peters, appointed Professor of Parasitology in 1954, shortly after Munro's retirement. In 1960 Shell funded a new laboratory at Ashurst to support Peters' work in plant nematology.[152] A connected appointment was that of M. J. Way as Reader in Applied Entomology in 1960. He came from Rothamsted and worked on the interaction of natural and chemical methods of pest control.[153] Munro, recognizing that ecological research was needed to parallel that on pests and parasites, encouraged also the work of botany lecturer A. J. Rutter.[154] To aid the ecologists Munro sought a large scale contour map of the Silwood/Ashurst area and consulted Professor Pippard who arranged for A. Stephenson, a distinguished surveyor and member of his staff, to prepare the map with the help of civil engineering students. This was the start of a tradition of regular surveying field trips to Silwood by the engineers. A further post-war development at Silwood was the construction of a small astronomical observatory at the request of one of the Imperial's Thames valley neighbours, Professor William McCrea of Royal Holloway College.[155] Other additions were a laboratory constructed by the Australian Scientific and Industrial Research Organization for the study of parasites of a wood wasp that had devastated conifer plantations in Australia, and a ninety-foot steel tower, with laboratory and radar scanning equipment atop. This was largely paid for by the United States Air Force and was operated by the department of meteorology. Silwood soon became more widely used, though still not by many botanists.

There was antipathy between Munro and the professor of botany William Brown. Brown did not cooperate in the development of Silwood or Ashurst and conducted much of his field research at Harlington alongside the agricultural chemists. One of Brown's wartime students R. K. S. Wood, a future head of department, joined the staff after the war. He, too, began field work at Harlington. In 1953 P. H. Gregory, assistant director of the plant pathology laboratory at Rothamsted, was appointed to the chair of botany which had been allowed to lapse since Farmer's retirement.[156] He conducted his research at the Chelsea Physic Garden, continuing there even after succeeding Brown as head of department in 1954.[157] Indeed, much botanical research was conducted away from the South Kensington campus.[158] Zoology and Applied Entomology had taken over one of the botany buildings after the war and though botany was promised a new building, near the corner of Queen's Gate and Imperial Institute Road, this never materialized. Instead some additions were made to the existing building. Gregory made the best of the limited space, modernized the department, brought in much needed new equipment, and paid attention to the library. He also gave the department a new name to reflect the fact that its research was principally in applied botany. It became the Department of Botany and Plant Technology in 1955. Gregory was unwell at the time and conducted much of the department's business from a hospital bed. He resigned in 1958 and returned to Rothamsted.

The mid to late 1950s saw many staff changes. Two long-serving women lecturers retired. One, Clara Pratt, was a specialist in plant geography and economic botany who had been on the staff for thirty-three years, and had conducted research around the world. The other, Margaret Lacy, was a lecturer in bacteriology. In 1953, the Nuffield Foundation funded an enzymology laboratory for Helen Porter.[159] But Porter had to wait until being elected FRS in 1956 before being offered a readership. An expert in plant carbohydrates, she brought her considerable chemical skills to botany and was among the first to use radio tracing (carbon 14) in the study of plant metabolism. In 1959, on the retirement of F. G. Gregory (not to be confused with P. H. Gregory), she was appointed to the chair in plant physiology and became Imperial's first woman professor.[160] At that point the ARC decided to break up the

Institute of Plant Physiology, located at Imperial since the First World War. In its place three separate units were created, at Rothamsted, East Malling and Imperial College. Porter headed the new Imperial unit. She retired in 1964 and was succeeded by C. P. Whittingham.[161]

Modernization continued under P. H. Gregory's successor as head of department, W. O. James, who came from Oxford in 1959.[162] James arrived during the expansionary period and was able to make a number of appointments in areas such as plant taxonomy, plant genetics, paleobotany, nitrogen fixation, and, an echo of earlier times, timber technology. He also brought in D. Greenwood's analytical cytology group from Leeds to introduce electron microscopy to the department. He clearly did not approve the earlier, more narrow, specialization and it is difficult to judge whether he was right to expand in the ways that he did. In retrospect we can see that the new areas did not truly flourish, but that work in the traditional departmental areas of bacterial and fungal disease, antibiotics, and virus control continued to be well funded by the DSIR. Connections were also made between these areas and the new biochemistry department. In 1954 the DSIR began its system of grants for postgraduates entering approved courses; botany's plant pathology and microbiology courses qualified.[163] James took an interest in the social relations of the department and introduced something that had not been there before, namely communal afternoon tea. When he retired in 1966 he was succeeded by Whittingham.

In 1967, on the retirement of its director, V. B. Wigglesworth, the Agricultural Research Council (ARC) Unit of Insect Physiology at Cambridge was closed down. Two of its senior scientists, J. S. Kennedy and A. D. Lees came to Imperial. Their salaries were paid by the ARC which founded an insect physiology group at Imperial,[164] and professorships were given to Kennedy in 1967, and to Lees two years later.[165] The move to Silwood gave Kennedy the opportunity of starting his own research group rather late in his career. He converted an empty building at Silwood, which had earlier been used by Munro for work on insect pests in stored grains, and turned it into an excellent laboratory. The list of speakers at his lunchtime seminar series read 'like a *Who's Who* of the world's leading behavioural scientists'.[166]

But despite successes such as these, there were serious problems in the biological sciences. The number of undergraduate students was relatively small and, despite much modernization, the research configurations of the two biological departments had not kept apace the integrative disciplinary trend of the period. Perhaps the enmity of Munro and Brown was to blame. Only after their retirement did botanists and zoologists begin to mix at Silwood and South Kensington. Ecological work, such as Rutter's, pointed to the future, but in the expansionary post-war period how to proceed was unclear. After the Robbins Report the Rector asked O. W. Richards, who had succeeded Munro as head of the zoology department in 1954, for plans to expand activity at Silwood.[167] But the Rector recognized that if students were not forthcoming any expenditure on these plans would defeat his purpose. Shortly before he died Linstead consulted Porter and others on what to do. Porter, by then retired in Edinburgh, recommended that the botany and zoology departments be combined into a department of biology.[168] This anticipated the major changes in the organization of the life sciences that were to occur in the 1970s.

Earth Sciences

After the war H. H. Read still headed the department of geology. Classical geology remained to the fore as a research discipline, though undergraduate courses in mining geology and oil technology drew more students. In 1950 Read resigned the headship (he became Pro-Rector in 1951) and was succeeded by the professor of mining geology, David Williams.[169] During Williams' tenure the balance between pure and applied geological research began to shift, a shift made easier due to the expansionary period. The discipline also changed as a result of a shift in theoretical emphasis towards the new science of plate tectonics. Relatedly, during the 1960s, as older uniformitarian ideas came under scrutiny, catastrophic events in the earth's history, including celestial impact, were given greater attention than before. Paleomagnetic work became more important and geophysical work was increasingly funded, much of it from defence budgets. Williams not only promoted his own field, he welcomed Bruckshaw's geophysics group into the department from physics, and encouraged also the development of geochemistry. Bruckshaw was given a professorship in 1958, the year he also became a consultant to the Channel Tunnel Study Group. He conducted a geophysical survey of the proposed route to reassess the findings of the survey of 1875–6, work which won him the George Stephenson Medal from the Institution of

Civil Engineers in 1962. Bruckshaw continued also his research in paleo-magnetism and developed instruments for geomagnetic surveying. He was helped in building his section by R. G. Mason, one of his students and a future professor.[170] Geochemistry was developed by J. S. Webb who built an excellent team in geochemical exploration.[171] Following the lead of the engineering professors, Williams also introduced a number of one-year MSc courses, the most successful being one in mineral exploration. Williams had earlier spent several years working in Southern Spain for the Rio Tinto Company and had written the most comprehensive account of the copper pyrites mines of the region. While at Imperial he took students to Spain on field trips and encouraged many to enter mining geology.

In the late 1940s staff in the oil technology section had given a number of short courses designed for oil industry personnel. G. D. Hobson, a reader in the section, introduced an MSc in petroleum reservoir engineering in 1955. In the same year, V. C. Illing who had built the oil technology section retired. But oil continued to be a major interest and when a joint Shell and Esso team discovered the Groningen Gas Field in 1959, heralding the start of North Sea gas exploration, the department developed interests there too. In 1961 W. D. Gill was appointed to the chair of oil technology (the title reverted to petroleum geology in 1974).[172] He introduced an MSc course in petroleum geology in 1963. His research complemented that of Webb and, in 1969, he established an organic geochemistry laboratory in which petroleum source rocks were studied. His work on petroleum sedimentology and his contributions to oil geology in the Middle East led to his being widely consulted by the oil industry.

Classical geology continued strong into the 1970s. New physical techniques in sedimentology and structural analysis had entered the field along with dating technologies associated with radioactivity. Imperial College geologists joined with others in giving modern structural explanations for formations around the globe. This was a lively activity during the 1950s and 60s, but these decades were to be an Indian summer for this kind of work at Imperial. Leading the structural work were three of Read's protégés, John Sutton, Janet Watson and John Ramsay, all of whom contributed much to the modern understanding of Pre-Cambrian rocks formations in the Highlands of Scotland.[173] Watson and Sutton were married in 1949. Just before their marriage Sutton wrote to his future wife telling her of an article by Violet Bonham Carter that he had read in the *Sunday Times*. Bonham Carter had noted the difficulty for young married women finding time to think intelligently when, as she put it, 'converted into resident charwomen'. In his letter Sutton wrote, 'I would not dream of asking you to marry me if it were at the expense of your geological career'.[174] From this distance it is difficult to know what to make of Watson's subsequent career, even when making allowances for the period in which she lived. As a research career it was highly successful but she remained Read's research assistant until his death in 1970, and was appointed to a chair only in 1974.[175] In a way what happened supports my earlier claim that while women were seen as fully capable of carrying out major research, they were not seen as suited to traditional academic careers. Read kept Watson as his assistant and promoted Sutton's career within the department, despite his believing Watson to be the more gifted geologist. He was probably acting in what he thought was the couple's best interests. In roughly the same period the department employed its first woman technician, Violet (Peggy) Hill. She began working after the war as a labwoman, became expert in ore polishing, and ended her career as chief technician from 1966–8.[176]

Sutton followed Williams as head of department in 1964. He was a good, albeit tough, administrator and very much a committee man.[177] The Robbins Report was a pointer to further expansion and the undergraduate course in geology was revised with growth in mind. Some of its more classical features were dropped, though structural geology remained central. There was new emphasis on sedimentology, and geochemistry and photogeology were added to the curriculum.[178] By the 1960s the undergraduate course took in about 30 students a year, but 'pure' geology still attracted fewer than did oil technology and mining geology. Sutton was able to appoint many new staff members but, in doing so, failed to consider the future unity of the department. The department attracted capable people but, as will be discussed in later chapters, their varied interests had a centrifugal effect over the longer term. In 1966 the geology department moved from the RCS to the RSM and its anomalous situation was resolved. Students taking the 'pure' geology course joined their fellow applied geology students in gaining also an ARSM on completion of their degrees.

Mining and Metallurgy

Professor Ritson continued as head of the department of mining until his retirement in 1953. Two years earlier the RSM had celebrated its centenary. At Commemoration Day in 1951 Sir Andrew McCance FRS, an eminent metallurgist and RSM graduate, was the Distinguished Visitor. Plans for the refurbishment of the RSM building and for the Bessemer Laboratory were on display as both the mining and metallurgy departments planned for growth, even before the 1953 expansion was announced. One new area was mineral technology, introduced at the end of the war when Ritson invited Marston Fleming, a Canadian who had spent the war years in Britain as a navigator with the RCAF, to join the department.[179] Related work in the metallurgy department was consolidated with that of Fleming in 1958, and his course in mineral technology became a model for mining departments at other universities. To reflect the new direction, the department was renamed the department of mining and mineral technology in 1961. When Ritson retired he was succeeded by another coal mining specialist, J. C. Mitcheson, who, like Ritson earlier, was sent on a trip to study the metalliferous mining industry in South Africa before taking up the chair.[180] While the department remained a major centre for training in metalliferous mining it had expanded its coal mining interests during the war. After the war, in addition to mineral technology, other new research groups including ones in rock mechanics and mine ventilation were added. Rock mechanics was the field of expertise of R. A. L. Black, appointed professor of mining in 1962.[181] He invited Evert Hoeck, a South African expert in rock mechanics, to come to Imperial as a visiting lecturer. On Black's death in 1967, Hoeck was given a readership and, two years later, a chair in rock mechanics. Hoeck introduced a new interdisciplinary MSc course, supported by Rio Tinto Zinc, which drew people from geology, mining, and civil engineering. Another Imperial graduate, R. N. Pryor, was appointed to succeed Black in the chair of mining. At the same time Fleming succeeded Mitcheson as head of the department. Fleming and Pryor reorganized the undergraduate courses. In the mining course more emphasis was placed on engineering science, the economics of mining, and on modern mine design and management. In the mineral technology course the engineering content was decreased and more emphasis was placed on applied chemistry.

The Royal School of Mines departments did not follow the lead of the departments in the other two colleges in eliminating the 'intermediate' year of the undergraduate course when the war ended. The four year course continued until 1953 when a three year course was adopted. As in the mining department, the metallurgy course was modernized after the war. C. W. Dannatt drastically reduced the amount of assaying required of students from 500 to 150 hours over the full course, and reduced also the number of hours of practical instruction in geology. Instead, students spent more time on mineral dressing, metallography and pyrometry, and were given instruction in physical chemistry and applied mechanics. Some argued that he had not gone far enough in the direction of physical metallurgy but, with an industrial lobby still arguing for practical instruction, he likely went as far as was then possible. Research in extraction metallurgy improved dramatically after 1950 when the Nuffield Foundation agreed to support a new research group. Under the leadership of F. D. Richardson it grew rapidly, and Imperial became a leading centre in the field.[182]

The expansion of the college during the 1950s allowed also for a new chair in physical metallurgy and J. G. Ball was appointed in 1956.[183] He succeeded Dannatt as head of department in 1957. At the same time F. D. Richardson was promoted to a new chair in extraction metallurgy. Richardson was tall, elegant (he wore a bowler hat to work) and slightly deaf. He ran a very strong section and had a major reputation bolstered also by an impressive record in wartime science. Ball was very different, both physically and in his personality. He was short and lively. The relationship between the two professors was not an easy one but they modernized the undergraduate course yet further, introducing a completely new programme in 1959. Basic science and mathematics were taught in the first year, applied mechanics was dropped, and geology instruction cut back yet further. More weight was given to fuels and refractory materials and, in the final year, students could specialize in engineering aspects of metallurgy, in process metallurgy, or in physical metallurgy. The overall shift in the postwar period represents a move from the more traditional metallurgy, tied to the needs of the resource industry, to one that met the needs of metallurgical industries more generally.

Student numbers, never very high before the war, fell significantly during the early 1950s — to about 20 a year. However by the late

1950s new industries were competing for too few graduates and began to advertise widely for employees. The appeal of the discipline grew as it came to be associated with a wide range of metals and composite materials, with gas turbine technology, supersonic aircraft, a budding space programme, and nuclear technology. The new course therefore coincided with widespread recognition that there was more to metallurgy than mining and refining. Student numbers began to climb — settling at about 40 per year by the late 1950s when one in three applicants were accepted. The recognition of metallurgy's possibilities drew many students to Professor Ball's research group in physical metallurgy. His group expanded considerably in the early 1960s and, as with Richardson's group, drew visitors from around the world to its colloquia and meetings. The physical metallurgy group had connections to the physics department and, in 1963, P. L. Pratt who held a readership in the group was appointed to a joint chair (with physics) in crystal physics, and a number of jointly taught courses were developed.[184] Also in 1963, A. V. Bradshaw was given a new chair in process metallurgy and C. B. Alcock a new chair in chemical metallurgy.[185] The Bessemer Laboratory was reconfigured during the 1950s and 60s and laboratories that had once been rather empty were, by the late 1960s, filled to overflowing.

Linstead's Evaluation

In 1963, when Linstead was thinking of how he would present Imperial College to the Robbins Committee, he sketched out his views of the various departments and gave them all grades from A to C. His notes were not intended for circulation but rather to sort out, in his own mind, what he should report. In his view the only solidly 'A' departments were civil engineering, especially in soil mechanics and transport, and the departments of meteorology, geology, mining, and metallurgy. Except for one, all other departments, including the new department of biochemistry, were graded 'B'. The history of science and technology was given a 'C'. Linstead noted that some other departments had sections worthy of an A. He listed the nuclear technology unit in chemical engineering, the nuclear power group in mechanical engineering, the computing and control section of electrical engineering, and the pest control centre in the department of zoology and applied entomology. Ten years later another Rector would see things very differently, especially with respect to the

RSM. This is a reminder both of how ten years can make a big differ-ence in the life of a department, and of the fact that such judgements are always highly personal, even when supposedly dispassionate. No doubt my own views can be read in the brief descriptions I have given above. They differ from those of Linstead, and from those of Flowers to be discussed in chapter thirteen — though not totally. But historians have the benefit of hindsight, and time reveals weaknesses and strengths not apparent earlier.

...

1. Lord Willis Jackson FRS, (1904–70), was born in Burnley and educated at Burnley Grammar School and the University of Manchester. After a brief period as a lecturer at Bradford Technical College, Jackson enrolled in a postgraduate apprenticeship programme at Metropolitan Vickers (MV), supported by an industrial bursary from the Royal Commission for the Exhibition of 1851. He did not complete the apprenticeship and accepted a lectureship in electrical engineering at the Manchester College of Technology. A paper that he read at the Institution of Electrical Engineers (IEE) drew the attention of E. B. Moullin, then a lecturer at Cambridge, who encouraged Jackson to apply to Cambridge and study with him for a PhD. Jackson refused the offer because of insuffi-cient funding. Instead he sought other jobs, including a readership under Fortescue at Imperial. When offered a lectureship instead, he turned it down. Shortly after, Moullin was appointed Reader in Engineering Science at Oxford and encouraged Jackson to apply for a DPhil research position there. This time Jackson accepted, but only after Moullin received a major grant from the DSIR and was able to support him well. Southwell, the professor of engineering science, appointed Jackson to a demonstratorship, and helped him to membership of Magdalen College. Jackson gained a DPhil from Oxford and a DSc from Manchester and, while Southwell wanted him to stay at Oxford, he left to become personal assistant to Sir Arthur Fleming and teach at MV in the college apprenticeship scheme from which he had earlier departed. In 1938 he was appointed to the chair of electrotechnics at the University of Manchester and, soon after, became involved in wartime work in signalling, microwaves, and radar. During the war he also thought seriously about what kind of technical education would be needed once the war was over and prepared a number of reports on this for the IEE. He was appointed head of the depart-ment at Imperial in 1946 but left in 1953 to succeed Fleming as Director of Research and Education for Associated Electrical Industries (which

included the former MV). He returned to Imperial in 1961, again as head of the department. Jackson was a major committee man both within the college and nationally. He was knighted in 1958, and made a life peer in 1967. See also D. Gabor and J. Brown, 'Willis Jackson: Baron Jackson of Burnley, 1904–70', *Biographical Memoirs of Fellows of the Royal Society*, vol. 17 (1971), 379–98. For Saunders see chapter 6.

2. See B/ Jackson. Jackson did not wish to offend people in Manchester and could not easily make up his mind to come to Imperial. The negotiations were up and down, but finally up. Jackson was offered a salary of £1800 and an allowance of £50 for each of his two children. Letters in the Imperial College archives, written to his mother, make clear why Jackson had rejected Southwell's earlier offer to stay at Oxford. He disliked the upper-class culture and the fact that all the people he met at Magdalen came from public schools. He was already an idealist wanting to upgrade the status of trades education in the country. Interestingly, a major influence in this was Miles Walker, one of his professors in Manchester. Walker was a close friend and mentor. In the final years of his life Walker was seriously ill. Jackson spent much time caring for him both in Menton and, after the war began, in Manchester. Walker died in 1941. Walker's mentor had been Silvanus Thompson with whom he worked for many years at Finsbury Technical College. Thompson believed it important to provide good scientific and technical education for young men who in earlier times would have been apprenticed in City guilds. Thompson's legacy can be seen in the careers of both Walker and Jackson.

3. See Gabor and Brown, *op. cit.* (1). Jackson was helped in building the department by John Ganley who had come to the college to work in Fortescue's department in 1924. During the war Ganley had been seconded to instruct RAF personnel in radio mechanics, but otherwise worked at the college for over fifty years. He was appointed Department Superintendent when Jackson returned to the college in 1961. Ganley oversaw the installation of a new telephone system in the college, TV studios and language laboratories, as well as carrying out the normal work in his department. He was known also for his party tricks, especially those involving exploding light bulbs.

4. See chapter 4 for discussion of pre-war four-year courses.

5. The reason for Jackson's departure at this time is not entirely clear. It would seem that he was frustrated by opposition to the demolition of the Imperial Institute and thought it would be a very long time before he saw a new building. He returned to Imperial shortly after the new building was completed.

6. Arnold Tustin (1899–1994) was apprenticed to the Parsons Company in Newcastle-upon-Tyne and won a scholarship to Armstrong College from where he went to Metropolitan Vickers (MV) as a postgraduate trainee. He began working on feedback control in connection with gunnery during the Second World War. He worked also on electric traction, and ended his time with MV at their Sheffield factory as Assistant Chief Engineer. A major figure in classical control work, he was Professor of Engineering at Birmingham University, 1947–55, before coming to Imperial. Several professors and other members of the academic staff who came to the college from the 1940s to the 1960s had earlier been either apprentices or post-graduate trainees at Vickers. Vickers was a major industrial presence in the armaments, shipbuilding, railway and aircraft industries. Its London head-quarters was for many years the Milbank Tower. See J. D. Scott, *Vickers: a history* (London, 1962).

7. That electrical engineering occupies the tallest building on the island site has to do with the redesigning of plans after Imperial was forced to keep the Colcutt Tower and to build less densely than it had earlier hoped.

8. Brian John Cory (1928–) became reader in electrical power systems in 1975. He, too, had apprenticed at MV.

9. Eric Laithwaite (1921–97) was educated at the University of Manchester and spent the war years with the RAE at Farnborough. After the war he returned to teach at Manchester where his rediscovery of the linear motor, in 1947, prompted him to work on the development of magnetically-levitated high-speed trains. The UK government put about £5m into testing these but then abandoned the project. Maglev trains were built in Japan and Germany. Laithwaite's work with gyroscopes, and the suggestion that they might contravene Newton's laws of motion, earned him notoriety in the profession and probably cost him election to the Royal Society. After Laithwaite retired from Imperial in 1986 he was asked to design a futuristic rocket launcher for NASA which entailed a long track inside a mountain along which a space capsule, powered by one of his linear motors, was to have been accelerated before hurtling out through the summit, and then on to space. He was working on this project when he died. Laithwaite was also a serious butterfly collector.

10. Colin Cherry (1914–79) worked as a laboratory assistant at the General Electric Co. from 1932–6, while attending evening classes at Northampton Polytechnic. He took an external BSc in electrical engineering at the University of London. During the war he worked at the Radar Research and Development Establishment at Malvern and, in 1945, was appointed

an assistant lecturer at the University of Manchester. Cherry appears to have had many of the characteristics of the autodidact, including voicing opinions on all kinds of things — but he was original. For example, he wrote an article in the *Sheffield Telegraph* (22 August 1963) in which he stated that the number of car accidents could be reduced if drivers were better able to communicate with each other. Not quite anticipating the mobile telephone, he recommended that the government legislate that all cars be fitted with radio transmitters and receivers with a range of 100 yards. Another of his ideas, promoted in the Press, was world government by television. He was also an early promoter of televising parliament, and thought that constituents should have two-way TV contact with their MPs. His book *On Human Communication* (Cambridge, MA, 1957) was very influential, especially among the young.

11. Including a future professor in the department, Igor Aleksander, who came to England intending to work with Cherry but, for technical reasons, did not. (Personal communication.) Steven Pinker, in his 1999 Colin Cherry Memorial Lecture, stated that when he read Cherry's *On Human Communication*, as a student in 1970s, it set him on his path.

12. *Topic* (December 1979); article on electrical engineering and Cherry.

13. Eric A. Ash, 'Dennis Gabor, 1900–79', *Nature* (2 August 1979), 431–3. Dennis Gabor FRS (1900–79) was born in Hungary and educated at the Technical University in Budapest (1918–20), and at the Technische Hochschule in Charlottenburg (1921–22) where he studied statistical mechanics under Albert Einstein. Gabor then worked for Siemens where he invented the high-pressure quartz mercury lamp. His contract was terminated when Hitler came to power and he returned to Hungary. Gabor came to England in 1934 and found work with the British Thomson Houston Company in Rugby. There he worked in the fields of electron optics (including work on the electron microscope), 3-D projection, and communication theory. Because of his status as an enemy-alien he was barred from working on radar, and on the all-important magnetron (transmitter), during the war. But Gabor was an inveterate and brilliant inventor and turned his attention to television for which he developed a flat screen, not then commercially viable. His invention of holography stemmed from work on the electron microscope. After the invention of lasers in the 1960s holography became a viable and widely used technology. English Heritage recently erected a blue plaque at 79 Queen's Gate where Gabor and his wife lived in a small flat, and where they frequently entertained his students and colleagues. See also T. E. Allibone, 'Dennis Gabor, 1900–

79', *Biographical Memoirs of Fellows of the Royal Society*, vol. 26 (1980), 107–47; obituary, *The Times*, 16 November 1979.

14. Ash states that he finds it hard to comprehend why recognition came so late. He thinks both the FRS (Gabor was elected in 1956) and the professorship should have come sooner. See Ash, *op. cit.* (13).

15. B/Gabor, box A. Gabor, quoted in *Think* (IBM house magazine, January/February 1970), 27.

16. See his lecture 'Inventing the Future', *Encounter* (14 May 1960), 3–16. This was later expanded into a book, *Inventing the Future* (London, 1963). Gabor led the discussion at a 1969 Touchstone weekend where he gave a talk entitled, 'The Future of Progress' in which he expressed the fear, common among his generation, that the physical sciences and technology had forged ahead of the social sciences and that society lacked the proper means to deal with the new. We can now see that this view, at least in its crude form, was misguided and, further, that it was inconsistent with Gabor's elitist attitude towards education.

17. Bruce McArthur Sayers FREng (1928–) graduated from the University of Melbourne in physics and electrical engineering. For several years he worked as an electronics design engineer on diagnostic instruments before coming to Imperial as a research student under Professor Cherry. Like some other Cherry students, Sayers was a pioneer in the area of signal processing. He joined the staff in 1962 and was appointed Professor of Electrical Engineering Applied to Medicine in 1968. The title was changed to Computing Applied to Medicine in 1984. Sayers was much involved in a range of Imperial and C&G activities.

18. Joseph Chapman Anderson (1922–2001), known as Andy, was educated at King's College, Durham and at the Regent Street Polytechnic where he took an external University of London BSc in electrical engineering. He served in the Royal Electrical and Mechanical Engineers, 1942–8, becoming Captain. Following this he spent short terms as a lecturer, first at the Northern Polytechnic and then at the University of the Witwatersrand. It was Anderson who largely fostered the field of electrical materials at Imperial. He was given a chair in 1965 and was the author of several books including the highly successful *Materials Science* (London, 1969), a first year text with a fifth edition published in the year of his death (additional authors brought out later editions). He had great rapport with research students, had many interests, and was a keen jazz pianist.

19. Among other thing this committee advised the government on when to introduce colour TV. In retrospect it is amusing to find a letter written to *The Times* (28 September 1965) by Sidney L. Bernstein, Chairman of

Granada TV Ltd. 'We in Granada …[thought the change to 625 lines was] an extravagant misdirection of national and private resources. … Can we afford another misdirection of resources which colour would undoubtedly be?' He opposed the findings of Jackson's committee which had reported to the Postmaster General, Tony Wedgwood Benn (as he was then known), stating that colour TV should be introduced in 1967. The committee also favoured the American system but recognized that the UK needed to go along with whatever was adopted by the major European countries. Final decisions were taken by Benn's successor, Reginald Bevins. The first (non-trial) colour BBC TV broadcast was of Wimbledon in 1967.

20. Peter Grootenhuis FREng (1924–) was educated at the Netherlands Lyceum in The Hague, and at Imperial College where he gained a BSc in mechanical engineering in 1944. From 1944–6 he was a graduate apprentice with Bristol Aeroengines, returning to Imperial as assistant lecturer in 1946. A specialist in vibration and acoustic technology, he also developed undergraduate teaching in dynamics. He rose through the ranks to a professorship in 1972. He has a great interest in sailing and is a past president of the Imperial College Sailing Club.

21. Walter Murgatroyd (1921–) was educated at Cambridge where he took a PhD in 1952. A specialist in reactor heat transfer, he worked for the UKAEA before joining Queen Mary College in 1956 as head of the department of nuclear engineering. Murgatroyd moved towards thermal power more generally when at Imperial. For the setting up of the nuclear power section see chapter 9.

22. Sir Hugh Ford FRS FREng (1913–), son of a freelance inventor, began his career as an apprentice at the Great Western Railway works in Swindon. He came to Imperial College on a Whitworth Scholarship. He was a research student in heat transfer under Owen Saunders, gaining a PhD in 1939. During the war he worked for the Paterson Engineering Company on the design of a high pressure polyethylene plant for ICI, and subsequently for the Iron and Steel Federation, helping to improve the operation of strip mills. Ford became Professor of Applied Mechanics in 1951 and Professor of Mechanical Engineering in 1969. He was knighted in 1975. See also chapter 13.

23. Dudley Brian Spalding FRS FREng (1923–), known as Brian, gained a BA in engineering science at Oxford before moving to Cambridge for his PhD. He was appointed Professor of Heat Transfer in 1958. He founded his own company, Concentration, Heat and Momentum (CHAM), in 1975 and retired from Imperial in 1988. See also chapter 13. For Eilon, see chapter 14.

24. John Malcolm Alexander FREng (1921–) was educated at the Ipswich School of Engineering and at Loughborough College. After an apprenticeship he came to Imperial and took a BSc in mechanical engineering in 1950. He became a research assistant with Hugh Ford (PhD, 1953) and, after working for British Electric, returned to Imperial in 1957. He received a conferred chair in engineering plasticity in 1963, and succeeded to the chair of applied mechanics in 1969. He resigned in 1978 and moved to head the department of mechanical engineering in Swansea.

25. Sir Derman Guy Christopherson FRS FREng (1919–2000) was educated at Sherborne School and Oxford where he took a degree in engineering science in 1937. He was a research student under Richard Southwell and took a DPhil in 1940. After war work on blasts and shock waves he became a lecturer in engineering at Cambridge. In 1949 he was appointed to a chair in Leeds. He was a specialist in lubrication but carried out little research at Imperial where he focussed largely on improving the undergraduate courses. He left Imperial in 1960 to become Vice Chancellor of the University of Durham.

26. Bernard George Neal FREng (1922–) was educated at Cambridge and worked there, aside from a short period in the USA, until being appointed Professor of Applied Engineering in Swansea in 1954. He came to Imperial in 1961 as Professor of Applied Science with reference to Engineering. In 1972, Neal moved to the chair of engineering structures in the civil engineering department. He became head of that department 1976–82, and held the chair of civil engineering 1981–2. Aside from engineering, Neal is known for his skill at croquet, and has been champion of the All England Club many times. Neal's successor at Swansea was Olgierd (Oleg) Cecil Zienkiewicz who was educated at Imperial and took a PhD under Southwell (Southwell had a small research group while he was Rector).

27. Saunders was Vice Chancellor of the University of London, 1967–9. He had long expressed concern over possible gas explosions in high-rise buildings, and joined the Ronan Point Inquiry in 1968 after the tragic collapse of that building.

28. Chapman brought much industrially related research to the department. He left in 1971 to become Director of the Constructional Steel Research and Development Organization. He later founded a company together with future Imperial professor, P. J. Dowling. For Sparkes, see chapter 6.

29. John Ian Munro (1928–85), known as Ian, was educated at the University of Glasgow. After army service, Munro worked with Ove Arup and Partners. In 1955–6 he attended a postgraduate structures DIC course in the department, and was appointed a lecturer in 1957. He specialized in

structural mechanics and became Professor of Civil Engineering Systems in 1980, and head of department in 1982.

30. Arthur Lempriere Lancey Baker (1905–86) was educated at the University of Manchester in civil engineering and, on graduation, worked on the Mersey Tunnel project. He then moved to Johannesburg as chief designer for the Reinforcing Steel Company where he designed a number of skyscrapers. He returned to England and worked as a consultant on reinforced buildings and other structures. During the Second World War, Baker was seconded to assist Reginald Stradling at the DSIR Buildings Research Station. It was Stradling who introduced Baker to Pippard, paving the way for his appointment at Imperial. Pippard very much enjoyed Baker's company and wrote to the Rector in support of his appointment, 'he has designed and carried out the most amazing jobs — large buildings on bad foundations, piers, jetties, a very beautiful domed church and so on … his designs are really brilliant'. During the war Baker designed a number of Mulberry harbours with shock absorbing fenders. ULP/Civil Engineering/5; Pippard to Rector (Southwell), 15 April 1945. Baker wrote a long paper (in ICA) on the proposed channel tunnel favouring part tunnel and part bridge. He believed that if proper shipping lanes were enforced there would be no danger. He defended this scheme in public on many occasions and thought that a tunnel posed major fire risks. See obituary, *The Times*, May 28, 1986.

31. Cedric Masey White (1898–1994) was educated at University College Nottingham and served in the Royal Tank Corps during the First World War. He was a research fellow, then lecturer, at King's College London before coming to Imperial as a reader in 1933. The hydraulics laboratory in the new building contained many of his design innovations. White had close connections to the hydraulics industry and worked on hydro-electric schemes in Scotland, Owen Falls, Uganda, and Dokan, Iraq. In Scotland he also worked on ways to allow salmon to migrate upstream around dams. See obituary, *The Independent* January 20, 1994. Alec Skempton noted that 'Pippard and White almost lived in the laboratories … [and] inspire[d] a love of scientific enquiry in the minds of those students willing and able to respond'. Judith Niechcial, *A Particle of Clay: the biography of Alec Skempton, civil engineer*, (Whittles, Caithness, 2002), 21–2.

32. The college had a long association with the London School because of its work in areas of botany and applied entomology preparing students for work in tropical countries. The Rockefeller Foundation came to Britain after the war looking for good public health research areas to support.

33. This field was developed later by Roger Perry; see chapter 13.

34. Peter O. Wolf FREng left Imperial in 1966 to become head of civil engineering at City University. Another person working in the hydraulics section in this period was Ralph Alger Bagnold (1896–1990). Bagnold was educated at the Royal Military Academy, Woolwich, and at Cambridge where he took the engineering sciences tripos. He had an interesting and exciting life as a soldier and desert explorer before coming to Imperial shortly before the Second World War to carry out research on the behaviour of blown sands. When war came he left the college and, as Brigadier Bagnold, founded the Long Range Desert Group, a small motorized reconnaissance unit which undertook raids deep into enemy-held territory. After the war he became Director of Research of the Shell Refining and Marketing Company, but resigned in 1949 to return to Imperial and a research position. He studied the transport of solids in streams of water as well as carrying on with his earlier work on blown sands. See R. A. Bagnold, *Sun, Wind, War and Water* (University of Arizona Press, 1991).

35. After the major flood at Lynmouth, Devon, in 1952, departmental advice was sought on the siting of new buildings and on the design of retaining walls and bridges.

36. John Robert Dark Francis (1920–79) took an external degree at the University of London before joining the Royal Navy. He was a lover of the sea and a keen sailor; wind, waves and water were part also of his research life at Imperial. He was much involved with the design of the Thames Barrier. In this he worked with Peter Cox who had earlier been a student in the department and who later became eminent in harbour and marine heavy engineering. As a student Cox was renowned for his Morphy Day activities, notably on the C&G tug-of-war team. He has served as President of the Old Centralians.

37. See B. G. Manton, 'History of the Civil Engineering Department, 1884–1956' (typed manuscript in ICA), 77. Terzaghi had developed his ideas while working in Istanbul.

38. Sir Alec Westley Skempton FRS FREng (1914–2001) was born in Northampton and educated at Imperial College. He was a research student in concrete technology but left before gaining a PhD when a scarce job opportunity arose with the Buildings Research Station in 1936. There he worked on reinforced concrete but, in a neighbouring laboratory, he met Leonard Cooling working in the new field of soil mechanics. He soon moved to join him. Soil mechanics was put on the map in Britain after the 1937 failure of the Chingford Reservoir embankment, a failure that Skempton helped analyse. His work caught the attention of Terzaghi who had been called in as a consultant. Skempton himself later became

a consultant on earth dams, dock walls, bridge foundations, road works, tall buildings, etc. to major engineering companies worldwide. A leader in the field, he was a prominent member of national and international professional bodies relating to soil mechanics. See Niechcial, *op. cit.* (31).

39. Alan Wilfred Bishop (1920–88) was educated at Cambridge and later worked as an engineering assistant at the Metropolitan Water Board. While Rector, Southwell had a research team at Imperial. Bishop joined the team and applied Southwell's relaxation ideas to the analysis of earth dams. On Southwell's retirement, Bishop moved to the Buildings Research Station where he met Skempton, before returning with him to Imperial in 1946.

    Bishop was given a chair in soil mechanics in 1965. He was consultant on many large projects and chaired the team of investigators at the 1966 Aberfan colliery tip slide disaster. Another member of the team was J. N. Hutchinson, a landslide specialist who became Professor of Geomorphology in 1977. Bishop remained a bachelor until he took early retirement. He then surprised his colleagues by marrying and moving to Scotland for part of the year.

40. Rudolph Glossop, who had been a geology student at Imperial 1920–24, had a major influence on Skempton and on the development of the course. Glossop turned to civil engineering after a slump in metalliferous mining during the 1930s. He worked on the excavations for the Leicester Square underground station and stayed in the field of earthworks and excavation. He was a British pioneer in soil mechanics and a director of John Mowlem and Company, until his retirement in 1968.

41. Quoted in Niechcial, *op. cit.* (31), 66.

42. Tidu Maini remembers Letitia Chitty, around 1964, carrying out longhand numerical analysis calculations (using Southwell's dynamic relaxation techniques to solve partial differential equations) while lying on the floor on top of paper cut outs of airplane wings. (Personal communication.)

43. See Niechcial, *op. cit.* (31), chapter 6. Also, (Sir) Alan Muir Wood, 'Geoffrey Morse Binnie 1908–1989', *Biographical Memoirs of Fellows of the Royal Society*, vol. 36 (1990), 45–57.

44. Eric Hugh Brown (1924–) was educated at University College Nottingham and then worked at the Royal Aircraft Establishment as a structural engineer. After the war he moved to the DSIR Building Research Station. He came to Imperial as a research student under Pippard in 1948 and was appointed a lecturer at the same time. He became Professor of Structural Analysis in 1971. For some of his musical activities see chapters 11 and 14.

Nicholas Neocles Ambraseys FREng (1929–) was educated at the National Technical University of Athens before coming to Imperial to study for a PhD in soil mechanics with Skempton. He joined the department as a lecturer in 1958 and was appointed Professor of Engineering Seismology in 1973. He founded and became the first chairman of the British National Committee for Earthquake Engineering.

45. B. G. Manton had been an engineer with the Port of London Authority before coming to the college in 1920. He wrote a history of the department up to 1956 (typed copy in ICA).

46. Sir Colin Douglas Buchanan (1907–2001) was educated at Imperial College. After a BSc in civil engineering he moved into town planning. He was Principal Inspector for the Ministry of Housing and Local Government, 1954–60, and a specialist in road traffic problems. He was the author of an influential report, *Traffic in Towns* (HMSO, 1963), prepared for Ernest Marples, Minister of Transport. Buchanan was against the building of car parks under London squares, thought Oxford Street a travesty of what a shopping street should be, and that Birmingham's Bull Ring was 'a sordid, ugly, brutal place'. He warned about traffic congestion in towns and insufficient funding for motorways. Buchanan was also a member of the Royal Commission on the site for London's third airport and wrote a minority report in favour of Foulness over Stansted.

47. The Committee under Lady (Evelyn) Sharp was set up in 1967. See 'Transport Planning: The Men for the Job' (HMSO, 1970).

48. The SRC selected Leeds, Newcastle, Cranfield, Loughborough and Liverpool (Marine) for new work in the transport area. It was also decided by the government that the new School of Advanced Urban Studies be sited outside London and Bristol was chosen.

49. See GB Minutes, 24 June 1966. For further detail on the civil engineering department in this and earlier periods, see Joyce Brown (ed.), *A Hundred Years of Civil Engineering at South Kensington*, (London, 1985). Buchanan's ideas were discussed again in 1994 when the various South Kensington institutions, in light of the coming millennium, put together the Millennium Consortium Project a long-term plan and architectural concept for the local area. This project is still under discussion.

50. In 1942 the department became known as the department of chemical engineering and applied chemistry, in 1954, the department of chemical engineering; and, in 1959, the department of chemical engineering and chemical technology.

51. K. E. Weale, 'Chemical Engineering and Technology' in Adrian Whitworth (ed.), *A Centenary History: a history of the City and Guilds*

College, 1885–1985 (London, 1985), 50. Weale was a long-time member of staff first appointed in 1948 as an assistant lecturer. As noted in chapter 11, Weale was active in the social life of the college.

52. J. M. Coulson and J. F. Richardson, *Chemical Engineering* (Oxford, 1964).

53. See 1989 citation for John Stanley Oakley when he was awarded an honorary associateship of Imperial College. Oakley wrote a handbook on safety for use in college workshops.

54. D. M. Newitt, 'opening remarks', *Journal of the Imperial College Chemical Engineering Society*, vol. 2 (1946).

55. Alfred Rene Jean Paul Ubbelohde FRS (1907–1988), known as Paul, was a polymath who made major contributions to chemical thermodynamics and to the science of explosions and detonation. He was born in Antwerp from where his family escaped at the beginning of the First World War to find refuge in England. Ubbelohde was educated at St. Paul's School and at Oxford where he was a research student under Alfred Egerton, working in a physical chemistry enclave of the Clarendon Laboratory. In a personal memoir Ubbelohde mentions that craftsmanship was highly esteemed at the Clarendon so he paid the glassblower to give him instruction and 'was rather vain of my glass blowing virtuosity' until Ernest Rutherford visited the lab and, on contemplating his apparatus asked, 'why all that gear?'. Ubbelohde studied further with Arnold Eucken at Göttingen (1931–2) and, in 1936, held a Dewar Fellowship at the Royal Institution under William Bragg. During the Second World War he worked under Sir Robert Robertson, then Director of Explosives at Woolwich. Robertson moved his team to Swansea for the duration of the war. After the war Ubbelohde was appointed head of the department of chemistry at Queen's University in Belfast. He came to Imperial as Professor of Thermodynamics in 1954. He lived nearby in Queen's Gate and had a fine collection of Chinese porcelain. In the 1960s he bought a pig farm in Sussex where he kept a flock of about 100 British Large Whites. He studied their energy efficiency with the same seriousness he paid to his academic research. Ubbelohde was also something of a wine connoisseur and chaired the college wine committee for many years. For further details see, F. J. Weinberg, 'Alfred Rene Jean Paul Ubbelohde (1907–1988)', *Biographical Memoirs of Fellows of the Royal Society* Vol. 35, 1990, 383–402. See also A. R. J. P. Ubbelohde, handwritten biographical lecture delivered at More House 7 February 1983. (B/Ubbelohde)

56. John Shipley Rowlinson FRS FREng (1926–) was educated at Oxford. Before coming to Imperial he was a senior lecturer in physical chemistry at the University of Manchester. He left Imperial in 1973 for the Dr.

Lee's Chair of Chemistry at Oxford in succession to Frederick Dainton. Rowlinson co-authored (with K. E. Bett and G. Saville, both lecturers at Imperial), *Thermodynamics for Chemical Engineers* (London, 1975), an important and much used text.

57. Peter Victor Danckwerts FRS FREng (1916–84) was educated at Winchester and Balliol College, Oxford where he took a degree in chemistry before moving to MIT for an MS in chemical engineering practice. He then worked as a research chemist with Fullers' Earth Union in Surrey, leaving to serve in the RNVR during war. He received the George Cross for outstanding gallantry in 1940. After the war he was appointed a demonstrator and lecturer in chemical engineering at Cambridge. In 1954 he left to become Deputy Director of Research and Development with the Industrial group of the UKAEA where he worked on the disposal of radioactive effluent at Windscale.

58. Kenneth George Denbigh FRS (1911–2004) was educated at the Universities of Leeds and Cambridge. He worked for ICI from 1934–8, and was a lecturer at the University of Southampton from 1938–41. During the war he was Chief Chemist at the Royal Ordnance Factory, Bridgewater. In 1945 he returned to ICI to head the physical chemistry laboratory at the Akers Division. From 1948–55 he was a lecturer at Cambridge before being appointed to the chair of chemical technology at the University of Edinburgh. He published some major work in thermo-dynamics. Denbigh left Imperial College to become Principal of Queen Elizabeth College in 1966.

59. Some of these concerns can be seen in a letter that Denbigh wrote to the College Registrar, E. M. Cutcliffe, 6 March 1963. See B/Linstead, Robbins Committee box (the letter was forwarded to the Robbins Committee by Linstead).

60. In 1962 the National Research Development Corporation (NRDC) published a 'Survey of Chemical Technology/Chemical Engineering', by N. McLeod, head of chemical engineering at the University of Edinburgh. Of Imperial, the report noted a 'conspicuous lack of research of a kind directed to the development of new chemical processes'. M. Zvegintzov of the NRDC wrote to the Rector (B/Linstead; 5 December 1962) that while there had been much progress in basic chemistry and physics, not enough had been translated into the chemical industry. Ubbelohde wrote a highly critical review of the report. He called it 'muddled and time wasting'.

61. Alfred Gordon Gaydon FRS (1911–2004), known as Dick, was educated at Imperial College where he was a research student under Alfred Fowler. After a brief period working at the British Cotton Industry Research

Association, he spent the rest of his career at Imperial, working first in the physics department before moving to chemical engineering. He was awarded a Warren Research Fellowship of the Royal Society in 1945 which extended through to his retirement in 1973. Despite poor vision, the result of a laboratory accident at the Research Association in 1936, he gained eminence as a spectroscopist and was awarded the Rumford Medal of the Royal Society in 1960, two years before being appointed to the chair in molecular spectroscopy. He was the co-author (with R. W. B. Pearse, also a member of staff at Imperial) of a standard text in his field, *Identification of Molecular Spectra* (London, 1940) which went into many editions.

62. For more on Sargent see chapter 13.

63. KCTI; chemical engineering correspondence file: statement by A. R. Ubbelohde to SRC 'Collaboration between Universities and Industries', (1967). While stressing research was fine, he was a bit dismissive of work carried out by chemical engineers.

64. Arnold Alexander (Lord) Hall FRS (1915–2000) was educated at Clare College, Cambridge where he took the mechanical sciences tripos in 1934, winning all but one of the prizes awarded. After a short period at the RAE, Hall returned to Cambridge to pursue research in aerodynamics under Bennett Melvill Jones and Geoffrey Taylor. In 1938 he rejoined the RAE where he worked on a range of problems including aircraft gunnery, high-speed wind tunnel design, jet propulsion, and the night time interception of aircraft. Hall, still under thirty years of age, succeeded Bairstow in the Zaharoff Chair in 1944. He came to Imperial with superlative letters of recommendation. Melvill Jones wrote that Hall was 'the most outstanding … man who has yet passed through my hands'. And, in the past few years, he has 'raised the gunnery department at the RAE from a place of negligible importance to what I regard as the main driving force in air gunnery at the present time'. (Letter to Southwell, 17 February 1944). Hall was remembered as a brilliant lecturer who used no notes, even when discussing complex material. Despite desperate attempts to accommodate him, including the planned new building which he helped design, Hall left Imperial to become Director of the RAE in 1951. He became Vice Chairman and Managing Director of the Hawker Siddeley Group, was knighted and later given a life peerage. Hall was a member of the Governing Body of Imperial College.

65. While the department taught aerodynamics and aerostructures, aircraft propulsion was taught (and research carried out) in the mechanical engineering department.

66. In the 1950s John Stollery built one of the first hypersonic gun tunnels in the country.

67. Pippard, with his experience in aeronautics, would perhaps have been well qualified to give the type of course Hall had in mind.

68. Sir Frederick Handley Page was a student in mechanical engineering (1911–13) and founded the country's first private aircraft firm. He was much enthused with ongoing glider activity among students and staff at the college. In 1955 Francis (Frank) G. Irving (1925–2005), a lecturer in the department, was navigator on the first 2-seater glider flight over the English channel, and A. D. Piggot, an instructor for the College Gliding Club, took a sail-plane to 23,200 ft. without oxygen, setting a height record. Irving was president of the Gliding Club 1969–99 and wrote *The Soaring Pilot* (1968) which was widely read in the gliding community. He became assistant director of the aeronautics department and worked on the construction of supersonic wind tunnels. When the larger of these had its two-minute run, the noise it created stopped all lectures in the vicinity. Irving was also Warden of Beit Hall for twenty years and was awarded an honorary associateship of the college in 1999.

69. Herbert Brian Squire FRS (1909–1961) was known as Brian. He was educated at Oxford and, after taking a BA in mathematics, became a research student under Southwell. He studied also in Göttingen (1932–3), returning to England to become a scientific officer at the RAE where he worked until the end of the war. After a brief period as a lecturer at Manchester, he returned to the RAE in 1947 as Principal Scientific Officer. Squire worked on aerodynamic problems related to the new high-speed age of jet propulsion. From 1949 he was on the staff of the aeronautics division of the NPL and in 1946 was made Chairman of the Helicopter Committee of the Aeronautics Research Council. Squire was against the use of cars in London and always rode a bicycle. He challenged the sartorial norms (dark suits) of professors of his day by dressing casually, and is described by his obituarists as having been 'an efficient but very laconic administrator', see S. B. Gates and A. D. Young, 'Herbert Brian Squire, 1909–1961)', *Biographical Memoirs of Fellows of the Royal Society*, vol. 8 (1962), 119–35.

70. Once the new building opened, the department took in 45 undergraduates per year.

71. John Hadji-Argyris FRS FREng (1913–2004) was educated at the National Technical University in Athens and at the Technical University in Munich where he gained a doctorate in engineering in 1936. He remained in Munich until the outbreak of war when he became private secretary to

the Greek ambassador in Berlin. After Germany declared war on Greece he was interned but his release was secured after several months of diplomatic negotiation. He left for Zurich where he took a postgraduate course in aeronautics at the ETH. He then moved to London and was a member of the technical staff at the Royal Aeronautical Society before joining Imperial. Argyris made a major contribution to the understanding of the mechanics of complex structures such as the aeroplane. Using a matrix formulation he reduced the overall problem to a large set of simultaneous equations which required high-speed computers for their convenient solution. This pioneering work in finite element analysis, and his continuing work on structures, was well received and he had much industrial support for his work.

72. Rectors Correspondence files; Argyris to Rector (Linstead), 21 March 1957.

73. Argyris was made an FIC in 1985.

74. Paul Robert Owen FRS (1920–90), known as Robert, was born in London and, after a grammar school education, won a scholarship to Queen Mary College London. At school he won the top prizes both in mathematics and English literature and his language skills are clearly to be seen in his letters and memoranda. He also had a gift for foreign languages and, as a younger man, was as an amateur actor of distinction. He graduated in engineering (specializing in aeronautics) in 1940 and, from 1941–53, he worked at the RAE where one of his colleagues was H. B. Squire. During a period in which the speed of aircraft increased dramatically, Owen worked on the stability and control of aircraft and on the design of wings. In the early 1950s he developed an interest in fluid dynamics and took this work first to Manchester, where he was professor and head of the department of the mechanics of fluids, and later to Imperial. At Manchester he worked with M. J. Lighthill who later also came to Imperial. See A. D. Young and Sir James Lighthill, 'Paul Robert Owen, 1920–1990', *Biographical Memoirs of Fellows of the Royal Society*, vol. 38 (1992), 269–85.

75. Symbolic of Britain's problem was the uneasy development of the de Havilland Comet, the world's first jet passenger aircraft. After it crashed in 1953, the American aircraft industry had a virtual monopoly.

76. He was appointed to the Commission of Enquiry after the collapse of the Ferrybridge cooling tower in 1965.

77. A laboratory was named for Donald Campbell which housed a large low-speed wind tunnel. G. W. (Jim) Cunningham, who joined the department in 1931 at the age of fourteen, working as a lathe operator for fourpence an hour, later constructed a three axis force balance for this laboratory.

Cunningham apprenticed with the aircraft industry during the Second World War, returning to the department as a technician in 1947. He was Superintendent, 1973–83 and was responsible for the design of a wide range of complex apparatus.

78. Michael James (Sir James) Lighthill FRS (1924–1998) was educated at Winchester and Trinity College, Cambridge where he read for the mathematical tripos. He moved to Manchester in 1946 as a senior lecturer and from 1950–9 was Beyer Professor of Applied Mathematics. In 1959 he was appointed Director of the Royal Aircraft Establishment at Farnborough where he remained until coming to Imperial in 1964. Lighthill was initially given space in the physics department but he soon moved to mechanical engineering where space was provided also for his Institute of Mathematics and Its Applications. Eminent in the fields of aerodynamics and fluid dynamics, Lighthill left Imperial to become Lucasian Professor at Cambridge in 1969. He was Provost of University College London, 1979-89. His administrative skills were legendary. He died in an attempt to swim around the island of Sark, something he had been the first to do as a younger man. Obituary, *The Times*, 20 July 1998.

79. Some laboratory space was provided also by the zoology department.

80. For more on this unit see C. G. Caro and K. H. Parker, 'Physiological Flow Studies Unit', in Whitworth (ed.), *op. cit.* (51). Caro was given the title Professor of Physiological Mechanics in 1977. More recently, Caro founded the spin-out company, Veryan Medical, which has developed synthetic tubes with helical twists. These replicate arterial geometry and generate a swirling blood flow that may have application in grafts and shunts by allowing blood flow to work as well as it does in natural arteries.

81. Rector's correspondence, Thomson to Linstead, Cambridge, 24 February 1958. Some work on plasmas continued at the college after Thomson's team left.

82. Devons left five years later for a chair in Manchester since Blackett did not wish to support his type of research. See below.

83. First quotation from letter; H. Allan to N. Barford, writing on nuclear physics under Thomson. See typed manuscript 'Departmental Histories', 73; (held in physics department). Second quotation from handwritten history of the physics department in the 1930s and 40s by Harold Allan (personal communication).

84. William David Wright (1906–1997) was known as David. He was educated at Imperial College gaining a BSc in 1928 and a PhD in 1930. Wright was a specialist in colour vision and his 1931 work on colorimetric

analysis formed the basis for an international system of colour measurement. Wright spent most of his career at the college, with brief periods also at Westinghouse and EMI, working in electron optics. In 1951 he succeeded L. C. Martin in the chair of technical optics. The title of the chair was changed to applied optics in 1963. Wright was the author of five books, and is remembered as an outstanding supervisor of research students.

85. Walter Thompson Welford FRS (1916–90) was educated at Hackney Technical School and worked as a hospital technician while taking an external degree at the University of London. He was Reader in Applied Optics, 1961-73, and Professor of Physics, 1973–83.

   Charles Gorrie Wynne FRS (1911–1999) was educated at Oxford before joining the Leicester firm of Taylor, Taylor and Hobson. He was a pioneer of computer-aided lens design and worked on lenses for astronomical telescopes. He was given a chair in optical design in 1969.

86. Harold Horace Hopkins FRS (1918–94) was educated at University College Leicester and came to Imperial after the war as a research fellow. He was appointed Reader in Optics in 1950 and left in 1967 to become Professor of Applied Optics at Reading. He regretted that his ideas in fibre optics were followed up by industry in the USA rather than in Britain. His flexible endoscope (to be followed also by a more rigid version) was first demonstrated in the USA in 1957. His work led to a range of new surgical techniques, including laparoscopy. See C. W. McCombie and J. C. Smith, 'Harold Horace Hopkins', *Biographical Memoirs of Fellows of the Royal Society*, 44 (1998), 238–63.

87. Anne Patricia Thorne (née Pery) (1928–), daughter of the 5th Earl of Limerick, was educated at St. Hugh's College, Oxford. She joined Imperial as a lecturer in the spectroscopy section in 1955 after two years as a research fellow at Harvard. Lady Anne served as Senior Tutor for women students for twenty years, was very supportive of women at the college, and took a leading part in the founding of the day nursery. For Gaydon, who joined the chemical engineering department in the early 1950s, see note 61 above.

88. Perkins examined the emulsions under a microscope back in South Kensington and reported the first 'star' disintegration of a Ag nucleus by a π-meson in 1947. This work paved the way for the bubble chamber work carried out at CERN by a later generation of Imperial physicists under C. C. Butler. Perkins later became professor of physics at Oxford.

89. Patrick Maynard Stuart (Lord) Blackett FRS (1897–1974) was educated at the Osborne and Dartmouth Naval Colleges. At sixteen he became a

midshipman intent on a naval career. He served in the Navy during the
First World War, seeing action both in the Falklands and at the Battle of
Jutland. An interest in aeroplanes prompted him to ask for a transfer to the
RNAS, but he was told that it took six years to train a naval officer and
only six weeks for an air officer and that his training could not be wasted.
Since war had interfered with the completion of his training at Dartmouth
he was sent to the Cavendish at Cambridge for a six month course in
1919. This experience prompted him to resign from the Navy and enroll
as a student. After gaining a degree in 1921 he became a research student
with Rutherford. At the Cavendish he learned about the cloud chamber
from C. R. T. Wilson, and its use in cosmic ray observations led him and
his co-worker, G. P. S. Occhialini, to discover the positron. This particle
was independently discovered at about the same time by C. D. Anderson,
working in the United States. Anderson, who published first, was awarded
the Nobel Prize in 1933. Many physicists believed that the prize should
have been shared with the Cambridge pair since they, unlike Anderson,
had identified the particle with Dirac's theoretical anti-electron. Blackett,
who used the cloud chamber to make many interesting discoveries, was
awarded the Nobel Prize in 1948 for his cosmic ray and cloud chamber
work. He was appointed to a professorship at Birkbeck College in 1933,
and to one at Manchester in 1937. He was elected President of the Royal
Society in 1965 and was created a life peer during the Wilson government.
For further details of his activities during the Second World War, and for
his role in British science politics after the war, see chapters 8 and 9. See
also, Sir Bernard Lovell, 'Patrick Maynard Stuart Blackett, Baron Blackett
of Chelsea, 1897–1974', *Biographical Memoirs of Fellows of the Royal Society*,
vol. 21 (1975), 1–115; and Mary Jo Nye, *Blackett: Physics, War, and Politics
in the Twentieth Century* (Cambridge, MA, 2004).

90.   Blackett was born in Kensington and had an attachment to the area.

91.   See Physics Department correspondence file; letter, Blackett to Hill,
29 December 1952. Blackett also brought along technician Robert F.
Wilkins who designed and built much of the equipment used both in rock
magnetism work, and in the bubble chamber/cosmic ray work. Wilkins
was also a regular contributor to art exhibits in the Consort Gallery. He
was made an honorary associate of the college in 1992.

92.   For correspondence, see note 91.

93.   Sir Clifford Charles Butler FRS (1922–99) was educated at a grammar
school in Reading and at Reading University. After gaining a PhD
he moved to Manchester as an assistant lecturer and began working in
Blackett's group with George Rochester. Together they discovered the

decay tracks of 'V-particles', later separated into two classes known as hyperons and K-mesons. For this work they used cloud chamber technology, both at Manchester and on the Pic-du-Midi in the Pyrenees. See G. D. Rochester and C. C. Butler, 'Evidence for the Existence of New Unstable Particles', *Nature*, 160 (1947), 855–57. The discovery of these relatively long-lived particles led theorist A. Pais to propose that their stability could be attributed to the separation of originally paired particles, and M. Gell-Mann and K. Nishijima to identify a new quantum number, 'strangeness'. Butler later turned to the accelerator as a source of particles. In a student article (*The Phoenix*, Winter 1959), Butler is described as having the looks of a rugby football player and the manner of an Upper Sixth history master. Butler left Imperial in 1970 to become Director of the Nuffield Foundation. From 1975 to 1985 he was Vice Chancellor of Loughborough University of Technology. For quotation, Clifford Butler, 'Recollections of Patrick Blackett 1945–70' (typed manuscript in ICA).

94. The Imperial College electromagnet and some cloud chamber work was soon transferred to an observatory on the Pic du Midi in the Pyrenees.

95. Derek G. Miller was Senior Experimental Officer (1958–78) and Senior Project Engineer (1978–98) with the High Energy Physics Group. He retired in 1998 and was awarded an honorary associateship of the college in 1999.

96. Harry Elliot FRS (1920–) was educated at the University of Manchester, returning there as a research student after serving in the RAF during the war. He joined the Manchester staff as a lecturer and came to Imperial in 1954. He was given a readership in 1957 and a professorship in 1960. An expert in cosmic rays in space, Elliot played a major role in both the American and European space programmes. His was expensive research. For example, in 1966 he was awarded £215,000 from the SRC to purchase a 'flying spot digitizer' and computer for bubble chamber work, and for analysis of the related interplanetary magnetic field work measurements gathered from space satellite instrumentation. From 1974–80 Elliot was Head of the Computing Centre. Much of the scanning of film in both Butler and Elliot's groups was carried out by women research assistants.

97. While at Birkbeck, Blackett had earlier carried out cosmic ray work in unused underground tunnels.

98. Paul Taunton Matthews FRS (1919–87) was born in India, the son of a missionary family. He won a scholarship to Mill Hill School and from there went up to Cambridge. A pacifist, Matthews served with the Friends' Ambulance Unit in China during the Second World War, returning to Cambridge after the war to complete his degree. He stayed

on as a research student in mathematical physics and was much influenced by Paul Dirac and by his doctoral supervisor, Nicholas Kemmer. Kemmer had been a Beit Scientific Research Fellow (1936–8), and demonstrator in the Imperial College mathematics department (1938–40) (see also chapter 8). Matthews worked on renormalization theory in quantum electrodynamics (work relating to elementary particles, notably mesons). His association with Salam began at Cambridge. Both men also spent a year together at Princeton before Matthews was appointed a lecturer at Birmingham University in 1952. In 1957 Salam was offered a chair in theoretical physics in the mathematics department at Imperial College and asked that Matthews be given a readership at the same time. Matthews is remembered as an outstanding lecturer and was much in demand as a conference speaker. He succeeded Butler as head of the physics department in 1970, but left in 1976 to become Vice Chancellor of the University of Bath. He died following a cycling accident in Cambridge. See T. W. B. Kibble, 'Paul Taunton Matthews, 1919–1987', *Biographical Memoirs of Fellows of the Royal Society*, vol. 34 (1988), 555–574.

Muhammed Abdus Salam FRS (1926–96) was born in the Punjab (in a part that later joined Pakistan). He was a prodigy and won a scholarship to Cambridge where he excelled in both the mathematical and physical sciences tripos. He then worked under Nicholas Kemmer, following closely the earlier work of Matthews in quantum electrodynamics. He and Matthews became close friends. Together they spent a year with Freeman Dyson at Princeton where Salam's work on renormalization theory brought him international recognition. Salam returned to Pakistan as professor of mathematics at Punjab University and Government College but continued his collaboration with Matthews. He found it difficult to advance his research career in Pakistan where he had to coach the college soccer team in addition to other duties. Given this, it is perhaps not surprising that he later disagreed with Blackett about the kinds of science to be supported in Third World countries. He wanted to see basic research encouraged in order to prevent brain drains, whereas Blackett favoured applied research, believing that 'pure' research was not something developing countries could well afford. Salam returned to a research fellowship in Cambridge before his appointment at Imperial in 1957. He succeeded to the chair held by Hyman Levy, becoming the first Asian to hold a chair in a science faculty in Britain. In 1979 he was awarded the Nobel Prize for physics (shared with Steven Weinberg and Sheldon Glashow), for his work on unifying theories of electromagnetism with those of the weak nuclear force. He was the founder and director of the International Centre for Theoretical

Physics in Trieste, and Chief Scientific Advisor to the Government of Pakistan. While a member of Pakistan's Atomic Energy Commission he was, like his friend Matthews, an active supporter of nuclear disarmament. A religious man, member of the Ahmadiyah sect of Islam, Salam sought harmony between his scientific work and his religious beliefs. For more on Salam see chapter 7; also T. W. B. Kibble, 'Muhammed Abdus Salam, 1926–96', *Biographical Memoirs of Fellows of the Royal Society*, vol. 44 (1988), 387–401; Alexis de Greiff, 'The International Centre for Theoretical Physics, 1960–79: Ideology and Practice in a United Nations Institution for Scientific Cooperation and Third World Development' (PhD thesis, Imperial College London, 2001).

99. Physics Department correspondence file, Blackett to Linstead, 9 March 1961.

100. Chapman resigned in 1946 to take up the Sedleian Chair of Natural Philosophy at Oxford. Interestingly he was later succeeded in that chair by one of his former Imperial College students, G. F. J. Temple (1901–92).

101. Harry Jones FRS (1905–86), was educated at the University of Leeds where he took a degree in physics before moving to Cambridge to work with R. H. Fowler and then to Bristol to work with J. E. Lennard Jones. At Bristol, together with Nevill Mott, he developed his interest in solid state physics, a field then being promoted by the DSIR. Their book, *The Theory and Properties of Metals and Alloys* (1958) became a classic. Jones was a lecturer in theoretical physics at Bristol from 1931–7, and was appointed Reader in Mathematics at Imperial in 1938. During the war he worked also at the Ministry of Home Security (mainly on detonation problems). When Levy retired as head of the department there was some question as to whether his successor should be W. G. Bickley or Jones. The college wanted Jones but the university favoured either Bickley or an outside candidate. Interestingly the college also had difficulties with the university over Jones's successor when he retired as head in 1970. Jones was a close friend of Dennis Gabor and was often a guest at Gabor's second home near Rome. See Sir Nevill Mott, 'Harry Jones', *Biographical Memoirs of Fellows of the Royal Society*, vol. 33 (1987), 327–42.

William Gee Bickley (1893–1969) was educated at University College Reading, and gained external degrees in physics and mathematics from the University of London in 1912 and 1913 respectively. After the First World War he lectured at Battersea Polytechnic, becoming assistant professor in mathematics at Imperial in 1930 and professor in 1946. Bickley, whose sight deteriorated seriously, created Braille notation for science and mathematics. See W. G. Bickley, 'Mathematics for the Blind', *Times Educational Supplement*, 22 February 1957.

102. George Alfred Barnard (1915–2002) was educated at Cambridge. During the war he worked on aircraft design, before transferring to the Advisory Unit at the Ministry of Supply where he worked on statistical methods and quality control. He joined Imperial College as a lecturer in 1945, became a Reader in 1948, and held the chair of statistics from 1955. In 1966 he left for a chair at the University of Essex. In his memoir, historian Eric Hobsbawm remembers Barnard as a fellow student and Communist Party member at Cambridge: 'the student Party's chief local commissar at the time was a lean-and-hungry looking mathematician from a working-class family, George Barnard of St. John's, who ended his career as President of the Royal Statistical Society'. For quotation see, Eric Hobsbawm, *Interesting Times: A Twentieth Century Life*, (London, 2003), 116. Barnard's younger sister, Dorothy Wedderburn, was also a professor at Imperial.

103. ULP/Maths/3, Jones to Linstead, 23 February 1956.

104. Walter Kurt Hayman FRS (1926–) was educated at Cambridge and appointed Reader at the University of Exeter in 1947, two years after gaining a BA in mathematics. Jones's objections were not personal and he welcomed Hayman, later expanding the presence of pure mathematics in the department.

105. Klaus Friedrich Roth FRS (1925–) was born in Breslau, Germany (now Wrocław, Poland) and came to England as a boy. Educated at St. Paul's School and the University of Cambridge, he taught at Gordonstoun School before moving to University College London where he took a PhD in 1950. A number theorist, Roth won the Fields Medal in 1958 and, in 1961, was appointed to a chair at UCL. He joined Imperial as Professor of Pure Mathematics in 1966. The college orator, who read Roth's 1998 FIC citation, amusingly portrayed him as the archetypical absent minded professor.

106. Sir Basil John Mason FRS (1923–), known as John, was educated at University College Nottingham but his studies were interrupted by the war. He was commissioned into the radar branch of the RAF and became Chief Instructor at Fighter Command Radar School. After the war he worked in telecommunications in the Netherlands East Indies before returning to Nottingham to complete his external University of London degree. He was appointed an assistant lecturer in the meteorology department of Imperial in 1948 and, nine years later, published his influential *Physics of Clouds* (Oxford, 1957). Linstead wanted to promote him to Reader in 1958 but the University of London objected. In 1960, Blackett,

Linstead and Sheppard (head of meteorology) managed to get a new chair to which Mason was appointed. Mason resigned the chair in 1965 to become Director General of the Meteorological Office. His chair was reassigned to the area of infrared spectroscopy and J. Ring was appointed.

107. Percival (Peter) Albert Sheppard FRS (1907–77) was educated at the University of Bristol. With a degree in physics he left to work at the Meteorological Office, which included work at Porton Down on the movement of poison gas clouds. He was a member of British Polar Year Expedition to the North West Territories, Canada in 1932–3 and was appointed Reader at Imperial College in 1939. During the war he was appointed Chief Meteorologist to the Ministry of Homeland Security and was responsible for area smoke screening over vital targets. Later he was in charge of meteorological aircraft and reconnaissance flights. He built up the department after the war and was also a major committee man beyond the college, including sitting as chairman of the space committee of the SRC.

    Frank Henry Ludlam (1920–77) joined Imperial as a lecturer in 1951 after working at the Meteorological Office. He was made Professor of Meteorology in 1965 and later headed the atmospheric physics section within the physics department.

108. Richard Segar Scorer (1919–) was educated at Cambridge University where he was a research student under G. I. Taylor. He worked for the Meteorological Office before coming to Imperial and was a notable early activist against air pollution. In 1972, he was a founder of a politically active group of scientists called Movement for Survival. A member of the Labour Party, he ran unsuccessfully for parliament, but successfully for an aldermanic seat in the London Borough of Merton. He is an expert on air flow and air pollution and attracted many young applied mathematicians to the department. When younger, Scorer was a keen cyclist and a marathon runner. Interestingly several members of the academic and non-academic staff in mathematics had political ambitions within the Labour Party in this period (E. P. Wohlfarth and H. Farebrother for example). For this reason they joined the ASTMS rather than the AUT since the former belonged to the TUC.

109. In addition to Jones, Hayman, Roth and Scorer already mentioned, they included Professors J. G. Clunie, D. R. Cox (who succeeded Barnard as professor of statistics), J. E. Ffowcs (Rolls Royce Professor of Theoretical Acoustics), C. W. Jones, G. E. H. Reuter, J. T. Stuart, and E. P. Wohlfarth.

110. James Dwyer McGee FRS (1903–87) was born in Canberra and educated at the Universities of Sydney and Cambridge. At Cambridge he was a research student under James Chadwick. Another of Chadwick's students, G. E. Condliffe, had set up a research lab for Electrical and Musical Industries (EMI) Ltd., then a new company, and invited McGee to join him. Because jobs were scarce, Chadwick advised he take it while stating, 'I don't think this business of television will come to very much'. But McGee and others proved him wrong. Indeed, it was the electronic Marconi-EMI system that was eventually adopted for the first regular TV service by the BBC in 1936. As part of the team, McGee worked on high-vacuum cathode ray tubes — one of his early transmitting tubes is in the Science Museum's television collection. He was also central to the development of the emitron camera. During the war McGee worked for the Admiralty, in conjunction with EMI, on signalling and infrared sensing devices. On coming to Imperial he remained a consultant for EMI. See also, B. L. Morgan, 'James Dwyer McGee, 1903–1987', *Biographical Memoirs of Fellows of the Royal Society*, vol. 34 (1988), 513–551.

111. Bryan Randell Coles FRS (1926–97) was born in Cardiff where he studied metallurgy at the University of Wales. This was the choice of a poor student who saw the booming South Wales steel industry as a possible source of employment. Not too happy with his course he gained inspiration from W. Hume-Rothery's *The Structure of Metals and Alloys* (1936) and, on graduating top of his class, thought of a career beyond Wales. He wrote to Hume-Rothery at Oxford asking if he could study further with him. At Oxford Coles worked on crystal structures, phase composition and phase boundaries, to infer the electronic structures of Ni-Mn alloys. With his DPhil he was appointed to a lectureship in metal physics jointly in Imperial's physics and metallurgy departments. While he carried out research in the physics department, at first he taught only metallurgy students. Only when Blackett came to Imperial did solid-state physics enter the undergraduate physics curriculum. Theoretical work in this field was being carried out by Harry Jones and his group in the mathematics department. Indeed, Jones was then at the cutting edge of the theory of the metallic state. At Jones's instigation, Coles spent two years at the Carnegie Institute of Technology in Pittsburgh where he learned many experimental techniques, notably those for low-temperature work. Coles developed techniques for studying the behaviour of impurity atoms carrying magnetic moments into different non-magnetic metals. A major experimentalist, he was also a major committee man, and editor of *Advances of Physics*. Coles was a good conversationalist and a lover of music, opera and the theatre.

According to his Royal Society obituarist, his mother read poetry to her children and Coles's remarkable memory was honed in this way. According to some of his colleagues, Coles regularly carried books of poetry under his arm. See David Caplin, 'Bryan Randell Coles', *Biographical Memoirs of Fellows of the Royal Society*, vol. 45 (1999), 53–6.

112. Sir Nevill Mott FRS (1905–1996), Nobel Laureate in Physics, 1977.

113. Pnina G. Abir-Am, 'The Molecular Transformation of Twentieth-Century Biology', in John Krige and Dominique Pestre (eds.), *Science in the Twentieth Century* (Amsterdam, 1997), 502.

114. ULOB/5 Gregory to Linstead, 7 March 1956.

115. B/Linstead; Bragg to Linstead, 5 July 1961.

116. Blackett, himself, briefly revived an old theory that had been kicked around in South Kensington during the nineteenth century. He suggested that there might be a connection between the earth's rotation and the generation of its magnetic field. William Ayrton and John Perry had the same idea as, later, did Arthur Schuster. See J. Perry and W. E. Ayrton, 'A New Theory of Terrestrial Magnetism', *Proceedings of the Physical Society of London*, iii (1879), 57–68. Blackett later demonstrated the falsity of this hypothesis.

117. The Tata Institute for Fundamental Research was under the directorship of Homi Bhabha, a Cambridge graduate, and close scientific friend of Blackett's. Bhabha also headed the Indian Atomic Energy Commission, founded in 1948. The Institute was funded by the Sir Dorab Tata Trust. The various Tata industries were, and remain, central to India's economy. Today the Tata industrial conglomerate is said to account for 2.8% of India's GDP.

118. Butler Recollections, *op. cit.* (93).

119. For a description of the new building see N. C. Barford, 'The Imperial College of Science and Technology: New Physics Department', *Nature*, vol. 188 (17 December 1960), 989–90.

120. By the end of his career times had changed and he could not totally avoid such meetings. Gossip has it that it was his wife who told him that the occasional meeting might be a good idea. As to his young staff, he would also advise them on how to set up their experiments and write up their results. Further, he told them that if a good research idea came to them while they were lecturing undergraduates, they were to leave the lecture hall immediately and return to their laboratories.

121. Quoted in Nye, *op. cit.* (89) 43, and inspired by Kipling's *If*. Date of Blackett's verse not known.

122. There is obviously some connection between Blackett's social democratic political stance and his attitude towards hierarchy at Imperial. But few of his staff remember him as a democrat; rather he is remembered as a far-sighted and largely benevolent autocrat who carried a strong whiff of his naval past. Even though he wanted the tea room to be open to all, like an admiral Blackett routinely went to the front of the queue (W. G. Jones, personal communication).

123. One sensible reason that Blackett gave for the many professorships was that by not offering deserving people chairs they would leave for better opportunities in the USA. The need to increase the number of senior academic posts in the UK was one of his public missions. See *The Times*, 17 March 1962, 5, for an article by Blackett on this subject.

124. Briscoe had worked on civil defence and had slept at the college during the war. His civil defence group was known as the seven dwarves and Briscoe was known as Snow White. He was good at keeping up morale during bombing raids and dealt with some fire bombs on his own. His other war work is discussed in chapter 8.

125. ULP/chemistry/org./2, Heilbron to Penson, 8 December 1948.

126. Arthur Herbert Cook FRS (1911–88) was educated at Imperial College and took a PhD under Jocelyn Thorpe. He developed the field of hetero-cyclic organic chemistry at the college. During the war he had worked with Heilbron on antimalarials and on the purification of penicillin. Funded by the Therapeutic Research Corporation of Great Britain Ltd., he led a team looking at structural analogues of penicillin, and attempts to synthesize some of these led to new and interesting chemical ideas. Cook was largely responsible for setting up the brewing industry laboratory at Nutfield, Surrey and succeeded Heilbron as its director in 1958.

127. Alexander Todd had not yet built up the Cambridge department but the departments at Oxford and Manchester were also strong at this time.

128. There is an interesting historical reason for the Rockefeller support. Sir Frederick Gowland Hopkins (1860–1947) was awarded the Nobel Prize in 1929 for his work on vitamins. When President of the Royal Society he gave a speech, at the British Association for the Advancement of Science in 1931, on the importance of biochemical research. The speech played a major role in the reorientation of funding by the Rockefeller Foundation which moved from funding the physical sciences to the molecular life sciences during the 1930s. After the war the Foundation funded much research in biomolecular and medical science in Britain; Linstead was one of the early beneficiaries.

129. See Board of Studies Minutes, 21 May 1954.

130. Richard Maling Barrer FRS (1910–1996) was born in Wellington, New Zealand and educated at Canterbury University College. While there he read Eric Rideal's book *An Introduction to Surface Chemistry* (Cambridge, 1926) and wrote to Rideal, Plummer Professor of Colloid Science, asking to join his research group in Cambridge. He was accepted and while there began to develop the idea of molecular sieves, gradually laying the groundwork for the industrial application of zeolites, especially in the petrochemical industry. He later published his influential *Zeolites and Clay Minerals as Sorbents and Molecular Sieves*, (London, 1978). Before coming to Imperial Barrer was head of the department at Aberdeen. He was a very good athlete, a cross-country champion and an excellent tennis player. On Rideal's retirement from Cambridge, Barrer invited him to be a senior research fellow in the department. Lovat V. C. Rees, 'Richard Maling Barrer, 1910–1996', *Biographical Memoirs of Fellows of the Royal Society*, vol. 44 (1998), 37–49.

131. Ernest Alexander Rudolph Braude (1922–56) was born in Germany and came to England as a refugee in 1937. He was educated at Birkbeck, and at Imperial College where he gained both his BSc and PhD. He was a research student under Heilbron before joining the staff as an assistant lecturer in 1945. See obituary by R. P. Linstead, *Proceedings of the Chemical Society*, (October, 1957), 297.

132. Sir Geoffrey Wilkinson FRS (1921–96) was born in Todmorden, Yorkshire where he was educated at the local grammar school before winning a Royal Scholarship to Imperial College in 1939. One of his fellow students in the chemistry department was Ralph Raphael, later to hold chairs of organic chemistry at Glasgow and Cambridge. One year ahead of them was Derek Barton. Wilkinson graduated at the top of his class and began research work under Briscoe. He was directed to work on a war-related project having to do with phosgene gas, later claiming that Briscoe directed this work 'from a safe distance'. As discussed in chapter 8, Wilkinson was recruited by Tube Alloys to work on atomic research at Chalk River in 1941. After the war he worked with Glen Seaborg at the Lawrence Livermore Laboratory at Berkeley and became, as he put it, 'the first successful alchemist' when he achieved the transmutation of platinum into gold. Wilkinson left nuclear chemistry on moving to Harvard. Working alongside R. B. Woodward he developed an interest in organometallic chemistry and began his study of ferrocene and other metallocene-transition metal compounds, work which was to bring him the Nobel Prize for chemistry (with E. O. Fischer) in 1973. His Imperial chair was renamed the Sir Edward Frankland Chair in Inorganic Chemistry in 1978. See M. L. H. Green and W. P. Griffith,

'Sir Geoffrey Wilkinson, 1921–1996', *Biographical Memoirs of Fellows of the Royal Society*, vol. 46 (2000), 593–606.

133. Sir Derek Harold Richard Barton FRS (1918–1998) was born in Gravesend and educated at Tonbridge School. On leaving school he worked for two years in his father's carpentry business before entering Imperial College in 1938. He made rapid progress, gaining a BSc in two years, and a PhD after a further two years working under Heilbron. During the war he worked also with Briscoe, later moving to Albright and Wilson (see chapter eight). Since this work was classified (it involved the separation of hydrogen isotopes), Barton was at first disadvantaged in seeking academic work. But Briscoe was a very good mentor and patron. After the war he invited Barton back to Imperial as assistant lecturer, while recommending he take also a junior visiting position at Harvard. There, like Wilkinson, Barton came under the influence of R. B. Woodward, and, in Barton's case, also Louis Fieser, but he was not then comfortable in the USA. As he wrote to A. A. Eldridge, everything was very efficient but 'the food is disgusting' and 'to be a professor in England is more attractive because it means so much more'. (B\Barton; Barton to Eldridge, 4 December 1949). Eldridge replied stating that professors were no longer what they used to be, 'it is a topsy turvy world … the plumber now wears a bowler hat and must be addressed 'Mr.' and the professor … wears no hat at all and doesn't much care how he is addressed'. (Eldridge to Barton, 8 December 1949). Briscoe then helped Barton to a readership at Birkbeck College. He wrote to the Academic Registrar of the University of London that Barton's 'breadth of interest and scholarship in the chemical field is quite extraordinary, and he is indefatigably skilful, original and ingenious in scientific enquiry'. (B\Barton; Briscoe to Henderson, 18 January 1950). All this was true and, by 1950, Barton had begun to publish widely, including papers on his path-breaking research on the conformation of steroid molecules. In 1955 he was appointed to the Regius professorship of chemistry at the University of Glasgow and, in 1957, returned to Imperial. For his conformational work, and insights into the spatial arrangement of organic molecules and reaction mechanisms, he was awarded the Nobel Prize for chemistry in 1969. In 1970 his chair was renamed the Hofmann Chair in Organic Chemistry. Barton retired from Imperial in 1972 but continued working, first as Director of the Institute for the Chemistry of Natural Substances at Gif-sur-Yvette (near Paris) and, from 1986, as Distinguished Professor of Chemistry at Texas A and M University. In 1995 he became the Dow Professor of Chemical Invention, a position he held till his death. See also, Steven V. Ley and Rebecca M. Myers, 'Sir Derek Harold Richard

Barton', *Biographical Memoirs of Fellows of the Royal Society*, vol. 48 (2002), 3–23.

134. Wilkinson hated wasting time in commuting and asked Linstead for the use of one of the college flats until he had the financial means to buy something close to the college. (B/Linstead; correspondence file; Wilkinson to Linstead, 12 November 1956). As discussed elsewhere, Linstead gave flats to Chain, the Halls and to Argyris, but he did not give one to Wilkinson. In his memoir, Wilkinson complained that he was not properly appreciated by the college administration. By contrast, when he won the Nobel Prize he was sent a case of Krug from the chemistry department at Harvard. Its members had been sorry to lose him and a case of champagne was the traditional gift of Harvard to its Nobel laureates. B/Wilkinson; handwritten memoir.

135. One catalyst he prepared is known as Wilkinson's catalyst. A rhodium complex, it was used commercially in the hydroformylation of olefins by Union Carbide, Johnson Matthey, and Davy Power Gas.

136. Green and Griffith, *op. cit.* (132).

137. Pratt and Evans had been students of Rex Richards at Oxford. Richards built an early solid-state nmr spectrometer.

Dennis Frederick Evans FRS (1928–90) was educated at Nottingham High School from where he won a scholarship to Oxford. After gaining his PhD under Rex Richards he spent two years with R. S. Mulliken at the University of Chicago before coming to Imperial with Wilkinson in 1955. Evans was promoted to a readership in 1964 and given a chair in 1981. He was an outstanding experimentalist and carried out work in a number of areas, though the magnetic properties of oxygen was an ongoing interest. His encyclopedic chemical knowledge made him a valuable resource both within and outside the department. For example, he was called as an expert witness at the inquest of Georgi Markov, murdered in 1978, when stabbed with the tip of an umbrella coated in ricin. (Markov, a writer, had defected to Britain from Bulgaria and had made a number of anti-Communist broadcasts over the BBC. Cold War politics and the KGB were behind this still mysterious death.) Evans had an interesting life away from the college. His politics were on the left (he appeared in a front-page photograph in the *Guardian* fending off a policeman during an anti-Suez rally), he moved in Chelsea Arts Club circles, and experimented with hallucinogenic and other drugs well before such activity became fashionable in the 1960s. The writer William S. Burroughs visited Evans' laboratory regularly in order to accurately weigh out his own daily heroin doses. The two were friends and Evans once rescued a very ill Burroughs in Tangier and brought him back to England for medical treatment. Evans kept exotic pets such as snakes, scorpions and spiders, and for a while shared his flat with Christine Keeler — to hide her and give her some privacy shortly after the Profumo affair became public knowledge. At the college he performed a range of stunts on social occasions, including 'swallowing' liquid oxygen which he exhaled through lighted cigarettes, inhaling low density gases to make his voice change, and appearing to eat wine glasses and similarly impossible objects. He was a brilliant scientist and much loved eccentric. See also M. L. H. Green and W. P. Griffith, 'Dennis Frederick Evans, 1928–90', *Biographical Memoirs of Fellows of the Royal Society*, vol. 46 (2000), 165–75.

138. B\Wilkinson, handwritten memoir. It was Les Pratt who then commissioned the spectrometer and defined its specifications. Wilkinson also appointed a crystallographer, (later Sir) Ronald Mason. He resigned in 1961, after just four years in the department, to take up a chair in Sheffield. He later moved to a chair at Sussex and was seconded to become Chief Science Advisor to the Minister of Defence at the time of the Falklands War.

139. Both David and Margaret Goodgame were college tutors and played a major role in undergraduate teaching and in mentoring students. Both are active members of the Imperial College Staff Christian Society. They were awarded honorary associateships of Imperial College in 2001. David Goodgame and William (Bill) Griffith, an expert in transition metals and Raman spectroscopy, later became professors.

140. Bernard Atkinson, a physical chemist in the department, has written a good history of the department for the years 1960–88. Copy on disk is in the ICA.

141. Frederick Clifford Tompkins FRS (1910–95) was educated at Bristol University before being appointed to a lectureship at King's College London in 1934. From 1937–46 he worked at the University of Natal, returning to King's in 1946. One year later he was appointed Reader in Physical Chemistry at Imperial and was given a chair in 1959. A distinguished surface chemist he was also the Editor and Secretary of the Faraday Society (later Faraday Division of the Chemical Society) from 1950–77 and President, 1978.

142. Michael Spiro was given a conferred chair in 1992. He is also a specialist in infusion processes and often gives talks about tea and coffee. One of Spiro's better known PhD students, known not just for chemistry, is Mary Archer. Dr Archer continued occasional work at the college, most recently in the Centre for Energy Policy and Technology.

143. Leonard Newton Owen (1914–99) was educated at the County School, Llangollen, and at the University of Birmingham. After gaining a BSc he became a research student under Norman Haworth and worked on starch and cellulose chemistry, work he continued when he moved to Imperial in 1943. During the war he worked on the nitrogen mustards and on ways to counter them. Later he worked on ways of chemically modifying these compounds, making them less toxic and thus effective chemotherapy agents. Owen lived in Beit Hall for many years. See also chapter 8.

144. There were many delays before chemistry acquired a new building, and then only in painfully slow stages. The first stage was completed only in 1993. Barton looked for space elsewhere and ran research teams both at Chelsea College and in the USA.

145. One of Barton's students, A. G. M. Barrett, today holds the Sir Derek Barton chair in organic chemistry. Barrett claims that Barton was a basically shy man who hid his shyness behind a gruff exterior. Barrett remembers him as kind and very generous towards his good students. (Personal communication.)

146. While the Colonial Office especially asked for more entomologists, the UGC argued against producing too many biologists at this time.

147. The Ministry bureaucrats argued that one of the original conditions of financial support for the purchase of Hurworth was that the field station would not be sold. They interpreted Imperial's move to purchase a new field station as indicative of intentions to sell the old one. Imperial argued that it was moving, not selling and closing, its field station. The Ministry, not wishing to move out, at first dug in its heels and refused to pay for Hurworth, but eventually did so.

148. The College purchased the main house at Silwood and 80 acres of land for £24,000 in 1947. The neighbouring Ashurst Lodge estate was purchased in 1948 (this was the former home of the Archer-Shee, 'Winslow Boy' family, and was later sold). In 1953 Silwood Park Farm and adjoining acres were purchased and, in 1961, some acres along Cheapside Rd. were added to the college holdings. Munro was given permission to convert the main building into laboratories, hostel, and a flat for himself and family. It was hard to get permits to buy furniture after the war, but Munro was once again resourceful. The High Wycombe Furniture Federation which had been helped by Munro's work on timber pests after the First World War came to the rescue and supplied furniture at wholesale rates. The person appointed to look after business matters was Captain John R. Barnes DSO, RN who was given a newly built residence on the estate — a further example of an ex-military man being given a senior administrative position

at Imperial. Munro displayed a paternalistic pattern of behaviour and his research students would dine with him each evening in the manor house. Munro's wife set up an infirmary at Silwood and a nurse was hired. A social club, now known as the Munro Club, was formed for technical staff.

149. KZFG; See J. W. Munro, 'Imperial College Field Station'; typed manuscript (1961), 30. Tizard, then Chief Scientific Advisor to the Ministry of Defence, shared Munro's grievance against the DSIR for having requisitioned the Hurworth field station during his tenure as Rector. In particular, he believed that the sidelining of Munro, and the taking over of Munro's work on pest control in food stores during the war, was misguided.

150. The move from Slough to Silwood was helped by James Stanley Porter who had joined the technical staff at Slough in 1930. He was appointed the clerk of works at Silwood and designed and carried through the alterations at Silwood Park House, and later at Ashurst Lodge, needed for scientific work to begin. He also later supervised construction of the Shell building at Ashurst and a number of other laboratories at Silwood. He was made an honorary associate of the college in 1977.

151. Later Rachel Carson was to write her famous book *Silent Spring* (1962), in part a response to the damage caused by the indiscriminate use of pesticides such as DDT, developed by applied entomologists in this period. Munro and his peers had to learn from their mistakes and could not have predicted all the consequences of their actions. Later, however, corporations were rather slow to learn.

152. Bernard George Peters (1903–67) was educated at Bristol Grammar School and Bristol University. After gaining a BSc in zoology in 1925 he moved to the Institute of Agricultural Parasitology at St Albans as a Grocers' Company Research Fellow. He gained a PhD from the University of London in 1928. From 1929–32 he was Deputy Director of the Imperial Bureau of Agricultural Parasitology at St. Albans before becoming a lecturer in helminthology at the London School of Hygiene and Tropical Medicine. During the war he was seconded to the Ministry of Aircraft Production (operational research for Bomber Command). After, he became Principal Scientific Officer at Rothamsted where, from 1952, he headed the new department of nematology. Peters was a pioneer in nematology as applied to plants. Until then nematology was seen largely as a medical-veterinary specialty within parasitology. For further details see John M. Webster, 'Nematology: From curiosity to space science in fifty years', *Annals of Applied Biology*, (1998), 132, 3–11.

153. Michael James Way, (1922–) was educated at Taunton Grammar School and at Cambridge University. He graduated in 1942 and moved directly into pest control, working first for the Ministry of Defence and then for the ARC at Rothamsted. Aside from a brief Colonial Office secondment he remained there until coming to Imperial. Way was given a chair in applied zoology in 1969.

154. Arthur John Rutter (1917–), known as Jack, was educated at Imperial College. He was the first research student of G. E. Blackman and gained a PhD in 1943. Together they studied bluebell growth in woodlands from an ecological perspective, developing new statistical methods of analysis for this study. During the war Rutter also worked with Blackman on both weed control and the selection of varieties of oil seed crops suitable for cultivation in UK. This work was not continued after the war when Rutter joined the staff as a demonstrator. But his ecological work was facilitated when the Forestry Commission invited university personnel to work on its lands. Rutter worked at Bramshill Forest in Hampshire and developed research programs on the ecology of wet-heaths, and on growth patterns in coniferous woodland. In this connection he demonstrated the extent of water loss from the forest canopy and raised questions relating to the conservation of water in soils. Rutter succeeded Whittingham as head of the department of botany and plant technology in 1971.

155. Stephenson was Chief Surveyor with the Polar Expeditions of 1930–31 and 1934–37. The observatory was actually intended for R. G. Waterfield, a consulting haematologist at Guy's Hospital. A descendent of John Herschel, he had inherited the family's astronomy gene and had been using the Royal Observatory until it moved to Herstmonceaux. Since his requirements were minimal, Munro recommended he be given a site at very low rent. Fowler had recently retired and astronomy at Imperial then ended. Waterfield agreed to allow college personnel to use his observatory, though few did. Some years later Blackett made Waterfield an honorary lecturer in the physics department.

156. Philip Herries Gregory FRS (1907–86) was educated at Imperial College where he gained a PhD for work in mycology and plant pathology. He then moved to Winnipeg where he worked under one of the world's leading mycologists, A. H. Reginald Buller FRS. He returned to England for work with the cut-flower industry in Cornwall, and at Hayne College in Devon on the diseases of flowering bulbs. When war broke out he moved to Rothamsted to work on viral diseases in potatoes. After the war he developed mycology work at Rothamsted before coming to Imperial in

1953. He was a specialist in the liberation and dispersal of fungal spores, and their relation both to plant, and human respiratory, disease. See J. M. Hirst, 'Philip Herries Gregory 1907–1986', *Biographical Memoirs of Fellows of the Royal Society*, vol. 35 (1990), 151–177.

157. When Professor Brown retired Gregory was the only professor of plant pathology in Britain. P. H. Gregory wrote a report in 1954 stating that Silwood was not ideal for botanical work because of poor soil and undulating land. He wanted expansion at Harlington but this did not happen (Botany Department files, Report 5 May 1954).

158. Some research was carried out at Kew. Field studies ranged throughout the country. Farmer had earlier taken students to Snowdonia but, while this continued for many years, the practice had largely died by the time war broke out. After the war it was decided to use the recently created centres of the National Council for the Preservation of Field Studies; first at Flatford Mill, Essex and later those at Dale Fort in Pembrokeshire and Malham Tarn in Yorkshire. By the 1960s the department increasingly used the facilities at Silwood Park.

159. Helen Kemp Porter (née Archbold) FRS (1899–1987) was educated at Clifton High School and Bedford College London where she studied chemistry. She came to Imperial in 1921 to further her studies in organic chemistry but turned instead to the department of botany. She was a research student under V. H. Blackman who, at the time, was concerned with spoilage in apple cargoes. To investigate this problem Blackman set up a team of women scientists which, besides Porter, included Elsie Widdowson FRS (1906–2000) and Dorothy Hayes, both of whom also had distinguished careers. Widdowson, who earlier had studied chemistry at Imperial, was to work at the MRC laboratory in Cambridge for over fifty years. She and Robert McCance are known for having compiled the first set of food tables in the United Kingdom, giving comprehensive nutritional information on many common foods: see, R. A. McCance and E. A. Widdowson, *The Chemical Composition of Foods* (Her Majesty's Stationery Office, 1940). The book is now in its fifth edition. Widdowson was awarded the FIC in 1994. The apple project prompted Porter to learn about carbohydrate metabolism in plants, a subject on which she became an expert. Learning was facilitated by the fact that the project entailed frequent trips to Cambridge where she met people working in the biochemistry department there. In 1931 Porter transferred to the staff of the Institute of Plant Physiology where she carried out further work on carbohydrate metabolism and enzymology. Much of her research was conducted at Rothamsted together with the director of the institute, F. G. Gregory.

Abandoning apples for barley, Porter corrected earlier views and gave what was accepted as the correct description of the formation of grain starch. In 1956 she was elected to the Royal Society. It had only been eleven years since women had become eligible for the Fellowship and, at the time, 10 of the 606 fellows were women.

160. For more on Gregory see chapter 6. F. G. Gregory was one of the 'old guard' and eminent in his field. When he died in 1961 he left his library to the college, and two-thirds of his residuary estate to form the Frederick Gregory Fund to benefit students in the botany department. Gregory had a serious interest in music and was a great admirer of Benjamin Britten. On his retirement the college, together with Imogen Holst, wrote to Britten asking whether he would compose a short piece to mark the occasion; money was collected for this purpose. Britten refused (see Rector's correspondence, Britten to Linstead, 17 November 1958). Instead the money was used for Gregory to attend composition classes at the Royal College of Music. (Linstead to Sir Ernest Bullock, Director of RCM; 11 December 1958.)

161. Charles Percival Whittingham (1922–) was educated at Cambridge and came to Imperial during the war to work on new herbicides under G. E. Blackman. When Blackman moved to Oxford, Whittingham went with him before returning to Cambridge to work for a PhD. There he lectured on plant physiology before his appointment to the chair of botany at Queen Mary College London in 1958. He came to Imperial in 1964, leaving in 1971 to head the botany department at Rothamsted. His research was on photosynthesis at the cellular level.

162. William Owen James FRS (1900–78) was educated at the Universities of Reading and Cambridge. At Cambridge he worked with F. F. Blackman, a leading plant physiologist, before moving to London to work with Blackman's brother, V. H. Blackman. In 1927 he moved to Oxford to work with A. G. Tansley. During the war he instituted the Oxford Medicinal Plant Scheme. James remained in Oxford until taking the chair at Imperial. Following in the footsteps of his two predecessors as head of department, he did not encourage work at Silwood.

163. S. E. Jacobs who had entered the College in 1923 as a student of chemistry moved to botany, via agricultural chemistry, and led the bacteriology section. He was appointed Reader in 1950. He was the editor of *Journal of Applied Bacteriology* which he built into a leading journal, and it was he who developed the microbiology course at Imperial.

164. The name Insect Physiology Group was used only because Kennedy believed all animal behaviour should be viewed physiologically. In fact people within the group studied insects from a range of perspectives.

165. John Stodart Kennedy FRS (1912–1993) was born in the USA, son of an itinerant Scottish railway engineer. The family returned to England after the First World War and Kennedy attended Westminster School before coming to Imperial as a student. He dropped out of the entomology course because he found it boring, and completed his degree at University College. During the Second World War he worked in the Colonial Office's Middle East Anti-Locust Unit organizing both RAF and Soviet crop-dusting aircraft. After the war he worked at Cambridge in the ARC Unit for 21 years. T. R. E. Southwood, who earlier had been helped by Kennedy to develop his own ideas on insect migration, was about to become head of Imperial's zoology department when the ARC unit closed. Despite being warned that Kennedy was a member of the Communist Party, something that had hindered his advance at Cambridge, Southwood invited him (and Lees) to join the department in 1967. So, thirty years after having dropped out of the college, Kennedy returned as Professor of Animal Behaviour. On Kennedy's retirement, Southwood, by then at Oxford, offered him a research associateship there. Kennedy returned to his earlier aphid research while maintaining research also at Silwood (with C. T. David) on moth flight. See John Brady, 'John Stodart Kennedy', *Biographical Memoirs of Fellows of the Royal Society*, vol. 41 (1995), 244–60.

     Anthony David Lees FRS (1917–92) was educated at Cambridge. During the war he carried out work with Wigglesworth on wire worms and sheep ticks, at the London School of Hygiene and Tropical Medicine. He moved with Wigglesworth to the ARC Unit at Cambridge in 1945. Lees became Professor of Insect Physiology in 1969. See D. S. Saunders, 'Anthony David Lees', *Biographical Memoirs of Fellows of the Royal Society*, vol. 40 (1994), 223–236.

166. Brady, *op. cit.* (165), 250.

167. The UGC paid a visit to Silwood shortly after Richards became head of department. This proved disastrous since the committee was unhappy with what they saw, and with Richards' plans. Linstead was not as supportive as he should have been. Despite earmarking the large sum of £200,000 for Silwood, it would appear that Linstead was uncertain how to proceed. Botany and zoology were, perhaps, his least favoured fields of endeavour and they did not receive sufficient attention at Imperial until Brian Flowers became Rector.

168. Rector's correspondence; Porter to Linstead, 26 June 1966.

169. David Williams (1898–1984) studied civil engineering at Liverpool and became interested in geology after his twin brother Howel, also a geologist, came under the influence of Percy Boswell at Leeds. Williams moved to

study with Boswell and his PhD research was on paleozoic volcanic rocks in Snowdonia. Later he worked in geophysical prospecting in Southern Africa and for Rio Tinto in Spain, before coming to Imperial as a lecturer in 1932. On the retirement of W. R. Jones in 1947 he became Professor of Mining Geology.

170. John McGarva Bruckshaw (1907–69) was educated at the University of Manchester. On graduating he worked under William Bragg on x-ray crystallography. In 1930 he joined A. B. Broughton Edge in an experimental geophysical survey of Australia. At Imperial he built the first school of applied geophysics in Britain. He was remembered as a modest man who encouraged those working under him, including his successor R. G. Mason.

    Ronald G. Mason (1916–) was educated at Imperial where he took a degree in physics before turning to geophysics under Bruckshaw. He studied for his PhD after a wartime spent working for the Ministry of Supply, largely on oil exploration in Nigeria. Mason developed some major theories on the spreading of the sea floor. He was professor of geophysics from 1967–84.

171. John Stuart Webb FREng (1920–) was educated at Imperial College, an outstanding student who managed to win just about every prize going while coxing both the RSM and Imperial College Eights, and becoming the college fly-weight boxing champion. Between his undergraduate and postgraduate degrees he served with the Royal Engineers and then worked for the Nigerian Geological Survey. Appointed lecturer in mining geology in 1947, he was given a chair in applied geochemistry in 1961. Webb was the major force behind the Imperial College *Wolfson Geochemical Atlas of England and Wales*, published by Oxford University Press in 1978. His sister, Mona Webb, was secretary both to Professor Read and Professor Williams.

172. William Daniel Gill (1916–1992), known as Dan, was educated at the University of Leeds. After gaining his BSc he began working for the Attock Oil Company and played a major role in the discovery of oil fields in the Himalayan foothills of Pakistan. By the late 1940s he was back in England, a lecturer at Nottingham, and an acknowledged expert on Himalayan geology. He worked towards his PhD which was good enough to win him a chair at Trinity College Dublin. He was an excellent field geologist and is described as having been 'larger than life'. He gave memorable parties at the end of the summer term where, in his Hampstead home, he accompanied himself at the piano and entertained members of the department while singing into the small hours. See, *GEOlogIC* (1992) 342.

173. John Sutton FRS (1919–92) came to Imperial before the war but his education was interrupted by army service. He returned to the college as a research student in 1946 and, at the suggestion of his supervisor, H. H. Read, began work on Lewisian Gneiss. This work was carried out alongside similar work by another of Read's students J. V. Watson. Indeed, they submitted joint work for their PhD degrees. Sutton and Watson later married and their geological work interconnected throughout their lives; many projects were conducted jointly. Sutton's obituarist stated that Watson was 'the careful fieldworker, painstaking recorder of detail, and it was she who wrote the greater part of their joint papers in which every sentence is carefully weighed and composed'. Claims of who did what in joint work are always contested, and this claim is no exception. Sutton built a research school with about 12 doctoral students by the end of the 1950s, though the rather routine work on Moine and Lewisian geology in which both Sutton and Watson's students were engaged earned their combined research teams the name 'Moine Machine'. Sutton was appointed Professor of Geology in 1958. See J. S. Spring, 'John Sutton, 1919–92', *Biographical Memoirs of Fellows of the Royal Society*, vol. 41 (1995), 440–56. In 1967 Marjorie Linstead gave Sutton her late husband's DSc gown which he had earlier inherited from Martha Whiteley.

Janet Vida Watson FRS (1923–85) was the daughter of D. M. S. Watson, professor of zoology and comparative anatomy at University College London. Her father taught Imperial College students during the 1930s when there was much student exchange between the two colleges. Her grandfather, David Watson, was a chemist/metallurgist who studied at the RSM under Edward Frankland and Andrew Ramsay. Watson was a student at Reading University, moving to Imperial after the war when she took a degree in geology. She came top of her class and was awarded the Watts Medal. As a research student she began to work on the migmatites of Sutherland, where Read had earlier worked, before moving to Lewisian rocks. Watson and Sutton married soon after gaining their PhDs in 1949. Watson then won a senior studentship from the Royal Commission for the Exhibition of 1851. At the time Read wrote, 'in my opinion Miss Watson is the most outstanding young metamorphic geologist in the Commonwealth'. See Royal Commission for the Exhibition of 1851 papers; Science Scholarships Committee; Meeting 83, 2 June 1949. Watson and Read co-authored a number of books, including a multi-volume *Introduction to Geology* (vol. 1, London, 1962), Watson was President of the Geological Society, 1982–4, and Vice President of the Royal Society,

1983–5. See D. J. Fettes and J. A. Plant, 'Janet Vida Watson, 1923–85', *Biographical Memoirs of Fellows of the Royal Society*, vol. 41 (1995), 501–13.

John Graham Ramsay FRS (1931–) was educated at Imperial College where he won both the Murchison Medal and Judd Prize. He stayed at Imperial as a lecturer, reader, and then professor (1966–73) before leaving to become head of the department of geology at Leeds. Early in his career he turned from Lewisian and Moine geology to work on the Alps. He also made the transition into plate tectonics. Ramsay was awarded the Wollaston Medal in 1986.

174. B/Sutton/ A/ 90; Sutton to Watson, 27 June 1948.

175. Other women of her generation had been appointed to the geology teaching staff; for example G. M. Wallace, also the wife of a member of staff (R. Shackleton), was a lecturer in the department for twenty years.

176. Hill's husband, Edward (Ted) John Hill, began as a lab boy. He came to the college before his wife, in 1925, and ended his career as Departmental Superintendent, 1961–70. Their son joined the technical staff in 1946. Ted Hill took an active part in college affairs and was a founding member of the Holland Club. He was a keen apiarist and kept hives, first on the roof of the RSM, later on the Botany building roof. They produced large quantities of honey.

177. One story from the 1960s is perhaps revealing. A South African mining company had presented a large block of Kimberlite to the department in the 1920s. Sutton thought it was taking up too much space and asked a technician to remove it. This was no easy task and the technician could only do so by attacking it with a sledge hammer and removing it one piece at a time. Two students, noting what was happening, asked to buy the rock pieces and the technician allowed them to do so, the money going to the department. The students began selling pieces to rock merchants. This proved profitable since, after the flooding of the Kimberley diamond mines, Kimberlite was no longer easily available. When Sutton found out, he did not praise the students' initiative. On the contrary, he was furious and severely reprimanded them and the technician. And, as noted by his obituarist (*The Times*, 15 September 1992): 'To be one of his students was like living on the slopes of a volcano. The soil was fertile, the view awe inspiring, but one knew that long periods of productive calm could suddenly be punctuated by an eruption'. See also chapter 13.

178. Sedimentology was the field of Douglas James Shearman (1918–2003) who was appointed assistant lecturer in 1949 and rose through the ranks to professor in 1978. He was a much loved figure and many tales were told

of his eccentricities. Like some of the civil engineers he was a consultant at dam sites throughout the world.

179. Marston Greig Fleming FREng (1913–1982) was educated at Queen's University, Ontario and graduated with a degree in mining and metallurgy in 1936. He worked for a few years in the Canadian goldmining industry before coming to Britain during the war. He was appointed Professor of Mineral Technology in 1961 and was head of the department of mining and mineral technology, 1967–74, and of the mineral resources engineering department, 1979–80. He was a major consultant to both governments and mining companies, and was a member of the Governing Body, 1968–80.

180. James Cecil Mitcheson (1898–1979), known as Cecil, came from an old mining family. His father, grandfather and great-grandfather were all mining engineers, the latter having worked with George Stephenson at the Old Killingworth Colliery. Mitcheson was educated in mining at the University of Birmingham and began his career there as a lecturer. In 1924 he left to work for the mining industry, in the Warwickshire Coalfield. When the mines were nationalized after the war he was appointed Mining Development Engineer in the West Midland Division but resigned shortly after to become a mining consultant.

181. Robert Alastair Lucien Black (1921–67) was known as Alastair. He was educated at the RSM and moved to South Africa after the war where he joined the staff of the Central Mining and Investment Corporation Ltd. in Johannesburg. In 1956 he became professor of mining at the University of the Witwatersrand. He succeeded Mitcheson as head of department but suffered from depression and died from a drug overdose.

182. Frederick Denys Richardson FRS (1913–83) was known as Denys. He was educated at University College London where he took a degree in chemistry, and carried out research under Charles Goodeve. Richardson spent two years as a postdoctoral fellow at Princeton before returning to England at the start of the war to work with Goodeve at the Admiralty's Department of Miscellaneous Weapons Development. He became Deputy Director with the rank of Commander RNVR and worked on a variety of projects from the protection of naval ships from magnetic mines, to flares, to anti-aircraft balloon barrages. After the war Richardson joined the chemical section of the British Iron and Steel Association, which Goodeve had founded, and began research in physical chemical problems related to the making of steel. At first his work on the properties of various silicate and iron slags was under the indirect supervision of two members of Imperial's chemistry department, J. A. Kitchener and J. O'M Bockris. Richardson moved to the metallurgy department at Imperial in 1950, as

Nuffield Fellow in Metallurgy, and was Professor of Extraction Metallurgy from 1957–76. Among his many publications is the two-volume classic, *The Physical Chemistry of Melts in Metallurgy* (London, 1974).

183. As noted in chapter 9, Ball came from Harwell where he had worked on the properties of plutonium. At Imperial he developed a wide range of work from metal creep to metal fracture and corrosion; and developed work also in semi-conductors.

184. Peter Lynn Pratt (1927–95) took a degree in metallurgy at the University of Birmingham, moving to the Cavendish for his PhD. After a research fellowship at Harwell he returned to Birmingham as a lecturer in 1953. At Imperial he set up research into the properties of ionic and ceramic crystals at high temperatures. He was given a chair in crystal physics in 1963 and was an important figure in the development of materials science at the college. He later became Director of the University of London Centre for Marine Technology.

185. Anthony Vernon Bradshaw (1923–) studied at the RSM and after obtaining his degree in 1944 worked at mines and smelters around the world. He returned to a readership at the college in 1958 where, with help from the Nuffield Foundation, Harry Oppenheimer, and the British Steel Corporation, he formed the John Percy Research Group in Process Metallurgy in 1965. Many overseas students, including several from Russia, came to work in his research group. He is said to have smoked about eight cigars per day. He left Imperial in 1975 to become Chief of the Division of Process Metallurgy in the Commonwealth Scientific and Industrial Research Organization of Australia.

Charles Benjamin Alcock (1923–2001) was educated at Imperial in chemistry and chemical engineering. After work with the British Iron and Steel Research Association he returned to Imperial as a lecturer in the Nuffield Research Group in Extraction Metallurgy. He rose through the ranks becoming professor of metallurgical chemistry in 1964. He left Imperial in 1969 and joined the University of Toronto as chairman of the department of metallurgy and materials science.

chapter eleven

Corporate and Social Life

Introduction

The Governing Body instructed the early Rectors to integrate the three older colleges into the Imperial College, and to take steps that would hasten the formation of a new corporate identity. In chapter seven we have seen how the college was promoted as meeting the needs of Empire, something members of all three colleges could agree on. But institutional identities are formed only slowly, by the passing down of stories and behaviours, and not usually by direction from above. For example, it took close to twenty years before all the professors agreed to adopt letter paper with the new Imperial College letterhead, and over thirty years before the names of the buildings were changed. Only in 1939 did A. J. S. Pippard, Dean of the City and Guilds College (C&G), remove the old plaque on the front of the building with the name 'The City and Guilds College, formerly Central Technical College' and replace it with one inscribed 'Imperial College: The City and Guilds College'. Much later, in the early twenty-first century, some of the new medical school staff were unhappy about similar losses of identity but, by then, those working in the older constituent colleges were happy to identify with Imperial. They raised little objection when the Rector, Sir Richard Sykes, introduced a faculty system which ended the governance role of the old colleges. By 2000 the name Imperial stood for excellence in science and engineering (and by then also medicine), and the names of the older colleges were no longer well recognized beyond the college.[1]

Corporate life at the college is bound up in the behaviours of both staff and students and it is difficult to disentangle the two. Before the Second World War close to half the students at Imperial came from Greater London, yet the student body was less locally bound than those at the major civic universities. The schools from which students came suggest that the college was a site for the mixing of young people from different classes and different geographical regions. In the early period, the Royal School of Mines (RSM) attracted proportionally more young men from public schools than did the other two colleges, in part a reflection of parental, and more generally upper-class, involvement in the mining and metallurgical industries worldwide. The staff, too, were a mix of class and geographical origin. Some undergraduates, and roughly thirty percent of the postgraduates, came from overseas — as was the case for much of the century. Further evidence, mainly anecdotal, suggests that the mix of class, region and, to a lesser degree, gender, worked well.[2] Some students faced racism, though less so towards the end of the century. In this, Imperial will have been a microcosm of the larger society.

As in the nineteenth century, science and engineering were pathways of upward social mobility. Imperial opened its doors to gifted young people, many from homes with little or no experience of higher education, and the training they received allowed the majority to find interesting work, and allowed more than a few to join national and international elites in science, engineering, business, and even government. Reflecting the larger society, students at the beginning of the century were, on average, poorer than their later counterparts in real terms, but with one important caveat. Accommodation near the college was relatively more affordable. Indeed, until the mid-1960s, most students could afford to rent rooms within walking distance, or a short bus ride away. Some, however, travelled long distances from homes in distant suburbs. By the end of the century about one-third of students lived in college accommodation but for the others, and for many of the staff, commuting in what had become much more crowded conditions was both tiring and time-consuming. Inevitably, patterns of conviviality changed.

While there was some early hostel accommodation, the great push to provide student residences came in the early 1960s when both the number of students, and the cost of living in central London, especially in the Kensington and Chelsea areas, increased dramatically. College authorities believed that without the provision of affordable accommodation Imperial would be unable to compete with other institutions for good students. Whether this assumption was correct remains open to question. Arguments for a more residential college, which were made increasingly after the Second World War, appear to have been based not solely on competition for students. They were heavily loaded with collegiate ideals associated with the ancient universities. It is an interesting question why those in authority did not look to universities in other great metropolitan centres, such as Paris, for their model. They appear to have held the view that some kind of Oxbridge polish would be conferred on students while in residence, that the huge expenditure would pay off in cultural terms, that social life within the college would improve, clubs would become livelier, and that Imperial would lose what some still saw as its demeaning trades school image.[3] But not everyone agreed. A student writing in *The Phoenix* in 1946 stated that the residential idea was 'just one more ridiculous attempt to ape Oxford and Cambridge. … My own impressions of the hostel are of bored inmates widely separated in the lounge studying *Punch* and the *Tatler* and of rowdy midnight coffees, and of occasional arguments on religion'. He stated that 'digs' were a far better preparation for life.[4] The question of what universities are for appears not to have been answered solely in terms of the lecture room, the field, the workshop, or the laboratory. Something more was needed to produce the ideal graduate, but what exactly this was remained a topic for debate.

There were those who pointed out that the location of Imperial College meant that a collegiate life made less sense than elsewhere, and that London provided a cultural backdrop unmatched in the provinces. Indeed, the history of twentieth-century student life at the college can be seen as one in which young people were pulled in different directions. Individuals had to negotiate their way among the demands of their studies, family and private life, social life within departments, local college activities, and the larger social and cultural life of London. In the more immediate college neighbourhood were diversions such as pubs, museums, and activities at the colleges of art and music. There were many individual routes through this maze and it is difficult to make generalizations. The array of possibilities became even more complex after the Second World War. National and international movements such as the Peace Movement of the 1950s and 60s and the student movements of the 1960s and 70s attracted many students and younger members of staff.[5]

Further, as in society more generally, by the 1950s students had become wealthier in real terms. Some could afford to replace bicycles with motor cycles, or motor cycles with cars. Almost all began to acquire new kinds of consumer goods such as radios, hi fi equipment and records, even television sets.[6] By the end of the century such acquisition had mushroomed to include CD players, personal computers, mobile phones, and a range of other electronic devices. Students became connected to a wider world with new choices and possibilities. Social life became attuned to a national and international popular culture, and the pull of the local diminished. Further, young people increasingly set the pace in the larger society, something that has only escalated since the 1960s.[7] Despite these changes some traditional behaviours, mainly associated with the college bars, local pubs, sporting events and rag days, persisted, albeit in new forms.

For members of staff similar choices existed and the record shows a range of commitment to college life. It would appear that there was a greater interest among the staff in clubs and social life at the college before the Second World War than later. Commitment to Imperial College was no less after the war, but it took different forms. The growth of the institution, changes in the demands made on the staff, and increased time spent on commuting to the college, changed the ways in which people interacted. This chapter will use historical illustration to make more explicit some of the above mentioned themes. Staff leave few, students even fewer, traces of their time at the college; but college publications, club minutes, and a few memoirs throw some light on a century of life at Imperial. What follows is a collection of small stories gleaned from these sources, and formed into a narrative designed to show something of how people identified with the college over a century. Change was inevitable but one thing remained constant. Students and staff were aware that they were part of a major institution known for excellence in science and technology, something that has always been a source of pride. This central aspect of Imperial's identity is the subject of other chapters. But those associated with the college also helped create a local culture beyond the lecture room and laboratory. It is this local culture that is discussed below.

Ceremonial and Image

One way that Imperial College promoted itself was through ceremonial. But this was not invented overnight. Annual graduation ceremonies, held in the University Great Hall, were low-key affairs until after the Second World War.[8] Before then the college promoted itself by other means, principally by the skilled use of a long-favoured social technology, the formal dinner. The three old colleges had developed their own traditional dinner culture for the celebration of anniversaries, and for the binding together of staff, current and former students. The Imperial College Annual Dinner was intended to do the same. It was a major event during the 1920s, but it declined in importance during the 1930s and did not survive the Second World War.[9] At first the dinners were held in large London restaurants, before moving to the smaller Goldsmiths Hall in 1930.[10] Much care was taken, especially in drawing up the guest list. People were chosen who were viewed as important to the promotion of the college, and members of the Governing Body and professors were asked personally to invite someone from the list as their guest. Unlike the constituent college dinners which were close to being all-male affairs,[11] these dinners included wives and a few honoured women guests. The dinners were intended to be sober and decorous occasions at which the college could display its serious side. Those making speeches noted the major accomplishments of the year and the achievements of old students. At one dinner, in 1924, the guests included government officials, representatives of the Empire, senior industrialists and distinguished academics. One of the after-dinner toasts to 'The Empire and Education' proposed by Sir James Allen, the High Commissioner for New Zealand, reinforced the identity of the college with empire. In the same year science students from across the country were invited to visit both the British Empire Exhibition and Imperial College where they were well entertained.

While the annual dinner was not a long-term success, the formal dinner continued as the favoured mechanism for binding people to the college, and as the ideal promotional form. For example, major dinners were given to honour retiring rectors and chairmen of the Governing Body and, in 1932, when it was decided to institute honorary fellowships, annual dinners celebrated the new fellows. The fellowships were to be awarded to graduates who had made names for themselves in the larger world, and dignitaries were invited to witness these college success stories. In the 1930s the fellowship dinners, attended by about 100–150 people, were held in the College Union which was heavily decorated with flowers for the occasion.[12] The large-scale formal dinner declined in importance after the Second World War, in part because the habit was lost during the long period of food shortages. By the late 1950s this Victorian invention had lost much of its appeal and large-scale dinners slowly gave way to more exclusive dinner parties.

After the Second World War the college decided to make its degree ceremony a major event and fellows were honoured at the same time as the graduating students, and at small dinner parties before or after the ceremony. With the events of 1945 still fresh in people's minds, it was decided to commemorate the Royal visit of that year by holding Imperial's convocation ceremony as close as possible to the 25th of October, the date of that visit. The first Commemoration Day ceremony was held in 1948 in the old Great Hall of the University. It began with a traditional academic procession, and after the degree ceremony there was a conversazione and ball. The Special Visitor that year was Deputy Prime Minister, Herbert Morrison. A few years later the 1953 Commemoration Day events were very festive, in part because it was the year of the coronation of Queen Elizabeth II, and in part because the government had announced a major expansion of the college. The Commemoration Day ceremony became the major social event of the year. It was the occasion for the staff orator to give witty, yet respectful, encomia for new honorary fellows, and for the Rector to give a well crafted speech outlining the year's major achievements. The speeches were issued as press releases. By the 1950s the timely press release was not limited to Commemoration Day speeches, but had become a routine publicity mechanism.

The college celebrated its Charter Day each July. This was a local event, often centred on an outdoor tea party, to which all members of the college, and many of its neighbours, were invited. By the 1950s the parties were held in Prince's Gardens and strawberries and cream were served. The opening of new buildings or programmes were sometimes tied to Charter Day. Such occasions went beyond the local, and eminent guests were invited. In 1957, in its silver jubilee year, the college celebrated the opening of the expanded Union, a newly planted garden in the Beit Quadrangle, and the opening of the Roderic Hill building by Queen Elizabeth the Queen Mother.[13] On this occasion the college hosted a traditional evening banquet at Mansion House, attended by the Minister of Education, Lord Hailsham. He gave a major speech in which he acknowledged engineering and technology as belonging within the university system, belated recognition of what, at Imperial, had long been a fact. But outside the college many people had been slow to accept that disciplines which once belonged in the area of trades schools and apprenticeships had become also major research fields. The practice of inviting members of the Royal Family to lay foundation stones and open new buildings has been a tradition since the early days of the three constituent colleges. An earlier grand affair was the 1926 opening of the Goldsmiths' extension to the C&G building. The Duke of York arrived in formal procession with the Prime Warden of the Goldsmiths Company and other City dignitaries, and a concert was given in the structures laboratory by a band of the Scots Guards.

Corporate spirit has also been reinforced by the commemoration of eminent historical figures. T. H. Huxley, a college icon, was the focus of more than one celebratory event in the twentieth century. In 1925 the centenary of his birth was marked with an exhibition of memorabilia, a memorial lecture given by Professor E. B. Poulton, a dinner presided over by H. G. Wells, and an open house which included a demonstration of search lights operated by students preparing to defend London 'from the air-raids of the next war'.[14] In 1995 the centenary of Huxley's death was marked by a major international conference on his life and legacy. A dinner was held at the college to which many of his descendants, including the conference guest of honour, Sir Andrew Huxley, were invited. The holding of commemorative conferences is more typical of the late century than earlier, and in part reflects the need of academic staff to connect with their peers worldwide, and to place their own institution in the spotlight.[15] Sir Andrew, himself a distinguished scientist, has been the honoured guest at a number of college occasions, including the 1981 centenary celebration of the Royal College of Science (RCS).[16]

By and large the celebratory efforts of the constituent colleges, such as the one in 1981, were more successful than those of Imperial College itself. This despite the fact that, by the middle of the century, people had begun to identify with Imperial. What appears to have happened is that the anniversaries of the constituent colleges became also anniversaries of Imperial. For example the 1951 celebration of the founding of the RSM was a college-wide affair. Two days of special events were organized to coincide with Commemoration Day. One attraction was an 'Aladdin's Cave', constructed in the RSM to display fluorescent minerals. The *Felix* reporter noted that visitor's false teeth, and old laundry marks on their clothes were, inadvertently, also on display.

Speeches at the 25 October Commemoration Day ceremony, then still held in the University Great Hall, also focussed on the anniversary. The special visitor was Sir Andrew McCance FRS, a former RSM student and past president of the Iron and Steel Institute, and the day ended with a banquet at Draper's Hall. The C&G College marked its first hundred years with similar celebrations and a more solid memorial, namely a book on its history.[17] The RSM 2001 anniversary proceedings were launched by the Princess Royal, Chancellor of the University of London. Once again there were displays and lectures. A gala dinner and ball were held in the Dinosaur Hall of the Natural History Museum, decorated for the occasion in the black, yellow, white and gold colours of the RSM. The Rector, Sir Richard Sykes, made some characteristically forward-looking remarks and noted that 'London is now the mining finance centre of the world, with three of the largest mining companies based here, and all the major petroleum companies represented'. In making these associations, Sykes was boosting the importance of Imperial College. But historian James Secord gave a speech which told of how historical identity is formed and embodied.

> How do we remember an institution like the Royal School of Mines, whose first century was so deeply rooted in heavy industry and high empire? ... We celebrate through stories passed down in common rooms, through memorabilia and portraits posted on walls, and at anniversaries like this one. Real history is not brief; it is not so much written down but embodied in what we do every day. To work in a laboratory, to go on a field class, to advise a company on a new product, to sit in a lecture hall; these are historically embodied actions, and to engage in them is to celebrate a particular vision of science that has come into being in the past two centuries.[18]

Ceremony and formal commemoration are aspects of this complex historical process of identity formation, but the day-to-day activities that Secord mentions are of even greater importance. Much of the day-to-day work is discussed in other chapters but what follows has to do with the cultural life of the college beyond ceremonial and beyond the strictly working environment.

Clubs and Collegiality among Students and Staff

The Early Years, 1907–1920

In the first two decades of the twentieth century poorer students were given a maintenance grant of about thirty shillings per week by the state.[19] This was enough to live on since landladies in Kensington, Chelsea, Fulham or Battersea would charge about fifteen shillings per week for a room with breakfast and evening meal. Typically students who lived in lodgings, and many living at home, walked or cycled to the college. One student who lived just south of Putney Bridge, and who cycled along the Fulham Road in the 1910s, recalled that traffic was so light that he was able to read a book while keeping just one hand on the handle bars. One of his peers had a motorcycle which was then still cause for wonderment.[20] It was rare for students to rent accommodation with kitchen facilities and prepare their own meals. Indeed, students were seen as young gentlemen who needed to be served meals and have their domestic needs cared for. Landladies were viewed as second best to servants in a residential college, and class assumptions about student needs were behind state maintenance grants in this period. Lunch at the college cost about one shilling, but a cheaper lunch could be had at the ABC restaurant near South Kensington station. A 'hearty college tea' cost five pence. For one penny one could travel a long distance on a bus, for two pence one could buy a pint of good beer, and for fourpence one could buy an ounce of tobacco. Cheap theatre tickets cost about one shilling. It would appear that many students of this period smoked, and many enjoyed the theatre, often attending in groups.[21] Some of the wealthier students would pay for coaching before examinations, a way for research students to earn extra money. Some students took jobs at the neighbouring museums. For example, H. H. Grainger remembered cleaning and tabulating fossils at the Natural History Museum.[22] Research students typically had larger grants or scholarships than undergraduates but less parental assistance. If they held demonstratorships they earned about £100 per year. They also appear to have had slightly different living arrangements from the undergraduates; occasionally sharing a flat, but more typically paying for bed and breakfast and joining their peers for evening meals at cheap restaurants.[23]

Both men and women students called each other by their surnames, a practice that did not fully fade until after the Second World War. Some exceptions were made to avoid confusion. For example, the Conrady sisters were called by their first names when more than one of them was present.[24] Most students wore ties or scarves displaying the colours of their college which, in the early period, meant those of the old constituent colleges, not Imperial.[25] Students also carried cushions with college colours, supposedly for sitting on in lecture theatres, but not infrequently used as missiles when lectures turned rowdy. The college authorities soon discouraged their use. Each morning when students arrived they had to sign their names in a book kept by one of the college porters, a practice that continued for some years ending first for postgraduates and, by 1920, also for undergraduates.[26] Before the founding of Imperial College, surprisingly little had been done for the corporate life of South Kensington students. Indeed, one of the ways of encouraging students to identify with the new institution was the creation of communal facilities within a student union. An Imperial College Union common room opened in the RCS building in 1909 and was decorated with a collection of pennants from American universities, a gift from the Rector, Henry Bovey. Women were not allowed in this room and were given a separate common room in the same building. Their room had been sectioned off from a large lavatory, had tiled walls, a dirty ceiling, and was reportedly an unpleasant place in which to spend time.[27] Common room space in the three old colleges was very limited, and what did exist was often reserved for the use of postgraduates and for third and fourth year students. Before the union building was constructed each college had a small lunch room and these kept going for a number of years after federation. The RCS lunchroom was in a large shed on the west side of Exhibition Road; it opened to all Imperial students after 1907. A description from 1911 is interesting in what it says about expected standards: 'our only accommodation for lunch and tea is an ill-ventilated and none too clean restaurant, where only half the tables possess table cloths, and a building which — we say with pain — is not much more than an old wooden shed.'[28] A neighbouring shed had a reading room which held a few newspapers and journals, and space for games such as chess and bridge.

A student union building opened in 1911. It was much smaller than the one in existence today; new floors and other extensions were added through the century. The idea behind the union was to encourage social intercourse between staff and students, past and present, from all three colleges. It was hoped that a new Imperial identity would result. But that took many years and, before the First World War, student and staff socializing was still very much bound to the unions and clubs in the old colleges. Some social areas and new catering facilities were included in the union building; but dining space was restricted to 150 at a time when there were 800 students enrolled and few cheap eating places nearby. Women students had their own lunch and tea room in the building. By 1919 two army huts were being used for additional dining space, but not until the mid-1930s when Henry Tizard separated the refectory services from the union did meal services come close to meeting the demand.[29] While it was agreed that the union by-laws would be written by the first elected officers, the governors decided that the union president was to be a current student. They also decided that no gambling would be allowed on the union premises and discussed whether card playing would be permitted — it was, provided that no money was exchanged. Another point of contention was whether the union should have a bar, something many professors opposed. But the majority view was that it was better that students drink on, rather than off, the college premises.[30] A bar was added after the war. At first a barmaid in traditional black silk dress presided, but after the first two barmaids married students it was decided to have male bar tenders. Ted Smith was hired and stayed for thirty-five years. Even before the bar came into existence complaints were made by neighbours. Some came from Queen Alexandra House, a hostel for women attending the art and music colleges in the area. In 1912 the Director wrote of 'the intolerable nuisance we have been suffering the last few weeks from the students in the new place opposite. I have called on the Secretary, and also written, but he does not seem able to do anything.' She complained of parties once or twice a week 'with a huge din and shouting and singing on the road' and that it was impossible to sleep.[31] An ongoing problem was who should control the union building and impart discipline. At first students were responsible for the building, and for organizing the catering, but these arrangements did not work well and the Union was soon in debt. Student power diminished during the 1920s and 30s as the college took over maintenance of the building, the running of the refectory, and overall discipline. New professional staff were hired and members of the academic staff were appointed honorary treasurers of catering and other union activities.

Sports were perhaps the main form of extra-curricular activity in this period, as was the case for much of the century. In the RSM there had been a tradition of sports on Wednesday afternoons and no classes had been scheduled at that time. In the other two colleges the idea of time off for afternoon sports was rejected as an Oxbridge practice, and as not part of the nose-to-the-grindstone ideal then held up as more noble than what was seen as a leisured student life in the ancient universities.[32] The RCS was often referred to as a 'cram shop' by students complaining of the work load. At first the Rector decided that there would be no period during weekday hours specifically designated for sports but, in 1910, the RSM students staged a revolt demanding their Wednesday afternoons back. After much discussion the new Rector, Sir Alfred Keogh, agreed. The Governing Body declared that no classes would be held anywhere in the college on Wednesday afternoons. Reluctant professors were forced to reschedule their lectures or laboratory sessions to meet the new routine. That it was principally RSM students who made this demand is due not only to a tradition lost, but to the higher proportion of public school boys in that college. They were then the main force behind both rugby and rowing, the most popular sporting activities at that time. But the college had no sports grounds of its own until after the First World War. Field games were played at other club sites, or informally in Hyde Park and Battersea Park. An annual sports day was held first at the Duke of York's regimental headquarters on the King's Road or at the Chelsea Football Club grounds at Stamford Bridge, before moving to the University of London grounds at Motspur Park in the 1930s. The tugs-of-war between the three colleges were the most anticipated events of the day, at least until the Second World War.

Those who lived at home, and had long commuting distances, tended to have a rather meagre social life at the college. Typically it centred on the lunch room, often with a student playing the piano, on sporting activities, and on the occasional visit to the theatre with friends. The weekly routine would sometimes be enlivened by an informal dance organized often in cooperation with students at the Royal College of Art, Queen Alexandra House, or with nursing students living in nearby hostels. Once a year a formal dance would be held at a venue such as the Empire Rooms. Tickets cost about five shillings and included a buffet.

At these events wives of staff and mothers of students chaperoned the young women. Finding congenial female company was a problem for students in an overwhelmingly male college and, for many, the informal dance was their main opportunity to do so. Within the separate colleges there were annual student smoking concerts where those with musical or comedic talent would entertain the others.[33] Staff members would join in and were expected to tolerate the occasional stinging joke or skit at their expense. Some professors were remembered years later for the wit, humour, or musical talent they displayed on these occasions. For example, Professor Witchell of the mechanical engineering department was a great favourite with students. He wrote his own songs and performed them well.

> In vectors and rotors, hydraulics and motors
> I'll tell you we're jolly hot stuff...[34]

For those who lived close to the college student life was richer. They could take part in the above activities and more. Evenings could be spent in local pubs talking with fellow students, and often with some of their teachers. They could attend evening events at the Royal College of Music (RCM) or Royal College of Art (RCA) and have a chance of meeting young women students.[35] They had good access to many of London's cultural activities and could more easily attend events at learned and professional societies. One student never forgot the excitement of going to a meeting at the Institution of Electrical Engineers at which Kelvin, J. J. Thomson, Evershed, Parsons, Marconi and Fleming were all present.[36] Percy Boswell enjoyed his life as a student, especially his postgraduate years, 1913–15. He lived fairly close to the college, in a bedsitter in Comeragh Road, and his friends included two other future geology professors, Vincent Illing and Harold Read. Boswell recorded a 1914 trip around Kent that he made with Read who had grown up in the county and knew it well. They and some other young men 'developed ... a kind of fraternity' which they called, and spelled, π or 'pie', not 'pi'. They met regularly for tea and spent time together in the evenings.[37] The more formal De la Beche Club was founded by this group in 1914. This was a student club, though staff were encouraged to join. Field trips and excursions were organized, as were sessions at which students read papers, before moving to a local pub for further discussion. The club went into abeyance during the war but was reinstituted by a later group of students in 1928. Later, in the 1950s, future professor Janet Watson was a member but was asked not to attend the annual dinners, even on the occasion when she had just delivered the anniversary lecture. Watson was no outspoken feminist and later described such rebuffs as mere 'pinpricks'.[38] The first honorary president of the club was Professor Watts who appears to have been a father figure much revered by the students. Watts was an early devotee of the motor car and students took much coveted turns driving with him on field study trips. He personally paid the expenses for students who otherwise could not have afforded to join others in the field. Watts's behaviour was not atypical of professors and other academic staff in this period. While the old colleges remained small, a family spirit existed within them and within departments.

Most departments had associations of the De la Beche type where students could exchange ideas, give papers, and even publish them in college journals. For example, the student Chemical Society, founded in 1896 by A. W. Tilden, was very lively and had its own journal until the 1920s. The journal was then merged with the *Scientific Journal of the Royal College of Science Association*. Students in the Chemical Society were mentored by the staff, notably Professors Baker and Philip. In 1913–14 nine papers were read at the society and five of them were published. One 'on the fixation of nitrogen' was by a future member of the chemistry staff, A. G. Pollard, and anticipated his later career as an agricultural chemist. Another was by C. H. Ingold. Something of a prodigy, Ingold published sophisticated papers some of which included new quantum theoretical ideas. The disciplinary clubs also organized outings; for example, chemists visited breweries, tar distilleries, pharmaceutical works etc., and the miners had an annual trip to the Kent coalfields. In 1913 the Chemical Society visited a factory where margarine was being manufactured using a new technology. The manager asked that 'now we had tasted Margarine he hoped we would remove the absurd prejudice which existed against it'.[39] This type of club activity helped bind students both to their future profession and to the college.

Students paid an annual union fee of two pounds, one of which went to the Imperial College Union and the other to the constituent college union. The old college unions welcomed incoming students with freshers dinners and supported a number of social clubs in addition to the sports clubs. For example there were chess clubs, literary and debating clubs, and, at the C&G, a radio club. Interestingly, the first clubs, other than for sports, to be founded in the Imperial College Union were political (not yet party political) and religious, closely followed by an international club, a society for music and drama, and a photographic society. The Imperial College Fabian Society appeared already in 1907 and George Bernard Shaw was one of the first to be invited to give a talk. A Rationalist Society was founded in 1909 and lively meetings were held on the then pressing topic of science and religion. A student group named the Wiffin Woffins was called upon to perform popular vocal music, comic turns, and magic tricks at college events. E. F. Relf organized a number of concerts together with students from the RCM.[40] The Imperial College rifle and revolver club was popular, and the college had a shooting range in one of its underground tunnels.[41] After the war, with the acquisition of a sports ground at Wembley, college-wide sports clubs came into their own and, while much sporting activity was still focussed on competition between the three old colleges, Imperial College teams began to compete with those from other London colleges and medical schools.

College journalism was lively in this period. In addition to journals devoted to publishing scientific and technical articles by students, there were a number of mostly short-lived literary and news journals. The most ambitious, and longest lasting, was *The Phoenix*, a descendent of the *Science Schools Journal* founded by H. G. Wells in 1886.[42] *The Phoenix*, at first the organ of students at the RSM and RCS, encompassed all three colleges from 1915. It was a lively journal before the First World War, with six to eight issues published annually. After the war the number of issues per year dropped to about four, and there were some lapses. The journal allowed students to write on a wide range of topics including college affairs, national and international politics, and the arts. Short fictional pieces and poetry were also published. Biographical items on the professors, student leaders, and senior administrators were a regular feature, as were book reviews. And each department had its correspondent expected to write on its affairs. H. V. A. Briscoe, a future professor of chemistry, was an active correspondent, and also the advertising secretary. Clearly there was an appreciation of humour, and interested students must have spent some time composing witty and humourous pieces about college life for publication. Some of this work shows a passing familiarity with serious literature:

For the Examiner

Oh Thou, who didst with Pitfall and with Gin
Best the road I was to wonder in...[43]

Examinations were a perennial topic. In a long article against over-examination one student wrote 'the future historian ... may not improbably refer to [the present age] as the Examination Age. The cult of the examination has risen to a height of popularity equalled only in China'.[44] Other perennial features include jokes about the nannies in Kensington Gardens, and tales of student pranks in the parks. By telling stories about each other, about their teachers, and about the college neighbourhood, the student journalists helped build a corporate spirit.

While the C&G College already had an old students association, the Old Centralians, the other two colleges associations appeared a few years after the founding of Imperial College.[45] One reason for this relatively late date is that the associations were envisaged as a means to ensure the continuity of the old colleges in light of the recent federation. For example, the Royal School of Mines Old Students Association was formed in 1913 with Professor Gowland as its first president. Its purpose was to 'foster the comradeship ... to advance the interests, and to express the opinion of old students and teachers in the RSM'.[46] And, as Professor Frecheville put it, 'to maintain the identity of the School'.[47] These associations, too, had their journals where old students could read about happenings in their colleges and could be roused into action when needed. For example, the RSM association journal was used to persuade old students, many of whom held senior positions in the metallurgical and mining industries, to help those who had fought in the war and were having difficulty finding work on returning home. The associations also held annual dinners to which student leaders and staff were invited. At the dinners old stories would be retold so as to keep alive the spirit of the colleges and, over the years, new lore was added. The first annual dinner of the RCS Association was held in 1908 at the Criterion Restaurant with H. G. Wells in the chair. Three years later the association dinner honoured T. W. Edgeworth David, an old student and member of Shackleton's expedition to the Antarctic.[48] *The Phoenix* reporter wrote, 'we could see his manliness in his face and hear it in his deep ringing voice', exemplifying a type of bonding hero worship still common before the 1914–18 war.[49] The Old Centralians had the advantage of being able to hold dinners in fine livery halls, an inducement for graduating students to join. To counter the centrifugal force of annual dinners of this kind, a favoured form of conviviality in the early period, Imperial organized an annual conversazione which, in 1909, was attended by over 700 people. Departments displayed their work, refreshments were served, and in the evening there was a concert and ball.

Then, as now, students held rag days to raise money for their favourite charities, and sometimes simply ran riot for no good reason. They also held the occasional political demonstration, and attended large ones with other students in Trafalgar Square. For example, women's suffrage drew students to rallies, both for and against the cause.[50] An annual excuse for bad behaviour was the Earls Court Exhibition. On closing night hordes of students from all three colleges would invade the premises and compete in trying to wreck the exhibits. Large numbers of policemen awaited them. Usually, but not always, the police kept the upper hand. On one occasion students placed a large number of balloons in the path of a royal carriage which was carrying Edward VII and the German Kaiser through Hyde Park. The sight of policemen chasing after the balloons was very amusing for the spectators. Rags in this period, while not as inventive as they came to be in the interwar period, helped unite the student body.

Before the war some of the RSM students belonged to a Royal Engineers OTC unit, led by mining professor William Merrett.[51] Merrett trained the unit for specialist tunnelling work but when war came there was much disappointment that the training was not used. Most of the young men dispersed and joined local infantry units. War brought an end to many of the old routines. *The Phoenix* closed in 1914 but was started up again in March 1915. One reason given for its reappearance was for it to be a 'forge between the College and the members of the College who have exchanged the pen for the sword'.[52] Wartime events were discussed in the journal, Belgian refugees were welcomed to the college; lists of those serving, of those injured, and of those who perished, were published. War related work of the professors was reported, and Keogh's wartime exploits were followed. But overall, the journalism was far less war-focussed than one might imagine, implying that much student life continued as before, though with a much reduced student body.

From the 1920s to the mid-1950s

After the war students returned to the college in large numbers, reaching a record high of 1305 in 1920–21. This was in part due to the many students returning from the war, but numbers grew more generally in the interwar period. In 1920 about 60% of the undergraduates registered also for external University of London degrees, a percentage that was to climb to about 75% by the end of the decade when Imperial formally joined the university. Ex-servicemen received grants from the Board of Education but, in general, grants were less generous than they had been earlier. The college instituted a loan fund for poorer students. Board and lodging climbed to about two pounds a week by 1930, and to about two pounds and ten shillings by the Second World War. Geoffrey Wilkinson recalled paying that amount for room and board at 4 Elvaston Place in the late 1930s, and that he spent a further shilling on chocolate bars and Saturday evening films at the Hammersmith Gaumont.[53]

Students returning from military service were given special attention. H. J. T. Ellingham, who joined the chemistry staff in 1919 after having fought in the war, was appointed liaison officer between students and staff and was very sympathetic to other ex-servicemen.[54] The war brought the three colleges closer together, and the sense of belonging to Imperial College became stronger. Money collected for a war memorial was used to purchase 22.5 acres of land in Wembley for a sports ground.[55] Tennis courts were built next to the Union, and the Tennis Club had about one hundred members by 1920. Other college-wide clubs began to flourish but finding space was difficult. The Badminton Club played in the Bessemer Laboratory. The Fencing Club was allowed to use the RSM examination hall but appears to have been frequently admonished for rearranging the furniture. During the Second World War this club was the only one of its kind still functioning in London. It opened its doors to other fencers, and was popular with allied services, such as the Czechoslovakian Army, stationed in the capital. Aside from mixed doubles in tennis, the only sport in which men and women played both together and against each other was ping pong.

Perhaps most successful in the 1920s and 30s was the Boat Club which, until 1938, used the Putney facilities of the Thames Rowing Club. D. W. Morphy, when honorary secretary, established a unified Imperial College team which he took to Henley for the first time in

1921. For much of the twentieth century Imperial College rowing teams were dominant within the University and did well at Henley. And, over the years, individual rowers have made a strong contribution to Britain's Olympic teams.[56] In 1920 Arthur Morphy presented a cup for an eights race between the three old colleges, and competition for the Morphy Cup became an important annual event in college life.[57] Competition for a place in the boats was so intense that, in 1945, the College Secretary, G. C. Lowry, presented a cup for a second eights race. On race day the towpath was the scene of further battles between team supporters. For many years the associated tug-of-war was viewed as being almost as important a test of old-college vigour as were the boat races. Old-college rivalry also took less organized form. Flour, soot, and rotten fruits and vegetables, collected in crates from Covent Garden market earlier in the day, were hurled at the enemy. Local residents looked forward to the annual Morphy Day and joined those from the college in watching events on both water and land. In 1938 Imperial College built its own boathouse next door to the Thames Rowing Club.[58] Included was an internal passage to the older boathouse bar. Tragically almost all who rowed in Imperial's first eight that year were killed during the war. New boats were named for them into the 1970s. The waterman hired by the college in 1939 was C. H. (Charlie) Newens. He helped coach both the men's and women's teams, became a fixture in the life of the club, and was a famous character on the Thames. Newens contributed to some major team successes already in the immediate postwar period, with the first eight winning the Danesfield Cup at Henley in the first peacetime regatta of 1945, and the Thames Cup in the following year.[59] He and his wife kept the boathouse and its equipment in good order until his retirement in 1980. Newens' successor as college waterman and boathouse manager was Bill Mason. Mason soon took on coaching duties and was made Director of Rowing. He has coached many successful teams and was chosen to be the head coach of Britain's women's Olympic rowing team at the 1996 games in Atlanta.[60]

Rugby was especially popular among RSM students and, for many of them, the major sporting event of the year was the match against the Camborne School of Mines. In alternate years it was played in Cornwall so that part of the fun was in the journey, with supporters in tow. The event was called the Bottle Match since the teams competed for a two-foot-high tin bottle.[61] The RSM rugby team also played in the Cartel des Mines against teams from Europe's top mining schools. Rugby teams began to compete for Imperial after the First World War but, by then, rugby practice was no longer permitted in Hyde Park since the park authorities deemed it dangerous to passers by. However, three pitches were still available for football practice. The Athletics Club was often successful in its competitive ventures and won the University of London's Rosebery Cup many times. Women's athletics teams performed remarkably well given the small number of women students at the college, and were occasional winners of the University Challenge Cup. Track and field and rowing, it would appear, were more welcoming to women in the early period than was swimming. The swimming club refused membership to women until forced to do so by the Rector after the Second World War. In the interwar period cross-country running became popular and students entered weekend competitions, often held at Wimbledon Common. Rambling and Youth Hostelling were also popular activities. Professor Finch was a great supporter of the Mountaineering and Sailing Clubs. Both before and after the Second World War he organized annual climbing trips to the Swiss Alps, and coached college crews sailing in the English Channel.[62] Until the late 1950s the various sporting events drew much support from college students and staff, many of whom routinely attended matches and boat races. After the First World War, while there was still much competition between the three old colleges, increasing attention was paid to external competition. This collective activity of watching and taking part in matches was a further important factor in the forging of an Imperial identity. People took pride in the increasing success of Imperial College teams.

An Imperial College hostel was built adjacent to the Union and opened for fifty students in 1926. There was an immediate waiting list and, a few years later, the hostel was enlarged. But still only few could be accommodated and most students continued to live at home or in digs. During the 1920s many students acquired motorcycles and rode them to college. For some, such as Sumantrao Moolgaokar who came to the college in 1926, they were a passion. He was remembered as a wild motorcyclist, fined five times for speeding in Exhibition Road. Later he rose through the ranks of Tata Industries to become managing director of its Telco Division.[63] The motorcycles soon became a problem; there was a small shed for parking motorcycles next to Holy Trinity Church, but this was soon filled to overflowing. Already by 1921 Professor Carpenter was complaining to the College Secretary that he found it impossible to enter the Bessemer Laboratory because of all the motorcycles parked in front of the door, most of them belonging not to metallurgy students but to engineers. Worse, some of the owners seemed to think the sump room a suitable place for carrying out repairs.[64] Other professors joined Carpenter in complaining about the motorcycles, especially about the noise. Indeed, the noise of motorcycles is something that many students of this period remembered years after leaving the college.

During the 1920s the Imperial College Union (ICU) began to come into its own. One on-going debate was whether to affiliate with the National Union of Students (NUS). The ICU did so first in 1933 but the affiliation was an on-off affair for many years. Most students were isolationist and objected to what they saw as the radical left-wing politics of the NUS. They wanted attention paid to more local issues. After the war two army huts were erected to provide extra dining space for the increased numbers. Food services improved dramatically after the 1925 appointment of Jimmy Peacock as Union Manager. Assisted by his wife, he took a special interest in the catering. Students appreciated their efforts and the Peacocks became great college favourites.[65] Their daughter, Queenie, helped also with the catering, continuing to work at the college after her parents' retirement. Even during the Second World War the Peacocks managed to produce decent meals. Standards declined after they retired in 1945, but recovered with the appointment of Victor Mooney as catering manager in 1953, a position he held until his retirement in 1985. Like his predecessor, he became a well-loved college figure. Already by the 1920s there was competition from new cheap restaurants in the area. One restaurant that features in a number of student memoirs of the interwar period is Kiski's Cupboard; it specialized in omelettes. The Good Humoured Ladies, on the King's Road, was a popular lunch place for botanists working at the Chelsea Physic Garden, though the lunches which cost about 1s 8d were not affordable on a regular basis by most students.[66] When the professors wanted to meet they would often do so over lunch at the restaurant in the Victoria and Albert Museum.[67] A favourite evening spot was the Blue Cockatoo Restaurant at the corner of Oakley Street and the Chelsea Embankment. According to Alec Skempton, in the late 1930s a good evening meal could be had there for two shillings and sixpence.[68] Some professors entertained students at dinner in their homes or hosted annual parties. For example, during the 1920s and 30s Professor and Mrs. Philip held 'good old Victorian' evenings with singing at their home in Bedford Park.[69]

Union and departmental clubs flourished. The RCS Mathematical and Physical Society appears to have had a very active programme through the 1920s. For example, in 1924, not only did students give papers, so did staff members A. N. Whitehead and Herbert Dingle, and visitors Oliver Lodge, Arthur Eddington and Richard Paget. And visits were arranged to the Royal Observatory, the National Physical Laboratory, the BBC, and the research laboratories of General Electric. Social clubs, too, resumed activities. The Musical and Dramatic Society put on plays at Christmas and Easter which always drew a crowd. In the 1920s the Rector and Mrs. Holland held dinner parties before opening nights and took their guests, mostly professors and their wives, to the performances. In 1923, Eric Ashby played the role of Alfred Doolittle in *Pygmalion*, and Patrick Linstead that of Henry Higgins. In 1924 Linstead played Napoleon in *Man of Destiny*.[70] Female roles were sometimes taken by Imperial students but, because there were so few women, also by students from the neighbouring colleges of art and music. After the final night's performance the Society typically held a midnight supper. In the 1940s and 50s plays by Terence Rattigan and J. B. Priestley were popular and, by the 1970s, plays by Tom Stoppard. In the 1920s Ashby was also active in the Literary and Debating Society which invited such luminaries as Edith Sitwell and Siegfried Sassoon to its meetings.

A student orchestra with about forty members was formed in the 1920s, conducted by Eric Speight. A full orchestral concert was given once a year, usually at Christmas; and, typically, there was an annual performance of a Gilbert and Sullivan opera.[71] In 1948 a college choir was founded by David Tombs, lecturer in electrical engineering, with much help from Bryan Thwaites, lecturer in aeronautics.[72] The choir was envisaged as one for both staff and students and, at first, had only male voices. Female voices were added later. In 1949 the choir performed a specially commissioned work at the Commemoration Day ceremonies: 'An Invocation to Science', conducted by the composer, Sir George Dyson, Director of the RCM. One early and enthusiastic choir member was Neville Blyth who enriched musical life at the college in a number of ways and, about three times a year, held what came to be known as 'Nevillaisian evenings'. He invited musically gifted students to dinner, and they would play chamber music and sing late into the night.[73] Lunch hour recitals began in 1950, the first given by Harold Allan and Eric Brown.[74] One positive consequence of bomb damage at the BBC was that the Promenade Concerts moved to the Royal Albert Hall during the war, and staff and students began to enjoy attending classical music concerts nearby. Folk music and jazz clubs were formed during the 1950s.[75] The strong musical roots put down from the 1920s to the 1950s have supported a lively musical culture within the college to this day. Musical activity has always been strong but other arts, too, have had a following. The Anonymous Club, founded in the 1920s, promoted art and sponsored an annual exhibition of work by Imperial College staff and students. And, in the early 1930s, 'talkies' were shown in the gymnasium, though the student projectionists appear to have had many difficulties. The 1930s also saw the opening of a circulation library with about 400 books that had accumulated in the hostel. Membership cost two shillings and sixpence a year, and this fee, together with an annual union grant of fifty pounds, allowed more books to be purchased. This was the start of what later became the Haldane Library. After the war it outgrew its original room in the Union and was moved to the Unwin Building and, by the mid 1950s, had about 4500 volumes.[76]

More exclusive clubs, associated with the old colleges, appeared in the 1920s. They were founded by young men who were fully engaged in college life, yet wanted something more. The 22 Club, founded in 1922 by T. B. Philip and W. Randerson, then president of the RCS union, was 'to provide a body consisting of those past and present students of the Royal College of Science, who have participated actively in the life of the College, and who can be relied upon to uphold the prestige and further the interests of the College'. Membership was by election and open only to male students. Staff and students from the other colleges were eligible for honorary membership. Twenty-two members were elected in the first year and membership was for life, provided fees were paid. By the 1970s the membership had stabilized at close to five hundred. The club, which still exists, held dinners, organized excursions, and invented a number of bonding rituals. For example, members had to wear club ties on Wednesdays, or be prepared to buy a drink for any other member noting the lapse. Like some other Imperial clubs, and the old college unions, this club had its own 'sacred' drinking tankard or 'pot' which was kept in the bar and which only members could use. At club dinners the pot became a communal drinking vessel, passed round along with a ceremonial snuff box. It was inscribed with the names of the membership and new pots were added as the membership grew. In 1951 an anthropologically minded student noted that the pots represented general camaraderie, 'tradition, high stools and revelry, bawdy songs and horse play'. He saw the pots as totems of the various college tribes.[77] Club dinners were usually rowdy affairs; bad enough that in some years the 22 club was banned from using the Union dining rooms. It marked its twenty-second anniversary in 1944 with a special dinner at the Union. H. J. T. Ellingham gave an amusing historical account of the club's activities while 'bombs were heard dropping outside'.[78] The club issued an annual booklet to inform its old student members of the main events of the year. It would appear that to be elected a member of the 22s it was necessary to be good at sports, convivial, a good raconteur, or a popular member of staff.[79]

Also founded in 1922 was the Chaps Club in the RSM. This club appears to have been well organized and its annual booklets were better printed than those of the 22 Club. They reported much news from abroad where many old RSM students found work. The Chaps organized an annual trip to the Derby, a tradition which lasted many years.

Derby Day 1930

…

The races over, back we started,
Pinching hats from sweating cyclists,
Signposts from along the roadside,
Singing, drinking, laughing, eating,
And amazed, the people round us,
Till we got back to the Union,
To the banquet and the bar.[80]

At a dinner in 1954, and with the expansion of Imperial about to get under way, Professor David Williams told a number of unprintable stories, and urged the membership to ensure that the RSM keep its individuality in the larger college.

The Links Club was founded in 1926 after some engineering students envied the success of the other two clubs. Its publications contain more inventive and humorous material than those of the other two clubs, but the humour is also cruder. One of the club's features was that new members were given nicknames during an induction ceremony at a club dinner. For example, one ICU president, Alan Kitchener, was named 'gnome'. Later honorary members, Peter Mee and Bernard Neal, were named Reggie and Ballboy respectively.[81] The clubs began taking annual photographs in the 1930s, and the Links members displayed themselves holding on to a linked chain, symbol of brotherhood and connectedness. The three clubs met for the occasional bit of friendly competition, such as raft races on the Serpentine and bowls matches.[82]

Some members of staff, often those who themselves had been students at the college, joined enthusiastically in student affairs, acted as honorary treasurers of clubs, became wardens in college residences, and joined in the activities of the old student associations. Tom Ellingham is a typical example. He clearly wished to retain something of the camaraderie he enjoyed as a student in his later life. He became something of a college historian, understood its constitution better than anyone at the time, and probably since, and was active in many clubs, including the 22 Club. He, and others like him, are emollients in the corporate life of the college. They keep things turning, organize events, encourage people to attend, pass on stories, and keep traditions alive.[83]

Ellingham was also active in the Freemasons. An Imperial College Lodge was founded in 1923 and its first Master was Sir Arthur Henry McMahon who had had a distinguished army career. McMahon was attached to the college only indirectly through his family, but he lived nearby and did much to promote Imperial's interests. Many well-known college figures were members of the Lodge but the membership, while exclusive, was representative of the entire spectrum of male participants in college life. Members included senior college administrators, professors, junior staff members, technical staff of all grades, some students and a few people living in the neighbourhood.[84] The Lodge was very active until the late 1960s and continues, somewhat diminished, to this day. At its peak it had a membership of about 150. While Masonry is seen as a secret brotherhood, in my view the main function of the Lodge at Imperial was to provide a space for meditation on life and morality for those who enjoyed doing so through ritual.[85] The Masonic technology of meditation, the 'ancient craft', will not be discussed here, but the Lodge was, and remains, a very different place of learning than other college

Illus. 47–49; top, King George VI and Queen Elizabeth, special guests at the 'centenary' celebrations at the Royal Albert Hall, October 1945; bottom left, Sir Roderic Hill at Silwood in 1952 and bottom right, Sir Patrick Linstead and Lady Falmouth at the opening of Falmouth Hall, Southside, in 1963.

Illus. 50–51; top, the Union Bar in 1954; note the rows of club and union pots. Bottom, the Union Lounge in 1958.

Illus. 62–65; top left, Helen Porter, Professor of Plant Physiology, the first woman to become a professor at the college; top right, Harold Read, Professor of Geology; bottom left, A. J. S. (Sutton) Pippard, Professor of Civil Engineering; bottom right, Mickey Davies, College Secretary.

Illus. 66–67. Field work: top, geologists D. Burgoyne, Janet Watson and John Sutton when research students; bottom, botanists at Dale Fort, Pembrokeshire, in 1949; front from left, John Levy, Mary Mayer, Clara Pratt, Jack Rutter (a future head of department) and Professor William Brown; at back, Fred Last.

Illus. 68–69; top, Imperial College Silver Jubilee, 1957: the Queen Mother in procession, leaving the Beit Quadrangle on her way to open the Roderic Hill Building; bottom, demolition of the C&G Waterhouse Building. Sir Owen Saunders, Dean of the C&G, opening a time casket found under a foundation pillar, and watched by members of the demolition crew, 1962.

Illus. 70–72. Three Nobel Laureates: top left, Dennis Gabor (awarded, 1971), top right, Sir Derek Barton (awarded, 1969) and bottom, Sir Geoffrey Wilkinson (awarded, 1973).

Illus. 73–74; top, Geology class under the supervision of Ted Hill, 1962; bottom, Wet-Grinding Mill in Mineral Dressing Laboratory, 1954.

Illus. 75–76; top, the civil engineering structures laboratory in the early 1960s with model of Verwoerd Dam (of the Orange River) at centre; bottom, Stanley Sparkes, Professor of Engineering Structures.

Illus. 77–79; three departmental heads of the 1960s: top left, Professor Lord Willis Jackson, electrical engineering; top right, Professor P. R (Robert) Owen, aeronautics; bottom, Professor Harry Jones, mathematics.

Illus. 80–83; top left, Professor Sir Hugh Ford, head of mechanical engineering; top right, Professor Roger Sargent, head of chemical engineering and chemical technology; bottom left, Professor Eric Laithwaite with his gyroscopes; bottom right, Professor Sir Richard Southwood, former head of zoology and applied entomology, planting a tree at Silwood to mark the opening of Southwood Hall in 1983.

Illus. 84–85; top, Manor House at Silwood with iconic bluebells. Bluebell ecology was studied in the early 1940s by G. E. Blackman and A. J. Rutter. Bottom, the spraying machinery workshop at Silwood in 1966.

Illus. 86–87; top, the official opening of the biochemistry building, 1965. From left, Professor Sir Ernst Chain, the Queen Mother, Lord Sherfield and Sir Isaac Wolfson. Bottom, the lower hall of the pilot plant in 1965, with a row of 300-litre fermenters on the right.

Illus. 88–89; top, the opening of the Computer Unit in 1964; seated, Lord Halsbury; standing from left, Professor John Westcott, Professor Stanley Gill, Sir Patrick Linstead (Rector), Mr. Hudson (IBM) and Lord Sherfield (Chairman of the Governing Body). Bottom, Professor Manny Lehman, Head of the Department of Computing (1979–83).

Illus. 90–91; computing, top, in the thermal laboratory (mechanical engineering) in the late 1960s and bottom, in the computing services centre in the late 1970s.

Illus. 92–93; top, three heads of civil engineering, from left, Professor Sir Alec Skempton, Professor Bernard Neal and Professor Ian Munro. Bottom, Colin Cherry, Henry Mark Pease Professor of Telecommunication.

Illus. 94–95; top, in 1964 promotional photographs were taken to attract more women students to the college. This one shows six women artificially posed around some equipment in the aeronautics department. (In 1964 the total number of undergraduate and postgraduate women students in aeronautics was seven.) Bottom, a model of a section of the Thames Barrier arrives for testing in the civil engineering department, 1973.

locations. It is interesting as a setting where more traditional hierarchies, religious affiliations, class, race and ethnicity did not, officially, count for anything. Several lodge members had Chinese and Indian names. Rector, professor, laboratory technician, and student were, in this particular venue, fraternal equals — which was, perhaps, part of its appeal. It was a site where friendships could be made across generational, class, ethnic, and religious boundaries, and where one could connect also to a much larger international brotherhood. The Masons are not a service organization, but they are charitable. The college lodge made regular donations not only to masonic charities, but supported also various college activities.

With the exclusive centrifugal activity of clubs such as the Links, Chaps and 22 Clubs, it was a good thing that the annual December conversazione, organized by Imperial, became a major, albeit very different, social event during the interwar period — one that brought people together. Many guests were invited and not only did departments exhibit their work, lectures of general interest were given, the Dramatic Society gave a performance of its Christmas play, buffets were set up in all the main buildings, and a bus moved continuously between the sites. The conversaziones were full-dress affairs and, despite this, drew many attendees. On one occasion some of the fine clothes were ruined when a lecturer, showing off the 'thermite reaction' in the old Huxley Building lecture theatre, caused an explosion that blew out the windows and filled the room with dense clouds of smoke.[86]

In the 1920s and 30s women students became more politically active than earlier and made demands for more equal treatment. They insisted on being called 'women' and not 'lady' students. In 1924 there were about fifty women students and the Rector decided, though not very imaginatively, to improve conditions for them. His first step was to engage the help of the honorary members of the Imperial College Women Students' Association (ICWSA) mainly the wives of professors, but also the redoubtable Martha Whiteley, President of ICWSA since its founding in 1912.[87] Behind her back the women students called her the 'Queen Bee'.[88] The older women helped in redecorating the dingy common room in the RCS building. As a finishing touch Lady Holland presented ICWSA with a fine silver tea service. Incongruously, water for the tea had to be boiled on a single gas ring, relic of an earlier laboratory, located in the women's lavatory. In 1930, after the construction of additional hostel and union space, a new common room was opened to both men and women.

ICWSA engaged in many of the same activities as the men. For example, rowing was popular and an eights team was soon organized.[89] During the Second World War, it was women students who did much to keep the social life of the college alive. While many British women took on traditional male roles during the war, the women students appear to have been pulled in two directions. On the one hand they were in the unusual position of being women studying science or technology at Imperial College; and most were also engaged in some kind of war-related work, at least during vacations. On the other, they became major social convenors. They organized a range of events, including a number of dances. For these they decorated halls with salvaged goods such as beer-bottle tops, and did their best to organize good buffets. At one dance, in aid of the Red Cross Prisoners of War Fund, a cabaret was put on by a group of American soldiers. For their efforts the women were rewarded once the war had ended by being allowed into the hostel for the first time. They were given twelve places, and a new common room in the Union building. But they were also expected to continue in their social convenor role, and to organize dances. The ICWA minute book reveals a membership keen to provide some comfort in the new common room despite post-war shortages. Old chair covers were taken to a dyer in Gloucester Road, Constance Sherwood donated £5 for the purchase of an electric fire, and it was decided to decorate the walls with reproductions of French Impressionist and Van Gogh paintings. The tables were to have 'some suitable vases and statuettes'. They debated which magazines and newspapers to order, deciding on *Vogue*, *Theatre World* and *Saturday Evening Post*, in addition to some dailies.[90] All of this was, perhaps, a gesture towards creating a feminine space in a dominantly male environment. In the post-war period ICWA held an annual dinner to which they invited successful women; for example, Dame Laura Knight, Kathleen Ferrier and Barbara Castle. But women's access to the Union was still limited. For example, the gymnasium was open to women only when being used for rehearsals by the Musical and Dramatic Society. The bar was the last all-male bastion to be stormed, but only in the 1970s.[91] A day nursery was set up in 1970 as part of an initiative to attract more women to the college, but with only fifty places it very soon failed to meet the demand. It was expanded only in 1992.

Interestingly mascotry was less a feature of the interwar period than it was in the thirty years between 1950 and 1980, though it certainly existed. For example, in 1924 some C&G students parked their mascot, an antique car named Boanerges, outside 10 Downing Street with an effigy of the Prime Minister sitting in it eating a biscuit. This was a comment on the fact that the new Prime Minister, Ramsay MacDonald, had accepted the gift of a car from a biscuit company. Since Boanerges was left without petrol in its tank the police had to organize a tow.[92] In the following year there was a major disturbance when a large number of students from University College caused damage in the Union and then stormed the C&G building looking for their mascot 'Finias' which had been stolen and hidden on the roof by RSM students. The Rector, Thomas Holland, showed up dressed in top hat and morning clothes, and accompanied by four policemen. But this did not impress, and all five were hosed down by engineering students armed with RSM hosepipes — 'most impressive', according to one witness. For this offence, and for damage caused, a fine of one shilling per student was levied from the unions of both colleges.[93] Rags became quite inventive in the 1920s. For example, a 'happening' that was inspired by a humourous article in *Punch* complaining that nobody had anything good to say about the Albert Memorial, and that it was time for some praise. Metallurgy student, A. J. C. Clasen, later Luxembourg's ambassador to Britain (and FIC), had the bright idea of organizing a service of worship. He set a time for 'the faithful to assemble outside the Albert Hall' and about 250 students, some dressed in surplices, followed their leader who was dressed as a bishop. Accompanied by an orchestra of mouth organs and combs, they advanced in slow procession to the Memorial where they 'worshipped' at its foot. Traffic was held up for some time.[94] Also inventive were the push-ball contests, in existence already early in the century. One between the RSM and the United Hospitals was played on a pitch that had Marble Arch as one of its goals and Hyde Park Corner as the other.[95] The ball was 6 ft in diameter and when the police intervened to deflate it the students decided to use a policeman as a substitute. Needless to say the police escorted the RSM team back to South Kensington. In the same period a group of students purchased tickets to a temperance meeting at the Royal Albert Hall. One of their number, disguised as an Anglican vicar, sat near the platform and, at the start, made loud comments in support of the speakers. He was then invited on to the platform to give his own views. He rose, drew a bottle of Guinness from his pocket and, to the loud cheers of his supporters, downed it in one gulp.

After the Second World War it became the custom to have a Guy Fawkes bonfire behind the Royal Albert Hall. To aid in the collection of money for charity, the bonfire was accompanied by fireworks and by other activities such as clockwork car racing on the Hall steps. In 1949 the evening ended in a pitched battle with the police in Hyde Park, after students burned an effigy of Queen Victoria on the steps of the Albert Memorial. The Fifth of November activities became even rowdier in the early 1950s, with fireworks being set off under cars and buses. The result was that the bonfires were banished first to the Rector's lawn, and then out of South Kensington altogether — to the college sports grounds at Harlington.[96] But the Royal Albert Hall, the Albert Memorial, the Serpentine, and the Round Pond have always been magnets for rag activity. On one occasion an American doctor rented the Royal Albert Hall to promote his wonder cure and a notice went up in the college for 'the halt, maimed and blind' to assemble at 6.30 pm. A large group then stumbled into the Hall doing their best to cause pandemonium — and succeeding. In 1953 a set of large white footprints mysteriously appeared leading from the Royal College of Music, across Prince Consort Road, through the Beit Quadrangle (which in those days meant up and down the wire netting and across the tennis courts), over the Union roof to the Albert Hall, and to the neighbouring Queen Alexandra House.

Students in the interwar years engaged also with the politics of the period. The 1926 General Strike saw most, though not all, supporting the government. Unlike a later generation, most students then still believed in demonstrating support for authority.[97] For example, H. Hewer and P. Linstead worked at the YMCA milk distribution centre in Hyde Park. They also took two days of training as railway signalmen, after which they were let loose on the Great Central Railway near Rugby.[98] A. B. Goggs, later a senior executive at ICI, was among the many temporary bus conductors. Others stoked boilers at the Lott's Road power station, and many joined the Special Constabulary Reserve. The Depression of the late 1920s and early 1930s made it difficult for many students to find work and led to an increased political awareness. In 1934 the editor of *The Phoenix* put together a collage of newspaper clippings in the journal in which descriptions of great poverty were juxtaposed with descriptions of great wealth. For example, a clipping on a poor man sentenced to hard labour for stealing a growing mushroom from a field, was positioned next to one on Lady Londonderry's parties with their conspicuous display of wealth, and tables piled high with food. In 1937 the Union saw its first service club, the Social Services Club, which began by opening an Imperial College Boys Club on a housing estate in North Kensington. Many left-wing students of the 1930s, and with more justification than their counterparts in the 1960s, believed that only revolution could bring a better future. Money was raised for the Republican cause in Spain and at least one student fought in the International Brigade. There were those who supported Chamberlain and appeasement, but antifascism was the stronger theme in student journalism of the period.[99] The journalism also suggests that a significant minority of students were pacifist and that others were concerned that the government was not doing enough for civilian defence, especially against possible air attacks and poison gases.

During the Second World War students were more caught up in the collective war effort than in traditional student fun. As a *Phoenix* contributor noted, 'the policemen around South Kensington must be losing interest in their job. It must be ages since a bevy of rollicking, tastefully decorated, buses was seen on Exhibition Road'.[100] While students undoubtedly missed the gaiety of the pre-war years, and the decorated buses, the war provided enough excitement for policemen and students alike.[101] The fact that Imperial was the only major London college not to be evacuated meant that students lived through the bombing and joined in the collective effort described in chapter eight. In 1941 the Governing Body took the position that students not engaged in military service had to serve in some other way, whether in the Home Guard, in civil defence, or in scientific or technical work. These duties took up much student time but they helped draw people together resulting, as was the case by the end of the first war, in Imperial College ties being strengthened.

Already in 1943, the Governing Body was looking ahead to the end of the war. It asked the new Rector, Richard Southwell, to live in the college and to think of ways to improve the corporate life of the college. The governors wanted more rounded graduates, people capable of taking part in public affairs. In 1944 Southwell moved into a refurbished flat in the hostel and, in a memorandum of that year entitled, 'The Social Life of the College', stated, 'we want to attain a fuller corporate life in order (1) that students (of all ages) may go out from the college with a richer background of experience (2) in order that our college may count more than it does at present in the outside world'. One of his conclusions was, that on the life of our college as it now exists, *and without disturbing any of its established customs*, we should superimpose something aimed at what it does not yet enjoy — a reputation for college hospitality based on an evening life that in its turn is centred on "Dinners in Hall". Those of us who at Oxford or at Cambridge took this evening life for granted like the air we breathed know now … how much it meant to enjoy … a corporate life.[102]

After the war, dinners were to be held once a fortnight, for about two hundred students. At dinner the students were to meet senior professors and distinguished visitors, and supposedly gain some social capital as a result. Southwell purchased silver to give the tables what he considered an appropriate appearance, and the occasions some extra shine. He encouraged staff to donate more. Perhaps all of this was nostalgia for a way of life he had enjoyed before the war, but he did his best to bring something of Oxbridge to Imperial.

In 1945 the college mounted an appeal for money to be used, principally, for further enhancing corporate life. Surprisingly, people were willing and able to donate at this time. Some of the money went towards new hostel accommodation. Southwell did not wish to see the existing hostel expanded. He held the view that, 'maturity comes from association with men of maturer age and where a student sleeps is not important'.[103] His ideal was the small hostel, with not more than twenty students, administered by a mature married couple in residence. One such hostel, named Selkirk Hall after its chief benefactor, was opened when a large house in Holland Park was converted into a residence for sixteen students.[104] Southwell appointed Lowry as its first 'tutor' and the Lowrys moved from being wardens of the Beit hostel to administering the new residence along Southwell's preferred lines. As Southwell put it, the amenities were 'for the first time in our history fairly comparable … with the amenities of an Oxford or Cambridge College'.[105] Southwell was also persuasive with the UGC which gave the college money to purchase another house for conversion along similar lines. Other members of the staff were asked to join the Rector and live in the main hostel. They were to help him create an Oxbridge style there too. One person who moved in during the war and stayed until her retirement was Constance Sherwood. She had come to the college in 1919 as a shorthand typist and had risen to become personal secretary to the Rector. Affectionately known as Sherry, she became the unofficial 'Director' of the halls of residence and 'official hostess' at the formal dinners.

The fuel crisis of 1947 put a major damper on dinners in hall, since their preparation was seen as a frivolous use of fuel. More seriously, the fuel crisis hampered research and left students and staff working in unpleasantly cold conditions through the winter months. This, together with other shortages, prompted Southwell to report that students had to 'endure… discomforts…and scarcities that would have been unthinkable before the war'.[106] But it was not just the fuel crisis which threatened Southwell's ideas. His imported vision was not without its critics. *The Phoenix* reported that at a student union meeting it was maintained by some 'that Dinner in Hall is merely an attempt to make Imperial College look like Oxford or Cambridge, by those who think all good universities should look like them'. And that such attempts were a 'poison' to avoid. But *The Phoenix* board supported the dinners provided that costs were kept low, and that informal dress was allowed.[107] Staff members, too, were sceptical. As W. G. Bickley, then a reader in mathematics, put it, 'London is too attractive, and full of all sorts of opportunity, that — short of a moat and a drawbridge — community life is bound to be far less dominant than, say, at Oxford'. Most people, he claimed, depended on public transport to reach their homes, not always dependable in the evening hours.[108] The tension displayed over the dinners and expanded hostel accommodation is interesting. What it implies is that many wanted a share in the kind of privilege they saw institutionalized at Oxford and Cambridge. But they did not wish to be seen as imitating the older universities. Some had a utopian vision, one in which the culture of privilege was all inclusive, or at least sufficiently expansive to include them. Others were more self-confident. They recognized that they were in the metropolis and, as a consequence, had advantages not fully shared by their Oxbridge counterparts.[109] While the 'dinners in hall' continued for many years, and other houses were purchased as intimate student residences, overall Southwell's vision failed. He would have been better served had he paid more attention to local history and college tradition. Later, Hill and Linstead did just that; their approach to producing the well turned out student is discussed below and in chapter fourteen.

During the 1920s and 30s there had been some organized field trips to Europe but such travel became more usual after the Second World War. Already in 1946 James Newby took twenty-three students on a tour of Swiss industry. It was not an easy journey because of war damage to European railways. For fourteen days the students were overwhelmed by the hospitality they received at the factories they visited. 'The lunches that were provided … surpass the imagination of a war-starved Briton.' Indeed, the students saw the two weeks as a 'lunch tour' as much as an industrial one.[110] Newby had been given the position of Vacation Work Clerk in 1934 and had built up a good organization for placing students in interesting workplaces during their summer vacations. This was interrupted by the war but, by 1949, when he was appointed Superintendent of Vacation Studies, he was already sending students abroad for training. In 1948 he organized a conference at Imperial which led to the establishment of the International Association for the Exchange of Students of Technical Experience (IAESTE). Newby became its General Secretary. By 1959 the Association was dealing with close to 6000 UK student exchanges involving 26 countries.[111] Through this and other means, such as organized field trips, continental Europe, and increasingly the rest of the world, opened up to Imperial students. Other types of opportunity also presented themselves. For example, many of the East European Communist countries held World Youth Festivals in the 1940s and 50s and invited Western students to attend. This was undoubtedly a propaganda exercise, but some students took the opportunity of a fully paid holiday, and will have learned something in the process. For example, some Imperial students attended a festival held in Bucharest in 1953 at which Emil Zapotek, the Olympic champion in the five and ten thousand metres, was the star attraction. But more important will have been the exchanges among young people. By the 1960s RSM students were organizing 'Foreign Students Week'. Students from other European mining schools were invited, entertained in South Kensington, and taken to industrial sites and sporting events such as the Oxford vs Cambridge Rugby match at Twickenham. This tradition continued into the later century and, on one occasion in the 1980s, was the subject of an amusing correction in the press, 'students who are providing an evening of pornographic films for the entertainment of visiting European students this week are from the Royal School of Mines, not the Royal School of Mime as inadvertently reported on Saturday'.[112] The Rector, Brian Flowers, who among other things was by then working hard to make the college a comfortable place for women students, was not amused and put pressure on the students to cancel that part of the week's arrangements.

But overall, the many exchanges, field studies, and other trips, often taken in groups, helped strengthen the bonds between college students and gave Imperial a profile also in Europe.

The Later Twentieth Century

The major college expansion announced in 1953 prompted people in the college to think more deeply about the kind of institution in which they wanted to work and study. As student numbers grew, many of the old assumptions had to give way. While some hoped to recreate a collegial way of life in the planned new residences, this largely failed. During the late 1950s and 1960s there was something of an Indian Summer within the old college unions since it was easier for students to identify with them than with the increasingly amorphous Imperial College Union. Indeed, it was in the 1950s that the RCS and RSM decided to acquire antique vehicles as mascots, to match the C&G's Boanerges. In 1955 the RSM took possession of a 1919 Aveling steam-driven vehicle to be named Clementine, and the RCS purchased a 1916 Dennis fire engine which was named Jezebel.[113] Other mascots were owned or acquired. For example, the RCS students guarded an oversize thermometer named 'theta' in the basement of their building, and the C&G students made a spanner which was cast in the Bessemer Laboratory from sixty pounds of nickel bronze.[114] And new union chants such as the Kangela were adopted.[115] Rags and intercollegiate rivalry continued, but despite all this union activity student participation slowly declined. From the 1960s to the 1990s the number of clubs increased, the Union building was expanded, and new dining rooms and common room space added. But the sheer number of students led to a certain anonymity. Within the Union students were no longer able to recognize, let alone know, most of their peers. The ICU came to be viewed by many simply as a provider of club facilities, and of space in which to relax or eat meals, no different in principle from a cheap restaurant or coffee bar. Indeed, already in the 1950s, those who valued old union conviviality saw the new coffee bars with their shiny Gaggia machines as a threat. Coffee bars were opening up all over London and two appeared on the King's Road, an increasingly trendy place in the late 1950s. A 1957 article in *The Phoenix*, somewhat sceptical, noted that 'each is equipped with a hideous Emettesque stainless steel expresso [sic] machine which, amid loud hissings and gurglings, gives forth a small cup of foam which bears the name of coffee.'[116] But coffee bars provided a space where both male and female students could meet informally, could chat, or even stay put for long periods reading or studying, and at little expense. More cheap restaurants appeared in the South Kensington area from the 1950s on and these, too, drew people away from communal dining in the union.[117]

Patrick Linstead, when Rector, recognized some of the problems associated with expansion and that it implied giving new kinds of people access to the social and cultural life of the college. He, himself, appears to have enjoyed the older fraternal and clubbish way of life. He was a regular attendee at old student dinners, the 22 Club, and other such fraternities. In 1957 he opened the new bar in the Union and enjoyed the associated ritual. At 6.00 pm the student orator, in ceremonial gown, read out a proclamation and Linstead drew the first pint. This was drunk by Alan Kitchener, the union president, even though it was full of sawdust because the pipes had not been properly cleaned out after the carpenters left. The room was filled with students and staff who were attached to the Union. Drinking and story-telling went on well past midnight and those who were left in the small hours had to be piled into taxis and taken home.[118] It did not occur to Linstead, or to the much younger union leaders of the period, that the women students who wanted to drink in the bar had a right to enter. When Lord Penney opened the refurbished bar in 1972 women were finally admitted.

But, while never making it explicit, Linstead must have sensed that the way forward in producing the ideal graduate entailed greater inclusion both of women and of non-white students in social activities. It entailed opening up to the wider world, a process that began already during Hill's Rectorship. 'Colour prejudice' became a major topic in student journals during the 1950s. Overseas students began to complain more loudly about their treatment both within and outside the college. African and West Indian students had serious difficulties in finding accommodation, especially during the period of major immigration from the West Indies. The 1958 riots in Notting Hill made matters even worse for some of them. As one African, who had trouble finding a room in 1960, put it 'the British think it is an offence to be coloured and a crime to be a Negro'.[119] Alongside his comments in *The Phoenix* was an editorial reminder to readers that most of the African students at the college came from privileged families, were likely to hold influential positions later in their careers, and that patronising attitudes were not only morally wrong but likely to be counterproductive. Clubs were told that they should help overseas students feel welcome. *The Observer* newspaper, aware of the internal discussion, generously wrote of Imperial 'it is a cosmopolitan and liberal society, with representatives of 66 countries among the students, and a kaleidoscope of skin-colours.' It also pointed out as something unusual that a recent president of the C&G union was a black Guyanian.[120]

In the 1950s and 60s student journalists still reported on day-to-day college life in the student papers. A new fortnightly student paper, *Felix* was founded in 1950 with the purpose of keeping students informed of college events and *The Phoenix* became a review journal. In 1958 it was edited by David Irving, the future controversialist historian and serious minimiser, if not denier, of the Holocaust.[121] Irving had come to the college in 1956 and had founded an organization called the National Guardian Movement which he claimed was an anti-Communist movement but, given its literature, was clearly inspired by the British Union of Fascists. It had some student members from Imperial and some from other colleges in the university. Irving soon became well known in the college. A gifted journalist, he helped in the running of *The Phoenix* after the editor had been injured in a mountaineering accident. He worked hard, contributed a number of drawings and short articles and, despite some misgivings among board members shaken by the content of some of his articles, was asked to take over the editorship in 1958. At first he appeared to have had a good idea, namely starting a series on old students, 'They Came to Imperial'. But his first subject was the German industrialist Fritz von Thyssen and it is clear that Irving was interested in him solely for his connections to Hitler. This and other questionable content led to Irving being fired from the editorship. But, for reasons not entirely clear, he was then asked to edit the 1959 carnival magazine for the University of London Union. That year the carnival was to raise money in support of the World University Service and its work with black students in South Africa. Irving produced a magazine but secretly inserted some extra inner pages. In these he is said to have defended Apartheid, included racist and sexist jokes and cartoons, claimed that Hitler's regime was 'the first great unifying force that Europe had known for six hundred years', and reprinted material from Oswald Mosley's journal *Action*.[122] News of what he was up to leaked out of the Imperial

College Union press room where the magazine was being set up, but Irving managed to get the copy as far as the printers before it was found and destroyed. People at the college were genuinely shocked by what had happened. The Rector, who expressed outrage, avoided the embarrassment of what to do next when Irving failed his exams and had to leave.[123]

While racist jokes were rare in student publications of this period, sexist jokes and articles were still acceptable. It would appear that the only people who objected to the Spring 1958 issue of *The Phoenix* (edited by Irving), which by today's standards was outright sexist, if not misogynist, were some women students. Judy Lemon wrote a spirited reply which was published in the following issue but her response was largely ignored. By the time Brian Flowers became Rector in 1973 blatant sexism was no longer acceptable. Flowers led from the top in combatting it and in encouraging women to come to the college.[124] By then times had changed, and students would have thought twice before making the kind of jokes in print that earlier had been unremarkable. However, life was not that comfortable for outspoken feminists.

Until the late 1960s one could still read much about the college in student publications, but gradually student journalists used *Felix* as a vehicle for honing their skills as reviewers of films and pop concerts, as reporters of national sporting events, and increasingly as commentators on matters such as design, international finance, and popular culture. Late in the century, instead of biographies of their professors, students were writing about software moguls such as Bill Gates and Larry Ellison. The focus was increasingly on a world beyond the college and *Felix* came to mirror mainstream journalism in its content.

After the war the World Government Society and the International Club were major venues. In 1948 the latter was addressed by the left-wing publisher Victor Gollancz, a number of MPs including Richard Crossman, and by Lord Pethick-Lawrence who gave an appreciation of Gandhi shortly after Gandhi's assassination. But, despite this and despite the internalization of new concepts such as Marshall McLuhan's 'global village', the trend in student societies was away from internation-alism.[125] One consequence of globalization is that differences are exported worldwide. By the late 1950s many new political, national, ethnic and religious clubs were being founded by overseas students, a reflection also of new nationalisms in the post-colonial era. For example, in the early 1960s, the Arab Society was attracting about one hundred people to its meetings and was promoting the Palestinian cause on the campus. The non-partisan Political Club bifurcated. In 1950 it became a Conservative Society and a Socialist Society. Well known politicians and journalists appeared regularly at college events during the 1950s and 60s. The sciences became increasingly fashionable and many people came to the college wanting a part of the action. Both the self-image of students and the profile of Imperial were boosted as a consequence. Students began to recognize that as engineers and scientists they had something to contribute to the political process and to national discussions.

C. P. Snow's 1959 essay on the 'Two Cultures' was opportunistic in the sense that it latched on to what was already a major discussion point, the cultural gap between those educated in the arts and sciences.[126] Linstead, very much aware of the wider discussion, encouraged politicians and well-known journalists to visit the college and give talks. He supported initiatives in which students were to look beyond their specialist education, and beyond the college, for ideas, and encouraged students to join in other activities. Staff, too, were encouraged to make outside connections. For example, the Consort Club was founded to bring staff from the college into contact with those at the RCA and RCM at regularly held informal dinners.[127] This led to a number of new joint initiatives. For example, in the 1950s many RCA students wanted to learn more about the sciences. To help in this and to promote contact between the students, Linstead and Professor Carel Weight from the RCA supported joint discussions, and created venues where art students could learn about the sciences and Imperial students could take art classes.[128] Another arrangement was made with the Royal Court Theatre for students to sit in on rehearsals, attend plays, and even act as consultants on matters relating to science or engineering in the scripts.[129] Students were encouraged to attend theatres, concerts at the Royal Albert Hall and the Festival Hall, and the college negotiated for unsold tickets to be made available on the day of performances at much reduced rates. Lunch time lectures on a wide range of topics, as well as the Touchstone weekends, were widely promoted. Much of this activity was organized by C. K. McDowall who had been appointed by Roderic Hill soon after the war to run the General Studies programme and the Touchstone weekends (see chapter fourteen). McDowall made many arrangements with outside cultural bodies allowing students to take advantage of what London had to offer. He also organized the annual Commemoration Day ceremonies, was someone who communicated well with all people in the college, and took good care of many overseas students.

The term 'brown bagger' was coined earlier in the century but came very much into use in the 1950s.[130] The following verse captures well the mood of the time with its implied criticism of the non-engaged student.

Brown Baggers

Are you coming tonight to the debate?
Oh, your dinner's at eight;
Aren't you interested in the Welfare State?
Oh, you mustn't be late.[131]

Whether internal college initiative, or the socio-cultural climate more generally, was responsible for the openness to activities beyond the college is open to debate. But Hill, Linstead, and many of the professors did much to promote such openness during the 1950s and 60s. One of the more successful clubs to emerge in this period was the H. G. Wells Society with its purpose of encouraging debate on problems relating to 'science and society'. Professor Willis Jackson was a major backer, and introduced Sir Barnes Wallis, the speaker at the first meeting in 1968. Jackson was himself a speaker on several occasions.[132] While a few students were political activists, most were not. The most popular clubs were still those for sports or conventional social activities. By the late 1950s there were over thirty social clubs in existence, a number that continued to grow in later years. For example, with the end of postwar austerity, a wine-tasting club was founded. Twenty years later, in a period of even greater prosperity, the Real Ale Society made its appearance.[133] The founding of many new clubs reflected a growing student population and was coincident with a decline in mainstream union activity.

With the new outward looking attitude came also more service oriented activity. Members of the Social Services Club set up a children's playground in Notting Hill, read to the blind, and visited old people's homes. In 1970 ICNightline began helping troubled students. Fifty Samaritan trained students took turns in receiving calls at 8 Prince's Gardens. ICNightline was the first such scheme in a British university and grew over the years, especially after Chelsea College and Queen Elizabeth College asked to join. Social conscience exhibited itself in a range of ways as, for example, the arrival of Amnesty International at the college in the 1960s. In 1975, Sinclair Goodlad, then a lecturer in the electrical engineering department, was awarded a grant by the Leverhulme Trust Fund to start a programme in which engineering students visited local schools to help in the science classes. This programme, which began in Pimlico, and is known as the Pimlico Connection, proved a great success and is still active.[134]

Outdoor recreational clubs diversified to include activities such as scuba diving, hot air ballooning, motor racing, caving and orienteering. Some of these activities clearly reflect an increase in student wealth. Student exploration was supported by an Exploration Board, founded in 1955–6, which began seeking ways to help fund expeditions. Outside agencies such as the Royal Geographical Society helped in this. The first expedition was ambitious. Sir Eric Shipton led a group of mountaineers to the Karakoram region of the Himalayas. The Mountaineering Club was prominent in the early years of the Exploration Board and organized further expeditions such as to the Caucasus, the High Atlas, the Hindu Kush, and the Andes.[135] While all expeditions were supposed to have a scientific aspect, more overtly scientific exploration by botanists, zoologists and geologists were especially supported. Some engineering students tested an ordinary production car on the roads of Kashmir. The Underwater Club mounted an interesting expedition to Malta in 1969 where nine students set up two underwater inflatable houses, and operated a two-person underwater laboratory. The Exploration Board was inundated with requests for support and, from the Arctic to Bolivia, from Ghana to Thailand, from Iran to the Solomon Islands, from Iceland to China, Imperial students set out to discover new parts of the world, bonding in new ways with like-minded adventurers.[136]

Not surprisingly, with more young people entering higher education, maintenance grants declined in real value later in the century. A new pattern of working through the long vacation began to emerge. Before the Second World War proportionately far fewer students needed to earn extra money in this way. But, with greater access to higher education came a change in the way students were viewed. They were no longer young gentlemen and ladies, they were simply students. And, while support was still relatively generous in the 1950s and early 60s, maintenance grants were no longer intended to support gentlemanly habits.[137] By the 1970s they were not even compatible with full-time study. Students without scholarships needed loans, parental assistance, or jobs. Nationally and internationally, students took on a new identity, one increasingly tinged with radicalism. And they became more attuned to a growing popular culture than to the older more gentlemanly one. How the college authorities reacted to this is discussed in chapter twelve but, with students and young members of staff seeking more democratic structures, much changed. Popular music with its seemingly rebellious stance was a good indicator of change in this period; students took an increasing interest in it and many of them formed bands.[138] Physics student Brian May started Smile, a band that did well and made its name by first giving a number of charity concerts with college support. For example, in October 1969, the band performed in aid of the National Council for the Unmarried Mother and her Child, a name that now seems very dated. May, Freddie Mercury, and others then went on to form the spectacularly successful Queen.[139] Some of its first gigs were at Imperial.

Classical music, too, continued to draw many. In 1975 the college received much good publicity when four members of its staff staged the first 'Sing and Play along Messiah'.[140] Agreeing to share in the loss, the four rented the Albert Hall, sold tickets in what they thought an appropriate distribution for instrumentalists and choir members, and hired professional soloists. But, instead of making a loss the event was an outstanding success, repeated and copied worldwide in subsequent years. Another success dating from this period is the Exmoor Singers, a group that formed at Imperial and later became one of the country's best amateur choirs.[141] The OpSoc (opera) was also very active in the 1970s, in part because of the enthusiasm of Mike Withers, a student who even after leaving the college took productions to the summer festival at Budleigh Salterton. Musical life was given a further boost by Eric Ash. A keen violinist, Ash appointed Richard Dickins in the 1990s to help revivify orchestral activity which had lapsed a little since the retirement of C. K. McDowall in 1974.[142] Dickins had earlier taken over the conducting of the college orchestra from Gavin Park. At the time the orchestra played second fiddle to the choir which, conducted by Eric Brown, was the most active musical ensemble at the college. Under Dickins, orchestral and chamber music picked up. The symphony orchestra grew in size and began tackling more difficult repertoire. Today there are a number of other thriving instrumental ensembles and the tradition of lunchtime concerts continues strong.

The student union president in 1969–70 was Piers Corbyn. His appearance was similar to that of many student radicals of the period in that he wore his hair long, and his dress, while not exactly that of the proletariat, was a statement against upper-class formality.[143] He wore the new uniform of youth as jeans and cords replaced grey flannel for men, and dresses and suits for women.[144] At the end of his term of office Corbyn reported, with some justification, 'a tremendous amount has been achieved this year'.[145] It was, perhaps, a peak year for student political activity. A 'sleep-in' was held at the hostel to protest a 'lodgings crisis', and students set up their own lodgings bureau. They had already been given observer status at Governing Body meetings but, in 1969, demanded it also at the Board of Studies. This demand culminated in a major demonstration when the Queen came to the college in November to open the new College Block (later Sherfield Building). While the bells on the newly restored Queen's Tower were ringing, and after the organist from Salisbury Cathedral had played the National Anthem on the organ in the new Great Hall, Corbyn, in front of an audience of nine hundred, begged leave to present a petition to the Queen in her role as Visitor to the College. The large envelope that he handed over included a letter, and a petition for the revision of the Charter to allow students to have a greater say in the running of Imperial College.[146] By and large the students were successful. They did gain access to more college boards and committees, though this was no unalloyed triumph. Further, Corbyn and his fellow union representatives successfully negotiated with the Governing Body for a 'sabbatical' for future union presidents.[147] In the spring term a refectory boycott was organized to protest rising prices. The union was, at the time, affiliated with the NUS and supported also a number of national issues related to student housing and maintenance grants.[148] The largest national demonstration of the year was one against Apartheid, and many students gathered at Imperial for a march to Senate House. The situation in South Africa was clearly oppressive and the demonstration attracted many students who were not, otherwise, activists.

One of the major internal issues was the curriculum, and the student leadership called for 'more diversification'. By this they meant a broader curriculum, one not restricted to science and engineering. David Christopher, a student witness at the Parliamentary Select Committee on

Education and Science in 1969, stated that Imperial was 'no "academic community" working towards "one knowledge". It is a factory producing human goods for the industrial oligarchy'.[149] This was no doubt the case, but hardly the whole story. As will be shown in chapter fourteen, the professors were ahead of the students in thinking of curricular reform. On a more practical level students painted a zebra crossing on Exhibition Road after their demands for one had not been met. Later, after a near fatal accident, an official crossing was installed. While the activism was part of a larger movement, there was surprisingly little reported in the student papers on the 1968 student revolts in Paris, Tokyo, and elsewhere, or of the earlier events at the University of California at Berkeley. Student radicalism at the LSE was reported, but in a way that suggests ignorance of the wider context.

Much else that was going on at the college suggests that the majority wanted nothing to do with political activism and countered it in many different ways. For example, in 1969 the Archbishop of Canterbury, A. M. Ramsey, visited Imperial for five days. The visit began on a Sunday with a special service at St. Augustine's on Queen's Gate. During the week the Archbishop addressed the H. G. Wells Society on 'The Meaning of Life', gave three other major addresses, and held several informal question and answer sessions. About two hundred people attended Eucharist with the Archbishop in the foyer of the mechanical engineering building, and the visit ended with a party in the union building. Whether the Archbishop believed it important to mingle with students in this period of unrest, or whether he simply accepted an invitation from the joint Christian societies is not clear. These societies were very active during the 1960s and, like Ramsey himself, increasingly ecumenical.[150] Teaching weeks, missionary appeals, pastoral work, carol singing, and walking weekends drew students together. The Anglican chaplain of the period, Ivor Smith-Cameron, was energetic and arranged regular mid-week discussions, and a pilgrimage to Iona. Roman Catholic students were similarly active, with Bible studies, weekend house parties, and pilgrimages to Rome. The Campus Crusade for Christ came to Imperial in 1974 but was not a success and soon moved its resources elsewhere. Other religions, too, had active societies, but with much smaller memberships.

One student, Philip Marshall, was more overt in his opposition to radicalism than those in the religious societies. He booked a large lecture hall to hold a semi-serious, semi-humorous counter-protest. The room

filled to overflowing and those present passed a motion pledging themselves to 'militant apathy'. Outsiders were seen as importers of dangerous
propaganda and Marshall urged the students to 'warn off Tariq Ali and
the rent-a-mob'.[151] He stated that he was 'sick to death of reading in the
Press about student militants. It is about time people realized that all
students are not pro-Moscow, pro-Mao, or any other title. I am here to
study engineering, not to change the world'.[152] But the world did change
and student life and attitudes with it. Indeed, it was largely students that
made the cultural revolution. They were no longer willing to be patronised
by their teachers and while this had many positive consequences, some
were negative. One result was that, in higher education, bureaucratic
procedures replaced discretionary ones, not necessarily with better results
at the human level.[153] Decision making became more open and democratic, but the more openly adversarial approach to problems led many to
withdraw from participation in college affairs. Students found new ways
of associating. Unions became less important as sites of social exchange;
residences, departments, and research groups more so.

The move toward social equality affected also non-teaching staff.
They, too, had a social club, the Holland Club, which had been founded
after the war and named after the third Rector. Professor Levy gave the
club some space in the basement of the old Huxley building and a small
refectory opened there. The club arranged a number of events including
Christmas dinners, dances, and annual outings to the seaside or coun-
tryside. These usually included some sporting event such as a bachelors
versus married men cricket match.[154] In 1962 the space in the Huxley
Building was needed for the college expansion and the club moved to
premises at 15 Prince's Gardens. There it ran its own dining room and
bar until the tumultuous year of 1969. It was perhaps fortuitous that the
College Block was close to completion in 1968. It made people think
about access to the spaces it opened up. For example, should there be
separate senior and junior common rooms? If so, who should be allowed
entry into the senior common room? The Holland Club food services
were losing money at the time and the college authorities decided to
step in. The question they faced was whether the non-academic staff
should continue to be segregated in their own dining room, join with
the students, or with the academic staff, in the new building. After much
discussion the Senior Common Room Committee agreed to 'some degree
of common usage'.[155] The College Block (now Sherfield Building) has

become the main centre for dining in the college.[156] Entry to the Senior Common Room is given to all academic staff and doctoral students and to many of the clerical, technical and secretarial staff.

ICWA folded in the late 1970s. Women students joined their male counterparts in the full range of student union activity and a separate union presence was redundant. An Imperial College Wives Club that had come into existence after the war continued to be active in the 1960s and 70s. It organized a range of social events. excursions, teas and luncheons, and helped the college in traditional ways with appeals and fund raising for different causes. These included helping the wives of overseas students, the day nursery, and the furnishing of student houses. But this type of voluntarism was already in decline by the 1960s, as wives increasingly found identities not so closely tied to those of their husbands. The club later changed into a loose association open also to all women staff at the college. Gradually the staff took over and wives participated less.

By the late 1970s about 35% of the students were living in college residences.[157] Earlier, selection for hostel places had been based heavily on a student's potential for contributing to Union activity. But, after the major construction of hostel space in Prince's Gardens, personal situation became a stronger determinant in selecting occupants for the new halls. Further, because the halls were used as recruiting tools, more spaces were reserved for freshers. By 1971 single rooms cost about £40 per term without meals. By the 1990s, students were expected to bring their own linen and clean their own rooms. The final trace of the 'gentleman' student who had earlier expected his meals to be prepared, bed to be made, room to be cleaned, and laundry to be cared for, had disappeared. The socialization of students had changed and they were expected to be more self-reliant. More of the responsibility for socialization fell on senior students within departments as staff took on other responsibilities. Senior students also took over the organization of freshers' dinners from the unions and the dinners became increasingly departmentally based. Some of the dinners soon faded into pub crawls for the hardy few. Student unions, once the medium for the formation of close and lasting ties between students and staff, became impersonal bureaucracies providing a range of much needed services.

None of this happened overnight and traces of older tradition can still be seen. Unions continued to organize rags, mascots were still 'stolen', and money still raised for charity. But these events, while often inventive, became less central to student life as the twentieth century came to a close. Some union activities were original enough to draw small student audiences. Raft races on the Serpentine and tugs-of-war across the Round Pond were popular in the 1960s. Tiddlywinks races along Oxford Street were popular in the 1970s. At the first race in 1969 about 50 students from the C&G and RCS winked their way in and out of Selfridges and the C&A and then down Oxford Street, while others from the RSM winked down Regent's Street. Later they all met in the Piccadilly Underground Station for the 'World Underground Winks Championship' before retiring to the Ennismore Arms.[158] Pushing beds around with students piled on top was a fad during the 1960s and 70s, and a small group took part in a London to Southampton race in 1969. In the 1990s students took to clowning outside Harrods, selling pies to passers by who could then throw them back. Some students then discarded their pie-ridden clothes and ran naked back to the college, a practice that wasn't strictly clamped down on until a woman student joined in. Traces of many of these activities still exist but ragging and the ways in which money is collected for charity has changed character. Typical of the later century were charity runs, or versions thereof. For example, a 1983 team of C&G students attempted to set a record by peddling non-stop around the coast of Britain on a specially designed tricycle, raising money for the Royal National Lifeboat Institution.[159] Also more typical is the individual gesture. For example, Ian Howgate, the union president in 1987–88, changed his name to Sydney Harbour Bridge to raise money for Comic Relief. As students became more attuned to a growing national and international popular culture, older behavioural forms declined in importance. But they still could be seen. For example, in the 1980s and 90s, the executive of the Royal College of Science Union were still being given a ritual initiation by being covered in a mix of rotting fruit, fish, milk and other such materials, before being ceremoniously dunked in the Round Pond.[160]

Special interest clubs continue to thrive and sports facilities have been much improved.[161] Outstanding sporting achievement is still admired, but routine matches which once had large followings are today largely ignored except by close friends of the participants.[162] Already by the 1970s it was clearly a struggle to get students out to sporting as well as to other union events.[163] By the 1990s many more will have followed the Imperial teams on University Challenge, a sporting event of a different kind.[164] Sitting at home in front of a television set was a more comfortable option than sitting in bleachers, or standing on towpaths, or on the sidelines of sports fields. Media success was becoming cooler than sporting success. The creating of television programmes, and taking part in the twice weekly transmission from the college studio became increasingly popular in the later part of the century. But there were those who missed the older forms of conviviality and sought ways to reinvent them. For example, one of the student residences in Evelyn Gardens is named Holbein House, for Arthur Holbein, a civil engineering student who came to the college after military service in the First World War. Despite being severely wounded Holbein became captain of rugby. His identity was very much bound up with Imperial, with the Links Club, and with the City and Guilds of London Institute. He was a successful civil engineer, an enthusiastic Old Centralian, and was on the Governing Body of the college for many years. He was convivial, a member of many clubs, returned often to drink in the Union bar, and was present at major college sporting events throughout his life. He was helpful to many young students entering his profession and exceptional, even among his generation, in his devotion to the college. To some of the C&G students living in Holbein House during the 1970s he was an ideal. They chose the name Bean, Holbein's nickname, for their club which was founded in 1976. Membership of Bean's Club was exclusive and by election. Only male residents of Holbein House were eligible and the stated aim was to 'further the spirit of Holbein in the past and present invited members of Holbein House'. Rules were laid out in 16 articles and 39 clauses and rituals invented in an attempt to recapture an earlier form of fraternal conviviality, and to encourage support for sports teams. But though once the norm for those with a need for this type of exclusivity, by the 1970s this kind of club behaviour was unusual.

A number of small dining clubs, of ranging longevity, existed in the later twentieth century. One, the Fourier Club, was founded by Richard Kitney in the early 1990s at a time when there was much tension between the academic and administrative staff.[165] Kitney invited authors of recent historical works to be after-dinner speakers and, by encouraging conversation in areas of common interest, did something to reduce tensions. Intended to bring the staff together, the club was a small but successful venture. For most people, however, social exchanges occurred in the college bars and in local pubs, rather than at dining clubs. The Southside bar was a major meeting place until its recent demolition,[166] and the popularity of local pubs has not diminished since 1907. The Queen's Arms, the Hoop and Toy, and the Zetland, the latter a great favourite in the 1960s when it pioneered pub television, are still much frequented.[167] Restaurants in the area do not have the same longevity as the pubs, but each generation of Imperial staff and students has had its favourites.

Conclusion

Those who fashioned the idea of the Victorian university had in mind the creation of a type of elite that, by the early twentieth century, no longer entirely fitted Britain's needs. The founders of Imperial College were aware of this and put in place a modern institution which, they believed, would offer an education that met the needs of industry and empire. But what this entailed for the corporate life of the college was unclear. This chapter has illustrated something of how the college saw itself through ceremonial, how community and conviviality were managed, and how the idea of the Imperial student was contested over the century. Understanding what happened is complicated by society's change in attitude towards students once their numbers began increasing after the Second World War. Perhaps the actions of Richard Southwell, and the way in which others responded to him, illustrates well the tension of the mid-century; a tension which, by the century's end, was largely resolved. Southwell was very keen to import something like Oxbridge collegiality into Imperial. His reasons for doing so were not so much elitist as they were communitarian and pedagogical. He believed, with some justification, that young students need to mix with successful older people in order to become properly socialized. He would have agreed with Cardinal Newman, though not approving his anti-technological imagery, that the university must be an *'Alma Mater*, knowing her children one by one, not a foundry, or a mint or a treadmill'.[168] But Southwell was not very imaginative when it came to conceiving how this might be achieved at Imperial. As a result he tried, with little success, to import commensal practices that he himself had enjoyed at Cambridge as a student, and at Oxford as a professor. Southwell seemed unaware of Imperial's pride in its distinctiveness, and of a home grown paternalist tradition that could, perhaps, have been institutionalized in new ways. By the late 1960s even that opportunity was lost. Older paternal patterns were no longer viable; student politics, increased numbers, and new demands on academic staff, led college life in a new direction. But Southwell's dilemma remains important. How should older professional and business people help in the socialization of students?

What we have seen is a change from student unions which were once social clubs for both staff and students, to student unions which today are bureaucracies for facilitating the increasingly disparate activities of the student body. In the early years of the college, the union was a site where students could mix with staff and with old students, many of whom stayed in touch with the college all their lives. Within departments staff took on paternal roles, both with respect to discipline and to general socialization. But this behaviour began to give way already in the 1950s. By then students had more disposable income than before and many looked beyond the college in constructing their social lives. During Linstead's Rectorship the college helped this more outward-looking trend, as did the increasing influence of women, highly critical of much of the fraternalism that surrounded them. New forms of student culture emerged but they were never entirely free of older tradition. One constant is the college location. It has allowed students and staff to enjoy something of the life of Kensington and Chelsea. The pubs, parks, restaurants, museums, the many other colleges, the Royal Albert Hall, and the King's Road, are perennial fixtures that have impacted the lives of those who pass through Imperial.

There have always been gifted and ambitious staff at the college, and many workaholics. But new bureaucratic forms of accountability have led to new forms of self-policing among the staff. Too much time spent hanging around with students is clearly a bad thing. Students, too, understand that it better serves their future prospects to make personal connections within specific academic areas than at union events. Gone are the days when Henry Tizard took good union men with him to Whitehall. Further, information that once was obtained from college-wide student journals is now obtained through official publication, much of it on the web, and from departmental or research group newsletters.[169] The student journals that remain are outward looking and reflect more on the larger world than on the college. One has the impression that today's student journalists are hoping to find work in the media. Departments have grown significantly, and postgraduate students now identify with Imperial almost exclusively through their research groups. Interestingly it is in research groups that one finds the strongest trace of the type of paternalism and mentorship that once existed also at the undergraduate level. Before the Second World War there were relatively few research students and they joined with undergraduates in union-focussed activity. A noticeable trend of the later century is that alongside the diversification of interests, and the atomization of the staff and student body, social service related activity increased. Voluntarism that earlier was more inwardly focussed is now focussed elsewhere. In some ways the college community is the loser but others, some in the college neighbourhood, may have gained. Those seeking community within the college may still find it at the departmental or section level, but growth and increasing specialization within the university sector have been bought at the cost of an earlier form of social cohesion.

...

1. The City and Guilds College is perhaps the exception. Its ties to the City were still valued, its alumni association was strong, and its loss within the college governance structure was felt more strongly than the loss of the other two colleges. However, by 2000 the presence of the medical school at Imperial and new interdisciplinary formations made the old governance structures increasingly impractical. The old college names and traditions survive in aspects of the social life of the college, and in the alumni associations.

2. The number of women students at the college was low but the proportion of women remained roughly constant, and below 2%, until the Second World War. Proportionately the number of women increased during the Second World War (to about 4%) because of a decrease in the number of men. However, after the war the number of male students shot up much faster than the number of female students. The numbers of both men and women did increase but the proportion of women declined through the 1950s (to about 2.5%) and only began to climb again in the mid 1960s. See appendix.

3. This latter view was sometimes voiced, perhaps by those who themselves had been educated at the ancient universities or held them as ideals. R. V. Southwell is the prime example.

4. R. P. Westwood, 'Education for the Specialist', *The Phoenix*, (December 1946).

5. After the Suez debacle, and after the Hungarian uprising, a new coalition of the Left emerged. One manifestation was the Campaign for Nuclear Disarmament (CND) protesting the proliferation of nuclear weapons. CND with its simple message 'Ban the Bomb' was popular with students and staff in the 1960s.

6. A. H. Halsey, in his *No Discouragement: An Autobiography* (London, 1996) 43, notes the difference between his own arrival at the LSE in 1947, with virtually no possessions, and his needing a VW mini bus to accommodate his son's 'hi fi and other gear' when taking him to Imperial College in 1976.

7. The term 'generation gap' first appeared around 1967.

8. The Great Hall, used mainly for examinations, was part of the Imperial Institute and belonged to the University of London. Shortly before the examination hall was demolished, in the late 1950s, Commemoration Day ceremonies moved to the Albert Hall where they are still held.

9. Within the college smaller groups were also boosting Imperial by means of dinners. For example, the short-lived Catalysts Club, founded in 1918, was to show off the college, and to help promote careers, by entertaining distinguished men of science visiting London.

10. GJ8/1 File on annual dinners.

11. Women staff and senior students were invited but spouses were not.

12. The first Fellows of Imperial College were: Sir Charles Vernon Boys FRS (physicist), William Frecheville (emeritus professor of mining), Ralph Freeman (designer/construction engineer of Sydney Harbour Bridge, which had just opened), Sir Thomas Holland FRS (former Rector, then Principal of the University of Edinburgh), Herbert Alfred Humphrey (chemical engineer, designer/construction engineer for Billingham Works, ICI, and a Telford Gold medallist), Richard Oldham FRS (Director of the Geological Survey of India), Sir William Pope FRS (Professor of Chemistry, Cambridge), James Whitehead (eminent patent lawyer) and Sir Herbert Wright (rubber botanist, industrialist and college benefactor). See *The Phoenix* (May 1932). The fellowships have been used ever since to honour eminent graduates and, since the mid-twentieth century, also others of distinction. Today, under revised university statutes, Imperial can give honorary degrees without needing ratification by the University, and a few have already been awarded.

13. The Beit Quadrangle tennis courts gave way to the garden. New tennis courts were then constructed at the eastern end of Prince's Gardens.

14. *The Phoenix* (1925), 130. There is a good description of these events in a diary kept by O. V. S. (Peter) Heath (privately owned by Robin Heath).

See 4 May 1925. Heath was the son of Sir Frank Heath, Secretary of the DSIR and a member of the Governing Body of Imperial College. Heath records there was also a conversazione in the evening at which formal dress was obligatory. E. B. Poulton was Professor of Zoology at Oxford and had known Huxley well.

15. The Huxley conference, and a similar one to commemorate Patrick Blackett in 1998 (he was born in 1897), were organized by the Centre for the History of Science, Technology and Medicine, and helped to raise the visibility of those working in the history of science at Imperial College. The British Society for the History of Science played a role in both conferences.

16. This was in fact the centenary of the founding of the Normal College of Science. The name RCS was used from 1890.

17. Adrian Whitworth (ed.), *A Centenary History: a history of the City and Guilds College, 1885–1985* (London, 1985). See also Joyce Brown (ed.), *A Hundred Years of Civil Engineering at South Kensington: the origins and history of the department of civil engineering of Imperial College, 1884–1984* (London, 1985).

18. James Secord is Professor of the History of Science at Cambridge. See *IC Reporter*, 5 June 2001 for both Sykes and Secord quotations. Secord was a senior lecturer at Imperial in the 1980s and played an important role in setting up the London Centre for the History of Science. See also chapter 14.

19. Such funding was formalized in the 1918 Education Act. Grants took the form of aid in paying approved fees, and in maintenance grants to not exceed £80 per annum. Grants were administered by local education authorities. Family income and the number of children in a family were taken into account, as was scholarship money from other sources. Fees were paid in full for those qualifying for maintenance grants. In the mid 1920s the average maintenance grant in the London area was £74. 1s. In 1920 close to 2500 maintenance awards were made to students in England and Wales, a number that rose to about 3700 by 1932. National Archives Kew PRO/ ED 24/1984; ED 12/266; ED 54/35; Education Act 1918, Grant Regulation, No. 26. See also 'Statutory Rules and Orders (1920), No. 1514, Board of Education (University Scholarships, England), Regulations.' (23 August, 1920); for maintenance allowances see also, *Board of Education Maintenance Allowance Awards at Secondary Schools and other Institutions of Higher Education: Statement of Expenditure incurred by Local Education Authorities (England and Wales) … for financial year, 1923–4* (HMSO, 1925).

20. KEE, Electrical Engineering correspondence file; Stanley Watkins to A. Tustin, 12 December 1959. Watkins wrote of his time at the college, 1905–

11. He also recalled that many road surfaces were still made of wooden blocks and that in wet weather buses would frequently skid, a road hazard for cyclists in the first decade of the century. Those who lived in more distant suburbs used public transport.

21. See, for example, KEC Civil Engineering correspondence file; F. Newhouse to D. W. Hopkin, 27 February 1958.

22. Notes on 'Physics in South Kensington' by Harold Allan mention a 1982 letter written by Grainger, a student in the RCS 1907–10. (Typed copy of Allan's notes are in the physics department.)

23. William Brown, a future professor of botany, and a research student at the college from 1912, shared a flat with some fellow students above a restaurant in the King's Road. See S. D. Garrett, 'William Brown, 1888–1975', *Biographical Memoirs of Fellows of the Royal Society*, vol. 21 (1975), 158.

24. Hilda, Doris and Renée Conrady were the daughters of Professor Conrady and all were students at the college. See Marion Gossett, 'ICWA Then and Now', *RCS Handbook* (1974–5).

25. In 1914 a student was expelled from the college for stealing another student's scarf, implying harsher disciplinary standards than later in the century.

26. The civil engineering department lagged behind the others in this. Professor Pippard ended the practice in his department only in 1939.

27. See *The Phoenix* (December, 1924), 45.

28. Quoted from a 1911 article in *The Phoenix*, in *Annual Report*, 1985–6. The RCS lunch room was later converted into a physical chemistry laboratory.

29. In his diary, O. V. S. (Peter) Heath noted that, in the 1920s, lunches at the V&A were cheaper than at the Union; *op. cit* (14).

30. See B/Lowry; G. C. Lowry typed memoir, 8.

31. B/ Sykes. Winifred Broome to J. C. C. Sykes, 29 March 1912. Sykes was a member of the Governing Body.

32. This view was widely expressed. It is interesting that I was unable to find any evidence of identification with the widespread practice of factories closing down on Wednesday afternoons to allow workers time to engage in, or attend, football matches. For the role of sports and athleticism in producing the ideal Cambridge graduate see Andrew Warwick, *Masters of Theory: Cambridge and the Rise of Mathematical Physics*, (Chicago, 2003), chapter 4. While at Cambridge the tag *mens sana in corpore sano* was taken seriously, at Imperial exercise was seen by many as a distraction until later in the century when fitness became fashionable. This is not to imply that sporting activities and success were not admired.

33. This form of entertainment survived, albeit in diminished form, into the 1980s.

34. *The Central* (June 1939) reminiscences. These lines from a song from 1908 by E. F. D. Witchell.

35. Women were also a minority in these colleges. There are several examples of Imperial students later marrying students from the art and music colleges. For example, Alec Skempton made many friends at the Royal College of Art, including his future wife Mary Nancy Wood. See Judith Niechcial, *A Particle of Clay: the biography of Alec Skempton, civil engineer* (Whittles, Caithness, 2002), chapter 2.

36. Watkins to Tustin, *op. cit.* (20).

37. Read graduated with an ARCS in geology in 1911 before becoming a research student under Watts. He was an editor of *The Phoenix*. Illing, a graduate of Cambridge, came to study with Watts but he soon moved into oil technology research. Boswell also remarked on the interesting fact that Read's wife was the champion lady boxer of Kent. B/Boswell; P. G. H. Boswell typed memoir. See chapter 6 for more on Read, Boswell, and Illing.

38. Watson, quoted in *GEOlogIC* (1985), 235.

39. *The Phoenix* (1912–13), 105. Report of visit to Otto Monsted and Co. Ltd., Southall.

40. Ernest Frederick Relf FRS (1888–1970) was a great musical force in the early years of the college. A physics and mechanical engineering student, he joined the National Physical Laboratory after leaving Imperial and later was appointed the first Principal of the RAF College at Cranfield, founded after the Second World War.

41. There are many underground tunnels and ducts around the South Kensington site. Most of the old buildings were connected by tunnels, including the Imperial Institute and the old Science Schools Building on Exhibition Road. These tunnels, some of which are still accessible, have been the site of many a student adventure.

42. Wells was later keen to remove any record of his early work and bought up as many copies of the *Journal* as he could, and destroyed them. One of his student pieces, 'The Chronic Argonauts' was later reworked as *The Time Machine*. *The Phoenix* faded away in the 1960s and remerged briefly in the 1980s.

43. *The Phoenix* (January 1908), 26.

44. *The Phoenix* (January 1909), 37. Exams were still written with quill pens in this period.

45. The name 'Old Centralians' recalls the first name of the college, The Central Technical College. The three associations were brought together in 1972 to form also an Imperial College Association, while not disbanding the old college associations.

46. *Mining World*, 26 July 1913.

47. Minute Book of the RSMOSA, 25 January 1916.

48. Sir Edgeworth David (1858–1934), a former geology student, joined the Geological Survey of Australia, and later became Professor of Geology at the University of Sydney. He had conducted a number of daring expeditions of his own, including earlier ones to the Antarctic. One of these was a remarkable three-man trek of over 2,000 km to the magnetic south pole, using hand-hauled sleds. While on the Shackelton expedition (1907–9), he was the first to climb Mount Erebus.

49. *The Phoenix* (1910–11), 117.

50. The Pankhurst family began their movement with meetings at their home in Camera Square, Chelsea (no longer in existence, but it was located to the south west of Elm Park Gardens). Several students lodged in the area and became involved in the movement.

51. Some other students belonged to a University of London OTC before the war. Later, because of Imperial's problems with the University, many students shunned the University OTC. When Imperial joined the University in 1929, a new University OTC unit (Infantry Company C) was formed under Major P. C. Bull DSO, a lecturer in the chemistry department.

52. *The Phoenix* (new series, 1915), 1.

53. B/Wilkinson, handwritten memoir.

54. Harold Johan Thomas Ellingham (1897–1975), known as Tom, was a chemistry student at the college before leaving in 1916 to join the army. He served in Mesopotamia, 1916–18, learned Arabic, and was to become something of an arabist. He was promoted to a readership in 1937 but left the college after the Second World War. He was a major figure in the recovery of old records after the college decided to open a muniments room. I have relied on his meticulous notes in many places in this book. See B/Ellingham.

55. This was later expropriated by Wembley Council and the college bought land in Harlington.

56. Until the 1970s Imperial boats routinely defeated even the University of London team. But the latter became a powerhouse in the late 1960s causing some Imperial rowers to prefer rowing for the university than for the college. It was a University of London first eight (with Imperial representation) that represented Britain at the Mexico City Olympics in 1968.

57. Donald W. Morphy was an electrical engineering student from 1917–20. Arthur Morphy was his father. D. W. Morphy left the college to become a graduate apprentice at Metropolitan Vickers and later co-founded the firm of Morphy Richards. A faithful Old Centralian, he died in 1975. See *The Central* (Autumn 1975), 18–19. Morphy Day declined in popularity in the 1970s but continued into the early 1990s.

58. The boathouse was opened by Viscount Falmouth who had donated a large sum towards its construction. It has gone through many improvements since 1938. In the late 1970s women rowers finally had access to their own changing room and showers and could leave the boiler room, which had served until then. Further improvements were made in 1999 with some help from alumni and the National Lottery Sports Fund.

59. There was much camaraderie among the rowers, and more generally among students who attended the college during wartime. Something of rowing in this period is captured in the brief memoir of Edgar Cove (Mechanical Engineering, 1945) which can be read on the Imperial College web page (Alumni News/Alumni Memories) where it was posted 24 April 2006. Cove mentions that in 1995 the stewards of the Henley Regatta allowed six members of the original crew that won the Danesfield Cup to have a ceremonial paddle along the course, fifty years on.

60. William (Bill) Graham Mason (1950–) continued to make the college a major force in rowing. He was made an Honorary Associate of Imperial College in 1993. In 1996 the college won the Grand Challenge Cup at Henley.

61. The bottle, a model beer bottle, was 'liberated' from a Bass Charington beer lorry in 1946. Matches between the RSM and Camborne, however, predate this event having begun in 1921. They continue to this day. The Camborne School of Mines has been amalgamated with other colleges within a Cornish campus of the University of Exeter. Today a number of parallel sporting events are held alongside the rugby match.

62. Finch was a famous alpinist and was honorary president of both the mountaineering and sailing clubs.

63. Tata Industries is India's largest industrial conglomerate. One who recalled his own passion for motorcycles was O. V. S. (Peter) Heath, a botany and geology student in the 1920s, lecturer in the college during the 1940s and 50s, and later Professor of Horticulture at Reading University. See Heath diary; *op. cit.* (14), and Terry A. Mansfield, 'Oscar Victor Sayer (Peter) Heath, 1903–97', *Biographical Memoirs of Fellows of the Royal Society*, vol. 44 (1998), 219–35.

64. K MET; metallurgy professors correspondence file; Carpenter to Gow, 17 June 1921. Similar complaints have been made more recently about skate borders blocking entry to college buildings.

65. Peacock began working at the college as assistant to his father who was Clerk of Works. In 1908 he succeeded his father. In 1925 financial and other problems in the Union prompted the Rector, Thomas Holland, to look around for new management. Peacock held a number of money raising events such as Derby Day parties and, with better management, put the Union on a sounder financial footing. He was a major supporter of sporting events at the college.

66. Professor Farmer and his demonstrators often had lunch there. See, for example, Heath diary, *op. cit.* (14), 27 November 1922.

67. This was a tradition going back to before 1907. Professor Watts remarked that the V&A dining room 'formed a pleasant meeting place where many matters were quietly and amicably settled without speeches or resolutions. Farmer, Callendar, Perry and I were pretty regular attenders.' See David Williams (ed.), 'History of the Department of Geology', 10 (typed manuscript in ICA).

68. See Judith Niechcial, *A Particle of Clay: the biography of Alec Skempton, civil engineer* (Caithness, 2002), chapter 2. The restaurant was destroyed by a bomb during the war. It stood where a sculpture of a boy and dolphin now stands. Many students lived in digs on Oakley Street before the war. Skempton lived in the hostel during his first year at the college and then in digs, first in Cromwell Place and then in Prince's Gardens.

69. *Record of the RCSA* (Winter 1983–4), memoir by W. J. Gooderham. Bedford Park was an experimental Victorian suburb built in Turnham Green where many of the ideals of John Ruskin and William Morris were adopted. Some of the houses were designed by R. Norman Shaw, the architect who also designed 170 Queen's Gate.

70. Eric (Lord) Ashby was educated at the City of London School before entering Imperial as a botany student in 1920. He joined the staff as a lecturer before moving to Sydney as Professor of Botany in 1938. During the war he was a science attaché with the Australian embassy in Moscow. In 1950 he returned to Britain and the chair of botany at Manchester. He became Vice Chancellor of Queen's University Belfast in 1954, Master of Clare College Cambridge in 1958, and Vice Chancellor of the University of Cambridge, 1967–9. He was knighted and later given a life peerage. Ashby was major voice in mid-twentieth-century discussions of higher education policy. Despite his many later experiences, his was still very much an Imperial College voice, reflecting also some of the earlier views

of H. G. Wells which Ashby likely imbibed as a student. Ashby emphasized the need for universities to embrace the practical, and to be inclusive of the technologically skilled. Linstead later became Rector of Imperial College.

71. The orchestra had its ups and downs during the 1930s. During Sir Roderic Hill's Rectorship, Lady Hill, herself an accomplished violinist, helped to revive it and the orchestra began to flourish after the war under the leadership of violinist Harold Allan, lecturer in physics, and outside conductors Dennis Fry and Frank Kennard. Once again the orchestra began performing Gilbert and Sullivan operas in cooperation with the Dramatic Society. Later the Operatic Society took over these performances. See Harold Allan, 'Fifty Years of Music at Imperial College: a personal reminiscence' (college pamphlet, 2000; copy in ICA).

72. Sir Brian Thwaites, later Principal of Westfield College, was a keen keyboard player and persuaded the Governing Body to purchase a cheap piano for choir accompaniment. Roderic Hill purchased a further good piano for the college in 1950. The first choir conductor was John Clements, a former conductor of the BBC Theatre Chorus. Other conductors included Imogen Holst and Reginald Jacques. In 1953 Eric Brown, who had been the principal accompanist, took over the conducting role and built up the choir over many years. (Brown was a member of the civil engineering department and became Professor of Structural Analysis in 1971.)

73. See Harold Allan, 'Neville Blyth and Music at Imperial College', *Icon*, vol. 19 (1980), 7. Blyth was educated at Oxford and joined the RSM as a lecturer in 1949 after a career in industry. He helped renovate the Bessemer Laboratory for the teaching of mineral dressing. Blyth made a beautiful carved mahogany music stand that he left to the college. In 1990 Mrs. K. M. P. Blyth left £240,000 in her husband's memory to enhance the arts at the college. In 1996 his family donated a further sum bringing a total of £472,000 towards the founding of the Blyth Music Centre. The Centre was opened in March 2001 with a concert by the Belcea Quartet.

74. See Allan (note 71). Both Brown and Allan were active in the musical life of the college for many years. Students, staff, and professional musicians have performed in the lunchtime concerts, ongoing today. Brown played a major role in setting the specifications for the organ in the Great Hall. On 27 November 1975, Albert Ferber gave the 600th lunch hour concert, a wide ranging piano recital.

75. London jazz clubs feature in a number of memoirs of the interwar period. Students often spent evenings together listening to jazz at a number of

different venues. But a college club was formed only in the 1950s. It would also appear that there is some connection between the interest in both jazz and folk music and left-wing political views — but this is just an impression, hard to back up with concrete evidence.

76. Peter Rowe, 'The Union Library', *The Phoenix*, (Autumn, 1956). Today the Haldane Library holds about 20,000 items including books, CDs, etc.

77. *The Phoenix* (Spring, 1951), 11.

78. 22 Club papers, typed leaflet on its history. The dinner was held on 4 November 1944. One might question the statement that bombs (doodle-bugs perhaps?) were heard falling, but that is what Ellingham claimed.

79. Members included well-known college figures such as H. J. T. Ellingham, R. P. Linstead, H. R. Hewer, J. F. Levy who were all elected as students, and P. C. Bull and G. C. Lowry who were given honorary membership.

80. Chaps Club booklet, 1930–1.

81. Peter Mee was educated at University College London and came to the college in 1959 after national service. He continued his military association serving in the 44th Independent Parachute Brigade of the Territorial Army, 1954–80, retiring with the rank of Major. He was a major supporter of sporting functions at the college. He became College Registrar, 1967–96, Clerk to the Governors, 1990–7 and College Secretary, 1996–7. On retirement he stayed at the college as head of the college alumni relations office. Bernard Neal was Professor of Engineering Structures and head of the civil engineering department from 1973–82. Neal's nickname is probably a double entendre, having a sexual connotation while at the same time referring to his remarkable skill at croquet. Alan Kitchener was president of the ICU 1957–8.

82. ICA holds the annual booklets and minute books for all three clubs.

83. A later example is Kenneth Weale (1923–98). As an Imperial student he was captain of rugby and also a keen cricketer. He joined the chemical technology and chemical engineering staff in 1948, and was appointed Reader in 1970. Weale retained strong links to student clubs and was a convivial college man. He was Warden of Falmouth Hall, 1963–71, and on the Governing Body, 1973–77.

84. Among the better known Masons were: Rectors T. Holland and H. Tizard; College Secretaries A. Gow, G. C. Lowry and J. Corin; Union Manager J. G. Peacock; Professors P. G. H. Boswell, H. V. A. Briscoe, G. I. Finch, W. D. Gill, H. R. Hewer, J. W. Hinchley, V. C. Illing, W. Jackson, W. R. Jones, J. F. Levy, W. H. Merrett, L. Owen, H. H. Read, D. Williams; academic staff members N. Blyth, A. Bramall, P. Bull, H. J. T. Ellingham; and student Herman Shaw, later Director of the Science Museum. At first,

membership was not cheap. In the 1920s the initiation fee for students was £10 and for others £20. Annual subscriptions were £17. But in order to attract new members the annual fees were dramatically reduced to two guineas in the 1930s.

85. A conclusion arrived at after reading the minutes of this society. Clearly those who invented new and meaningful rituals were much admired.

86. See C. W. Dannatt, 'The Royal School of Mines: some records and recollections', speech given at 100th annual dinner and ball of the RSMA, 23 November 1984. The thermite reaction is one between aluminium and iron oxide which produces aluminium oxide and iron plus much heat.

87. ICWSA was later renamed the Imperial College Women's Association (ICWA), acknowledging also staff membership.

88. Martha Whiteley was an assistant professor (reader) in the chemistry department. Lady Holland, Lady Thorpe, Mrs. Philip and Mrs. Baker were all active supporters of ICWSA. See Marion Gossett, 'ICWA Then and Now', *RCS Handbook* (1974–5). Gossett was President of ICWSA in 1923.

89. Women's rugby teams appeared only in the 1980s.

90. ICWA minutes, 8 May and 10 October, 1945.

91. ICWA was formally abolished in June 1981. Women then joined the ICU on equal terms.

92. This Boanerges was a 1904 Rover. In 1933, beyond repair, it was replaced by a 1902 James and Brown car. Both these cars participated in the annual Brighton run and in a wide range of college activities. On 9 July 2002 the second Boanerges spent its 100th birthday being driven along the back roads of France from Boulogne to Paris. A back-up team was needed to rebuild the engine, but the car made it to Paris. A letter was delivered from the Rector, Sir Richard Sykes, conveying greetings to the École Polytechnique.

93. B/Lowry; see G. C. Lowry memoir. See also Heath diary, *op. cit.* (14), 19 March 1925.

94. B/Lowry; See G. C. Lowry memoir for mascotry and student rags. The Albert Memorial procession was filmed by Phillip Bull, a lecturer in the chemistry department.

95. The first record of this game that I have found was for Armistice Day, 1918. On that occasion the pitch was Prince Consort Road and the goals were at Exhibition Road and Queen's Gate. But the game was probably played even earlier.

96. The Rector's lawn was the future site of the Roderic Hill Building, opened

in 1957.

97. The General Strike was not general but selective. The TUC brought out key workers in the power, transportation and communication industries in support of the miners. The strike lasted nine days. By the 1960s and 70s even those students who did not openly demonstrate against traditional authority were largely unsympathetic towards it.

98. B/Linstead; notes by Humphrey Hewer

99. These several claims are based on a reading of student journalism of the period. A 1934 issue of the student magazine, *The Phoenix* (vol. 43, No. 1, 1934), has an interesting cartoon depicting factory-like working conditions at Imperial with the caption 'The Royal College of Science is wondering why Hitler thinks IC is a kind of concentration camp.' Himmler established Dachau under his SS in the summer of 1933, the model for later camps. By 1934 there were others which, at that time, were used mainly for the incarceration of communists and left-wing dissidents. While most Imperial students were not political, opposition to Hitler was early expressed in student publications, and clearly the cartoonist was aware of what was happening.

100. *The Phoenix* (March 1940), 4. In his diary, O. V. S. (Peter) Heath records watching a crowd leave the college to watch a C&G vs RSM rugby match on a Wednesday afternoon in 1923. They departed in fancy dress in four double-decker buses. About twenty people accompanied them on motor bikes all dressed as members of the Ku Klux Klan, 'most effective!'. There was 'much shouting, police rattling, throwing of flour bags and paper streamers, and setting off of fireworks'. Heath diary, *op. cit.* (14), 7 March 1923.

101. Some students took the excitement in their stride. Alan Woodworth Johnson FRS, a research student during the war, recalled sitting in a fifth-floor flat just off Queen's Gate during bombing raids, listening to music oblivious to the din and danger all around. (See Johnson's FIC citation, 1972.)

102. GJ/2 Southwell Memorandum, 15 September 1944. Subsequent Rectors did not wish to live in the Beit Hostel and the college later purchased a house in Pembroke Gardens which was used by the Rectors until the college acquired 170 Queen's Gate back from its leaseholders in the late 1960s. Before that the building was occupied by a secretarial college, predictably known as the 'House of Sex'. For quotation see *The Phoenix*, vol. 53 (1945), emphasis in original. Southwell contacted W. N. Stocker at Brasenose College asking whether he could buy some Port for Imperial's

cellar. GJ/2/1 Southwell to Stocker, 11 December 1944.

103. GB minutes (1944). Southwell, *op, cit.* (102).

104. Selkirk Hall opened on 24 October 1946. It was made possible by a donation of £17,200 from mining engineer William Selkirk who had been a student at the RSM. The name of the hall was later transferred to one in Prince's Gardens. Selkirk donated a further £12,000 for student scholarships in mining. See Rector's Summary Reports for 1945 and 1946.

105. Annual Report, July 1945. Southwell noted that this was achieved at the cost of £200 per year per student as compared to £200 per twenty-four weeks at Oxbridge.

106. Rector's Report 1947.

107. *The Phoenix* (December 1946), 5–11.

108. Rector's correspondence; Bickley to Southwell, 24 January 1945.

109. Cambridge, especially, was very parochial until fairly late in the century. The modern university with its science research parks, global business links, and entrepreneurial activity would have been unthinkable not so long ago.

110. *The Phoenix* (December, 1946), 17–18.

111. James Newby (1892–1978) joined the college in 1913 as a clerk in the C&G Registry. He continued his work on vacation training until his retirement in 1960, and was awarded the FIC in 1967, See *Topic* (15 May, 1978). IAESTE continued to be run from Imperial College after his retirement but was always short of money. Both William Penney and Brian Flowers were very supportive and both became presidents of IAESTE during their Rectorships.

112. *Daily Telegraph,* 10 December, 1984.

113. For a few years male RCS students chose a 'Queen of Jezebel' from among the female students. But already in the later 1950s feminist sensibilities worked against this. Most women students stood back from entering anything like a beauty contest. To keep it going men entered the contest during the 1960s, often in drag. The contest continued as farce into the 1970s.

114. One story is that it was cast from spent cartridges from the C&G Rifle Club, but this is debated.

115. The Kangela was reputed to be a Swahili fertility chant and was adopted in the mid 1950s. It had been used already in the 1940s by the RCS Boat Club. The C&G had a chant known as the Boomerlaka and the RSM its Miner's Song.

116. *The Phoenix* (Summer, 1957) 41–2.

117. A number of 'continental' restaurants had opened during the war to serve

displaced communities; for example, the Daquise, a Polish restaurant still in existence. Asian restaurants also began to appear. The Rice Bowl (no longer in existence) at South Kensington Station was a major draw in the 1950s and 60s.

118. B/Black; Alan Kitchener to Martin C. Black, 27 September 1972. Black had invited Kitchener to return for the opening of the newly refurbished bar but Kitchener was unable to attend. In his letter Kitchener recalls the earlier opening.

119. Quoted in *The Phoenix* (December 1960). By the time of the 1976 Notting Hill riot, the accommodations situation for non-white students had improved considerably.

120. Quote from *The Observer*, 16 October 1960. The Guyanian referred to here should not be confused with Trevor Phillips, also from Guyana, and an Imperial College Union president (1975–7). A chemistry graduate, Phillips was the first black president of the NUS (1978). He later became editor and presenter of London Weekend Television's (LWT) current affairs programme 'Eyewitness'; then head of current affairs at LWT before becoming an independent maker of TV documentaries. In 2003 he was appointed by the government to chair the Commission for Racial Equality and has since been given a knighthood.

121. Irving had come to the college as part of a new venture to attract good arts students into the sciences. These students were given instruction during their first year at the college designed to allow them to enter regular first year science courses in their second year. This experiment was abandoned after a few years.

122. Quotation in *Felix*, 6 May 1959. Information on the content of the missing pages is taken from issues of *Felix* in the same period. Irving's actions were a hot topic for several weeks.

123. Irving's love affair with Hitler was evident already when he was a student, though at the time he appears to have been viewed simply as a crank, albeit an intelligent one. On failing his exams, Irving did not immediately leave the area but continued working as a labourer on the new physics building site, something he had already been doing for about a year.

124. In 1975 about 10% of the student body of approximately 4000 students were women.

125. In 1952 the Sir Arthur Acland essay prize was awarded to A. Conway for his essay, 'An Excursion in Ethics'. It was about the need to unify humankind and for a planetary ethos to be created. Political utopianism of this kind reflected postwar anxieties but was no longer fashionable by the

1960s.

126. C. P. Snow, *The Two Cultures and the Scientific Revolution*, (Rede lecture, Cambridge, 1959). See also chapter 14.

127. This club was very much something of its time and did not survive as an active site of exchange. It became something of 'old boys' group later in the century.

128. The late 1950s and early 1960s was a heyday for the RCA with graduates such as David Hockney and Peter Blake causing a sensation at the 1963 Venice Biennale.

129. This was the idea of Sir Tyrone Guthrie, Director of the Royal Court Theatre, who held the view that contemporary theatre had to take note of the fact that it existed within, and was reflective of, a scientific and technological age.

130. Maurice Marples, in his *University Slang* (London, 1950) credits Imperial College with the origin of the following terms: brown bagger, amber (light ale), dep (department) and soc (society). A 'brown bagger' was someone who came to the college only during working hours and then returned home. Brown baggers not only carried brown brief cases, they carried their lunches in brown paper bags, symbol of a lack of engagement in college life. The term first appeared in *The Phoenix* in 1918 and was later taken up by H. G. Wells.

131. First verse of a poem by Ben Jacob, *The Phoenix* (Spring 1954), 21.

132. Jackson gave many talks to student societies at this time, especially to those with an engineering interest such as the Radio Club. This was a technical club; a campus radio station run by the Union opened in the mid 1970s. STOIC (Student Television of Imperial College) began in 1971, at first using a studio available to all educational institutions in the Greater London area. Barnes Wallis was a frequent visitor to the college in this period and a major booster of British technology. He appears to have held the unrealistic hope of Britain regaining technological supremacy over the USA and the rest of the world.

133. The Wine Tasting Club continues strong. Its teams were winners in several of the Peter Dominic Inter-University wine tasting competitions in the 1990s.

134. See also chapter 14.

135. The expeditions were often dangerous and in 1961 four students lost their lives in a drowning accident off Jan Mayen Island, and three in a climbing accident in the Swiss Alps (reports in *The Times* 28 June and 15 July, 1961). The 1961 expedition was the second to Jan Mayen Island. An earlier one, organized in 1938 by chemistry lecturer Alexander King, included also

botanists, geologists, meteorologists and geologists. Earlier King had been the editor of the short-lived irreverent student magazine the *Muckraker*. Henry Tizard was so impressed by King's organizational abilities that he lured him away from academic work to a role in wartime Whitehall. In 1968 King was one of the founders of the Club of Rome (which had much representation from Imperial College) and was later made its honorary president. He was made a FIC in 1992.

136. Reports of the expeditions have been published and can be found in the ICA. Robert C. Schroter was a long-time chairman of the Exploration Board. Schroter was a student in the chemical engineering department where he gained both a BSc and PhD before joining the new Physiological Flow Unit in 1966. He later became Professor of Biological Mechanics in the Department of Bioengineering.

137. In 1958 the government set up a committee under Sir Colin Anderson to examine and clarify existing financial support for students. The committee report recommended a new centralized system of support to replace the old system of county and state scholarships. This committee, not anticipating the huge increase in student numbers, suggested that all students accepted at university should be eligible for a maintenance grant (dependent on parental income). It also proposed that the then modest university fees be lumped in with the awards package; the proposals were accepted. Later it proved very difficult to disentangle fees from maintenance provision. Anderson Committee Report, 'Grants to Students' (Cmnd 1051, 1960).

138. Perhaps the start of the 1960s band craze was over the San Francisco acid rock group, Jefferson Airplane, and the homegrown Beatles. Their 'all you need is love' struck a chord with a generation opposed to the Vietnam War and in favour of both sexual and social revolution. Students did not yet see the irony in their support of a burgeoning mass marketing industry. Tidu Maini proudly remembers organizing the 'kidnapping' of three of the Rolling Stones, including Mick Jagger, during Rag Week in the late 1960s. They were 'ransomed' for £100 which was donated to charity. Maini was a civil engineering student at the time. He graduated in 1966, returning later to take a PhD in rock mechanics. One of his professors delegated him to solve the problem of a leaky wine cellar at Lord Penney's home near Oxford. Penney became something of a mentor to Maini who maintained close ties to the college during his subsequent career in management consultancy. Maini returned to Imperial as Pro-Rector, Development and Corporate Affairs in 2001. He has been a president of the C&G Association (personal communication).

139. Only May was an Imperial student.

140. The four were: Don Monro from the electrical engineering department, David Burgess, Reg Garton and Gavin Park from the physics department. Park was the conductor.

141. The name Exmoor was a play on the fact that Alan Moore, who began by organizing a barbershop style quartet, reorganized the group's various singers after leaving the college.

142. At first Dickins was given only a half-day per week appointment. Dickins was a postgraduate student at the Royal College of Music at the time. He now holds a half-time position at Imperial and is engaged in a wide range of musical activities.

143. Twelve years earlier the editor of *The Phoenix* was Anthony Hodgson who applied to the University of Birmingham to do postgraduate work in chemistry. He received a letter from Professor Stacey, the head of department, stating that he did not like beards on young men and 'expects his postgraduate students to express their personalities through their inorganic chemistry and not through their unorthodox Bohemian-type appearance'. See, *Phoenix Review* (1957–8). By 1968 such a letter would have been unacceptable, though some academics would still have held Stacey's point of view.

144. While women increasingly wore jeans, mini-skirts and long skirts were also popular in this period. While the beatnik culture of the 1950s hardly penetrated college life, the hippie culture of the 1960s had a major effect on the appearance and tastes of students.

145. Imperial College Union, President's Annual Report 1969–70.

146. The Union had the aim of 'creating a socially relevant co-adult academic community ... necessitat[ing] student participation and representation at all levels in the college'. The Queen passed the petition to the Secretary of State for Education and Science. For a report of this event see *The Times*, 28 November 1969.

147. This allowed the union president to be registered at the college for the year without paying fees and without attending classes. Corbyn also negotiated a retroactive sabbatical for himself. The college paid union presidents a grant of £150 and the union paid a grant of £200. Later other union officials were also given sabbaticals.

148. In 1972 Alastair Cameron, then Reader in Lubrication, and a warden in Prince's Gardens, held a tea party at his flat to which a number of students were invited to meet Margaret Thatcher, Minister of Education. She wanted to hear about the NUS and its affiliates and was concerned that their business affairs be conducted well. A new NUS constitution was

being considered at the time. Cameron reported to the Rector that the students 'had been completely charmed by her'. Mechanical Engineering Department correspondence file; Cameron to Rector, 7 January 1972.

149. Select Committee on Education and Science. Sub-Committee C. Minutes for session 1968–9, 14 May 1969.

150. For example, Jeremy Sammes, of the Staff Christian Association, successfully lobbied to have the annual Commemoration Day service moved from Holy Trinity Church to the Great Hall. The student Christian societies, too, wanted a say in who was to give the sermon/address. While, in 1971, the combined Christian societies agreed to ask the Rector to invite Bishop Mervyn Stockwood to address what was the first service to be held in the Great Hall, the Anglican link was loosened and people of other denominations were increasingly invited to conduct or address services.

151. Tariq Ali, Pakistani, Trotskyite, controversialist and editor of the lively and unorthodox magazine *Black Dwarf*, ('Paris, London, Rome, Berlin: we shall fight, we shall win').

152. 579 voted for the motion, 50 against and 21 abstained. For a report of this event see *Daily Telegraph*, 1 February 1969. In 1969 the Communist Society had a membership of twenty.

153. Early pointers to the concern with over bureaucratization can be seen already in student publications of the late 1950s where fun is made of new terminology: 'accommodation units' instead of 'rooms' or 'houses', 'rodent operative' instead of rat catcher etc. See, for example, *The Phoenix* (Spring, 1957).

154. GSGH/ Holland Club papers.

155. See FC and EC minutes, 4 October 1968. The Holland Club bar opened its doors to the larger community, moving first to the west of the College Block and then, in 1977, to the basement of the Block, renamed Sherfield Building. Further space was found in the basement of the new Huxley building for a coffee and snack bar, and for billiards, ping pong, and darts.

156. Refectory services are open for weekday lunches. In the evenings, and at weekends, they open only for special occasions.

157. Later, a major project of the 1990s was the purchase of the Clayponds estate in South Ealing for £9,855,000. It was refurbished for student residences. The 138 units of one and two-bedroom flats brought the number of student residency places to 2245. See FC and EC minutes, 10 May 1991.

158. See *RCS Broadsheet*, 31 October 1969. About £60 was raised for charity. One former student recalled the tiddlywinks, and also helping to knit a scarf which was unfurled the full length of the Piccadilly Line escalator at Piccadilly Circus to draw attention to a group of students who were carol

singing there. See Mary Freeman, Imperial College Alumni News website. Tiddlywinks along Oxford Street was later stopped by the Metropolitan Police on the grounds of it being too disruptive.

159. In the 1970s some students in the C&G Motor Club captured the Three Peaks record by completing a journey from sea level at Fort William to sea level at Caernarvon via the summits of Ben Nevis, Scafell Pike and Snowdon using motor transport between the mountains. Their time of 14 hours and 6 minutes was 89 minutes shorter than the previous record.

160. See photograph in *Felix*, 18 May 1984.

161. The Sports Centre in Prince's Gardens was opened in 1968. A new and enlarged centre, named *Ethos*, opened in 2006.

162. Much was made of the Imperial Eight winning the Grand Challenge Cup at Henley in 1996 and that three of the men rowed also in the boat that won the gold medal for Britain at the Sydney Olympics in 2000. One of them, Louis Attrill, is said to have learned to row only after coming to Imperial.

163. Martin Black, secretary of the ICU in the early 1970s, has left an interesting file in the archives. A convivial man, he belonged to many student clubs but his papers reveal frustration that others did not commit themselves, as he did, to college events.

164. Imperial teams were successful and, in 2001, won the Challenge for the second time in five years.

165. The club was named for J-B. J. Fourier (1768–1830). The temporary suspension of the College Secretary position and the appointment of Angus Fraser as Managing Director was seen as an attempt to redefine the college. It caused much distress and was the impetus for the formation of this club. The distress had little to do with Fraser personally but was a consequence of difficult economic times and increasing commercialization of the college. Harsh measures were taken to save money, including the contracting out of work. Some long-term employees lost their jobs.

166. The Southside residences in Prince's Gardens had construction problems almost from the start, and they were very expensive to maintain. They have recently been demolished and will be replaced with new residences. The plan is to have the new buildings more consistent in style with the remaining Victorian buildings, and with the idea of a garden square.

167. The Ennismore Arms was another favourite until its recent closure. Student drunkenness is a perennial phenomena. But the behaviour of young people, even in this, has changed. Today, with an anti-establishment popculture having become mainstream, we see the extension of youthful behaviour patterns into near middle-age and, accompanying this, increased drunken-

ness.

168. Newman quoted in W. Moberley, *Crisis in the University* (London, 1949), 33.

169. The college community is now largely dependant on a newsletter (*Reporter*) put out by the administration.

chapter twelve

The Making of the Modern College, 1967–85
Part One: Governance in a New Political Climate

Introduction

One way of seeing the Rectorship of William Penney is in terms of his response both to government directives and pressure from the young. The government, post-Robbins, was spending increasingly large sums of money on higher education and wanted more control over university expenditures. At the same time, a new postwar youth culture that had been growing already in the 1950s spread rapidly through university campuses during the 1960s. Many young people joined national and international causes such as the peace movement, including support for nuclear disarmament and opposition to the Vietnam War, and the civil rights movement, especially support for the anti-apartheid cause in South Africa. Students and young members of staff appointed during the expansionary years also sought radical reform in university governance. By 1968 students had become big news. While events at Imperial were a pale reflection of those at the LSE, let alone those in Japan, Paris or, earlier, at Berkeley, Penney was taken aback by much that happened.[1] He was basically sympathetic towards the more liberal aims of students and young staff members, but disliked their methods, appearance, and noisy behaviour.[2] Indeed, new behaviours and lifestyles were a shock to many of his generation. Students appeared to have changed overnight. Penney's term of office was marked by a number of *ad hoc* responses to campus events, only some of which had lasting effect. Yet he did initiate collective thinking on possible future directions for the college, the fruits of which were to be of great use to his successor.

The 1960s was also the decade which brought a clearer, and for many a not altogether comfortable, understanding of Britain's post-imperial status and loss of power. Further, while many Britons were emigrating to the 'old' Commonwealth, others from the 'new' Commonwealth were immigrating to Britain and challenging traditional ideas of what it meant to be British. British identity was challenged also by moves to make Britain a member of the EEC, moves which succeeded finally in 1973, the year that Brian Flowers became Rector. One could perhaps argue that 1973 marks the end of the postwar period and the start of something new. This appears to be true of Imperial College where Flowers, very much a realist, brought in reforms that gave the college a more modern outlook. This was no easy task since, in addition to the social and cultural pressures just mentioned, 1973–4 saw the economy dive into a recession unequalled since the early 1930s.[3] Inflation was a serious problem already in the 1960s and Harold Wilson, having tried a number of other measures, devalued the pound in 1966. But deflationary policies were hard to sustain in a climate of trades union militancy, and at a time when old non-competitive industries were clearly failing. Even Rolls-Royce, an icon of British quality, went into receivership in 1971 and was taken over by the government. Further, having joined the EEC in 1973, Britain was obliged to phase out both domestic food subsidies and preferential tariffs for Commonwealth goods.[4] 1973 was also the year of the Yom Kippur War which led to a temporary oil embargo against the United States, followed by major increases in the price of oil imposed by OPEC (Organization of Petroleum Exporting Countries).[5] All of this, together with serious problems in ongoing government negotiations with the National Union of Miners, made the recession in Britain particularly severe. When Harold Wilson returned as Prime Minister in 1974 he was faced with a social and economic picture far worse than in his previous term. In 1975 inflation came close to 30%. Partly as a result of economic uncertainty, the old quinquennial university grant system ended in 1976 to be replaced by annual grants and higher tuition fees.[6] Even tougher measures came with the first Thatcher government. The recession did not fully bottom out until the early 1980s, by which time Britain's investment in North Sea gas and oil was finally paying off.[7]

In addition to dealing with a recession, the Wilson government was under pressure to improve working conditions and benefits for employees. Two new bills were introduced which were to affect the college and make its financial problems even more acute — though its working conditions better. The Health and Safety at Work Act (1975) meant, for example, that chemicals had to be stored under more secure conditions, and new ventilation systems and fume cupboards installed. The Employment Protection Act (1976) introduced maternity leave and benefits also for part-time employees. In 1976 the VAT was introduced at 8% and, one year later, national insurance payments were increased by 23%.[8] All the added expense, which included having to prepare invoices in sextuple, together with more general inflation (fuel bills more than doubled in the period 1972–5, and refectory prices jumped by 40%,[9]) led to a hiring freeze and a stop to routine building maintenance.[10] Given the inflationary times, and the fact that the college spent about 88% of its annual income on salaries, there was no alternative to these measures.[11] A programme called SLICE (save lighting and Imperial College energy) was introduced but made little impression on climbing deficits.[12]

Hiring freezes were in place elsewhere, not just at Imperial. Universities were unable to recruit young staff needed for rejuvenation of the system and a generation of academic talent had to find alternate means of employment. While the recession began to ease in the early 1980s, the Treasury demanded still more cuts in the 1980–81 spending round. There was some alleviation with so-called 'new blood' money for the appointment of young lecturers in 1982. Because there was especial concern over science and technology, universities also received an extra £47m from the Department of Education and Science (DES) to be used in those areas.[13] But the UGC was under stress and its members were being asked to resign in protest. As money became tighter the UGC became more intimately involved in the running of individual universities, having to take decisions on departmental closures and so on. While the overall situation was grim, Imperial received preferential treatment and escaped the worst; but, from the early 1970s to the late 1980s, times were tough. Already in the late 1960s monies for completion of the Jubilee expansion were not forthcoming, and new buildings planned for mathematics and chemistry were left in abeyance.[14] By the end of the decade, after difficult negotiations, and with much cost-cutting, construction on the mathematics building was allowed to begin.[15] But the chemistry building was not completed until the 1990s. The late 1960s saw also the National Incomes Commission looking closely at academic pay, and with negative consequences. No wonder the period also saw an increase in union activity on UK campuses.

These difficulties were compounded by doubts over expansion. During the 1970s the demand for university and polytechnic places in science and engineering had not kept up with what had been projected by the Robbins committee.[16] Further, from the mid-1960s to the late 1970s, many graduates emigrated to find well paid work. There was a 'brain drain', mainly to the USA.[17] Newspapers published angry articles and letters about it, and about graduates driving minicabs. Some graduates were finding work in financial institutions, a trend that grew in the 1980s. This was justified *post hoc* by Brian Flowers, who stated that it was a good thing to have numerate and technically trained people in the City.[18] Clearly there had been some miscalculation by the Robbins Committee, since good jobs for highly trained scientists and engineers were in short supply. Robbins had imagined that much of the planned university expansion would be in science and engineering, and that graduates would find work. In fact the major expansion of university places during the 1960s and 70s turned out to be in the arts and social sciences. The Imperial College Appointments Board, taking seriously the problem of graduate employment, began collecting statistics and working also to achieve better liaison with both departments and prospective employers. The 'brain drain' as it affected the academic world was further exacerbated by an old rule that had been introduced by the UGC after the Second World War to help prevent universities from raiding each other's staff. The ratio of senior (professors, readers and senior lecturers) to junior staff had to be one to two. For many years this rule had been ignored but, given the financial exigencies of the period, the UGC decided to enforce it. In 1968 the senior staff at Imperial comprised close to 40% of the total and the college agreed to a target of 37.2% by 1972.[19] Not surprisingly, many junior staff saw the writing on the wall and looked to North America and elsewhere for advancement. Given that the college had recently undergone a major expansion and had begun to see itself more clearly as a major research institution — one central to the life of the nation — the loss of younger staff was especially galling.

In the 1970s academics were further discouraged by government planning on research funding. A 1972 Government Green Paper, 'A Framework for Government Research and Development' contained two reports, one authored by Lord Rothschild and the other by Sir Frederick Dainton.[20] Rothschild recommended that some of the research paid for by the government should be on a 'customer-contractor' principle.

He suggested that there be a 25% decrease in funding to the research councils, and that the money withheld be redirected towards government departments for contract research.[21] Dainton had to take on board recent criticism of the research councils and, while supportive of them, and against government interference, he made some suggestions as to how they could be made more efficient. For example, to counter a move by the Ministry of Agriculture and Fisheries to take control of the ARC, he suggested instead that there be a new chartered body, an overall Board of Research Councils, which would be answerable to the Ministry of Education and Science, and would keep the broad policies of the councils under review.

There was one thing on which close to everyone at Imperial was agreed, namely opposition to the report by Lord Murray on governance within the University of London.[22] The Murray Report went nowhere, but much time was expended in the early 1970s replying to its various recommendations. Had they been implemented, Imperial would have had to give up some of its independence won in earlier struggles. Murray was overly guided by the interests of the smaller colleges within the University and was not enthusiastic about Imperial and other large colleges having seats on the CVCP. He recommended far more central control and consolidation of resources than then existed, and even suggested withdrawing Imperial's direct access to the UGC.[23] Even before Murray had been asked to look into the governance of the University, there had been discussion of a 'University of South Kensington'. While in many ways this was the continuation of an old debate, renewed interest in merging with some of the other local colleges, such as Queen Elizabeth College, the Royal College of Art, and the Royal College of Music was related also to the recession and declining enrolments.[24] It was thought that greater size would be advantageous in a shifting economy. But some people were motivated by a sincere dislike of the University of London's federal system which to them appeared to confer few, if any, advantages. For example, R. K. S. Wood wrote to the Rector,

> as the head of the largest university group in plant pathology in the UK … I get no advantage and I confer no advantage in belonging to the University that I could not get or confer if Imperial College were independent.

He wanted 'release' from what he called a time-wasting 'bureaucratic machine'.[25] Professor Ubbelohde, too, wrote several letters to the Rector stating that he wanted Imperial to have the kind of independence enjoyed by Loughborough or Brunel,

> our membership of London University does a vast deal of harm and has atrophied our own muscular development. I do not think we can ensure communication at all levels at Imperial College without the *full infrastructure enjoyed by any modern university, on our own campus*.[26]

But others, such as R. M. Barrer, head of the chemistry department, and R. N. Pryor, professor of mining, wanted to remain within the University.

The Board of Studies in its submission to the University stated that if Murray's proposals were accepted it would urge the college to seek status as a separate university. After the shelving of the Murray Report the issue of Imperial's independence disappeared for a couple of years, eclipsed by more pressing issues. But, in 1975, it erupted once more, and serious discussions began within the college and the university on what changes were needed to the university statutes. H. E. Cohen, Professor of Mining, and Chairman of the Academic Staff Assembly, was asked to head a working group to consider the interests of Imperial. Cohen received many letters, including one from T. R. E. Southwood. Critical of the 'Byzantine complexity' of university governance, Southwood's letter was itself somewhat byzantine; but it summed up the frustration that many felt with the existing system. Southwood recalled that Owen Saunders had earlier recommended that the other schools in the university be given some of the independence that had already been won by Imperial. But, while some independence had been given, the university began

> immediately weaving nets to pull everyone in again and Imperial College (in science at least) got partially caught along with the rest. Since then countless committees have met at great cost in both time and money, and we are largely where we were fifteen years ago. [The university is] too disparate to be governed effectively and [the present governance] is a web of Byzantine complexity designed to trap the smallest common denominator.

Southwood went through all the usual arguments made in support of being part of the university such as shared resources, but noted that many resources were shared also by outside institutions, and that Imperial could well become one of those without ill effect. He also made the point that in his field, zoology, as opposed to Cohen's field of mining, there were over twenty schools and medical schools involved in decision making, and that this was absurdly time consuming. He wanted 'root and branch' reform and that the larger schools needed 'dominion not colonial' status.[27] The University Senate agreed on a private bill to go before parliament in 1976. Even though the bill contained no amendments to the university statutes, and was designed simply to allow the university to make changes to its constitution internally, the AUT petitioned the House of Lords to block passage.[28] The AUT stated that its London membership wanted more protection for the independence of the Schools, and better representation of its membership on Senate, since it was there that future decisions on the statutes would be made. But the bill passed, and a new University of London Act came into law in 1978. Major statute reforms were then discussed but not fully achieved until the end of the century.[29]

In 1970 Jack Straw, a future Labour government minister, was president of the NUS. He declared higher education to be a 'right' and did not focus on the defects of the Robbins projections. Instead he criticized the failure of the government and institutions of higher education to meet the projections with respect both to access to higher education and student housing. Straw's paper was in some ways prescient, and interestingly anticipatory of the views of Kenneth Clarke, Minister of Education in John Major's government. Straw advocated breaking down the barriers between the universities, polytechnics and other institutions of higher education, and the creation of a single national funding agency. His ideal was the 'polyversity', the higher education equivalent of the comprehensive school.[30] But, by 1970, the expansionary mood had passed. As Harold Perkin has argued, the zenith of professional society occurred in the years after the Second World War when a new consensus on the nature of society and the welfare state provided work for increasing numbers of civil servants, doctors, social workers, teachers and academics. By the 1970s the consensus was at an end, partly the result of the events mentioned above. In addition, according to Perkin, there was a backlash within the private sector towards what was increasingly seen as the complacent and arrogant public professional.[31] The backlash reached its peak during the Thatcher years, but it was clear already earlier that state funded institutions could no longer function as before.[32]

In 1983, and at the end of his term as chairman of the CVCP, Sir Edward Parkes claimed that the Thatcher government had not come in with any developed philosophy of higher education.[33] This was true; however it did have a clear commitment to reducing public expenditure despite recent improvements in the nation's economic situation. Universities were not immune from cuts and they arrived soon after the 1979 election. There were cuts to Imperial's budget of about 2% for each of the years 1979–80 and 1980–81, at a time when student enrolment increased by about 3.7%. Staff numbers were cut by about 10%.[34] By the Spring of 1981 there had been an overall cut of 8.5% in the recurrent grant. Part of the reduction was due to the fact that the government began to withdraw that part of the grant seen as a subsidy for foreign students.[35] The idea was for universities to charge foreign students whatever they considered reasonable up to the full recovery of costs. At Imperial the number of foreign students declined by about 30% in 1981 and by a further 15% in the following year. Further, universities were not compensated in full for properly negotiated pay settlements recently completed, something unthinkable until it happened. The consequence of the many cost cutting measures was that universities had to contract even faster than by natural attrition. To achieve this the UGC asked the government for money to buy out staff willing to take early retirement. But this was problematic since it was not always those that the universities wanted to retire who did so.[36] One of Imperial's governors, Caroline Benn, was concerned that many more technical than academic staff were being made redundant. She was an activist governor and had a number of causes at this time, including improving both access for women and working conditions for technical staff.[37]

Shrinkage also gave government the opportunity to push the question of tenure, asking universities to consider its abolition, or at least limiting its term. In 1981 a Day of Action was organized by the three major university unions (AUT, ASTMS and NALGO). MPs were lobbied on the cuts and rumoured changes to tenure rules, and a large rally was held. But the protest did little to change the government's mind. Since none of these issues invoked much response from the public or the media, vice-chancellors collectively realized that they had a serious public relations problem on their hands.[38] The Minister of Education, Sir Keith Joseph, was blunt. In a letter to all vice-chancellors he stated that he was concerned over their reluctance to move on tenure, and that the government was prepared to bring in legislation if nothing was achieved voluntarily. He was of the opinion that such legislation should be welcomed by universities since it would provide them with the means of getting rid of unproductive people. Shrinking the university sector, he believed, would lead to improvement in quality and efficiency.[39] Needless to say those working in universities disagreed, as did middle class parents wanting higher education for their children. The late 1960s to the mid 1980s were difficult times, not only for universities but for the country as a whole.[40] At Imperial College it was the task of Penney and Flowers to lead in those times.

The Penney Years, 1967–73

Fortunately William Penney was good with people and managed to navigate the college through the shoals of the late 1960s and early 1970s without any major upset.[41] When he became Rector one of his first decisions was to appoint fellow mathematician, Harry Jones, as Pro-Rector. When Jones died in 1986, Penney wrote to his widow recalling that in the late 1960s 'the students were sometimes uncontrollable, the heads of departments were bewildered and angry, the rest of the staff, including the technicians, wanted a say in how everything was run, Harry had the respect of everyone'.[42] The same could be said of Penney. As an old student of the college, and with a distinguished career behind him, Penney was highly respected despite his association with nuclear weaponry and an increasingly unpopular nuclear industry.[43] In his letter to Mrs. Jones, Penney captured the earlier mood well. He had much administrative experience but, on coming to Imperial, was faced with a new situation in having to deal with angry students and a demoralized staff. His quiet approach and light administrative touch were likely an advantage. Shortly after his arrival, Penney, with the prompting of Lord Sherfield, Chairman of the Governing Body, set in motion a College Appeal. This was launched in 1968 under the direction of Air Vice Marshall A. A. Adams. As with the 1945 Appeal, it was targeted mainly at improving student services and housing. It proved difficult to raise much money, and while much of what was raised was spent on additional student housing, this did little to reduce student discontent.[44] Linked to the appeal was the first of a series of attempts by Penney and subsequent Rectors to improve ties with the alumni.[45] The college had a very small endowment fund and Penney was hoping to increase it. Even today, income from endowments is a small fraction of the total. At the same time as seeking ways to raise new money, Penney sought ways to save. He set up a working party, under Professor J. G. Ball, to look into how the college deployed its financial resources, and set a savings target of £200,000 for the 1968–9 year.[46]

Despite all the financial difficulties, Penney clearly wanted to make some changes. He soon asked the Board of Studies to suggest ways to improve college organization, stating repeatedly that he found the college somewhat old-fashioned, and its structures of governance antiquated. While treading carefully he managed to suspend the Delegacy that had been put in place to oversee the affairs of the City and Guilds College when Imperial was founded.[47] Another problem that Penney identified was departmental isolation. He blamed it on the fact that departments were located in separate buildings, and that departmental heads had complete control over the use of space within those buildings.[48] He ordered a review of how college space was being used and slowly moved control over space to the central administration.

Penney overstated the degree of departmental isolation, Imperial had encouraged interdisciplinary research from the start. However, that there should be more interdisciplinary work became a rallying cry in the 1970s. To address this, and the academic future more generally, Penney put in place a Growth Points Steering Committee which focussed first on two academic areas, materials science and bioengineering.[49] As is discussed in chapter thirteen, materials science was already a major research area in the college, pursued in a number of departments. The time seemed right to push for an undergraduate course in the field. Bioengineering was less well established and the aim was to see what more could be done to bring departments together on joint research projects. The committee also noted that some existing work was being carried out in collaboration with people at a number of London medical schools, and that in light of the 1968 report of the Royal Commission on Medical Education this was something to be encouraged. The Commission had recommended forging closer ties between undergraduate medical schools and universities so as to give the schools, 'access to the expanding world of scientific knowledge'.[50] Penney saw this as a way for a medical faculty to come to Imperial in the future. The Growth Points Steering Committee recommended that exploratory talks begin with officials at the Royal Postgraduate Medical School in Hammersmith, the Brompton Hospital, and the National Heart Hospital to promote common interests and more interdisciplinary research. These discussions were the start of reorganization of work both in engineering and in the life sciences, and anticipated later mergers with a number of medical schools. Later, in 1979, Brian Flowers headed a University of London Working Party which looked at medical and dental teaching resources in the university in light of the Todd Report. By 1979 Imperial was considering more seriously the bringing of a medical school to the college. But in a time of financial strain people were wary of taking on medical students who were seen as being even more expensive to train than engineers. As Flowers then put it, the time was not yet right.[51]

A little earlier there had been lengthy discussions with the Architects Association on a proposed merger with its school. The plan, put forward already by Linstead, was to include a new building for architecture students in South Kensington. Much private money was raised and the UGC had agreed to contribute further funds once a merger agreement had been reached. But negotiations were poorly handled, foundering mainly on the very different pedagogical philosophies of the two institutions. For example, Imperial wanted the architecture students to take some form of standard degree examinations. The architects, on the other hand, wanted a greater emphasis on the evaluation of architectural drawings, site planning, and intern work. The teachers were nearly all practising architects who worked at the school on a part-time basis. Many of them took students into their firms as interns. The cultural gap proved too wide but, in retrospect, this appears to have been a missed opportunity. A few years later, when Imperial very much wanted to diversify its offerings, the interaction of architects, civil engineers, and materials experts could, for example, have led to interesting new developments.

Failure of the negotiations is interesting also in light of the report of the 1970 Committee, a report which Penney had to take on board. This committee, under the chairmanship of Hugh Ford, had been set up by Linstead in 1962 to plan for the 1970s. Aside from Ford, who was older, its members were youngish professors mostly under the age of forty in 1962. They were unanimous in wanting more breadth in undergraduate teaching, something a merger with architecture could have aided. The engineering departments, especially, were under pressure to broaden their undergraduate curricula after the publication of two major government reports on engineering education. The Dainton and Swann Reports both recommended the inclusion of subjects such as sociology and economics, and that students have more practical working experience during their studies.[52] Overall, the college responded well to the changing needs of British industry and introduced a range of new courses which attracted many applicants once the pendulum began to swing back in favour of science and engineering studies in the late 1970s.[53]

Penney welcomed suggestions for college reform coming directly to him. This may have been a mistake since he was bombarded with letters from a wide range of college constituencies. Some were from angry individuals, others from those wanting to contribute their own ideas on how to address the various problems faced. There was much correspondence from the ICAUT which, as one might expect, was especially active in this period.[54] Under the leadership of Bryan Coles and Brian Spalding, professors in the physics and mechanical engineering departments respectively, and presidents of the ICAUT in the late 1960s and early 1970s, all kinds of initiatives were proposed. Both men were receptive also to student demands, and appear to have been advocates for the students as well as for the academic staff. In 1969 the ICAUT instructed Spalding 'to collect information about the degree of satisfaction felt by academic staff with all aspects of College activities and procedures'.[55] Spalding collected information, albeit not in a disinterested way, something he acknowledged. He wrote confidential letters to twenty-six members of staff, one professor and one lecturer from each of the larger departments. He chose people that he 'thought might answer' a questionnaire that he had composed. That he chose to look just at large departments is understandable given that it was from these that most complaints came. People cited lack of consultation on issues that affected them, heads of department who never held department meetings, no machinery for complaints, and so on. All but two of his twenty-six correspondents wanted heads of department to become chairmen with fixed terms. The responses led Spalding and others to the creation of an Academic Staff Assembly, formed in 1971. All this activity had results. For example, non-professorial staff were admitted to the Board of Studies and, more generally, were given an increasing say in how things were run. Further, while the ICAUT demand that the heads of department be replaced by chairmen with limited tenure was not met, the term of office for heads of department was limited to five years, though renewable. Spalding also made what at the time was a radical suggestion, namely that students be allowed to evaluate their courses and that some good feedback on teaching might be the result.[56] At the same time the technical and clerical staff wanted more say in their working lives. The former were represented by the Association of Scientific, Technical and Managerial Staff (ASTMS) and its highly vocal Secretary, Harry Fairbrother.[57] A national Consultative Committee on Clerical Services was formed at this time and, in 1970, the CVCP and the TUC came to an agreement on a joint negotiating body to cover the conditions of service of clerical staff in English universities.

Increased power sharing was something on many people's minds. For example, Professor Southwood wrote to Penney that he wanted not just teaching staff, but also research staff and demonstrators to have more say in the running of departments.[58] Penney's replies to the many letters he received are interesting for their tone — polite, mildly encouraging, and legalistic. He did not acknowledge the political and cultural context within which he was working but, advised by H. J. T. Ellingham, looked carefully through the provisions of the charter to see where, legally, students and junior staff members might have some input denied them by custom.[59] He wanted liberal reform, but within existing frameworks. Students and some members of staff responded by demanding changes to the charter. Constitutional change was first among the many demands of the short-lived Imperial College Representative Council, the brain-child of Piers Corbyn the ICU President and Harry Fairbrother.[60] Seats on this Council were assigned according to numbers in the college. Thus students, being the largest body, had fifteen seats, just short of a majority. The ICAUT refused to participate in this student-led initiative, seen by many as a step too far. In considering how to deal with radical students Peter Mee, the College Registrar, was somewhat Machiavellian. Not fully appreciating the tenor of the times, he wanted to isolate what he saw as a small group of activist students 'intent on making the pace on various matters', and that the way to do so was by the time-honoured technique of divide and rule. He suggested that steps be taken to strengthen the old college unions, something he believed would dilute student activism in the ICU.[61] Professor Francis, on the other hand, wanted to abolish the old colleges altogether, 'except for sporting and legal purposes' and organize departments in faculties according to 'genuine academic links'.[62] Similarly, Professor Pratt wanted an American style administration with Pro-Rectors with well defined roles.[63] Both suggestions were later adopted. The Pro-Rectors in the 1980s, and faculties in the early twenty-first century.

Clearly this was a period for rethinking old procedures, and one of Penney's initiatives was to try and improve, as he put it, 'the machinery of communicating'. In this he was much aided by the College Secretary, M. J. Davies, who understood the college well and provided Penney with good briefing notes on a wide range of issues.[64] At Davies' suggestion, Penney organized a meeting of the Governing Body, the Board of Studies, and the ICU. By the middle of 1968 they had hammered out the terms of reference for a new body, the Joint Union Rector Governors Committee (JURGO).[65] This committee was disbanded in 1972 but its formation was, perhaps, a necessary response to pressures of the period. Its members soon agreed that it was a good idea to have student observers at the meetings of the Governing Body, but also turned their attention to the matter of student discipline. Student discipline was a national concern and, in 1968, the CVCP and the NUS issued a joint statement regarding the need for an agreed code of procedures, something slowly put in place over the next two decades.[66]

The rate at which new directives were being issued by government departments, the UGC and the research councils, increased dramatically from the late 1960s onwards, a consequence of the hugely expanded higher-education sector. These directives could not be ignored were the college to remain competitive. Indeed, the history of the college during the last quarter of the twentieth century could be written in terms of its response to policies and directives from government departments and funding agencies. But institutions never follow exactly what is being asked of them; rather they respond in the ways they see best, often with unexpected results. During Penney's tenure, the administration began to modernize, took on board some lessons in industrial relations, and began thinking how to reorganize the academic units so as to make the college attractive to both industry and funding councils. It was recognized that the college needed to become more competitive and businesslike. It was in furthering these steps towards modernization that Penney's successor, Brian Flowers, was to make a major contribution.

The Flowers Years, 1973–85

Brian Flowers came to the college after a period as Chairman of the Science Research Council.[67] He therefore had a good understanding of how Imperial College departments measured up nationally. He later recalled that while Imperial was a very good institution it was not as good as it thought it was, and that he wanted to make it better.[68] Flowers was an activist Rector and it is difficult to know where to begin in describing the many changes that occurred during his twelve-year term. He wanted people at the college to face up to political, economic, and industrial realities, and his persistent prodding moved the college to modernize on a number of fronts. His work as it relates to already existing departments is discussed in chapter thirteen, his desire to have the college diversify its offerings so as to include also the social sciences and humanities is discussed within chapter fourteen. To be discussed here are some matters of governance, student affairs, and the creation of new interdisciplinary research centres.

Just as earlier, the governing body played a central role in decision making during this period. Lord Sherfield's term as chairman of the governing body ended in 1974. To mark his retirement and many years of service, the College Block was renamed the Sherfield Building.[69] Flowers worked mainly with Sherfield's successor Sir Henry Fisher[70] and was assisted by two very able College Secretaries, Mickey Davies who retired in 1979, and his successor John Smith.[71] Both were skilled in many ways, including in their ability to provide pertinent and informative briefing notes. Fisher at first questioned Flowers' commitment, not wanting him to accept an invitation to become the first president of the European Science Foundation. This entailed Flowers spending two days a month in Strasbourg as well as having some brief time commitments in London.[72] Flowers was already spending half a day per month to retain membership of the UKAEA, and was Chairman of the Royal Commission on Environmental Pollution. However, since Lord Sherfield supported Flowers' request, Fisher agreed. Later he questioned Flowers' taking on the chairmanship of the Commission on Energy and the Environment. Overall, however, Fisher was supportive of the Rector who served the college well.[73] Flowers was extremely energetic and well able to handle a number of external duties which, he believed, reflected well on the college. In speaking to the governing body in 1976, he stated his concern that the college was not taken sufficiently seriously in the life of the nation, and that the government did not consult it enough. He stated that if there was one institution which could address itself to the important needs of an increasingly technological society, it was Imperial.[74] He believed, rightly in my view, that more public exposure, paying attention to issues of national concern, and the forging of links to Europe, would all pay dividends. Flowers was good at bringing attention to the college and raising its profile in the media. For example, the Silver Jubilee of Queen Elizabeth's accession to the throne was marked by the inauguration of a set of well publicized public lectures, the first given in 1978 by Sir Ieuan Maddock, former Chief Scientist at the Department of Industry and Secretary of the BAAS.[75]

In 1974 four members of staff, with Orwell's 1984 in mind, made some predictions as to what Imperial College would be like ten years later. Three of them appear to have had rather cloudy crystal balls. Bryan Coles thought too much attention was being paid to teaching at the expense of research, and that this would lead to a decline in serious scientific scholarship. Harry Fairbrother believed that computing would become ever more centralized with associated big-brother consequences, and Henry Cohen believed that 'delusional notions of equality' would have retreated and 'proper elitism' would prevail. Roger Sargent saw things more clearly. He anticipated the emphasis on maintaining standards and speculated that 'perhaps those in the RSM will one day cease to be miners … and those in the C&G will grow out of their technical college past'.[76] Standards were on many people's mind at this time. Brian Flowers was worried about entry standards at the college though, by and large, they remained high. He was also concerned to increase the number of women students which stood at about 10% in the mid 1970s. In 1975 women came first in their graduating year in botany, materials, metallurgy, mathematics, and mechanical engineering. Flowers used these successes in a number of speeches aimed at attracting women students. Leaving aside the ethical problem of equality of access, it was clear that one way of keeping enrolment figures high while maintaining high standards was by attracting more able women to the college.

The 1970s and 80s saw a rapid rise in feminist activity. One consequence was the founding of the Women in Science and Engineering (WISE) campaign. Dorothy Griffiths, then a lecturer in industrial sociology, challenged gendered language at the college and championed the use of the title Ms. She and others in the Women in Science group won little respect for their feminist activities in the still very masculine world of the 1970s.[77] There was some hostile reaction to what were seen as pushy women. However, both Flowers and his successor, Eric Ash, did their best to combat the hostility. Both were supportive of women in the college, and in some cases intervened to aid the promotion of women staff whom they believed deserving.[78] The position of the Rectors' wives, however, is interesting. Flowers often stated that his wife shared his job.[79] This was true. Mary Flowers worked in many ways to improve the situation of students, especially foreign students. She helped in the refurbishment of 170 Queen's Gate,[80] and was an important social convenor. Flowers had stated that one of his goals in taking the Rectorship was to build closer ties to the student body and he did work closely with student union leaders. Mary Flowers organized regular 'beer and bangers' evenings for students and staff to meet informally. She contributed also in bringing the different college constituencies together, improving relations with college neighbours, and by inviting important people from outside to the college. No doubt this was all done willingly, but Mary Flowers played the traditional role of the wife of a public man. Clare Ash did much the same. Indeed Eric and Clare Ash were very convivial and held many dinner parties in which they mixed up people from different areas of the college, and invited also civil servants and politicians from Westminster. Clare Ash worked also in support of the college crèche, and on the housing of important college visitors, renovating some flats for this purpose. Both women, who volunteered so much of their time to the college, were among the last generation of wives willing to subjugate their own lives and careers to those of their husbands. One negative consequence of this largely positive modern socio-cultural development has been a decline in voluntarism of this kind. But, up until the early 1980s, the wives of college personnel did much to domesticate the male environment. Their contributions made college life far more comfortable and pleasant than it might otherwise have been, not just for women but for all.

In 1979 the Advisory Board of the Research Councils set up a working party under Peter Swinnerton Dyer to advise the Secretary of State for Education and Science on the future of postgraduate education.[81] Swinnerton Dyer claimed that, on balance, the country was getting value for its money, but recommended there be more serious review of expenditures in the future. He also recommended that there be better supervision of students, closer monitoring of thesis completion, and that there be sanctions against institutions with poor completion records.[82] Further, postgraduate numbers should be kept up, grants increased, and interdisciplinary work encouraged. Before coming to these conclusions the working party sent a questionnaire to all colleges and universities. As was typical in such circumstances, Imperial sent one response to the University of London and another directly to Swinnerton Dyer. Some of the questions asked related to collaboration with other schools and so raised the question of Imperial's position in the University. Excitement over the Murray report had only recently died down but, once again, the Rector had to consider all the pros and cons of Imperial remaining a school within the university.[83] Flowers appreciated Imperial's association with the name and worldwide reputation of the University, and acknowledged the utility of the library, and of computing and other services. Also positive was Imperial's share of student places in university-run hostels, and some bureaucratic conveniences such as having the university verify the A-level results of incoming students. But much of the bureaucracy was seen as burdensome; for example, getting approval for new programmes, for senior appointments, and the need to comply on a multitude of academic procedures. Overall Flowers was ambivalent, but thought that with new statutes and further devolution of powers there would be little incentive to separate.[84] The Swinnerton Dyer report prompted smaller colleges within the university to consider mergers. The political mood was for the formation of just a few large colleges which would lead, it was assumed, to smoother university governance. For economic reasons merger fever grew even stronger later in the 1980s, but already in 1981 Imperial engaged in what proved to be unsuccessful negotiations with Queen Elizabeth College (QEC).[85] This small Kensington college, originally a women's college, had eight departments. Largely science based, it was well known for its work in food science and nutrition. But there was opposition to the merger among Imperial staff who claimed there would be a 'dilution' of standards were they forced to absorb the smaller and supposedly weaker departments from QEC within their own. This was very shortsighted. In the larger scheme of things people's working lives are short, the addition of a few extra staff would hardly have rocked the boat, and the addition of buildings, laboratories, and residences within walking distance of Imperial would have been a great asset.

Flowers was especially concerned with the financial situation of students, recognizing that maintenance grants had dropped to a level that forced those with no other source of support to seek work even during term time. He was critical of Sir Keith Joseph for increasing science spending while doing nothing for student grants. The government arrangements, he claimed, were *ad hoc* and decisions were being taken without proper forward planning.[86] Existing grants were too low and did not reflect the contribution graduates make to society during their working lives. Students, he believed, were justified in thinking that their problems were not being recognized. However, Flowers was supportive of the government move to provide a broader education for sixth formers and approved the introduction of the AS level examinations.[87] In this he did not represent college opinion. Most departments wanted the kind of students who did well in A levels and not the kind who would need what was then seen as remedial work.[88]

Foreign students were an especial concern, though not because of entry standards. As Flowers stressed on a number of occasions, entry standards were the same for foreign and domestic students. Since its founding, Imperial College had seen itself as international in outlook; indeed this was, and still is, very much part of its identity. When Flowers arrived there were about one thousand overseas students from eighty-five countries. About 40% of postgraduates and 23% of the total student body came from abroad. As noted above, even before Margaret Thatcher came to power, the government policy had been to increase the ratio of home to overseas students. Higher education institutions had been asked to increase home students overall by 3%, and foreign students were to be charged higher fees. Flowers opposed both measures but to no avail. In October 1980 the Thatcher government announced the withdrawal over three years of what remained of that part of UGC grant which supported overseas students. The college was forced to raise foreign student fees even further and, by 1983, very high annual fees of £2000 for mathematics and £3000 for science and engineering, were in place at the undergraduate level. A new £4 million government bursary scheme for foreign students was introduced but this did little to alleviate the situation. Despite the temporary drop in applicants from overseas, there was still sufficient demand for undergraduate places. Research units, however, were very dependent on foreign students to function well and it was feared that increased fees would cause students to look elsewhere. This prompted the Rector to begin what became routine visits to South East Asia and Japan to promote recruitment and make contact with alumni who could help the college.[89] There was a small drop in the number of overseas students for a few years, but it was only temporary. A new wave of students soon began arriving, especially from Hong Kong, Malaysia and Singapore. The overall student body changed also in the 1980s and 90s as more students, including postgraduates, came from the European Union to study in London.[90] There was also an increase in the number of student exchange programmes with European universities, something much encouraged by Flowers.[91]

Penney had brought in a five-year term for department heads but Rectors retained the constitutional right to choose the heads. Flowers had several such appointments to make shortly after coming to the college and wanted to find people both outstanding in their fields, and with some loyalty to Imperial, not just to their departments. This was not assured. For example, in 1980 the letterhead of the physics department still did not acknowledge Imperial College. The address of the Blackett Laboratory was given as Royal College of Science, Prince Consort Road; and requests to change this had been ignored. Further, some of the engineering professors placed far more emphasis on their outside connections than was good for the college. Their lives were wrapped up in professional societies, and in consultancy for engineering firms and industry.

Details such as these will have sent a signal to Flowers that it was important to choose future heads carefully. Perhaps if they recognized the major role he played in getting them their jobs, the college and not just their departments or business associates would be repaid with their loyalty.[92] Flowers had spent much of his career in the civil service, and in public service relating to science and technology. He had a clear vision both of the rights and duties of a leader. When he came to Imperial he took seriously his responsibility to the Governing Body that the college be run efficiently. With respect to department heads, he decided on a process of selection that entailed broad consultation both within the college and without. He promoted his approach as progressive, and more democratic than what had preceded it, stressing the fact that comments were invited from all members of the department, including technical staff and research student leaders. Indeed, his approach was an improvement on earlier practice and was seen as progressive by those who, for the first time, were being asked their opinion on a future head of department. Flowers was wary of going so far as to allow any internal (even informal) departmental voting. The idea that people could be trusted to vote rationally, and not engage in a simple popularity contest was foreign. American universities might engage in such elections, but in Britain old ruling-class assumptions about leaders having natural authority were still apparent.[93] None of this is to say that poor decisions were made. People were invited to report directly to Flowers who then weighed what he had learned. He was an autocrat, but a well informed and well meaning one. However the process, by very dint of its efficiency in information gathering, was highly effective in concentrating power at the centre. It was the antithesis of democracy. Flowers came to know the departments well. He gathered much information on the kinds of work being carried out, who was doing excellent or poor work, internal relationships, feuds and other problems, friendships and dependencies, and so on. As a consequence he was able not only to make well-informed choices for headships, but was also better informed in how to manipulate future departmental activity. Power moved from the periphery to the centre during his period of office; but it did so, by and large, consensually. People accepted that the Rector had made an effort to understand their concerns and so were accepting of his decisions. This contrasted with the methods of earlier Rectors. Typically they would make decisions over headships by consulting just a few senior departmental and college figures, and would depend more heavily on outside advice.[94] Once in place a department head was responsible to the Rector but, in effect, not much could be done were departments poorly administered. The five-year term allowed for new direction where necessary, and gave the Rector a quinquennial look into the internal workings of each department. But headship under these new conditions proved less attractive than it once was. In some cases arm twisting became necessary to ensure a good succession.

Flowers was strategic in his choices. Among other things, he wanted to encourage interdisciplinary work within and across departments.[95] The financial difficulties of the 1970s made clear that the potential for change lay not so much in the growth of departments but rather in cross linkages between them.[96] Flowers' Rectorship is marked by the appearance of several new interdisciplinary centres, setting a trend that continued beyond the end of the century. Perhaps a good example to illustrate much of what went on is the founding of the Centre for Environmental Technology. As mentioned, Flowers was Chairman of the

Royal Commission on Environmental Pollution from 1973–6 and was therefore knowledgeable in this area. In the late 1960s there had been attempts to bring people together from across the RSM to work on environmental technology, especially in the area of mine site recovery. And, in the biological sciences, ecological and environmental work was being actively promoted by T. R. E. Southwood. By the early 1970s environmental management was a hot topic. It was an area in which something could be done even in a time of financial stringency. Flowers set up a broadly based working party under John Sutton, head of the geology department, to explore possibilities in the area of environmental technology. In 1975 Sutton reported that environmental science courses existed already in many universities and were attracting high enrolments. Approaching things from the technological angle would allow Imperial to occupy a niche for which it was well suited. Further, Sutton was able to report that work relevant to environmental technology was already being carried out in almost all departments in the college, and that funding for such work exceeded £1 million.[97] Such research included the effects of air pollutants on grasslands, noise pollution, pest management, metal recycling, and hydrological and other problems related to mining and quarrying. Sutton suggested that the way forward was to introduce a new interdepartmental MSc course and, at the same time, build a number of interdisciplinary research teams that could be loosely administered in a small centre. These teams would help define concentration areas for the course, and help rationalize environmental research at the college. Having a centre would help in the liaison also with outside bodies and funding agencies. The Centre for Environmental Technology (ICCET) was opened in 1976 by Peter Shore MP, Secretary of State for the Environment. The administrative and teaching centre was located at 48 Prince's Gardens but later it outgrew these premises and moved to the RSM building. Over 120 students applied for the new MSc course and 24 were accepted in the first year.[98] The course was approved by NERC, SRC and SSRC and 14 students were given funding. Eight areas of specialization were identified, though only four were offered at the start.[99] When the Centre opened over fifty academic staff were associated with it, eight of whom were professors. The academic director was Gordon Conway, who earlier had directed the Environmental Management Unit and had lobbied Flowers for the new centre.[100] The new initiative got off to a good start and monies soon came in; for example, a 1978 contract from the Department of the Environment for ecological work worth £450,000.[101] This success encouraged the creation of other centres. By the early 1980s there were several more, not all of which survived. Ten years later ICCET was still flourishing and centres existed also in such areas as robotics and automated systems, semiconductor materials, population biology, and biological and medical systems.[102]

Towards the end of Flowers' term at Imperial, the CVCP set up a committee under the chairmanship of Sir Alex Jarratt to look into university management and to initiate a series of efficiency studies.[103] This was in response to a government White Paper suggesting, among other things, that universities find some way to measure excellence and performance. As is discussed in chapter fifteen, it was Flowers' successors who had to deal with the resulting bureaucratic demands of the Research Assessment Exercise and new forms of external teaching evaluation. Flowers may well have been relieved not to have these extra administrative duties, though no doubt he could have handled them. His administrative skills were highly respected and he left Imperial to become Vice Chancellor of the University of London. The students gave Flowers a special sendoff. After he had given his farewell speech they drove him home in style. While the constabulary redirected traffic, Flowers and the students rode in Jezebel.[104] From the Queen's Lawn, via the perimeter of Hyde Park and Kensington Gardens, Flowers was taken back to 170 Queen's Gate, the home he was soon to leave.

...

1. For more on the student revolt of this period see Colin Crouch, *The Student Revolt* (London, 1970). Crouch sees the revolt partly in terms of alienation, and partly as a reaction to increasing technological growth and economic planning. He associates the latter with the coming to power of a Labour government in 1964. Perhaps it was because students at Imperial identified with the technocratic modernizers that their reaction to new developments was more muted than elsewhere.

2. Penney belonged to a generation that had served the country during the Second World War and, coming from a military family, will likely have had some admiration for military tradition. He must have been bemused by the fact that the young, while ostensibly anti-war, chose military men such as Mao, Che Guevara, and the aspirant militarist, Trotsky, as their heroes. The basically liberal ideals of the students were sometimes pursued by adopting some of the far from liberal methods of these revolutionaries.

3. At the end of 1974 the *Financial Times Index* was at its lowest level since May 1954. The 1972 national salary increases for academics were the last for many years. In that year the total salary bill was increased by 7.5%, minimum salaries were brought in, and the new average salary for professors climbed from £5610 to £6003. See 'University Salaries: More than Increases', *Nature*, vol. 236 (23 March 1972), 5.

4. Shortly after assuming office in 1979, Margaret Thatcher negotiated better terms for Britain in the EEC.

5. The price jumped in increments, from roughly $4.00 US to $12.00 and then to $30.00; after the Iranian revolution of 1979 the price reached $40.00 a barrel. The OPEC decision was related to President Nixon having earlier severed the link between gold and the US dollar. As the price of gold climbed, oil producers recognized they were being paid with devalued dollars. Much of the oil money came to London since the OPEC countries could not absorb it all, but foreign investors were not interested in the plight of universities.

6. Ironically some of the UGC problems in this period were the result of an act of political correctness in which race trumped class. Earlier, the Department of Education and Science had suggested that there be differential fees for home and foreign students. Tony Crosland, the Minister, had tried unsuccessfully to bring in higher fees for foreign students in 1966. Ten years later Shirley Williams got around the problem by raising fees for domestic as well as foreign students, believing it unfair to discriminate against foreigners, especially those from poor countries. However, the government was being hypocritical since it covered the increase for all, not just poor, domestic students. This amounted to a major hand-out to better-off families many of which could continue sending their children to public schools knowing that university education would cost no more than before (only maintenance grants were still means tested). Sir Keith Joseph later attempted to bring in differential fees for more affluent students but failed because of a middle class backlash. A further result of Williams's policy was increased centralised control over universities. Students became the new unit of resource, and quinquennial planning by the UGC became increasingly irrelevant. Making sensible distinctions between universities for granting purposes became almost impossible before new assessment procedures were adopted.

7. Universities did not see any benefit until the end of the decade. This was due to the policies of the Thatcher government discussed in this chapter and in chapter fifteen. Cuts in university expenditures during the early 1980s were severe. Sir Alec Merrison, chairman of the CVCP in 1981,

described government plans as 'madness'. See *Nature*, vol. 290 (19 March 1981), 177.

8. At Imperial all departments had to take a 4% cut in income that year. VAT increased by a further 7% in 1979 when inflation was running at about 20%.

9. Trevor Phillips, the ICU President in 1973–4, made a spirited attack on increased refectory and other prices. He spoke out clearly on the ways in which increases in the cost of living were affecting students.

10. By the early 1980s there was much concern over Britain's 'crumbling universities'. One major expense at Imperial was the removal of asbestos from the RSM building. See for example, 'Worry over safety standards in crumbling universities', *THES* (30 December 1983), 1–3.

11. One area where savings were made was in the rationalization and closure of some interdepartmental service laboratories. In their struggle to survive, these laboratories took in increasing amounts of work from outside the college but were still not paying their way. One, the Analytical Services Laboratory, was located in the metallurgy and materials science department. It had a staff of ten and handled about sixty samples a month of which about twenty came from the college. But the service brought in only about £10,000 a year from outside and it was questionable whether the ten staff members should be retained. Similar problems existed in connection with the Crystal Growth Laboratory and Thin Film Microelectronics Laboratory, both in electrical engineering, a Mechanical Testing Laboratory in mechanical engineering, and a number of others.

12. The miners' strike of 1974 forced the entire nation to save electricity. Despite this, in 1974–5 the college paid the London Electricity Board over £200,000, and roughly the same amount to the Department of the Environment for heating. This, in addition to £129,000 for oil for the boilers. All this represented a doubling of costs since the start of the decade and was about 3% of annual income. A working group was set up under Professor John Brown, head of electrical engineering, to see how the college could save money in this area. See *Topic*, 10 February 1975.

13. The Report of the Select Committee on Science and Technology (1976) recommended that resources be concentrated in universities best suited to promoting industry-related postgraduate study. It also recommended encouraging industries to become more involved in higher education. This kind of thinking helped Imperial during the difficult years.

14. James Callaghan was Chancellor of the Exchequer in 1965 and, though more sympathetic to universities than many of his university-educated ministerial colleagues, indefinitely deferred building money for univer-

sities. About one-fifth of the quinquennium money for this purpose was lost.

15. The college plans included the demolition of 180 Queen's Gate (like number 170, it had been designed by Norman Shaw). This was opposed by the Victorian Society. Without the new buildings for chemistry and mathematics the college claimed it would be unable to meet its Robbins target of 5275 full-time students by 1976. The fact that the mathematics (Huxley) building was built during this cost-cutting period had negative consequences for its overall quality.

16. At Imperial College there were increasing concerns over enrolment goals. For example in 1970 thirty-seven of those offered places in the chemistry department declined and went to Oxbridge. In 1971 the figure was eighty. Nationally the number of applicants for chemistry degrees was down by 18%. Physics was similarly hurt. In 1971 the department made 450 offers with an expectation of only 260-80 acceptances. Imperial, as a whole, took in 200 below its target figure. Most departments were still only accepting people with As and Bs at A-level. But the mining department was accepting some students with Cs. In the late 1960s there was also pressure from the UGC for the University of London to do something about its seemingly high drop-out and failure rates, labelled 'wastage'. As it turned out Imperial's 'wastage' was relatively low, about 12% in science and 15% in engineering compared to national levels of 15% in science and 21% in engineering, but this was only determined after much discussion and research into national and local statistics. The coining of the term 'wastage', and its use by the UGC, is an interesting reflection of the times.

17. For an article on the 'brain drain' see 'Brain Drain Persists', *Nature*, vol. 267 (5 May 1977), 8.

18. Rector's Report, 1974. The move to the City continued through the 1980s at a time when industrial and other opportunities were shrinking. Good economic growth in the late 1980s and 1990s was accompanied by more jobs opening in new high tech industries.

19. At this time Owen Saunders, still technically a professor at Imperial, was Vice-Chancellor of the University of London. He insisted Imperial make an effort to meet the ratio, but allowed it some leeway given the high level of postgraduate work at the college. See GDK/Saunders to Rector, 26 March 1968. Difficulty over promotions lasted into the 1980s. Flowers was very conscious of the need for more promotions but his hands were tied. For example, in 1978 he was able to appoint or promote only eight people into the senior ranks.

20. Lord Rothschild was then Head of Central Policy Review Staff and Dainton chaired the Council for Scientific Policy Working Group. For Green Paper see Cmnd 4814 (1972). For contemporary commentary on the Rothschild Report see Eric Ashby, Harold Orlans, John Ziman and V. C. Wynne-Edwards, 'The Choice and Formulation of Research Problems: Four Comments on the Rothschild Report', *Minerva*, vol. 10 (1972), 192–208. Rothschild was the Special Visitor at the college Commemoration Day ceremony in 1975 and spoke of the difficulty in predicting the future.

21. By the later 1970s the college was increasingly turning outwards towards contract research. However its main source of research funding was still the research councils. The Science and Engineering Research Council (SERC), founded by the 1965 Science and Technology Act, was the successor to the Science Research Council (SRC). The Act also created the Natural Environmental Research Council (NERC), the new Agricultural Research Council (ARC) and the new Medical Research Council (MRC). The Act brought all the councils under the Department of Education and Science, and brought to an end the Department of Scientific and Industrial Research (DSIR). Until then, the DSIR had responsibility for both research councils and a number of government research laboratories. The latter were put under the new Ministry of Technology.

22. Lord Murray, formerly Sir Keith Murray, chairman of the UGC. A preliminary version of the report was available for discussion purposes in 1972; the final report appeared in 1973. For Imperial College response see Annual Report, 1972–3. See also GB minutes, 19 December 1975.

23. Imperial was the only college in the university that had direct access to the UGC at that time but other large colleges were seeking the same. It should have been clear to anyone with the slightest historical sense that Murray's main recommendations would go nowhere.

24. At Imperial applications to undergraduate programmes declined (by approximately 9% in the early 1970s), but there were still enough applicants to meet college intake projections without seriously compromising standards. See F and EC minutes, 20 November 1981.

25. Rector's Correspondence; R. K. S. Wood to Penney, 28 September 1970.

26. Rector's Correspondence; Ubbelohde to Penney, 28 July and 20 November 1969. Emphasis in the original.

27. University of London/Imperial College Relations file. Southwood to Cohen, 14 January 1976.

28. The 1926 University of London Act had given the power to change statutes to the Board of Education.

29. For discussion of proposals in the early 1980s see GB minutes, 17 December 1982. See also chapter fifteen. Eric Ash, when Rector, fought hard for Imperial, and for greater devolution of power. The late 20th century reform of the University of London statutes reflected his efforts.

30. Jack Straw's paper (NUS, 1970) was in response to CVCP paper 'University Development in the 1970s' (1969). The ICU paid attention to Straw and advocated the creation of a polyversity in South Kensington in which the college facilities would be pooled with those of the other major colleges in the area. The Union communicated its views to the CVCP.

31. Harold Perkin, *The Rise of Professional Society in England since 1880* (London, 1989).

32. The future was anticipated by Margaret Thatcher when she was Secretary of State for Education and Science. Her department published a report 'A Framework for Expansion' (1972) which covered education from nursery schools to universities. Thatcher was appointed to the Governing Body of Imperial College in 1974. This was perhaps a good move since she had little love for universities in general. She also had little love for the civil servants in her ministry who, she believed, were far too sympathetic to teachers' unions and held left-wing philosophies of education. See Margaret Thatcher, *Road to Power* (London, 1995).

33. Parkes speech; copy in ICA.

34. See EC and FC minutes, 20 November 1981. Most cuts were in the non-academic staff area. By 1984 the number of academic staff in UK universities had shrunk by one-sixth and 20,000 student places had been lost. For an angry article about well qualified school leavers not finding university places see *The Guardian*, 8 February 1984. See also *The Economist*, 15 September 1984 for a sympathetic but more conservative take on the problems then faced by universities.

35. This was something urged on the government by the International Monetary Fund during the economic turmoil of the mid 1970s when the UK needed to be bailed out. As noted in note (6), Shirley Williams did increase foreign (and domestic) fees, but despite these increases universities were not then charging foreign students the full cost of their education.

36. This scheme began in 1978. While only 9% of Imperial's eligible staff took up these offers, many more did so elsewhere. The first compensation package which was seen as generous was not repeated. Some universities found themselves giving post retirement contracts to those who had taken compensation, hardly a cost saving measure. One way that Imperial tried to bring in new people and ideas in this period was by increasing the number

of visiting professors on one or two-year appointments (renewable to five years). This scheme, which began in 1972, helped to stimulate research.

37. For Benn on attrition see F and EC minutes, 20 November 1981.

38. On 12 December 1983 the CVCP had a luncheon meeting with editorial staff at *The Times* and complained about the poor coverage of university affairs.

39. Rector's Correspondence; Sir Keith Joseph to Flowers, 9 May 1984. Joseph believed that university expansion had gone too far and that standards had declined. He wanted fewer students and fewer academics, and never pushed the Treasury for university funding during his tenure as Minister of Education.

40. For a critical overview of the politics of higher education in the 1960s and 70s see, John Carswell, *Government and the Universities in Britain: Programme and Performance, 1960–80* (Cambridge, 1985). Carswell claimed that 'the tragedy with which this book is concerned' was the deterioration in the traditional university ethos. For an interesting personal viewpoint see, Lord Annan, 'British Higher Education, 1960–1980: A Personal Perspective, *Minerva* vol. 20 (1982), 1–24.

41. William George (Lord) Penney OM FRS (1909–91) was known as Bill. He was born in Gibraltar where his father was in the Royal Ordnance Corps. When the First World War broke out, his mother took the children back to England and settled in Sheerness. Penney was educated in the local schools, including the Sheerness Junior Technical College. He was a mathematics prodigy, and entered Imperial College in 1927 with exemption from the first year course. A star athlete at school, he played football for the RCS team. In 1931 he was awarded a PhD and then left for two years on a Commonwealth Fund Fellowship, gaining an MA under J. H. van Vleck at the University of Wisconsin. He then entered Trinity College, Cambridge with an 1851 Exhibition Senior Studentship, and gained a second PhD before returning to Imperial College as assistant professor in 1937. Penney's work on the Manhattan Project during the Second World War is discussed in chapter eight. He was one of two British observers to accompany the flight of the bomb dropped on Nagasaki. (The other was Group Captain Leonard Cheshire.) After the war and the failure of the UN to reach any agreement on the control of nuclear weapons, Penney supported Britain's going ahead and making its own. He was appointed Chief Superintendent of Armaments Research under the Ministry of Supply, working at Fort Halstead in Kent. When, in April 1947, the Attlee government decided that Britain would develop its own atomic bomb, Penney was asked to head the team which, in 1951, moved

to Aldermaston. From there Penney directed work on the production of plutonium in the Windscale reactors. Britain's first atomic bomb was ready for testing in 1952 and, after its successful test in the Montebello islands off the north-west coast of Australia, Penney was made a KBE. In 1954 the UKAEA was formed and took over responsibility for atomic weapons research from the Ministry of Supply. Also in 1954, Penney was asked to direct work on the H-bomb. In 1959 he succeeded John Cockcroft as Director of the UKAEA's Harwell establishment with responsibility for the Authority's entire research programme. In 1964 he succeeded Lord Sherfield as chairman of the UKAEA. Sherfield became Chairman of the Governing Body at Imperial and it was he who invited Penney to return to Imperial as Rector in 1967. In the same year Penney was made a life peer.

42. Penney, quoted in Lord Sherfield, 'William George Penney, OM KBE, 1909-1991', *Biographical Memoirs of Fellows of the Royal Society*, vol. 39 (1904), 296.

    For Jones, see chapter 10. The position of Pro-Rector was akin to the later position of Deputy Rector. Pro-Rectorships as currently understood were introduced in the 1980s.

43. Penney displayed ambivalence towards nuclear weapons and worked hard for the Test Ban Treaties of the 1950s and 60s.

44. The Bernard Sunley Charitable Foundation, the Ford Foundation, and a number of City companies were the chief donors to this appeal. The Sunley Foundation financed the renovation of four of the seventeen houses acquired (leased) in Evelyn Gardens, and the Ford Foundation helped modernize facilities at Silwood. But the college was disappointed by the poor response to the Appeal from industry. By 1972, at a time when the annual income of the college was about £10.5m, the appeal had reached about £650,000. This was a period of enormous rise in London property values. For example, the UGC was unable to help the college purchase some Queen's Gate and Jay Mews properties which had been conditionally allocated to the college by the Royal Commission for the Exhibition of 1851. For more on the appeal see FC and EC minutes, 7 February 1969.

45. Penney stressed this as one of his most important achievements in his first 'Annual Report' (1969). He also began the modernization of Imperial's archaic system of student prizes, seeking alumni donations to increase the value of prizes, some of which had become pitifully low by the early 1970s. He also sought a fairer distribution of prizes among the departments.

46. In 1972 the college began renting out student accommodation in the summer, something that later proved very profitable.

47. This involved negotiation with the CGLI, the Goldsmiths and the Clothworkers, the three principal City parties to the Delegacy. The original suspension was for three years but the Delegacy never regained its control. For the origin of the Delegacy see chapter 4. For its suspension see GB minutes, 21 March 1969.

48. GCB 1/6 Board of Study minutes, 1967–9. Perhaps he agreed with a letter from S. Eilon making the same point. See Rector's correspondence; Eilon to Penney, 13 November 1969.

49. Professor Ubbelohde was so incensed by the hiring freeze and a 6% reduction in his departmental vote in 1968–9 that he refused to cooperate with the Growth Points Committee. He had not been able to equip the new ACE building, or hire new staff, and refused to discuss growth points until his department had 'caught up'. (Rector's Correspondence; Ubbelohde to Penney, 1 and 2 July 1968.) He was not alone in opposing the formation of this committee. Professor Skempton also wrote to Penney stating that he saw no grounds for a Growth Points Committee when he could hardly keep existing programmes going in civil engineering. (Skempton to Penney, 20 June 1968.) Perhaps this is a lesson in the importance of naming committees carefully.

50. Royal Commission on Medical Education (Cmnd 3569, 1968). This Commission, chaired by Lord Todd, also recommended the grouping of London medical schools and that some should be associated with Imperial College. While the schools identified in 1968 were not exactly the same as those that later joined Imperial, the Commission's recommendations were largely followed, albeit many years later. Interestingly the Commission made no recommendation concerning the Royal Postgraduate Medical School in Hammersmith. This must have bothered the School since some of its officials approached Penney stating that they, too, wanted to cultivate links with Imperial. (See FC and EC minutes, 7 February 1969.) Bruce Sayers already had links to Hammersmith at this time. See chapter 13.

51. Flowers' decision was based on the findings of Sir Hugh Ford who he had earlier asked to chair a committee to look again at the Todd Report from the point of view of Imperial College. Ford advised a 'go slow' until the funding situation improved. See F and EC minutes, 16 November 1979.

52. In 1965 the Council for Science Policy, under the chairmanship of Sir Frederick Dainton, established the Enquiry into the Flow of Candidates in Science and Technology into Higher Education. Its report was published in 1968. Dainton was also a member of the Working Group on Manpower Parameters for Scientific Growth under the chairmanship of Lord Swann, which also reported in 1968. Both of these reports should be understood in

the context of the decline in applicants for places in science and technology in higher education institutions. Much attention was paid to what industry, especially manufacturing industry, needed in its graduate recruits.

53. See chapters 13 and 14 for how this broadening was accomplished.

54. The ICAUT was given formal union recognition in 1971. The ICASTMS was given recognition in 1974 for consultation and negotiation of conditions of employment.

55. See D. B. Spalding, 'AUT Review of College Procedures' (1969).

56. Some department heads agreed to student evaluation at this time. By national standards the ICAUT was not radical. For example, it did not support the AUT marking boycott in 1978, called off only when the government agreed to rectify some salary anomalies.

57. Fairbrother, on the staff in mathematics, was an antagonist, though he made some reasonable demands on behalf of union members. Flowers later lost patience with his tone and wrote that he was not prepared to conduct business with Fairbrother unless he changed his approach, 'almost every communication [from the ICASTMS] was characterized by intemperate language ... deliberate misrepresentation and crude threats'. Rector's correspondence, Flowers to Fairbrother, 21 December 1973.

58. Southwood to Penney, 14 November 1969. Southwood was a moderate, not wanting to go so far as to give demonstrators and research staff full voting rights.

59. For Ellingham see chapter 11.

60. For more on the activities of Corbyn and the ICU in this period see chapter 11.

61. SRC/Peter Mee to Rector, 26 November 1969. The letter was written one day before the Queen opened the new College Block (Sherfield Building) and Piers Corbyn presented his petition. See chapter 11.

62. SRC/J. R. D. Francis, Professor of Hydraulics, to Penney, 17 November 1969.

63. SRC/Pratt to Penney, 14 November 1969. P. L. Pratt, Professor of Crystal Physics, jointly appointed between the departments of metallurgy and physics.

64. Michael John Davies (1918–1984) was known as Mickey. Davies was educated at the University of Cape Town before going to Oxford as a Rhodes Scholar. He joined the Colonial Service in Tanganyika in 1940 and served as private secretary to a succession of governors. He also became Minister of Constitutional Affairs. He was appointed College Secretary and Clerk to the Governing Body in 1962 and held the post until his retirement in 1979. He helped put in place the new committee and admin-

istrative structure introduced by Linstead during the major expansion of the 1960s. See obituary, *The Times*, 14 July 1984. Davies was among a generation of colonial administrators who joined British universities from the 1950s to 1970s. The Empire was contracting just as universities were expanding and looking for experienced staff.

65. This committee included the president and eleven other members of the ICU, three members of the GB, the Rector, Pro-Rector and Deans of the constituent colleges, the Registrar and College Secretary. Agenda items were the privilege of the Rector and the ICU President. The committee met once a term. The composition was soon changed, with fewer student members and greater representation of staff, such as hostel wardens, who had much student contact. See GB minutes 15 March 1968 and 21 March 1969.

66. For discussion of student discipline see GB minutes, 17 December 1971.

67. Brian Hilton (Lord) Flowers FRS (1924–), son of a Baptist minister, was born in Blackburn and grew up in Swansea. He read physics at Cambridge, moving from there to join the wartime Anglo-Canadian Atomic Energy Project (Tube Alloys) in Montreal and Chalk River. In 1946 he joined the Nuclear Physics Division at the Atomic Energy Research Establishment at Harwell, transferring to the Theoretical Physics Division in 1948. After a short period at Birmingham University he returned to Harwell in 1952 as Head of the Theoretical Physics Division where he succeeded Klaus Fuchs. Flowers pioneered computing methods at Harwell and applied them to the solution of problems relating to nuclear structure. He was appointed Chief Research Scientist at Harwell in 1958. In the same year he moved to Manchester as Professor of Theoretical Physics, becoming Langworthy Professor of Physics and Head of Department in 1961. From the early 1960s Flowers was increasingly drawn to work in areas of public policy related to science and higher education. In 1964 he succeeded William Penney as Chairman of the National Institute for Research in Nuclear Science (NIRNS) Atlas Computer Committee, and its successor the SRC Atlas Computer Committee. He was a member of the Council for Scientific Policy, 1965–7; Chairman of the Computer Board for Universities and Research Councils, 1966–70; Chairman of the Science Research Council, 1967–73; President of the European Science Foundation, 1974–80; Chairman of the Royal Commission on Environmental Pollution, 1973–6 (in succession to Sir Eric Ashby, a former Imperial student); Chairman of the Standing Commission on Energy and the Environment, 1978–81; Chairman of the CVCP, 1983–5; Vice Chancellor of the University of London, 1985–90; and Chairman of the Nuffield Foundation, 1987–98.

He was knighted in 1969 and made a life peer in 1979. He was Chairman of the House of Lords Select Committee on Science and Technology, 1989–93. Flowers was a founder member of the Social Democratic Party in 1981.

68. Flowers, personal communication.

69. The College Block, part of the Jubilee expansion, was completed in 1969. Another governor from this period who worked hard on behalf of the college was businessman, David Elmer Woodbine Parish, chairman of the CGLI.

70. Sir Henry Arthur Pears Fisher (1918–) is the eldest son of a former Archbishop of Canterbury. He was educated at Marlborough College and Oxford, returning to a fellowship at All Souls after service in the Second World War. He was a barrister and QC who became a High Court judge on the Queen's Bench. He resigned from the bench to become President of Wolfson College, Oxford.

71. John H. Smith (1928–) was educated at University College London and entered the Colonial Service in 1951. Before his appointment at Imperial College in 1979, Smith had spent six years as Governor of the Gilbert and Ellice (later Gilbert) Islands.

72. One result of these European activities was the setting up of the European Academy of Science to offer independent and authoritative advice on European scientific affairs.

73. M. J. Davies to Sir Henry Fisher; 2 October 1974. See also B/Flowers, Box 3–5, Fisher correspondence, 1978–9.

74. GB minutes, 25 June 1976. Flowers was also defending the broadening of the curriculum in this context. He understated Imperial's role in advising governments in that the tradition of staff giving advice to various government bodies continued strong even in this period.

75. In 1984 the Jubilee Lecture was given by an important college neighbour, Sir Bruce Williams, Director of the new Technical Change Centre, located on Cromwell Road and supported by the Leverhulme Trust, SERC and SSRC.

76. *Icon*, October 1974.

77. Griffiths (personal communication).

78. This was necessary because hostility typically took the form of ignoring what women were doing. When the 'new blood' lectureships were introduced in the early 1980s women did poorly in winning positions. This was in part because the positions were open only to people in the 27–35 age bracket, peak years for child raising. On this subject see comments by

Professor Daphne Jackson, president of WISE, *New Scientist* (16 February 1984), 4.

79. See, for example, speech to RCS Association dinner, 9 November 1973. Also his farewell speech, 2 July 1985.

80. Flowers formed a small committee to refurbish the lower floor of the house so as to make it suitable for the entertainment of VIPs. Professor Ubbelohde joined Mary Flowers on the committee.

81. Sir Peter Swinnerton Dyer succeeded Sir Edward Parkes as chairman of the UGC. A few years after the report discussed here he authored an even more influential document. 'A Strategy for Higher Education into the 1990s' (HMS Stationary Office, 1984). It is discussed in chapter 15.

82. An earlier article on slow completion rates quoted Sir Geoffrey Allen, chairman of the SRC and former Imperial professor, as stating that there would be a penalty for slow completion rates. In a list of SRC funded PhDs, on average those at Birmingham finished in the shortest time. Imperial students came thirty-first out of thirty-five universities listed. See *The Times*, 14 January 1981, 4.

83. See GB minutes, 12 December 1980.

84. The details of statute reform were finally worked out during Eric Ash's rectorship. For Flowers' response, see Flowers to Swinnerton Dyer, 11 July 1980; letter reprinted in F and EC minutes, 21 November 1980.

85. See F and EC minutes, 20 November 1981.

86. See, for example, letter to editor, *The Times*, 11 December 1984. This was written in Flowers' capacity as chairman of the CVCP. The cuts prompted the CVCP to 'stop being purely reactive' and take initiative in debates concerning higher education. See also *The Times*, 20 December 1983, 3.

87. Flowers' ideal student was someone with two A levels in, say, mathematics and physics, and two AS levels in English literature and economics. But the new AS syllabus was difficult for schools to introduce and cover well in one year.

88. This was especially the case as the government had dismissed the idea of four year courses. Later this decision was reversed. Another scheme which did not find favour in the college was proposed by the Society for Research into Higher Education (Leverhulme Enquiry, 1983). It was for a two year degree course supported by mandatory grants. Those achieving a satisfactory standard could continue for a further two years supported by discretionary awards. The Board of Studies stated that such a scheme was not suited to the high flyers that came to Imperial.

89. See F and EC, 22 May 1981. Flowers was followed by professors, and staff from the Registry, in the recruitment drive in S. E. Asia. Other universities were doing the same.

90. From the point of view of fees, such students counted as home students.

91. Exchange programmes were set up in this period with Chalmers University of Technology in Sweden, with the Delft Institute in The Netherlands, and with the ETH in Zurich.

92. I think this worked well in the life sciences where Flowers was able to get his way over Silwood Park. Silwood was losing money and Flowers wanted either to sell the land or build a research park and have more work move there, so as to relieve the space problems in South Kensington. Both options were opposed but Flowers was able to persuade some of his new heads to accept the second option.

93. The turn to elected and/or rotating chairs in American universities appears to have taken place in the early 1950s and was widespread by the end of the decade. See Robert M. MacIver, *Academic Freedom in Our Time: A Study Prepared for the American Academic Freedom Project at Columbia* (New York, 1955).

94. Typically heads of similar departments elsewhere in the country and senior industrialists would be consulted.

95. See also chapter 13.

96. The genesis of interdisciplinary centres is a complex issue. Some scholars, looking at the emergence of centres in the United States, claim that the impetus was internal to science and technology; and that wartime cooperation, such as in the Manhattan Project, showed scientists the advantages of having multidisciplinary teams when working on complex problems. Intense competition for resources also drives the formation of centres. See, for example, Henry Etzkowitz and Carol Kemelgor, 'The Role of Research Centres in the Collectivisation of Academic Science', *Minerva*, vol. 36 (1998), 271–88.

97. Some of the largest grants were in civil engineering and metallurgy.

98. Flowers was not entirely happy with admission standards at first. But by the 1980s the number admitted had risen to forty and entry standards had risen also.

99. The first four offered were ecological management, environmental physics, pollution and mineral resources. To be added later were water resources management, land use and transport planning, pollution control, and safety engineering and energy policy.

100. Sir Gordon Richard Conway FRS (1938–) was educated at the University of Wales (Bangor), Cambridge, Berkeley, and the University of the West

Indies. He joined Imperial as a research fellow in the department of zoology in 1970. He rose quickly through the ranks and was given a chair in 1980. He early carried out work on yellow fever vectors in East Africa and, following Southwood's example, took on many projects around the world. He became the Ford Foundation's representative for India, and has been a major figure in helping to establish ecologically sound farming practices globally. In 1992 he was appointed Vice Chancellor of the University of Sussex and, in 1997, President of the Rockefeller Foundation. He retired from the Foundation in 2004, returning to Imperial as a part-time professor in international development. He is a member of Imperial College's development advisory board, was made FIC in 2003 and was knighted in 2005. For some of his ideas, see his *The Doubly Green Revolution: Food for all in the Twenty-First Century* (Penguin, 1998).

101. For more on the early period of ICCET see 'Annual Report of ICCET' (1978). ICCET became a department with John Beddington as head in 1997. In 2004, 158 students took the MSc course with its various options. The MSc course was long under the direction of J. N. B. (Nigel) Bell who followed Conway as director of the centre. See also chapter 13.

102. One shorter lived centre was for remote sensing. This was set up in 1973 to make use of data collected by satellites such as LANDSAT and SEASAT. For more on some of the centres, see chapter 16.

103. For more on Jarratt see chapter 15.

104. Jezebel is a 1917 fire engine, mascot of the RCS.

chapter thirteen

The Making of the Modern College, 1967–85
Part Two: Academic Restructuring

Introduction

Few changes in the academic programme occurred during William
Penney's tenure as Rector. He focussed mainly on modernizing the
governance of the college and on student affairs, something that
occupied vice-chancellors and principals throughout the world in this
period. His successor, Brian Flowers, saw his task as the reorganization
of the academic side so as to bring in modern degree offerings, improve
academic standards, and encourage excellence in research. Flowers was
keen to see departments develop, each with a number of lively research
sections, and with interdisciplinary links where possible. He envisaged
the college both along departmental lines and in terms of a linking
'horizontal grid'. As he put it in a speech to the Old Centralians, he
wanted to set up a two-dimensional structure at Imperial to face up to
'the totality of real technological activity'.[1] The move to more interdisciplinary work contributed to increasing tension between the needs of
research and those of teaching, a theme that is discussed below and in
the chapters that follow. In the 1970s the two most important areas for
interdisciplinary work were the environment and materials science, but
other areas soon followed. By the 1990s there was a dramatic increase
in interdisciplinary work in the medical field. As we have seen in earlier
chapters, interdisciplinarity was not new at Imperial but during Flowers'
tenure it began to take on its modern form.

One development associated with the new interdisciplinary centres has been an increase in the number of MSc courses. Understanding this development, and the rise of MSc courses more generally, would benefit from more analysis than I can give it here. After the Second World War the UGC, and later the research councils, encouraged the growth of the MSc. The courses were seen as good vehicles for the development of vocational specialisms, and for the training of a highly skilled workforce. However, university motives in offering the courses were mixed. In some cases there was genuine enthusiasm to develop new expertise. This was undoubtedly the case, for example, with Professor Baker and the MSc in concrete technology. This course was a great success and recently celebrated its sixtieth anniversary. In other cases, financial support from the UGC was a greater incentive than perceived vocational need. In 1974 Flowers ordered a critical reappraisal of all MSc courses to consider true demand, national interest, and vocational value. Many courses had low enrolments and it was clear that the MSc courses needed a shake-up. Where demand had declined entry standards had slipped, something ignored by departments increasingly dependent on student fees. For example, there was little demand for the MSc courses in biochemistry, a matter of some concern. Flowers saw a solution in their closure and, to keep the department solvent, the introduction of an undergraduate course which would likely attract many more students. MSc courses were also suspect in that they demanded much staff time which many believed would be better spent on research. Flowers did close down a number of the older MSc courses, but encouraged the founding of others within new interdisciplinary research centres. His motivation appears to have been mixed.[2] He genuinely believed that research needed to be more interdisciplinary and that people needed to come together in new forms of association. But he also recognized that in the harsh economic climate of the 1970s new money was more likely to come to fields, such as the environment or materials, that required input from many different experts. And one way to financially support new clusters was to offer attractive MSc courses and enroll students, something that remains true today.

While some of the major themes just mentioned are discussed below, the richness and breadth of activity in this period can only be suggested in a chapter of this length.[3]

The Life Sciences

The Division of Life Sciences was formed in 1973, at the end of Penney's rectorship, with T. R. E. Southwood as its first chairman.[4] At this stage the idea was simply to facilitate cooperation and interdisciplinary work between departments, not the abolition of departmental structures. Given the imminent retirement of Professor Chain, one of the last problems faced by Penney before his own retirement was what to do about biochemistry. He set up a committee, chaired by Albert Neuberger, Professor of Chemical Pathology at St. Mary's Medical School, to consider whether Imperial should continue running the biochemistry department on the microbiological and physiological biochemical basis that had been developed under Chain.[5] The committee recommended that existing lines of research be continued, but stated that Chain's successor need not be a 'copy-cat' researcher.[6] Some committee members, however, believed that it was important to appoint a molecular biologist who could turn the department around. Given the impetus molecular biology had received in the postwar period due to new methods in x-ray crystallography, electron microscopy, associated methods of computation, and the discovery of the structure of DNA, this desire for a new direction is understandable.[7] Flowers, who was consulted before his arrival at Imperial, was ambivalent. Not inclined to give up on what Chain had achieved, he saw a solution to the problem in a second chair of biochemistry, and the appointment of someone who was capable of taking on the pilot plant. This would allow for a more general biochemist to be appointed as Chain's successor and head of department. In a letter to the Rank Group Trustees, who were approached to finance a chair in physiological biochemistry, Flowers wrote that there was a national need to maintain the expertise developed at Imperial 'because of the importance of collaboration between industry and the whole broad academic field of functional biochemistry'.[8] As mentioned the two MSc courses offered by the department were poorly attended and did not attract the best students. Flowers suggested closing them down and their replacement with an undergraduate course linked to chemistry and to the biological sciences. Flowers was aware that a department occupying as much space as biochemistry, and with as few students, would no longer find favour with the UGC.

The creation of a Division of Life Sciences was far from plain sailing.[9] The idea had originally come from some of the botanists who wanted a more 'biological' approach to the life sciences and thought that the drawing together of the various departments was a way to bring this about.[10] In this they had the support of Southwood who wanted to see traditional disciplinary boundaries disappear. Anne Beloff Chain, however, was wary of the idea of a Division, wanting to preserve the biochemistry department's independence and ensure that its research carried on much as before. Fears for the department's future were exacerbated when Penney, understandably, decided to leave all the major decisions to his successor and appointed Southwood as acting head of the department. During the interregnum there was much debate over what sort of research orientation the department should have. In the event, Brian Hartley, a structural biochemist, was appointed to head the department in 1974.[11] He immediately began reorganizing, beginning with the setting up of a group to carry out research on the structures of proteins and enzymes, and made some good appointments.[12] Among them Howard Morris, a specialist in biopolymer mass spectrometry, who came with him from Cambridge and became a central figure in the department.[13] Morris brought his former doctoral student, Anne Dell, from Cambridge. She held a postdoctoral position, later becoming a professor in the department. Morris was the first to identify the structures of some of the opioid peptides (enkephalins and endorphins) using mass spectrometry, and people from around the world came to rely on his expertise in determining the structures of biological polymers.[14]

Other appointments were made in the areas of molecular biology and genetic engineering.[15] Among them were David Lane and Jean Beggs, both to have major careers.[16] Returning to England from a position in the USA, Eric Barnard was appointed to the second chair in biochemistry, the Rank Chair in Physiological Biochemistry, in 1975.[17] He brought along his colleague, Oliver Dolly, and together they worked in the area of neurological biochemistry, notably on problems related to muscular dystrophy.[18] Chain's legacy continued in the work of Harry Bradford and others engaged in physiological biochemistry.[19] As it turned out, Hartley and Barnard found it difficult to cooperate and it was Hartley, not Barnard, who did his best to make use of the pilot plant. He began by thinking of ways of producing new micro-organisms for industrial fermentation processes. One of his suggestions was that physiological biochemistry be hived off into a new cell biology and physiology department. What was left would then have formed a new department of molecular biology and applied microbiology with links to chemistry and the new biophysics unit. It was to be a new department at the forefront also of genetic engineering. These plans were opposed not only by the botanists but also from within his own department; they did not materialize. Nonetheless links were made to other departments and work in the areas Hartley envisaged increased. Brian Flowers helped out by allowing Hartley to set up a new Centre for Biotechnology in the pilot plant, an interdisciplinary centre with a focus on genetic manipulation, and on fermentation of industrial interest. But the plant proved too costly to run without external support and by 1994, when industrial funds and industrial contracts dried up, it was closed and dismantled.[20]

With respect to teaching, Hartley began with closure of the MSc courses and the setting up of a broadly based BSc course. A new MSc course was added later. Converting what had been largely a research institute into a teaching and research department was no easy task, one requiring that people rethink their roles. Hartley had to think also in terms of Flowers' metaphorical 'horizontal grid', both in the design of the undergraduate course and in the development of research. As to the course, he collaborated with others in the Division on a common first year programme for undergraduates in the life sciences.[21] But this was to have a shaky history and was eventually discontinued. The Division of Life Sciences had five undergraduate streams, botany, zoology and entomology, biology, microbiology and biochemistry, and student numbers began to rise across the Division in the 1970s.[22] In part this was because Flowers pressured the departments to think seriously about how to address environmental concerns in both teaching and research, knowing that this would attract good students. On environmental matters, Flowers' horizontal thinking stretched beyond the life sciences. He wanted links also to those concerned with energy studies, namely physical scientists and engineers.[23] In this he was supported by Southwood and A. J. Rutter, head of botany and plant technology, as well as by people in the RSM. At the time of Flowers' arrival, there was little environmental work within the Division, but much ecological activity. Southwood trained many people who have kept Silwood at the forefront of ecological research to this day. At the time, Flowers pushed to have leadership in ecology move a little more towards Rutter and the botanists and, to some degree, this occurred. He also pressed for the Directorship of the Silwood Park field station to be given to the botanists, but this proved more difficult.

Rutter had assumed the headship of the botany and plant technology department after some difficult years. His predecessor, C. P. Whittingham, had taken over from W. O. James who, as discussed in chapter ten, tried to diversify the departmental interests, but with little success. When Whittingham inherited the chair of plant physiology, and the headship, the only vigorous research areas were in plant pathology under R. K. S. Wood,[24] and in environmental ecology under Rutter. When Whittingham retired, the chair in plant physiology, founded in 1918, and which in earlier times had brought fame to the department, was allowed to lapse. Nonetheless, research in the field was kept alive under J. Barber and J. M. Palmer, both later professors. Timber technology, another old Imperial specialty, was doing well under J. F. Levy.[25] Rutter was a good and very sociable head, and departmental morale improved under his leadership. He encouraged much field work and some of his legacy was seen later in the teaching of his former research assistant, J. N. B. Bell.[26]

In the postwar years the Zoology and Applied Entomology undergraduate course remained fairly classical, perhaps a little old-fashioned by the 1960s, with students working their way through the taxonomic groups — invertebrates, vertebrates etc.[27] Roy Anderson, who was a student in the later 1960s, appreciated his more 'classical' education and credits the many field trips to places such as Jersey, Slapton Lee in Devon, and Dale Fort in Pembrokeshire, with keeping students interested and engaged.[28] The department remained more focussed in its traditional areas of research than did Botany, and this focus served it well. Work continued in systematics under O. W. Richards though after his retirement this field received less attention.[29] Research in the environmental and ecological aspects of entomology increased under Southwood, work in parasitology continued under J. D. Smyth, and work in the behavioural and physiological aspects of entomology continued under Professors Kennedy and Lees.[30] While Munro had earlier brought these various sets of expertise to the college, his own line of work continued also in the applied entomology of M. J. Way, and at the Overseas Spraying Machinery Centre at Silwood where courses were given in pest management.[31] A new Environmental Management Unit under Gordon Conway, supported by the Ford Foundation, carried out computer modelling and economic analysis with the intent of improving pest control decisions and promoting ecologically sound agricultural practices around the world.

Added to the life sciences mix was biophysics under David Blow. He joined the physics department in 1977.[32] Blow, like his old Cambridge colleague Hartley, wanted to see more work in molecular and cell biology at the college. Flowers faced a difficult problem in rationalizing all the various activities. He had to note the professors' preferences while keeping a clear eye on student numbers, as well as trying to predict national needs. By and large, the botany and zoology and entomology departments were content with the way the Division of Life Sciences was working, but the biochemists and biophysicists were not. Further, Flowers needed to sort out what work should be carried out in South Kensington, and what at Silwood. For example, should 'whole organism' work be totally shifted to Silwood? The biological sciences, both at Silwood and at South Kensington, needed more modern facilities and there were debates as to where they should go. Aside from biochemistry, the life science departments had not been part of the redevelopment undertaken during the Jubilee expansion. To complicate, or possible simplify, matters, several people left in 1979. Hartley resigned as head of biochemistry,[33] though not as a professor; Rutter retired, and Southwood, who had been Flowers' chief guide through the maze, left for the Linacre chair of zoology at Oxford.[34] Already in 1978 Flowers set up an advisory committee under Professor Barnard to plan the restructuring of the life sciences.

Because the horizontal connections were not yet as effective as Flowers might have wished, there was still much duplication of effort. Biochemists were going their separate ways in the biochemistry and botany departments, and neuroscientists were going theirs in biochemistry and zoology. Matters were further complicated by changes to Britain's high school syllabus. The new Nuffield Syllabus introduced students to cell and molecular biology, and reinforced the idea that these areas represented modern trends. While true, this was far from being the whole truth. But this is what good high school graduates believed, and Imperial wanted its share of them. Further, it was biology, rather than zoology or botany, that was then being taught in schools, and it was seen as high time to unite botany and zoology at the college.[35] Taking advice from Barnard's committee, Flowers decided to move on this and put the matter to a departmental vote. In 1981 botanists and zoologists, fully aware that a negative vote would leave them out of favour during Flowers' rectorship, voted to unite in a new department of pure and applied biology. Botanist R. K. S. Wood was appointed as its head.[36] It was also agreed to strengthen work in cell biology and ecology, and to end the formal association of the Directorship of Silwood with zoology. This did not prevent a zoologist from being in charge at Silwood and, indeed, M. J. Way was appointed to succeed Southwood as director. It was no easy task keeping Silwood going in what were difficult economic times.[37] Way asked for more student residences and proposed the formation of a Centre for Pest Management which came into existence in 1984. The Overseas Spraying Machine Centre changed into the International Pesticide Application Research Centre (IPARC) under Dr (later Professor) Graham Matthews.[38] Ecological work also continued strong.[39] By the time of Wood's resignation in 1984, there were eight other professors in the new department.[40] In 1989 it became the Department of Biology, returning to a name that had originally been used by T. H. Huxley in 1881 when the Normal School of Science opened in South Kensington.

The Physical Sciences

In this area the priorities for Flowers were the reorganization of the chemistry department and the further development of computing science. His years with the SRC had persuaded him that while Imperial's chemistry department had a proud history, by the early 1970s its standing had declined relative to other departments in the UK.[41] In 1973 the department had seven professors, five of whom were FRSs. It could be compared favourably to its old rival University College London which, at that time, had nine professors, none of whom was an FRS. Flowers did not care for the formal division of chemistry into the three seemingly outdated sections, inorganic, organic and physical. Nor did he care for the departmental constitution, drawn up by Linstead, which perpetuated this division. Professor Barrer was due to retire as head of department in 1977; Professors Barton, Tompkins and Owen were also soon retiring, and Professor West resigned in 1975.[42] West's appointment as reader in analytical chemistry in 1963 had brought a new instrumental approach to the subject, ending the heavy emphasis on gravimetric analysis that had long been taught by L. S. Theobald. The nature of the undergraduate programme, especially in inorganic chemistry, changed radically as a result.[43] Aside from Wilkinson, the only other professor from the 1960s remaining by the later 1970s was crystallographer D. Rogers.[44] With relatively few young people on the staff, the age profile of the department was also a concern. Organic chemistry was the exception in having two recently appointed future stars in place, A. G. M. Barrett and S. V. Ley.[45] In Barrer's field of surface chemistry, L. V. C. Rees had an independent reputation in zeolite research, but physical chemistry was, on the whole, rather weak.[46] Wilkinson still led a lively inorganic section, his own group working on transition metal alkyl complexes.

Flowers decided that the tripartite structure would have to go but this was not something fully accomplished until after his tenure — though he did get rid of the Linstead constitution. He wanted more sections under a single head of department. And, always thinking of horizontal links, he wanted more connections to chemical engineering, biochemistry, materials science, geochemistry and public health engineering. He persuaded Barrer to relinquish the headship two years before his retirement and appointed Geoffrey Wilkinson to succeed him.[47] Wilkinson had the same problems as other heads of department in getting his staff promoted during this period but, with the appointment of two new section heads, C. W. Rees as Hofmann professor of organic chemistry,[48] and W. J. Albery as professor of physical chemistry, as well as the appointment of A. R. Fersht as professor of biological chemistry, it was hoped the department would prosper.[49] In Albery, Flowers had an ally in his determination to break down the traditional barriers between organic, physical and inorganic chemistry, no easy task given the strength and self-sufficiency of the organic chemistry section. Flowers was keen to see a theoretical chemistry section and was supported in this by Wilkinson. But the new professors were largely against the idea which led to some internal tension, and to a degree of dismissal of the department by the Rector. Flowers was also concerned because he had received a series of complaints from students about their treatment in the department. He insisted on more open assessment procedures, that laboratory courses cease being open ended, and that there be a mechanism for student voices to be heard.[50] Perhaps the most serious problem was one of space. In the mid-1970s there was no foreseeable money for the second phase of the new chemistry building.[51] This, together with financial stringency more generally, contributed to a decline in morale. But gradually the old tripartite structure was eroded and chemistry, like physics, became a multi-section, multi-professorial, modern department.[52]

Mathematics and physics in this period continued much in the manner envisaged by the former heads of department, Professors Jones and Blackett respectively. When Jones retired in 1970 there were ten professors remaining which made it difficult to argue for further promotions in the financially tight years that followed. But Jones was succeeded as head of department by David Cox who worked in an area where growth was possible, namely statistics.[53] Cox is a statistician of international repute who trained many doctoral students and set up a one year MSc course in his field. This was timely since, already in the late 1960s, governmental and other bureaucracies, under increasing pressure to account for monies spent, were seeking statistical advice. Further, the Government wished to promote research in a number of areas that demanded statistical expertise, such as environmental protection, social welfare, and medical services and outcomes. Universities were asked to train more statisticians. The financial sector, too, was seeking statisticians. Indeed, business demanded mathematical skills more generally, and the department moved to expand also in the areas of numerical analysis and operational research. Applied mathematics, too, continued strong, notably in fluid mechanics. Two future heads of department, J. T. Stuart and F. Leppington, worked in this area introduced earlier by Professor Scorer.[54] When Cox left the department, work in statistics temporarily declined but recovered with the appointment of Adrian Smith to the chair in statistics a few years later. Jones had been among the last of the feudal heads. Under Cox and his successors, J. T. Stuart and G. E. H. Reuter, more democratic departmental procedures were introduced.[55]

In 1963 Blackett was elected President of the Royal Society and gave up the headship of the physics department. He was succeeded by C. C. Butler who began with a major reorganization of the departmental governance structure. Butler set up some well-defined committees and, unlike Blackett, held regular meetings with the heads of the research groups. In planning for the 1972–77 quinquennium he decided to allow Professor McGee's chair to lapse and redirect some of McGee's group to join James Ring, recently appointed to a chair in infra-red astronomy.[56] He also decided to close down acoustics. Butler's aim was to strengthen astrophysics, applied optics, and solid-state physics, and to provide room for biophysics. Professors Salam and Matthews wanted to create an institute for theoretical high energy physics, but this did not receive widespread support in the college.[57] Butler's successor, P. T. Matthews, served just one term and paid much needed attention to the undergraduate course which underwent complete revision.[58] He, himself, gave a survey course for first-year students. He also reestablished the post of Director of Undergraduate Studies and introduced student evaluation of courses. This turned out to be a more positive experience than its opponents had feared. Matthews extended the optics section to take advantage of new developments in lasers. D. J. Bradley was appointed as professor of laser optics in 1973, becoming head of department one year later.[59]

In 1974 Norman March was appointed to a chair in theoretical solid-state physics, but he left for Oxford in 1977 and the chair was not filled until 1981 and the appointment of John Pendry.[60] Pendry had to carve a space for himself between the mathematicians, who had owned theoretical work on the solid state since the early days of Harry Jones, and the theoreticians in physics engaged in particle theory. Under Pendry, the surfaces of solids became a major field of theoretical study. Bryan Coles had built up the experimental side in solid state (condensed matter) physics and, in 1984, with the appointment of R. A. S. Stradling, a specialist in semiconductor physics, the field became even stronger.[61] The biophysics section under David Blow (see above) was formally opened in 1978 by Max Perutz. The section offered a third year option for undergraduates and a postgraduate programme. In 1977 the department grew physically larger when the atmospheric physics group built an observation tower on the roof. This housed a weather detection radar system to study the development of serious rain storms across the Home Counties. In the same year, sixty years of optics was celebrated with an open house. But the older applied optics was to reconfigure in the 1980s, coming together with electron optics, lasers, and photonics. These changes were supported by Pilkington Brothers Ltd. who funded the appointment of J. C. Dainty as professor of applied optics in 1984.[62]

After the resignation of Bradley in 1980, Ian Butterworth became head of department.[63] An expert in bubble chamber work, he resigned in 1982 to become a director of CERN and was succeeded as head of department by Tom Kibble.[64] While much new research was carried out in this period, it appears that the research orientation of the department did not change radically in the years after Butler. The power that he and Blackett enjoyed as long-term heads had allowed them to structure the department much as it was to remain until the end of the century.

Mining, Mineral Resources, Metallurgy, and Earth Sciences

The late 1960s and early 1970s were tough times for mining departments in UK universities. The department in Birmingham closed, the undergraduate course at Sheffield was terminated and a number of chairs around the country remained unfilled. For many years mining departments had trained people to work overseas, largely within the empire. But in the post-colonial world, work once carried out by British graduates was being done by others, many of whom were trained within their own countries. When Flowers came to Imperial he was determined that the mining and mineral technology department either reinvent itself or close. Metallurgy and geology, too, needed restructuring in his view, but less drastically. Marston Fleming, who had been head of mining and mineral technology, moved to become Pro-Rector with the responsibility for forging contacts with industry at the highest level. He also mediated discussions with his old department. Succeeding him as department head was the professor of mining, R. N. Pryor.[65] Flowers was worried about declining enrolments and was critical of the MSc courses in mining and mineral technology. The courses had low admission standards and low enrolment, especially from the UK. Pryor defended the MSc courses. Writing to Flowers he stated, 'regarding your query as to whether the staff effort and resources involved in mounting the courses justify the return, one must, I think, revert to history'. Pryor claimed that the MSc courses made only a low demand on staffing and that 50% of the students were in fact British. He defended all the courses but in particular the vulnerable mineral process design course where only 10% of the students were British, 'for the industry and the profession it would be a sad day indeed if Imperial College were to decide to pack [it] up'. He noted that the course was well run by Henry Cohen, that it had won worldwide acclaim in the mineral industry, and that 120 people had graduated over the past ten years.[66]

Flowers asked Fleming to report to him on what could be done in the area of mineral resources. Fleming described the subject as covering the whole spectrum of mineral supply and demand; namely, resources, reserves, extraction, processing, marketing and finance. This was a much broader definition of the subject than had been earlier envisaged. He suggested that one possibility would be for the RSM to establish itself as an independent mineral resources institute, but that this would require both external funding and some form of institutionalized connections between different sections within the RSM. Noting the tenor of the times, Pryor decided that a name change might give his department a new sense of direction, and it became the Department of Mineral Resources Engineering in 1976. But a name change was not enough. Pryor did his best to improve morale. He sent a circular to members of the department asking for suggestions. Noting that the academic staff room and coffee club were not generating the kinds of discussion he wanted to see, he suggested that people gather at an informal Friday lunch.[67] But most of the senior staff were overworked with many commitments both within the college and outside. They were not fully prepared to engage in serious discussion of new initiatives and, at that stage, few were as far-sighted as Flowers on the likely outcome for a department such as theirs, unprepared to change.

One initiative marking the way forward was the Environmental Research Unit, set up by Pryor in 1972. It was to examine environmental problems associated with large scale, open cast, non-ferrous metalliferous mines and stone quarries. Fleming, too, had been considering setting up a group to work on ways of restoring open pit mines, the utilisation of underground excavations, and the hydrological problems associated with such sites. Flowers, always conscious of what he called 'horizontal links', saw a possible connection to environmental work elsewhere in the college. As discussed in chapter twelve, he set in motion discussions that were to lead to the Centre for Environmental Technology in 1976.

Another issue came to the fore in 1974 when the UGC decided that Imperial, Heriot-Watt College, and the University of Aberdeen would receive support for training people for the off-shore oil and gas industry. Flowers brought the mining and mineral technology and geology departments together and asked them to collaborate on new degree courses suited to the industry. He told them that it was necessary for the outside world to be reminded that the college had a long history in petroleum studies, and stated that he would seek funding for a new chair in petroleum engineering. Fleming wanted all the petroleum work to be moved into the mining and mineral technology department, but the geology department was opposed.[68] In the event, it was decided that a course in petroleum geology would remain in the geology department, but that petroleum reservoir engineering would go to mining and mineral technology. C. G. Wall was appointed to a new chair in petroleum engineering and introduced an MSc course in the field.[69]

Sadly, Pryor died prematurely in 1979. He was succeeded as head of department by C. T. Shaw.[70] Already under Pryor there had been much revision of the undergraduate courses, and revision continued under Shaw. Students were given more options, courses diversified and became more flexible, and subjects such as mineral economics and management were added to the curriculum. In 1984 the department saw the temporary end of instruction in the Tywarnhaile mine; computer simulated underground surveying was for a while seen as providing sufficient background knowledge for a new generation of engineers, but some actual underground surveying returned in the 1990s.[71]

The departments in the RSM became more integrated both as a result of the oil and gas developments in the North Sea, and because the economic situation in the 1970s forced them to be more conscious of what others were doing. In this connection the geology department had an identity problem not properly resolved in this period. Geology, justifiably, saw itself as a good scientific department. The record of 'pure' geological research going back to the Victorian period was excellent, upheld during the 1960s by people such as Professors Sutton and Watson, and continued into the 1980s by N. Price, and by Watson's student M. P. Coward, appointed to the H. H. Read chair in 1984.[72] But the departmental strength was principally in applied areas of geology. When John Sutton became head of department in 1964, not only did he oversee the physical remodelling of the department, which included a system of interchangeable specimen drawers and the opening of a new Watts Library, he also pushed the department to increase its range of applied, exploration, and engineering work.[73] In the early 1960s this was justifiable but, by the 1970s, it had become a questionable strategy. The department had to struggle to maintain its share in a diminishing market.

Professor Gill retired in 1978 and was succeeded in the chair of petroleum geology by Robert Stoneley.[74] When Stoneley came to Imperial the MSc course in petroleum geology attracted only about 12 students each year. The numbers were very volatile, but Stoneley managed to revive the course and, in addition, designed some short courses for oil company personnel.[75] But, in the longer term, petroleum geology was to merge with petroleum engineering in mineral resources engineering. G. R. Davis modernized the mining geology course but this, too, was a contested field of study. Further, engineering geology was similarly betwixt and between and the professor, J. L. Knill, had difficulty justifying his field's continued existence in the geology department — a difficulty that only increased when he succeeded to the headship in 1979.[76] In part this was because new mega-projects such as the oil terminal at Sullon Voe in the Shetlands, the hydro-electric pumped storage station at Dinorwig in Snowdonia and the Channel Tunnel, all needed engineering geology expertise, but they needed it within a larger engineering context. Projects such as these suggested a different future for the training of engineering geologists. Petroleum studies and engineering geology were not the only contested areas. Turf wars over what should be taught where, and in which departments research groups should be situated, grew fiercer during the 1980s. The competition was not confined to Imperial College; earth sciences, engineering and mining departments across the country were competing for diminishing resources.

While diversification may have seemed a good idea to Sutton in the late 1960s, in retrospect his expansionary moves appear not to have been clearly thought through. When Flowers became Rector he found a department in serious financial difficulty. He believed many of the departmental projects would be better situated in other departments, or within interdisciplinary research centres. To help improve the financial situation, Sutton's successor, G. R. Davis, revived an idea that he had discussed with the Saudi Arabian government in the 1960s.[77] He suggested that oil and mineral exploration in that country extend to the sedimentary cover of the Precambrian shield. The shield had been the main focus of exploration to that date, but Davis believed that rocks in the cover would contain mineral deposits as well as oil and gas. In 1974 he reopened negotiations and, in 1978, a contract was signed between the college and the Saudi government to allow the geology department to conduct an 'industrial-style field operation' over four years.[78] While fruitful in terms of pedagogy and research, this was a stop-gap measure from the financial point of view.

The department was understandably anxious about its future under a Rector who saw much of its work as fragmentary and in need of rationalization. Flowers wanted to merge geology with the other RSM departments into something he believed, and perhaps rightly, was better suited to the modern world. Offended by this, someone in the geology department designed a Stegosaurus stamp to represent what were held to be the antediluvian attitudes of the college administration. This particular dinosaur was said to have had more nerve cells in its rear than in its head.

Curiously, the department adopted the Stegosaurus as its own logo and, in later years, underwent a near similar fate. The department, which had been one of the largest earth sciences department in the country in the mid-1970s, was closed down in 1998. But, as will be discussed in chapter sixteen, geology was merged with earth resources engineering and the discipline has recovered well within a new departmental structure.

Geology is a basic science and geological expertise is needed for all kinds of activities such as mineral exploration and seismology, for large civil engineering projects such as dams and tunnels, and for more academic pursuits such as archeology. Clearly the nation needs to train large numbers of geologists. But the question of where research activity in geology's borderlands should be located was debated in this period. At Imperial, the problems were, perhaps, more acute than elsewhere. The college had always encouraged work of industrial importance and the department had built itself into a good applied geology department before overextending. Ironically William Watts, Imperial's first professor of geology, wanted a department that carried out research and provided training in the basic discipline. Graduates, he believed, would then be able to use their skills in a wide variety of work. Despite his introducing some applied courses he warned about going too far in that direction, and it was against his wishes that the department began its gradual migration from the RCS to the RSM. Had the department remained more focussed it could, perhaps, have weathered the storms that were to come its way in the 1990s.

Already in the early 1960s, people at the college recognized the importance of setting up courses and research in the area of materials and, by 1963, a postgraduate course was offered in 'The Science of Materials particularly as applied to the Chemical, Electrical, Mechanical and Metallurgical Industries'. Centred in the metallurgy department, it drew on expertise in the departments of chemical engineering and chemical technology, electrical engineering, chemistry, and mathematics. In the following year the course was awarded nine DSIR studentships and eleven people were enrolled. It soon evolved into two MSc courses, one in the science of materials and one in materials technology. By 1970, the college had gone some distance in developing the field of materials science across a range of departments and, in that year, the metallurgy department introduced an undergraduate course in materials science. By the end of the decade more undergraduates were enrolled in it than in the metallurgy course. Research in the department had turned in the same direction leading to a name change in 1973 when it became the department of metallurgy and materials science. In 1971, Brian Flowers, then chairman of the SRC, opened Europe's first million volt electron microscope at Imperial, a machine principally used by Professor Swann and his group to study a range of chemical reactions and phase transformations in metals and non-metals.[79] The microscope was large, occupying three floors of the building. Swann was highly skilled in designing peripheral equipment and marketed a number of his designs.

When Flowers came to Imperial as Rector he was keen to expand materials science and technology yet further. But, in 1977, the SRC refused to support the MSc courses any further believing, with some justification, that by then it was the duty of industry to take over. Flowers then set up a working party, which he himself chaired, to make recommendations on the department's future.[80] This was timely also because of a number of retirements and resignations, notably the retirement of Professor Richardson from the chair of extraction metallurgy and the resignation of Professor Bradshaw who headed a process metallurgy group.[81] In 1977 only five students were enrolled in the extraction metallurgy MSc course. Industrial expertise in extraction metallurgy had by then reached such a high level that its importance as an area of research went into decline, though some research continued at the college. Increasingly, extraction metallurgists made links to the mining and mineral technology (later mineral resources engineering) department. Just as with some of the borderland areas of geology so, too, metallurgical extraction work, both pyro- and hydro-, moved over to mineral resources engineering.[82]

The new situation favoured those working in the general fields of physical and chemical metallurgy, and the study of a range of inorganic materials extending beyond metals and alloys to ionic crystals, ceramics, glass, glass-ceramics, solid-state electrolytes, semiconductors, and polymers. The working party noted that while the undergraduate course in materials science was increasingly popular, its curriculum was still too tied to metallurgy. Chemists, for example, claimed that too much time was spent on teaching atomic rather than molecular science. Part of the problem was that in a field as vast as materials, it was difficult to organize a sensible range of teaching input from other departments. One suggestion was for a Division of Materials Science to be formed rather along the lines of the Division of Life Sciences. While this did not occur, better links to chemical engineering, physics, and other departments were made, more interdepartmental colloquia were organized, and the undergraduate course evolved in a more satisfactory direction.

In 1979 D. W. Pashley was appointed to succeed J. G. Ball as head of department.[83] Flowers was able to attract funding from the Wolfson Foundation for a new chair and B. C. H. Steele was appointed professor of materials science in 1981.[84] Steele was a ceramicist working in the ceramics/electronics field and headed a group working in solid-state ionics and in the development of solid oxide fuel cells. Metallurgy, once dominant, became just one aspect of research and teaching in the department. The older BSc in metallurgy was folded into a new and revised four-year BSc course in materials and, in 1986, the department was renamed the Department of Materials.

Engineering

New undergraduate courses

Perhaps the most important development in this period was the introduction of a range of new engineering courses at the undergraduate level. Hugh Ford, the head of mechanical engineering, was among the first to recognize that the training of engineers needed to be modernized so as to produce graduates more in line with the demands of manufacturing industry. Ford had a vision for his own department, outlined in a 1971 paper. In the past, he wrote, the department 'turned out men with a style, just as Cambridge had its style, and I believe that the engineers that we turned out ... were well suited to the kind of work then available.' But the engineering discipline had changed; heat engines had become thermodynamics, heat transfer and combustion had become engineering dynamics, and the strengths of materials had fractured into stress analysis and the science of materials.[85] Ford acknowledged that the new emphasis on engineering science was necessary but, at the same time, believed that many students were failing the BSc (Eng) course because of it. Students, he claimed, could see no connection between what they were studying and the work they were later to do in the 'real' world. He wanted the more practical aspects of engineering not to be forgotten in the move towards engineering science, and that students be exposed to 'real mechanical engineering situations'. He drew attention to an article in *The Times* by Minister of Education, Shirley Williams, in which she stated that 'the party is over for scientists' because public expenditure by research councils seemed to bear little connection to the GNP.[86] Ford was early sensitive to the new political mood and, thinking along more utilitarian lines than was then fashionable in university circles, gave much thought to the kind of practical education that might better fit new graduates to the needs of industry.[87]

Ford invited D. P. Huddie to join the mechanical engineering department in 1971, asking him to plan a new model course to run alongside the one already in place.[88] Ford called the new course 'total technology'.[89] In it students were asked to think of engineering in its totality, that is in terms of research, design, development, planning, manufacture, project management, marketing, industrial relations, sociological and environmental concerns, and finance. Traditionally, engineering graduates were expected to learn about the broader aspects of engineering on the job. Many who joined industry on graduation spent two years as advanced apprentices before being given permanent employment. In the total technology course, a three year programme at the college was sandwiched between two years in industry. Students thus had some industrial experience before entering the college, and contact with their sponsors was maintained throughout their studies. The course was first offered in 1974 with eighteen students and eight industrial sponsors. Ten years later the annual intake of students was forty-five and there were about one hundred industrial sponsors.[90] In addition to taking more traditional subjects, total technology students were able to select also from offerings in industrial sociology, economics, business, communication, and foreign languages.[91]

Ford was ahead of the national trend in setting up this course. That similar courses would be funded nationally was announced in the House of Commons on 16 May 1976 by the Secretary of State for Education, Shirley Williams. The new courses were to include instruction in subjects such as economics, industrial relations, accounting and management, and were to be aimed at a small number of high flyers. In 1977 the UGC invited proposals from universities for four year courses (five years in polytechnics) to train future leaders for the manufacturing industries. Imperial College was among the few successful institutions bidding for these courses — to be offered in the mechanical, electrical, and chemical engineering departments.[92] The Rector asked Professor Brown, head of electrical engineering, to draw up some plans. Known as Dainton courses after Sir Frederick Dainton, chairman of the UGC, they were to be of two kinds: engineering enhanced and technology enhanced. As earlier for the total technology course, it was decided that Dainton students, too, would be sponsored by an engineering firm or manufacturing company. Professor Swanson headed a curriculum committee to plan for the courses and, after its proposals were approved, the UGC provided funding for new members of staff. By the end of the 1970s, there were about forty students a year in both the Dainton and Total Technology courses.

A further development came as a consequence of a 1980 report, 'Engineering our Future' by Sir Montague Finniston who headed the Government Enquiry into the Engineering Profession.[93] One of its proposals was that able school leavers (Finniston suggested the top 25%) be given a broad engineering course leading to an MEng degree after four years. This degree, too, was to be aimed primarily at manufacturing industry, and the course was to include industrial projects and plant design. Once again the emphasis was on the practical, and was seen as putting a further brake on the move towards engineering science that had taken place in undergraduate education since the Second World War. To sum up, three new types of degree, all with a bias towards manufacturing, entered Imperial during the 1970s and 80s. In retrospect, given the global economy and the decline in traditional manufacturing in Britain, that all these new courses were oriented towards manufacturing, appears questionable.[94] But Ford had no crystal ball and was not alone in wanting to give British manufacturing a lift by training graduates attuned to the realities of working in industry, and to the social consequences of both engineering projects and technological change. He was supported by the funding councils and the courses were a brave attempt to provide what was then deemed relevant. A certain amount of fusion among the various courses took place, but the four-year degree, with some broadening of the curriculum, and leading to an MEng, gradually became the norm.[95] This was true also beyond the three departments which first engaged with these new programmes; other disciplines, too, joined the trend towards four-year degrees.

While undergraduate courses were undergoing revision, Flowers was also concerned with how to organize the various sets of expertise within the RSM and C&G so as to generate a lively research culture. He was thinking of a combined school of engineering with groupings based on perceived useful subject concentrations. But the old college and departmental traditions were too strong for such radical change. Instead, Flowers had to work with existing departments and nudge them in the directions he thought best. Some of what happened is described below.

Aeronautics

Aeronautics experienced a difficult decade in the 1970s. The protracted problems related to Professor Argyris's position in the department (see chapter ten) were resolved with his resignation in 1975. Flowers then considered what to do next. The department was one of the largest of its kind in the UK with about 130 undergraduates and about 30 postgraduates, but entry standards were dropping, despite there being four applicants for each undergraduate place. In 1977 Professor Owen, still head of department, asked that he be allowed to let entry numbers drop without financial penalty. He wanted a few years in which to raise standards and achieve a new equilibrium.[96] However Flowers, sceptical about any long-term future for the British aircraft industry, was not sure whether an aeronautics department still made sense. He much appreciated Owen's efforts at diversification but thought that the kind of work set in motion could readily be integrated into the other engineering departments.[97] When rumours of this possibility began circulating, Flowers received a deluge of mail. For example, the Director of the Royal Aircraft Establishment, R. P. Probert, wrote stating that national considerations should outweigh any local financial problems that Imperial might be experiencing. In reply, Flowers stated that aeronautical work would not be closed down. Rather, what was being considered was the removal of the aerodynamical work to mechanical engineering and the removal of structures work to civil engineering. This idea prompted a long and well-argued letter of protest from the former Zaharoff professor, Sir Arnold Hall. Hall, then Chairman and Managing Director of the Hawker Siddeley Group, wrote that when he arrived in the department at the end of the war it was 'little more than a pile of dust after the ravages of the previous five years' but that it had been built into something well worth preserving. He noted that new technical fields especially dedicated to aircraft design and performance had developed out of both the general subject of gas dynamics and the general theory of structures. Even though aeronautical engineering is related to other branches of engineering, he stated that there was much

merit in the coupling of aeronautical dynamics with aeronautical structures, and the combined subject deserved its own department. Flowers,
he wrote, should scrutinize with great care any possible move to reduce
aeronautics to more general engineering.[98]

Flowers acknowledged that the department had a fine reputation
in gas dynamics and accepted that the department gave undergraduates good instruction in structures and in aerodynamics. But few of its
students actually joined the aircraft or aerospace industries.[99] What,
Flowers asked, was the department for? In the face of Owen's organized
opposition to disbanding,[100] and despite misgivings, Flowers decided to
leave the department alone and see where it might go under new leadership. He looked for a new head of department, though this proved
difficult to find. Clearly few will have wanted to take over the headship
with Owen still holding the Zaharoff chair.[101] In the event, Flowers
appointed a senior lecturer, G. A. O. Davies, as department head, a
good choice as it later proved. But much of the work carried out in this
period was a continuation of what Owen had begun earlier, and much
was related to industries other than aircraft and aerospace.

Chemical Engineering and Chemical Technology

Aeronautics was not the only engineering department that gave concern.
Flowers identified problems in chemical engineering and chemical technology, problems that had been apparent already to Penney. In the
late 1960s and early 1970s the department lost a large number of its
staff: junior people who recognized their chances of advancement were
poor during the period of financial stringency, and more senior people
who saw problems with the department's direction under Professor
Ubbelohde. As noted in chapter ten, the department had been criticized by the National Research Development Corporation (NRDC)
in 1962 for not paying enough attention to the needs of the chemical
industry, but Ubbelohde had dismissed such criticism. J. R. A. Pearson,
who succeeded Denbigh as professor of chemical engineering in 1972,
found little improvement ten years later and began complaining to
Penney almost immediately after his arrival.[102] Part of the problem was
the old divide between chemical engineering and chemical technology.
The legacy of the earlier decision to fuse an undergraduate chemical engineering programme on to a postgraduate research institute in chemical

technology remained problematic. Ubbelohde's attitude towards the engineers was dismissive. When complaining to Penney about having to maintain a high ratio of junior to senior staff, he stated that while he could use junior staff for undergraduate teaching in chemical engineering, he needed more senior staff in chemical technology where research was important.[103] Flowers addressed this problem head-on. Why, he asked, was chemical engineering seen as 'hack teaching', why was it assumed that good research existed only in chemical technology, and why should all the senior posts go to chemical technology?[104] Flowers would have liked to fill the vacancy left by Professor Rowlinson with a chemical engineer. But he also wanted to find an internal successor to Ubbelohde as soon as possible, and was determined that that person be a chemical engineer. In the end he compromised. He appointed Roger Sargent, a chemical engineer, as head of department but agreed that Rowlinson's successor would be a chemical technologist.[105] Geoffrey Allen was appointed to found a group in polymer science. Among those he brought in was a future Dean of the C&G, Julia Higgins, appointed lecturer in 1976. But one year later Allen moved to become chairman of the SRC, leaving Higgins and others in his group to make their own way.[106] At roughly the same time another chemical technologist, Anita Bailey, came to the department when Kodak endowed a chair in interface science.[107] A condition of the endowment was that the college support the new section and fund an additional lecturer in the field. The support was not fully forthcoming.

Both Sargent and Pearson rode the wave of the computer revolution in developing models for chemical process engineering.[108] Some of Sargent's colleagues, critical of the attention being paid to the mathematical and computing end of things, complained that under his leadership the experimental side was being neglected. But, overall, Sargent's tenure brought the department together as it had not been earlier. Professor Denbigh's warning that the department needed to take more note of industrial needs and practices was finally heeded, with good longer-term consequences for the undergraduate course, for its graduates, and for research funding within the department. Staff were appointed with a better eye on the teaching needs of the chemical engineering course, and on the needs of the process industries. Sargent's own field blossomed and helped make the department, once again, the leading chemical engineering department in the country. One of Sargent's students from this period, John

Perkins, later became head of department and a future leader in the process area. Another future head of department, Stephen Richardson joined Alan Cornish as a research student in 1972. Richardson's career is interesting in that it illustrates both the coming together of chemical engineering and chemical technology, and the external influence of the discovery of North Sea oil and gas.[109]

In the early 1970s the five main sections in the department, chemical engineering, combustion, high pressure, thermodynamics, nuclear technology, and a few smaller groups focussing largely on materials, began to interact. In this they were encouraged by Flowers, forever seeking horizontal connections also within departments. But not all the five major areas remained prominent. Rowlinson's replacement by Allen marked the beginning of a decline in high pressure work. G. R. Hall the professor of nuclear technology left in 1970 and the chair was filled by G. N. Walton.[110] However, by the late 1970s few students were enrolled in the nuclear technology MSc course. Flowers, believing that the future of the nuclear technology industry looked bleak, encouraged Sargent to let the section shrink.[111] Combustion, the section with the oldest roots in the department, then headed by Professor Weinberg, also began a slow, perhaps temporary, decline.[112] As in the old metallurgy department, materials science, and not just in the area of polymers, grew in importance; interactions between materials science and process engineering proved fruitful.

Civil Engineering

Flowers had to intervene also in civil engineering which needed attention with the impending retirement of the department's long-term head, Professor Skempton. Earlier, Penney had asked Bernard Neal to move from mechanical engineering to head the structures section in civil engineering on the retirement of Professor Sparkes.[113] This was because Penney thought the department needed propping up at a time which saw also the retirement of Professor Baker and the resignation of Professor Buchanan. A. J. Harris was appointed to succeed Baker but the transport chair remained unfilled.[114] In 1976 Flowers chose Neal to take on the headship on Skempton's retirement. Unlike Pippard, Skempton had not been an interventionist head and had not prepared for his succession. His nonchalant attitude towards administration was never seriously challenged, in part because of his high reputation in soil mechanics, his charisma and large physical presence, and in part because the various sections were happy to go their independent ways. The under-graduate course had not noticeably changed since Pippard's time and, according to Professor Brown, it produced a 'perennial slump in student morale in the second year, and its intellectual demands in the third year [were] disputed'.[115] Neal was seen as someone who could be trusted to put things right, and as having the authority to mediate between the different research sections in planning the department's future. Another problem was that civil engineering, reflecting the wide range of depart-mental research interests, offered a large number of MSc courses, not all of which functioned well. In the early 1970s there were courses in concrete, hydraulics, hydrology, public health, seismology, soil mechanics, structures, timber, and transport. Concrete and soil mechanics were the most heavily enrolled, though the demand for transport, which only had space and staff for fifteen students, was proportionately higher. Just about everything in the department needed attention and Neal was likely a good choice as head, except that his own retirement was looming. Given the financial exigencies of the period, he did not have enough time to turn things around in the few years he had left. In 1982 John Burland stepped in briefly as head, but was not keen to take on a major adminis-trative load so soon after coming to the college.[116] J. I. Munro was soon appointed and was doing well, but sadly he died of a heart attack after less than three years in office, and before the department had reached a steady course.[117] After Munro's death John Burland stepped in once again for a short period until the appointment of Patrick Dowling in 1985.[118]

The various difficulties were compounded by a number of retire-ments, and the death of J. R. D. Francis in 1979. His chair remained vacant until the appointment of P. Holmes in 1983.[119] Holmes was a specialist in offshore and coastal engineering which fitted well Neal's ambition, articulated already in the late 1960s, to bring aspects of what he then called 'ocean studies' to Imperial. Neal envisaged this as including a range of work: meteorological, oceanographical, geophysical, biological and engineering.[120] The opportunity he was seeking came in 1975 with an invitation from the Engineering Board of the SRC . Universities were asked to submit research proposals which would be of use to the new North Sea oil and gas industry. Imperial and University College were selected to set up what became the London Centre for Marine Structures and Materials.[121] Neal encouraged work on the underwater environment, on the best materials for underwater construction, and the design of marine structures and underwater cables. But work of this kind had only a limited period of support. After the North Sea industry had learned what it needed, and as the oil reserves diminished, industrial and government funding dried up and the marine technology work was run down.

When F. E. Bruce retired from the readership in public health engineering, Neal decided to unite work in public health with that in hydrology to form a new section, Public Health and Water Resource Engineering, under Roger Perry, a chemist and specialist in water pollution.[122] Perry had connections to the industrial waste management industry which helped fund the new section. He brought in a number of other non-engineers to the department, including more chemists and some microbiologists. Also joining Perry was hydrologist H. S. Wheater, a future head of the section. The new section began to grow in importance as environmental concerns such as waste management and water conservation came to the fore. Civil engineers added words such as 'sustainability', 'climate change' and 'pollution' to their vocabulary.

Professor Harris retired in 1981. While he had been in the department for only eight years, his departure was viewed with regret. A towering figure in the civil engineering industry, Harris was a great popularizer of engineering and a brilliant raconteur. Meetings in the concrete structures section began with Harris pouring a glass of wine for everyone present. Clearly a hard act to follow, he was succeeded in the chair by J. W. Dougill.[123] After the departure of Colin Buchanan, the transport section shrank. When Dowling became head of department he considered closing it down. But, based on earlier consultations, Flowers was not sure that this was a good idea. Among those he consulted was Buchanan, then head of the School of Advanced Urban Studies in Bristol. Buchanan referred back to his earlier problems with Skempton over space, and his not being allowed to compete for Imperial to become a major centre in transport studies.[124] He advised Flowers that with the retirement of Skempton a new chair in transport studies could revive what was left of his old section, but he was not in favour of it moving further toward the social and economic understanding of transport issues. This was not because he believed those areas unimportant, on the contrary. Rather, he thought they did not mesh well with Imperial's mainstream engineering interests. Perhaps he feared a repeat of his own rather negative experience. Flowers decided to revive the section by appointing a major figure as its head. Dowling agreed to this, but it took a while to find someone suitable. T. M. Ridley, already a visiting professor, was eventually persuaded to become permanent head of the transport section, but this occurred only in 1990.[125]

As it turned out this section did develop in the direction that Buchanan had warned against. Many of Ridley's appointees were not engineers, but worked in areas such as transport economics. For example, Stephen Glaister, who headed the section in the early twenty-first century, is an economist. The diversification of staffing within the department, notably in transport and public health, posed a problem for the organization of teaching at the undergraduate level. The problem was not unlike that faced earlier by the chemical engineering and chemical technology department. There, the staffing favoured research work in chemical technology, which led to demoralization among chemical engineers who carried an unfair share of the undergraduate teaching load. Similarly, the civil engineering department was appointing people unable to carry out basic undergraduate teaching. This led to some tension among the more traditional engineers. But, overall, the department came through a major transformation very well. It had excellent support from industry, and continued to be the best department of its kind in the country.[126]

Electrical Engineering

John Brown was appointed to head the electrical engineering department when Willis Jackson became Pro-Rector in 1968.[127] Like Jackson, Brown had strong pedagogical views but his were different. Brown disliked the common first year course and, despite Jackson's opposition, won support for a complete revision of the undergraduate programme. Brown's course was much more focussed on electrical engineering, including in the first year. As to research, the department was then moving in three principal directions, namely electrical engineering as applied to medicine, communication, and materials. Brown was not totally convinced that these directions were sensible. A traditionalist, he saw much of what was happening, albeit lively and interesting, as being on the fringes of mainstream heavy and light electrical engineering as he understood them. Not that the mainstream was ignored and some work continued in electrical power systems. Nonetheless, in recognition of current developments, Brown sensibly wanted his department renamed so that 'electronic' became part of its name, something that surprisingly did not happen until 1991.

In 1970 the electrical engineering department had about 270 undergraduates, 60 research students and 40 taking advanced MSc courses. Brown wanted to close down some of the MSc and short post-experience courses, believing them too demanding of staff time, and that failure rates were too high. Aside from Bruce Sayers' courses in medical technology, which he supported, he saw little advantage in them. Sayers had founded the engineering in medicine laboratory in 1962. He was an interdisciplinarian, bringing together people working in medical electronics, computer modelling, physiology, and epidemiology.[128] The group specialized in the analysis of biomedical signals, and was notable for its work in cardiology, audiology and respiratory neurophysiology. The short intensive courses of which Brown approved were for people working at the Royal Postgraduate Medical School (RPMS) in Hammersmith. Sayers soon expanded these offerings, attracting students also from other medical schools to a DIC course which was offered until 1971 — at which point an MSc course was introduced. By the early 1970s, Penney and Sayers were discussing increased research collaboration with the RPMS, Charing Cross, and Westminster Medical Schools.

The materials section was still being led by Professor Anderson and included a number of chemists and physicists as well as electrical engineers. The section disbanded in 1981 as materials science was restructured throughout the college. Anderson then formed a new section in solid-state electronics, retaining some of his old staff while others moved to join materials groups elsewhere. The communication section was still being led by Professor Cherry. Uncharacteristically, by the 1970s Cherry had difficulty in seeing what the future held for his field. That it had a major future was not in doubt. Cherry recognized that since the Second World War the field had spawned computers, control engineering, cybernetics, information theory, and much else. Businesses had undergone major reorganization as a result, and would continue to do so. But much depended on investment in internationally compatible communication networks, and Cherry could not predict how or when these would emerge. Within the UK, the Post Office was the largest single network controller, but Cherry believed major competition was just around the corner. He recognized there would be an integrated system of cables, radio and satellite communication, and that international communication was a fast growing area, but was uncertain what to predict or how to prepare for the future.[129] Sayers, who had one foot in the communication field, was prepared to predict. Not entirely correctly, he foresaw the end of business offices, of bricks and mortar, and a worldwide conversion to digital electronic communication. He claimed that within a few years much university instruction would be on-line.[130] The problem for others, especially for Flowers and Brown, was how to evaluate the many ideas then current.

In 1976 Brown was given the chair of electrical engineering that had remained unfilled since the death of Willis Jackson. His own chair in light electrical engineering was to be redesignated so as to expand the communication section. The plan was for the headship of the department to move to communication after his retirement in 1979. But finding someone suitable proved difficult.[131] In the event, Flowers appointed Sayers. Sayers envisaged a department focussed largely on information engineering, covering the 'acquisition, transmission, storage, display, processing, enhancement, analysis and interpretation of data'.[132] This pretty well defined the field of informatics as it came to be known. One of the problems during the 1970s was that Brown had been unable to promote younger people to professorships. This had resulted in the migration of talent and a period of decline in the department, something Sayers wished to stop. Despite continuing financial problems, two people were soon given chairs. L. F. Turner became Professor of Digital Communication and E. M. Freeman, Professor of Electromagnetics.[133] Further promotions followed later in the decade but, by then, Flowers, who appreciated Sayers' organizational ability, asked him to move over to head the department of computing.

Computing and Control

A major area of concern for Flowers was computing. As discussed in chapter nine, computing had made a start in both the mathematics and electrical engineering departments. At first a Unit, then a Centre in Computing and Control, had been set up. Professor Jackson was unwilling to let Westcott's control section leave electrical engineering, delaying departmental status which was achieved only after Jackson's death. It is understandable that Jackson wanted to keep the control engineers in his department. The group flourished during the 1960s and included several good lecturers, among them David Mayne who was to become a leading control theorist. The Department of Computing and Control was founded in the C&G in 1970. One year later Penney wrote to Flowers, then chairman of the SRC, outlining the state of affairs in the department. Professor Gill headed the computing section and Professor Westcott the control section. Gill, perhaps not the best of organizers, had trouble within his section. He also had some disagreements with Penney and resigned shortly after Penney became Rector. Finding a good successor was not easy, something that Penney believed was due to the generally low standard of senior computing staff in UK universities. He wrote to Flowers that while there were some promising junior staff at Imperial, he was concerned about promoting them too soon.[134] He mentioned S. J. Goldsack, a reader in the physics department as a possible chair. Indeed, later that year Goldsack was appointed to a new chair of computing science, with the financial support of Control Data Ltd.[135] Aside from research, Goldsack's mandate was to promote computer instruction among all students at the college. This fitted well the vision of Brian Flowers who was to encourage people to think further of ways in which computer skills could be taught across all departments.

Penney had set up a committee, chaired by Hugh Ford, to consider how much of the college resources should go into computing, and how they should be spent. Among other things, Ford's committee recommended the separation of the college computing services from the department, one of the first things arranged by Flowers after he arrived. Professor Harry Elliot from the physics department was put in charge of a new computing services centre.[136] Flowers also set up a computing committee, chaired by Professor Sargent, to advise him on long-term computer planning.[137] This was followed by a computer policy committee, chaired by Professor J. T. Stuart, to advise on overall computing needs in the college. By the early 1980s it was recognized that personal computers were the way of the future and Stuart's committee had to consider how best to decentralize both general computing instruction, and computing services.[138]

Planning for computing science and for computing facilities for the college was very difficult at this early stage. This was in part because of the rapid change and development within the computer industry, and in part because of the huge expense involved. Given today's situation, it is difficult to recollect the enormous cost of the early computers, their slow speeds and small memories. Without the funding of chairs, donations of equipment, and assistance with operating costs, by a computer industry concerned to get things going, universities could not have afforded to develop in this area. When Flowers took over from Penney, Professor Westcott still headed the department of computing and control and ran also the control section. M. M. Lehman had been appointed to succeed Gill and had returned to the college in 1972 as head of the computing section.[139] At the time, the department had an MSc course in computing science, the successor to a DIC course developed earlier by W. S. Elliot. Gill and Westcott had been in the midst of designing an undergraduate course. But Lehman claimed that it would have included too much mathematics and not enough software engineering which, he believed, should be central to any computing science course. Both Lehman and Westcott were interested in developing software skills. Westcott was by then working on economic modelling and wanted his control section to be known not just in the engineering and manufacturing worlds, but also by economists and people in government.[140] Lehman, in turn, asked Flowers for permission to set up what became the Imperial Software Technology Company. This had three major investors, the NatWest Bank, Plessey Engineering, and P. A. International.[141] Students began designing software of interest to the investors and for some other customers, including British Telecommunication and Motorola. While students gained experience this way, the business arrangement soon foundered because the contractual bias of business did not always fit the pedagogical needs of the department.

In a seeming departure from his interdisciplinary bias, Flowers also wanted to separate computing scientists from control engineers, and have the computing department become more mainstream. As a department of computing and control it was somewhat imbalanced with the computing scientists doing most of the teaching and, as yet, little research, and with the control specialists carrying out research, but little teaching. In 1979 Lehman succeeded Westcott as head of department and, one year later, the control section was hived off and returned to electrical engineering. The new department of computing then had about 250 undergraduates and about 60 postgraduates. There were two undergraduate courses, one in computing science, and a four-year course in software systems engineering. Lehman brought a lot of energy to the computing department and used a democratic, more American, style of management in dealing with the many new young lecturers. His was a department run by committees, a somewhat alien idea at Imperial College at the time. The courses he introduced flourished well into the 1980s as demand for information technology specialists grew. Lehman also brought new research funding to the department but was, perhaps, too inward looking and did not take enough care to position the department well within the college. Flowers decided on new leadership and, in 1983, appointed Bruce Sayers to head the department.[142] At roughly the same time Igor Aleksander was appointed to the new Kobler Chair of Management Information Technology.[143] Sayers had been head of electrical engineering and perhaps thought he was being sent to bring order to a weaker department. But, by the time he arrived, computing had made great strides and fared better in the first RAE than did his old department which was still in a bit of a slump.[144] Given Sayers' bias towards informatics, the division of work between the two departments remained somewhat fluid in this period. For example, in 1985 Schlumberger Management and Control endowed a chair in engineering software in the electrical engineering department to promote computer science in the fields of electronic, electrical, mechanical and civil engineering.

Mechanical Engineering

In the late 1960s Hugh Ford was seeking to expand work in the areas of computer assisted design, manufacturing technology, nuclear power, turbo machinery, biomechanics, and materials. Biomechanics, a relatively new field, was supported by a readership funded in 1969 by the Arthritis and Rheumatism Society for S. A. V. Swanson.[145] In 1964 Swanson had formed a research partnership with M. A. R. Freeman, a consultant orthopaedic surgeon at the London Hospital. A small unit was founded, with the support of the Ministry of Health, to work on orthopaedic implants and the mechanics of synovial joints. A range of new implants for the repair of major bone fractures and for hip joint replacements was developed. But this unit was not funded by the UGC and had little support from the research councils. While some funds continued from the Ministry, from the Arthritis and Rheumatism Society, and from the sale of implant designs patented through the NRDC, money was very tight. By the late 1970s the two principals had fallen out and research funding largely collapsed.[146] Ford found funding also for a second readership in 1970, this one in polymers and composite materials. It went to another of his former students, J. G. Williams.[147] While Penney was supportive of expansion into these new areas, he wanted Ford to close down some other work so that he would be able to support the two new readers when their grants ended. It is not clear that any closure occurred. The department had some independence due to good industrial funding for much of its work.

While many Imperial staff took on consulting and contract work, it appears to have been the norm among mechanical engineers. Ford, himself a major consultant and on the boards of several companies, had no objection to his staff spending one day a week on consulting.[148] But a number of staff members were going further and, by the late 1960s, had founded their own companies. This meant that keeping account of time spent on college and external activity became more complicated. Not only time, but also space allocation and the use of college equipment became serious issues.[149] Further, there were some new tax advantages in forming umbrella companies to manage a range of consulting activity. The first person to request forming such a company was Professor Spalding. The idea behind it was to pool all incoming money from consulting, contracts and sales, and come to terms with the college on what was owed for time, space, and equipment. The company then paid all the outgoings, including fees to Spalding and some of his staff, and covered student research grants and expenses.[150] Spalding was carrying out so much external work that Ford had wanted to make his professorship half-time, but in the end did not. The two men, both very gifted, had an abrasive relationship.[151] Another member of the department who brought in many contracts during this period was Professor Cameron who had founded the Lubrication Laboratory in the 1950s and was an expert not only on lubricants but also in the optical analysis of bearings.[152] Locally the 'Lube Lab' was better known as a hotbed of rowing. Cameron designed racing boats and attracted oarsmen also to his tribological line of work.

The fastest growing areas for new contracts, and for a range of entrepreneurial activity, involved the use of computers. There was much industrial demand for computer aided stress analysis and computer aided design. For example, college personnel were using computers to aid the precision forging of gas turbine blades and the design of pressure vessel components. Professor Grant and C. B. Besant both set up design companies in the early 1970s, and others followed.[153] Grant and Besant wanted to rent college space for their companies' computers and drawing offices. This presented Ford with a problem, though there were college precedents. For example, IC Optical Systems paid the physics department £500 per annum for 100 sq. ft of space in one of the laboratories. Given the cost of rental space in South Kensington this was a bargain, and questionable also on legal grounds. Strictly speaking such arrangements were against UGC regulations which provided university buildings for teaching and research, not for commercial sub-letting. The term 'entrepreneur' did not yet have the positive connotation it was to acquire later and Penney was not sure that the college should permit so much business activity. While there was a contracts office within the college administration, it had never had to deal with internal entrepreneurial activity on this scale before. The College Secretary, Mickey Davies, suggested that an industrial liaison officer be appointed within the mechanical engineering department. Ford agreed to this so as to 'make an effective and more professional job of maintaining and developing industrial contracts'. He was told to appoint someone who was not to be paid by results, but was to be given a fixed salary taken from contract profits.[154] The benefits of such business activity to the college as a whole were not yet recognized, and the various ethical matters it raised were not properly resolved. Only after new businesses had mushroomed across the college was the administration of such activity centralized.

Several people have remarked that Ford's own office could have been taken for that of a chief executive of a large corporation. Ford moved in business circles and was well connected also to the Ministry of Defence. His many connections were of great benefit to the department and the monies that flowed in as a consequence kept many a research project afloat. The fine office may well have helped in this.[155] In 1969 when the SRC created science and engineering boards, Ford was asked to chair the new engineering board. He was among the last of the long-term heads of department and hated outside interference in departmental affairs. But the Rector began to gather more power in the central administration, at first related to the control of college space. Penney also wanted more control over hiring procedures, and that there be less discretionary power in the hands of heads of department, not all of whom used it well. Such moves disturbed Ford. As he wrote 'either the head of department has the responsibility for the leadership and is accountable … or he is a chairman, a pawn in the hands of powerful professorial colleagues'.[156]

As discussed above, by the time Ford retired he had contributed much to engineering pedagogy. His successor as head of department, Alan Swanson, continued to think in terms of serving the needs of manufacturing industry. When J. M. Alexander, professor of applied mechanics, resigned in 1978, Swanson redesignated the chair. T. M. Husband was appointed to a chair in engineering manufacture in 1979. In 1981 Husband founded the Centre for Robotics and Automation.[157] But the move towards robotics didn't work out as envisaged since research in this field attracted computer scientists rather more than engineers interested in manufacturing. Swanson raised money also for a chair in design. In 1980 a joint MSc course was launched together with the Royal College of Art in engineering design. It was to be given first on a trial basis but has continued to this day.

Conclusion

The period covered in this and the previous chapter saw the beginning of major restructuring leading to the modern college; that is, to Imperial College as it has become a century after its foundation. New ways of teaching were attempted, the 'feudal' system and the old departmental structures began to break down, disciplinary boundaries were challenged, and a wide range of interdisciplinary work was introduced. The three most notable research areas in which people found common ground during this period were materials, computing applications, and environmental science. Links to medicine were also made, though these flourished more in the following decades. New industrial connections were also apparent, harbingers of an increasing shift in the financial support for research from government to industry. But it was also a period of mega projects, such as the development of North Sea oil and gas, for which both government and industrial resources were forthcoming.

Interdisciplinary work, however, brought new problems. For example, we have seen how it led to a new staff profile in the civil engineering department, a profile not entirely suited to the teaching requirements of the basic undergraduate programme. Indeed, tension between the needs of undergraduate teaching and the needs of research remains problematic not just in civil engineering. As Flowers recognized, it was (and remains) important for Imperial's overall position that there be value added for undergraduates — over and above what a student would gain by attending another institution. While added value comes from studying at a college where entry levels are high, and one's fellow students gifted, it is also important that teachers be motivated to teach the basic disciplines. But, in this period, what to teach undergraduates was highly contested, and finding people to cover basic areas was not always easy.

Consulting work has long been the norm among college staff. Entrepreneurial activity is largely a more recent phenomenon though some members of staff conducted businesses already early in the century. Ever since its founding the college has encouraged entrepreneurial behaviour even if not for personal gain. Those who recognize new opportunities, take risks, and build their own research teams by bringing in outside funding have always been rewarded — financially and socioculturally. But the 1970s saw a different level of activity, with college based businesses opening on a new and increasing scale. This was largely the consequence of advances in computing. Software development for a range of applications provided a relatively low capital-intensive area within which new kinds of expertise could flourish and find a market. Theoreticians had opportunities to construct and test models as never before and, as we have seen, applied their models to the process industries, stress analysis, epidemiological problems, banking, the economy, and more. It took time before all this business enterprise found wide approval within the college, and time for it to be seriously regulated. But the economic problems of the 1970s changed people's attitudes and, with them, the idea of the university. The older gentlemanly ethos in which knowledge was pursued for its own sake, and then shared by means of publication, while never dominant at Imperial, increasingly gave way to a business ethos. In importing new kinds of business values also in the running of departments, Imperial was ahead of many other institutions, and may account for its increasing success in a range of league tables later in the century. While recognizing new realities is never easy, facing them is usually a winning strategy.

One aspect of the turn towards a business ethos has been the demand for greater accountability. Staff have to show how they use their time and the resources and opportunities offered them. New projects have to be rationalized as never before. This has both positive and negative consequences. Many people once had much discretionary power and never saw the need to account for themselves. That they now have to do so is a good thing. But some of the problems associated with the new ethos are only too obvious. The time needed to justify one's own actions and to judge those of others is considerable. Keeping up with new regulations and the filling of seemingly endless forms takes time away from teaching and research. It also engenders some nostalgia for 'the good old days' which were, in fact, not so good for the majority. However, daring research, research that might conceivably not produce dividends, or at least not quickly enough, will likely be rejected in favour of projects that have a better chance of providing short-term results.

The late 1970s saw a shift in importance away from departments and towards departmental sections and new research centres, a topic to be discussed more in later chapters. As Professor Ford had predicted, heads of departments became in effect chairmen, though still with some discretionary power. They mediated between the heads of sections while taking note also of activities in the expanding number of free standing and informal research centres. They also had the increasingly difficult role of figuring out what undergraduates needed to learn. It was a difficult period, but Flowers provided good leadership. He was very much aware of general industrial trends, of the situation of universities nationally, of the national political mood, of where research monies might come from, and where graduates might find jobs. He tried to encourage departments to rethink both their research directions and their course offerings. He was perhaps the most interventionist Rector in the history of the college. But his interventions, while often painful, appear in retrospect to have been for the most part positive.

...

1. B/Flowers; box 1, speech to Old Centralians, 13 December 1973.
2. On the matter of MSc courses, Flowers was ambivalent at times. While critical of many MSc courses he was, for example, also critical of the chemistry department for not joining the MSc bandwagon. The department offered no vocationally oriented MSc courses though one such course, in analytical chemistry, had existed until the resignation of Professor West. Professor Wilkinson's position was that chemistry was a basic science and that chemical expertise was needed in a wide range of industries, in medical research, and elsewhere. He saw the department's

role as preparing good graduates at the BSc and PhD level, able to turn their skills in a number of different directions. Further, he believed that research within the department, while having industrial and other applications, should remain basically fundamental. He told Flowers that senior executives in the chemical and pharmaceutical industries agreed with him, and that they preferred hiring PhD graduates over those with vocationally directed MScs.

3. The creation of interdisciplinary research centres, is discussed in chapters 12 and 16. For a good account of research developments internationally in this period see John Krige and Dominique Pestre (eds.), *Science in the Twentieth Century* (Amsterdam, 1997).

4. Sir Thomas Richard (Dick) Edmund Southwood FRS (1931–2005) was educated at Imperial College and at the Rothamsted Experimental Station where he was a research student in applied entomology under C. B. Williams. He returned to Imperial in 1955, was soon appointed to a chair, becoming both head of the department of zoology and applied entomology, and director of the field station at Silwood Park in 1967. He moved to Oxford University as Linacre Professor of Zoology in 1977, and was Vice Chancellor of the University, 1989–93. Southwood's work at Imperial focussed on ecology, notably on the migration of insects, and the dynamics of natural populations. Much of his research work was carried out overseas. After leaving Imperial he was appointed Chairman of the Royal Commission on Environmental Pollution (1981–5); and of a Working Party on Bovine Spongiform Encephalopathy (BSE) set up in 1984 by the Ministry of Agriculture, Fisheries and Food and the Department of Health and Social Security. His report on BSE was criticized for being too optimistic in its forecast. As a child on his family's dairy farm in Kent, Southwood already displayed a prodigious memory for the names of plants and animals, notably for those of his life-long love, insects. He also displayed a phenomenal memory for people's names, something for which he was renowned in the college. He is remembered also as an excellent organizer and research leader. He was the author of many articles and books including the influential, *Ecological Methods with Particular Reference to the Study of Insect Populations*, (London, 1966). See also obituary, *The Times*, 1 November 2005.

5. Albert Neuberger FRS, editor of the *Biochemical Journal*, was a distinguished biochemist who specialized in protein metabolism. Before becoming head of pathology at St. Mary's he had headed the biochemistry section of the MRC National Institute of Medical Research at Mill Hill.

6. Rector's Correspondence; report of Neuberger Committee, 29 June 1972.

7. The first protein x-ray photograph was made in 1934 and from then on protein structure was the holy grail of structural biochemists. This research programme allowed chemists and physicists into the biological sciences. The discovery of recombinant DNA in the early 1970s suggested that proteins be viewed also as products, or as objects to be engineered. See Pnina G. Abir-Am, 'The Molecular Transformation of Twentieth Century Biology' in Krige and Pestre (eds.), *op. cit.* (3), chapter 26. See also, Robert E. Kohler, *From Medical Chemistry to Biochemistry: The Making of a Biomedical Discipline* (Cambridge, 1982).

8. Rector's Correspondence: Flowers statement to Rank Group Trustees, 21 May 1973. Chain was a friend of the Rank family and had helped lay the path for this approach. In 1967 the Rank Group had offered to fund a chair in some area of food research but nothing came of this. The same is true of a Southwood led initiative to obtain funding for a chair in agricultural ecology in 1968.

9. For example, people engaged in physiological flow within aeronautics did not want biologists making decisions affecting their field and feared a united front.

10. For an early suggestion along these lines see KB/ Memorandum from R. K. S. Wood, 25 September 1970.

11. Brian Selby Hartley FRS (1926–) was educated at Queen's College, Cambridge and the University of Leeds. After gaining a PhD he returned to Cambridge as ICI Fellow in the department of biochemistry. He came to Imperial after eighteen years at the MRC Molecular Biology Unit at Cambridge. Hartley was part of Fred Sanger's group in the Unit. One interesting Imperial connection is that Sanger was one of Professor Chibnall's students at Cambridge.

12. While molecular biology's star was rising this does not mean that, in principle, it might not have been a good idea for some biochemistry departments to continue working, as Chain had done, namely from biology down to chemistry. The younger generation of structural/molecular biochemists who came to the college with Brian Hartley were, perhaps, too dismissive of the older biochemistry. But Rectors need to be realists and to think about where grant monies are likely to come from over the longer term. Colleges such as Imperial with no money of their own cannot easily afford to counter fashion.

13. Howard Morris FRS (1946–) was educated at the University of Leeds in chemistry and biophysics. He later joined the MRC Molecular Biology

Unit at Cambridge, and came to Imperial as a lecturer in biochemistry in 1975. He was appointed to a chair in biological chemistry in 1980.

14. The work of Hartley and Morris had much to do with attracting biophysicist David Blow and biological chemist Alan Fersht to Imperial in 1977 and 1978 respectively. Morris's work depended on the very sensitive bioassaying technique developed in Aberdeen by Professor Hans Kosterlitz. Morris constructed much of his own apparatus and later set up a company in the USA manufacturing mass spectrometers to his specifications. More recently the Centre for Structural Biology has built on this heritage. Research at Imperial has helped in the design and construction of a novel geometry mass spectrometer (Q-TOF) with attomolar sensitivity. In the early twenty-first century this is having a major impact on biopolymer analysis, especially in the fields of proteomics and glycobiology. For more on Dell see chapter 16.

15. In 1978 Hartley joined the Scientific Board of Biogen Ltd. which took over the financing of genetic work at the college earlier supported by the Wolfson Foundation. One of the first secure 'biohazard' laboratories in the United Kingdom was built at the college with Biogen's support.

16. Sir David Lane FRS FMed Sci (1952–) was educated at University College London and was appointed a lecturer in the zoology department at Imperial in 1977. He moved to the biochemistry department in 1981 where he set up a laboratory with Cancer Research Campaign funding. Lane discovered the p53 protein SV40T antigen complex and has continued to work in the area of human tumour suppressor gene function. He moved to the University of Dundee in 1990 as Professor of Molecular Oncology. There he founded Cycladel, a biotechnology company for the development of new drugs for cancer treatment.

Jean Duthie Beggs FRS (1950–) was educated at the University of Edinburgh and was appointed lecturer in biochemistry at Imperial in 1979. Her field is yeast genetics and she discovered a cloning vector for gene manipulation. Beggs returned to Edinburgh in 1985 as a Royal Society University Research Fellow and, in 1999, was appointed Professor of Molecular Biology.

17. Eric Albert Barnard FRS (1927–) was educated at King's College London where he gained both a BSc and PhD in biology. Chain was opposed to the appointments of both Hartley and Barnard. He believed that they would take the department in the direction of molecular biology, undo his good work, and contravene the terms both of the original Wolfson, and the more recent Rank, endowments. He even threatened legal action, including over

the £70,000 which remained in the Fleming Memorial Fund. As it turned out Barnard's work fitted the terms of the Rank endowment well.

18. In 1978 Barnard was given a major grant from the Muscular Dystrophy Association of America. A team of six, including his wife Dr. P. (Penny) J. Barnard, worked with a colony of specially bred chickens in the Eastwood Muscular Dystrophy Research Facility, named for another major donor, W. and J. B. Eastwood Ltd., one of the largest industrial producers of poultry and eggs in Europe. Dolly later worked with botulism neurotoxin (botox) in a special containment laboratory which opened in 1996. While both cosmetic companies and the military may have been enthusiastic to discover new uses for botox, the work proved too toxic for Imperial. Dolly left to work in Dublin taking much funding with him.

19. Henry (Harry) Francis Bradford (1938–) was educated at University College London, the University of Birmingham, and the Institute of Psychiatry, University of London. He joined Imperial as a lecturer in 1965 and was made Professor of Neurochemistry in 1979.

20. This was despite Hartley having increased external contract income from about £6,000 when he took over to about £65,000 in the 1980s. In 1982 the plant became a commercial entity called Imperial Biotechnology. The college invested, as did some venture capitalists, raising a total of £400,000. The idea was to provide for what appeared to be a growing need for purified enzymes. The Centre for Biotechnology drew on expertise from a number of departments working on enzymology, genetic engineering, microbiology and the conversion of biomass into food and methane production. For pilot plant, see GB Minutes, 30 June 1995 and 26 June 1998. After the dismantling of the pilot plant, the refurbished building was named for Lord Flowers.

21. Botany and Zoology and Entomology were already cooperating on a first year biology course for their students. Botany and biochemistry were cooperating on the microbiology course.

22. 67th Annual Report, 1973–4.

23. Flowers understood where monies were likely to come from. In the early 1980s the Department of the Environment was one of the few government sources for large grants. The Department gave over £500,000 to Imperial in this period. For the Centre of Environmental Technology see chapter 12.

24. Ronald Karslake Starr Wood FRS (1919–) entered Imperial on a Royal Scholarship and, on graduating, became research assistant to William Brown. During the war he worked at the Ministry of Aircraft Production and joined the staff of Imperial in 1945, completing his PhD studies in

plant pathology in 1948. He specialized in mycology and fungal physiology. He succeeded Rutter as head of the department of botany and plant technology and was the first head of the department of pure and applied biology.

25. John Francis Levy (1921–2005) was educated at Imperial College where he studied chemistry before moving to botany. As a student he coxed a number of Imperial eights crews, despite having only one leg, and was captain of the RCS Boat Club and, later, of the Thames Rowing Club. After retirement he became president of the Imperial College Boat Club. His interests in both wood and boats (he worked on the 16th century warship the Mary Rose) drew him to the work of Alastair Cameron (see below) in the design and construction of racing boats, though carbon fibre was by then the chosen material. Levy was a Warden of Weeks Hall and served on the College Refectory Committee for 26 years. He was a great college booster. See obituary, *The Guardian*, 29 September 2005.

26. John Nigel Berridge Bell (1943–), known as Nigel, was educated at the University of Manchester and the University of Waterloo. He came to Imperial as a research assistant in 1970 and, in 1989, was made Professor of Environmental Pollution. Bell is a specialist in the effects of low-level atmospheric pollutants on plant performance. He was Director of ICCET, 1984–94.

27. Students specialized in parasitology or entomology in their third year.

28. Anderson, personal communication. The biological departments took advantage of the centres founded by the National Council for the Preservation of Field Studies after the war. The department head, O. W. Richards, made a big impression on the young Roy Anderson who clearly remembers the interesting questions posed by Richards when he was interviewed before coming to the college. For more on Anderson see chapters 15 and 16.

29. Richards retired in 1967 but remained active in the department as a senior research fellow until 1979. He published much, including *The Social Wasps of the Americas* (1978). R. G. Davies, another systematist who retired in the 1980s, has written a good historical account of entomology at Imperial (typescript in ICA).

30. By the late 1970s the department also had some dynamic readers: Anderson, Brady, Conway, Hassell, and Mordue. Professor Peters had founded a good parasitology division, then under Professor Smyth. Protozoology was given a boost by the arrival of P. C. C. Garnham as a senior research fellow in 1968. Cyril Garnham FRS (1901–94) had been Professor of Medical Protozoology at the London School of Hygiene and

Tropical Medicine and was a leading figure in malaria research. Research on the malaria parasite was furthered by Robert Sinden who joined Imperial in 1971 and by Elizabeth Canning, a former student. Both were to become professors at the college (see also chapter 16). For Kennedy, Lees and the ARC unit, see chapter 10. Interestingly W. D. Hamilton who was developing his ideas on kin selection and altruistic behaviour at this time, and who became renowned as an evolution theorist, had a relatively low profile in the department. William Donald Hamilton FRS (1936–2000) was a lecturer, 1964–77, before moving to the USA. When Richard Southwood moved to Oxford he invited Hamilton to join him there in 1984. See Richard Dawkins' obituary of this remarkable man, *The Independent*, 3 October 2000.

31. The machinery centre was supported by grants from the Overseas Development Agency as was an Immunology Unit which provided a testing service for blood meal identification in connection with some disease vectors such as tsetse flies and mosquitoes. See also note 30 above.

    Michael James Way (1922–) was educated at Taunton Grammar School and at Cambridge where he took a BSc in zoology. He worked in pest control research at Rothamsted, and was seconded by the Colonial Office for work in East Africa. He received a conferred chair in applied zoology in 1969. In 1981, when Way was Director of Silwood, he persuaded the Commonwealth Agricultural Bureaux (CAB) to move its biological control headquarters to Imperial. When he retired in 1985, a joint CAB and Imperial College Library at Silwood was dedicated to him.

32. David Mervyn Blow FRS (1931–2004) was educated at Cambridge and was a research student under Max Perutz at the Cavendish Laboratory. When Blow joined him in 1954, Perutz headed a small MRC unit for the study of the molecular structure of biological systems. He was then beginning work on haemoglobin and Blow began his research career trying to purify and crystallize haemoglobin taken from the blood of a variety of animals. Later he turned to more theoretical problems including data analysis and, in this connection, pioneered computer methods for molecular biology and helped lay the foundations of protein crystallography. After completing his PhD, Blow spent two years working in the USA before returning to a post in the MRC unit which by then had joined with Fred Sanger's unit to form a larger MRC unit (later MRC Laboratory). Brian Hartley was part of Sanger's group and collaborated with Blow in his research on the digestive enzyme chymotrypsin. Blow led the team that determined its structure in 1967, only the third or fourth protein structure to be known.

33. Hartley was frustrated in not being able to take his department forward as he wished. Flowers perhaps held out hopes which he could not later deliver due to financial exigencies. For example, some staff salaries paid for under Hartley's MRC grant were not later taken over by the college when the grant expired in 1979.

34. Southwood was succeeded briefly in the headship by J. D. Smyth, Professor of Parasitology, before the department merged with botany.

35. In some ways botany and zoology were being compared unfairly with biochemistry at this time. Biochemistry, in its molecular work, was a fashionable field which brought in research funding and, at Imperial, had more Fellows of the Royal Society than the other life sciences. There was strong pressure on biologists to become more laboratory oriented, more molecular in focus, and more quantitative in approach. The lowering costs of computation in this period led many into mathematical modelling, perhaps at some cost to laboratory, whole organism, and field studies.

36. Some disaffected botanists erected a poster, 'Botany Killed By Flowers' which must have puzzled people unaware of college politics. Professor Smyth took early retirement in 1982.

37. There was some discussion of closing down activities at Silwood since they were a drain on college resources. But the Governing Body recognized that, were Silwood to be sold, the college would never be able to regain such an asset in the future.

38. Graham Anthony Matthews (1936–) was educated at Imperial College before joining the Colonial Office in 1958 as a research entomologist; he was stationed in the Rhodesias and Nyasaland. He returned to Imperial as a lecturer in the zoology department in 1967. One year later he was seconded to Malawi, returning to join the country's newly established Agricultural Research Council. Back at Imperial in 1972, he rose through the ranks becoming Professor of Pest Management in 1993.

39. Some of the ecological work was perhaps given a jolt by serious tree losses at Silwood due to Dutch Elm Disease.

40. Wood remained professor of plant pathology. The other professors were M. J. Way, R. M. Anderson (parasite ecology and mathematical modelling), J. Barber (molecular biology), E. Canning (parasitology, immunology), G. R. Conway (applied ecology, pest management, application of economics), R. G. Davies (application of computing techniques to insect taxonomy), J. F. Levy (ecology and nutrition of microorganisms invading wood, timber technology) and M. P. Hassell (insect population ecology and mathematical modelling).

41. For earlier work in this department see chapter 10. Barton received the Nobel Prize in 1969 and Wilkinson received it in 1973, the year Flowers came to Imperial. Flowers recognized that the department had flourished in the 1950s and 60s, but believed it needed reconfiguring.

42. Thomas Summers West FRS (1927–) was educated at the University of Aberdeen and the University of Birmingham and was on the staff at Birmingham before being appointed to the new readership in analytical chemistry at Imperial in 1963. He was promoted to a professorship two years later. While one of West's specialties was organic analysis on a sub-micro scale, the chair was listed in the inorganic section which led to problems in the 1970s when money was tight. Wilkinson resented having to share resources with West and relations between them deteriorated. West, a vocal proponent of Imperial becoming an independent university, resigned in 1975 and moved to become Director of the Macaulay Institute at Aberdeen. When Flowers asked Wilkinson whether he wanted to keep the chair in analytical chemistry he replied that he would prefer to redirect the funds elsewhere. Later, he complained to Eric Ash that the chair had been taken from the department against his wishes. See Chemistry Department correspondence; Wilkinson to Flowers, 12 March 1975 and Wilkinson to Ash, 28 September 1988. Barrer, on the other hand, strongly advised Flowers to maintain the chair in analytical chemistry. See Barrer to Flowers, 27 February 1975.

    Barton resigned shortly before he was due to retire, and did so without proper notification. The Rector first heard about it through the grapevine and an announcement in *Chemical and Engineering News*. A further loss to the department at this time occurred with the death of E. R. (Eric) Roberts who, after many years of service, and increasingly handicapped, died of complications due to Parkinson's disease in 1975.

43. Students were introduced to physical methods such as atomic spectroscopy, polarography, molecular luminescence, microwave spectroscopy etc. The department had earlier been designated the University of London Centre for X-ray diffraction and Raman spectroscopy services. West's MSc course in analytical methods was heavily oversubscribed in the 1970s, and about twelve undergraduates a year opted to specialize in analytical chemistry. Further, about one-third of the non-academic chemistry positions advertised in this period asked for expertise in analysis. In retrospect, it seems wrong to have allowed this chair to be terminated.

44. Donald Rogers (1921–99) was educated at King's College London. Before coming to Imperial in 1961, to the new Readership in Crystallography, he had been a lecturer at the University of Wales (Cardiff). He worked on

molecules of biological interest and was given a conferred chair in 1965. Rogers took over the teaching duties of Archibald John Edmund Welch (1915–88), a crystallographer and physical chemist with a forty-two year association with the department, as both student and member of staff. Welch became Assistant Director of the department in 1960 and retired in 1981.

45. Anthony Gerald Martin Barrett FRS FMedSci (1952–) received his BSc in chemistry from Imperial and became a research student under D. H. R. Barton. He was appointed lecturer in the department in 1975. In 1983 he left for a chair at Northwestern University in the USA and from there moved to Colorado State University. He returned to Imperial in 1993 as Glaxo Professor of Organic Chemistry and Director of the Wolfson Centre for Organic Chemistry. In 1999 he was appointed Sir Derek Barton Professor of Organic Chemistry. In 2000 he won the GlaxoWellcome Award for Outstanding Achievement in Organic Chemistry.

   Steven Victor Ley FRS (1945–) was educated at Loughborough University of Technology where he gained both a BSc and PhD. He joined Imperial in 1974 after spending two years as a research fellow at Ohio State University. He was appointed Professor of Organic Chemistry in 1983 and head of department in 1989. He left for a chair at Cambridge in 1992.

46. Lovat Victor Charles Rees (1927–2006) was educated at Aberdeen where he gained a PhD under Barrer. He began his career at Aldermaston becoming a section leader in nuclear chemistry before joining Imperial as a lecturer in 1958. He rose through the ranks to become Professor of Physical Chemistry in 1987 and moved to Edinburgh in 1993.

47. When Wilkinson became head of department, he promoted the first woman to become a Departmental Superintendent at Imperial College. Christine Wright succeeded F. W. L. (Les) Croker in the position.

48. Flowers, anticipating problems relating to status, wanted the Hofmann chair to lose its name and for Rees to become simply the professor of organic chemistry. But Rees was reluctant to relinquish the title and suggested that, to maintain parity, the inorganic chair be named for Edward Frankland, which it was. Later the principal organic chemistry chair was named for D. H. R. Barton and the Hofmann chair became associated with the headship of the department.

49. Charles Wayne Rees FRS (1927–2006) was educated at University College, Southampton. He was lecturer in organic chemistry at Birkbeck College, 1955–7, reader at King's College London, 1963–5, Professor of Organic Chemistry at the University of Leicester, 1965–9, and Heath Harrison

Professor of Organic Chemistry at the University of Liverpool before coming to Imperial in 1978.

Wyndham John Albery FRS (1936–), known as John, was educated at Winchester and Balliol College, Oxford. On graduation he won a fellowship at University College, Oxford and worked there until his appointment at Imperial. Albery brought an old strength, electrochemistry, back to the department and applied it to some biological problems. He returned to University College as Master in 1989. In the theatrical tradition of his family, Albery has also written sketches for review and TV performance.

Sir Alan Roy Fersht FRS (1943–) was appointed Wolfson Research Professor of the Royal Society in Biological Chemistry in 1978. As with Professor Blow (biophysics) and Professor Hartley (biochemistry), Fersht came from the MRC molecular biology laboratory at Cambridge. At Imperial Fersht continued to work with his former colleagues, work that encompassed the mechanism, energetics and specificity of enzyme catalysis. A pioneer in protein engineering, Fersht left for a chair at Cambridge in 1988.

50. At this time chemistry was the only department without a set minimum amount of work required. Students felt compelled to complete as much laboratory work as possible and complained that quantity was valued over quality. In the 1970s the department had approximately 200 undergraduates (250 by 1984) and about 160 postgraduates. Chemistry also had a high number of postdoctoral research workers, with about twelve per year funded by the UGC (funding later taken over by the research councils).

51. The history of funding for this building is complex. It would appear that Linstead persuaded the UGC to redirect some monies intended for the second chemistry building towards the new biochemistry building. Linstead claimed that the UGC had agreed that further monies for chemistry would be forthcoming in the near future, but that turned out not to be the case. The college tried to force the UGC's hand by getting it to fund the demolition of the central part of the old RCS building in 1971. But, as Professor Wilkinson put it, 'this only left a hole in the ground' and the UGC's bluff was not called. The timing could not have been worse given the recession and the freezing of UGC funds. Not until 1989 were expansion proposals approved allowing the second phase of the building to be completed in 1993. See G. Wilkinson, 'Report on the Chemistry Department' (November, 1976). See also B. Atkinson, 'Department of Chemistry: recent history, 1960–89', (copy on disc in ICA). Atkinson's history gives a good account of the department in this period.

52. The end of the tripartite structure was not fully achieved until the tenure of David Phillips as head of department. See chapter 15.

53. Sir David Cox FRS (1924–) was educated at the universities of Cambridge and Leeds. During the war he was a statistician at the RAE and later became Professor of Statistics at Birkbeck College (1955–66). Very versatile, he applied statistical methods in many ways: in scientific and medical research, in helping government departments and the pharmaceutical industry organize and evaluate data, and in the epidemiology of HIV AIDS. His proportional hazards regression model is a major contribution to modern statistics. He was Professor of Statistics from 1966, leaving Imperial in 1988 to become Warden of Nuffield College, Oxford.

54. John Trevor Stuart FRS (1929–), known as Trevor, was educated at Imperial College and returned to join the staff after a few years working at the aeronautics division of the National Physical Laboratory. He was appointed Professor of Theoretical Fluid Mechanics in 1966. He succeeded Cox as head of department in 1974.

    Frank Leppington (1939–) came to Imperial in 1964 after gaining a PhD at the University of Manchester. Leppington was head of department 1986–92, and was principal of the faculty of physical sciences, 2002–04.

55. Gerd Edzard Harry Reuter (1921–92) was educated at Cambridge. He taught at Manchester and at Durham where he was Professor of Pure Mathematics (1959–65) before joining Imperial as Professor of Mathematics in 1965. He was head of department, 1979–83, and carried out work in the dynamics of fluid flow, the dynamics of epidemics, and of biological populations.

56. James Ring (1927–) was educated at the University of Manchester. He joined the staff there before moving to become Professor of Applied Physics at the University of Hull in 1962. He joined Imperial as Professor of Physics in 1967.

57. This proposal, for a national institute, was discussed by the Growth Points Steering Committee in 1968. It was opposed by several people, among them Hugh Ford, who wrote to Penney arguing that such an institute would drain funds from smaller projects and that space at the college was scarce and needed for the expansion of computing facilities. KEM/16/7, Ford to Penney, 29 February 1968. See also FC and EC minutes, 4 October 1968. One unfortunate result was that Salam, wishing to spend more time at the international centre for theoretical physics that he founded in Trieste, became a half-time professor at Imperial from 1973.

58. Undergraduate numbers increased considerably during Butler and Matthew's tenure, reaching about 165 (12% women) per year in 1976,

the year Matthews left the college. For more on the work of Butler and Matthews. see chapter 10.

59. Daniel Joseph Bradley FRS (1928–) was educated at Birkbeck College and at Royal Holloway College where he was a student of Samuel Tolansky. In 1960 he was appointed to a lectureship in the physics department at Imperial where he set up a research programme in UV solar spectroscopy using rocket technology to reach high altitudes. In 1963 he began work in laser physics but returned to Royal Holloway College as a reader one year later. In 1966 he was appointed professor and head of department at Queen's University, Belfast. He returned to a chair in laser optics at Imperial in 1973 and headed a group in optical physics, laser physics and space optics. Bradley, like others in this period, was frustrated by cut backs and the UGC rule governing the ratio of senior to junior positions. For example, he was unable to maintain the chair in optical design which had a proud history at the college. He was also justifiably critical of the college administration's handling of some departmental SRC grant applications. He resigned in 1980 and moved to a chair at Trinity College Dublin.

60. Norman Henry March (1927–) was not a major presence at Imperial. While he published many books and papers he did most of his work close to his home, in the library at Egham. He left in 1977 for the chair of physical chemistry at Oxford. Like Charles Coulson, his predecessor in that chair, March was a Methodist minister.

Sir John Brian Pendry FRS (1943–) was educated at Cambridge and began his career at the Cavendish Laboratory. Before coming to Imperial, he headed the theoretical group at the Daresbury Laboratory 1975–81. He currently works on new optical materials (metamaterials) and, in 2005, was awarded (together with his five-member international team) the European Commission's Descartes Prize. Pendry was the first Principal of the Faculty of Physical Sciences, 2001–2.

61. Richard Anthony Stradling (1937–2002), known as Tony, was educated at Oxford and was appointed a lecturer and teaching fellow (at Christ Church) in 1968. In 1978 he was appointed to the chair of natural philosophy at St. Andrew's University, moving to Imperial in 1984 as Professor of Semiconductor Physics.

62. James Christopher Dainty (1947–) was educated at the Polytechnic of Central London and Imperial College. He held positions at Queen Elizabeth College and the University of Rochester N. Y. before returning to Imperial in 1984 as Pilkington Professor of Applied Optics. There is an interesting historical connection between the University of Rochester and Imperial College. Professor Conrady's daughter Hilda and her husband

Rudolph Kingslake moved to the University of Rochester in the 1920s to set up the technical optics department. The technical optics department at Imperial, where they had been students, was their model. Rochester is situated at the heart of the American optical industry (Eastman/Kodak, Bausch and Lomb etc. are located there).

63. Ian Butterworth FRS (1930–) was educated at the University of Manchester and worked as a Scientific Officer at Harwell before coming to Imperial in 1958 to work with C. C. Butler. He became involved in the design and construction of bubble chambers and carried out important work on particle detection. In his 1988 FIC citation for Butterworth, John Smith, the staff orator, likened him to 'an unmade bed'. Whatever this implies, it was clearly no obstacle to advancement. Butterworth took over leadership of the high energy physics group from Butler before moving to CERN in 1982. He left CERN in 1986 to become Principal of Queen Mary College.

64. Thomas (Tom) Walter Bannerman Kibble FRS (1932–) came to Imperial in 1959 as a research student under Abdus Salam. Kibble joined the staff and made important contributions to particle theory with his idea of the cosmic string. He was appointed professor of theoretical physics in 1975. In 2005 he was given the *Nature*/NESTA award for lifetime achievement in creative mentoring in science. The award was in part based on reports of his outstanding supervision by former research students. See *Nature*, 434 (24 March 2005), 421.

65. Robert Nelson Pryor (1921–79) was born in India where his father was a mining engineer. He was educated at the RSM but his studies were interrupted by the war when he returned to India with the Royal Engineers. After completing his degree in 1948, he worked for the Rio Tinto Company in Spain, becoming Chief Mining Engineer. He moved back to London in 1960 where he worked as a consultant for the company. Pryor served as President of the Institution of Mining and Metallurgy. He was appointed to the chair of mining in 1968, but had to be seriously wooed before accepting the job. At Imperial he taught production management, the design of large open-cast operations, and the environmental recovery of spent mines. See obituary by M. G. Fleming, *Nature*, 282 (December 1979), 890.

66. Fleming correspondence file; Pryor to Flowers, 17 October 1977.

67. Mineral Resources Engineering Department correspondence file; Circular, 27 May 1976.

68. R. G. Mason, professor of geophysics, wanted petroleum studies, geochemistry and geology to remain together. He and Professor Gill, professor

of petroleum geology, wanted an institute or school of petroleum studies, rather than having the field split between geology and mineral technology. See Geology PHS/37; minutes of meeting of heads of section in geology, 29 April 1974.

69. Professor Gill's oil technology chair was renamed chair in petroleum geology. Colin Gulliver Wall (1927–98), was educated in petroleum engineering at the University of Birmingham where he also took a PhD in chemical engineering before working for Shell International as a petroleum engineer. He had experience in offshore drilling and reservoir engineering and, like many senior staff in the RSM, spent much time in consulting work.

70. Charles Timothy Shaw (1934–) was educated at the University of the Witwatersrand and McGill University. A specialist in mineral exploration, he worked in the South African mining industry before moving to Virginia Polytechnic Institute where he was associate professor of mining until his appointment at Imperial.

71. The college did not divest its full interest in the mine until later.

72. Neville Price (1926–2005) came to Imperial in 1964 from the Mining Research Establishment at Isleworth where he had been Head of Geology and Strata Control. He joined John Ramsay's group and was instrumental in setting up the MSc in structural geology and rock mechanics. He succeeded Ramsay as head of the structural geology section, leaving for a chair at University College in 1984.

    Michael Peter Coward (1945–2003) worked in the mining industry in South Africa after completing his PhD under Watson. He returned to academic life at the University of Leeds and was appointed to the H. H. Read chair in 1984. Coward entertained the department with his singing in Gaelic. He was an enthusiastic dancer and a lover of the outdoors and mountains. See *GEOlogIC* (1985) 235.

73. Geologists, it seems, wanted to occupy all the borderlands of their subject. For example, according to the quinquennial statement for 1972–77, the following were just some of the research areas in the geochemistry section alone: prospecting in desert and glaciated terrain, interpretation of geochemical anomalies, geochemistry in mineral exploration, remote sensing techniques, compilation of geochemical maps, agricultural applications of the geochemical survey, significance of trace element distribution in relation to animal and plant nutrition and disease, applied marine geochemistry, and geochemical reconnaissance in pollution surveys. An example of the latter work, which could equally well have been carried out in the chemistry department, is Iain Thornton's work on the study of

heavy metal and radioactive pollutants in dusts and soils. All were legitimate areas of work, but this number, when added to the many projects in all the other sections of the department, represented a problem.

74. Robert Stoneley (1929–) was educated at Cambridge and, after graduating in 1951, worked on geological mapping in Antarctica. He joined British Petroleum in 1953, engaged in exploration activity around the world, and was appointed Chief Geologist to the Oil Service Co. of Iran 1974–8. He came to Imperial at a difficult time but managed to turn the MSc petroleum geology course around and hire new staff. He introduced petroleum geophysics in 1982 and reservoir geology in 1987. Some important industrial contracts came to the department as a result.

75. By the early 1980s the oil industry was supporting much research and teaching across the college. For example British Petroleum funded work in geology, work carried out by Professor Pendry in physics, and research work in the Physiological Flow Unit. Shell funded a readership in petroleum engineering.

76. Sir John Lawrence Knill FREng (1934–) was a student in the department and graduated with the Watts Medal in 1955 and two years later with a PhD (under Sutton) and the Judd Medal. For some years he worked with the civil engineering firm Sir Alexander Gibb and Partners, on groundwater resources around Tehran, and on the water tightness and seismic conditions of oil reservoirs. He returned to the college as a lecturer in 1959, and was appointed Professor of Engineering Geology in 1973. Knill resigned from the college in 1988 to become Chairman of the Natural Environment Research Council.

77. Grosvenor Rex Davis FREng (1922–) was a graduate of the University of South Africa (Rhodes) and was appointed professor of mining geology at Imperial in 1966. He had much experience in the mining industry and as Chief Geologist in Southern Rhodesia (later Zimbabwe). Davis was succeeded in the chair of mining geology by Dennis Langston Buchanan (1947–). Buchanan was educated at The University of South Africa (Rhodes) and came to Imperial as a research student in 1973 after working as a geologist in the South African mining industry. An expert in economic geology, he was appointed lecturer in 1980 and was given the Mineral Industries Chair in Mining Geology in 1984.

78. See Mary Pugh, 'The History of the Geology Department. Imperial College, 1958–88' (typescript, 1995), 135. The Saudi government funded the project with a budget of £400,000 and a fee to the college of £100,000.

79. Peter Ronald Swann (1935–) was educated at University College Swansea and Cambridge. For a few years he worked for the US Steel Corporation

in Pittsburgh and while there set up a company with his brother manufacturing parts for electron microscopes. He joined Imperial as a senior
lecturer in the metallurgy department in 1966. He was appointed to a
chair in 1974 with support from the Central Electricity Generating Board.
Swann, a leading corrosion expert, returned to the USA in 1977.

80. Vice-Chancellors and people in Flowers' position naturally seek out areas
of possible growth. In the mid 1970s growth was a problem in the university sector, but materials science was seen as something that could still
attract outside funding.

81. Paul Grieveson (1934–) was appointed to succeed F. D. Richardson.
Grieveson was educated at King's College, Newcastle and Imperial College.
He then joined the faculty of the Carnegie Institute of Technology before
leaving to work for the US Steel Corporation. He returned to Britain as a
lecturer in extraction metallurgy at the University of Newcastle. A specialist
in ferrous metallurgy, he came to Imperial in 1978 from Strathclyde where
he held a chair in extraction metallurgy.

82. Hydrometallurgy was developed at Imperial by Alfred Richard Burkin
(1923–). Burkin joined the RAF after leaving school and later took an
external BSc in chemistry at the University of London. He then worked
for Ilford Ltd. while carrying out research also for a PhD. After a few
years as a lecturer at University College Southampton, he joined the
Department of Mining in 1953, moving to metallurgy in 1960. He was
promoted to a chair in hydrometallurgy in 1980. Professor A. J. (John)
Monhemius who came to Imperial in 1966 as a PhD student worked with
Burkin and joined the staff on graduation. Hydrometallurgy moved to the
department of mineral resources engineering in 1986 but research in this
field gradually wound down. Monhemius became Dean of the RSM in
2000 (Dean of Engineering when the new faculty system came into effect
in 2001).

83. Donald William Pashley FRS (1927–) was a physics student at Imperial,
gaining a BSc, and a PhD as a research student in Morris Blackman's
electron diffraction group. He stayed on at the college as an ICI Research
Fellow, continuing work in the area of electron diffraction, before joining
the research laboratories of Tube Investments in 1955. He became Director
of the Laboratories in 1968, returning to the college as Professor of
Materials in 1979.

84. Brian Charles Hilton Steele (1929–2003) studied chemistry at the University of Birmingham. He worked for Morgan Refractories before coming to
Imperial in 1958 as a research student. For his PhD he carried out path-
breaking research in solid-state electrochemistry. He stayed at the college,

becoming a reader in 1975 and Professor of Materials Science in 1981. In 2001 Steele, together with some of his former students and colleagues, John Kilner, Alan Atkinson and Nigel Brandon, founded CERES, a company developing fuel cell technology and products — a spin-out from their research at Imperial. It raised £16m when floated on the London Stock Exchange in 2004 and is a leading company in its field.

85. KEMA/1/24 H. Ford, 'Mechanical Engineering' (typescript paper, 1971).

86. Williams quoted in Ford, *op. cit.* (85). According to Ford, while research council monies in science and engineering rose by 14% per annum over the period 1960–66 and about 12% from 1966–71, the GNP rose on average only about 3%. While one should not expect any direct correlation, part of the problem may have been due to overly huge sums being invested in risky research projects such as controlled nuclear fusion and new supersonic aircraft during the 1950s and 60s. Knowledge of the lack of connection between research expenditure and the GDP was hardly new in the 1970s. It was known within the Ministry of Technology and led to much disillusionment. As Edgerton has put it, the lack of connection 'fatally undermined ... the analyses that led to [the Ministry's] creation'. David Edgerton, *Warfare State: Britain, 1920–70* (Cambridge, 2006), 231. Williams may have been responsible for popularizing the idea that there should be a strong link between GDP and higher education and, as a consequence, gave future governments the perfect excuse for cutbacks.

87. What does seem clear is that other university people reacted less intelligently to the political mood than Ford and began, more and more, to justify themselves almost solely in economic terms. Even in arts faculties people began to see themselves as no longer providing a liberal education for an elite, but a mass education for the good of the national economy. Because of its charter, Imperial College had always been sensitive to the economic needs of the country and so had less need to adjust the ways in which it justified itself to the outside world.

88. Sir David Patrick Huddie (1914–98) had been Chairman and Managing Director of Rolls Royce Aero Engines. He resigned from this position when Rolls Royce went into receivership in 1971. See obituary by Hugh Ford, *IC Reporter* 23 June 1998. Huddie's successor at Rolls Royce Aero Engines was Sir Ralph Robins, a mechanical engineering student at Imperial in the 1950s who entered Rolls Royce as a graduate apprentice.

89. For a brief discussion of this see David Dickson, 'A move away from "academic" engineering degree courses', *THES* (8 November 1974), 4.

90. For further details see J. Barnes and P. Grootenhuis, 'Mechanical Engineering' in Adrian Whitworth (ed.), *A Centenary History: A History of the City and Guilds College, 1885–1985* (C&G, 1985).

91. For the development of these non-technical areas of study at Imperial see chapter 14.

92. See letter from Sir Frederick Dainton to the Vice Chancellor of the University of London, 16 May 1977 (Rector's correspondence file). In this letter Dainton stated that there had been a good response to his invitation but at that stage the only proposal from London that the UGC was considering was the one from Imperial College. The other successful institutions were the University of Manchester, Manchester Institute of Technology together with the Manchester Business School, and the University of Strathclyde. Later some other institutions were included.

93. Sir Harold Montague Finniston FRS (1912–91) was a leading industrial metallurgist. In 1933–5 he had been a lecturer in metallurgy at Imperial. He was Chief Metallurgist at the UKAEA, 1948–58, and became Chairman of the British Steel Corporation, having earlier been part of the government committee to renationalise the steel industry. In 1979, the year prior to publication of his report, Finniston was the Special Visitor at Commemoration Day.

94. One plausible response to this might be that Imperial College students should be prepared to help advance manufacturing industry worldwide.

95. At first UGC rules did not permit the MEng as a first degree, and the four-year degrees started as BEng. By the 1990s all the BEng courses at Imperial had been abandoned. The longer MEng courses also allowed better integration with European universities and colleges of technology where a five-year training of engineers was not unusual. Student exchanges became more manageable.

96. Rector's correspondence; Owen to Flowers, 20 October 1977.

97. Another direction Flowers considered for the department was aviation management since he did see the air travel industry as secure and growing. Flowers approached David Nicholson, chairman of the board of British Airways about supporting a chair in the field. B/Flowers; Flowers to Nicholson, 2 January 1974. For Owen's diversification of departmental research see chapter 10.

98. B/Flowers; Probert to Flowers, 18 December 1979; Hall to Flowers, 10 January 1980.

99. Students joined other engineering industries, the automobile industry for example. Increasingly students unable to find engineering positions moved to the City.

100. I think it likely that some of the younger staff members supported Flowers and may have seen greater opportunities for career advancement within the civil or mechanical engineering departments. But Owen was among the last of the 'barons' at Imperial and expected fealty. He appears to have received it. See chapter 9 for the 'feudal system'.

101. The only other professor in the department, P. Bradshaw, refused the headship. Flowers also tried to bring in someone from another department but without success. Owen was to retire in 1985 but agreed to give up the headship in 1982.

102. John Richard Anthony Pearson FRS (1939–) known as Anthony was educated at Trinity College, Cambridge where he studied mathematics. His postgraduate work was in fluid mechanics. After a few years in industry he returned to Cambridge as Assistant Director of Research in the Chemical Engineering Department. His book *Mechanical Principles of Polymer Processing* (Oxford, 1966) became a standard work. He and Ubbelohde had one thing in common, an interest in food and wine. Pearson was an active member of the Imperial College Wine Committee and helped reorganize the dining facilities in the Senior Common Room (SCR), including establishing a SCR bar. Pearson insisted that the TOEFL test, for students for whom English is a foreign language, be taken by foreign students taking his MSc course in polymer science and engineering. He was the first at Imperial to do so and the practice was soon copied. Pearson joined Schlumberger after leaving Imperial in 1982.

103. KCT14 Ubbelohde to Penney, 28 July 1969.

104. See correspondence between Flowers and Ubbelohde in KCT 18. To give the department something concrete to think about, Flowers established a working party to look into how safety in the installation and operation of plant was taught at the college. He was influenced in this by a spate of industrial accidents, notably the explosion and fire at the Flixborough chemical plant.

105. Roger William Herbert Sargent FREng. (1926–) was educated at Imperial College and, after seven years experience with Air Liquide in France, he returned to the college as a senior lecturer. He was appointed to the chair of chemical engineering in 1962 when Denbigh succeeded to the Courtaulds chair. Sargent is widely seen as the father of process system engineering and his many students can be found in important positions around the world. His SPEEDUP design package brought considerable royalties to the college.

106. Sir Geoffrey Allen later became Director of Research at Unilever. Higgins (personal communication) claims that, in retrospect, being left to think

for herself was good for the future development of her career. For more on Higgins see chapters 15 and 16.

107. Anita Irene Bailey (1926–) was educated at the University of the Witwatersrand and at Cambridge University. On graduating she continued to work at Cambridge until joining the MRC Biophysics Research Unit at King's College London. In 1970 she moved to head a research group at the Frauenhofer Gesellschaft in Stuttgart, and joined Imperial as Kodak Professor of Interface Science in 1976.

108. In the early 1980s the IBM Institute project came to the department of chemical engineering and chemical technology. The Institute was designed by IBM to help universities introduce computing into their teaching, and to give students access to computing equipment. In this case, the Institute donated equipment relating to process control and three pilot plants were placed under the control of the new system. IBM donated process control computers also to Bath, Birmingham, Exeter, Heriot-Watt and Strathclyde.

109. Richardson, a chemical engineer, was later appointed to a lectureship and carried out research with Graham Saville, a chemical technologist. Their collaboration was encouraged by Sargent and was supported by North Sea gas interests. By the end of the century Richardson had become a world expert on the safe delivery and use of oil and gas, and his work was funded by international oil interests. For more on Richardson see chapter 16. Cornish later became head of department at University College London.

Graham Saville (1935–) was educated at the University of Oxford and joined the department in 1963 as a research fellow. He was appointed Reader in Thermodynamics in 1995. He taught thermodynamics for many years, including to his future co-researcher Stephen Richardson.

110. Gilbert Northcott Walton (1915–91) was educated at Oxford where he took a degree in chemistry. He served in the Royal Army Ordnance Corps during the Second World War and in 1948 joined the Atomic Energy Research Establishment at Harwell.

111. Flowers could see a future in reprocessing plants and in anti-proliferation fuel cycles but not much else. See Walton to Flowers, 24 February 1978 and Flowers to Walton, 28 February 1978. Walton retired in 1980 and the UKAEA (Harwell) removed all the plutonium and americium from the department's laboratories.

112. Felix Weinberg FRS (1928–) arrived in Britain after the war. He was a survivor of Auschwitz. He took an external BSc at the University of London and came to Imperial as a research student. He was appointed assistant lecturer in 1954 and was given a chair in combustion physics

in 1967. His work has been in one of Imperial's traditional fields, the structure of flames. He has added significantly to the combustion legacy of William Bone and Alfred Egerton, and continues to carry out research as an emeritus professor.

Henryk Sawistowski (1925–84) was another professor who had been uprooted by the war. After serving with the Polish forces in Italy he attended the Polish University College in London, arriving at Imperial as a research student in 1952. Known as an excellent lecturer, he stayed in the department and was appointed Professor of Chemical Engineering in 1976.

113. There was concern over the structures section which had three strong readers all with different ideas about what needed to be done. Shortly after, in 1972, this section was divided into structural mechanics under E. H. Brown and structural engineering under B. G. Neal.

114. Sir Alan James Harris FREng (1916–2000) was an evening student at the Northampton College of Technology (now City University) where he took an external University of London degree. After distinguished war service (he became a colonel in the Sappers, engaged in a number of diving exploits, and won the Croix de Guerre), Harris worked in Paris on prestressed concrete with one of the world's experts in the field, Eugène Freyssinet. He married into the Freyssinet family and remained a lifelong Francophile. Furthering the Freyssinet interests in the UK he became a director of the Prestressed Concrete Company before setting up his own engineering firm, Harris and Sutherland, in 1955. The firm designed a wide range of structures, from schools to nuclear reactor buildings. Harris was an expert in the use of a wide range of materials from concrete to laminated wood. He was also a keen and competitive sailor.

115. Rector's correspondence; E. H. Brown to Flowers, 18 March 1976.

116. John Boscawen Burland FRS FREng (1936–) was educated at the University of the Witwatersrand and at Cambridge. He came to Imperial in 1980 from the Building Research Institute where he was head of the Geotechnics Division. For more on Burland's work see chapter 16.

117. While head of department, Munro set in place good computing facilities. For more on Munro see chapter 10.

118. Patrick Joseph Dowling FRS FREng (1939–) was educated at University College, Dublin and came to Imperial as a research student to work on steel bridge decks. He joined the department on gaining his PhD in 1968. A specialist in steel and steel concrete composite structures, he was given the British Steel Corporation Readership in Structural Steelwork in 1974. He was appointed Professor of Steel Structures in 1979 and was

head of the structural engineering section from 1981. Dowling founded a consulting firm, Chapman and Dowling, together with J. C. Chapman who had left the college in 1971 (returning later as a visiting professor). Dowling used his expertise in a range of bridge work, and for work on North Sea oil platforms and the Thames Barrage. He was awarded a CBE for his role in strengthening the links between higher education and industry. He resigned from Imperial in 1994 to become Vice Chancellor of the University of Surrey.

119. Patrick Holmes (1939–), was educated at University College, Swansea. Before coming to Imperial he held the chair of maritime civil engineering at Liverpool where he was also head of department.

120. KEM/ B. G. Neal paper on Ocean Engineering (1968).

121. This was one of four centres established in the light of North Sea oil and gas exploration. Four Imperial departments were involved in the London Centre: aeronautics, civil engineering, mechanical engineering and metallurgy and materials.

122. Roger Perry FREng (1940–95) studied chemistry at the University of Birmingham and joined the staff there on gaining his PhD. At Birmingham he developed some analytical techniques for use in the detection of water pollutants. He joined the public health engineering section as a lecturer in 1970 and was given a conferred chair in public health and water technology in 1981. When he became Director of the Centre for Environmental Control and Waste Management, his title changed to Professor of Environmental Control and Waste Management. A new environmental engineering laboratory has recently been named for Perry.

123. John Wilson Dougill FREng (1934–) was educated at King's College London and at Imperial College where he was a research students with A. L. L. Baker. He became Professor of Engineering Science at King's College in 1976. He rejoined Imperial as Professor of Concrete Structures and Technology in 1981 but left in 1987 to join the Institution of Structural Engineers, becoming Chief Executive and Secretary in 1994.

124. Rector's correspondence; Buchanan to Flowers, 24 February 1977. See chapter 10 for Buchanan's earlier problems.

125. Tony Melville Ridley FREng (1933–) was educated at the University of Newcastle, Northwestern University, the University of California, and Stanford University. He was Director General of Tyne and Wearside Passenger Transport, 1969–75, and Managing Director of Hong Kong Mass Transit Railway Corporation, 1975–8. During the 1980s he was Chairman and Chief Executive of London Underground Ltd., resigning

after the King's Cross underground station fire. He was Managing Director of Eurotunnel 1987–9. He first came to Imperial as a visiting professor in 1981 and was appointed to the Rees Jeffreys Chair of Transport in 1990. He was head of the civil engineering department, 1997–9. (From 1998 renamed civil and environmental engineering.)

126. Over the period discussed in this chapter the share of government funding of research in civil engineering at Imperial dropped from about 75% to about 50%. By the early twenty-first century it had dropped to about 28%. The department has scored well in many league tables and has consistently received 5* in the RAE. For more on civil engineering in this period see Joyce Brown (ed.), *A Hundred Years of Civil Engineering at South Kensington* (C&G, London, 1985).

127. John Brown FREng (1923–) was educated at the University of Edinburgh where he studied mathematics under E. T. Whittaker. He worked at the Radar Research and Development Establishment (1944–51) and joined Imperial as a lecturer in 1951. While on the staff he also gained a PhD in physics. Brown, a major theoretician in applied electromagnetism, moved to University College London as a lecturer in 1954 and was given a chair in 1964. He returned to Imperial as head of department in 1967. On retiring from Imperial he became Dean of Technology at Brunel (1988–91). He is the uncle of Chancellor of the Exchequer Gordon Brown.

128. In connection with epidemiology, Sayers was a consultant to the World Health Organization and tracked both the spread of a measles epidemic in South America, and the spread of Rabies in European foxes. His laboratory later became part of the Centre for Biological and Medical Systems.

129. Colin Cherry 'Long Range Thinking' (Typescript, October, 1977).

130. Rector's correspondence; B. McA Sayers to Flowers; 19 April (no year given).

131. Among the external candidates being considered was Eric Ash, but he declined.

132. Rectors Correspondence; Sayers to Flowers (undated).

133. Ernest Michael Freeman FREng (1937–) was educated at Kings College London and was working at Brighton Polytechnic when invited to join Eric Laithwaite at Imperial for one day a week on a joint research project. Freeman moved to Imperial as a reader in 1973. He works in the area of interactive computer graphics and their application in the analysis of a range of electromagnetic devices.

Laurence Frank Turner (1935–) began his career as an apprentice with British Thomson Houston (later part of AEI) in Rugby. After part-time study towards a BSc, he became a research student at Birmingham, gaining a PhD. He joined Imperial as a lecturer in 1963.

134. Computing department correspondence file; Penney to Flowers, 5 July 1971.

135. Stephen James Goldsack (1926–) was educated at King's College, Newcastle and was a research student at the University of Manchester under P. M. S. Blackett. After postdoctoral work in Brussels and the University of California (Berkeley), he joined the University of Birmingham as a lecturer. He came to Imperial in 1960 as a senior lecturer in physics where he developed the computing side of the bubble chamber work.

136. GB minutes, 21 June 1974.

137. See Annual Report 1973–4.

138. In thinking about dispersed computing, Elliot and Stuart had to consider the future in light of the 'Report on Microprocessors', published by the Computer Board for Universities and Research Councils in 1979. In this period reports on computing were being published almost every month.

139. Meir M. (Manny) Lehman FREng (1925–) was born in Germany and came to England with his family in 1931. He left school at the age of sixteen and worked for Murphy Radio while taking evening classes in electronics. Lehman sat the college entrance examinations and entered as a mathematics student in 1950. He stayed on as a research student, worked on the college computing engine, and gained a PhD in 1957. Lehman worked briefly for Ferranti before leaving for Israel where he was engaged in the construction of computers in the scientific department of the Ministry of Defence. In 1964 he moved to the United States to work with IBM. He returned to Imperial in 1972. Lehman has a lively mind. Interested in software design he has continued to be active in research into his eighties. For his work as a research student see chapter 9.

140. Westcott used the following example in explaining the kinds of questions he was then interested in and, by implication, the kind of software that he wanted students to learn how to design: 'How hard and how soon do you put on the brakes in an overheating economy without causing a skid into catastrophic recession?'. Westcott to Flowers, 11 January 1974 (computing and control correspondence file). Westcott was a founder member of the Club of Rome (as were some others from Imperial) and, in the 1980s, headed a 'Programme of Research into Optimal Policy Evaluation' (PROPE) to apply control and optimisation methods to macroeconomic policy design. The work was supported by the SSRC.

141. Plessey, the principal investor, was then engaged in defence electronics, and P. A. International was a firm of systems consultants. According to

Professor Lehman, Flowers encouraged the formation of college businesses after returning from a trip to MIT where he saw much business activity on the campus. (Personal communication.)

142. In connection with the departmental isolation, one problem may have been that Lehman, an orthodox Jew, did not attend the Friday meetings of heads of departments. In an on-line interview, Lehman claims that Flowers told him that the fledgling department needed a stronger 'mafiosi' type to lead the department in its fight to get established *vis à vis* the stronger departments in the college. (See Meir M. Lehman, Electrical Engineer, an oral history conducted in 1993 by William Aspray, IEEE History Center, Rutgers University. Copy on IEEE www pages.) Lehman took early retirement but remained at the college as a senior research fellow, funded by external contracts for many years before moving to Middlesex University in the early twenty-first century.

143. Igor Aleksander FREng (1937–) was educated at the University of the Witwatersrand and came to England intending to become a research student under Colin Cherry. However, while waiting for the start of the session, he took a job with Standard Telephone and Cable and decided, instead, to stay with the company (personal communication). He later joined Queen Mary College as a lecturer, moving to the University of Kent as Reader, and to Brunel as Professor of Electrical Engineering. He joined the computing department at Imperial in 1984, moving to become Professor of Neural Systems Engineering in the electrical engineering department in 1988. Aleksander is an expert in artificial intelligence and neural networks and designed the world's first neural pattern recognition system in the 1980s. Aleksander's group received financial support from the Department of Trade and Industry which helped fund the new unit of management information technology with four lectureships and support staff.

144. In 1986, number scores were not used in the same way as later. Computing was named 'outstanding' and electrical engineering 'better than average'. In 1989 both departments scored 5.

145. Biomechanics came to Imperial because of the interest of Owen Saunders who had connections to the Ministry of Health Research Department at Roehampton. Sydney Alan Vasey Swanson FREng (1931–), known as Alan, was educated at Imperial College and was a research student under Hugh Ford. After a few years working for Bristol Aircraft he returned to the college as a lecturer in 1958. Swanson married the daughter of F. Howarth, a former assistant director in the botany department. Mr. Howarth's son, D. J. Howarth was a professor in the department of

computing and control. In 1978 Swanson was appointed head of department in succession to Ford.

146. Freeman was one of the foremost surgeons in this field and founded his own company for the manufacture of joint replacements.

147. This was the result of a successful application to the SRC. James Gordon Williams FRS FREng (1938–), known as Gordon, was educated at Imperial College. After a few years working for the RAE he returned to the college as an assistant lecturer in 1962. He was given a chair in polymer engineering in 1975. In 1990 he was appointed Professor of Mechanical Engineering and head of department.

148. Consulting fees in the late 1960s were still reported in archaic guinea units. Typical consultancy charges by engineers were about ten to fifteen guineas per day. Eric Laithwaite, in electrical engineering, charged what was then considered an astronomically high fee, namely seventy-five guineas per day. The professorial salary range in 1969 was £3,649–5,090 plus £100 London allowance. In 1970 the minimum was raised to £4120 and in 1971 to £4,533 with a permitted university average of £5,610.

149. Linstead sent for a document on how MIT dealt with individuals holding consultancies and directorships; and how it dealt with entrepreneurs.

150. In 1969 Spalding set up Combustion, Heat and Mass Transfer Ltd (CHAM). The company's role has changed, but it is still in existence today and is run by Spalding's son. Among the things that Spalding and his students were engaged in during this earlier period was the design of a centrifuge for isotope separation, for the UKAEA. Spalding ran the heat transfer laboratory which was later subsumed within the Computational Fluid Dynamics Unit.

151. Spalding was one of several at Imperial who had contacts with scientists in the Soviet Union. There were a number of scientific and technical exchanges in this period, but little is mentioned of them in the college records. I had expected to write more about the Cold War, especially in this chapter, but found few relevant sources. However, the Governing Body minutes of 1976 record an interesting discussion as to whether a FIC should be awarded to Benjamin Grigorievich Levich, an eminent Jewish electrochemist who had been refused permission to leave the USSR. He was nominated by Professors Spalding and Barton, but there was opposition from some professors and governors who stated their concern that the college not involve itself in Cold War politics. Professor Sutton was a vocal opponent of the nomination. Sir Edward Playfair told his fellow governors they 'should not be pusillanimous about the word 'political'', since many decisions taken over fellowships were in fact political. Playfair argued

that provided Levich was worthy of a fellowship on scientific grounds he should be supported as, in the end, he was (GB Minutes, 17 December 1976). Spalding, or perhaps his wife, paid a small price for his support in that he was denied a visa to deliver a paper at a conference in Minsk. Mrs. Spalding presented the paper for him. See *The Times*, 28 October 1977, 4. Shortly after, several Imperial physicists joined others in boycotting an international conference on laser-plasma interactions held in Moscow. This was to protest human rights violations related to Soviet Jews. See *The Times*, 25 November 1978, 13.

Sir Edward Playfair (1909–99), a member of the Governing Body, had a distinguished career in Whitehall and was Permanent Secretary at the Ministry of Defence 1960–61.

152. Alastair Cameron (1917–2000) was a student at Imperial College and obtained both a BSc and PhD in the chemistry department. He rejoined Imperial in 1954, after war service and periods working for PAMATRADA (Parsons and Marine Engineering Research and Development) and at Cambridge with Eric Rideal. He became Warden of Tizard Hall on the opening of the new residences on the south side of Prince's Gardens. Cameron was appointed to a new chair in lubrication funded by Mobil Oil in 1972.

153. Grant was involved in the design of a range of nuclear power installations in the UK and elsewhere. In 1974 GEC gave him a contract worth £1 million over two years in connection with a fuel handling system for a Romanian nuclear power installation. For more on Grant and nuclear power see chapter 9.

Colin Bowden Besant FREng (1936–) was educated at Plymouth College of Technology and Imperial College. On graduation he worked for the UKAEA for a few years before returning to Imperial in 1964. He rose through the ranks becoming Professor of Computer-Aided Manufacture in 1989.

154. Rector's correspondence; Ford quoted in letter from Davies to Penney, 28 September 1972.

155. Ford is said to have held lively parties at his Chelsea home to which industrialists were invited. These may have also indirectly contributed to departmental research income.

156. Rector's correspondence; Ford to Penney, 7 April 1973. See also, Ford to Penney 18 March 1968 for complaint over someone having been sent to examine use of space in his department.

157. Thomas Mutrie Husband FREng (1936–) was educated at the University of Strathclyde. On graduating with a PhD he worked in industry for

several years before returning to academic work at Strathclyde. He was Professor of Manufacturing Organization at Loughborough University before coming to Imperial in 1981. When Swanson became Pro-Rector in 1983, Husband succeeded him as head of department. He left Imperial in 1990 to become Vice Chancellor of Salford University.

chapter fourteen

Diversifying the Curriculum

Arts, Humanities, and Modern Languages

The Second World War was a watershed in thinking about the broader education of Imperial College students. Before the war people took pride in the focussed scientific and technological education offered at the college, believing that, in many respects, it was superior to that offered at other universities, notably at Oxford and Cambridge. In part this was a defensive reaction to the lofty position held by Oxbridge in the socio-political life of the country. But the utility of engineering and science education was valued, and scientific study was seen, justifiably, as being no less liberal than that in the humanities. However, this view was not always distinguishable from the smug belief that Imperial's education was superior to anything the arts had to offer, and that it better matched the needs of the new scientific and technological age. Some even claimed that formal education in the humanities was unnecessary, and that one could acquire such knowledge in one's spare time.[1] It was assumed that the smartest people would naturally gravitate towards a scientific or technological education and absorb whatever else was needed for a rounded life.[2] Many of Imperial's professors were very well educated and had extensive knowledge in fields other than their own — art, music, literature, languages, or politics. Hyman Levy and Alfred Egerton, for example, were held up as ideals in this regard, though they, themselves, are unlikely to have been among the smug.

The horrific experiences of war changed this way of thinking. Nazi science, the Holocaust, and the devastation of Hiroshima and Nagasaki came as major shocks. The postwar generation had enormous difficulty in coming to terms with what had happened, and with the presence of nuclear weapons in its midst. The atomic bomb, and the nuclear arms race that followed the war, had a major effect also on educational philosophy. The gap between what C. P. Snow was to label the 'two cultures' in 1959 was a major talking point already in the late 1940s.[3] Scientists and engineers were put on the defensive, and blamed for not thinking about the consequences of their actions and discoveries. They were criticized for being too narrowly educated, and accused of not being able to see the larger picture, of not knowing how to think ahead, and of lacking humanistic values. The typical response from scientists was that they were simply engaged in pushing back the frontiers of knowledge. How discoveries were used, they claimed, was strictly political and out of their hands.[4]

These short-sighted and limited ways of thinking, on both sides of the cultural divide, were challenged by many people in the academic community. The UGC, concerned with the kinds of graduates that British universities were producing, wanted specialists who were also well-rounded.[5] Its views were shared at Imperial, notably by Willis Jackson who, beginning already in the late 1940s, gave many talks, to both student and professional audiences, on engineering pedagogy, the need to diversify curricula, and that arts students, too, might benefit from a better understanding of the sciences. Even before the end of the war the Governing Body was determined to broaden the education of students at the college and instructed the new Rector, Richard Southwell, to find ways to do so. Southwell, who had been educated at Cambridge and had spent much of his working life at Oxford, brought a typically Oxbridge vision to Imperial, with mixed results. His largely unsuccessful attempt at developing a more collegial and commensal culture is discussed in chapter eleven. But he also wanted Imperial students to be better educated in politics and the arts, and it is here that his efforts had more lasting effect. Some of the money raised by the 1945 college appeal was directed towards non-technical education, namely towards weekend cultural activities and lunchtime lectures.

Southwell's original plan was for Imperial to have a cultural centre somewhere outside London, where students could attend weekend and vacation courses. But it was his successor, Roderic Hill, who succeeded in bringing this about. Supported by Jackson, Hill set up a committee which first met in July 1949 to plan for a centre.[6] It was agreed to name the new project 'Touchstone', and that a centre be located at the recently purchased Silwood Park.[7] Touchstone weekends began at Silwood in March 1950. At the first of these Hill explained that in earlier times the touchstone was used to measure the purity of gold and silver objects, and that the name had been chosen to symbolize the value that would come from the project. A number of people interested in bridging the cultural gap were invited to Silwood during March and April to help those at Imperial think of ways to integrate new disciplines into the curriculum and to encourage students to attend short extra-curricular courses during vacations and weekends. These sessions coincided with a series of BBC talks given by Sir Walter Moberly on 'The Idea of a University' which helped move the discussions forward.[8] Touchstone weekends became a great success, helped by the fact that Professor and Mrs. Munro were excellent hosts.[9] The first weekend session to which students from both Imperial and the Royal College of Art (RCA) were invited was held in May 1950. Led by David Pye of the RCA those present, including Hill, discussed 'Design in Everyday Life', a topic close to the Rector's heart. Hill asked C. K. McDowall to run Touchstone and, over the years, many distinguished guests were invited to lead discussions at Silwood.[10] Early guests included art historian Ernst Gombrich, trades union leader George Woodcock, the chairman of the Sexual Law Reform Society, John Robinson,[11] economist Joan Robinson, and philosopher Alasdair MacIntyre. In November 1956 psychologist Margaret Knight conducted a Touchstone weekend on the topic 'Scientific Humanism'.[12] This was a great success and led to the formation of the Huxley Society, set up by students for ongoing discussion of problems relating to religion, atheism, agnosticism, and ethical behaviour. McDowall had a good rapport with students and encouraged many to attend the weekend discussion groups.

The other of Southwell's aims was to have lunchtime lectures at the college. These, too, were established by Hill. He decided to lengthen the college day on Tuesdays and Thursdays and asked McDowall to arrange both single lectures and a few lecture series in the now extended lunch break.[13] The resulting programme, known as General Studies, began in 1952. The list of people who came to the college to give lectures during the 1950s and 60s is very impressive. Student societies enthusiastically joined in, arranging talks both within and without the General Studies framework. Noteworthy were the lectures on music given in the first two years by Reginald Jacques, and later by Anthony Hopkins.[14] Well attended were lectures in the history of art by Ernst Gombrich,[15] on atomic weapons and military planning by Professor Blackett, a short series on employment and the law by K. W. Wedderburn; series on political parties by Robert McKenzie, on drama by Ivor Brown, on literature by M. R. Ridley, on art by Barnett Freedman and on 'shaping the postwar world' by historian Michael Howard. Students in the 1950s and 60s were exposed to a wide range of opinion and to accomplished people working in many disciplines.[16] This kind of exposure was also what industry was demanding, believing that it would help produce 'graduates who can think and act for themselves, capable of choosing between truth and falsehood, wisdom and folly, beauty and ugliness'.[17] It was held that curricula should be broader and should prepare students for the needs of the modern world. Diversification became almost a mantra.

In 1961 Linstead appointed Professor Rupert Hall to open a small department in the history of science and technology, but surprisingly Hall and his colleagues did not involve themselves in discussions on curriculum diversification.[18] Once again Willis Jackson took the lead. In 1961 he wrote to the Rector that he wanted more than the lunchtime lectures, and was seeking ways of introducing a compulsory component of 'non-technical' work into the undergraduate course in his department.[19] He was prepared to sacrifice time spent on electrical engineering to do so, and to use departmental funds to pay the fees of visiting lecturers.[20] Jackson was active also nationally, and, in 1964, was the force behind a major national symposium 'Liberal Studies in Technological Education' held by the British Association for Commercial and Industrial Education (BACIE), the year he was president of the Association. The BACIE journal also published many articles on the subject. Former Imperial student, Sir Eric (later Lord) Ashby, was a major contributor. In his 1961 letter to the Rector, Jackson stated that Professor Tustin was concerned about the poor writing skills of the students and had asked that someone be appointed to help improve them, and help more generally in promoting communication skills. Jackson already knew who he wanted to appoint, namely Sinclair Goodlad who had had some experience working with engineers at MIT.[21] Linstead was reluctant to grant this request since, to that time, such teaching had been carried out by visiting staff. But he relented, and Goodlad was appointed to a lectureship in the electrical engineering department. Goodlad helped not only to improve writing skills but, by 1963, when the department gave over a further three hours a week to humanistic, social, and economic studies for second and third year students, he, and others, began giving lectures in new areas.[22] From an article Goodlad wrote in a student paper, it would appear that he was stimulated by the more popular writing of Professors Gabor and Cherry both of whom were much engaged with the 'new industrial revolution' in the area of computers, control, communication, and engineering in medicine.[23]

In 1965 Goodlad put forward a plan for a department of social sciences, humanities and languages.[24] One year later, Steven Rose, a lecturer in biochemistry, proposed the creation of a sociology of science unit, something for which he had the support of Professor Hall. These proposals came at a time when it was hoped that 400 architecture students would soon be joining the college.[25] By 1966 Goodlad's plan had expanded into one for an institute for human sciences, with a staff of about forty, covering the history and philosophy of science, economics, sociology, communication, psychology, modern languages and politics. A 1969 iteration was for a School of Languages, Social Sciences and Humanities.[26] But by then William Penney was Rector, and he was more sceptical than Linstead about moving in the direction envisaged by Jackson and Goodlad.[27] Penney was conservative in his approach and, as discussed in chapter twelve, was having to deal with the campus politics of the late 1960s. He will have noticed that in the larger academic world many radical activists came from the social sciences. Whatever the reason, none of the proposals mentioned was accepted, but access to non-technical fields of study did increase.

Professor Coles, chairman of the General Studies Committee in the late 1960s, was in favour of coordinating all the non-technical areas at the college within a single administrative unit. A committee, chaired by Jackson, was set up to review the situation. Coordination was problematic since it was unclear what 'non-technical' meant. For example, by this time an industrial sociology unit existed in the mechanical engineering department and its head, Joan Woodward, understandably did not wish to be lumped in with humanists and language teachers, many of whom were part-time, and most of whom were not carrying out research.[28]

Others did not wish to give up control of departmental curricula. For example, Professor Pryor did not want the economics and management courses that had been established in the mining department to be administered by a new unit. Despite opposition, some amalgamation occurred in 1973. The old General Studies courses joined others in language instruction, psychology, industrial psychology, sociology (not industrial sociology), and twentieth century history, under a new loosely structured unit called Associated Studies.[29] Professor D. D. Raphael was appointed to direct the unit, and to give instruction also in philosophy.[30] In 1980 Professor Hall and M. B. Hall were to retire and the Rector decided to merge what remained of the history of science and technology department with Associated Studies to form a new Department of Humanities, with Professor Raphael as its head.[31] The idea was for Raphael, a distinguished academic, to build a solid department with staff engaged, as he was, in serious research and publication. But he was, perhaps, too close to retirement for that to become a reality. While the department had some good lecturers such as Eric Stables who taught French, and John Thole who taught modern history, serious research did not take off.[32] Not only were academic interests and language instruction represented in this department, but loosely associated with it were also the college choir, orchestra, the chamber music society, the operatic society, the Haldane Library, the Consort Gallery where art work was displayed, and remnants of the lunchtime lecture programme begun in General Studies.

During the 1970s and 80s a permanent full and part-time Humanities staff gradually took over from the interesting visitors that had come to the college in the 1950s and 60s. It is not clear that students benefited overall from this turn to in-house instruction. But the college had little choice but to turn to more permanent staff as it expanded. There was no other way to provide students with the necessary number of courses in the humanities, languages, and social sciences. By the late 1970s, students were being awarded some academic credit for work in these areas. However, the less formal lectures continued, organized by McDowall until his retirement in 1974. Musical and orchestral activity declined in the few years following his departure, but picked up again when the orchestra invited Richard Dickins, a postgraduate student at the Royal College of Music (RCM) to be its conductor. A few years later Eric Ash gave Dickins a half-day position as Musician in Residence within the Humanities Department, a position that later expanded into a half-time appointment.[33] In the early 1990s a new scheme allowed five talented undergraduates to have free instrumental tuition at the RCM.[34] By the late 1990s a physics and musical performance degree was offered.

The university politics of the 1980s affected small academic groups throughout the country. Even before the Research Assessment Exercise (RAE) became fully established, the UGC decided that, with a few exceptions, it would no longer support small departments and isolated individuals. It encouraged consolidation, and supported the movement of staff between universities. Two of Imperial's better scholars in the humanities moved at that time.[35] However, James Secord, then a senior lecturer in the history of science, sensibly wanted to maintain a history of science and technology presence at Britain's largest science and technology college, and sought the means to do so. Recognizing the need for some consolidation, he negotiated with other small units within the University of London and helped to form the University of London Centre for the History of Science, Medicine and Technology in 1986.[36] More interested in academic research than most of their remaining colleagues, the historians of science wanted to associate with, and be judged by, others in their field. Shortly before he moved to Cambridge in 1992, Secord attempted to enlarge the area of science studies at Imperial and, perhaps thinking of adding some sociologists and philosophers of science and technology to the historians, pushed for the formation of a new section, to be named Science and Technology Studies (STS). In this he was supported by John Durant, Director of the Science Museum, who helped also to set up a second MSc course in science communication under the direction of Ros Herman.[37] STS was intended to grow further and had the support of both Brian Flowers and his successor Eric Ash.[38] The unit was still linked to Humanities which by then had been downgraded from a department to a programme.[39] But STS proved an awkward marriage and divorce soon followed. Science Communication remained within Humanities and the historians were given their own Centre for the History of Science and Technology (CHOST).[40] David Edgerton and Andrew Warwick were appointed in 1992, specialists in the history of twentieth-century technology, and in modern physics and mathematics, respectively.[41] Edgerton headed the new Centre which began to flourish with the appointment of new staff.[42] In 1993 Goodlad succeeded Eric Stables as Director of the Humanities Programme which continued to provide useful service courses at the undergraduate level.[43] In 1994 the

Board of Studies agreed to allow students to take up to one-eighth of their undergraduate course in humanities or social sciences instruction and the demand for such courses increased. By the late 1990s about 800 students a year were taking courses in modern languages, and many others were enrolled in a range of humanities courses. Many of these were being taught by people from CHOSTM.[44] At the MSc level, the Humanities Programme now offers courses in Science Media Production, Scientific Translation, and Creative Non-Fiction Writing, as well as the older course in Science Communication.[45] A few of the staff have taken on PhD students working in areas related to the MSc courses.

Social Sciences, Management, and Business

Willis Jackson was a leader not only in bringing a broader liberal education to Imperial College, he was interested also in managerial education.[46] But he was not alone in experimenting along these lines at Imperial. For example, in the 1950s Owen Saunders encouraged a number of mechanical engineering students, mostly postgraduates, to take courses in economics and industrial management at the LSE.[47] With strong backing from Linstead who was seeking ways to expand postgraduate education, Saunders decided to launch a new MSc course in production engineering and management.[48] The term 'management' was something of a red flag and many people at Imperial objected that management could not be taught, at least not unless those taking the course had some previous experience in industry.[49] Despite such objections, Nicol Gross was appointed Reader in Production Engineering and asked to design a new MSc course to begin in 1955.[50] One of Saunders' doctoral students, Sam Eilon, who completed his PhD in 1955, was given a temporary lectureship to help Gross set up the new course.[51] Following Saunders' advice, Eilon had earlier taken management courses at the LSE. Further, he had a serious interest in operational research, something to which he had been introduced during army service in Israel. Eilon's job was to design the management side of the course. Arrangements were close to completion when Gross decided to leave Imperial and take a senior position with British Oxygen.[52] The course had not yet been advertised, but since five students had registered it was decided not to turn them away. Eilon was placed temporarily in charge of the new Production Engineering section, housed at 14 Prince's Gardens. He taught from his strength, namely operational research and management.[53] John Alexander, seconded from the applied mechanics section, was given responsibility for the production engineering side. Alexander was assisted by Roy Brewer who later took over.[54] The new MSc course was one of the first courses of its kind in the country.[55] Other people were soon added to the teaching staff, including Joan Woodward who first came to Imperial as a part-time lecturer in industrial sociology in 1957 (full-time in 1962).[56] There was also a reciprocal agreement with the LSE by which instructors came to Imperial to give lectures in subjects such as accounting, finance, and industrial relations.

In 1957 Eilon left for work in Israel but, when a new readership in the production engineering section was established in 1959, he successfully applied and returned to Imperial as head of the section. By then there were about fourteen students taking the MSc course, and a review was undertaken to plan for its future. It was decided to expand, to build a new industrial engineering laboratory, take in more students, and give them more options on the management side. All this activity marked the start of a parting of the ways. Eilon introduced a new MSc in Operational Research and Management Studies (ORMS) in 1961 at which point the section was renamed Production Engineering and Management Studies. Two years later Eilon was given a chair.[57] By 1965, because of growth on the management side, the section outgrew its Prince's Gardens premises and moved to the mechanical engineering building. One year later the reciprocal agreement with the LSE was terminated when Eilon decided to have his own staff teach the MSc students. The result was a further expansion of staff in areas new to the college. New short courses for industry and business managers were launched as the section moved slowly in the direction of more conventional business studies.[58] Even statistics, which had been taught in the mathematics department, was now taught by staff in the section. All this was possible because Eilon was very successful in bringing in contracts and research funds.

During the 1960s management studies at Imperial became caught up in the national politics of business education.[59] The success of American business schools had convinced many that similar institutions were needed also in Britain. Oliver (Lord) Franks was invited by a group of business and governmental organizations to advise on what should be done.[60] His report, published in 1963, stated that there should, at first, be two major business schools in the country, one in London and the other in

Manchester. Since the LSE and Imperial were already engaged in related areas, the two colleges acted as sponsors and cooperated in the planning of a school for London.[61] What was proposed was a London Graduate School of Business Studies. While the government was willing, even wishing, that either Imperial or the LSE would take on the new School, curiously both colleges declined. Linstead wanted to take it on but was having enough trouble managing the Jubilee expansion.[62] Further, there was opposition within Imperial to a curriculum that, to many, seemed to be expanding too quickly in non-technological directions. A possible merger with the Architects Association School was also under discussion and there were those who saw that, and expansion in the area of management, as threatening the identity of the college.[63] Clearly not taking on the business school was a missed opportunity, but that Linstead bowed to pressure is understandable. The new London Graduate School of Business Studies (later London Business School, LBS) opened in 1965, first as a college neighbour on Exhibition Road, and then in quarters on Northumberland Avenue.[64] Once opened, it forced Imperial to think more clearly about its own role in management education.[65]

In 1965 when the topic of business studies was on everyone's mind, Imperial received 229 applications for the forty places in the ORMS course. Eilon had wanted Imperial to take on the new business school and move ORMS into it. He even considered leaving Imperial and moving ORMS to the new school. Nothing came of this idea, though at least one of Eilon's staff did move.[66] By the mid-1960s both Eilon and Joan Woodward had been so successful in attracting funds to the section that it became increasingly unbalanced.[67] ORMS began to overshadow production engineering which saw its enrolment decline. A split was inevitable and, in 1966, production engineering moved to join the applied mechanics section under Professor Alexander. There, a new MSc course in production technology was introduced. At the same time, Jackson's committee was reviewing non-technical teaching in the college. Jackson looked at ORMS and saw that while there was good work being carried out in operational research, it was a small island surrounded by a sea of social sciences and business studies. By the early 1960s Woodward had become very well known for her field work among factory workers in East London. Before becoming a full-time lecturer at Imperial she had moved her operational base from Oxford to a technical college in Essex so as to be closer to the factories where she carried out her work. She was an excellent lecturer and attracted good students. She also attracted funding that allowed her area to expand and she wanted her own MSc course in industrial sociology. In this she was supported by Jackson. Needless to say there was tension between Eilon and Woodward, each with a different vison of how the section should grow once production engineering had left. The situation was not helped by a serious space problem in the mechanical engineering building.

Jackson recommended the establishment of a separate Industrial Sociology Unit and Woodward and her colleagues migrated back to Prince's Gardens. Unlike Linstead and Jackson, Penney was not enthusiastic about the expanding social sciences. While he agreed to the creation of the new unit he also initiated discussions as to whether it should continue in the longer term. Interest in the proposed industrial sociology MSc course was, perhaps, the deciding factor in keeping the unit going. By 1969 the course was approved and Woodward was given a chair. Sadly she died in 1971 and, once again, questions were raised about allowing work in industrial sociology to continue.[68] This time politics more clearly entered the discussion. One of Woodward's possible successors, Dorothy Wedderburn, was openly left-wing. Wedderburn, working very much in the Woodward mould, was interested in industrial behaviours such as strikes and absenteeism.[69] She received good grants from companies such as ICI but was nonetheless seen by some in the college as being dangerously on the side of the workers. However by 1973, and before any decision on the future of the Unit had been taken, Imperial had a new Rector. Brian Flowers had more sympathy for industrial sociology than his predecessor and appointed Wedderburn as Director of the Industrial Sociology Unit in 1973.[70] Also in 1973, Sir David Huddie, who had been asked to think about the new engineering courses, suggested that there be a department of social and economic studies.[71] In this he had the support of a number of major industrialists who wanted to see engineering graduates have more instruction in these areas. The department that he envisaged was one that would provide service teaching mainly for engineers. He did not envisage a department that would engage seriously in research, something that proved to be problematic in the longer run. Eilon was opposed to any new department, arguing that his department and the ISU were already providing all the service teaching that undergraduates needed. However, other staff approved of the idea and Eilon was overruled. The new Department of Social and Economic Studies was established in 1978. Wedderburn was given a chair and became head of department. Z. A. Silberston, who had been appointed to a chair in economics in 1977, became head of the economics section in the new department.[72] Unfortunately this department did not take off and Imperial failed to establish a good foothold in the social and economic study of science.[73]

The creation of the ISU in 1967 prompted discussion over what to do with the remains of Eilon's section. Eilon wanted independence for management science, but Hugh Ford was only willing to support an independent department if new space were found for it. He did not wish to compete with another head of department for lecture room and other space in what he considered to be his domain, namely the mechanical engineering building. But some agreement over space and lecture scheduling must have been reached since, in 1971, Eilon was given his own department of management science. By then his section had fifteen academic staff, and the MSc course had about sixty students (300 applicants in 1971) working on quantitative, analytical, approaches to management. The department also had some research students. That Eilon wanted independence is understandable, especially in light of what was happening elsewhere. Ten other British universities had set up management departments or schools, including Oxford, Warwick, Birmingham, Strathclyde and Cranfield.

While Eilon's MSc course was still heavily tilted towards operational research it was moving more and more in the direction of other areas of business studies. In the longer run this proved to be a serious problem, indeed a fatal weakness. Turf wars, not just within Imperial College, became fierce during the 1970s when university funding declined. But the longer-term problems were not immediately recognized at Imperial since the course was doing well. It drew most of its students from engineering departments, and its graduates found good jobs. Operational Research as applied to industrial and management problems is a technical field which requires a high level of numeracy. Had Eilon remained focussed on producing people specialized in this way he would have had little competition. But, in attempting to fuse the highly technical work with work more usually associated with traditional business schools, and in setting up joint undergraduate courses with other departments, he overextended.[74] When he began asking for new staff and facilities to cover the new courses, he was largely refused.[75] Eilon was in expansionary mode and appears to have been somewhat oblivious to the larger political and economic climate of the 1970s. By the late 1970s he wanted to start an undergraduate course in management science, something Ford was very much against.[76]

By the time Eric Ash was appointed Rector, it was clear that some rationalization of resources was needed. By the 1980s Eilon's department, once well funded, was in financial difficulty due to overextension during the recession. Further, there was considerable redundancy within the departments of management science and social and economic studies, redundancy that needed to be viewed also within the context of what was being offered elsewhere in the University of London. While there was bitterness over his decision, it is understandable that Ash merged the two departments into what became a new management school. Since Eilon and Silberston were close to retirement, they had less say than might otherwise have been the case.[77] Wedderburn left in 1985 to become Principal of Bedford College. The new School of Management opened in 1987 with David Norburn as Director.[78] Norburn insisted on more salubrious quarters than those offered in the mechanical engineering department building. He was given new premises in a house on Exhibition Road, remodelled for the purpose.[79] At the time, he was quoted as saying of Imperial 'the place is wall to wall brains' and that the Management School planned to teach about 120 in a full-time MBA program, about 150 part-time MBA students, and about 150 undergraduates in a business degree course — all in addition to the existing courses the new School had inherited.[80]

The idea behind the departmental merger was for Imperial to have a business/management school to compete with the best business schools in the country. Indeed, the new Management School began to advertise itself accordingly, 'the primary aim of the Management School is to become a top international business school' focussed on 'the integration of new technology with managerial skills'.[81] The turn towards business was opportune given that growth, and attracting a wide range of students, appear to have been important to the college. Few people have the mathematical or technical skills needed for the kind of work that began under Eilon in the old Production Engineering section. At first something close to the two old MSc courses continued; namely Management Science and, in place of industrial sociology, a course called Technology and Industrial Organization. But the two were soon merged into a single MSc in Management.[82] In addition to the newly defined MSc course, a new three-year part-time Executive MBA was introduced for managers with some years of working experience.

Aside from its masters degree courses, the Management School was still very much a service teaching centre for the whole college. This was true also at the postgraduate level, where staff provided segments in other courses such as the MSc in Environmental Technology, Mineral Resources Management, and Manufacturing Automation. Norburn's wife, Susan Birley, an expert in entrepreneurship, was appointed to help run the MBA programme. Together, they and their colleagues were able to ride the MBA wave of the 1980s and 90s.[83] But the heavy emphasis on teaching brought problems and, in 1992, the School scored only a 3 in the Research Assessment Exercise (RAE). Given Imperial's sensitivity to its placement in league tables, and given the financial consequences of a low RAE score, it was agreed among the senior administrators that something needed to be done. When Richard Sykes became Rector in 2001 he was prepared to close the school down unless it could be transformed into a major research body. Fortune was on its side, however, since an alumnus, Gary Tanaka, promised a major donation to the college in 2000.[84] The donation, together with other resources, was sufficient for a new building and it provided the occasion to rethink where the college should be heading in the management area. The Governing Body (about to become the Council) decided on the business school approach. The name was changed from 'management' to 'business' and, when the new building opened in 2003, the school was named the Tanaka Business School.[85] Norburn retired and a new Director, David Begg, was appointed.[86] Begg was given a mandate to restructure the School. It was to become a high powered research centre, and the research was to be more rigorous and quantitative in approach than had been the case earlier.[87] Some of the staff were offered retirement/redundancy packages, as the School moved to adopt a more typically Imperial College research ethos.[88] The Tanaka Business School was to fill a niche, or rather several niches, and to focus on the management of innovation, entrepreneurship, health care management, and new high technologies applied to finance.[89] Given that, by the early twenty-first century, the MBA bubble had deflated, this more focussed approach appears to be a good strategy. Accordingly, Begg has begun to build the business school anew, appointing staff in the designated growth areas.

Conclusion

Looking back, the 1950s and 60s appear to have been something of a golden age with respect to the arts at Imperial. Widespread public awareness of what C. P. Snow was to call the 'Two Cultures' prompted many working in the arts and humanities to take an interest in the sciences, and to wish to be associated with them in some way. Highly gifted people, including many who were engaged also in public affairs, came to the college and were welcomed by students and staff alike. Aspects of the cultural mixing that occurred at that time are discussed in chapter eleven, but in this chapter we have seen how Imperial College gave up these early experiments in broadening the outlook of its students, and gradually moved to incorporate non-technical education into its curriculum. Those who came to lecture at the college during the 1950s and 60s saw themselves as doing their part in bridging the cultural divide, and on both sides people's lives were enriched. But once the college decided to institutionalize courses on a more permanent basis problems arose. The in-house turn of the 1970s and 80s resulted in a new generation of students being offered courses of uneven quality. With hindsight we can see that this was a consequence of inherent structural problems. Some of the people appointed to teach in the social sciences and humanities were outstanding. The academic standing of sociologist Joan Woodward and of philosopher David Raphael, for example, more than matched that of many of their professorial colleagues in engineering and science. By the time Raphael came to Imperial he had already published widely and was well recognized in his field. Woodward was able to carry out major sociological research while at Imperial because she was well funded by industry. But for most of those who came to teach humanities, modern languages, and social sciences, there were few, if any, such opportunities. Many were part-timers. Indeed, people with academic ambition in the humanities or social sciences would more probably have sought work elsewhere. In order to flourish in an academic environment people need a limited teaching schedule, research opportunities, close colleagues, and to be judged by others working in similar fields. Goodlad's proposal of an institute providing service teaching over a wide range of topics would probably not have worked well. By comparison, the small history of science and technology department, set up by Linstead, was a model that could have worked had more attention been paid to it. Professor

Hall and Marie Boas Hall supervised many PhD students who later went on to good academic positions. But the Halls were not interested in undergraduate education and failed to build the department to meet those needs. They failed also to bring in staff with new research interests, interests more in line with those of the college, and when they retired the department fell apart.[90] The more recent Centre for the History of Science, Technology and Medicine has been better nurtured and, while small, is already larger than the Hall's department ever was. It may prove more viable since its members not only carry out serious research but provide also undergraduate teaching to a high level.

In the earlier period, the overall result of poor decision making was a Humanities Department that became home to an assortment of people who, with a few exceptions, were service teachers with little hope of professional advancement. The same is true of its successor the Humanities Programme, though some attempt is being made to introduce research activity. Despite all these difficulties, with respect to providing a rounded undergraduate education for scientists and engineers, Imperial compares well with science and engineering faculties in most other British universities. It compares less well with some of the world's major technological institutions such as MIT. The difference between MIT and Imperial is both structural and financial. By the end of the Second World War, five schools were established at MIT: science, engineering, humanities and social sciences, management, and architecture and planning. While the engineering school is by far the largest there has been a strong financial commitment to all five, each with separate departments. The economics department, in the humanities and social sciences school, was especially strong in the postwar period; and the Sloan Foundation gave major support to business education within MIT's management school. Academic staff in the humanities and social sciences school teach undergraduate service courses, but they also belong to lively research schools able to attract good scholars.[91]

Professor Eilon was correct in assuming that it would be a good thing for MSc students in Production Engineering to be given some information about life on the factory floor. He was lucky to find Joan Woodward. But her success proved problematic. Her excellent reputation attracted others to the college and, given the support she received from industry, the field of industrial sociology grew to a point where it caused imbalance within the section. She was given independence but after her untimely death the momentum ceased, and there was little interest in providing any longer-term stability for the industrial sociology unit. This was because the college had difficulty in coming to terms with its main problem in the area of arts and social science education, namely how to attract and keep high-quality permanent staff in non-mainstream fields deemed to be educationally important for scientists and engineers. How can the college ensure that its students are given a full and rounded education? This problem is still not fully resolved.

In the 1970s and 80s there were discussions with other institutions over possible mergers. Imperial could have merged with Westfield College which would have brought in expertise in modern languages and literature.[92] A further possibility was a merger with Royal Holloway and New Bedford which would have resulted in a multi-faculty institution. In the early twenty-first century a merger with University College was even considered. Such mergers would no doubt have brought other problems and I am not suggesting that they should have occurred. Indeed, my own preference is for academic institutions only slightly larger than the size of Imperial today. But had the mergers occurred, they would have given Imperial College students access to many good professionals in the arts, humanities and social sciences.

The hybrid legacy inherited by the Tanaka Business School is interesting. It includes the technical management legacy of Professor Eilon, the industrial sociology (now organizational behaviour) legacy of Professors Woodward and Wedderburn,[93] and the MBA legacy of Professor Norburn.[94] Collectively the legacy tells much of the aspirations of Imperial College in the later twentieth century. Its graduates were to be well rounded and ready to take their place as future leaders in the academy, industry, government, and commerce. But the new Business School is set to become a major research institution and, as such, is less likely to provide as much service teaching in the future.[95] And, the teaching it offers is likely to be highly technical. The staff in the Humanities Programme offer mainly language courses[96] and CHOSTM, while good, is small relative both to other history departments and to other departments at Imperial. It can offer only a limited range of lecture courses. All this leaves open the question of the future of undergraduate pedagogy in 'non-technical' fields. Will today's market economy value the arts, humanities and social sciences as forming an integral part of the education of its future leaders in new industrial, technological or medical fields? Will well-rounded graduates be better rewarded in their careers? These are difficult questions to answer, but upon them depends the future of non-technical education at institutions such as Imperial College.

...

1. There is a parallel tale told of Frederick Lindemann (Lord Cherwell) who, when asked by the wife of the Warden of All Souls what he did, replied that he was a scientist. On hearing this she stated that according to her husband (B. H. Sumner) anyone with a first in Greats could 'get up' science in a fortnight.

2. Among the generation that attended university in the period following the First World War were scientists engaged in perhaps the most exciting period of nuclear science. Some, such as P. M. S. Blackett, were associated with Rutherford at the Cavendish Laboratory. But science flourished also in other areas. Noel Annan has written about this, his own, generation, and claims that its greatest triumphs were indeed in the sciences. See Noel Annan, *Our Age: Portrait of a Generation* (London, 1990). These scientific success stories were impressed on the young and will have influenced the way the generation at university after the Second World War thought about their world.

3. C. P. Snow, *The Two Cultures and the Scientific Revolution*, (Rede Lecture, Cambridge, 1959). See special issue on the 'Two Cultures': *History of Science*, vol. 43 (June 2005). An article in this issue by David Edgerton, 'C. P. Snow as Anti-Historian of British Science: Revisiting the Technocratic Moment, 1959-64' (pp. 187–208) is of especial relevance in that it examines the polemic against the 'narrow-minded' specialist in the light of politics of the period. See also David Edgerton, *Warfare State: Britain, 1920–1970* (Cambridge, 2006), chapter 5.

4. This compartmentalized way of thinking has been discussed by Forman in the context of German science after the First World War. See Paul Forman, 'Weimar Culture, Causality and Quantum Mechanics, 1918–27', *Historical Studies in the Physical Sciences* vol. 3 (1971), pp. 1–115.

5. UGC Report, 'University Development, 1952–57' (1958). In 1961 Patrick Linstead, who had headed a group sponsored by the British Study Association looking into secondary school education, published a report advising that having students in grammar schools specialize in science and technology subjects already for A-level examinations was misguided. He recommended a broader sixth form school curriculum. See report in *The Times*, 19 July 1961, 7.

6. See EC minutes, 26 July 1949.

7. This site was opposed by those on the committee who believed it would result in the centre being monopolized by biology students. But they were outvoted.

8. Sir Walter Moberly, educated at Oxford, had been Vice Chancellor of the University of Manchester before becoming the first full-time chairman of the UGC. His *Crisis in the Universities* (1949) provoked much discussion over what might be the ideal university education. The debate invoked earlier ideas, including those of Cardinal Newman, Thorsten Veblen and Abraham Flexner. Moberly's views appear to have been informed by both Christian and humanist traditions. But he was no expansionist. He saw universities as a training ground for the professions which he believed to be already overfilled. In Britain, perhaps the greatest experiment in bringing curricular breadth to the university was carried out in the late 1950s at the University College of North Staffordshire (later the University of Keele), by A. D. (Lord) Lindsay, socialist and Master of Balliol.

9. J. W. Munro, Director of the Field Station at Silwood.

10. Charles Kenneth McDowall (1909–1997), known as Kenneth, was educated at Eton and Balliol College, Oxford. He became a schoolteacher and taught French and German before working in military intelligence during the Second World War. He rose to the rank of Major and, after the war, helped reorganize the German education system. He came to Imperial in 1949 as Administrative Secretary and was an important figure in the life of the college, responsible for many things including the annual Commemoration Day ceremonies. He also led the second violins section of the college orchestra for many years. McDowall was a historian and guide to Westminster Abbey and continued in this role after his retirement from Imperial in 1974. See profile of McDowall by Mervyn Jones, *Felix* (19 February 1964). See also chapter eleven.

11. It was in his capacity as chairman of that Society that J. A. T. Robinson, Bishop of Woolwich, was invited to Silwood. Later he testified in favour of the publication of *Lady Chatterley's Lover* in the 1960 trial against Penguin Books, and caused a sensation within church circles with the publication of *Honest to God* (1963). Another Touchstone guest was Mary Whitehouse who founded the National Viewers and Listeners Council in 1964 to combat what she saw as a loosening of moral standards in public broadcasting.

12. Margaret Knight, a psychology lecturer at the University of Aberdeen, caused a major furore with a short series of BBC lectures given in 1955. She advocated that moral education in schools should be divorced from religious education and was an important figure in secular humanist circles.

13. Later, Eric Ash lengthened the college day yet more in order to further accommodate humanities education. In 1991 the college day was to start at 9.00 am and end at 5.50 pm.

14. Jacques came to Imperial first in 1951 as the conductor of the college choir. He was much loved for his brilliance and warmth. Hopkins taught at the Royal College of Music and was an excellent broadcaster on musical topics.

15. I attended one of Gombrich's lecture series in the early 1960s and remember an interesting experiment that he conducted. One of the larger lecture halls was filled with students, few of whom knew much about modern art before attending. Gombrich brought in a large collection of slides. Some showed the work of Jackson Pollock, others those of his imitators. Without knowing which were which, we were asked to evaluate the paintings and give them a mark out of ten. The next week Gombrich announced the results. The class had consistently scored Pollock's paintings higher than the others, despite it being unclear to most of us exactly what the differences were. This gave Gombrich the opportunity for an interesting lecture on aesthetics and psychology. Ernst Gombrich was at the Warburg Institute (which until close to this period was located in the Imperial Institute) and later became Professor of the History of Art at University College London.

16. Among those giving single lectures in the 1950s and 60s are the following, chosen to illustrate the range of expertise: Eric Ashby, Isaiah Berlin, Anthony Blunt, D. W. Brogan, Hugh Casson, Cecil Day Lewis, Edward Heath, Dilys Powell, G. H. N. Seton Watson, Eric Shipton, John Skeaping, Dudley Stamp, Richard Titmuss, Barnes Wallis, Huw Weldon and John Wolfenden. Imperial staff members also gave lectures. among them H. Levy, H. B. Allan, G. J. Whitrow, A. W. Skempton, and K. Roth.

17. 'Education for Modern Needs', *Nature*, 180 (1957), 1151.

18. See chapter 9. Hall focussed on teaching at the postgraduate level. He introduced a good MSc course and trained some professional historians of science and technology. The new department was given financial support by Imperial Chemical Industries.

19. Electrical Engineering correspondence file; Jackson to Linstead, 16 January 1961. Jackson had been thinking along these lines even earlier when he had come to Imperial after the war. But he left Imperial in 1953 and began pushing for diversification of the curriculum more seriously only after his return in 1961.

20. This pattern became the norm throughout the college until there was consolidation and centralisation of resources. The result was the eventual creation of the humanities department and the development of credit courses. In the 1960s students were expected to attend a departmentally set minimum

of lectures on non-technical subjects but were not normally examined on them.

21. Electrical Engineering correspondence file; Jackson to Rector, 6 November 1961. John Sinclair Robertson Goodlad (1938–), known as Sinclair, was educated at Cambridge where he took a BA in English. He had taken science subjects at A level when a student at Marlborough School. He taught for one year in Delhi and one year at MIT before coming to Imperial. One of his most successful innovations at Imperial was the Pimlico Connection. Each Wednesday afternoon about twelve undergraduates from electrical engineering visited Pimlico School as unpaid tutorial assistants in science classes and acted also as guides for groups of children visiting the Science Museum and other venues. This programme expanded in later years to include about twenty schools and many more students, also from other departments. It continues to be successful and has inspired similar schemes around the country.

22. While Goodlad taught more than writing skills, including English literature and drama, students were also encouraged to take economics, industrial management, and some other courses at the LSE.

23. Sinclair Goodlad, 'Pieces in a Jigsaw', *The Spanner* (1965), 27–9. Goodlad gives a good account of the courses offered to electrical engineering students at that time.

24. The idea was to provide service teaching and to carry out some research. There was no plan to offer degree courses in these fields.

25. See chapter 12. The merger with the Architects Association School did not occur.

26. In 1969 Goodlad and some students carried out a survey; 1060 students responded to a questionnaire. 43% of respondents stated that they wanted about 15% of their degree to be devoted to non-technical studies. See J. S. R. Goodlad, R. Mohan, J. Shields and D. Wield, 'The Demand for Non-Technical Studies at Imperial College', *Liberal Education* (Spring, 1970), 32–7.

27. Rector's correspondence file. See Penney's memorandum, 18 November 1969. Penney was worried about all the new disciplines and having things spin out of control.

28. Instruction in scientific German and French had been given in many departments since before the founding of Imperial in 1907. Scientific Russian was added later. In some departments students were expected to take examinations in two languages. This usually entailed carrying out a translation, with dictionary, within some designated time period. In the 1960s more general

language instruction came to the college and more emphasis was placed on oral skills; but language instructors were part-timers.

29. Associated Studies and General Studies coexisted for a while until the retirement of C. K. McDowall. General Studies still drew on experts from outside the college. For example in 1971 a lecture series 'The English Tradition of Dissent', was given by a good group of historians: R. H. Hilton, Christopher Hill, Gwyn Williams and K. R. Minogue. In 1973 E. F. Schumacher gave a lecture series, 'Economics as if People Mattered'. In Associated Studies, lectures were given in modern languages, history, modern literature and drama, sociology and politics. For a report on the formation of Associated Studies see Michael Binyon, 'A Philosopher among the Scientists at Imperial', *THES* (1 September 1972), 4.

30. David Daiches Raphael (1916–) was educated at Oxford. Before coming to Imperial he was Professor of Philosophy at Reading University, but had spent much of his professional life at the University of Glasgow where he had been Edward Caird Professor of Political and Social Philosophy. He has written extensively in the areas of moral and political philosophy, and on Scottish Enlightenment thinkers.

31. The Rector, Brian Flowers, considered closing down work in the history of science since the Halls had not built up the department and had not provided for a succession. Instead he promoted lecturer Norman Smith to reader and moved him to the new Humanities Department. Smith, a specialist in the history of technology, was joined in 1981 by Simon Schaffer, appointed lecturer in the history of science. When Schaffer left to take a position at Cambridge, James Secord came in his place. Secord left in 1992, also for Cambridge. Both Schaffer and Secord later became professors in the history of science at Cambridge.

32. Eric Stables, (1931–) was educated at King's College London and taught there and at Goldsmith's College before joining Imperial in 1974. He became Director of the Humanities Programme in 1987. John Bernard Thole (1930–91) was educated at Cambridge. He was appointed to teach history and politics in the Department of Chemical Engineering and Chemical Technology in 1966, and later transferred to Associated Studies where he organized the lunch hour lecture programme. Thole created a Victorian-style music room at 53 Prince's Gate.

33. See chapter 11 for further details on musical life at the college.

34. According to Dickins, the standard was so high that most of the students chosen for such instruction could have entered the RCM in their own right. See Annual Review, 1990–91.

35. Professor Raphael attempted to encourage research and publication but a serious academic culture in the humanities was difficult to build up. Aside from the historians of science and technology, philosopher John Milton and historian Kathleen Burk were exceptional in having a serious record in academic publication. They moved to larger departments at King's College and University College respectively.

36. The London Centre was formalized in 1987 and the other participating institutions were University College and the Wellcome Institute for the History of Medicine. This was acceptable to the UGC which allowed the separate institutions to accept both masters and doctoral students in the history of science and technology.

37. Science Communication was a programme that began in 1991 with a major donation from the Leverhulme Trust. Ros Herman had studied computer science and was a journalist with the *New Scientist* before coming to Imperial.

38. In 1992, the year that Secord left for Cambridge, Norman Smith, reader in the history of technology, also retired. This meant that the Rector had to make a decision either to close this area or bring in new people to revive it.

39. Humanities was downgraded in 1987. The STS unit was the only part of humanities to be submitted in the 1992 RAE. It did poorly, gaining a 3 (as did the management school).

40. The Centre later included also the history of medicine and now has the acronym CHOSTM. Its staff supervise doctoral students and teach both in the MSc course and in the undergraduate humanities programme.

41. David E. H. Edgerton was educated at Oxford University and Imperial College. He taught economics of science and technology at the University of Manchester, 1984–5 and, from 1985–92, was a lecturer in the history of science and technology at UMIST and the Manchester Centre for the History of Science, Technology and Medicine. He was appointed to head the Centre of History of Science and Technology at Imperial in 1993, and was promoted to a chair in 1998. A major donation from the Rausing family led to the creation of the Hans Rausing Chair to which Edgerton was appointed in 2002. Andrew C. Warwick was educated at Cambridge University and held a research fellowship at St. John's College before coming to Imperial in 1992. He was appointed to a chair in 2002 and succeeded Edgerton as head of the Centre in the same year.

42. One of the major ongoing research efforts in CHOSTM is the Newton Project under the direction of Rob Iliffe. Many of Isaac Newton's original theological manuscripts are being transcribed and, together with scholarly apparatus, have been made available electronically on the world wide web.

Iliffe was first appointed to a governor's lectureship, under a scheme initiated by Lord Oxburgh when he became Rector. Serafina Cuomo and Andrew Mendelsohn were similarly appointed a few years later.

43. Goodlad was later succeeded by Charmian Brinson, a Germanist who was given a chair in 2004 — one of few in the Humanities Programme to have achieved that rank.

44. A new language laboratory was opened in 1995 intended to help both native English speakers learn foreign languages and non-native English speakers improve their English. The Humanities Programme has some staff on the academic scale who teach at the MSc level, but most are on the professional services scale, providing service teaching mostly in languages. CHOSTM is the only humanities unit at Imperial to be RAE rated. For a small, relatively isolated, history department the 5A rating in both 1996 and 2001 was a notable success. In addition to postgraduate teaching, the staff offer a range of undergraduate humanities courses in the history and philosophy of science, technology and medicine, European history, and politics.

45. The scientific translation course attracts the most students with about 50 in 2005. The course is devoted mainly to the theory of translation and students can opt for work in the languages of their choice. Mandarin, Spanish and Japanese are the most popular.

46. Jackson will have been influenced by his mentor at Metropolitan Vickers, Sir Arthur Fleming, who was an advocate of such education.

47. King's College engineers were attending industrial management courses at the LSE even before the war.

48. This was made possible in part because of postwar American Counterpart Funds made available to Imperial College. The first UK chair in production engineering was at the new College of Aeronautics at Cranfield.

49. See Minutes of Production Engineering Exploratory Committee, 22 November 1957. This meeting was held after the course had run for a year. Part of the problem was that the technical meaning of 'management' in this new context was not widely understood.

50. Nicol Gross (1910–1969) was educated at Cambridge, had much industrial experience before coming to Imperial, and was a world authority in welding technology. He became managing director of British Oxygen in 1967.

51. Samuel Eilon FREng (1923–) studied electrical engineering at the Israel Institute of Technology, Haifa, and at Imperial College. His own research covered production and inventory control, distribution systems, and management control.

52. British Universities had difficulty in filling senior positions in production engineering at this time. Courses existed at Birmingham, Manchester and

Durham Universities. The chair in Birmingham was demoted to a reader-ship, and at Nottingham where they wanted to build in this area, and had advertised a chair, they could not find a suitable person. (KEMM; internal memorandum.)

53. This caused some friction between Eilon and some of the mathematicians who also taught operational research methods at that time. Interestingly, one of the best known operational research departments in the country was at United Steel. It was headed by K. D. Tocher who earlier had worked in the mathematics department at Imperial, but was asked to leave when work on the college computing engine was terminated. See chapter 9.

54. For J. M. Alexander see chapter 10. Roy Brewer had taught in a similar course at Salford Technical College and was appointed lecturer at Imperial in 1957. He died in 1965 just after having accepted a chair in production engineering at the University of Bradford.

55. In the USA similar courses were being offered at MIT and the Carnegie Technical Institute already in the early 1950s. Indeed, the legacy of war time operational research was taken up enthusiastically in management science, business, and other areas. See M. Fortun and S. S. Schweber, 'Scientists and the Legacy of World War II: The Case of Operations Research (OR)', *Social Studies of Science* 23 (1993), 595–642.

56. Joan Woodward (1916–71) was educated at Oxford and Durham in classics. During the Second World War she became a senior labour manager at the Royal Ordnance Factory in Bridgewater. It was there that her interest in industrial sociology began. She had several academic posts before coming to Imperial. Her *Industrial Organization: Theory and Practice* (Oxford, 1965) was highly influential and attracted staff and research students to Imperial. Among them were Dorothy Wedderburn and Sandra Dawson, both of whom later had major careers. See 'Professor Joan Woodward: Industrial Sociology at Imperial College', *The Times*, 19 May 1971, 19; and obituary, *The Times*, 21 May 1971, 16.

57. Eilon was at first appointed Professor of Industrial and Management Engineering (later shortened to Industrial Management). He had wanted to change the title to Management Science, but this was not permitted by Hugh Ford who claimed the title would imply a change in emphasis away from engineering. But the title was changed in 1971 when Eilon's section became a separate department.

58. Students were given tasks such as analyzing the performance and strate-gies of different companies. Business policy seminars first instituted in 1967 expanded during the 1970s and lasted until the department was closed down and such activity moved to the new Management School.

59. The early history of business education in the UK is very complex. The Imperial College archives contain much correspondence on this topic. In the 1950s and 60s the major Imperial College players were Linstead, Saunders and Jackson. There is much correspondence between them, Sir Keith Murray, chairman of the UGC, and J. W. Platt and Sir John Wolfenden who were both associated with the Foundation of Management Education which was helping to finance management studies at the college. Platt, a senior executive with Shell, was later in favour of the proposed business school for London coming to Imperial. Also much engaged with this issue was the CGLI which, since the 1950s, had supported the C&G College moving in the direction of some forms of business education.

60. Lord Franks was asked to prepare a report by the Federation of British Industries, the British Institute of Management, the Foundation for Management Education and the National Economic Development Council. The idea was for one or two business schools to be opened along the lines of the best such schools in the USA. It was agreed that business education would be supported by a mix of government and industrial funding.

61. Linstead and Sir Sydney Caine, Director of the LSE, had many meetings and exchanged much correspondence on the new school.

62. Willis Jackson supported Linstead in wanting the business school at Imperial.

63. Evidence for these claims can be gleaned from Linstead's correspondence in this period.

64. It now occupies part of the old Bedford College site in Regent's Park.

65. Penney was under pressure from Arthur Earle, the first Principal of the LBS not to invade his turf. Earle was understandably worried about the spread of resources, as many universities were then trying to jump on the business bandwagon. 'As you will know there is a considerable ferment of activity in many British Universities'. He was concerned about competition in business and management studies generally, and in particular about some of the business policy and finance areas in which Imperial was proposing to expand. Earle to Penney, 12 November 1968 (KEMM). See also minutes of meeting on 'The Future Organization of Industrial and Management Engineering' (KEMM, no date, 1965).

66. Roger Hall, a lecturer in the section, moved to become a founder member of the academic staff at the new London Business School.

67. The section was also funded by the Foundation for Management Education and the UGC.

68. According to Dorothy Griffiths (personal communication), who was a young research assistant under Woodward at the time of Woodward's

death, Penney met the staff and assured them that the Unit would not be closed within the following two years. During that time its future would be discussed. In the interim, the staff were to decide on their own form of internal governance, but they had to choose someone to represent the unit in the college. In the event, they set up a management committee and elected Keith Alan Smith as its chairman. It was he who represented the Unit. Smith left the college in 1973. Griffiths is now Deputy Principal of the Tanaka Business School.

69. Dorothy Enid Cole Wedderburn (1925–) was educated at the University of Cambridge. On graduating she stayed at Cambridge as a Research Officer in the Applied Economics Unit. She made her name with studies on the impact of redundancy, and in challenging what she saw as a complacent welfare state. Wedderburn came to Imperial as a lecturer in 1965, became a reader in 1970 and professor in 1978. She was a member of the Royal Commission on the Distribution of Income and Wealth (1974–5). On leaving Imperial in 1981 she became Principal of Bedford College, later Principal of Royal Holloway and Bedford New College. As a college principal Wedderburn played a major role in the restructuring of the University of London and was Pro-Vice-Chancellor, 1986–8.

70. Flowers wrote to Southwood that he wanted to bring more of the social sciences to Imperial provided they were of the 'hard-headed variety … [and] if it were initially a post-graduate activity accepting only science-based students. There are not many science-based Maoists! … [I have] not yet plucked up the courage to talk to Dorothy Wedderburn'. Rector's Correspondence; Flowers to T. R. E. Southwood, 2 July 1974.

71. For Huddie and the new engineering courses see chapter 13. For the later Huddie Report on subjects other than science and technology, see GB minutes, 25 June 1976.

72. Zangwill Aubrey Silberston (1922–), known as Aubrey, was educated at the University of Cambridge where he studied history. While a prisoner of war during the Second World War, he began to study economics. He returned to Cambridge after the war and took a second degree in economics. After a short period working with Courtaulds he returned to Cambridge as a lecturer. From 1971 he was a Fellow of Nuffield College, Oxford, coming to Imperial in 1978. In 1981 he succeeded Wedderburn as head of the department of social and economic studies.

73. One lecturer of this period was Gary Werskey, author of *The Visible College* (London, 1978). This work is a collective biography of left-wing scientists of the interwar period, including Hyman Levy and Lancelot Hogben who both worked at Imperial College.

74. The ISU, later integrated in the Department of Social and Economic Studies, contributed much more to the overall general undergraduate education at the college than did Management Science. Eilon was seeking to expand into the undergraduate field by introducing joint degrees with departments such as chemistry and biochemistry. Some of these joint courses survived for a few years but were later redesigned in the Management School.

75. In 1977 Eilon was unhappy because the UGC thought he should be using more visiting staff — as, for example, from the LBS. See Eilon to Rector 7 July 1977. (KEMM). As discussed in chapter 12, Flowers was unable to support any staff increases in this period.

76. KEMM; see Ford to Flowers 18 June 1979.

77. Eilon applied for sabbatical leave in 1987, the year that the new Management School opened. He did not return to teach at Imperial and retired in 1989. He became a senior research fellow in the department of mechanical engineering. Silberston became a senior research fellow in the new management school.

78. David Norburn (1941–) was educated at the London School of Economics and City University. Norburn worked for the Burroughs Corporation and Price Waterhouse before returning to City University as a Senior Research Fellow. He joined the London Business School as a lecturer in 1972, becoming Director of the MBA programme. In 1982 he was appointed Franklin D. Schurz Professor in Strategic Management at the University of Notre Dame, Indiana. He returned to England as Professor of Strategic Management at Cranfield Institute of Technology and joined Imperial in 1987.

79. This Victorian house was originally home to the Cunard family. It had earlier been occupied by people from the mathematics department and is, today, home of the new Institute for Mathematical Science.

80. *The Independent*, 27 August, 1987. Not all of these plans materialized. No specialized undergraduate course was offered though management was taught in a number of joint degree courses.

81. Pamphlet advertising the MSc in Management, 1989–90.

82. Management science became just one option in the new MSc degree run by Roger Betts. Industrial sociology virtually disappeared. Other options were the management of innovation, the management of new ventures, project management and finance. A review of individual projects submitted for the degree during the early 1990s indicate they were not that different from those one might expect in any other business school. For example, 'Financial Analysis of North Sea Oil Extraction'; 'Feasibility of a Pulp and Paper Mill

on the Ivory Coast'; 'Feasibility of a London-based Helicopter Ambulance Service'.

83. Many new appointments were made to meet the MBA demand. For example, Sandra Vandermerwe, Professor of International Marketing and Services in 1995; David Miles, Professor of Finance in 1996; and Sumit Kumar Majumdar, Professor of Strategic Management in 1998.

84. Gary Tanaka promised £25m in October 2000 for the relocation of the management school and £2m to support internet technology (it is not clear how much of this money has been received). Tanaka graduated with a PhD in mathematics in 1970 and, in 1979, became Director and President of Amerindo Investment Advisors. This firm became preeminent in the field of investment advising, especially in the management of emerging technologies stock portfolios for institutional investors.

85. Perhaps this was a poor decision. The building should have been named for Tanaka, but not the School. I make this comment in light of the emphasis on branding and the high repute and recognition of the Imperial College name.

86. David Knox Houston Begg (1950–) was educated at the Universities of Cambridge and Oxford, and at MIT where he took a PhD. He returned to Oxford as a teaching fellow at Worcester College and was appointed Professor of Economics at Birkbeck in 1986. Begg is also a research fellow at the Centre for Economic Policy Research.

87. In fact the management school was improving even before any Tanaka money came to the college. In 1996 it received a 4 in the RAE and, in 2001, a 5.

88. For example, Professor Begg, noting which journals counted for most with the various ranking bodies, insisted any member of staff who wished to stay on at Imperial should be prepared to publish at least one article a year in one of the highly rated journals (personal communication).

89. Professor Christofides had earlier worked with Eilon in areas of operational research. He later moved towards finance from his earlier work in scheduling and vehicle routing. Nicos Christofides (1942–) was born in Cyprus and was educated at Imperial where he took a BSc and PhD in electrical engineering. After working briefly with Associated Electrical Industries he returned to Imperial, and the Management Section of Mechanical Engineering. He rose through the ranks to become Professor of Operational Research in 1984 and is now head of the Centre for Quantitative Finance. This area of the business school is very successful and graduates about 10–15 PhDs a year, many finding high-paying jobs in the City. (The salaries of recent graduates is an important criterion in the ranking of business schools.)

Financial mathematics is also a strength in the mathematics department and there are some turf wars in this area.

90. It should be recognized that the kind of industrial support for academic positions in science and engineering does not exist in the humanities so the Halls will not have had an easy time, especially in the 1970s. However, government and foundation monies could have been more actively sought. The Halls were primarily interested in 17th and early 18th century science. It would perhaps have been good to bring in specialists in areas such as modern physics or evolutionary biology.

91. See, Howard Wesley Johnson, *Holding the Centre: Memoirs of a Life in Higher Education* (Cambridge, MA, 1999). Johnson was President of MIT. Humanist and social science scholars at MIT were, and remain, sensitive to their environment and have typically undertaken research projects that connect in some way to the larger world of science and engineering. Perhaps this is another lesson for those at Imperial. But MIT has the advantage of being very much wealthier, and is able to provide far higher levels of funding in the humanities.

92. There was also much defunct science space at Westfield which could have been utilized. See GB 5 December 1986.

93. It is interesting that at least three very successful women were engaged in sociological work at Imperial: Woodward, Wedderburn, and Dawson. This pattern is not atypical. In male-dominated institutions, such as Imperial College, it is easier for women to succeed at the margins than at the centre of traditional fields. By carving out a new space for themselves, in a field that was well chosen, they increased their chances of success.

Dame Sandra June Noble Dawson (1946–) was educated at the University of Keele and the University of Cambridge. She joined Imperial as a research assistant in the Industrial Sociology Unit in 1969 and rose through the ranks to become Professor of Organizational Behaviour in the Management School in 1990. She is a specialist in the organization of health services and moved to Cambridge in 1995 where she is Director of the Judge Institute of Management.

94. The economics legacy of Professor Silberston is less marked since there is little work in this area in the new business school, despite its current head being an economist.

95. The plan is to integrate some business education in all of the undergraduate engineering courses. There will be no further joint management and engineering (or science) courses.

96. The Humanities Programme is now attempting to diversify its course offerings. This, presumably, is to attract more students and preserve jobs.

Leaving aside the valuable language instruction, if humanities and social sciences education is to be a serious adjunct to an education in science, technology or medicine, if the point is to help students achieve a more historical, cultural, social, political and economic understanding of their principal areas of study, a more focussed approach than the one currently adopted will be necessary.

chapter fifteen

The Expanding College, 1985–2001…
Part One: Governance and the Medical School Mergers

The Politics of Higher Education

The history of Imperial College at the end of the twentieth century cannot be understood without some knowledge of government policies as they related to higher education. The earlier decisions that led to greater access in the 1960s and 70s resulted in sweeping changes to the ways in which higher education was funded and managed. By the late twentieth century we see a wide range of interested parties attempting to work out a new social contract — one in which society agrees to continue its support for universities, but only in exchange for some recognizable contribution to the public good. What such a contribution might be is part of an ongoing debate. This section summarizes some of what occurred in the 1980s and 90s.[1]

In 1984, after discussion with the government, the University Grants Committee (UGC) issued a document titled 'A Strategy for Higher Education in the 1990s' and asked all vice chancellors and principals to respond.[2] According to the UGC, the basic aims of the Robbins Report were still in place but, in future, higher education was to encompass a wider range of skills and disciplines than earlier envisaged. Further, since people were now expected to have to retrain during the course of their working lives, universities were asked to think of ways to provide remedial and short courses with life-long education in mind. Continuing Education was already a major talking point, including at Imperial. Brian Flowers had appointed Professor Pratt as Director of Continuing

Education in 1981 and, in 1985, Flowers held discussions with the Vice Chancellor of the Open University on possible collaboration on distance learning.[3] Universities were told that they also needed to better prepare students for the information technology and biotechnology fields. Student numbers were to grow, and a shift towards science and engineering was to be encouraged by the use of directed funding. Research activity was to increase but not at the expense of teaching. To accommodate this expansionary vision, and support greater access, economies would have to be made. The unit of resource per student was to shrink and universities were asked for their views on how best to plan for the future in the new political and economic climate.

Not surprisingly academics objected to being asked to do more with less. Imperial College responded by stressing that, in accordance with its charter, it was carrying out work that was important to industry, and that all subjects taught at the college were 'directly relevant to the wealth-increasing capacity of the country'. Further, the college had joined with financial institutions in founding companies concerned with technology transfer and was doing its best to add to its UGC income. Most of the staff, however, wanted to be teachers and researchers, not entrepreneurs. Therefore, if the UGC were to see all the college business activity as an excuse for offering less, any incentive to seek outside funding would be lost. If government funding were to be seriously diminished then, by the 1990s, Imperial would be forced to retreat into teaching and contract research only, a very unhealthy situation. If very deep cuts were to be made, it would close down departments, including some that were among the best in the country. This would be a serious national loss. Indeed, the college was already turning down many qualified applicants and had not even begun to tap seriously the potential of women.[4]

Already in the 1970s the government attitude towards funding had begun to change. Until the major expansion of university education, UGC monies were seen as 'deficiency' grants in that they were intended to bridge the gap between what universities claimed to need and the income they received from other sources. The various needs were reported to the UGC which then decided on priorities and made its case to the government. After more than fifty years of working within this system, universities were accustomed to bridging grants and had a well developed sense of entitlement. However, university expansion in the 1960s and 70s put an end to entitlement and universities had to make do with what the

Treasury thought it could afford, or find other sources of funding.[5] As discussed in chapter twelve, in the early 1980s the education minister, Sir Keith Joseph, began to envisage university funding in a radically new way. In so far as research was concerned, the government wished to stop providing all of its grants through research councils, and was thinking instead of contracting for some research.[6] But this idea was opposed both by the research councils and the UGC and, for a while these bodies retained their independence.[7] By the mid 1980s, however, government policy had matured. Government was to become a purchaser of both teaching and research services, and universities were to compete for its custom. The government also envisaged that many university departments would be funded solely for teaching, and not at all for research.

In this uncertain climate the internal management, financial arrangements, and the teaching and research carried out in universities, all came under review. With respect to university management the Committee of Vice Chancellors and Principals (CVCP) set up a committee in 1984 chaired by Sir Alex Jarratt. With respect to subject areas, the UGC began with a review of the earth sciences looking to increase efficiency in the national delivery of both teaching and research in this area. Reviews of physics and chemistry were intended to follow. As is discussed in chapter sixteen, the earth sciences review was to have devastating consequences for Imperial's geology department. The Jarratt Report, for its part, spelled out a long list of managerial failures, making clear that universities were far from being well managed.[8] Given the prevailing business climate it is perhaps not surprising that Jarratt advised university vice chancellors to think of themselves as chief executives, and governing bodies to see themselves as boards of directors. One of the report's principal recommendations was that sound management data needed to be systematically gathered and, in line with government intent, that universities should collectively set up a system of quantitative performance indicators covering management and the delivery of services. The UGC and the CVCP set up a working party in 1986 which invited the views of all universities on what the performance indicators should be and how they should be weighted. Needless to say this exercise, involving academic opinion from across the country, was fraught with difficulty.[9]

At Imperial, the idea of performance indicators, and the principle of selectivity, were treated warily given that selectivity, at least in the academic area, might possibly act in the college's favour. The only serious objection made was to the increased bureaucratic cost in having to demonstrate research, teaching, and administrative efficiency. The CVCP also asked universities to list the research fields they were most keen to promote. Those listed by Imperial were: composite materials, ceramics, electrical materials, plant biochemistry, neuroscience and pharmacology, the biological effects of trace elements, and the broader aspects of information technology. The college also noted that it had recently moved towards more undergraduate education. Namely it had moved from a balance of undergraduates to postgraduates of 60:40 to one of 68:32, but wanted to return to the earlier ratio by increasing postgraduate numbers, not decreasing undergraduates. Further, the college did not wish to see the retirement age lowered to sixty as had been suggested. And, fearing loss of tenure, it stated that staff must be protected from dismissal except for essential reasons spelled out in advance.[10] The college was also critical of the UGC which was seen as no longer playing a neutral role, but rather one of imposing government decisions on universities.

For some reason, not entirely clear, in the midst of these deliberations the CVCP recommended that the UGC's constitutional position be examined. The government willingly set up the Croham Committee to look into this and, together with the Jarratt Report, it gave the minister what he needed to close the UGC down.[11] Croham recommended reconstituting the UGC, albeit still as an independent body. The new committee, Croham stated, should be smaller and half its membership should come from outside the academic community. It should be drawn from all regions of the country; and industry, commerce, the professions and public service should be represented. The chairman could be an academic but, if so, was to be someone with considerable experience also outside the academic world. The chief executive officer was to be a distinguished academic.[12] Further, the new committee should no longer simply hand over block grants to the universities, but should monitor performance and future planning.[13] Both Croham and Jarratt stated that universities needed to be more accountable, and that the government should have more control over tax monies spent on higher education.

Sir Keith Joseph's successor was Kenneth Baker. Despite belonging to a cabinet in love with the free market, Baker was a centraliser keen to take control of university expenditure. It was his task to implement the new Tory vision.[14] Baker shared Thatcher's view that universities were 'pushing out poison' and that it was time to take them on. The growth in higher education and the fall-out from the student revolution had led many to question the role of universities. Thatcher appears to have been among the many and set out to curtail both their funding and independence. Taking note of both the Jarratt and Croham reports, her government decided on some major changes. Its new position was spelled out in a 1987 White Paper 'Meeting the Challenge'.[15] Whitehall was to decide what the educational needs of the economy were, and to 'make adjustments where needed'. Student demand and faculty interest would no longer determine which courses were offered. Instead, courses would depend on 'the demands for highly qualified manpower, stimulated in part by the success of the Government's own economic and social policies.' The government and its central funding agencies '[would] bring higher education institutions closer to the world of business'.[16] This was all vaguely threatening and not entirely clear. As Simon Jenkins put it, 'the world of business was seen by ministers, few of whom had any experience of it, in a golden haze: it was a vague amalgam of the free market, hostility to unions and sympathy to the Tory Party'.[17] The White Paper was not so much about saving money as about taking control. Whitehall would now decide what was needed and universities would come under a form of *dirigism* inspired by practices in the private sector. The government also set in motion a discussion of student fees. It wanted to shift university funding more toward fees so as to make institutions more competitive, and to test their ability to attract students.[18]

The 1988 Education Reform Act did not put in place all the proposals listed in the White Paper. Some were dropped after strong opposition in the House of Lords. But universities did lose much of their independence. The University Grants Committee was terminated and its place taken by a new University Funding Council (UFC), the name itself a mark of attitudinal change.[19] The new council was not independent, as Croham had suggested, but answerable to government. In accord with Croham, just over half its members came from outside the universities. The first chairman of the UFC was Lord Chilver of the Cranfield Institute of Technology who shared the government's concern that higher education produce graduates to serve the needs of industry.[20] In adopting the Jarratt principle of performance indicators, new subject sub-committees were set up under the UFC for the centralized quality assessment of all universities expecting research funding.[21] Put in commercial terms, the government and the taxpayers needed to know what they were purchasing.

In a yet further move to centralize, the 1988 Act took the polytechnics away from local government control and created a new funding agency, the Polytechnic and Colleges Funding Council (PCFC).[22] In 1989 it was announced that maintenance grants for students would be abolished and replaced by a loan system.[23] To cap it all tenure was attacked along the lines suggested earlier by Sir Keith Joseph.[24] Tenure provisions for all new appointees and for existing staff looking for promotion were ended. This blow was only slightly softened when the government accepted an amendment to the 1988 Act (proposed in the House of Lords by Lord Jenkins) which protected traditional academic freedom of expression.[25]

During the government of John Major the most significant event was the abolition of the binary funding system and the merger of the polytechnic and university sectors. A single funding agency, the Higher Education Funding Council (HEFC), was introduced.[26] Over the previous years, the polytechnics had gained some degree giving powers. Kenneth Clarke, the Education Secretary, saw the merger as a way of improving Britain's university graduate figures by allowing the polytechnics to give yet more degrees, even to rename themselves universities if they so wished. Until this merger the universities had roughly three times per capita more government money to spend than had the polytechnics. But, with the end of the binary budget line, and with research funding not significantly increased, universities were seriously pressed. Universities, colleges and polytechnics all competed for roughly the same amount of research money that, earlier, had gone almost exclusively to universities.[27] This was in part a consequence of Thatcher's view that all should have the chance to compete, and that the spirit of national competitiveness needed a boost. While many of the polytechnics will have been delighted with their new status, and the opportunities now provided, the legislation created problems elsewhere. Britain could claim to have doubled the number of students attending university, but at a cost. By the early 1990s, universities were receiving about half the amount per student they had received in the pre-Thatcher years. Given the enlarged university sector, taxpayers could no longer be expected to shoulder the burden of the traditionally high staff to student ratios. The average faculty to student ratio in the older universities dropped from 1:9 to 1:17. The staff to student ratio at Imperial was much higher, namely 1:10.8 in 1988–9, and reflected the research intensive nature of the college. Imperial College had for some time been under pressure from the UFC to decrease its ratio, but justifiably responded by stating that the council should be more discriminatory and treat centres of research excellence differently from the rest.[28]

Labour may have killed the grammar school, but it was the Tories who killed both the traditional university and the polytechnic. The polytechnics, like Imperial, were good models for the linkage of higher education to Britain's industrial and commercial sectors. In killing them, the government may well have left the country vulnerable in some areas of trades education. In retrospect it is surprising that it was under the Conservatives that a mass system came into being. Not that equality trumped merit in all respects. For example, against the wishes of the AUT, the national system of salaries was freed up at higher levels in 1988. And, in 1989, the move to separate research from teaching for funding purposes allowed merit back into the system and led to the preservation of some centres of excellence such as Imperial College. In a 1991 speech given to the CVCP, Kenneth Clarke called for yet greater efficiency in the universities. He wanted modular courses to be introduced to allow students to access the system at different times of year, and at different points in their lives.[29] He asked for more distance learning, more part-time programmes, and more teaching-only contracts. In 1994 the HEFCE spawned the Higher Education Quality Council (later Quality Assurance Agency) to monitor teaching quality. But while the results of its assessment are made public, teaching grants are awarded on a per capita basis, something that later was much criticized by Imperial College Rector, Sir Richard Sykes. In 1995 a cabinet reshuffle resulted in the transfer of the Office of Science and Technology from the Cabinet Office to the Department of Trade and Industry (DTI), and the merger of the Department of Education with the Department of Employment. Imperial staff expressed their concern over this seeing the DTI as not caring for any long-term research. As Professor Adrian Smith put it with respect to his own discipline, 'the best mathematical research tends to have an impact longer downstream.... The DTI mentality may be more geared to supporting work with shorter lead time to industrial impact.'[30]

In 1996 the Secretary of State for Education and Employment set up a further committee to look at the longer-term development of higher education. The committee was chaired by Sir Ronald (later Lord) Dearing.[31] It published a huge, somewhat platitudinous, report with ninety-three recommendations. Once again the Robbins philosophy of greater access was reiterated, but the major discussion was over university funding. The best news coming out of this committee was the recommendation that spending cuts come to an end and that universities needed more, not less, funding. One result was the formation of the Joint Infrastructure Fund and the Joint Research Equipment Initiative, which have provided Imperial and other universities with some much needed money for the upgrading of facilities. Dearing recommended that funding levels be linked to the GDP, and that students contribute more to the cost of their education. It was suggested that they pay about twenty-five percent of the full cost, and that the current loan system which was not working well should be ended.[32] Dearing recommended that careers services be better integrated into university governance and that students be expected to pay back into the system when well established in the workplace.[33] Further, it was suggested that government should support entrepreneurship at universities by helping to fund start-up companies. The report placed much emphasis on information technology, recommending that all students have access to the internet, and that they own portable computers by 2005.[34] Perhaps most contentious was the idea that university teachers be certified as to their teaching ability.[35] While this committee was appointed by a Conservative government, a Labour government swept into office two months before Dearing published his report.[36] The new education minister, David Blunkett, at first refused to accept much of it, but gradually changed his mind.

The question of topping up fees, suggested by the CVCP in light of dire projections of government funding, and having students pay back some of the costs of their education when in employment, as suggested by Dearing, became major political issues in the new century.[37] Blunkett's junior minister, Tessa (Baroness) Blackstone, Minister of Higher Education and a former Principal of Birkbeck College, expended much of her energy fighting Oxbridge over the matter of separate college fees. Other institutions were caught in the cross fire when she ended the right of all universities to charge fees, precipitating the long-drawn out debate over top-up fees.[38] Sir Richard Sykes, noting the absurdity of the situation, stated that for Imperial to prosper it would have to charge fees of about £10,500 or, if it were to redistribute fee income to cover the cost of poor students, about £15,000 would be needed from those who could afford it.[39]

Looking back over these various changes one can understand the Treasury wanting to shed more light on rapidly increasing university expenditures. There had to be accountability and greater efficiency in the use of resources, but this could have been achieved without the attempt to micro-manage internal university policy. One can understand also the termination of the UGC which, clubbish, exclusive, and gentlemanly, had become something of an anachronism by the late twentieth century. In England it was replaced by the HEFCE set up to ensure that university expenditure was no longer exempt from government control. The loss of independence was a major blow and learning how to be accountable took time. The early teaching assessment exercises, for example, were something of a farce. In the first instance, institutions were asked to describe their goals and evaluate how well they had performed in light of them. A consequence was that institutions that set low goals for themselves, and achieved them, were relatively better rewarded than those which set higher goals not fully reached. Gradually, however, assessment procedures both in research and teaching have improved. A negative aspect of the current system is that it encourages rapid publication, and thus short-term over longer-term, and possibly more thoughtful, work. Audits that demand quantity will get what they deserve; further refinement of the RAE is needed. But perhaps even more troubling is that the country is losing its teaching capacity in technical areas because good departments are being forced to redirect their efforts towards research or close down. At a time when graduates in science, engineering, and mathematics are seen as necessary for the future well-being of the country, departments that produce good graduates (including future school teachers), if not first-rate research results, are closing for lack of funds. The recent closure of several chemistry departments is a case in point. For example, the department at King's College was unable to survive in the funding climate of the 1990s and, while some of its good research staff came to Imperial or moved elsewhere, good teachers and a major teaching facility were lost.

While one expected New Labour to recognize that we live in a capitalist society, it was surprising to see Blair's government so keen to adopt Thatcher's free market economic ideology, only to apply it so inconsistently. Higher education ministers have behaved like business executives in situations where they should not have done so, and like socialists in situations where the country can ill afford it. Despite all the problems, Imperial College's history will probably stand it well in today's business-inspired cultural climate. The long tradition of applied work, and the possession of a charter directing the college to think of industrial needs, has given it an edge over many of its competitors. Further, its strong overseas connections have proved invaluable since many foreign students still seek good scientific and technical education at Imperial, despite the very high fees.[40] The RAE, with all its faults, has helped also in the building of a new corporate ethos. People have had to learn how to compete while pulling together internally, and the college administration has been forced to improve. Success in the RAE, and Imperial's high placing in various league tables, has had some positive feedback. The standard of successful applicants for both staff positions and student places has risen. Lord Oxburgh's move to bring in new young lecturers at college expense has had positive feedback also. Oxburgh made a business case to the Governing Body that the college should spend money on a major hiring programme.[41] While very costly, the scheme paid off in college rejuvenation, in subsequent success in the RAE, and in the improved government funding that followed.[42] In many respects Imperial College today is better than it has ever been, and people at the college are happy to be associated with its successes. Morale improved considerably during the 1990s. Where the college is likely to be less good, however, is in providing space for spontaneity in research, and for a type of scientific and technical genius, examples of which can be found in earlier chapters of this book. This is not to say that genius is a thing of the past, but rather that individualists with little patience for the bureaucratic demand for accountability are unlikely to fare well in Imperial's new academic climate.[43]

Governance: the Ash and Oxburgh Years

Lord Flowers had done much to help the college adjust to new economic and political realities following the end of the postwar boom. His successor, Sir Eric Ash, took over at a time when the economy was beginning to grow, but with a government still seeking cuts in the university sector.[44] Staff morale was low since most had suffered a decline in real income and many had experienced blockages in promotion. By the time Lord Oxburgh took over from Ash, new government monies were slowly flowing into the college but, given the dislocations of the previous twenty years, he had much healing to do.[45] Besides, the problems for universities were far from over.

Government policy in the 1980s and 90s encouraged debate over the optimal size of the college and whether mergers with other institutions were desirable. As we shall see, Imperial did grow by negotiating successful mergers with a number of medical schools and with Wye College.[46] Also merged for economic reasons were the Imperial College and the Science Museum Libraries, though only after long and difficult negotiations.[47] Government policy also encouraged further expansion in the area of information technology (IT). In this connection, one major experiment of the period was the Alvey programme (1983–8). This was a direct response to the fact that in 1981 the Japanese Ministry of International Trade and Industry (MITI) announced the development of so-called Fifth Generation computing which, it (wrongly) believed, would allow the Japanese computer industry to overtake IBM within ten years. Japan invited some European cooperation, but Britain declined and the government set up a committee under John Alvey, Director of Research at British Telecom, to advise it on how to proceed. The Science Research Council also set up some academic committees to advise on what was needed. There was clearly much concern over Britain's competitiveness in the IT sector, seemingly confirmed when Alvey's report to the government predicted a shortage of about 5000 university-trained IT personnel by 1985.[48] It suggested a programme, to be funded jointly by government and industry, designed to foster cooperation between universities and computer technology companies, and to promote also integration with computer technology used in other European countries.

Under this scheme Imperial received more contracts than any other university but the results were mixed. From the financial point of view Alvey was a failure. As Sir Eric Ash noted in his presidential address to the IEE, Imperial College found its Alvey contracts 'a reliable path to ruin' since overheads were far greater than anticipated.[49] But this short-lived programme set longer-term patterns for collaboration between the college and industry in a highly competitive area of industrial research. One of its weaknesses was that the marketing of the various project successes was poorly planned, reflective of a period when entrepreneurship was still viewed with suspicion at universities. But, as Ash stated, academia was underused by both government and industry. With understanding on all sides, especially with regards to intellectual property, he claimed, the situation could change — as indeed it did.

Ash had come to Imperial from University College where he headed a department that was well funded by industry. When at Imperial, he chivvied departments to follow his example. Despite the terms of the Imperial College Charter, this did not always go down well, especially with departments that wanted to maintain research programmes in pure rather than applied science. For example, Geoffrey Wilkinson, head of the chemistry department, felt strongly on this issue and was very critical of the way in which UGC monies were distributed at Imperial. He believed that the recurrent vote formula was unfair, and that it was biased towards those departments bringing in major funding from industry, resulting in engineering being favoured over science.[50] He wanted a stronger weighting in the formula towards the numbers of students winning grants from research councils. Chemistry and physics were running deficits at this time and rightly saw their standing as pure research departments to be in jeopardy. In practice, however, it was impossible to carry on as before. Increasingly departments needed to bring in outside funding if they were to survive. The direction taken by the Thatcher government with respect to universities could not be reversed, something learned only with difficulty. Ash took the view that it was unfair for departments making an effort to bring in outside funds to be expected to bail out the others. This is not to say that he was unsympathetic towards pure science; on the contrary. He continually urged the government to put more money into the university science budget and repeatedly mentioned the vulnerability of pure research in the existing political climate. But facing budget deficits there was little he could do. Without more industrial funding jobs would have to go.[51] Ash was also concerned with the declining real value of academic salaries. As the author of a report on this topic, published by the CVCP, he claimed that the increasing affluence of the 1980s had completely by-passed universities, leading to an imbalance that needed to be addressed.[52]

To help bring more outside funding to the college, Ash took the opportunity of the 1989 resignation of the College Secretary, John Smith, to make some changes to the college administration. Instead of appointing a new Secretary, he decided on a new post, namely Managing Director and Angus Fraser was appointed.[53] The new name was in line with suggestions in the Jarratt report which stressed that universities management become more businesslike. Ash appears to have also had in mind that Fraser be responsible for generating greater income for the college. In retrospect the new position appears to have been a hybrid of the traditional College Secretary's position and that of an American style development officer. At the time only about £5m of the college annual income of £70m was coming from industry, an indication that the new government policies had not yet fully taken hold.[54] The very term 'managing director' was enough to cause distress among academics who were fearful of the college evolving into some kind of giant commercial enterprise. The distress even spilled into the national press. For example, after Fraser announced the contracting out of some service jobs, including those of the security staff, Professor Bruce Sayers spoke out publicly against him, and against the introduction of 'factory floor' and 'commercial' attitudes.[55] Within the college Sayers was even more outspoken. He was not alone. As management planning became a mantra, people seeing a succession of private sector business practices invade the college voiced their disapproval. As it turned out, the experiment of having a Managing Director was short-lived and ended when Oxburgh became Rector. He reinstituted the College Secretary and, after a brief interregnum, appointed Tony Mitcheson to the position.[56] The college also moved to form a separate development office to engage in fund raising activities.[57]

Ash was very concerned to attract good students to Imperial and found the college far too complacent on this score. The belief that good students would naturally gravitate towards South Kensington was widespread. However, in 1985, the proposed enrolment figures were undershot by 10% and Ash was determined this would not recur.[58] He focussed on recruitment, especially of women and overseas students, and did much to improve liaison with schools. He insisted that a more professional approach be taken toward recruitment, and that both college and departmental literature become more attractive. One does indeed see a change in the appearance of annual reports and departmental prospectuses dating from Ash's rectorship. Up to the early 1980s college literature retained something of an Edwardian character. Annual reports, for example, were very long and were still addressed to the Queen. Under Ash, college publications were modernized; they became more glossy, had fewer words and more colour illustrations, and more attention was paid to student services.[59] Student numbers and high entry standards were, by and large, maintained, and the college was able to compete for additional student places.

Student morale was low in the mid 1980s. The reasons for this are complex. In part it had to do with low and then vanishing grants, and the high cost of living in London. In part it was due to poor conditions in the student residences, refectories, and student union buildings, all of which had been allowed to run down during the recession. And in part it was due to poor job prospects. Low morale among PhD students was a national problem, not confined to Imperial. In 1988 Ash chaired a working group of vice chancellors and principals looking into the problem. They supported the American idea of a taught component to the PhD degree, that there be tighter selection of entrants, that a four year completion rate be the general rule, and that very good students be passed 'with distinction'.[60] Ash was worried about poor completion rates also at the college, and by complaints from students about poor teaching and supervision. The Science and Engineering Council (SERC), too, complained about the college's PhD completion rates, especially in the computing and aeronautics departments. The departments responded by saying that the poor figures reflected the fact that their best students were being wooed away by industry before completion. When the UGC visited Imperial in 1985 it reported negatively also on undergraduate teaching. Ash set up a committee under Professor Blow to look into the situation, but this simply opened a debate that continued well into the 1990s over how to weigh teaching and research responsibilities. The fundamental problem was hardly new, but had come to the fore with recent government policy. Academics have always been rewarded more for research than for teaching. But with an increased emphasis on research output, even those researchers who were good teachers could ill afford the time to teach and some began resenting having to do so.[61]

Imperial needs to attract good students and so needs to support good teaching. Blow suggested that student evaluation, which some instructors had already introduced, be more widely used in the college. He introduced computer readable forms for the purpose.[62] There was some pressure to separate teaching staff from those carrying out research but Ash stood against this. The same problem concerned also Oxburgh who sought new ways to reward teaching excellence.[63] One initiative to boost student morale was to encourage more study abroad, principally in Europe. In 1990 the Board of Studies put forward proposals for six courses to include one year at a European institution. This initiative was successful and brought also European students to spend a year at Imperial.[64]

Ash was very much a modernizer and ambivalent towards Imperial's traditions. He did not like to wear academic gowns on formal occasions, such as when addressing the freshers, and gave up the practice. He also objected to the tradition of having people in the administration prepare the Rector's speeches. A good orator with a good memory, he was able to speak *ad lib* on occasion, and hold his audience's attention. He pushed the college to adopt e-mail as the normal mode of communication, something for which there was surprisingly much opposition still in the early 1990s. Like Lord Flowers and Sir Richard Sykes, he saw the three old colleges as increasingly anachronistic, as sacred cows that needed slaughtering, something later achieved by Sykes.[65] Surprisingly for an old student, Ash had little attachment to the old colleges which he saw as impediments to progress.[66] Many shared his point of view; but others, believing that tradition gives authority to the new, disagreed. I am with the traditionalists on this point and, viewing educational institutions as organic, believe that status is to be gained in the modern world by having strong roots, and celebrated connections to the old.[67] This may seem irrational, but the claim is well supported by the status accorded an Oxbridge education, and Oxbridge dons, regardless of merit.[68] But Ash was correct in claiming that the college's administrative structures needed modernizing and that new disciplinary groupings were desirable. Flowers' earlier move towards the creation of research centres was one way to go in meeting this need. Indeed, encouraged by patterns in government funding, many new centres were founded also in this later period.

Ash's contribution to medical education at the college is discussed below, but another of his major achievements was to forge a new settlement with the University of London. The old antagonisms between the college and the university, discussed in several of the earlier chapters, had surfaced yet again during Flowers's Rectorship. As mentioned in chapter twelve, in 1980 the chairman of the UGC, Sir Peter Swinnerton Dyer, had asked Imperial and the other large London colleges to submit their views on the future of the university. This prompted much discussion within the college, and prompted Flowers to write a paper listing the advantages and disadvantages of remaining within the university. He was, however, reluctant to take matters further.[69] Ash decided that changes were needed and succeeded in pushing the university to a new settlement. In his speech at the 1992 Commemoration Day ceremony, he stated,

> our relationship to the Federal University is not an issue which we would wish to address, were it trouble free and productive. Unhappily it is not always so. While we have been directly funded by the UFC for over a quarter of a century, there remain some significant financial streams which flow through the university. Our faith in the ability of the Federal University to deal efficiently and equitably with Imperial College business has been stretched, sometimes to near breaking point and occasionally beyond that point.

He further objected to the additional layer of management involved when appointing senior staff at the college, and when examining students. The university administration, he claimed, provided no discernible added value to Imperial and entailed much additional cost.

By 1992 there were eight new universities in London, all with greater independence than Imperial in matters relating to appointments and student regulations. As discussed in chapter twelve, the 1978 University of London Act had given the university the right to change its own statutes and there had been ongoing discussion of possible changes since that date. By 1993 this had led to a set of provisional statutes written by Graham Zellick, a lawyer and Principal of Queen Mary College (later Vice Chancellor of the University of London). But the question of whether the new council, which was to replace the university senate, would have the legal right to grant the kind of changes that Imperial, and some of the other large colleges within the university, wanted was unclear. Ash took the view that if Imperial did not get its way it should seek independent university status. He told the Vice Chancellor, Stewart (later Lord) Sutherland, that he wanted some of the key powers that had been given to the vice chancellors of the new London universities such as the University of Westminster. Sutherland responded by inviting debate on three options:

a) The devolution of examining powers provided there be some mechanism of continuing comparability of standards.
b) Statutory changes that would give degree-awarding powers to separate colleges. Any school which took this route would become a univer-

sity *per se*, with just a historical link to the university. An alternative, but in same spirit, would be to get constitutional reform to make the University of London a federation of universities rather than of colleges.

c) The University should agree to devolve to some colleges the right to award degrees and some other powers.[70]

Discussion among the major schools showed that there was the most support for option C, but there were many on the university senate representing smaller colleges who were against the changes that this implied. In a speech to the senate, Ash stated that he believed the university was worth preserving; but, if Imperial was not given powers to award degrees as it saw fit, and appoint its own professors and readers, and if certain financial issues were not resolved, the college would seek permission from the Privy Council to go it alone.[71] In the end option C was agreed on. Revised statutes were drawn up and, by 1998, Imperial was given the power to award degrees, including honorary degrees, and to appoint senior staff without having to consult the university.

In line with the revision of the university statutes, Imperial College worked also on the revision of its own charter, something that was in any case needed after the medical school mergers.[72] The main aims were to reduce the size of the governing body (to be renamed council) from about 55 to about 25,[73] to give *ex officio* status on the new council to more college administrators and the ICU president, to define new terms of office for governors, and to set up a body (later court) to advise the governing body (later council). In its revisions, Imperial had to conform with new rules set out by the University Commissioners and the Nolan Report on Standards in Public Life.[74] The new Council became the governing and executive body of the college. While there is a range in the number of years that council members are allowed to serve, the maximum for those from outside the college is two terms of four years.[75] In the long run this may have some negative consequences. It was always possible to remove governors who did little to help the college; but under the old system some of the governors served very long terms, were extremely loyal, and were often of great help.

The new Court is a large body (160 members at the start) with representatives from a wide range of constituencies. New kinds of people, for example local high school teachers, were brought in to the governance circle. Some overseas representatives, and others who had earlier served on the governing bodies of Imperial and the newly merged medical schools, now joined the Court. The Council is bound to report to the Court which, according to the statutes, need meet only once a year.[76] In 1998 the former chairman of the Governing Body, Lord Vincent, became the first chairman of both Council and Court.[77] The new statutes also provided for a Senate to replace the old Board of Studies. And the Queen reserved 'for herself and her successors the role of Visitor acting through the Privy Council'.[78] The Visitor's jurisdiction is final.

While the new charter was being written, the day-to-day college administration was also under review. In a paper delivered to the Governing Body in 1993, Ash described his own job as 'overloaded' and recommended that changes be made so that the next Rector would not have over thirty people reporting to him.[79] He asked the governors to consider introducing a North American style faculty system and that there be a set of recommendations in place for his successor, both on internal governance and on relations to the university. Lord Oxburgh was less bothered than Ash by the horizontal administrative structure that he inherited, but his very success in founding the Imperial College School of Medicine made the old internal forms of governance no longer viable. Oxburgh was fortunate in having had a workaholic Deputy Rector, W. A. Wakeham, who helped to keep the administration afloat during this transitional period. Wakeham played a major role in bringing new research monies to the college in the late 1990s, and in rationalizing the financial management of the college.[80] A faculty system was introduced by Oxburgh's successor, Sir Richard Sykes. It replaced a structure still largely defined in terms of the old colleges. Thus, at the start of the new century the college had in place a new charter, a new governing council, a senate, and a court, and had also restructured its day-to-day administrative procedures. The revised charter contains also a clause defining the new mission.[81] Central to this was the following.

> The objects of the College shall be to provide the highest specialized instruction and the most advanced training, education, research and scholarship in science, technology and medicine, especially in their application to industry.

There is also the directive to work in cooperation with other bodies, something that had already proved important in the later twentieth century. It is to be expected that this mission statement, not unlike the one enshrined in the 1907 charter, will continue to guide work at the college in the twenty-first century.

When Lord Oxburgh came to the college in 1993 the major recession was well in the past and money had become available for some much needed refurbishment of college facilities; research funding, too, picked up. The long-awaited second phase of the chemistry department was under construction already by 1991. In the same year Sir Norman Foster was appointed consultant architect and began helping the college formulate an estate development plan.[82] With some exceptions, the college did well in the 1992 RAE, but the restructuring of the academic programmes still had a long way to go and needed good direction. After all the cutbacks, and measures taken to increase efficiency, Oxburgh needed to extend a hand to both students and staff and rebuild morale. In this he was largely successful. He held lunches for heads of department, and created a new Pro-Rectorship for student affairs to which he appointed Professor Alan Swanson. Student problems began to be dealt with more systematically than before, and personnel management, too, generally improved. Swanson was also directed to think of ways to monitor teaching standards and to assure teaching quality in line with new government directives.[83]

By the mid 1990s the college had grown such that separate graduation ceremonies were conducted for undergraduate and postgraduate students. In 1996 the first postgraduate ceremony was held in the Royal Albert Hall. The speaker on behalf of the honorary graduands was Lewis Wolpert who, interestingly, had begun by studying civil engineering at Imperial, and had taken an MSc in soil mechanics, before turning to cell biology. By the early twenty-first century the number of MBAs awarded, and the number of MSc degrees given in areas such as history of science and science communication, had become proportionately greater than before. Imperial was becoming more diversified in its postgraduate offerings.

As will be discussed, the need for growth in the direction of new biotechnology and medical science was met by the founding of a new medical school. But there was another merger during Oxburgh's Rectorship, namely with Wye College. Wye, an agricultural college in the University of London, was a victim of the new RAE system and close to bankruptcy by the mid 1990s.[84] Its grants had been severely cut after poor research assessments in the 1980s, and again in 1992. Wye approached Imperial about a possible merger and Oxburgh set up a committee to consider the request. The committee, seeing many possibilities at Wye, noted,

> as world population doubles again in the next thirty or so years, the demand for natural resources, energy, food, and water, will lead to pressure on the environment on a scale hard to imagine. There will certainly be major social consequences, and the opportunities for excellent science that will also have practical considerations are considerable.[85]

Oxburgh understood that Imperial was already positioning itself for this kind of research, and that Wye, with its body of agricultural knowledge, could help in new areas of biotechnology. Besides, the food industry is a perennial industry, and could be expected to generate research problems well into the future. Wye was also 'the counterpoint to overworked urban environments', and with its 350 hectares in an area designated as being of 'outstanding natural beauty', would allow people to focus on problems related to the environment and 'rural development'.[86] The only problems were seen to be academic, namely Wye's poor results in the RAE, and its low A-level entry levels. The seven-person committee asked to look into the merger voted 6-1 in favour.[87] A merger took place in 1999–2000 though, as it turned out, the problems were not simply academic but also financial.[88] Since the merger undergraduate teaching at Wye has been largely closed down and the remaining staff work as part of a new Centre for Environmental Policy in the Faculty of Natural Sciences. A range of MSc courses are offered, including in areas related to agricultural economics, agri-business, and rural development.[89]

Oxburgh had a yet more ambitious merger in mind and began discussing a possible union with the principal officers of University College (UCL) and the London School of Economics. These discussions occurred before the final settlement with the University of London mentioned above. The idea was to consider a possible new model for the university. Oxburgh was not convinced that a merger would have added value overall and believed that, while the science departments at UCL and Imperial could be merged, the engineering departments would prove difficult.[90] The discussions, only exploratory, collapsed shortly after Sir Derek Roberts, the Principal of UCL, was replaced by Sir Christopher Llewellyn Smith. Later, when Roberts returned to the principalship, Sykes reopened negotiations with UCL. But these, too, ended when Sykes recognized that the government would not contribute the money needed for a smooth merger.[91]

While both Flowers and Ash worked to increase access for women students, the situation for women staff at the college was far from ideal.[92] Oxburgh, turning his attention to the problem, set up the Academic Opportunities 2000 Committee under Professor Julia Higgins. It was to give the college some guidance in how to improve working conditions for women.[93] Higgins was also the President of the Association of Women in Science and Engineering (AWISE) and chaired the national steering committee of the Athena Project.[94] The latter project, arising from a suggestion made by the CVCP and sponsored by the Royal Society, was to foster the mentoring of young women in postdoctoral and junior academic positions.[95] The Academic Opportunities Committee hired an outside consultant with a view to making recommendations on the creation of 'a level playing field for women academics at Imperial College by removing any barriers that may exist in appointment or career advancement for highly qualified women, and to ensure that the numbers of such women in the college are as high as possible'.[96] A questionnaire was sent out to academic women staff and showed that the major complaint was with the lack of a career structure consistent with family life and the raising of children. The response rate was 37%. About 15% of those who responded complained of sexual discrimination in advancement, about 2% complained of sexual harassment, 1% of racism, and about 26% of 'an unsupportive working environment'. While sexual discrimination was on the decline, and equal-opportunity legislation in place, what the report clearly showed was the need for a better working environment for women, better maternity (and paternity) provisions, improved child-care provisions and, where the family situation warranted, that people be given more time to show their academic worth. In other words the college should become more inclusive and that there should not be just one set formula for career advancement. While desirable, the latter was difficult to put into practice, especially in the highly competitive university system of the early twenty-first century. The level playing field will have to be achieved in the home before it is fully achievable in the academy and, even then, the childless (men and women) will probably have some early career advantages over others. Nonetheless an improved working environment is possible and it appears that the college is working towards this.[97]

Also in 2000 Imperial formed a partnership with the British Association for the Advancement of Science (BAAS) and Creativity in Science and Technology in developing a scheme to attract good students into science and technology. Good lecturers, such as chemistry professor David Phillips, gave lectures to young people in Britain and overseas as part of the scheme. Relatedly, Oxburgh encouraged the BA to hold its millennium meeting at Imperial and make it into a festival of arts and sciences. The festival named 'Creating Sparks', was billed as the largest festival of its kind to be held in Kensington since the Great Exhibition. While not reaching such heights, it was a great success and allowed Imperial to showcase some of its work.

Entrepreneurship

It was not simply the economic situation and fall-out from the Jarratt Report that encouraged entrepreneurship within universities.[98] There was a major cultural shift during the 1980s and 90s and people began to engage in entrepreneurial activity more generally. Whether this was a consequence of Thatcherism or whether Lady Thatcher's ideas were simply a vivid reflection of their period is impossible to say. But Thatcher did encourage entrepreneurs, enobled a few, and gave them political voice. And her government encouraged universities to develop spin-out companies, something later reinforced in response to Dearing. The dot-com phenomenon, too, was a thing of wonder.[99] It is not surprising, therefore, that young people increasingly viewed the starting of their own businesses as a noble option. The rise in the number of academic spin-out companies is a reflection, thus, not simply of universities attempting to increase their sources of income, but also of a wider cultural shift. Paradoxically, Imperial's long history of ties to industry at first worked against its adapting well to this new cultural turn. There had always been some entrepreneurial activity at Imperial, even in the early twentieth century, but most of the staff interpreted the charter in terms of supporting existing industries. Thus, in seeking increased industrial and commercial involvement during the 1980s, there was a tendency to carry on as before. In the early 1980s the college saw an increase in the number of its staff joining boards of directors, becoming consultants, taking out patents, and issuing licenses. These activities were far more common than the creation of new companies. However there were exceptions.[100] One of the successful companies of the late 1970s was Queensgate Instruments, founded by members of the physics department. It specialized in optical and infra-red instrumentation for remote sensing, had links to both NASA and ESA and, in 1988, won a Queen's Award for technological and export achievement.[101]

The creation of new companies began to accelerate in the mid-1980s. In 1986 Imperial College entered a joint venture with Investors in Industry plc (3i) and Research Corporation Ltd. and formed a new company, Imperial College Exploitation Ltd (IMPEL) to assist the transfer of technology from the college to industry and commerce. In 1986 Imperial and 3i built a science park at Silwood where the Technology Transfer Centre, owned exclusively by the college, was sited.[102] The total area was about 65,000 sq. ft. and space in the buildings was rented out to college entrepreneurs by yet another Imperial College company, Imperial Activities Ltd. (IMPACT) which also provided services to the tenants.[103] The original idea for IMPEL was that it undertake an audit of research activity at the college and draw up a list of projects with commercial possibilities. Further, there was to be a register of academics who wished to be specialist consultants. Sir Frank Cooper, deputy chairman of the governing body, became chairman of the board of directors.[104] Other Imperial College directors included Professor Howard Morris, head of biochemistry, and Dr. David Thomas, Director of Industrial Liaison.[105] A further company, Imperial Biotechnology, was founded jointly by Imperial and 3i at this time. It was to develop and manufacture materials in biochemistry's pilot plant, then under Professor Hartley.[106]

Thomas had earlier worked with the Alvey scheme and, in his new position, was to bring more research contracts to the college and encourage also entrepreneurial activity. At first he was not that supportive of spin-outs, but he helped people take out patents and sell licenses. However, it turned out that this was not very profitable for the college since close to the same amount of money was spent in establishing patents as was gained from the sale of licenses. Given poor accounting practices, and the fact that full overheads were not properly included, this fact did not emerge until discovered by Wakeham in the mid-1990s. But, given the increased entrepreneurial spirit of the times, things changed. In 1990 IMPEL was split into two: a technology transfer company which retained the old name, and Imperial Consultants Ltd. (ICON) which acted as a consultancy service, and oversaw the commercial use of college facilities. Thomas proposed that offices be opened in some European cities, and in Japan, to make contact with industries that might make use of Imperial talent.[107] By the late 1990s, and with the medical mergers underway, IMPEL gave way to IC Innovations which had roots also in RPMS Technology Ltd.[108] This new company made it easier for Imperial staff to set up new companies provided all interests were declared. Under IMPEL only about two new companies a year were formed. However, by 1992 the company had brought in about £750,000 in royalty and licensing income. Under IC Innovations the number of new companies jumped to about two per month though not all survived. The college sold Imperial Innovations, as it is now known, but holds about 70% of its shares. Imperial Innovations continues to seek venture capital from institutional and other investors for Imperial spin-out companies. These companies have generated revenues of about £30 million for the college since 1997.[109]

It is impossible to name all the companies that were spun out during this period but a few will be mentioned in order to give a sense of the type of activities involved. Among the earlier companies were several which created computer software with business or industrial applications in mind. Many of the more recent companies have medical interests. An example from the early 1980s is Imperial Software Technology which was formed jointly by Imperial College, NatWest, PACTEL and Plessey.[110] Students in the computing science department worked on projects of interest to the company shareholders. Further, the company sought external contracts; for example, from British Telecom, and from the United States military. One interesting project involved the development of Chinese character input. But, overall, this company did not work well since the department's pedagogical ideals did not mesh with the contractual arrangements. In the Annual Report of 1998, Wakeham is quoted as stating 'we estimate that around 750 staff are now employed in 37 companies that originated in technologies developed at Imperial College. We hold an equity stake in 10 of these companies, with holdings valued at about £5m'.[111] A successful company of the early 1990s was Turbo

Genset, set up by Professor Colin Besant and some of his colleagues in the mechanical engineering department. The company exploited technology that had been developed for small-scale electricity generation. It was listed on the London Stock Exchange in 2000 and Imperial made a profit of about £10m on the sale of its shares that year.[112] The value of the company's shares plummeted in the dot-com crash of 2002 — despite it not being a dot-com — but the company survived and is still in business.

In 1998 the college introduced a new entrepreneurship course in the Management School under Professor Sue Birley. The course was aimed at encouraging students and young faculty members to become involved with spin-out companies. Three years later the Entrepreneurship Centre was launched to foster business innovation among students. In addition to providing skills training, the centre promotes competition in the creation of business plans, and organizes a number of extra-curricular activities for striving entrepreneurs. The Centre's motto is 'one day you will need more than just a degree'. Entrepreneurship was catching on and, by 2003, the college held a stake in fifty-five spin-out companies covering a wide range of technologies and markets. The companies had created also about 1000 new jobs in the previous ten years.[113] Among those founded in the more recent period, and perhaps representative, are: Acrobot, a company that develops surgical robotic systems and associated software;[114] Implyx, for the preparation and purification of large proteins capable of crossing cell membranes to deliver drugs directly to individual cells; Geotechnical Observations for the delivery of high quality geotechnical services; Toumaz Technology, working in the field of analogue and mixed chip design for application in wireless communication (including novel wireless hearing aids); Ceres Power, developing novel solid oxide fuel cells; Hydroventuri a renewable energy company with applications in the area of tidal power; Gene Expression Technologies, developing gene therapies for cancer; and Casect, developing plasma chips and miniaturised gas chromatography devices for biological and environmental applications. The creation of spin-out companies is not without a range of potential problems,[115] and whether any of the existing companies will survive over the longer term remains to be seen. By 2005, over sixty spin-outs, the majority with medical applications, were listed on the college website. Clearly Imperial College was well launched into some of the new industrial concerns of the early twenty-first century.

The Imperial College School of Medicine
(Faculty of Medicine from 2001)

The 1968 report of the Royal Commission on Medical Education (Todd Report) concluded that there were too many medical schools in London and that some of them, including St. Mary's, were too small.[116] The report suggested that there be some amalgamation but little was done at the time. However, as stated in chapter twelve, Lord Penney took note of the Todd Report and foresaw the possibility of bringing a medical school to Imperial. The recession of the 1970s, and the spending cuts that followed, prompted the Vice Chancellor, in 1979, to ask Lord Flowers to head a working party to look again at the Todd Report. Flowers was to review the management and financing of medical and dental education in London, and make some new recommendations.[117] Like Todd, Flowers recommended a number of mergers and suggested that there be greater collaboration between those carrying out research in medical schools and those working in university science and engineering departments. He viewed this as especially necessary for small independent schools such as St. Mary's. Flowers suggested that it amalgamate with the Middlesex Hospital Medical School and that there be greater academic collaboration with Imperial College, and with the Royal Postgraduate Medical School in Hammersmith, the leading centre of clinical research in the country. The Vice-Chancellor of the University, Noel (later Lord) Annan, largely agreed with Flowers's findings, though not in the case of St. Mary's which he recommended should be closed rather than amalgamated. This was in part because University College had agreed to absorb the Middlesex school but not St. Mary's. There was much opposition to closure and, for a while, St Mary's retained its independence. However, by the mid-1980s when funding became dependent on results in the RSE/RAE, St. Mary's came into serious financial difficulty.[118] While St. Mary's had expanded its research in the 1970s, and its biochemistry department had tentatively jumped on the molecular biology bandwagon, it was difficult for a small school to compete with university research in this area. In 1986 the Dean of St. Mary's, Peter Richards, approached Sir Eric Ash and asked whether he would be interested in a medical school. Ash replied, 'Yes, I think that would be rather a good idea.'[119]

Ash, believing that the twenty-first century belonged to medical science, saw that it would be wise for Imperial to reposition itself. A medical school would provide a good platform for new kinds of increasingly important research. It was also a good way to bring in more women students, something Ash very much wanted.[120] At the time, Imperial College had about 2000 members of staff and St. Mary's had about 250. St. Mary's feared being swamped and losing its identity. But people at the medical school also recognized that their financial situation gave them little hope of surviving, let alone retaining their independence. Selling the merger was therefore easier at St. Mary's than at Imperial. At Imperial there was concern that the medical school would be a drain on college funds and that its low research standards would pull the college down. But perhaps more problematic was that many shared Ash's view that medical science was becoming increasingly important. They were worried about the future. Could it be the case that Imperial's identity was also at risk? Was St. Mary's the thin end of the wedge? Would traditional science and engineering become swamped by medical science?

As discussed in earlier chapters, Imperial had a long history of cooperative ventures with various London medical schools in areas such as new surgical technology, electronics, artificial joints, medical computer software, and biochemistry. Ash asked Pro-Rector, Alan Swanson, who had earlier worked on the biomechanics of joint replacement, to head a working party to look into the academic consequences of a merger.[121] After a largely favourable report, Ash decided to take the matter to the Board of Studies and put it to a vote. He claimed not to know in advance what the outcome would be, and that he had never before asked for a vote in such circumstances.[122] Bruce Sayers made an impassioned speech against the merger but was countered by others, notably by Roy Anderson.[123] The Board voted overwhelmingly for the merger, as did the Governing Body. Things then moved quickly. The smoothness of the transition owed much to the diplomatic skills of the College Secretary, John Smith. Smith and Ash did everything they could to make people from St. Mary's feel welcome. They did not interfere in the running of the medical school and, since funding issues remained contentious, allowed the budget lines to remain largely separate for three years.[124] The revised statutes were legalized by Act of Parliament in 1988, as was the new college name, Imperial College of Science, Technology and Medicine. St. Mary's became the fourth constituent college at Imperial. As had been the case when the City and Guilds College joined Imperial in 1907, St Mary's was given a delegacy to oversee its affairs, though subject to the Governing Body of Imperial. The Chairman of the Delegacy was Sir Evelyn de Rothschild, a former governor of St. Mary's who had long fought for the school's survival. Peter Richards remained as Dean of St. Mary's.

At first St Mary's identity remained largely unchanged. Medical students continued to be taught in Paddington and day-to-day affairs continued to be managed largely from there. But the St. Mary's merger was only the first. In 1992 a further report, the Tomlinson Report, recommended the establishment of postgraduate medical education within multi-faculty institutions that had graduate schools covering a wide range of scientific and technological disciplines.[125] One year later Imperial had a new Rector whose principal task was to negotiate further medical mergers and create an enlarged medical school, with Tomlinson's report in mind. Lord Oxburgh began thinking about how to proceed even before taking office.[126] One merger being considered was with the Charing Cross and Westminster Medical School (CXWMS). But this school was in poor shape both financially and academically.[127] Oxburgh believed it needed to put its house in order were it to come to Imperial and, if it could not, then Imperial should go it alone and expand its own medical school based at St. Mary's. Indeed, that is what Richards, who opposed Imperial approaching other schools, wanted. Oxburgh invited discussion also with three postgraduate medical schools all of which were in good financial and academic shape: The National Heart and Lung Institute, The Institute of Cancer Research, and The Royal Postgraduate Medical School (RPMS).[128] Oxburgh had to keep an eye also on National Health Service plans for hospitals in the area. There were six main hospitals offering a wide range of services, and a number of specialist hospitals. If it were decided to consolidate or move them (as was then under consideration), this would have an affect on future clinical teaching at the college.[129] This, together with the financial and academic consequences of the various mergers, were the major things to be weighed.

Oxburgh wanted more than simply a merger. He wanted Imperial to bring something new to medical education at both the undergraduate and postgraduate level. This was timely since the General Medical Council was much concerned with improving existing teaching methods and wanted to see reform. In order to plan for a new academic structure for the enlarged medical school, Oxburgh asked Peter Richards to resign as Dean of St. Mary's. The new Dean, John Caldwell, Professor of Pharmacology, was the first non-clinician to head St. Mary's. At the same time, Oxburgh appointed Professor Christopher Edwards to make some organizational and academic plans for the newly enlarged school, and to become its first Principal.[130] Edwards had some preliminary organizational plans in place by November 1996, envisaging a Medical Academic Council, drawn from all the constituencies, to advise in a general way, a smaller advisory group for day-to-day decision making, and an institutionalized reporting mechanism for the chairs of the different medical departments.

But before any of this could come about Oxburgh had to spend much time talking to the leaders of the various institutions, explaining how the new system would work. He also had to calm the fears of people at St. Mary's and the fears of some at Imperial. Many at Imperial were unhappy that the St. Mary's merger had pulled the college down in the ranking tables after the 1992 RAE, and feared that things would only become worse were there to be a merger with CXWMS. But Oxburgh wisely took a longer-term view, believing that with some resignations and retirements, and with new appointments, something positive would emerge. He recognized that 'Imperial is jealous of its reputation for academic excellence and any institution with which it might merge should have similar aims and reasonable prospects for realizing them'.[131] He believed that bringing together a number of small units in a major new concentration of medicine, science and engineering would result in synergy at the postgraduate and research level, and that Imperial could become a powerhouse in medical research. Oxburgh's plans were ambitious, but were they to be realized many things needed to fall into place. First, it was necessary to persuade the government and its funding agencies to meet the immediate needs of Imperial College. The biology department needed new accommodation before consideration could be given to bringing in new biomedical departments. Ideally there would be a completely new building, or set of buildings, to house the medical school, the biological and biomedical departments. Further, the prospective postgraduate schools needed to be persuaded that the mergers were in their interests. The plans then had to be approved by HEFCE before the mergers could go ahead.

Both the strategic and the human problems were considerable. How to persuade the government to fund a new and costly building? How to merge the new medical schools in practice? St. Mary's would have to be persuaded to give way to a single united medical school, with consequent loss of identity. Collectively the medical schools had many more governors than had Imperial, many of whom did not wish to lose their positions. Imperial's governing body, on the other hand, refused to be swamped. These and many other issues had to be negotiated. It was in part to provide a home for people dropped from existing governing bodies that the Council/Court solution described above was proposed. Staff jobs, too, had to be cut. In working through all these problems, Oxburgh had much help from Sir Frank Cooper, chairman of Imperial's governing body. Cooper was an experienced Whitehall man and proved invaluable in the many negotiations leading to the formation of the medical school, in persuading people to accept a Court, and in persuading the government to fund the new building.[132] While CXWMS had to be forced to modernize, the already modern RPMS had to be persuaded that it was in its interests to be associated with the CXWMS and to join Imperial. The Principal of the RPMS, Sir Colin Dollery, was reluctant and had to be convinced of the benefits of being part of a larger body.

Oxburgh wanted to bring all pre-clinical teaching to South Kensington, another major bone of contention. Perhaps the key to the merger was the promise of a new state-of-the-art building for non-clinical teaching, and for medical, biomedical and biological research. The promise of better facilities for all the players helped persuade them to come on side. Many of these problems were discussed with officials at the Ministry of Education, and with the Minister of Health, Virginia Bottomley, who was supportive.[133] The money was approved but the negotiations continued. Oxburgh spent much time 'eating for the college', discussing the numerous details over endless breakfasts, lunches and dinners.[134] Everything was up for discussion, not simply financial and academic matters. For example, Oxburgh wanted some cultural changes. One such was that there be an end to the St. Mary's beer-drinking orgies, an integral part of the initiation of freshers at the start of each academic year. Surgeons and doctors joined in, and some students became near paralytic in order to demonstrate new loyalties. Oxburgh did not wish to see such behaviour in the new medical school.[135] He recognized that new medical students have insecurities over and above those of other freshers, and tried to encourage new forms of acculturation.

After much effort by many people, the Imperial College Bill was placed before Parliament and approved in January 1997. The new Imperial College School of Medicine was comprised of nine divisions over four sites. It held 49% of the college staff, 24% of the student body, 48% of the total income, and 51% of research income. Imperial college was still not large. The total student body was 8,824 full-time students of which about one-third were postgraduates and about one-fifth were from overseas. The new building was designed by Norman Foster and Partners and cost £65m. The building committee was chaired by Stuart (later Sir Stuart) Lipton, a member of the governing body. Lipton is a major builder whose earlier projects included the Sainsbury wing of the National Gallery, the Royal Opera House renovations, and the library extension at Imperial.[136] The building, named for one of St. Mary's most famous medical researchers, Sir Alexander Fleming, was opened by the Queen on 21 October 1998. She also presented the college with its new charter.[137] Three generations of the Fleming family were present at the opening ceremony. Richard Dickins conducted the college Brass Ensemble in the first performance of 'Imperial Fanfare'.[138]

There was still much to be worked out. Even before the 1997 Act, Oxburgh had begun discussions on how to rationalize the various areas of expertise that were to come together at Imperial. The new medical school operates at four main sites, raising questions of the degree to which expertise should be allowed to overlap. For example, there were good kidney transplant teams at both St. Mary's and at Hammersmith, and associated immunological and other research was carried out at both sites. In the United States transplant units were not generally considered viable unless at least one hundred such operations were performed each year. Neither St. Mary's nor the Hammersmith Hospital performed that many at the time. The NHS wanted similar standards to those in the United States and Imperial wanted to coordinate the research work at a single site. This exemplifies the new kinds of problems and negotiations that the Rector faced. But in resolving them he received much support from the North West London NHS.[139] In this particular case, St. Mary's agreed to give over transplants to Hammersmith Hospital, and research in the field to the RPMS site of the new medical school. In exchange St. Mary's Hospital became the major diagnostic/scanning centre for the district, and research in this field consolidated in Paddington. Oxburgh, concerned about the poor medical school showing in the 1992 RAE and the improved, but still weak, showing in 1996, had also to worry about the assessments looming in 2001.[140] In this connection he had to persuade the new medical school to make over two hundred people redundant, and to insist on higher research performance, especially from those who had come from the CXWMS. By 2001 the RAE results were greatly improved and the medical school was beginning to be recognized as a major success.

Much of the credit for this must go to Lord Oxburgh. Sir Eric Ash had the vision to recognize that a medical school would be a great asset in the twenty-first century and took the first steps at a time when many of his contemporaries at Imperial were hostile to the idea. But the merger with St. Mary's was far less complex than what followed. For a short period, St. Mary's continued to function much as before but the new medical school required that all participating medical institutions merge their identities with Imperial and become working units of the college.[141] This was a great deal to ask of people proud of their associations with the individual medical schools, each with their separate historical traditions. Further, many people had to face up to negative assessments of their research standards and were resentful. Others who did not carry out research, notably some of the pre-clinical/clinical teachers, many with considerable skills and achievements, complained of coming under mindless attack. There was some justification to their complaints. That Oxburgh was able to come to largely amicable settlements with so many different interest groups, that he persuaded the government to provide generously for the mergers, and for a new building, speaks to his administrative and diplomatic abilities. There seems little doubt that the move into medicine was timely and will serve the college well in the years to come.

...

1. For a fuller treatment of these themes see, Michael Shattock, *The University Grants Committee* (Open University, 1994); A. H. Halsey, *The Decline of Donnish Dominion: The British Academic Professions in the Twentieth Century* (Oxford, 1992); Robert Stevens, *University to Uni: The Politics of Higher Education in England since 1944* (London, 2004); Brian Salter and Ted Tapper, *The State and Higher Education* (Ilford, 1994); Harold Silver, *Higher Education and Opinion Making in Twentieth-Century England* (London, 2003); Elie Kedourie, 'Diamonds into Glass: The Government

and the Universities' and 'Perestroika in the Universities', *Minerva* 31 (1993), 57–105. One academic who, in mid-century, anticipated much of what came later, thought clearly about the role of science and technology in university education, and recognized their centrality to the modern state, was Eric (Lord) Ashby, a former student and governor of Imperial College. Ashby receives a good treatment in Silver's book (chapter 8). See also Eric Ashby, *Technology and the Academics: An Essay on Universities and the Technological Revolution* (London, 1958). See also chapter 12.

2. 'A Strategy for Higher Education into the 1990s: The University Grants Committee's Advice' (HMSO, 1984). This document contains a summary of points raised in earlier discussions between the Secretary of State for Education, Sir Keith Joseph, and the chairman of the UGC, Sir Edward Parkes. It was written by Parkes's successor, Sir Peter Swinnerton Dyer. An earlier document 'Higher Education into the 1990s' had been published by the Department of Education and Science in 1978. This was concerned mainly with ways of accommodating more students. In 1985 Keith Joseph issued a Green Paper, 'The Development of Higher Education into the 1990s' which began by citing foreign competition and Britain's poor showing since 1945 as his main reason for action. Higher education institutions should, he stated, 'beware of "anti-business" snobbery ... [and in future there will be] a distinct emphasis on technological and directly vocational courses at all levels'.

3. Peter Pratt, Professor of Crystal Physics; see GB minutes, 6 December 1985.

4. Imperial College response to circular 'Strategy for Higher Education into the 1990s'. See GB minutes, 16 March 1984.

5. In the mid 1980s about 75% of university income came from block grants and fees. About 8% came from the research councils. In 1984 universities came under the Ministry of Education and a 'universities branch' handled the UGC. Ironically the civil servants in the Ministry of Education were less sympathetic towards universities than those in the Treasury, perhaps because of their long established ties to Local Education Authorities and teachers unions.

6. Some of these ideas are to be found in the 1972 Rothschild Report discussed in chapter 12. It was suggested about 25% of funding be taken from the research councils and redirected towards contract work.

7. The government soon set up an Advisory Board for the Research Councils (ABRC) as had been suggested by Sir Frederick Dainton (see chapter 12). Its role was to devise strategy for the handing out of research funding. For example, in its report of May 1987 the ABRC advised that greater selec-

tivity was needed, that research councils should concentrate on areas of national importance, and that they should channel more of their support into interdisciplinary research centres. No wonder such centres began to mushroom in this period. Support was also to be given to research undertaken in collaboration with industrial or other users. The Agriculture and Food Research Council was the first to come round to the government way of thinking, allowing that excellence alone was not a sufficient criterion for assessing research proposals. More concrete returns from research were needed. Other Councils soon followed in setting up new funding criteria to show they were 'getting value for money'. In 1993 the government published its White Paper, 'Realising Our Potential: A Strategy for Science, Engineering and Technology' (Cmnd 2250) announcing the creation of a new Director General of the Research Councils and the Office of Science and Technology, thus formalizing earlier policies of government intervention in research funding.

8. Committee of Vice-Chancellors and Principals Report of the Steering Committee for Efficiency Studies in Universities (The Jarratt Report, 1985).

9. Some performance indicators were already in use, such as the number of applicants per undergraduate place, the A-level grades of entering students, research grants and contracts, publication records, staff to student ratios etc.

10. College discussion at the time favoured instead buy-outs for people in the 35–50 age group so as to make room for more active younger researchers. Strictly speaking, Imperial College staff did not have tenure but in practice the system was not unlike that in other institutions. The college received a copy of a letter sent from the Secretary of State for Education and Science to the CVCP, in which tenure came under attack. 'I continue to believe that tenure in the strongest form in which it is enjoyed in some universities cannot be justified ... I therefore propose that tenure should be limited'. See copy in FC and EC minutes, 2 November 1984. See also GB minutes, 15 March 1985.

11. Lord Croham, 'Review of University Grants Committee' (Cmnd 81, 1987).

12. Sir Peter Swinnerton Dyer continued as the chief executive officer in this new body. In the 1990s the chief executive of the by then renamed HEFCE was Brian Fender who had been a chemistry student at Imperial during the 1950s. His earlier career included being Director of the Institut Laue-Langevin in Grenoble and Vice Chancellor of Keele University.

13. Croham recommended a three year funding period, but this was rejected Funding was promised on a four-year basis, but the level was reviewed annually as before.

14. Joseph was a cerebral politician with many ideas but poor political instincts. *Private Eye* gave him the name Mad Monk which stuck. Baker was a more able politician and put in place many of the measures Joseph had dreamed up, as well as some of his own, with mixed results.

15. Cmnd 114, 1987.

16. In 1987 the National Audit Office (NAO) sent representatives to visit all universities to make an inventory of industrial collaboration with universities. They wanted to know how much money was coming in from industry and 'the extent to which it is necessary for the university to tailor its work to what industry wants in order to generate income'. They looked also at what efforts universities were making to market the results of publicly funded research. The NAO labelled this exercise the 'Value for Money Investigation'. See letter from UGC to Clerk of the Court, University of London, 11 November 1987; (copy in Heads of Department Meetings File, 1987).

17. Simon Jenkins, *Accountable to None: The Tory Nationalization of Britain*, (London, 1995), 143.

18. 'Public Funding and Fees' (Department of Education and Science, 1988).

19. See White Paper, 'Higher Education: Meeting the Challenge', (Cmnd 114, 1987). The new UFC was in line with the recommendations of the Croham Committee. For the decline and fall of the UGC see Shattock, *op. cit.* (1).

20. Lord Henry Chilver FRS.

21. The Research Selectivity Exercise (RSE) preceded the Research Assessment Exercise (RAE). The results of the first RSE, published in 1986, prompted newspapers to begin publishing league tables, a practice that seems to have mushroomed.

22. This was formal recognition of what had long been understood, namely that the national need for technically trained people required more state control over technological education. It was a major step along the road leading to increased state control over all areas of higher education.

23. Formalised in the Education (student loans) Act 1990. It was the Blair government, in 1998, that fully ended traditional maintenance grants and introduced a flat-rate fee for all students entering higher education.

24. The Education Reform Act (1988) introduced a new body, the University Commissioners. It was charged with ensuring that the statutes of all universities include provisions for dismissal due to redundancy or other

good cause. Provision was also to be included for both disciplinary, grievance, and appeal procedures for academic staff. In 1990 the University Commissioners proposed some model statutes which they asked all universities to seriously consider. These posed problems for Imperial which never had any statutes of this kind and was governed according to its charter, and various procedures formally approved over the years. These included non-statutory grievance procedures that the college believed had stood the test of time.

25. For discussion of these issues see Conrad Russell, *Academic Freedom*, (London, 1993). See also, *Minerva* 32 (1994), for a series of articles on relations between British universities and the state in this period.

26. 'Higher Education: a new framework' (Cmnd 1541, 1991). The Further and Higher Education Act (1992) led to the splintering of the HEFC into separate regional funding agencies for England, Scotland, Wales and Northern Ireland; hence the Higher Education Funding Council for England (HEFCE).

27. In 1992–3, the year that the binary system ended, the UGC gave out £547.5m for research and the PCFC only £7m.

28. See GB minutes, 10 March 1989.

29. Speech by Kenneth Clarke, Secretary of State for Education, given at the annual conference of the CVCP at Warwick University, 25 September 1991.
    Imperial had been moving slowly in this direction and had some modular options as, for example, in its MSc in environmental assessment and analysis. There were some distance courses and a range of summer programmes.

30. Smith quoted in *IC Reporter*, Issue 10 (1995). In fact, the Chief Scientific Advisor, who headed the Office of Science and Technology, retained direct access to the Prime Minister and the budget continued to be negotiated with the Treasury.

31. 'Higher Education in the Learning Society' (Report of the National Committee of Inquiry into Higher Education, 1997). Both Sir Ronald (later Lord) Oxburgh and Sir Richard Sykes sat on this committee. Dearing had served as Chairman both of the Post Office Corporation and of the UFC. According to many cynical observers, the committee was struck not so much to resolve the serious problems facing universities, but rather as a way of finding a route to charging students higher fees. This is plausible given the fact that the formation of the committee was prompted by the CVCP reaction to the budget cuts of 1995, and its threat to levy

£300 per student. See also Paper D, 'Report of the National Committee of Inquiry into Higher Education', GB minutes, 17 October 1997.

32. The funding of students was a very complicated issue, made more so as students from Europe began to come to British universities in large numbers. The Treaty of Rome requires Britain to treat EU students as though they were British. Provision for students was traditionally far better in the UK than elsewhere in Europe. Determining parental income as was necessary under the older grant scheme was close to impossible for foreign students.

33. In the late 1980s the University of London was pressuring its colleges to upgrade their career advisory services and to take over more of the work that was then still done centrally. Imperial College took a while to figure out how to deal with this. A row broke out in 1988–9 when Clare Ash hired a secretary to help with projects she was engaged in to help improve student life, and conditions for visiting scholars. The hiring was coincident with a cut in £10,000 in the career advisory services and her action annoyed many people. Imperial had a wholly independent career service in place by 1992. After Dearing, the Russell Group commissioned the Greenaway Report on Higher Education which stated that during their lifetime university graduates, on average, could expect about £400,000 in income over and above those without higher education. In light of this a graduate tax dedicated to education seems just, though perhaps impractical. A fee system, even if paid in stages, is likely to be more manageable and was the approach taken by the government.

34. Earlier Dearing had spent some years trying to modernize the Post Office and, persuaded by the benefits of modern technology in that arena, was perhaps keen to apply it elsewhere. He appears to have been an enthusiast for distance education delivered on-line, not perhaps realizing that the labour needed to keep such systems up-to-date in the university sector is considerable. It can take only a year for on-line courses to become ossified. Besides, it is questionable whether people learn as well without the personal contact of both instructors and fellow students.

35. It was suggested there be an Institute for Learning and Teaching in Higher Education, and a Quality Assurance Agency to ensure courses had the proper educational content and that standards were maintained across the sector. Further, students were to be given a Progress File and their performance monitored by the state.

36. The Dearing Committee's terms of reference and its membership were set up with cross-party agreement.

37. The tensions within Universities UK (as the CVCP was renamed in 2001) can well be seen in the discussion surrounding the earlier appointment of Diana Warwick, a former General Secretary of the AUT, as its chief executive in 1995. Despite having accepted a peerage, she was firmly committed to an egalitarian approach among universities and was not supportive of institutions such as Imperial demanding special privileges. Those who teach weaker students in underfunded institutions that are now part of the university system do have serious needs. But, with limited government funding, major research institutions such as Imperial need to be able to charge sensible top-up fees to maintain high-level teaching and research. Universities UK became so large and unmanageable that the leading research universities formed their own group, the Russell Group (so-named because the vice-chancellors first met in the Russell Hotel in London). The Coalition of Modern Universities represents the old polytechnics.

38. As a later Higher Education Minister, Margaret Hodge, put it in 2002, the government wanted both excellence and equality, a seemingly impossible goal. Not all universities wanted top-up fees, and lengthy discussions contributed to a delay in the publication of the Government White Paper (2003). By then Charles Clarke was Minister of Education and was trying to work out a solution. In so far as Imperial is concerned, perhaps the best news to come out of this was a revised RAE system which should funnel more of the available research money its way. While universities can still not set their own fees, let alone collect them, the government planned to allow fees for home students to be set up to £3000 (up from £1,100). The fees were to be collected by the government in accordance with the tax system, some of it after graduation. The White Paper returned some independence, at least in principle, to the universities, but only time will tell whether the *dirigism* of the previous years will decline.

39. See *The Times*, 18 October 2002. This article is based on a leaked document, namely Paper M, 'The payment of Home/EU undergraduate fees which reflect the true economic cost of undergraduate education', that Sykes presented to Council, 18 October 2002. In this paper, Sykes stated what he considered to be the true cost of undergraduate education. This was in anticipation of the 2003 Government White Paper.

40. After a drop in the number of foreign students in the late 1970s and early 1980s due to sudden fee increases, numbers began to increase again in the later 1980s. Many students came to Imperial from Hong Kong and other places in South East Asia. In 1990 the college won a Queen's Award for Export Achievement for bringing in £11m in foreign student fees and

foreign contracts. In 1990–91 Imperial had 23.1% foreign students (not including those from EU), exceeded in the major UK universities only by the LSE with 38.6%. By 1995 the fees for overseas students at Imperial were: medicine, £15,000; science and engineering, £9,450, and mathematics, £7,450. Higher education for foreign students has become a big business in the UK; though whether for much longer remains to be seen.

41. See EC minutes, 'Governor's Lectureships', 6 May 1994. Some of the money for the lectureships came from savings made already during Ash's rectorship, and targeted for new appointments.

42. See EC Minutes, 6 May 1994. In that year HEFCE changed its funding methodology which had been based more on staff to student ratios, to one with more emphasis on staff being active in research. To compete, Oxburgh wanted the staff to student ratio at Imperial to be more in line with that at Cambridge. The cost of adding thirty new lecturers was approximately £900,000 per year. But the added HEFCE grant could be as much as £750,000 and the new staff were expected to bring further external funding. More new hiring followed the original thirty lectureships.

43. It is worth asking why all of Imperial's Nobel Prize winners belong to an earlier period of the college's history, a generalization that can be roughly extended to Britain as a whole.

44. Sir Eric Albert Ash FRS FREng (1928–) was born in Germany and came to England at the age of ten. He was educated at University College School and Imperial College where he took a first class honours degree in electrical engineering. He was a PhD student under Dennis Gabor and, on graduating, went to Stanford as a Fulbright Fellow. He returned to London in 1955 and worked as a research engineer with Standard Telecommunications. In 1963 he accepted a senior lectureship at University College London and was appointed to a chair in 1967. He was Pender Professor and Head of the Department of Electronics and Electrical Engineering, 1980–85. His own research is in electron optics and ultrasonics, as applied to imaging. Ash was awarded a Marconi Fellowship in 1984. An active member of the Institution of Electrical Engineers, he was awarded the Institution's Faraday Medal (1980), and was elected President 1987–88. He has been Secretary of the Royal Institution, a Vice-President and Treasurer of the Royal Society, and was awarded the Royal Medal in 1986. In the citation for Ash's Fellowship of Imperial College his 'unflagging dedication to the College … in a time of great financial stress' was rightly acknowledged.

45. Ernest Ronald (Lord) Oxburgh FRS (1934–) was educated at Oxford, and at Princeton where he took a PhD in geology with Harry Hess. A

specialist in geophysics and tectonics, Oxburgh later taught at Cambridge where he was appointed Professor of Mineralogy and Petrology in 1978, Head of the Department of Earth Sciences 1980–88, and President of Queen's College, 1982–89. In 1988 he was appointed Chief Scientific Advisor at the Ministry of Defence. Oxburgh was knighted in 1992 and made a life peer in 1999. He sits on the Cross Benches and has served as Chairman of the House of Lords Select Committee on Science and Technology. He also served as non-executive Chairman of Shell from 2004–5. On his appointment at Imperial Oxburgh listed his interests as including orienteering, mountaineering, reading, and repairing old cars and houses. He also breeds Burmese cats.

46. Ash had a number of merger discussions with other University of London colleges, including Royal Holloway and Bedford New College.

47. The two libraries were both situated in the Sherfield Building from 1969 but existed as separate units until 1992. The unification is an uneasy one, not made easier by the high rent that the Science Museum is charged by the college; separation is under discussion.

48. The Alvey Report (1983) called for a five-year programme to cost about £350m of which the government was to provide roughly two-thirds and industry one-third. The money was to be used to fund two hundred projects, each typically involving about two industrial firms and one or two universities; this in addition to boosting undergraduate numbers in computing science.

49. Eric Ash, 'Higher Education, Industry and Government: new rules for a ménage à trois', Presidential Inaugural Lecture, IEE, 1 October 1987. Ash also headed a SERC working party on the Alvey scheme and its report, in 1986, urged the government to put more funds into the programme. See discussion in *Financial Times*, 1 May 1986. It is worth remembering that in 1985 Imperial College was turning away six applicants for every place in its computing science courses. And it had only one computer terminal for every six students in this field. See special report, 'City and Guilds at Imperial College', *The Times*, 27 February 1985. With hindsight, it also seems incomprehensible that the college installed a new telephone system in 1984 which had the capacity to handle only 69 incoming calls at the same time. The system was upgraded in 1989 at which time the college was receiving about 4,700 calls a day. The college installed its first fax facility in 1986.

50. Flowers had introduced a new departmental vote system and Ash tried to make it yet fairer, but criticism continued.

51. See, for example, report in *Times Higher Education Supplement*, 1 November 1985. In 1987 the deficit was running at about £1m and Ash planned for a cut of 100 positions across all categories. Another serious problem in this period was building maintenance, despite an earlier UGC directive that universities increase their building maintenance budgets by 20%.

52. For Ash's report, see *Daily Telegraph*, 7 June 1991. Ash claimed that academic pay had fallen 36% behind non-manual earnings nationally. (The recession had bottomed out in the early 1980s.)

53. According to Ash the title was suggested by the appointee, and he agreed to it (personal communication).

54. By 1997–8 the college research income was £120m. Less than half of this amount came from the research councils. About 15% came from industry and about 30% from charitable foundations. The large amount from charitable donations is a reflection of the medical school mergers.

55. *The Guardian*, 26 November 1991, 4. The outsourcing of jobs was then a relatively new phenomenon. By the mid-1990s the college was engaging in this practice more regularly.

56. If Fraser had been more successful in bringing funds to the college, perhaps he could have survived. But in fact some of his schemes cost the college money. For example, trying to set up links with Japan and plans to open a Japan office did not result in any serious Japanese investment in the college. A major industrial research park in Newport, Wales, a scheme, proposed by David Thomas, Director of Industrial Liaison at the college, and approved by Fraser, soon faltered. By contrast a simple scheme, the Harlington Gravel Trust, set up earlier by John Smith and Peter Mee, made a profit. The purpose was to oversee the extraction and sale of gravel from part of the college's Harlington site, and to use the income to help maintain and improve sports and athletics facilities for students. Founded in 1985, by 1994 it had generated about £2.5m in royalties.

    Peter Mee acted as College Secretary in the brief interval between Fraser and Mitcheson. Kenneth Anthony (Tony) Mitcheson was educated at the Royal Military Academy at Sandhurst and had a career in the Royal Artillery before coming to Imperial.

57. Ash had begun work to improve alumni records with fundraising in mind.

58. Recruitment concerns continued until the end of the decade. A survey carried out by the Registrar, Peter Mee, of applicants who had turned down placements in the years 1987 and 1988, showed that the expense of living in London was a major factor in their decisions.

59. For the historian, the annual reports became less useful. Much of the earlier detail was stripped away. Publicity material followed more general

trends and became relatively lightweight compared to what had been produced earlier. But it served a purpose and gave prospective students something attractive and easy to read. In 1990 Imperial won the *Times Higher Education Supplement* award for the best annual report published by a higher education institution.

60. *THES*, 8 April 1988, 1.

61. Given that teaching still counts towards promotion, heads of department have some hold on their staff. But the fact that teaching loads matter more to heads of department, concerned about overall departmental income, than to their staff often more concerned with personal research output, has led to tension.

62. See David Jobbins, 'Imperial's students take consumer test', *THES*, 24 February 1989, 5. According to Jobbins, Imperial already had the most comprehensive lecture assessment programme in the country.

63. See Oxburgh's discussion paper, 'Recognition of Teaching Excellence within the College', 25 April 1995. Oxburgh instituted a prize system for good teachers.

64. The college could tap into the European Union ERASMUS scheme (replaced by SOCRATES in 1997). These schemes were to support European exchanges between universities by providing a set of guidelines, including a system for the recognition of academic work carried out in other participating institutions. Funding came from the EU to participating institutions. Imperial and the major technical colleges in Delft, Zurich and Aachen formed the IDEA League and negotiated common policies to allow the movement of students between them. In 1989 a Europe-wide survey of engineering colleges and universities by the journal *Liberation* placed Imperial College at the top. The college also placed top in the Halsey survey for the *THES* in 1990.

65. Oxburgh also saw the old colleges as inconvenient, albeit worthwhile preserving. He wanted to find a new, less intrusive, place for them within the college structure (personal communication). Alumni continue to identify with the old colleges since students still graduate with associateships. But in the day-to-day running of Imperial, the old colleges no longer play a role.

66. Personal communication. Ash believed the old colleges were also an impediment to the creation of first rate Imperial College sports teams, with the exception of rowing.

67. In some ways the college understood this. For example, in 1990 the CGLI stated that it wished to give its Associateship to students other than those graduating from Imperial College. This was a highly contested issue and,

in the end, Imperial was able to retain exclusive rights, at least for now. See GB minutes, 16 March 1990.

68. Until the late 1970s it was virtually impossible to leave Oxford or Cambridge without a degree. While first class degrees were difficult to achieve, so was failure. In a posthumous publication, Lord Ashby noted that many Oxbridge dons were good college men who would 'not be considered good enough scholars to be offered tenured posts in the less distinguished universities on Britain's academic periphery'. See Lord Ashby, 'Centre and Periphery in Academe', *Minerva* 34 (1996), 95–101. Quotation, 97.

69. For the college reply to Swinnerton Dyer, see FC and EC minutes, 21 November 1980. For Flowers paper, see, GB minutes, 12 December 1980.

70. Stewart Sutherland, 'The University of London and the Award of Degrees', 24 June 1993.

71. Speech by Ash to University Senate, 7 July 1993.

72. See, for example, GB minutes, 19 December 1990, Paper D. The old Governing Body did not function well by the late twentieth century. Meetings were often attended by fifty people and so it was hard to come to any final decisions, especially since members did not simply rubber stamp the decisions of the Finance and Executive Committees. Further, as John Smith, who was appointed College Secretary in 1979, recalls, before Flowers and his successors modernized the college administration few important matters came on to the agenda from departments.

73. In fact the first Council had thirty-two members. This number was decreased to nineteen voting members in February 2005. The role of Council became more strategic leaving all management to the administrative staff.

74. See GB minutes, 19 December 1990. The old Governing Body was unusual among university councils of the time in having a large number of its members nominated by external bodies such as the professional engineering institutions, the Commonwealth High Commissions, and the Greater London Council.

75. In compliance with the Nolan recommendations.

76. The Council retained the power to appoint the Rector. The Chairman of the Council, someone external to the college, is formally appointed by the Court on recommendation of the Council. In theory this means the Court could veto the Council's first choice, though this seems unlikely.

77. Field Marshall Sir Richard (later Lord) Vincent (1931–) was educated at the Royal Military College of Science at Shrivenham and became Commandant at the college in 1980. Vincent had a distinguished military career, was appointed Chief of the Defence Staff in 1991–2, and Chairman

of the Military Committee of NATO. He is a specialist in electronics and guided weaponry.

The Deputy Chairman of the Governing Body (later Council) was Sir Peter Baxendell, Chairman of Shell Transport and Trading. He was a 1946 Imperial graduate in geology (oil technology). He and his wife, Rosemary, a graduate in biology, met when students at the college. Baxendell was shortly to be replaced as Deputy Chairman by The Hon. Sara Morrison. Morrison is a prominent member of the Tory Party who had served on the Governing Body since 1986 and was a director of GEC. Earlier she had a major career in local government, working especially with young people and the disabled. Morrison is an example of a good governor later lost by the new term rules.

Lord Vincent remained chairman of the Council until 26 March 2004. Eileen Buttle was then acting chairman until the appointment of Lord Kerr on 1 January 2005.

78. Statute 22, Imperial College Charter and Statutes (1998).

79. 'Academic Structure', paper delivered by Ash to GB. See GB minutes, 24 June 1993. As Rector, Ash received reports from 23 heads of department, the managing director, the Pro-Rectors, the finance director, and the registrar.

80. For more on Wakeham see chapter 16.

81. In planning its submission to the UFC in 1989, the college had created a somewhat less elegant mission statement, not enshrined in the charter. This was for the college to be judged 'one of the premier institutions in Europe and each academic department to be ranked among the top three in the UK' and that 'the learning process be stretching and exhilarating' and, with a nod to humanities, that it extend 'beyond the confines' of a student's chosen subject.

82. By 1994 the Norman Foster and Partners team had come to the conclusion that Imperial had a poor visual and pedestrian environment. It was suggested that there be a new heart to the college created around the Queen's Lawn, with improved refectory and other services. Further, that there be a clearly defined entrance to the college, and that there be more raised pedestrian walkways between buildings. Suggestions were also made for the redevelopment of the RSM and Bessemer buildings, the Beit Quadrangle, and the North side of Prince's Gardens. New floors were to be added to the Library and plans for a new biomedical building prepared. In 1996 plans for a new sports centre on the north side of Prince's Gardens were added. These various plans have all since largely materialized. A Waterstones bookstore opened in a ground-floor extension of the library in 1997.

83. Swanson, a former head of the department of mechanical engineering, had held a Pro-Rectorship earlier. Work on teaching quality continues under Rees Rawlings, the current Pro-Rector. There is now a new Centre for Education Development which conducts teaching workshops for new staff. Some instruction is mandatory, but staff can also opt for further education courses and take Advanced Studies in Teaching and Learning leading to an MA.

84. Wye had joined the University of London in 1900. A college has existed on the site, near Ashford in Kent, since the fifteenth century. In the mid 1990s Wye had about 500 resident undergraduate students and about 200 postgraduates. A high proportion were from overseas. It also had a fairly large distance education programme with about 1200 students. There were twenty different degree options. Its teaching assessments were very good, a fact which sheds poor light on its forced decline into bankruptcy. Little national attention had been paid to what the country actually needed in the way of teaching and research in the rural development/agricultural area. Wye had an annual budget of about £12.5m which was roughly the budget of an average Imperial College department. This sum was clearly insufficient to keep both the college and its associated agricultural enterprise going. One of its problems was that being small it had to submit to the RAE as a single unit with the result that its good departments were pulled down by its weak ones.

85. Quoted in 'Report from Rector', *IC Reporter,* 23 June 1998. See also GB minutes, 26 June 1998.

86. Rector's Report recommending the merger, 23 June 1998. 'Rural development' was a catch-phrase taken from the revised EU common agricultural policy.

87. Discussion Paper on Wye prepared by the Deputy Rector for the Management and Planning Group, 19 January 1998.

88. In part these were related. To help with assessment/league table problems, Imperial insisted that Wye increase its admission standards. But this led to a sharp drop in the number of students and a major loss in fee income and government support.

89. A committee, chaired by Peter Bearman, proposed a new business plan for Wye in 2005 which was accepted by the Imperial College Council. It proposes the construction of new industrially-related research facilities, convention space, housing, and a range of cooperative ventures with input also from local government. The 'Wye Concordat' was signed on 6 December 2005 by the Rector, leaders of Ashford Borough Council and of Kent County Council, paving the way for a new '£1 billion world-class

science research and manufacturing facility', *Reporter,* 8 December 2005, 1.

90. Oxburgh, personal communication.

91. Sykes, personal communication. Perhaps just as important was major opposition to the merger within both colleges — almost total opposition at UCL. Sykes and his supporters had in mind a great metropolitan university that could compete with the best in the world. His vision was not unlike that of Lord Haldane's early in the twentieth century. Both Roberts and Sykes had experience with business mergers, but the merger they suggested in this instance was probably not in the best interests of either college, or of higher education more generally.

92. Statistics on the use of college student counselling services in the late 1980s and early 1990s show that the services were more heavily used by women than by men. An interesting exception were students in computing science. The per capita heaviest users were women in mineral resources engineering.

93. Earlier Ash had had been under some pressure from the ICAUT to do something about the growing number of short-term contracts many of which were given to women. He asked Professor W. A. Wakeham to look into the problem. Wakeham issued a report, 'Fixed-Term Contract Staff' (1991) which showed that among the people on short-term contracts, some had been employed for over thirty years. Clearly many such people had been exploited by the college. But the college was not alone. According to the CVCP there were about 17,000 short-term contract employees in UK universities in 1990. At Imperial fixed-contract staff were about 40% of the total academic staff (not including research assistants). Oxburgh worked to put some new policies in place, including better salaries, and the creation of new status-enhancing job titles.

94. AWISE was founded in 1984. Starting in 1985, Imperial put on a two-day event with short science and engineering courses for high school girls in collaboration with AWISE. This lasted for several years with about 380 girls attending each year, about 200 being accommodated in the residences. One interesting observation is that the performance of women students in the sciences in the 1990s was less good than that of men at most British universities. At Imperial, however, in science and engineering there was no significant difference. (Based on a study by Jane Mellanby of students at the University of Oxford that was reported in *The Independent,* 14 August 2000, and claimed to be the norm; also, on statistics collected at Imperial College. For the year 1998–1999, excluding medical students, 25% of

women and 25.5% of men gained first class degrees; and 95% women and 92% men gained either first or second class degrees.)

95. This has been a successful initiative. At Imperial younger women on the academic staff are now well represented among those winning good research grants and awards, and in gaining international recognition.

96. Report on the findings of the Academic Opportunities 2000 Committee, *IC Reporter*, 16 March 1999.

97. The college received an Athena Silver SWAN (Scientific Women's Academic Network) award from the Royal Society in 2006 for improvements in the working conditions for academic women.

98. See, for example, Henry Etzkowitz, 'Entrepreneurial Scientists and Entrepreneurial Universities in American Academic Science', *Minerva*, 21 (1983), 198–233; B. Clark, *Creating Entrepreneurial Universities* (New York, 1998); Henry Etzkowitz, Andrew Webster and Peter Healy, *Capitalizing Knowledge: New Intersections of Industry and Academia* (New York, 1998). Etzkowitz is the author of the idea of the 'triple helix' a model for understanding university/government/industry relations. Etzkowitz discusses the link between economic development and innovation within universities, something that occurred earlier in the USA than in Britain.

99. Three students who built a successful dot-com in this period were Wendy Tan (computer science), Gareth Davies (physics) and Erik Pettersen (physics). All were students in the early 1990s. Their company was called Moonfruit and it provided people with the tools to build their own websites. About 35,000 of these had been created before the 2000–2001 crash, and the company was said to have been worth about £40m on paper.

100. See also the section on mechanical engineering in chapter 16.

101. The financial success of this company did not mean any money came to Imperial. The fact that Imperial College employees could profit so well from their college association without anything coming to the college was clearly something that needed rectification. Under Ash's direction, David Thomas began to put in place procedures for the college to be given some equity (typically about 20%) in spin-out companies. For Thomas see below.

102. As mentioned in chapter 12, it was Lord Flowers who pushed for a research park, and suggested that Silwood might otherwise have to be sold because it was losing money.

103. See FC and EC minutes, 7 November 1986. A variety of research projects moved into these buildings: work on virus resistant tomatoes, the safe disposal of radioactive waste, the protection of beaches from wave erosion, sweeter and cheaper sugars, birth control drugs, research following the Chernobyl disaster, work on semiconductors, and more.

104. Sir Frank Cooper became Chairman of the Governing Body in 1988. He had a distinguished career in banking, industry, and in Whitehall where he was Permanent Secretary at the Ministry of Defence. See obituary, *The Times*, 30 January 2002.

105. Thomas had a difficult task rationalizing Imperial's many contracts with industry and to ensure that the college received some return from staff interactions with outside industry and with the rising number of spin-out companies.

106. The first product was peptidase, an enzyme used in the cheese industry to speed up the cheese maturation process. But the pilot plant projects failed to make enough money to justify its continued existence. See chapter 10.

107. This had the support of Ash and Angus Fraser. Thomas became chairman of the new company and Paul Docx was its chief executive. The offices in Europe and Japan did not pay off.

108. Royal Postgraduate Medical School (RPMS), Hammersmith.

109. In the mid to late 1990s there was much discussion, led by Bill Wakeham, on the role of spin-out companies. It was not envisaged that these companies would make a noticeable difference to college finances, but it was recognized that for reasons of staff motivation and retention it was necessary to support entrepreneurial activity. As it turned out, some money did come to the college as a result.

110. Imperial College held about 20%˙of the shares with a total initial investment of £20,000. See also chapter 13.

111. See also 'Spin Out Companies' (paper D), GB minutes, 13 November 1998.

112. According to the college 1999-2000 accounts, the college sold 25% of its shares in Turbo Genset that year for £9,924,000.

113. Council minutes, 28 March 2003. In 2003 depressed capital markets put a damper on the growth of spin-out companies. Further, the college was running out of space and thinking of ways to acquire a new site for the incubation of companies. But spin-outs began increasing in number again by 2005.

114. The original acrobot was a robot for knee surgery developed by Brian Lawrence Davies FREng (1939–). Davies was educated at University College London and after an apprenticeship with GEC Ltd. he returned to UCL as a lecturer. He joined Imperial as a senior lecturer in 1983 and became Professor of Medical Robotics in 2000. A specialist in robotic and computer aided surgery, Davies has created a number of medical robots including PROBOT that was the first to be used to remove tissue from a patient in a clinical setting — during prostate surgery.

115. Potential problems include staff devoting college time to their business interests, misuse of the research time or work of research students and, lured by prospective venture capital, the forsaking of a disinterested stance towards the results of research. For some American experience with these problems see Jennifer Washburn, *University Inc.: The Corporate Corruption of American Higher Education* (New York, 2005). While Washburn's book focusses more on the effects of major funding by large corporations and the consequences for disinterested research in that connection, spin-out companies are also discussed. Her book is salutary, even for those who hold a utilitarian view of education, and support entrepreneurial activity within academic settings. I mention the potential problems only because it has been difficult to find anyone at Imperial willing to acknowledge the existence of any actual problems.

116. Royal Commission on Medical Education (Todd Report, Cmnd 3569, 1968). For a more detailed account of the merger of St. Mary's Hospital Medical School with Imperial College as seen from the perspective of St. Mary's, see E. A. Heaman, *St. Mary's: The History of a London Teaching Hospital* (Montreal and Liverpool, 2003), chapter 14. This merger was the first of several medical school mergers which led to the creation of the Imperial College School of Medicine.

117. An internal Imperial College committee, chaired by Hugh Ford, also looked again at the Todd Report. It concluded that the college should consider becoming a centre for preclinical training but that it was not something to rush into. Ford's report includes a list of some major areas of collaboration between Imperial and London medical schools. At the time, thirteen departments were involved with eighteen different medical schools or hospitals. See FC and EC minutes, 16 November 1979.

118. In 1989 St. Mary's scored only an average of 3.00 in the RAE.

119. Quoted in Heaman, *op. cit.* (116), 399. For Ash's proposal see FC and EC minutes, 16 May 1986.

120. After the St. Mary's merger the number of women students rose to 22%. The number rose to about 30% after the later medical school mergers.

121. Other members of the working committee were Professors Anderson, Blow and Morris, as well as representatives from St. Mary's.

122. Ash, personal communication.

123. It is curious that one of the chief opponents of the merger was Bruce Sayers, then head of the electrical engineering department, and himself a major researcher in medical electronics and signal processing. Roy Anderson was then head of the department of biology; today he is Director of the Wellcome Research Centre for the Epidemiology of Infectious

Diseases situated at the St. Mary's campus of Imperial College. See Board of Studies minutes, 18 February 1987. See also GB minutes, 19 June 1987.

124. By several accounts, St. Mary's was not transparent in its finances. Several irregularities were uncovered after Imperial had access to the books. The keeping of separate books had been as much to protect Imperial's interests as St. Mary's.

125. Sir Bernard Tomlinson, 'Report of the Inquiry into London's Health Service, Medical Education and Research' (HMSO, 1992).

126. See Oxburgh, 'Medical Strategy for Imperial College', EC minutes, 12 November 1993. See also GB minutes 1 July 1994.

127. EC minutes, 20 March 1992 and 12 November 1993. The Charing Cross and Westminster Hospital medical schools had merged in 1984. The CXWMS had about 160 medical students and about 170 research-active staff in 1993. In the 1992 RAE, CXWMS scored only 3.00. This was the same score as St. Mary's received, but the CXWMS had submitted only 80% of its academic staff for evaluation. As a consequence it received no further research income from the state. Its budget was about £4m and it was running a deficit of about £1m.

128. See EC minutes, 12 November 1993 and Oxburgh, 'Medical Strategy', *op. cit.* (126). The first to merge was the National Heart and Lung Institute which joined Imperial in 1995. The RPMS agreed to join in late 1996 but the Institute of Cancer Research later joined UCL. The Kennedy Institute of Rheumatology, an independent unit located on the CXWMS campus, later joined Imperial.

129. In 2001 the government approved a major project that had been under consideration for many years, namely to consolidate many of the area's tertiary health services in Paddington Basin (renamed Paddington Waterfront). In 2005 this development came under question and, as of writing, it is unclear what will happen.

130. Christopher Richard Watkin Edwards (1942–) was educated at Cambridge and began his career at St. Bartholomew's Hospital. He was appointed Professor of Clinical Medicine at the University of Edinburgh in 1980, becoming Dean of the Faculty in 1991. He was appointed a Governor of the Wellcome Trust in 1994. He joined Imperial in 1995, leaving in 2001 to become Vice-Chancellor of the University of Newcastle-upon-Tyne.

131. EC minutes, 12 November 1993.

132. See Eric Ash, 'Sir Frank Cooper, an appreciation', *IC Reporter*, 12 February 2002. Cooper's obituary in *The Times* (see note 104) had focussed more on his earlier career. Ash describes Cooper's many contributions to Imperial

College. Cooper had also been chairman of the board of King's College School of Medicine.

133. See EC Minutes, 6 May 1994. There is some correspondence between Oxburgh and Bottomley in the ICA. See for example, Oxburgh to Bottomley, 15 March 1994, on bringing pre-clinical teaching to South Kensington.

134. Oxburgh, personal communication. Typically it is easier to get agreement over a dining table than over an office desk.

135. It is not clear whether the participating surgeons were on duty during freshers week. Oxburgh was largely successful; freshers no longer feel compelled to join in such rituals. Oxburgh especially had in mind the increasing number of women medical students some of whom had already voiced their disapproval of having to demonstrate loyalty in the old ways.

136. In addition to the library extension, Lipton's company converted the vacated east and west sides of the Beit quadrangle for student accommodation. The greenhouses on the roof were dismantled, and a new storey added. Later the north and south wings were refurbished. But much of the student union space remained as it was, though further renovations are planned.

137. The Queen reopened the restored Albert Memorial on the same day. For good coverage of the college at this time see Special Report, published on the occasion of the opening of the Sir Alexander Fleming Building, *The Times*, 22 October 1998, 40–43. Sir Alexander Fleming (1881–1955), the discoverer of penicillin, was Professor of Bacteriology at St. Mary's Medical School, 1928–48.

138. Composed for the occasion by John Madden, Director of Music at Radley College.

139. Most hospitals depend on people paid from academic sources in order to provide a full range of specialist services. Since a united medical school would have the ability to move such people from one hospital to another, the independent hospital trusts were placed in a difficult position. The NHS regional authority was often impatient with the independence of the trusts and was keen to use the unification of the medical schools as an opportunity to rationalize resources in the region's hospitals. It agreed with Imperial in wanting to have sensible concentrations of expertise at the different hospitals rather than allowing each hospital trust to decide on the services it offered. Kidney transplants were a case in point where the dispersal of expertise could lead to less than optimal results for patients, and to greater expense overall.

140. In 1996 the NHLI scored 5*, the RPMS scored 5* for some of its units and 5 for others, but the CXWMS rated only 3 in each of its three units. St. Mary's received a 5 for basic medical science and a 4 for clinical science. St. Mary's scores were an improvement on the 1992 ratings.

141. At a college-wide meeting in 2001, I witnessed a doctor from St. Mary's telling Sir Richard Sykes that he very much resented the loss of identity. Sykes replied that the doctor should be proud to be part of Imperial College which also had a fine history and was today a world leader in higher education.

chapter sixteen

The Expanding College, 1985–2001…
Part Two: Some Academic Developments

Introduction

Restructuring of departmental and college governance is ongoing and it will probably take some time before a relatively stable configuration is reached. The reasons for this have less to do with the expansion into medical science and technology, though that is certainly a factor, and more to do with the larger issues discussed in chapter fifteen. The past quarter century has seen a radical shift in how university funding is viewed, and attitudes have yet to stabilize. Nationally, the number of undergraduates increased by about sixty percent over the final quarter of the twentieth century, with over half a million students enrolled by 2001.[1] Unless the public can come to terms with greater selectivity in the funding of undergraduate programmes, these may well cease to flourish in major research institutions. Perhaps this is what the future holds, namely two kinds of institution: undergraduate colleges and research universities.

For the present, the college must continue to offer quality undergraduate education. Thus, when Sir Richard Sykes became Rector in 2001, he was faced with the problem of how to restructure the college so that it could continue to be competitive in both the undergraduate and research areas.[2] Change was made a little easier by the recent medical school mergers and the foundation of the Imperial College School of Medicine. The college was in any case seen as needing new forms of governance over and above its new charter, council and court. One of Sykes' major innovations was the introduction of a faculty system, something that Eric Ash had called for earlier.[3] This brought to an end the already diminished role of the old colleges in Imperial's governance. From a managerial point of view faculties make sense and have possibly helped enhance the college profile with respect to attracting industrial and other funding. But the advantages have been bought at the cost of a new and expensive bureaucracy and, perhaps more seriously, at the expense of departmental autonomy.[4] Research sections, with their inter-disciplinary linkages, are slowly becoming more dependent on faculty than on departmental decisions for their well-being. Further, research sections, often with connections to one or more of the centres, have their own agendas, not always commensurate with those of the various depart-ments. Undergraduate teaching remains the responsibility of departments, and some heads of departments have had an increasingly difficult task in persuading people to dedicate much time and effort to it.[5]

We have seen that an early response to new forms of govern-ment funding, and to increased competition for resources, was a rise in the number of interdisciplinary research centres. In this chapter it will be shown that such centres grew and proliferated during the final years of the twentieth century. Many new MSc courses were introduced, built around the combined research interests of staff associated in new groupings. PhD students, while nominally attached to departments, began to associate with the centres in increasing numbers. For now, however, departments remain the unit of assessment, and so are rivals for internal funding. Faculties and interdisciplinary centres, however, appear better to represent the outward-looking face of the college. For the foresee-able future, the newly founded faculties will need to fund some shared departmental projects from above, and support basic disciplines when needed.[6]

In this chapter aspects of the history of interdisciplinary work in the period 1985–2001 will be discussed and, in light of this, also the recent history of the departments. The college plan for 1994–9 listed four areas where interdisciplinary work would be encouraged: medicine and the life sciences, the environment, information technology, and materials. This plan, already in action by 1994, has been largely successful. A fifth interdisciplinary area of major significance has been that of energy, one which has attracted research in renewable sources, in cleaner energy sources, in carbon capture, and in a range of other related areas. In 1998 the Graduate School of Environment was renamed the Environment

Office (later the Energy and Environment Office) and, more recently, a college-wide *Energy Futures* initiative was launched to assist the many cross-disciplinary research groups.[7]

Already in the 1960s, Patrick Blackett was an advocate for the multi-professorial department. By the 1990s this had become the norm in all large departments, several of which had over twenty professors on their staff by the end of the century. It is no longer possible, in a chapter of this length, to even briefly mention the work of all the professors. Further, since the period covered was one of great transition, the chapter may give a false impression of Imperial's overall position in the national and international context. Some departments are shown struggling to compete under difficult circumstances and the chapter outlines many of the problems they faced. It is well to remember that similar problems affected departments across the country. Looking back from today's vantage point, one can see that Imperial came through the 1980s and 90s very well and is today ranked one of the best teaching and research institutions in the world. Even knowing this, it is difficult for the historian to have much perspective on recent events and so this chapter remains tentative and very sketchy. Time may well prove that my selection of what to discuss and what to omit, was misguided.

Interdisciplinary Centres

The founding of the Imperial College Centre for Environmental Technology (ICCET) was discussed in chapter twelve. By the late 1980s its links to other areas of the college had increased in number. For example there were new links to the Management School, to research in the areas of the environment and transport in civil engineering, to the environmental geochemistry group, and to other centres such as the Reactor Centre (later the Centre for Analytical Research in the Environment, CARE),[8] the Centre for Remote Sensing, and the Global Environment Research Centre.[9] Indeed ICCET had grown so large that it was given quasi-departmental status in 1989. It moved to more spacious accommodation in the RSM building and, in 1997, became a full department. In addition to its successful MSc course the centre had about forty PhD students by 1990. In 1992 ICCET teamed up with thirteen other European institutions to offer a European MSc in environmental management.

The mid-1990s was a high point for this centre/department. In 1994 a committee chaired by Sir John Mason recommended the founding of a Graduate School of the Environment to coordinate teaching and research in this field college-wide.[10] At the time, the environment was high on the agendas of both government and industry and Imperial was seeking to tap into available research monies, and to position itself as a national and international centre of expertise across all aspects of environmental work: scientific, technical, medical, management, economic and legal.[11] A later experiment was the creation of the T. H. Huxley School for the Environment, Earth Sciences and Engineering by Ronald Oxburgh in 1997. This was to subsume the work of the Graduate School of the Environment and was an attempt by Oxburgh to further rationalize the many course offerings and research units in the environmental area. He had a new taxonomic vision with respect to the grouping of the associated disciplines, one which was, perhaps, a little before its time. He envisaged a school that would provide for 'coherent study and research into the evolution of the Earth, the exploitation of its natural resources, and for understanding the consequences of anthropogenic environmental inter-ventions in natural systems'.[12] Oxburgh brought together the Department of Earth Resources Engineering, the Department of Geology and the new Department of Environmental Technology in the School, and appointed John Beddington as Director.[13] There was some imbalance in that the geology and earth resources engineering staff numbered about 80 and the environmental science and technology staff about 30. With goodwill this experiment could, perhaps, have succeeded, but there were financial problems within all sections which impeded cooperation. The Department of Environmental Science and Technology had become too diverse and somewhat overextended. Its finances were problematic despite Beddington having expanded the MSc course which was taking in about 150 students by the late 1990s. The other departments had additional problems related to staff who were not active in research, thus bringing down RAE scores and further reducing income. When Wye College merged with Imperial many of its staff were also placed within the Huxley School. The already existing fissures and financial anxieties became yet more pronounced. Beddington attempted to turn things around, to increase efficiency, and to replace non-research-active staff.[14] But, when Sykes introduced the faculty system he decided to end the experiment and closed down the Huxley School. He arranged for the Department of Environmental Science and Technology, and those working at Wye, to form separate units within the new Faculty of Life Sciences; and he merged the Departments of Geology and Earth Resources Engineering in a new Department of Earth Science and Engineering within the Faculty of Engineering. These moves were supported by most of those involved, but further rationalization was to follow.

In 2005 the Faculty of Life Sciences followed the example of the Faculty of Medicine and created divisions.[15] The Department of Environmental Science and Technology and the Department of Agricultural Sciences (Wye) were closed and some of their staff merged to form a new Centre for Environmental Policy within the Division of Biology. Scientists previously employed in the departments either joined other science or engineering departments, or left the college. The new Centre, under the direction of Jeff Waage, is relatively small.[16] It includes a few economists, environmental lawyers, and social scientists. Scientists and engineers in other departments still associate with the centre, and they still teach on the MSc course which remains large. However, in many respects the Centre has come full circle and is not unlike its leaner, younger (ICCET) self of the 1970s.

Several centres were sponsored by the University of London in order to encourage inter-collegiate collaboration. Just three examples will be given. First, the London Centre for Marine Technology which continued to function in this period. During the 1980s, civil and mechanical engineers, geologists and materials scientists helped design Conoco's Hutton oil platform in the North Sea, and were involved in other large off-shore projects. By the mid-1990s, however, North Sea oil and gas development was winding down and work at this centre declined. This pattern of people coming together to address relatively short-term problems is not unusual. Several other centres have shown a similar lifespan. Second is the Interdisciplinary Research Centre for Semiconductors, led first by physicist Bruce Joyce.[17] This centre has proven to be longer-lived and was restructured in 1997 as the Centre for Electronic Materials and Devices. From the start it attracted many talented people.[18] The third, and youngest, example is the ongoing London Centre for Process Systems Engineering, founded in 1994. This centre has a strong representation from the department of chemical engineering and chemical technology where Roger Sargent and John Perkins had earlier built up the field, and a smaller representation from the department of electrical engineering. These and the following more local examples demonstrate that centres have life patterns that are responsive to the demands of the larger economy.

A new interdisciplinary space was created when the William Penney Laboratory opened in 1987. It brought together a number of information technology related research groups. The largest of these was the Imperial College Planning and Resource Control Research Centre (ICPARC), founded in 1993 and discussed further below. This centre, now closed, focussed on large-scale optimization problems in the internet and transport sector. For a while it had much industrial support, including from British Airways, but this collapsed early in the new century. Several other centres were founded in the 1990s, some with relatively short life spans and some that are still in existence. Among the former were the Centre for Toxic Waste Management, and the Wellcome Centre for Parasitic Infections. The first of these, under the direction of Roger Perry brought together chemists, chemical engineers, biotechnologists and civil engineers. It was sponsored by the Wolfson Foundation and by one of Britain's leading waste disposal contractors, Cleanaway Ltd. Those within this centre have regrouped as the Centre for Environmental Control and Waste Management. The initial focus of the Wellcome Centre, under Roy Anderson, was on parasitic worms, said to affect roughly one-third of the world's population. Like the other centres it was multidisciplinary and interdisciplinary. It included molecular biologists, immunologists, epidemiologists and mathematicians, working together also with people in the Caribbean, Africa and India. Chemists, too, were involved. For example, Steven Ley synthesized a molecule known as avermectin B1a which proved to be a useful parasiticide.[19]

The Centre for Biological and Medical Systems emerged from the old Physiological Flow Unit and the Engineering in Medicine Laboratory when experts in biomedical fluid mechanics came together with specialists in biomedical signal and image processing, computing, biophysics, and applied mathematics. It opened in 1989 and Colin Caro, the former head of the Physiological Flow Unit, was appointed Director. A donation from the Leon Bagrit Foundation enabled new premises to be built in space that had earlier been occupied by the steam laboratories of the mechanical engineering department. The new premises were opened in 1991 and the centre was named for Leon Bagrit.[20] By then Caro had retired (though still research active) and Richard Kitney took over the directorship.[21] The centre offered an MSc in Engineering, Physical Science and Medicine. Further donations from Lady Bagrit helped its later development, and it became the Department of Biological and Medical Systems in 1997. By 2001 Christofer Toumazou headed the department, renamed the Department of Bioengineering.[22] An undergraduate course was introduced which soon enrolled about forty-five students.

In order to facilitate yet further exchanges between those working in the new Imperial College School (later Faculty) of Medicine and those working in the life sciences, the Graduate School of Life Sciences and Medicine was created in 1999. A range of interdisciplinary programmes are associated with this School. For example, MSc courses are offered in computational genetics and bioinformatics. Also associated are the new Centre for Integrated Genomic and Microchemical Analysis and Synthesis, the Centre for Molecular Microbiology and Infection, the Department for Biological and Medical Systems, the Centre for Cognitive Systems (focused on auditory and visual perception), and the Genetics and Genomics Research Institute — among others.[23]

Not all the new centres and groupings that emerged in this period have been mentioned, and not all were successful. Indeed, with the decline of manufacturing industry in the United Kingdom, and with little government funding, there was a decline in MSc courses in some areas. But, in general, centres and institutes continue to thrive as research and teaching umbrellas under which people from a range of disciplines congregate, work together, and exchange ideas. These few examples illustrate that people at Imperial are continually seeking new ways to associate beyond the traditional departments. When Brian Flowers helped to create the first interdisciplinary centres he envisaged them as more briefly transitory than has turned out to be the case, as simply a means of bringing people together across disciplines. While centres tend to have limited life spans, some have been able to grow and gain full departmental status and some, like bioengineering, have offered new undergraduate courses. But most are ultimately dependent on short-term funding, or on fees from MSc courses, and are thus highly vulnerable to market forces. Despite this, interdisciplinary work is here to stay as can be seen also in the recent history of the mainstream departments. It is to this history that we turn next, something that cannot be discussed without yet further mention of centres.

Recent Developments in the Departments

Life Sciences

Biochemistry which had reinvented itself under Brian Hartley ran into difficulties after he resigned as head of department in 1979. Part of the problem was that Hartley and his two successors, Eric Barnard and Howard Morris, each had a different vision of the department's future. Brian Flowers had tried to help by creating the Centre for Biotechnology for Hartley but, as mentioned elsewhere, this did not prove successful in the longer term. Barnard, who succeeded Hartley as head of department in 1979, left Imperial in 1985 to direct the MRC Molecular Neurobiology Unit at Cambridge. Morris, who succeeded Barnard, was a reluctant head and resigned in 1989. Ash then appointed David Glover, a fruit fly geneticist, to take over. Glover was one of Hartley's appointees, but being head of a fractious department was not to his liking, and he left shortly after for a chair in Dundee.[24] It would appear that the department lost its way for a few years, something that is reflected also in its low RAE rating of 4 in 1992. Later in the decade research contracts were lost and the department recorded deficits.[25] When Glover resigned, Ash was determined to appoint an outsider as head of department. This brought the biochemists together since they feared being lumped in with the biochemistry department at St. Mary's, and with people who did not share the same molecular biology interests. They suggested to Ash that he invite James Barber to come over from Biology to head the department, and Ash agreed.[26] Barber, professor of plant physiology, had recently created the Centre for Photomolecular Science and had invited George Porter, a specialist in photochemical kinetics, to form an association with it.[27] Barber stayed for two terms as head of biochemistry but, in the end he, too, had problems with his staff, as did his successor Anne Dell.[28] She was head for only a short period before Sykes decided to merge the department with biology, which had its own problems of a largely financial nature, and to create a Department of Biological Sciences in the new Faculty of Life Sciences.[29] Just before the merger, biochemistry received a 5* rating in the 2001 RAE, suggesting that, despite problems, Barber and Dell had helped the department to regroup and regain its reputation as one of the best in the country. The Department of Biological Sciences was short-lived. As mentioned above, the Faculty of Life Sciences was formed into Divisions in 2005 resulting in departmental expertise being realigned yet again.

How things will turn out in the post-genomic age is uncertain. The Genetics and Genomics Research Institute should encourage links across a wide range of disciplines. The Centre for Molecular Microbiology and Infection has already demonstrated this kind of linkage. Occupying three floors of the Flowers Building (the refurbished Pilot Plant) it was set up by G. Dougan and D. B. Young.[30] Dougan, a specialist in microbial pathogenesis, was appointed to the Rank Chair in 1991 and, together with Young, attracted funding from the Wellcome Foundation and a number of other sources for the centre which opened in 1999. Biochemistry was further strengthened by the return to Imperial of Colin Hopkins, an earlier Rank Professor, who was appointed Professor of Molecular Cell Biology in 2000.[31] The biochemistry department of the 1990s was militantly molecular.[32] While molecular medicine, new nanotechnologies and synthetic biology appear to have exciting futures, perhaps being part of a larger faculty will encourage some biochemists to work also at other levels, something that Ernst Chain earlier believed was being wrongly neglected.

As discussed in chapter thirteen, the botanists and zoologists were joined in a new Department of Pure and Applied Biology in 1981, renamed the Department of Biology in 1989. Roy Anderson succeeded R. K. S. Wood as head of department in 1984 and, in 1989, became Director of the Wellcome Research Centre for Parasitic Infections, mentioned above.[33] Anderson, together with his deputy Michael Hassell, continued the modernization of Silwood begun under Professor Way. Ashurst Lodge was sold to raise money for new laboratories and the science park was enlarged to help put operations at Silwood on a more secure financial footing.[34]

When Anderson left for Oxford in 1993 it was agreed that he could take his team of epidemiologists with him; but the distinction between epidemiologists and parasitologists was unclear and a large number of people left at the time. Anderson was succeeded as head of a diminished department by ecologist Michael Hassell. Hassell was also appointed Director of the Silwood Park campus, a position that had been in abeyance since Way's retirement.[35] During the 1990s work in the biology department continued much as before: in ecology, plant pathology, microbiology, applied entomology, animal physiology and neurophysiology.[36]

Despite the migration to Oxford work in parasitology continued.[37] One area that has flourished is research into the malaria parasite. As noted in chapter thirteen, this work was started at the college by Professor Garnham and continued under Professors Canning and Sinden. More recently Andrea Crisanti's team, working together with scientists at the European Molecular Biology Laboratory at Heidelberg, have created the first transgenic malaria mosquito.[38]

Environmental work with links to ICCET (as it then still was), remained strong, notably as applied to global agriculture and fisheries. Biologist John Beddington, appointed Director of ICCET in 1994, applied economic and population biology theories to natural resource management — especially fisheries management. For example, his research team has worked on the Falkland Islands squid fishery, helping to determine sustainable levels in light of international competition for the resource. A new addition in this period was the Centre for Population Biology funded by NERC and located at Silwood. This successful unit had about fifty people working under the direction of John Lawton.[39] Lawton's interests in issues of diversity and threatened species brought the Centre much press attention during the 1990s when conservation and species extinction were almost permanently in the news. One of Lawton's major projects was the Ecotron. This consists of sixteen small environmentally controlled chambers in which ecosystems (microcosms) are exposed to different conditions (soil, light, water and wind) over long periods. The purpose is to carry out controlled experiments which can be replicated and the results subjected to statistical analysis.[40]

Anderson, Hassell, Beddington and Lawton are all model builders who were able to ride the new computing wave at a point when fast and powerful computation became relatively cheap. Further, they were all, in varying degrees, influenced by the biomathematical work of Robert May.[41] It is, perhaps, this group legacy of ecological and epidemiological modelling that most clearly marks the biological sciences at Imperial in the later twentieth century — the era of theoretical ecology.[42] By the early twenty-first century, however, the momentum had shifted to molecular biology, an area that is growing remarkably quickly in the post-genomic age.[43] The explosion of new information has led also to the rise in importance of bioinformatics, a field led at Imperial by Michael Sternberg.[44]

The Physical Sciences

The departments of chemistry, physics and mathematics were all affected in varying degrees by the economic problems of the 1970s and 80s. When Professor Wilkinson retired as head of department in 1988, chemistry was running a deficit which later accumulated to about £3m. This was largely a consequence of changes to the way in which funding was managed — changes discussed in chapter fifteen. The old UGC had eleven different bands for the funding of students; chemistry was in the top band since it costs more to run undergraduate chemistry courses than most others. During the cutbacks, and with the introduction of HEFCE, the number of bands was reduced to four. While chemistry remained in the top band, the funding differential was much reduced. Without a radical change in undergraduate teaching patterns, a deficit was almost inevitable. Indeed, nationally, some chemistry departments were closed as a consequence. At Imperial, chemistry students spend many hours in the laboratory. This has long been a distinguishing feature of the department, a legacy from the days of Hofmann and the Royal College of Chemistry, and one much appreciated by employers in the chemical industry.[45] It was something that Wilkinson, and those who succeeded him as head of department, were unwilling to give up. There was some pressure to use new computer technology and to install virtual laboratories, but this was dismissed as a bad idea. Eric Ash allowed the department to run a deficit but nonetheless encouraged reform. Organic chemist Steven Ley became head of department in 1989 but moved to a chair at Cambridge in 1992.[46] At roughly the same time Professor Mingos, who had succeeded Wilkinson as head of the inorganic chemistry section, left for Oxford. It was therefore a good time to restructure the department and to think both about research and undergraduate education. Together with Tony Barrett, head of the organic chemistry section, and Vernon Gibson, head of the inorganic section, the head of department, David Phillips, finally brought the old tripartite system to an end.[47] New groupings were formed in synthesis, catalysis and advanced materials, analytical chemistry, interfacial science, theoretical/computational chemistry, and biological chemistry.[48]

This reorganization which brought more section heads to the department had positive consequences also for funding. New links were made to departments in the medical school and to biochemistry and chemical engineering. Chemists are associated also with many of the college and university centres. For example, the Zeneca/SmithKline Beecham Centre for Analytical Sciences where Andrew De Mello has been working to develop 'lab-on-a-chip' analytical methods with medical applications. Tim Jones, head of the electronic materials group, became Director of the Centre for Electronic Materials and Devices, and Co-Director of the London Centre for Nanotechnology with which De Mello is also associated.[49] Andrew Miller is Director of the Genetics Therapy Centre, supported in part by the Mitsubishi Chemical Corporation.[50] From these few examples, it should be clear that industry responded well to the new departmental groupings. The hope that the new arrangements would help bring the department's 5 rating in the 1996 RAE up to a 5* in 2001 was realized. People who did not fit the new scheme, or who were not productive in research, had by then left the department. The older divisions of inorganic, organic and physical chemistry still made some sense at the undergraduate level where basic ideas are taught but, there too, major restructuring has occurred. The undergraduate course became a four-year course, in part to compensate for a de-emphasis on mathematics and the quantitative aspects of chemistry in secondary schools.[51]

Physics, too, experienced serious financial difficulties in the 1980s. The department had a high staff to student ratio and its staff was seen as not bringing in sufficient research funds.[52] Despite this the department had great reserves of strength and, with several distinguished scientists on its staff, was able to weather the financial storms. Head of Department, Tom Kibble, showed good leadership, and schemes such as the advanced fellowships set up by the research councils allowed some new talent to enter.[53] The economy improved in the 1990s though there was still some difficulty attracting good physics students. The ratio of acceptances to applicants was close to one early in the decade.[54] In 1990 David Blow was appointed head of department (The Blackett Laboratory) in succession to Professor Kibble.[55] He was followed in fairly quick succession by David Southwood, John Pendry, and Peter Knight.

Several major intellectual lineages remained visible in the department during this period. As we have seen in earlier chapters, optics has a long and proud history at the college reaching back to the spectroscopic work of Norman Lockyer in the RCS, and to technical optics introduced during the First World War. By the late 1980s there were two large optics groups: applied optics, and laser optics and spectroscopy.[56] Between them they included theorists and experimentalists in the areas of electron optics, lasers and photonics. The laser group was, and remains, the largest laser related physics group in a British university. It has had many connections to similar groups in the European Union, from where much of its funding has come. Peter Knight joined the applied optics section in 1979 where he worked in the area of theoretical quantum optics and strong field physics.[57] He is also part of a larger group at Imperial interested in quantum computing and information processing. A legacy from the days of George Thomson is the continuing work in plasma physics, led for many years by Malcolm Haines.[58] If controlled nuclear fusion one day becomes a viable energy technology, the efforts of this group will, in part, be responsible. One of its members, David Potter, left to found the successful company, PSION.[59] Butler's legacy in the area of high energy physics can be seen in continuing work on particle detectors.[60] Another section, the space and atmospheric physics group, emerged largely from Professor Elliot's old cosmic ray group, but can trace some of its history also to earlier work in meteorology.[61] Overall, this section has been heavily involved in space instrumentation and, for much of this period, was headed by David Southwood. Southwood was seconded from Imperial and became Director of the ESA space science programme in 2001. He has played an important role in the joint NASA/ESA Cassini-Huyghens mission to Saturn and Triton.[62]

Two interests of the solid-state group founded earlier by Professor Coles, superconductivity and magnetism, came together in response to the 1986 discovery of high-temperature superconductors. David Caplin gained support for an interdepartmental centre for high-temperature superconductivity which was set up in 1987.[63] In part because of its tradition in condensed matter work, and in part because of growth in this new field, Imperial was a natural choice for the SERC Interdisciplinary Research Centre in Semiconductor Growth and Characterisation which opened in 1988. This University of London Centre, mentioned above, had major input also from the departments of chemistry and materials. Its first director, Bruce Joyce, developed the method of molecular beam epitaxy, used for growing thin molecular, metal, or semi-conductor layers on surfaces, and with application to the construction of a range of electronic devices. Highly successful, the centre connected also to other departments and, within physics, to the work of Professors Pendry and Stradling.[64] In 1997 the physics department, together with the departments of chemistry and electrical engineering, set up a new Centre for

Electronic Materials and Devices which replaced the one in semiconductor growth and characterization. Gareth Parry came from Oxford as Director.[65] Donal Bradley, son of a former head of department, joined the Centre when he returned to the department in 2000 as Professor of Experimental Solid State Physics.[66] While biophysics under David Blow and K. H. Ruddock was another successful area, the kind of expertise needed for research in biophysics was no longer commensurate with that needed in basic physics teaching.[67] The field, which had largely begun with crystallographic work, was becoming more biological in nature by the late twentieth century. The department attempted to keep biophysics, but the section moved to the new department of biological sciences in 2001.

In 1998 a bust of Patrick Blackett was unveiled in the physics department by the President of the Royal Society, Sir Aaron Klug.[68] As the above discussion illustrates, despite the many innovations since his time, Blackett's legacy in shaping the department could still be seen at the turn of the century.

When Eric Ash became Rector he introduced a new formula for the departmental vote which probably affected mathematics more than any other department. At the time, the department was taking in about sixty students a year and had close to that many staff members. According to Ash's formula, staff to student ratios weighed heavily, as did the ability to bring in outside funding, something the mathematicians, working largely individually, were then rather poor at. The result was that internal funding for the department declined dramatically. The department was unable to replace its staff, renovate its space, or even pay for basic supplies. The problem was only resolved when it began to market itself in schools, and started to take in more students. Within a few years the annual intake had almost doubled. By the early 1990s, the department had done so well in attracting students that it was considering raising entry standards, while at the same time extending its search for students elsewhere in the European Union. By the early twenty-first century the mathematics department had an annual intake of over 200 students and was offering nine different BSc courses. It had recovered from its slump.

To achieve this recovery, not only was an increase in the number of students necessary, mathematical training had to be rethought with an eye to the job market. One major success has been in the area of financial mathematics which has attracted funding also for a range of interesting research projects. When statistician Adrian Smith became head of department in 1990, he continued the promotion of financial mathematics begun earlier by Professor Cox.[69] Links were made to what became the Centre for Quantitative Finance under Gerald Salkin and Nicos Christofides.[70] Another who has worked increasingly in this field in recent years is applied mathematician Colin Atkinson.[71] Work in the area of stochastic analysis increased, with some applications also in the financial mathematical area.[72] Terry Lyons, working in real (including stochastic) analysis, attracted many research students before leaving for a chair at Oxford in 2000.[73] When Smith resigned to become Principal of Queen Mary College, he was succeeded as professor of statistics by David Hand, another who works in areas related to finance.[74]

In applied mathematics, too, a new generation came to the fore. Philip Hall, a specialist in fluid dynamics who collaborates also with people working in bioengineering, became head of the department in 1998.[75] His successor, John Elgin, is also from the applied mathematics section.[76] Among those working in numerical analysis are Jeff Cash, Gerald Moore and John Barrett.[77] The mathematical physics legacy of Harry Jones and E. P. Wohlfarth in solid-state (condensed matter) physics continued under David Edwards until his retirement.[78] Today, however, the mathematical physics group is more diverse and less concentrated than earlier in condensed matter work.

As mentioned in chapter ten, Harry Jones was at first reluctant to allow pure mathematics into the department. But, after the appointment of W. K. Hayman, research in pure mathematics gained ground and several other appointments were made. Pure mathematics, with an emphasis on complex variable theory, became very strong. By the mid 1980s, with the retirements of Professors Hayman, Clunie and Reuter, the field needed new leadership. A new generation of algebraists, including Gordon James and Martin Liebeck, soon made its way.[79] So, too, did analysis specialists, Alexander Grigor'yan and Boguslav Zegarlinski.[80] Joining the pure mathematicians, Simon Donaldson became head of the geometry and topology section in 1999.[81] Early in 2005 the new Institute for Mathematical Science was launched with Professor Donaldson as President. Situated in old mathematics territory on Exhibition Road (in a building vacated by the Management School but earlier occupied by mathematicians), the Institute will provide facilities for a number of interdisciplinary teams applying mathematics to areas such as biological systems, laser optics, medical imaging, efficient hospital management, stock exchange computing, and financial management.[82] Within the mathematics department the old individualism has given way to a much more collaborative stance.

Geology, Earth Resources Engineering, Materials

In 1985 these three departments were in the Royal School of Mines; by 2001 they had become part of the new Faculty of Engineering. As mentioned in chapter thirteen, geology was already experiencing some difficulties in the 1970s. The department did not have a strong central core that could, perhaps, have allowed it to weather the storms and cutbacks of the 1980s. There were turf wars in the areas that geologists occupied, and J. L. Knill, who became head of department in 1979, stated that because of the diverse interests there was also a lack of communication within the department.[83] Indeed, the range of specialization built up earlier by Professor Sutton within the mining geology, petroleum geology, engineering geology, geophysics and geochemistry sections, had a centrifugal effect. People in these various fields were often closer to those working in other departments, or in industry, than they were to each other. The department faced two external reviews. The first, in 1982, was internal to the University of London and resulted in several mergers, and the closure of some geology departments. However, at that point Imperial's department was left alone. In 1986–7 there was a national review conducted by the Earth Sciences Review Committee of the UGC. Its first chairman was Ronald Oxburgh, then still at Cambridge. The Committee's report recommended that there be about ten large earth sciences departments in the country, with some minimal training being available at all universities. Some of the larger departments were to be well funded for teaching and research at all levels, others for more limited research, and for teaching also at the MSc level. The UGC largely accepted the report and Imperial's department had to compete with others wishing to be designated a major research department.[84]

Imperial's bid was unsuccessful and the department was told that it should become a medium-sized department, though with some PhD work supported. The UGC (by then UFC) decided on a 30% cut in the departmental budget, entailing the loss of about ten academic and fifteen technical and clerical staff.[85] This was a major blow to a department with a proud history, and one which had seen itself as among the best in the country. Eric Ash did his best to counter the decision since not being in the top rank of research departments was unacceptable at Imperial. Ash wrote to Sir Peter Swinnerton Dyer, chairman of the UFC, complaining that a review committee composed entirely of pure geologists, albeit very distinguished, had wrongly assessed a department renowned for its work in applied geology.[86] But Swinnerton Dyer backed the committee, claiming that Imperial's department did too much contract work (its contract and external research income was then the third highest in the country) and too much private consulting. There was some logic to the UFC claim that its role was to support academic research and teaching, and that if Imperial wanted to undertake contract research then industry should pay.[87] Ash was later quoted as saying that the Earth Sciences Review was a 'total disgrace', that the committee had made up its mind about the Imperial department before visiting, and had then spent only two hours at the college.[88] This was true, but what then occurred was less a disgrace than a consequence of differing philosophies of education and, on this occasion, possibly wrongly, Imperial's tradition worked against it. Knill resigned in 1988 to become chairman of NERC and Ash did his best to prevent the department from shrinking too far. Together with new department head, Richard Selley, he sought industrial support.[89] To a degree they were successful and some new appointments were made.[90] But the department lost six of its professors to retirement or resignation in this period and, worse, lost good younger people who moved to positions elsewhere.

Selley and his successor, Michael Worthington, did their best to rebuild but were perhaps demoralized by what had occurred.[91] By 1990 the department had trouble with falling enrolments, and its 'pure' geological core had shrunk so far that it looked to Natural History Museum personnel for help in teaching its courses.[92] It was only a matter of time before some major restructuring would be needed. Worthington sensibly wanted a merger with earth resources engineering but, in the mid-1990s, the engineers refused. Reflecting geology's woes was a 4 rating in the 1996 RAE. In 1997 the department was relocated, along with the department of earth resources engineering and a new Centre for Petroleum Studies, in the short-lived T. H. Huxley School for the Environment, Earth Science and Engineering where, as mentioned above, further cuts were made internally by the Director, John Beddington. Despite the earlier shrinkage, the geology department was hard pressed to avoid running a deficit. Even after the major national reorganization in the earth sciences there were perhaps still too many departments seeking the available funding.[93] The many problems could not easily have been anticipated. As discussed also in chapter thirteen, they are only too clear in retrospect. In 1998 the decision was finally taken to merge the geology department with earth resources engineering and the newly associated departments were administered by David Sanderson, the H. H. Read Professor.[94] Three years later the T. H. Huxley School was broken up and the new Department of Earth Science and Engineering joined the Faculty of Engineering.[95]

A parallel tale can be told of the department of mineral resources engineering, renamed earth resources engineering in 1995. It was the descendant of the old department of mining and it, too, was in difficulty during the 1980s and 90s. Brian Flowers had managed to save it during his rectorship, and its earlier name change was symbolic of that. In the late 1980s the department listed its research interests as being in oil and natural gas engineering, mine design, rock mechanics, geostatics and computer assisted design, hydrometal and surface science, and quarry/ environmental engineering. The second name change symbolizes the restructuring that continued into the 1990s when the department was focussed in both the environmental and oil and gas engineering fields. The department served the increasing demand for environmentally sensitive practice in mining, oil and gas extraction, and in quarrying.[96] Given the world demand for energy, it is not surprising that there was some renewed interest also in coal deposits, especially those in China and India where some opportunity for environmentally informed work might exist. Some older related areas, such as rock mechanics, also continued.[97]

Back in 1985 E. T. Brown was head of department.[98] He was succeeded in 1987 by J. S. Archer a specialist in petroleum reservoir engineering who oversaw much of the departmental restructuring.[99] Among the problems Archer faced was the difficult decision of whether to close down the undergraduate course in mining engineering. The possibility of cooperation with a number of other European partners was considered; but the BSc mining engineering course was, in the end, closed and a short-lived MEng course in environmental and earth resources engineering opened in 1993. The MEng in petroleum engineering continued and

Imperial maintained its traditional strength in this field. As mentioned above, those with related interests across departments came together in the new Centre for Petroleum Engineering which opened in 1996.

The department ran a deficit in the mid-1990s and its research income fell, despite a relatively good 5 rating in the 1996 RAE. Contributing to the financial difficulties was a decline in undergraduate enrolment. There was a problem in finding a permanent head after John Archer became Deputy Rector in 1994 and J. D. Woods was brought in from outside.[100] But the deficit continued and the department plan for 1996–7 was not accepted by the college administration. Dean of the RSM, Rees Rawlings, came in as Acting Head in 1996.[101] He was asked to rationalize the staffing level and put the department on a reasonable academic and financial footing. Morale was low also because of heavy teaching loads which were seen to interfere with research and the possibility of departmental recovery. Rawlings helped to restructure the courses so as to alleviate some of the problems. He promoted cooperation between the departments in the RSM and supported the creation of the T. H. Huxley School. As noted above this was only a temporary solution. How the recently combined Department of Earth Science and Engineering will fare in the new Faculty of Engineering remains to be seen. But it appears that things are looking up; the first head of the merged department, geophysicist M. R. Warner worked to rebuild geological and engineering strength.[102] Undergraduate degrees are offered in a range of geological and geophysical options, and masters degrees are offered in petroleum engineering, petroleum geoscience, and metals and energy finance.[103] Further evidence of recovery is that the merged department received a 5* in the 2001 RAE.

In chapter thirteen it was shown how the old metallurgy department evolved into the department of materials. In 1985 the head of department was D. W. Pashley.[104] Work continued along the lines developed earlier by Professors Ball, Swann and Steele. Swann left for the United States in 1977, but work in electron microscopy continued under H. M. Flower and F. J. Humphreys.[105] Steele continued working in the ceramics/fuel cells area. Working with him was J. A. Kilner, a future head of department.[106] As elsewhere in the college, materials scientists turned increasingly to theoretical work when money was tight in the 1980s. There was much modelling of materials microstructures, such as in polymers and metal and ceramic matrix mixtures. However, this work was not yet sufficiently sophisticated to obviate the need for experiment, and actual industrial testing continued. In 1983 the Centre for Composite Materials was founded, and by the end of the decade much departmental research was associated with it. Good industrial links resulted in some large projects coming to the centre. For example, one on the prediction of fatigue in new materials, funded by British Aerospace, Du Pont, Rolls Royce, Shell and Westland Helicopters. A range of short post-experience courses was offered, designed to help people from industry update their skills and knowledge, and an MSc course was offered first in 1988. Following the example of Professor Ball, the centre ran a highly successful seminar series which attracted many visitors. Also associated were people working in aeronautical structures and materials and F. L. Matthews, professor of composite materials in aeronautics, later took over directorship of the centre. By the late 1990s the centre was doing less well than earlier since the market for its MSc courses declined.

Malcolm McLean succeeded Pashley as head of department in 1990, coming to Imperial after many years at the National Physical Laboratory.[107] 1990 was also the year in which a new four-year BEng course, including a year at a continental European materials department, was introduced. By the end of the decade the department moved towards four-year undergraduate degree courses for all students. A new MSc course in corrosion of engineering materials was offered in 1993. In the College Plan of 1991–5, the department listed its interests as: new alloy preparations, technical ceramics, metal ceramics matrix composites, and semiconductor materials. Joining the department in 1995, as Professor of Ceramic Materials, was Larry Hench. Hench is a leader in the development of prosthetics materials and is Co-Director (with Professor Julia Polak) of the Tissue Engineering and Regenerative Medicine Centre.[108] Tissue engineering was an exciting new field, but the department was running a deficit in the 1990s and needed more space so as to be able to expand into new areas. Early in the twenty-first century expansion became possible. The department left the Bessemer Laboratory, its home for close to a century, and moved to new premises in the refurbished Goldsmiths' Extension.[109] Reflective of new directions, an undergraduate degree course in biomaterials and tissue engineering was first offered in 2003, with input also from the Faculty of Medicine and the Department of Chemical Engineering and Chemical Technology.

Engineering and Computing

Aeronautics

Brian Flowers appointed G. A. O. Davies to succeed P. R. Owen as head of the department of aeronautics in 1982.[110] Owen's legacy was long-lasting in that Davies and the two successive heads, P. W. Bearman and J. M. R. Graham, were all his protégés.[111] The department continued with its two major fields, aerodynamics and aerostructures but, within these, a number of specialist sections arose. As noted in chapter ten, aero-structures was left somewhat in disarray after the departure of Professor Argyris and it fared poorly in the tough economic times of the 1980s. Things began to turn around with the 1989 appointment of Michael Crisfield to the chair of computational mechanics.[112] Crisfield, an expert in finite element methods, rejuvenated the section before his death in 2002. On the theoretical side, new computing facilities, including high-speed parallel computers, enabled complex modelling work to be carried out. On the experimental side, new testing equipment was acquired, including a major water flume facility for experiments on current and wave action, structural load and fatigue testing machines, impact rigs, and a gas gun for high speed impact studies. There was a new composite materials laboratory and, as elsewhere in the college, such materials became a focus of study.

Aerodynamics fared better than aerostructures during the 1980s. Early in the decade the Honda Automobile Company approached the college offering to pay for the construction of a large wind tunnel in exchange for the testing of its models, and the training of its staff in wind tunnel technology. Honda had been unable to design cars with the petrol efficiency of its competitors; hence one of the largest contracts ever, between the Japanese automobile industry and an academic institu-tion, came to Imperial. Peter Bearman led the design and construction of what was then one of the world's largest wind tunnels.[113] It opened in 1985 and had a moving floor used to simulate automobile motion. Model cars of one-third size could be tested in it. In part because of the work carried out at Imperial, Honda began to flourish in the 1990s and some of the young engineers who were trained at Imperial later held senior positions in the company. Aerodynamic work continued also in other areas such as wind turbines, a major interest of Michael Graham. While the department remained not much focussed on the aircraft or aerospace industries, a hypersonics group was founded by J. K. Harvey, a specialist in gas dynamics and high-speed flow. This group, partially funded by the defence industries, continues to work on rocketry and related topics.[114] The present (2006) head of department, Richard Hillier, the first since Owen not to have been one of his protégés, is a member of the hypersonics group.[115]

British Aerospace representatives came to the college in the early 1990s wanting Imperial to become a centre of excellence in aerospace. But since the industry had in mind a contract model in which money would flow to the department only for specific projects, the offer was refused. At the start of the period under discussion here the department attracted fewer women than the other engineering departments but, by the end of the century, it had about 14.7% women (undergraduates and postgraduates) placing it marginally higher than computing and mechanical engineering.[116] The department has many overseas students, including some of China's best aeronautics graduates sent to Imperial for work at the PhD level. The department has scored consistently well in the RAE, gaining a 5* rating in the last two rounds.

Civil Engineering

As we have seen in chapter thirteen, civil engineering underwent some major restructuring after the retirement of Professor Skempton and, as a result, was relatively well positioned by the mid 1980s. The department has performed consistently well in the RAE and has been placed at the top, or near the top, in a range of league tables. The older fields of structures (notably steel and concrete[117]) and soil/geo mechanics remained strong, and there was continuing interest in the areas of geotechnical and seismic engineering.[118] In the late 1980s and early 90s much attention came to the college because of the work of John Burland and his team in soil mechanics. In the late 1980s Burland helped with the construction of the Queen Elizabeth II Conference Centre in Westminster, built on complicated foundations due to differing rates of settlement around the site. Burland's expertise was needed also in connection with the construction of a large underground car park at the Palace of Westminster. There was concern over what effect the excavations might have on the stability of the Parliamentary Clock Tower (Big Ben). Monitoring by Burland's team showed a seasonal tilt, but demonstrated that the tower was not in danger. By the early 1990s the team was involved with tunnelling for the Jubilee Line. This time there were problems in connection with the tower, and tunnelling had to stop until Burland's team put the proper measures in place.[119] This work is reminiscent of that carried out by college-trained engineers on tunnelling for the London Underground in the 1920s and 30s.[120] Burland is perhaps best known to the general public for his work in helping to save the Leaning Tower of Pisa from collapse.[121] Unlike Professor Bishop, his predecessor in the chair, Burland is a theorist and his work bridged into the computer age from older methods of calculation. His successor as head of the soil/geo mechanics section, David Potts, is fully immersed in the computer modelling era.[122]

The channel tunnel provided a range of work, not just for civil engineers but for a number of other college specialists. In addition to providing some tunnelling and materials expertise, civil engineers were funded by the engineering firm Wimpey to carry out research into air pollution problems in the tunnel. As discussed in chapter thirteen, environmental problems more generally came to the fore in civil engineering and the department was renamed Civil and Environmental Engineering in 1998. In the early 1980s Professor Perry set up the Centre for Environmental Control and Waste Management. Later in the decade civil engineers were helping Oxfam in the provision of simple transportable clean water supply systems to be used in parts of Central America and in parts of Africa.[123] Gradually hydrology took over from hydraulics in having more importance in the department.[124] After Perry's death in 1995, the field was developed further by his successor as head of the environment and water resource engineering section, Howard Wheater.[125] The section's focus on the water cycle, sustainability and toxic waste has served it well. Many contracts, and much research funding, have come its way.[126] An Imperial team will be working on water conservation and reuse with the developers of the site for the 2012 London Olympics.

The new emphasis on environmental engineering with the ensuing appointment of chemists and other specialists caused some problems for undergraduate teaching and, as a consequence, some tension between new and more traditional engineers. Similar problems, now largely resolved, arose in connection with the transport and design research areas. In 1990, when the Rees Jeffreys Chair of Transport Engineering was established, Professor Ridley joined the department on a permanent basis. In

1992 he founded the Railway Technology Strategy Centre which drew new types of specialists to the department.[127] For example, in 1998 economist Stephen Glaister was appointed Professor of Transport and Infrastructure and later succeeded Ridley as Director of the Centre.[128] Glaister knows the transport area well and has generated support for a range of research projects. In addition to his work on railway infrastructure, he has worked in the area of road charging and has supported the congestion charge in London. His team has followed the consequences of the charge for commuting times and noise levels on London streets, and found them both much improved. Recent emphasis on design and materials has similarly brought in new types of expertise. A chair to promote the role of design in civil engineering was funded by the Ove Arup Foundation, and Chris Wise was appointed in 1998.[129] Wise, a brilliant designer, is bringing new and creative ideas to the department, and is playing an important role also in undergraduate teaching, and in developing a twenty-first century curriculum for civil engineering. Civil engineers are paying more attention than earlier also to how buildings are used. They have moved into areas such as the management of the occupancy of large and tall buildings, from both the safety and functional angles. Thinking along these lines was stimulated already by the Ronan Point disaster of 1968, but was renewed by the more recent 9/11 (2001) terrorist attack on the Twin Towers in New York.[130]

Patrick Dowling, head of the steel structures section, was appointed head of department in 1985.[131] After he left to become Vice-Chancellor of the University of Surrey in 1994, the headship went through several changes before the current head, David Nethercot, another steel structures specialist, was appointed in 1999.[132] More generally, the department had some difficulty retaining its staff during this period since many were offered well-paying jobs elsewhere. It adopted the practice of giving bonuses to those bringing in large contracts or major research funding.[133] By the late 1990s the department was taking in about 65–70 undergraduate students a year, and placing much emphasis also on its one-year MSc courses. It appears to have given more of these than any other department. Late in the century about 150 MSc students were enrolled each year.[134] Recently, however, the EPSRC has withdrawn some funding for these students, stating that more support should come from industry. Overall, civil engineering has been a very successful department. It is, however, very much a collection of sections with seemingly little contact between them. For some reason, unlike for geology, a decentralized structure has worked well for this department.

Mechanical Engineering

Owen Saunders restructured the mechanical engineering department after the Second World War. His legacy was still visible at the end of the century.[135] Saunders' own areas of combustion, heat transfer and applied thermodynamics continue in importance today. After his retirement, work in related areas was carried forward by Professors Spalding and Whitelaw until their respective retirements in 1988 and 1999.[136] Spalding's legacy was also in the computational area. Work in fluid mechanics continues across several departmental sections, as does work in the areas of plasticity and lubrication/tribology, though all much transformed since Saunders' time. The tribology section is now under Hugh Spikes[137]; and plasticity, Hugh Ford's early area of interest, has become part of the strengths of materials section. This was headed for fifteen years by one of Ford's former students, Gordon Williams, until he became head of department in 1990. Williams is a theoretician with interests in metal plasticity and polymer fracture mechanics, and is representative of the turn to theory and computer modelling in this period.[138] The department has also moved more generally into materials science and has a good adhesives section under Tony Kinloch.[139] The only major loss in expertise has been in the area of nuclear power; the remnants of this group were moved to CARE and ICCET. When, at the end of the century, Williams and several others professors retired, there was no clear internal successor as head of department. Oxburgh brought in R. A. Smith as the new head in 2000.[140]

As discussed in chapter thirteen, Ford was concerned that the growth of engineering science, albeit necessary for research, was having a negative effect on undergraduate pedagogy. He held what was, perhaps, a dated view on the kind of work Imperial graduates might expect. But, like many others at the time, he believed that students needed more practical experience and a broader education than was then being offered. The tension between theoretical and practical interests is still detectable in today's pedagogical debates, despite everyone knowing that a grounding in both theory and practice is essential. As to practical pedagogy, it is interesting that where once mechanical engineering students would test new theoretical ideas in the laboratory, increasingly they do so on site — in factories, on oil rigs, and so on. Computer simulations have replaced much of the testing machinery seen earlier in the department. Ford was concerned with educating people to help revitalise Britain's manufacturing industry. Today's students are taught to think about adding value to industry at a global level, and that while selling added value is important to Britain's economy, where it is sold is less so.

Mechanical engineering comprises a set of core disciplines and the sections within the department reflect this. But the department has spawned new interests that cross traditional disciplinary boundaries as, for example, biomechanics discussed in chapter thirteen. More recently the mechatronics section was created which, while still relatively small, brings together mechanical engineers with electronics specialists. The section's main interests appear to be in medical applications, and in hybrid powered vehicles. Given the booming sales in electronic devices with mechanical parts, mechatronics should grow. While mechanical engineering staff are associated with a number of centres created elsewhere in the college, few centres appear to have been created within this department — perhaps because its research has always been well funded. There are one or two exceptions, such as the Centre for Vibration Engineering which opened in 1990 under the direction of David Ewins, head of the dynamics section.[141] It is interesting to find some members of this department voicing support for the further collapse of departmental boundaries, and for the creation of a large school of engineering. This was an ideal of Brian Flowers, but one that during his period as Rector was impossible to achieve.

As to teaching, the department is now seen as having a good balance between engineering science, theory and practice. Students are introduced to design problems already in their first year. Recently there has been a strong interest in racing car design, and student teams have won a number of international competitions. But, as was earlier the case in civil engineering, ideas of conservation and sustainability are slowly making their way into mechanical engineering pedagogy. As a consequence, student activity of this type may well become less fashionable in the future. There is a greater focus on providing undergraduate education than in the department of civil engineering, and about 125 students now enter each year. But the department offers fewer MSc courses, taking in only about 50 students a year. While the department appears conservative in its tendency to build on its traditional strengths, the conservatism appears to have served it well. Perhaps there was a lesson in the earlier turn to robotics — an attempt to be fashionable that largely failed. Overall, however, the research sections have done very well and the department has received a 5* rating in recent RAEs.

Electrical and Electronic Engineering

The electrical engineering department was renamed electrical and electronic engineering in 1991. By then the old section of heavy electrical engineering was much diminished. During the 1970s the department still had strength in electrical machines, power systems and power distribution. Eric Laithwaite had led in some of these areas but, by the 1980s, these fields were less fashionable than they had been in the 1960s. When Bruce Sayers became head of department he wanted to move the department more in the direction of informatics and, to a degree, was successful. In pursuing the path he did, Sayers was in line with international trends. However, late in the century a small section associated with the control and power group still dealt with 'heavy' work, and recently there appears to have been a small resurgence in the field.[142] The legacy of light electrical engineering, which in the early days had been largely focussed on wireless and telephone communication, has been more lasting.[143] Professor Westcott brought his control group back to electrical engineering from the computing department in 1980. There he was joined by D. Q. Mayne, Professor of Control Theory, a major theorist who was to succeed Sayers as head in 1984. David Limebeer, a future head of department, joined Mayne's control group in the same year.[144] By the late 1980s Westcott and Mayne had built a fine international reputation for theoretical control at Imperial. Mayne and some others were associated also with chemical engineers in the London Centre for Process Systems Engineering.

Professor Cherry's influence was still alive during this period, though his group split apart after his death and a younger generation began work in a range of information technology related areas. His former student, Robert Spence, led a group on human-machine communication and analog circuit design.[145] A. G. Constantinides set up a digital signal processing group and now heads a combined communication and signalling processing group.[146] Another person much influenced

Illus. 96–97; top, physics professors in 1964; front row from left, M. A. Salam, C. C. Butler, P. M. S. Blackett, W. D. Wright, H. Elliot; back row from left, P. T. Matthews, M. Blackman, J. D. McGee, B. J. Mason. Bottom, Professor Harry Elliot (left) and Professor David Wright looking at a model of the first international space satellite, Ariel 1, in 1962, the year in which it was launched. Elliot contributed to the British instruments on board.

Illus. 98–99; top; the construction of new buildings ongoing in 1963. The main section of the mechanical engineering building can be seen in the centre of the picture; bottom, the Steam Plant Laboratory, 1961, located in the first new section of the Mechanical Engineering Department. In 1989 this space was converted for use by the Leon Bagrit Centre for Biological and Medical Systems (later, Department of Bioengineering).

Illus. 100 – 101; top, Southside in 1963 shortly after opening; bottom, Southside Dining Room in 1963.

Illus. 102–103; top, the opening of the College Block (later Sherfield Building) in 1969 by Queen Elizabeth II. On her left the Rector, Lord Penney, and on her right ICU President, Piers Corbyn; bottom, the Great Hall with freshers being welcomed by the Rector (not in picture) in 1984.

Illus. 104–105; top, Professor David Mayne showing Prime Minister Margaret Thatcher around the Department of Electrical Engineering during Tech 2000 (held in 1985); standing behind Thatcher is Professor E. M. Freeman. In the background is an illustration of an invention of Professors R. Spence and M. Apperley in the field of human-computer interaction, the 'fisheye lens'; bottom, view of the Queen's Tower and the Electrical Engineering Department (when its tiles were being replaced) in 1979. The picture was taken from the top level of the mechanical engineering building.

Illus. 106–107; top, engineering professors on their way to the annual dinner for new Fellows of the Royal Academy of Engineering, 1993. Posing with C&G mascot Boanerges, from left, Roger Sargent, John Perkins, Bob Spence, John Archer, and Igor Aleksander; below, the Chaps Club after their Derby Day outing, 1983 (Royal Albert Hall in background).

Illus. 108–109; top from left, HRH the Princess Royal (Chancellor of the University of London), Professor Bryan Coles, Lord Flowers (Vice-Chancellor of the University of London), John Smith (College Secretary), and John Robinson (Chairman of Imperial Activities Ltd. (IMPACT)) at the opening of the Technology Transfer Building at Silwood in 1988. Bottom from left, Professor Patrick Dowling, Sir Eric Ash and Professor John Archer at the Royal Academy of Engineering Soirée Exhibition held at Imperial College, 1993.

Illus. 110–111; four physics professors, top from left, Tom Kibble and Malcolm Haines; bottom from left, Sir John Pendry (first Principal of the Faculty of Physical Sciences) and Sir Peter Knight (first Principal of the Faculty of Natural Sciences).

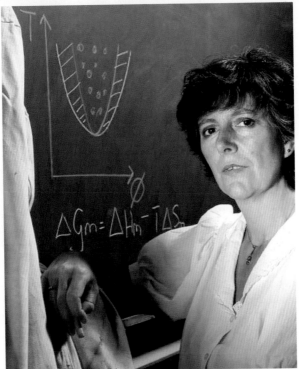

Illus. 112–113; top, Chemical Engineering and Chemical Technology Christmas Party, 1990; from left, Professor Bill Wakeham and Professor Geoffrey Hewitt; bottom, Professor Dame Julia Higgins in her laboratory in the early 1990s. The diagram and equation on the blackboard relate to her work on polymer mixtures.

Illus. 114–116; top, biologists: left, Professor Sir Roy Anderson and right, Professor Michael Hassell. Bottom from left, Lord Flowers, Lady Flowers, Sir Gordon Conway, Professor Nigel Bell and Professor John Beddington on the occasion of the 25th anniversary of the Centre for Environmental Technology (ICCET) in 2002.

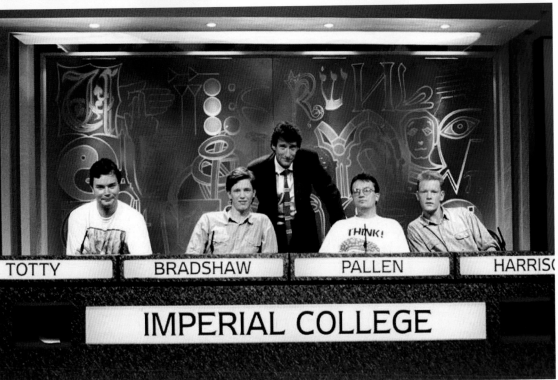

Illus. 117–118; top, the RSM Rugby team victorious in the Bottle Match, 2002; bottom, Imperial College team wins University Challenge, 1996. With Jeremy Paxman from left, Jim Totty, Nicholas Bradshaw, Mark Pallen and Chris Harrison.

Illus. 119–120; top, the opening of the Sir Alexander Fleming Building in 1998 by Queen Elizabeth II with Lord Oxburgh, centre, and members of the Fleming family at right; bottom, the front of the Sir Alexander Fleming Building with reflection of the Queen's Tower.

Illus. 121–123; top left, student band on Queen's Lawn, 1988; top right, Commemoration Day, October 2005; bottom, engineering students with C&G spanner mascot join the Lord Mayor's Parade, 2005.

Illus. 124–125; top, Governing Body Meeting, 1995, held in Board Room, 170 Queen's Gate. Note the portraits of former Rectors around the walls. Bottom, three rectors in 2000: Lord Oxburgh (centre) with his predecessor, Sir Eric Ash, on his left and his successor, Sir Richard Sykes, on his right.

Illus. 126; photograph taken of those present at the last meeting of the 32-strong Council, together with some non-voting senior administrators, February 2005.

Front row from left, Professor Geoffrey New, Professor Sean Hughes, Mr. Donald Hearn, Sir Richard Sykes (Rector), Dr. Eileen Buttle (Deputy Chairman), Lord Kerr (Chairman), Mr. Ram Gidoomal, Professor Nigel Bell, Sir Peter Williams.

Middle row from left, Dr. Martin Knight (Chief Finance Officer), Dr. David Wilbraham, Dr. Caroline Vaughan, Mr. Jeremy Newsum, Dr. George Gray, Mr. Tony Mitcheson (College Secretary and Clerk to the Court and Council), Mr. Chris Gosling (Director of Human Resources).

Back row from left, Dame Rosemary Spencer, Dr. Tidu Maini (Pro-Rector, Development and Corporate Affairs), Professor Dame Julia Higgins, Mr. Richard Walker (Deputy President, ICU), Mr. Mustafa Arif (President, ICU), Professor Rees Rawlings, Mr. David Brooks Wilson (Director of Estates), Professor Julia King (Principal, Faculty of Engineering), Mr. Vernon McClure (Academic Registrar), Dr. Rodney Eastwood (Director of Strategy and Planning).

Illus. 127–128; top, demolition of Southside, December 2005; bottom, new entry to the College from Exhibition Road. The Department of Mechanical Engineering is to the left and the Tanaka Business School is to the right of the entrance. The Goldsmiths Extension is on the far right of the picture.

by Cherry's work is Igor Aleksander who moved to electrical engineering from the computing department in 1988.[147] There is some continuity also in the work of Erol Gelenbe who holds the Dennis Gabor Chair.[148] The older applied physics and materials science of Professors Gabor and Anderson continued in the work of Mino Green who formed a Device Science Research Group in the mid 1980s. Work in this field continued under Richard Syms.[149] In the years since Anderson's death there have been multiple applications of his and other earlier work, including the development of thin film transistors, new fibre optics devices and liquid crystal displays. The engineering in medicine laboratory founded by Bruce Sayers moved to join the Leon Bagrit Centre and, as mentioned above, his legacy is now with bioengineering.

By the end of the century, the department was taking in about 150 undergraduates a year and about 65 MSc students, but it was having difficulty in attracting all the PhD students it wanted.[150] Pedagogy, which has been debated within this department ever since the days of Jackson, Gabor, Cherry, and Brown, continues to be a lively topic. The department takes pride in emphasising language skills among its students (the legacy of Sinclair Goodlad), and in interesting its students in areas beyond the strictly curricular. It arranges lectures and seminars on a wide range of subjects. For example the Colin Cherry Memorial Lectures have attracted some well-known speakers over the years, including Ernst Gombrich, Douglas Adams, and Steven Pinker, and have attracted also capacity audiences. The same is true of the more recently founded Dennis Gabor lectures.

Chemical Engineering and Chemical Technology

In 1985 Roger Sargent was still head of the chemical engineering and chemical technology department. During his long tenure he built the department well, resolving many of the difficulties that had existed earlier between chemical engineers and chemical technologists. Sargent was succeeded as head of department by Bill Wakeham.[151] Under his leadership the department continued to thrive; so well, in fact, that it was said to be 'untouchable' in a 1995 article in the *Times Higher Education Supplement*.[152] Given the economic problems of the 1980s, Sargent had some difficulty with staff retention. To help with this problem, he introduced a system of staff exchanges with industry. One result was that Geoffrey Hewitt came to the department, at first for only two days a week.[153] For a while Hewitt retained his position as Director of Research at Harwell but, in 1990, joined the department full-time in the Courtaulds Chair. He appears to have been something of a powerhouse, with a wide range of projects always on the go. Like many chemical engineers he is interested in multi-phase systems and in the many problems they pose for the chemical industry. In that connection one of his group's inventions, which relates also to more general college interests in oil technology, was a device for the measurement of flow rates of different phases (oil, water, gas etc.) from oilfields. Others with interests in the oil industry include Stephen Richardson who became head of department in 2001.[154]

Multiphase problems are of interest also to Brian Briscoe, professor of interface engineering, who developed the materials engineering technology section. Briscoe, a specialist also in tribology, works with complex materials including foodstuffs.[155] Julia Higgins runs a strong research group and is one of many who was drawn to materials science in the 1980s.[156] Her group studies a range of complex materials, notably polymer mixtures. The process systems engineering section remained a major departmental presence in this period. It was run by John Perkins after Sargent's retirement.[157]

John Perkins succeeded Wakeham as head of department in 1996. By then the department was taking in about 75 students a year and was running a four year MEng course.[158] It had two small MSc courses, one associated with the Centre for Process Engineering. Some of the older departmental research areas were running down, including combustion and nuclear technology.[159] Work in nuclear chemistry was finally closed in the late 1990s, allowing space to be redesignated for growing fields such as biotechnology and environmental control. Clean energy production became a major concern. For example, David Chadwick's group in catalytic engineering processes carries out research on new catalysts with application in this and other environmental areas.[160] Biotechnology, too, has spawned a new range of chemical engineering problems. At the turn of the century, the department was planning to enter the field of tissue engineering, making further links to the department of materials and new ones to the faculty of medicine.

Computing

As mentioned in chapter thirteen, Brian Flowers had asked Bruce Sayers to move from electrical engineering to head the computing department in 1983. Sayers, acculturated in a more hierarchical tradition than Professor Lehman, tried to bring a more vertical line of command to the department. This was probably a mistake since computer people worldwide seem to work best in looser, more horizontal, structures. Sayers arrived at a difficult time, in part because of a number of unsettling consequences stemming from the 1981 announcement that the Japanese were embarking on a so-called Fifth Generation of computers. The Japanese predicted that their computer industry would overtake IBM within ten years. In retrospect we can see that little came of this ambition to rule the world of computing. But the claims caused some political panic at the time. The Alvey Committee, discussed in chapter fifteen, recommended a huge increase in the number of IT graduates and that a range of research projects be funded by both government and industry.

As a consequence, things appeared to be looking up for the computing department which, since its foundation, had seen itself as a poor relation at Imperial College. But the prospect of generous new funding led to infighting over what type of research the department should engage in. Sayers was not a computer scientist, though he had some good ideas on how to use computers. As had been the case in his old department, he wanted to see application driven research; but not everyone agreed. Further, the applied work he favoured was not entirely in the spirit of the computer revolution envisaged by the Japanese. A debate over which way the department should move ensued, as those favouring strengthening software engineering, those who had more theoretical interests, and those who were in Robert Kowalski's logic programming section, put forward their ideas.[161] The debates related also to the unfolding history of the discipline.[162] By the 1960s time-sharing operating systems allowed new areas such as computer graphics and artificial intelligence to come to the fore. Attention was directed towards abstract models of computation and the study of computers as a phenomenon. This opened the door for logicians, mathematicians, engineers, and philosophers to approach computing from a number of different angles. Not surprisingly there were disputes over research agendas, and over the best approaches to programming, algorithms and information processing. At Imperial, with its tradition of working also to apply new technologies in industrial and business settings, the situation was yet more complex. The department would have needed a first-rate diplomat at its head to navigate the shoals of the Alvey and Fifth Generation period without internal dispute.

There was a further complication in the later 1980s when the Alvey projects were already underway. Fujitsu, one of the main partners in the Fifth Generation project, offered to set up a High Performance Computing Centre at Imperial.[163] The idea was for work to be carried out on parallel computer applications and to achieve faster speeds for large scale calculations, such as those carried out in weather forecasting and various forms of finite element analysis. John Darlington, a specialist in functional programming, helped design the parallel machine with the idea of grid computing as a future goal.[164] The Japanese were attracted also by the success of the logic programming group under Professor Kowalski, since at the time it was fashionable to approach computing from the artificial intelligence angle that he favoured. With Fifth Generation support Kowalski's group expanded during the 1980s. The rapid rise of what was seen by some as a rather esoteric, albeit fashionable, field did not help the internal departmental dynamic. Another complicating element was the arrival of Dov Gabbay who came to work with Kowalski during his sabbatical leave. Gabbay stayed at Imperial and was given a chair.[165] A brilliant logician, he soon formed his own group, though not without some strife. Together with Barry Richards he also formed a new centre, IC-PARC (see above), which focussed on constraint logic programming.[166] For a while this functioned well, supported by a range of industrial grants, notably from British Airways interested in programming related to routing and scheduling problems.

Tom Maibaum, and later Samson Abramsky, joined Imperial from Queen Mary College. Together they built up the theoretical area that had been introduced earlier by John Florentin.[167] But the largest group in the department was composed of software engineers and much Alvey money came to that area. One research student in this field was Jeff Kramer who was to become a leader in the field of distributed computing and software architecture, and was later head of department.[168] But the department had little space for all this growth and there was increasing tension also over the space taken by IC-PARC in the William Penney Laboratory. Sayers had done well to find funding for this building and, more generally, had put the department on a sound financial footing.

But continuing feuds over future directions made his position untenable. Eric Ash saw the need to replace him when his term as head ended and chose Tom Maibaum, then still a reader, as his successor. At the time, the department described its interests as being in logic and functional programming, engineering software, signal processing and parallel architectures.[169] Research was going well, but internal problems did not go away under Maibaum. Departmental finances became more tenuous during the 1990s, exacerbated by competition with electrical engineering for information technology work. Maibaum resigned in 1996. It was then rumoured that Oxburgh was considering breaking up the department and distributing its staff between mathematics and electrical engineering.[170] In retrospect this seems unlikely to have been the case since the department, despite its problems, was viewed as one of the best in the country. Oxburgh appointed Kowalski as head of department. Kowalski attempted to heal wounds but, unable to manage the department any better than his predecessor, resigned and took early retirement in 1999.[171] He was succeeded by Jeff Kramer and the department moved into a more stable phase. By then the economy had improved and Kramer was able to secure infrastructure funding for a complete refurbishment of the department, and for an increase in staff numbers. IC-PARC, however, began to wind down. Centres which rely on short-term money are vulnerable and need lively MSc courses to tide them over difficult times. Perhaps a course in constraint logic programming could have helped in this regard; but no such course was offered. Today the space once occupied by the centre is being reconfigured for new use. The department has widened its interest and now has five main research clusters. Newer fields include interactive media and bioinformatics.[172] Looking back on the history of this department one can see that it was a child of its time. Its culture is reflective of the 1960s and in many ways it appears more democratic than other departments.[173] It also has something of a 1960's anarchical streak. Nonetheless, the department has always performed well in external ratings such as the RAE.

Conclusion

It is risky to attempt making any serious generalizations on recent historical events but a few tentative comments can be made. This chapter has shown that the merger with St. Mary's, and the later founding of the Imperial College School of Medicine, had a major impact on the kind of funding that came to the college late in the century, and on the kind of research undertaken. New groupings of expertise have emerged, and much medically related work is being carried out in a range of new research centres. Today there are biologists, biochemists, chemists, physicists, chemical engineers, materials scientists, aeronautics and computing specialists, mathematicians, mechanical, civil, and electrical engineers who have research connections to people in the faculty of medicine. New medical research is reliant on knowledge and skills that have been built up at Imperial over many years. Electronic medical devices, medical robots, prosthetics, and molecular medicine all depend on basic research in mathematics, physics and chemistry, as well as on engineering science and technology. Further, new theories and models of disease, infection, and epidemics depend not only on medical, but also on mathematical, biological and biochemical expertise. Synthetic biology, too, is multidisciplinary, drawing on experts from many fields. Imperial is fortunate in having so many specialists in one place and medical research can only profit from its association with Imperial's long tradition in applied science and engineering.

The turn to medicine does not mean that other areas of work are not continuing. Imperial's traditional strengths in many areas of science and engineering are still very evident. As earlier, the computer continues to make its mark, affecting everything that happens in the college. Computer modelling is now central to work carried out in all departments, from the biosciences to engineering. The computer is central also to communication among researchers and to the rise of interdisciplinary work.

A highly volatile external funding climate has, over the past quarter century, prompted much internal restructuring in an attempt to fit in with national and international trends, and to appear attractive to industrial, governmental, charitable and other funding agencies. The earlier transformation of the mathematics department, and of the earth science and engineering fields, are clear examples of this. More recent changes have affected departmental autonomy, a traditional strength at Imperial. In moving forward the college should remember that it has been well served for much of its history by the lack of any well-defined vertical hierarchy, and should perhaps pause to evaluate recent administrative changes before any further action is taken. Whether the new faculties will prove their financial worth remains to be seen.[174]

Not only new faculties, but also graduate schools, new centres and institutes have emerged during this period. All these new groupings have put pressure on traditional departmental activity in the competition for resources. Since the college wants the best students to come to Imperial, it must provide excellent undergraduate courses. To date it has managed to do so, but only just. The college depends heavily on HEFCE funding for its undergraduate programme, but this funding is insufficient, even with higher student fees, and teaching has had to be supported also from other sources. Departments need to be successful in the RAE, which means that virtually all academic staff have to be research active for there to be even a chance of a 5A let alone a 5* rating. Good ratings have allowed the college to tap into new funding, and to rejuvenate the academic staff after a long and difficult period. A younger staff has clearly benefited the undergraduate programme and there has been some feedback from research success to teaching, but not enough. I do not subscribe to the view that research and teaching skills are incommensurate, and would not like to see yet further separation. But recent change has placed academic staff in something of a dilemma when it comes to fulfilling their various responsibilities.

Naturally there are periods of rise and decline in different areas of research. Physics, which was dominant at the end of the Second World War, has been overtaken in recent years by work in the biomedical area. Loss of expertise is sometimes problematic since proper judgement can often be made only retrospectively. For example, had the college closed down the biological sciences after the Second World War, as was seriously considered, the loss would have been enormous. Less clear is the case of aeronautics. What would have been the result had Brian Flowers closed the aeronautics department, as he was inclined to do, and moved the staff to civil and mechanical engineering? In the period covered in this chapter a major case in point is nuclear technology research. It was allowed to decline in two departments, but it may need to be revived. Further, and relatedly, environmental research in science and technology, which saw a rapid growth from the 1970s to the mid 1990s, while still active appears to have declined in recent years. Materials science has been a major growth area. But, with interests in new materials not only within the department of materials, but in just about every other department in the college, future turf wars seem likely. For now, however, with new nanotechnologies, new materials being used in electronic devices, in major building structures and in transportation vehicles, and with new areas such as tissue engineering becoming important, there is plenty of work to go around.

Throughout the century, Imperial has attracted students from overseas. In the late twentieth century it would appear that the academic staff, too, are increasingly international. In this, Imperial College is a microcosm of the larger academic world, and is reflective of the globalization of higher education. The college has responded positively to historical developments beyond its control, and has relied on its long tradition of welcome to overseas students, and its tradition of overseas work. It has become a livelier place as a result. But, while a gradual withdrawal from dependence on governmental sources of income is no doubt a good thing, putting too many eggs in the globalization basket is dangerous, especially for an institution with no major endowment to cushion the blow that could result from a serious downturn in the global economy.

...

1. At the beginning of the twentieth century there were about 20,000 university students.

2. Sir Richard Brook Sykes FRS FMedSci (1942–), was educated at Queen Elizabeth College London and the University of Bristol. On gaining his PhD he joined Glaxo, becoming Head of the Antibiotic Research Unit, 1972–7. In 1977 he moved to the Squibb Institute for Medical Research in New Jersey, where he became Vice President for Infectious and Metabolic Diseases, 1983–6. He returned to Glaxo as Deputy Chief Executive 1986–7, and Chairman and Chief Executive Officer, 1987–93. After working to merge Glaxo with Wellcome, he became Deputy Chairman of Glaxo Wellcome, 1993–7, and Chairman in 1997. A further merger saw him become Chairman of GlaxoSmithKline, a position he held until 2002. Sykes was given a knighthood in 1994 for services to the pharmaceutical industry. As Rector he retains strong links to industry and serves also on a number of government and higher education committees. In 2004 he was made an honorary citizen of Singapore for his services to that country's biomedical sciences industry.

3. Four faculties were formed in 2001. The first faculty principals were: John Perkins FREng, Faculty of Engineering; Michael Hassell FRS, Faculty of Life Sciences; Sir Leszek Borysiewicz FRCP FMedSci, Faculty of

Medicine; and Sir John Pendry FRS, Faculty of Physical Sciences. The official line was that the faculty structure would foster integration and collaborative work, the effective management of change, and that it would ensure accountability. (The Tanaka Business School and some other units are outside this structure.) At the time that the faculties were founded, Sykes was quoted as saying, 'scientists and engineers can no longer sit in a "silo" of narrow interest because the biggest opportunities for discovery are between specialities'. As we have seen, few were still sitting in 'silos'. But the rhetoric was used to further boost interdisciplinary work which, Sykes believed, would be better fostered in faculties. Quotation, *IC Reporter*, 5 June 2001, 1.

4. How to restructure is a difficult question and depends on how the future is envisaged. Will the college be able to charge realistic fees? If not, will it become solely a postgraduate institution? To what degree will it be dependent on government funding? Will a future government be far more selective in how it funds universities also at the undergraduate level? Choosing a good governance structure and a good 'business' model in these uncertain times is problematic.

5. This, despite teaching being a factor in promotion and salary determinations. The problem is a consequence of government policies that have placed enormous pressure on academic staff to carry out research and publish. Indeed, departmental funding depends on their doing so, just at it depends on how much teaching gets done. In the 1990s, for a department with an academic staff of 32, not more than one could be non-research active for a 5A or 5* to be awarded in the RAE. For RAE see chapters 12 and 15.

6. This will remain the case unless undergraduate teaching is securely funded. Informal internal taxation has gone on before. Rectors have allowed departments in difficulties to run deficits and have forgiven debts at the expense of others. See discussion of chemistry and mathematics below.

7. There have been a number of these, including the Centre for Environmental Policy, the Natural Energy Research Centre, and the Centre for Energy Policy and Technology. In connection with the latter also the Urban Energy Futures project co-directed by Nilay Shah of the department of chemical engineering and chemical technology, and the work of David Fisk who holds the BP/Royal Academy of Engineering Chair in Engineering for Sustainable Development.

8. The Reactor Centre was losing money, hence its transformation into CARE under Professor A. (Tony) Goddard who later became a part-time Professor of Environmental Safety, with Special Reference to Nuclear Power, in the Department of Earth Science and Engineering. In the 1990s

Goddard helped train young Russian scientists and engineers in nuclear safety. In light of concerns over oil supplies and global warming, people with skills in nuclear power may need to regroup in the future.

9. This Centre was set up by Iain Thornton in 1990. Iain Thornton (1934–) was educated at the University of Durham where he studied agriculture. After work in Gambia he came to Imperial as a research student, gaining a PhD in applied geochemistry in 1968. He joined the staff at Imperial, working in the Applied Geochemistry Research Group and ICCET, and was appointed Professor of Environmental Geochemistry in 1988. Thornton has studied urban and rural dusts to determine levels of radioactive and heavy metal content; and the pathways of trace heavy metals through both animal and human food chains. He and Tony Goddard carried out a study of radioactivity in domestic dusts near UKAEA installations in Cumbria in 1984–5.

10. For B. J. Mason see chapter 10.

11. In 1995 Imperial appointed Richard Brabazon Macrory to the new Denton Hall Chair in Environmental Law, funded by international law firm Denton Hall Burgin and Warrens. Macrory was associated both with the Management School and with ICCET until 1999 when he joined the law department at UCL. Two environmental lawyers were associated with the Centre for Environmental Policy when it was founded in 2005.

12. Rector's statement on T. H. Huxley School, *IC Reporter*, 18 November 1997.

13. John Rex Beddington FRS (1945–) was educated at the London School of Economics where he took a degree in economics, and at the University of Edinburgh where he carried out research in population biology. Before coming to Imperial as a lecturer in 1984 he held a lectureship at York University. In 1991 he was appointed Professor of Applied Population Biology. Beddington was Director of ICCET (1994–97). In 2005 Beddington took over as Chairman of the Science Advisory Council of the Department of Environment, Food and Rural Affairs (DEFRA) from Roy Anderson when Anderson moved to become Scientific Advisor at the Ministry of Defence.

14. Needless to say, this was at some human cost. Many good teachers and people who had served the college well for many years lost their jobs at this time.

15. Basically the name change reinforced authority at the faculty level, but the divisions were not that different in structure from the older departments. The medical faculty has both divisions and departments within its divisions.

16. Jeffrey King Waage (1953–) was educated at Princeton and Imperial College. At Princeton one of his professors was Robert May who, recognizing his interest in entomology, advised him to enrol at Imperial for postgraduate work. On gaining his PhD in 1978, Waage became a lecturer in the ecology section of the biology department. In 1986 he was appointed Director of the International Institute of Biological Control at Silwood, before becoming the chief executive of CABI Bioscience (Commonwealth Agricultural Bureaux), its parent organization. This allowed him to retain close links with scientists working at Silwood. He rejoined Imperial in 2001 as Professor of Applied Ecology and Head of the new Department of Agricultural Sciences at Wye.

17. Bruce Arthur Joyce FRS (1934–) was educated at Birmingham University and worked in industry becoming Senior Principal Scientist at Phillips Research Laboratories in Redhill. From there he joined Imperial in 1988 as Professor of Semiconductor Materials. See also below.

18. For example, one of the earlier people to join was Ronald Newman, a leading expert on impurities and defects in semiconductors. Ronald Charles Newman FRS (1931–) was educated at Imperial College. On gaining his PhD he worked for Associated Electrical Industries before joining the University of Reading as a lecturer in physics. He was appointed to a chair in Reading in 1975, and returned to Imperial as Professor of Physics in 1989.

19. In 1982 Steven Ley and Tony Barrett won a £200,000 award from the Wolfson Foundation for their work on a new class of parasiticides, the avermectins and milbemycins, which have low toxicity for mammalian hosts. The avermectins proved successful in combatting River-Blindness, a serious problem in parts of Africa.

20. Eric Ash had connections to the Bagrit Foundation which provided generously for the Centre.

21. Richard Ian Kitney FREng (1945–) was educated at the University of Surrey and at Imperial College where he was a research student under Bruce Sayers, gaining a PhD in electrical engineering in 1972. He left to work for EMI and was a lecturer at Chelsea College, before returning to Imperial in 1985. He was appointed Professor of Biomedical Systems Engineering in 1989. Kitney has carried out computer simulations on the human thermo-regulatory system and works also in the area of biomedical imaging, including 3-D visualization techniques, with application in the study of arterial disease and cardio-respiratory control. He is a Co-Director of the Imperial College–MIT International Consortium for Medical

Imaging Technology, and founder of a spin-out company, ComMedica, specializing in diagnostic imaging, and related technology.

22. Christofer (Chris) Toumazou (1961–) received a PhD from Oxford-Brookes University (in collaboration with UMIST). He joined Imperial in 1986 as a post-doctoral fellow and, just eight years later, in 1994, was appointed Mahanakorn Professor of Analog Circuit Design in the Department of Electrical and Electronic Engineering. In 2003 he founded the Institute of Biomedical Engineering (which will move into the refurbished Bessemer Laboratory area in 2007). A specialist in microchip technology with wireless medical application, some of his recent work has been in the development of low-powered analogue circuitry used in the design (with Canadian company Epic Biosonics and surgeon Alan Lupin) of a fully implantable cochlea to overcome childhood deafness. This work has received much attention, with clinical trials ongoing in Canada and the UK. Toumazou helped introduce electronic master classes at Imperial. One of the first, in 2001, was given by Sir Ara Darzi in cybersurgery (microsurgery using robotics) enabling interactive distance learning in this new technology.

Sir Ara Darzi FMedSci (1960–) studied medicine and surgery at Trinity College Dublin. He is the Paul Hamlyn Professor of Surgery, Head of the Division of Surgery, Oncology, Reproductive Biology and Anaesthetics, and Head of the Department of Surgical Oncology and Technology. He is a pioneer in non-invasive micro-surgery, and in the use of robotic surgical tools.

23. The Genetics and Genomics Institute in the Faculty of Medicine brings together a wide range of experts working in many areas: disease gene discovery and investigation, gene therapy, pathogen genomes in plants and animals, transcriptomics, proteomics and bioinformatics — to name just some of the latest descriptors in these fast evolving fields.

24. David Moore Glover (1948–) was educated at Cambridge and University College London. After a post-doctoral research fellowship at Stanford University he joined Imperial in 1975 as a lecturer in biochemistry. He was appointed Professor of Molecular Genetics in 1986. Glover later left Dundee to become Arthur Balfour Professor and Head of the Department of Genetics at Cambridge.

25. There were also some new additions. For example, the Wellcome Trust funded new molecular parasitology laboratories in 1998 for Professor Deborah Smith. In 2000 Smith's work on *Leishmania* won an award from the Athena Project, a project designed to promote the academic careers

of young women in science and engineering. For the Athena Project see chapter 15.

26. James Barber FRS (1940–) was educated at the University of Wales (Swansea) in chemistry, and at the University of East Anglia where he gained a PhD in biophysics. He joined Imperial as a lecturer in 1968 and was appointed Professor of Plant Physiology in 1979. He built a large group working on the molecular processes of photosynthesis. When Barber moved to biochemistry he headed a section called Photo Bioenergetics and Molecular Dynamics.

27. George (Lord) Porter OM FRS (1920–2002) became Chairman of the Centre for Photomolecular Sciences and Professor of Photochemistry when he joined Imperial in 1987. A Nobel Laureate in chemistry (1967), Porter had been Fullerian Professor at the Royal Institution (1966–87), and Director (1966–85). He was President of the Royal Society (1985–90). Porter was also a major scientific communicator, a great champion of science, and well known to many through the BBC (including the Dimbleby Lecture of 1988).

   Two other photochemists who left the Royal Institution for Imperial at the same time were David Klug and David Phillips. Phillips joined the chemistry department as head of the physical chemistry section. Klug later became Professor of Chemical Biophysics.

28. Anne Dell FRS (1950–) was educated at the University of Western Australia and at Cambridge where she was an 1851 Exhibition Scholar. She began her doctoral studies under D. H. Williams, but after he left Cambridge she worked with Howard Morris. She came to Imperial with Morris as a postdoctoral fellow, was appointed lecturer in 1979, and Professor of Carbohydrate Biochemistry in 1991. She is a specialist in glycomics and is associated with the Centre for Structural Biology (set up largely with BBSRC funding) and the new (2004) Glycobiology Training, Research and Infrastructure Centre (GlycoTRIC). This centre is reflective of multidisciplinary interest in the function of sugars in life processes, and in the development of new therapies for sugar-related metabolic disorders.

29. Biophysics also became part of the new Department of Biological Sciences, headed by Martin Buck, Professor of Molecular Microbiology. When founded, in 2001, the department had 97 academic staff, 250 research staff and 350 postgraduate students. But its financial situation was poor and putting biologists and biochemists together in a single department was probably a bad idea. Within the divisions some more natural alignments, some also between biologists and biochemists, may form. The Faculty of

Life Sciences merged with the Faculty of Physical Sciences to form a new Faculty of Natural Sciences in 2006.

30. Gordon Dougan (1952–) was educated at the University of Sussex and the University of Washington. Dougan left Imperial in 2004 for the Sanger Institute in Cambridge.

   Douglas Brownlie Young FMedSci (1953–) was educated at the University of Edinburgh and at Oxford University. After research posts in India, Seattle, and with the MRC at Hammersmith, he was appointed Fleming Professor of Medical Microbiology at Imperial (St. Mary's) in 1991. Today he works in the Division of Investigative Science in the Faculty of Medicine. He has carried out work on leprosy and tuberculosis and is on the coordinating board of the Global Partnership to Stop TB. He leads an international consortium to develop drugs for the treatment of latent TB (since the genetic sequencing of the TB bacterium there has been much new research in this field). His work is generously supported by the Bill and Melinda Gates Foundation and the Wellcome Trust. He has also received support from the BBSRC and the EPSRC for a Centre for Integrative Systems in Biology to be operational by 2007.

31. Colin Russell Hopkins (1939–) was educated at the University of Wales (Swansea) and for much of his career worked at the University of Liverpool Medical School, becoming Professor and Head of the Department of Medical Cell Biology in 1975. He came to Imperial as Rank Professor of Physiological Biochemistry in 1986. Hopkins moved to UCL in 1991 as Director of the MRC Laboratory for Molecular Biology. New emphasis on molecular cell biology has attracted a number of good scientists to Imperial. For example, Anthony Magee must have seen possibilities at Imperial when he moved from the National Institute of Medical Research at Mill Hill to a chair in membrane biology in the Division of Biomedical Sciences in the Faculty of Medicine.

32. The department profiled itself as follows in the 1991–5 College Plan: biomolecular structure and function with focus on medical problems; plant biochemistry and molecular biology; molecular genetics and gene regulation; molecular parasitology; neurochemistry and molecular neurobiology; molecular cell biology. Strictly speaking biochemistry is by definition a molecular science, but this department included also some non-molecular areas of biotechnology in its early days.

33. Sir Roy Malcolm Anderson FRS FMedSci (1947–) was educated at Hertford Grammar School and Imperial College. He was a research student under June Mahon and received a PhD in zoology (parasitology) in 1971. On graduation he won an IBM scholarship in biomathematics

which allowed him to visit a number of laboratories in the USA. After brief periods at Oxford and King's College London, he returned to Imperial as a lecturer in 1977, and to run the MSc course in ecological management. Anderson is a theoretician and uses models and mathematical analysis in his approach to biological problems. He increasingly moved towards epidemiology and his team took on problems related to parasites, to HIV AIDS, and later BSE and SARS. Anderson was rapidly promoted, becoming professor of parasite ecology in 1982. In 1993 he followed Southwood in the Linacre chair at Oxford, and as head of the department of zoology. In 2001 he returned to Imperial's new faculty of medicine as Director of the Wellcome Research Centre for the Epidemiology of Infectious Diseases (which replaced the earlier centre in parasitology). Anderson's work has been very influential and is widely accepted. But it is also controversial. This was evident in connection with his modelling of the progress of the Foot and Mouth epidemic during the late 1990s. It was largely on his advice that the widespread culling of animals occurred. It is interesting that government agencies believed, possibly correctly, that theoretical epidemiologists were better able to make informed decisions on these matters than the veterinary profession — perhaps because the veterinarians were unable to come up with a united front.

June Mahon (1928–78) gained a BSc in zoology from Imperial College and a PhD from the University of Neuchatel. She was appointed a lecturer in parasitology (helminthology) in 1959.

34. See also chapter 12. The Garden Wood Laboratories were built at this time.

35. Michael Patrick Hassell FRS (1942–) was educated at the University of Oxford and came to Imperial as a lecturer in 1970. He was appointed Professor of Insect Ecology in 1979. Hassell works in the field of population ecology and, relatedly, on insect parasites and their possible use as biological control agents. He has developed mathematical models for both field and laboratory work. See Michael P. Hassell, *The Spatial and Temporal Dynamics of Host-Parasitoid Interactions* (Oxford, 2000). Hassell became the first Principal of the Faculty of Life Sciences in 2001.

36. Those working in microbial physiology worked alongside biochemists in the short-lived Centre for Biotechnology.

37. Many parasitologists migrated back to Imperial when Anderson returned in 2001.

38. Crisanti is Professor of Molecular Parasitology. The idea is to control transmission of the parasite by gene manipulation. Another malaria scientist at the college, Austin Burt, Professor of Evolutionary Genetics, recently

received US$8.8m from the Grand Challenges in Global Health Initiative (funded by the Bill and Melinda Gates Foundation, the Wellcome Trust and the Canadian Institute of Health Research) for his work on genetic strategies to prevent the spread of malaria. Imperial's malaria group was joined in 2005 by Fotis Kafatos, Professor of Immunogenomics, who is a pioneer in the field of insect molecular biology and genomics.

39. Sir John Hartley Lawton FRS (1943–) was educated at the University of Durham, gaining a PhD in 1969. He worked at the Universities of Oxford and York before coming to Imperial to set up the Centre in 1989. Its new premises at Silwood were opened in 1990 by Margaret Thatcher. Lawton carried out pioneering work in theoretical ecology, population dynamics, and on global environmental change. He was a founder and chairman of the Royal Society for the Protection of Birds. He left Imperial in 1999 to become Chief Executive and Deputy Chairman of the Natural Environment Research Council (NERC).

   Lawton was succeeded as Director of the Centre for Population Biology by Charles Godfray. Hugh Charles Jonathan Godfray FRS (1958–), was educated at the University of Oxford and Imperial College where he gained a PhD in 1983. On graduation he stayed at Imperial as a NERC post-doctoral fellow, before moving to Oxford as a demonstrator in ecology in 1985. He returned to Imperial in 1987 and became Professor of Evolutionary Biology in 1995 (Godfray has strength in Imperial's traditional area of entomology). In 2005 he became Head of the Division of Biology in the Faculty of Life Sciences.

40. This facility opened in 1991 and is now operated as a NERC service open to the entire scientific community in the UK.

41. The nexus around May is of much interest in connection with ecological work at the college. Robert (Bob) McCredie May FRS (1936–) had earlier been a professor of theoretical physics at the University of Sydney. He became interested in the dynamics of animal behaviour, and in the relation between stability and complexity in the dynamics of animal populations. This work was in part stimulated by that of Robert McArthur, a specialist in island biogeography at Princeton. May was appointed to a zoology professorship at Princeton in 1973. His own work on island biogeography, and his recognition of chaos in biological systems, brought him great recognition. He was also to lay the mathematical foundation for modern studies in the epidemiology of infectious diseases. May visited Silwood in the early 1970s since he was interested in the theoretical ecological work being carried out there by Roy Anderson, Michael Hassell and others. Professor Southwood, a good networker and organizer, encouraged May to become

a visiting professor at Imperial (1975–88). Later May was appointed Royal Society Research Professor (1988–2001) — jointly appointed at Oxford. Many younger biologists of this period, wanting to acquire new mathematical skills, were keen to associate with him. At Imperial, in addition to Hassell and Anderson, they included John Beddington, Michael Crawley (later professor of plant ecology), John Lawton and, later, Charles Godfray. Beddington, and later Lawton, came to Imperial from York University, another good centre for biomathematics, also visited earlier by May. These scientists, together with two others working elsewhere, formed a close-knit group. Its pattern was not unlike that of the nineteenth-century X-Club in that the group was largely London based, often dined together, and went on walking holidays to the Lake District, the Alps and other locations. They work/ed together closely in various combinations, publish/ed many joint papers, and support/ed each other's careers. Like the X-Club members most (all of those mentioned above) have become FRS, and one of their number (May) was elected President of the Royal Society. May has received a life peerage.

42. While the modellers use computers for their simulations, Robert May is said to be a reluctant computer user. He is reputed to carry much in his head and to theorize 'on the backs of envelopes' etc. Nonetheless May is a leader and helped people like Hassell put their biological/ecological ideas into rigorous mathematical form, enabling them to begin building serious models.

43. One major development is a switch from the reading to the writing of genetic code.

44. Michael Joseph Ezra Sternberg (1951–) studied natural sciences at Cambridge before coming to Imperial for an MSc in computing. He then moved to Oxford for a DPhil in molecular biophysics. He returned to Imperial as Professor of Structural Biophysics in 2001, having been Head of the Laboratory of Biomolecular Modelling at the Imperial Cancer Research Fund. He works also with Professors Darlington and Muggleton, of the computing department.

45. Until the 1980s about 60% of the department's graduates were employed by large companies such as ICI or the big pharmaceutical companies. In the early twenty-first century about 50% have joined small or medium-sized companies.

46. Ley succeeded W. J. Albery who briefly headed the department (1988–9) before leaving for Oxford. Wilkinson remained at the college as a Senior Research Fellow, continuing his research in a laboratory funded by Johnson

Matthey plc. (A founder of this company, George Matthey, was a student of Hofmann's at the Royal College of Chemistry.)

47. David Phillips (1939–) was educated at the Universities of Birmingham and Southampton and came to Imperial in 1989 from the Royal Institution (RI) where he was deputy to the director, George Porter. Like Porter, Phillips is a photochemist. At the RI Phillips worked on gas phase photo-chemistry where he developed supersonic jet spectroscopy to study systems at low temperatures. At Imperial, where he held the Hofmann Chair, Phillips turned to the study of fluorescence in biological systems, using time resolved Raman and Infra Red spectroscopy. He also works in the area of photodynamic therapy, work supported by the EPSRC. Phillips is renowned as a lecturer, especially for his demonstration lectures given in the tradition introduced at the RI by Humphry Davy and Michael Faraday. In 1997 he was awarded the Royal Society Michael Faraday Award, for the scientist who has done most to further the public understanding of science in the UK.

In 2002 Phillips was succeeded as head of department by Richard Templer. Richard Templer (1958–) studied physics at Bristol before moving to Oxford to study archeology where he specialized in solid state luminescence. He later changed fields to work on the physical basis of membrane-protein interactions, joining Imperial College in 1990 as Royal Society University Research Fellow. He was appointed Professor of Biophysical Chemistry in 2001.

Vernon Charles Gibson FRS (1958–) was educated at the University of Sheffield and at Balliol College, Oxford. Following a NATO postdoc-toral fellowship at the California Institute of Technology he was appointed to a lectureship in inorganic chemistry at Durham University, becoming Professor of Chemistry in 1993. He joined Imperial College in 1995 as Professor of Polymer Synthesis and Catalysis, and was given the title Sir Geoffrey Wilkinson Chair of Inorganic Chemistry in 1999. Since 2001 he has been the Sir Edward Frankland BP Professor of Inorganic Chemistry. Gibson works in the area of catalysis, designing new systems for the controlled synthesis of industrially important polymers, work that has been well funded by British Petroleum.

For Barrett see chapter 13. Eric Ash and Professor Charles Rees wooed Barrett back to Imperial from the USA in 1993 by setting up the Wolfson Centre for Organic Chemistry, with support also from Glaxo.

48. These later changed. By 2003 the groupings were: biological and biophys-ical; catalysis and advanced materials, computational, theoretical and structural; electronic materials, interfacial and analytical; and synthesis. As

in the life sciences, so in chemistry, modelling work grew in importance with the advent of cheap electronic computers. Nick Quirke headed a large theoretical section with strong support from industry and the EPSRC. The head of the biological chemistry section, Robin Leatherbarrow, worked with Imperial's first specialist in this area, Sir Alan Fersht. Leatherbarrow's research is concerned with molecular recognition between proteins and their ligands, with protein engineering, and with the design of enzyme inhibitors. He is associated with the Chemical Biology Centre, which co-ordinates activities between Imperial College, the Institute for Cancer Research and the London Research Institute of Cancer Research UK; and was responsible for establishing the first EU Marie Curie Centre in Chemical Biology in the UK.

49. Timothy Simon Jones (1962–) was educated at the University of Liverpool, gaining a PhD in 1988. He was appointed a lecturer at Imperial in 1991, and Professor of Chemical Physics in 1998. In 2000 he was appointed STS/Sumitomo Professor of Electronic Materials.

    Andrew De Mello (1970–) was educated at Imperial and returned to join the staff from a lectureship at the University of East Anglia. He is Professor of Chemical Nanosciences.

50. Andrew David Miller (1963–) was educated at the Universities of Bristol and Cambridge. He came to Imperial in 1990 and is Professor of Organic Chemistry and Chemical Biology.

51. The percentage of students taking A level examinations in chemistry has declined in recent years, though the absolute numbers have remained fairly steady. In 1997 the three year course was finally phased out and the four year MSc course became the norm, the BSc being given only for non-performance at the MSc level.

52. In 1987 the average research income of the academic staff was about £31,000. This compared unfavourably with the Oxbridge average of £55,000. See EC and FC minutes, 6 November 1987.

53. These fellowships introduced a competitive element in that universities had to make themselves attractive to those who had been awarded research funding. As it turned out, several who came on advanced fellowships in the 1970s and 80s later became section leaders in the department. Professor New was able to attract two future department heads to join his group at Imperial, namely Peter Knight and John Elgin, head of physics and mathematics respectively.

54. In the 1996 RAE, the department received a disappointing 5 (the only physics departments to receive a 5* that year were at Oxford and Cambridge). In 2001 the department was rated 5*.

55. Kibble remained active after his retirement in 1998 and was Director of the new Centre for Astroparticle Physics and Cosmology. For more on Kibble and Blow see chapter 13.

56. Professor D. J. Bradley brought laser optics to the department in 1973. In the late 1980s the two optics sections were headed by J. C. Dainty and G. H. C. New, Professor of Non-Linear Optics. For Dainty (applied optics) see chapter 13. Geoffrey Herbert Charles New (1942–) was educated at the University of Oxford. After gaining a DPhil in 1967 he worked with Professor Bradley at Queen's University, Belfast, before coming to Imperial with Bradley in 1973. Bradley and New's work can be seen as one branch of an optical tradition that has its roots in Lockyer's spectroscopic work. In the early twentieth century the tradition continued in the work of A. Fowler, his student H. Dingle and, in the 1950s and 60s, in the work of A. G. Gaydon and W. R. S. Garton. In the late twentieth century Imperial also housed the Laser Consortium where work using large laser systems was carried out. Lasers have been revolutionary in a wide range of work; for example in research carried out by Professor E. A. Hinds FRS in the Centre for Cold Matter.

57. Sir Peter Leonard Knight FRS (1947–) was educated at the University of Sussex, gaining a DPhil in 1972. After work in the USA, at the University of Sussex, and at Royal Holloway College, he joined Imperial in 1979 and was appointed Professor of Quantum Optics in 1986. He became head of department in 2001. Knight received a knighthood in 2005 after heading a successful SRC initiative for the promotion of non-linear optics in the UK. He has supervised many doctoral students who have become leaders in the field. Knight is also the first non-American to be elected President of the Optical Society of America (the society is recognised as the international body in optics).

58. Malcolm Golby Haines (1936–) was educated at Imperial College where he spent all of his professional life. He was appointed Professor of Physics in 1975 and headed the plasma physics group 1975–2002.

59. David Potter FREng was educated at Cambridge and Imperial College. After gaining his PhD in 1970 he stayed in the physics department as a lecturer. He early saw a way to use microprocessors to build hand-held organizers and founded the company (Potter's Scientific Instruments Or Nothing) in 1980 which launched the Psion organizer in 1984, the first of its kind.

60. In this period much of this work was carried out by Peter John Dornan FRS (1939–). Dornan was educated at the University of Cambridge and joined Imperial after a few years working at the Brookhaven National

Laboratory, N.Y. In 2002 he was awarded the Rutherford Medal of the Institute of Physics, jointly with Professor David Plane of CERN and Dr Wilber Venus of the Rutherford Appleton Laboratory, for major contributions to the development of the detectors for use with the Large Electron Positron Collider.

61. For example, John Harries, who came to Imperial from the Rutherford Laboratory, works in the area of global warming and carried out the first good measure of the radiation budget (heat in and out of the planet).

62. David Southwood (1945–) studied mathematics at Queen Mary College before coming to Imperial as a PhD student under Jim Dungey in 1966. For an interesting article by Southwood on his education in space physics, and of the contributions made by Dungey to the understanding of the magnetosphere see, 'An Education in Space Physics' in Stuart Gillmor (ed.), *Discovery of the Magnetosphere*; vol. 7 in *History of Geophysics* (Washington, 1997), 185. After a postdoctoral appointment in the USA, Southwood returned to a lectureship at Imperial in 1971 and was appointed Professor of Space and Atmospheric Physics in 1986. He was Head of the Blackett Laboratory 1994–7. From 1997–2000 he was Head of Earth Observation Strategy for the ESA. He has been associated with a number of Russian, NASA and ESA missions, and has been seconded to the ESA for several years. Among other things he has been responsible for the dual technique magnetometer on the Cassini-Huyghens mission. Professor Michele Dougherty, who joined Imperial in 1991, became the lead scientist for the magnetometer instrument on this mission. She helped persuade the Cassini team to take a closer look at one of Saturn's moons, Enceladus — a good move since the results were surprising and revealed recent geological activity, and the presence of water at the moon's south pole.

63. This Centre brought people together from physics, chemistry, materials, electrical engineering and mathematics. It was absorbed within the Centre for Electronic Materials and Devices on Professor Caplin's retirement in 2002. Anthony David Caplin (1937–) was educated at Cambridge University and at Yale University. After gaining his PhD from Cambridge in 1964, he joined Imperial College as a lecturer and was appointed Professor of Physics in 1995. Caplin was also active in Scientists for Labour.

64. For more on Pendry and Stradling see chapter 13.

65. Gareth Parry FREng (1950–) was educated at Imperial College. He was Rank Foundation Professor of Electro-optic Engineering at Oxford, before

returning to Imperial as Professor of Applied Physics and Director of the Centre for Electronic Materials and Devices in 1997.

66. Donal Donat Connor Bradley FRS (1962–) was an undergraduate at Imperial before moving to the Cavendish Laboratory for his PhD. He returned to Imperial from a professorship in the department of physics and astronomy at Sheffield. He works on molecular electronic materials and, in 2003, was a co-recipient of the European Union Descartes Prize for the development of polymer light emitting diode displays. In 2005 he won the Latsis Prize for his work in nano-engineering. Bradley is a co-holder of the fundamental patent on polymer electroluminescence which led to the founding of Cambridge Display Technology Ltd (1990). In 2001 he founded a new start-up company, Molecular Vision Ltd, launched to develop polymer detection systems for 'lab-on-a-chip' microanalysis. Bradley succeeded Knight as head of department in 2006.

67. Keith Harrhy Ruddock (1939–96) was educated at Imperial College and joined the staff on gaining his PhD. He was influenced by Professor David Wright to pursue work in the area of vision. He was appointed Professor of Biophysics in 1988 but was killed in an automobile accident eight years later. Professor Blow died in 2004, after biophysics had moved to join the new but short-lived Department of Biological Sciences.

68. It is a copy of a bust made by Jacob Epstein in the 1950s.

69. Adrian Frederick Melhuish Smith FRS (1946–) was educated at Cambridge and University College London. A specialist in Bayesian statistics, he was appointed Professor of Mathematical Statistics at the University of Nottingham in 1977, and Professor of Statistics at Imperial in 1990. A good administrator, he helped also with reorganizing the management structure at the college. Smith left Imperial to become Principal of Queen Mary College in 1998. In the same year he was asked by the government to review post-14 mathematics education in the country and issued a much debated report in 2002.

70. Gerald Richard Salkin (1933–) was appointed a lecturer in mechanical engineering in 1966 and made the journey towards management science with Professor Eilon. Salkin was appointed Reader in Management Science in 1974 and founded the Centre for Quantitative Finance with Nicos Christofides in 1992. Located in the Management School it later became part of the Tanaka Business School. The Centre has strong links to the banking industry and is well supported also by a number of multinational companies. See chapter 14 for Eilon and Christofides.

71. Colin Atkinson FRS (1941–) was educated at the University of Leeds. He was a lecturer in the theory of materials at the University of Sheffield before

coming to Imperial as a lecturer in 1971. He has carried out research in a number of areas including non-linear diffusion (phase transformations), dislocation theory and fracture, solid mechanics and problems of drilling and oil recovery. He was appointed Professor of Applied Mathematics in 1985.

72. Work in this field has been carried out by Mark Herbert Ainsworth Davis (1945–). Davis was educated at Cambridge and the University of California (Berkeley). He studied electrical sciences before turning to mathematics. He was appointed to a lectureship at Imperial in 1971 and in 1984 became Professor of System Theory. He left Imperial in 1995 to become Head of Research and Product Development at Tokyo-Mitsubishi International, returning to Imperial in 2000. He is the author of three books on stochastic analysis.

73. Terence John Lyons FRS (1953–) was educated at Cambridge and Oxford Universities. Lyons came to Imperial as a lecturer in 1981, leaving for a chair at the University of Edinburgh in 1985. He returned to Imperial as Professor of Mathematics in 1993.

74. David John Hand FBA (1950–) came to Imperial in 1999 from the Open University where he was Head of the Department of Statistics. Before that he held a chair at King's College London. His research interests include data mining, classification methods, and the interface between statistics and computing. He first applied his statistical knowledge in medicine and psychology before turning to problems in banking and the management of consumer data. He has helped financial institutions fight credit card fraud and calculate credit risks.

75. Philip Hall (1950–) was educated at Imperial College and held professorships at Exeter and Manchester before returning to Imperial in 1996 as Professor of Applied Mathematics. During his headship work in biological fluid mechanics increased.

76. John Nicholson Elgin (1946–) left school aged fourteen and became an apprentice with Parsons Peebles. After taking evening classes and gaining an ONC in mechanical engineering he entered Heriot-Watt University, taking a BSc in physics in 1970. He then moved to Cambridge and, after taking Part 3 of the Mathematics Tripos, gained a PhD in applied mathematics. He joined Imperial as a post-doctoral fellow with Professor New in the physics department, and later received an SRC Advanced Fellowship. He joined the mathematics department as a lecturer in 1983. His research interests, with application in the telecommunication industry, are in the study of integrable systems (notably the vector Nonlinear Schrödinger

Equation) and in the study of chaotic systems through the properties of (unstable) periodic orbits supported by such systems.

77. Jeffrey Ronald Cash (1947–) joined Imperial in 1972 and was appointed Professor of Numerical Analysis in 1993. Gerald Moore (1951–) was appointed Reader in Numerical Analysis in 1996, and John William Barrett (1955–) was promoted to Professor of Numerical Analysis in 1999.

78. David Murray Edwards (1933–) was educated at Imperial College and joined the staff as a lecturer in 1965. He was appointed Professor of Applied Mathematics in 1987.

79. Gordon Douglas James (1945–) was educated at Cambridge University where he worked as a lecturer before coming to Imperial in 1985. He was appointed Professor of Pure Mathematics in 1989. Martin Walter Liebeck (1954–) joined Imperial as a lecturer in 1985 and was appointed Professor of Pure Mathematics in 1991. He now heads the pure mathematics section.

80. Alexander (Sacha) Grigor'yan (1957–) was educated at Moscow University and came to Imperial as a Reader in 1994. He was appointed to a chair in 1998. Jozef Boguslav Zegarlinski (1955–) was educated at the Silesian University, Katowice, and at Wroclaw University. He joined Imperial as a lecturer in 1993 and was appointed to a chair in 1996.

81. Simon Kirwan Donaldson FRS (1957–) was educated at Cambridge and Oxford. He impressed the mathematical world already as a research student with his 'Self-dual connections and the topology of smooth 4-manifolds', *Bulletin of the American Mathematical Society* (1983). After a year at the Institute for Advanced Study in Princeton, he returned to Oxford as Wallis Professor of Mathematics in 1985, the year he was elected FRS. Donaldson received the Fields Medal in 1986 and joined Imperial in 1999.

82. The official mandate of the Institute is to bring together mathematicians and researchers in the physical sciences, engineering, the Tanaka Business School, life sciences and medicine to tackle fundamental problems needing significant mathematical input; to train outstanding young mathematicians in their chosen field and/or in interdisciplinary research; to encourage the interaction of young researchers with more senior internationally-leading researchers from the UK and elsewhere; to provide a national resource for the UK mathematics community and those fields dependent on high level mathematics; to provide a forum for medium-term research in programmes related to emerging areas of mathematics and its applications in all disciplines. A cursory glance at what is going on suggests that there is currently much interest in medical imaging within the institute.

83. See departmental report, GB minutes, 15 March 1985. Knill stated that he helped found the newsletter *GEOlogIC* in order to foster communication between the sections. For Knill see also chapter 13.

84. Oxburgh was by then no longer chairman of the committee which continued as arbiter in the competition. This review was not set up to save money but to rationalize resources. About £30m of new money was given to geology nationally, used both to relocate staff and to provide new equipment.

85. It would appear that the department was a bit top heavy with technicians in the 1980s.

86. Rector's correspondence; Ash to Swinnerton Dyer, 14 March 1988. See also Mary Pugh, 'The History of the Geology Department, Imperial College, 1958–88' (typescript, 1995).

87. This is a difficult issue. Clearly Imperial's mandate is to work in support of industry, but the type of work carried out must also have academic merit and be publishable. Good geological work was being carried out at the time, but the committee found Imperial's overall research output lacking and that much of the applied work was in the form of private consulting with little or no financial benefit to the department. The department also appeared a bit old-fashioned in its reluctance to fully embrace the computer age in its research, and in its reluctance to use e-mail.

88. Interview with Ash on his being knighted, in *Record of RCS Association* (Winter 1989).

89. Ash invited a large number of industrialists to Imperial for a seminar at which he explained the situation. Industry responded well in helping to rebuild research, if not teaching, capacity.

    Richard (Dick) Curtis Selley (1939–) came to Imperial in 1961 as a research student under the joint supervision of John Sutton and Douglas Shearman. Aside from three years working in petroleum exploration in the North Sea and abroad, his career was made at Imperial College. He was head of the department, 1988–93, and became Professor of Applied Sedimentology in 1989.

90. For example, Howard Johnson was appointed Enterprise Oil Chair of Petroleum Geology in 1993. Howard David Johnson (1950–) was educated at the Universities of Liverpool and Oxford. He joined the Shell Group in 1978 and worked in a number of locations on oil exploration, production, and research. Johnson was the seventh industrially-funded member of the academic staff to be appointed in this period and became Head of the Sedimentary Basins, Structures and Fluids Group. His chair was renamed the Shell Chair when Shell purchased Enterprise.

91. Michael Hugh Worthington (1946–) was educated at Durham University where he took a degree in applied physics, and at the Australian National University where he took a PhD in geophysics. He was a lecturer at Oxford before coming to Imperial as Professor of Geophysics in 1985. He was Chairman of the Centre for Remote Sensing 1988–90 and is a specialist in electromagnetic prospecting methods in exploration seismology. He was head of the department from 1993–98.

92. See Management and Planning Group Papers 1989–90. There were still some core geologists on the staff, such as Jake M. Hancock, Professor of Geology 1986–93. Cambridge educated, he came to Imperial from King's College London and was a specialist in Cretaceous geology and paleontology.

93. This was the reason for Oxburgh (by then Rector) and Beddington's action. Letting people go who have devoted many years to the college cannot have been easy. But, as mentioned elsewhere, the older more gentlemanly ethos of the university had given way to a business ethos where productivity, in this case research productivity, is what mattered. There is little doubt that the college lost a number of good teachers during this period, and not just from the geology and earth resources engineering departments. Given the new political realities, teaching ability on its own was a luxury the college could no longer afford. Perhaps there are lessons to be learned from this period in how to handle job termination. On the positive side, space was opened up for some good new appointees.

94. David J. Sanderson (1947–) was educated at the University of Newcastle-upon-Tyne. Before coming to Imperial College he held chairs at Queen's University Belfast, and at the University of Southampton where he was head of department. In 1998 he became one of two assistant directors of the Huxley School. He is head of the fractures, faulting and fluid flow group in the earth science and engineering department.

95. Engineering geology, Knill's former field, was moved to civil engineering at this time.

96. Mine and quarry reclamation work has since wound down.

97. The joint petroleum engineering and rock mechanics section has been headed by Martin Blunt since 1999 when he was appointed to the chair of petroleum geology. Martin Blunt (1963–) was educated at the University of Cambridge in theoretical physics where he gained a PhD in 1988. After work with British Petroleum as a research reservoir engineer, he taught petroleum engineering at Stanford University before coming to Imperial. He is a specialist in aspects of flow and transport in porous systems, and is interested in carbon capture and its long-term storage.

98. Edwin (Ted) Thomas Brown FREng (1938–) is a graduate in civil engineering from the University of Melbourne. He came to Imperial as Professor of Rock Mechanics in 1975. He returned to Australia to become Dean of Engineering at Queensland University in 1987. He has published widely in his field, notably on fracturing around wellbores.

99. John Stuart Archer FREng (1943–) was educated at City University, sponsored by ICI for a four-year sandwich course in industrial chemistry, and at Imperial College where he gained a PhD in chemical engineering in 1968. Working at the interface between geosciences, engineering, and physics, Archer is a specialist in the characterization of petroleum reservoirs and has been a consultant worldwide. He was Deputy Rector of Imperial College, leaving in 1997 to become Vice Chancellor of Heriot-Watt University.

100. John David Woods (1939–) was educated at Imperial College where he took a BSc and PhD in physics. He then worked at the Meteorological Office before moving to the chair of physical oceanography at the University of Southampton in 1972. He later moved to a chair at the University of Kiel, returning in 1986 to become Director of Marine and Atmospheric Sciences at NERC. He was head of Earth Resources Engineering, and Dean of the Graduate School of Environment until both became subsumed in the T. H. Huxley School in 1997. Woods continued as Professor of Oceanography in Earth Science and Engineering.

101. Woods was against the merger with geology, only delaying the inevitable. Rees David Rawlings (1942–) came to Imperial as a metallurgy student in 1961. His entire career has been at the college where he is currently (2006) head of the structural materials group in the department of materials, and Pro-Rector for Educational Quality.

102. Michael Robert Warner (1954–) was educated at the University of York and, after a NERC Fellowship at the University of Edinburgh and work with the Bullard Laboratories in Cambridge, he joined Imperial as a lecturer in 1989. He was appointed Professor of Geophysics in 2000.

103. The metals and energy finance field was developed by Dennis Buchanan who earlier led the mining geology section. In 2006 he introduced a new MSc course in the field and brought the department more in touch with the financial aspects of mining operations. Geochemistry was a victim of the cuts. It is an expensive field to maintain but there appear to be moves afoot to find funding for new laboratories in this area.

104. Donald William Pashley FRS (1927–) was educated at Imperial where he was a research student under Morris Blackman, gaining a PhD in physics with a thesis on electron diffraction. He was ICI Research Fellow

in the department until 1955 when he joined Tube Investments Research, becoming Director of the Research Laboratories in 1968. He returned to Imperial in 1979, succeeding Ball as Professor of Physical Metallurgy (later Professor of Materials).

105. Harvey Milard Flower (1945–2003) was educated at Cambridge and at Imperial, joining the department as a PhD student in 1967. He was appointed lecturer in 1972 and Professor of Materials Science in 1992. He was a world expert in materials for the aerospace industry, notably in the metallurgy of titanium, aluminium and their alloys. Sadly he died in an accident in 2003.

Frederick John Humphreys (1941–) was educated at Oxford University and came to Imperial in 1978. He became Professor of Physical metallurgy in 1988 but left two years later for a chair in materials science at Manchester.

A new and powerful electron microscope came to the department in 2004. To be used in nanotechnology research, it has a resolution of 0.14 nanometers. Leading use of the new instrument is David McComb. Imperial is one of only three universities in Europe and the US with a microscope of this power.

106. Together they were founders of the spin-out company Ceres mentioned in chapter 15. John Anthony Kilner (1946–) was educated at the University of Birmingham where he took a PhD in physical metallurgy. He came to Imperial in 1979 as Wolfson Fellow in the Wolfson Unit for Solid State Ionics and worked there with Professor Steele. He joined the academic staff in 1987 and became Professor of Materials Science in 1995. He became head of the department of materials in 2000. His research has been focused on mass transport studies in oxide and semiconductor materials. He founded and became Director of the Centre for Ion Conducting Membranes in 1998.

107. Malcolm McLean FREng (1939–2005) was educated in physics at Glasgow University where he gained a PhD in 1963. A well respected metallurgist, McLean worked on a variety of topics, including the processing and properties of advanced high temperature materials.

108. Larry Leroy Hench (1938–) was educated at Ohio State University and came to Imperial in 1995 following thirty-two years at the University of Florida where he was Graduate Research Professor of Materials Science and Engineering, and Director of the Bioglass® Research Center. Bioglass®, invented in 1969 in response to the injuries of the Vietnam War, was the first manufactured material to bond to living tissues and is used clinically throughout the world for the repair of bones, joints and teeth. Professor

Hench's work has led to the development of a new generation of gel-silica optical components (Gelsil®) for environmental sensors, tissue engineering, and solid-state dye lasers. Hench has also made something of a splash with his books for children featuring Boing-Boing, the bionic cat. Dame Julia M. Polak FMedSci is Professor of Clinical Histochemistry in the Division of Investigative Science, Faculty of Medicine. Hench and Polack's work in the Centre is focussed on ways of growing lung and bone tissue.

109. The Bessemer Laboratory had been rebuilt earlier as part of the postwar reconstruction. As of writing the laboratory site is undergoing another complete renovation. The new Bessemer Building will house the Institute for Biomedical Engineering, including the Bio-Nanotechnology Centre and the Imperial Bio-Incubator. The latter will be a facility to house biomedical spin-out companies in their early stages.

110. Glyn Arthur Owen Davies (1933–) was educated at the University of Liverpool and the Cranfield Institute of Technology. He worked for British Aerospace (1957–9) before joining the aeronautics department at the University of Sydney. He joined Imperial as a lecturer in 1966. He followed Professor Argyris as head of the aerostructures section and was head of department 1982–9. He was appointed Professor of Aeronautical Structures in 1985.

111. Peter William Bearman FREng (1938–) was educated at Cambridge University and worked at the National Physical Laboratory before coming to Imperial as a lecturer in 1969. He became Professor of Experimental Aerodynamics in 1986 and head of department in 1989–99. He refused the title Zaharoff Professor, and the title has since fallen out of use. Bearman was Deputy Rector 2001–4.

John Michael Russell Graham (1942–), known as Michael, was educated at Cambridge University and Imperial College. Most of his professional life has been spent at Imperial and he became Professor of Unsteady Aerodynamics in 1990. He was head of department 1999–2003.

112. Michael A. Crisfield (1942–2002) came to Imperial from the Transport and Road Research Laboratory in Bracknell. His chair was endowed by Finite Element Analysis Ltd.

113. (Bearman, personal communication). See also 'Force 12 gales at the flick of a switch', *The Times*, 27 February 1985, 18. Honda's grant was initially £700,000. Bearman was joined by J. K. Harvey in directing the work, and Honda increased its investment in 1988.

114. John Kenneth Harvey (1935–) was educated at Imperial College. On gaining his PhD in 1960 he worked for a short period at Princeton

University before returning to Imperial as a lecturer in 1962. He was made Professor of Gas Dynamics in 1989.

Fluid mechanics, the modelling of turbulence and related topics, require good computing facilities which have come to the department during the past twenty-five years. Staff numbers in the computational area have also increased. For example, Michael Albert Leschziner FREng (1946–) was appointed to a chair in computational aerodynamics. Educated at City University and at Imperial where he gained a PhD in 1976, Leschziner returned to the college in 2001 after holding chairs at UMIST and Queen Mary and Westfield College.

115. Richard Hillier (1946–) joined Imperial as a lecturer in 1973. He was appointed Professor of Compressible Flow in 1995 and Head of Department in 2003. He has worked on the recently launched US scramjet. Hypersonics was pioneered earlier at Imperial by John Leslie Stollery FREng (1930–). Stollery was educated at Imperial where he was a student under Professor Squire. On graduating he left to work for the De Havilland Aircraft Company for a few years before returning to Imperial as a lecturer in 1956. He became a Reader in 1962 but left for the chair of aerodynamics at Cranfield University in 1973. The hypersonics group is well funded and carries out research in shock-wave physics, boundary-layer investigations, unsteady aerodynamics, supersonic fuel-injection and mixing, supersonic combustion and plasma flows.

116. By contrast, the civil engineering department had about 28% and chemical engineering about 45% women at the end of the century. In 1987 civil engineering gave seven of its thirteen first class degrees to women. *ICPR*, 26 October 1987.

117. The concrete section continued to flourish after the retirement of Professor Harris in 1981 and the resignation of Professor Dougill in 1987. In 1996 the MSc concrete structures course celebrated its 50th anniversary and people came from all over world for a gala dinner. The present head of the concrete structures section is Milija Pavlovic (1950–) who was educated at the Universities of Melbourne and Cambridge and joined Imperial as a lecturer in 1978. He was appointed Professor of Structural Engineering and Mechanics in 1996.

Nicholas Buenfeld, Professor of Concrete Structures, a specialist in concrete durability, heads the Concrete Durability Group, a multidisciplinary group which includes also chemists, physicists and materials scientists.

118. Two people associated with this field are: Richard (Dick) John Chandler FREng (1939–) and Amr Salah-Eldin Elnashai FREng (1954–). Chandler

was educated at Loughborough University of Technology and the University of Birmingham. He joined Imperial as a lecturer in 1978 and was made Professor of Geotechnical Engineering in 1990; Elnashai was educated at Cairo University and Imperial College. He joined the staff at Imperial in 1985 and was appointed Professor of Earthquake Engineering in 1992. In 2001 he resigned to take a chair at the University of Illinois.

119. Burland, personal communication. Since his retirement Burland has turned to more fundamental research.

120. For example, such work was carried out by former students Sir Harold Harding and Rudolph Glossop, founders of the firm Soil Mechanics Ltd. Later their company merged with John Mowlem Ltd., a company with which the department has had a long association.

121. Many engineers worked on this problem. Having determined that the lean was due to greater soil subsidence below one side of the tower, Burland and his team gradually withdrew soil from below the other side so as to almost level the ground — not completely, since Pisa still wanted a leaning tower. An interesting TV programme was made about Burland's work at Pisa.

122. David Malcolm Potts FREng (1952–) was educated at King's College London and at Cambridge. He worked for some years at the Shell Research Laboratories in Rijswijk on problems relating to marine and offshore structures. He joined Imperial in 1979 and was appointed Professor of Analytical Soil Mechanics in 1994.

123. The provision of mobile water purifying units is something the chemistry department was doing early in the century. Indeed, A. Wilhelm Hofmann and Edward Frankland pioneered modern water purification methods at the RCC and RSM already in the nineteenth century. For Imperial's role in supplying water purification units during the First World War, see chapter 5.

124. Hydrology used to be associated with sanitary engineering for local councils etc. and was viewed as far less glamorous than other areas of civil engineering. Times have changed; though, recently, with the appointment of Christopher Swan as Professor of Hydrodynamics, there has been a renewal of interest in hydraulics.

125. Howard Simon Wheater FREng (1949–) was educated at Cambridge and worked in the aeronautical industry before becoming a research student in hydrology at the University of Bristol. He joined Imperial in 1978 as a lecturer and was appointed Professor of Hydrology for Environmental Management in 1993. He is a specialist in flood hydrology, water resources,

water quality, and waste management. He has a special interest in arid zone hydrology and has worked extensively as a consultant in the Middle East.

126. Other professorial staff in this section include David Butler (1959–), Professor of Water Engineering, who heads the Urban Water Research Group. Butler was educated at Imperial and returned to the college after working for Ove Arup and teaching at South Bank University; also Nigel Jonathan Douglas Graham (1953–), Professor of Environmental Engineering, who was educated at Cambridge and Imperial. Graham directs the course in environmental engineering and is a specialist in water and wastewater treatment.

127. This Centre, supported by British Rail and the Rees Jeffreys Road Fund, was part of the larger University of London Centre for Transport Studies established jointly by University College and Imperial. The current head of the University of London Centre is John W. Polack, Professor of Transport Demand. He joined Imperial as a senior lecturer in 1995. For more on Ridley see chapter 13.

128. Stephen Glaister (1946–) was educated at the University of Essex (mathematical economics) and the London School of Economics. Before coming to Imperial in 1998 he was Cassel Reader in Economic Geography at the LSE. Glaister has served on a number of transport advisory boards and was a non-executive board member of London Transport, 1984–93.

129. Christopher Mark Wise FREng (1956–) was educated at the University of Southampton. After work in the UK, Australia, and the USA, he joined Ove Arup and Partners, becoming a director in 1992. He co-founded Expedition Engineering Ltd and co-designed a number of major projects such as the American Air Museum in Duxford, the new Commerzbank headquarters in Frankfurt, and Channel 4's new headquarters in London, before coming to Imperial as Ove Arup Foundational Professor of Civil Engineering Design in 1998. Wise co-designed the Millennium Bridge over the Thames (with Norman Foster and Partners and sculptor, Sir Anthony Caro). Because of its initial problems, he experienced some friendly teasing from some of his more traditional colleagues who noted that by paying a little more attention to engineering science he could have avoided the infamous wobble. Wise is a good communicator and has made several broadcasts. In BBC TV appearances he has, among other things, 'recreated' Caesar's crossing of the Rhine by building a replica bridge over the Tyne without the use of modern technology, and demonstrated the construction and use of a large Roman catapult.

130. Ronan Point was a 22-storey building in Newham which was brought down by a gas explosion. Owen Saunders was on the committee that looked into the causes of the collapse.

131. For Dowling see chapter 13. His immediate successor as head was Roger Edwin Hobbs (1943–) who was educated at Imperial and spent his professional life at the college. A specialist in undersea pipelines, he became Professor of Engineering Structures in 1990 and was head of department 1994–7.

132. David Arthur Nethercot FREng (1946–) was educated at the University of Wales (Cardiff) and began his career there as a lecturer before moving to the University of Sheffield. In 1989 he was appointed Professor of Civil Engineering at the University of Nottingham, where he was also head of department.

133. This practice, not unique to civil engineering, seems to have been more necessary in this than in other departments. In the early twenty-first century, about 30% of the civil engineering department funding came from government agencies, the rest largely from civil engineering firms and industry.

134. Richard James Jardine FREng (1953–), Professor of Geomechanics, is the current Director of the MSc programme. Jardine was educated at Imperial College and joined the staff as a lecturer in 1984. He was given a chair in 1998.

135. The longest serving appointee from the Saunders period is probably John Picket who joined the department as a junior technician in 1953 when he was fifteen years old. In the late 1990s he was running the carpenters shop. His son Kevin also came to work in the department as a technician.

136. James Hunter Whitelaw FRS FREng (1936–) was educated at the University of Glasgow and spent two years in the USA at Brown University before his appointment as a lecturer at Imperial in 1964. He became Professor of Convective Heat Transfer in 1974 and was head of the thermofluids section for twenty-five years. He brought many important research contracts to the college, and has been a major presence within the department. He continues to work as a Senior Research Investigator. Also working in this area were Frederick Charles Lockwood (1936–) who joined Imperial as a research assistant in 1964 and was Professor of Combustion 1990–2002; and William Phillip Jones (1943–), Professor of Combustion, who was educated at Imperial College and worked for Rolls Royce before returning to join the staff in 1977. For Spalding see chapters 10 and 13.

137. Hugh Alexander Spikes (1945–) was educated at the University of Cambridge and joined Imperial as a lecturer in 1972. He was promoted to

Professor of Lubrication in 1996 and won the Tribology Trust Gold Medal in 2004.

138. For J. G. Williams see also chapter 13. The chair in plasticity, held by J. Alexander, was converted to a chair in robotics on his retirement (see chapter 13). But robotics did not take off in mechanical engineering and was more attractive to computing specialists. The robotics group was later disbanded and those with mechanical engineering interests, such as A. Amis and B. Davies, joined the biomechanics section. Andrew Amis is now Professor of Orthopaedic Biomechanics and head of the section. For Davies see chapter 15.

139. Anthony (Tony) James Kinloch FREng (1946–) was educated at Queen Mary College. He headed a research group on advanced materials at the Ministry of Defence before joining Imperial in 1984 as Reader in Engineering Adhesives. He became Professor of Adhesion in 1990. He leads research in the areas of adhesion, adhesives and polymer and fibre composites in the mechanics of materials section, and is a specialist in how to join new thermoplastic composite materials. In 1997 he was awarded the Hawksley Gold Medal from the Institution of Mechanical Engineers.

140. Roderick (Rod) Arthur Smith FREng (1947–) was educated at the Universities of Oxford and Cambridge. He was a Fellow of Queen's College, Cambridge when Oxburgh was Master, and was Professor of Mechanical Engineering and head of the department at Sheffield before coming to Imperial. He was also Professor of Advanced Railway Engineering at the Royal Academy of Engineering and Chairman of the Advanced Railway Research Centre. Smith's earlier field was the fatigue/fracture of metals. He has applied this more recently in the area of railway engineering, the field which he brought to Imperial.

141. David John Ewins FRS FREng (1942–) was educated at Imperial College and the University of Cambridge. He returned to Imperial as a lecturer in 1967 and became Professor of Vibration Engineering in 1983. The Centre for Vibration Engineering includes now also the Rolls Royce University Technology Centre. Also associated with the Centre is Nicholas (Nick) Alexander Cumpsty FREng (1943–). Cumpsty was educated at Imperial College and the University of Cambridge. He was Professor of Aerothermal Technology at the University of Cambridge before joining Rolls Royce as Chief Technologist in 2000. He returned to Imperial as Professor of Mechanical Engineering, and as head of department, in 2005.

142. Today power systems and distribution problems are linked to control system theory. People are working with energy/power systems, with grids,

and with integrating many new power inputs from small to large scale, including wind, solar, tidal, as well as more conventional sources.

143. By the early twenty-first century there were research groups in: Control and Power, Circuits and Systems, Communications and Signal Processing, Intelligent and Interactive Systems, and Optical and Semiconductor Devices.

144. Sayers moved to become head of computing in 1983 (see below). David Quinn Mayne FRS FREng (1930–) was educated at the University of the Witwatersrand and Imperial College. He returned to the Witwatersrand, rejoining Imperial as a lecturer in 1959. He was Professor of Control Theory 1971–89 before leaving for a position at the University of California (Davis). He returned to Imperial as a Senior Research Fellow in 1996.

    David John Noel Limebeer FREng (1952–) was educated at the University of the Witwatersrand and the University of Natal. After working as an engineer in South Africa he became a research associate at the University of Cambridge before joining Imperial as a lecturer in 1984. He became Professor of Control Engineering in 1993.

    Another future head of section, Richard Bertrand Vintner (1948–), was educated at the Universities of Oxford and Cambridge and joined Imperial in 1974. A theoretician in control systems, he was appointed Professor of Control Engineering in 1991.

145. Robert Spence FREng (1933–) was educated at the Hull College of Technology and Imperial College. He worked for General Dynamics/ Electronics in Rochester, N.Y. before joining Imperial as a lecturer in 1962. He was appointed Professor of Information Engineering in 1984 and succeeded Professor Aleksander as head of department in 1997. Spence works in the area of computer assisted circuit design. He founded the spin-out company Interactive Solutions Ltd. which markets interactive graphics programmes and engineering design programmes.

146. Anthony George Constantinides FREng (1943–) joined Imperial College as a lecturer in 1971 from a position in the research department of the Post Office. Constantinides is a pioneer in digital signal processing, a generic technology with applications in many areas from the medical to the defence industries.

147. Igor Aleksander, Professor of Neural Systems Engineering, has an interest also in robotics and automaton theory. (One of the first modern automaton theorists in the UK was J. J. Florentin who earlier worked at Imperial in electrical engineering and in computing.) Aleksander succeeded Mayne as head of department. For more on Aleksander, see chapter 13.

148. Erol Gelenbe (1945–) was educated at the Technical University in Ankara and at the Université Pierre et Marie Curie in Paris. Professor Gelenbe is also Head of the Intelligent Systems and Networks Research Group.

     Work on pattern recognition and artificial intelligence is also carried out by Professor Ebrahim Mamdani FREng (1942–). Mamdani went to school in Tanzania before entering the University of Poona (Pune). He was a research student at Queen Mary College and joined the staff there on gaining his PhD in electrical engineering. He held the chair in electronics (1984–95) before joining Imperial in 1995 as the Nortel/Royal Academy of Engineering Professor in Telecommunications Strategy and Service. Now an emeritus professor, Mamdani continues to carry out research in the department.

149. Mino Green (1927–) was educated at Durham University before moving to MIT where he headed a group in the Solid State Research division. After working both at universities and in industry in the USA, Green joined Imperial as a lecturer in 1972 and became Professor of Electrical Device Science in 1983.

     Richard R. A. Syms FREng (1958–) came to Imperial in 1987 and was appointed Professor of Microsystems Technology in 1996. He has been Head of the Optical and Semiconductor Devices section since 1992.

150. By 2005 the annual intake of undergraduates was about 170. They entered one of two streams, electrical engineering or electrical engineering and management. This is likely to change as some management and business education will soon be integral to all engineering degrees; engineering and management degrees will no longer be offered.

151. William (Bill) Arnot Wakeham FREng (1944–) was educated at the University of Exeter in physics. A specialist in thermodynamics, the measurement of fluid properties, and multiphase fluid systems, he joined the department as a lecturer in 1971 and became Professor of Chemical Physics in 1985. He was appointed Deputy Rector in 1997, holding also the Pro-Rectorships in research and resources. He was an extremely hardworking administrator during the transitional period when the School of Medicine was founded. He left in 2001 to become Vice-Chancellor of the University of Southampton, though retaining a tie to his old department as a visiting professor.

152. *THES* (16 January 1995); article reporting on a US survey of chemical engineering departments in the UK. Needless to say the department consistently gained a 5* rating in the RAE.

153. Geoffrey Frederick Hewitt FRS FREng (1934–) was educated at the University of Manchester Institute of Science and Technology before

working for the UKAEA. Hewitt was appointed Courtaulds Professor of Chemical Engineering in 1990 when he came to Imperial full-time. Hewitt is not only a major researcher, but also a fine lecturer on a wide range of topics including heat and mass transfer and nuclear technology. On retiring in 1999 he continued to be active in research and in the supervision of postgraduate students.

154. Stephen Michael Richardson FREng (1951–) was educated at Imperial College and returned to the college as a lecturer in 1976, after a few years at Cambridge as a research fellow. He was appointed Professor of Chemical Engineering in 1992. Richardson and Graham Saville developed a computer programme called Blowdown which simulates actual blowdowns (sudden depressurisation in oil pipelines and containers) and has been used successfully on offshore oil and gas platforms and at gas terminals and petrochemical plants. For more on Richardson and Saville see chapter 13.

155. Brian James Briscoe (1945–) was educated at Cambridge and came to the college as Reader in Interface Science in 1978. He was made Professor of Interface Engineering in 1992.

156. Dame Julia Stretton Higgins FRS FREng (1942–) was educated at Oxford and, after working for a few years as a research fellow in Manchester and in France, came to the college with Sir Geoffrey Allan in 1973. She was appointed Professor of Polymer Science in 1989 and Principal of the Faculty of Engineering in 2006. Higgins is Foreign Secretary and Vice-President of the Royal Society. For some of her other contributions see chapter fifteen.

157. John Douglas Perkins FREng (1950–) was educated at Imperial College and was a doctoral student under Roger Sargent, gaining a PhD in 1973. After a short period at Cambridge, he returned to the department as a lecturer in 1977. Perkins was head of department 1996–2001. He became Courtaulds Professor of Chemical Engineering in 1999, in succession to G. Hewitt. Perkins was the first Principal of the new Faculty of Engineering, but left shortly after his appointment to become Dean of the Faculty of Engineering and Physical Sciences at the University of Manchester. As mentioned above, a University of London Centre for Process Engineering was founded in 1994. After Perkins' resignation, the directorship was taken over by E. N. Pistikopoulos.

158. The Director of the undergraduate programme in the late 1990s was L. S. Kershenbaum, Professor of Chemical Engineering, and a specialist in applied catalysis and process system engineering. Kershenbaum was educated at The Cooper Union and the University of Michigan and joined Imperial in 1970 after working for Dupont in Wilmington, Delaware.

159. Plans were afoot in the early twenty-first century to build up the combustion section with new appointments.

160. David Chadwick (1945–) was educated at University College London and joined the department in 1969. He was appointed Professor of Applied Catalysis in 1974. The department bid successfully for an Institute of Applied Catalysis in 1997.

161. Robert Anthony (Bob) Kowalski (1941–) was educated at the University of Bridgeport, and at Stanford and Edinburgh Universities. After further study at the University of Warsaw, renowned for its school of logicians, he worked in the new field of logic programming while holding post-doctoral research fellowships in Edinburgh. He was appointed Reader in Computing at Imperial in 1974, and Professor in 1982. Together with Alain Colmerauer in Marseille, he developed *Prolog*, a language for computer programming using a rule-based approach, and based on human modes of logical reasoning. Kowalski has posted an interesting autobiographical essay on the computing department website.

162. See Michael S. Mahoney, 'Computer Science: The Search for a Mathematical Theory', in John Krige and Dominique Pestre (eds.), *Science in the Twentieth Century* (Amsterdam, 1997), ch. 31.

163. In the early 1990s Fujitsu gave the college a 'massively parallel' supercomputer, a 128 processor AP1000, the only one of its kind in Europe. It was to be used for complex modelling problems and non-numeric computation. The London e-Science Centre is successor to the Fujitsu Centre.

164. John Darlington (1947–) was educated at the LSE and the University of Edinburgh. He joined Imperial as a lecturer in 1977 and was appointed to a chair in 1985. His work on parallel computing contributed to the 1992 commercial ICL Goldrush machine.

165. Dov Gabbay (1945–) gained his PhD in non-classical logics from Hebrew University in 1969. After work at Stanford University he returned to Israel, and a professorship at Bar-Ilan University. He came to Imperial on a sabbatical in 1982 before being offered a chair. He now holds the Augustus De Morgan Chair of Logic at King's College London. At Imperial Gabbay, together with Maibaum and Abramsky, began publishing handbooks. See S. Abramsky, D. Gabbay and T. S. E. Maibaum, *Handbook of Logic in Computing Science* (1992) and D. Gabbay *et. al.*, *Handbook of Logic in AI* (1993).

166. E. Barry Richards (1942–) came to Imperial from a chair at the University of Edinburgh. He was appointed Professor of Computing Science in 1989.

167. Thomas (Tom) Stephen Edward Maibaum (1947–) was educated at the University of Toronto, at Queen Mary College and Royal Holloway

College. On gaining his PhD he took an assistant professorship at the University of Waterloo. He joined Imperial as a lecturer in 1981 and was appointed Professor of the Foundations of Software Engineering in 1990. In 1999 he moved to King's College London, returning to Canada in 2004, and to a Canada Research Chair at McMaster University.

Samson Abramsky FRS (1953–) was educated at Cambridge where he took a degree in philosophy, and at Queen Mary College London. On graduating he worked as a programmer for GEC Computers Ltd before taking a lectureship at QMCL in 1980. He came to Imperial as a lecturer in 1983 and became Professor of Computing in 1990. In 1993 he resigned to take a chair in Edinburgh. Today he is Christopher Strachey Professor of Computing at the University of Oxford.

John Joseph Florentin, a research student under John Westcott, was appointed to a lectureship in electrical engineering in 1960. He left in 1964 for a period at Brown University, returning to Imperial in 1966 as Reader in the Theory of Computing. He left shortly after for a chair at Birkbeck College.

Another theoretician is Christopher Leslie Hankin (1954–). Educated at City University, he was a lecturer at Westfield College before joining Imperial as a lecturer in 1984. He became Professor of Computing Science in 1995, and later headed the section on programming language theory. He was appointed Pro-Rector for Research in 2004.

168. Jeffrey Kramer (1949–) was educated at the University of Natal and at Imperial where he took the MSc course in computing and control in 1971. He joined the staff as a lecturer in 1973 while carrying out research for a PhD with Jim Cunningham. He was appointed Professor of Distributed Computing in 1995. Other specialists in software engineering during this period include Morris Sloman and Jeff Magee. Magee, who succeeded Kramer as head of department in 2004, is concerned with areas such as database distribution, management, and security.

169. See College Plan, 1991–5.

170. For rumour, see Kowalski's autobiography on the department's website.

171. Logic programming turned out not to have the utility as a model for computation hoped for at the start. But Kowalski found applications for his work in new areas. For example, together with one of his former PhD students Marek J. Sergot, later a professor in the department, he worked on a European Community research project exploring some applications of logic programming to air traffic flow management. Another former student and close associate was Keith L. Clark, later also a professor in the department. Aspects of logic programming are still alive at Imperial also in the

work of another Edinburgh graduate, Stephen Howard Muggleton (1959–), who worked at Oxford after graduating with his DPhil. He was appointed Professor of Machine Learning at the University of York in 1997, joining Imperial in 2001 as the Joint Research Council Chair of Computational Inference and Bioinformatics. Sergot, Clarke, and Muggleton are today part of a strong Artificial Intelligence presence at the college.

172. Guang-ZhongYang, working in the interactive media section, is a specialist in the area of medical imaging and has close research ties to the medical school. Bioinformatics is part of a larger informatics section in the department. It includes also a group, under Berc Rustem, concerned with computational management more generally.

173. There are sartorial characteristics also. Walking through this department, including professorial offices, one sees far more casual clothing than in the mechanical or electrical engineering departments, both of which are Victorian foundations with more formal elements to their cultures.

174. As of now staff must be associated with departments for the RAE, and for internal assessment leading to promotion. But this could change. For example, teaching could be delivered increasingly by non-research active staff, large research groups within centres could become detached from departments for the purposes of the RAE, and people holding research professorships could be given the same rights as other professors, a motivation for others to leave teaching. Moves such as these could further detract from departmental autonomy and diminish the quality of undergraduate instruction.

chapter seventeen

Conclusion

The way in which an institution emerges can have profound long-term consequences for its future development. This is true of Imperial College where the ideals that motivated the founders of the three earlier colleges, and those that motivated the federation of the colleges in 1907, remain embedded within it. The college has what could be termed a mimetic, or even genetic, memory, in that many early ideals and behavioural patterns have replicated over time. This is not to say that the future of the college was predetermined, simply that its distinctive character owes much to its foundational history. The same is true of the academic departments each of which has its own characteristic history similarly influenced by the way and time in which it emerged.[1]

The three colleges that joined to form Imperial in 1907 were pioneering institutions that offered advanced level education in science and technology with a pedagogical emphasis on the practical. Even before it became widely understood that skilled scientists and technologists were important to the nation's economic well-being, the Royal School of Mines, The Royal College of Science (earlier also the Royal College of Chemistry), and the City and Guilds Central Technical College were producing graduates who were to play their part in government, academic and industrial science, in the mining and metallurgical industries, in engineering, and in science education. In chapter two we have seen something of the ideals that existed within the older colleges; that they were open to both men and women, and that their students came from a wide range of class backgrounds. Service and commitment to the state, to industry, to the advancement of knowledge, and to entrepreneurial ideals, were all valued, but weighted differently in each. Further, all three were responsive to the needs of a modern political economy for new kinds of technical expertise.

By the early twentieth century the need for more well trained scientists and engineers was widely voiced. Imperial College was founded by the state in response to lobbying by industry, academic scientists and engineers, those managing resource industries in the Empire, and the London County Council which was concerned that local industries were unable to find suitably trained employees. We have seen how the decision to unite the three South Kensington colleges in a new enterprise was made, and how it was highly contested. Given the mix of ideals existing in the older colleges, a way forward was possible only because people showed respect for what others were doing. Pure and applied research, teaching science and engineering from both pure and applied perspectives, taking out patents, publishing in academic journals, industrial and government contract work, working together with large engineering firms, being entrepreneurial — all were seen as noble options. The college charter with its directive to serve the needs of industry was taken seriously. The utilitarian approach, together with a determination to carry out both teaching and research at the highest level, gave Imperial a unique position among institutions of higher education in Britain. One hundred years later the college is especially well placed to help Britain in an increasingly global economy. There will be even more reliance than in the past on the skills and ingenuity of scientists, doctors and engineers if a high standard of living is to be maintained.

Not only did Imperial College early support a wide range of activity, until recently it had a horizontal governance structure which gave people much independence. This was necessary since the college had virtually no endowment and little income, and had to rely on the initiative of its staff. The principal responsibility of the Rector was to appoint highly motivated and capable heads of department, and to allow them freedom from as much central interference as possible. As was noted by Professor Chibnall, for example, his experience at Imperial in this regard compared favourably with his later experience at Cambridge. In return, the heads, while given much power, were to encourage independent research groups within their departments and build solid undergraduate programmes. Responsibility for funding was devolved from the centre of the college.[2] It was for the departmental heads to seek outside sources of support, to encourage their staff to do the same, and to build their departments as they saw fit. Even in the early days of the college the Dean of the City and Guilds College and, from the Second World War the Deans of all three colleges, had little say in departmental affairs.[3] They represented the interests of the older colleges on important college-wide committees, and ensured overall standards by keeping an eye on student admissions, and on staff appointments and promotions. They could not override departmental heads — that was the prerogative of the Rector to whom the heads were accountable, and to whom they had direct access. Imperial College was lucky in that many of the early heads set high standards for those that followed. For example, John Farmer, Professor of Botany, began with a department that occupied just two large rooms in the Science Schools building on Exhibition Road, and some greenhouse space on the roof. By the time he retired twenty-two years later, the department had two large buildings, four professors, two assistant professors (readers) and six junior academic staff members. That the department had four professors is illustrative of another characteristic of Imperial College in its early years, namely a willingness to promote talented people when still relatively young. This pattern was tempered after the college joined the University of London in 1929, but regained strength when Imperial became more independent from the university later in the century. The fact that departmental heads were left to their own devices could lead to problems, but overall it was a system that served Imperial well, at least until the Jubilee Expansion that began in 1957. Like Farmer, other strong heads from the early period left major traces in the departments they served. For example, some of the pedagogical ideals that A. Wilhelm Hofmann brought from Liebig's laboratory in Giessen to the Royal College of Chemistry are still observable in the chemistry department today. William Unwin, too, left an engineering legacy that was detectable many years later in the kinds of expertise found in both the civil and mechanical engineering departments. William Bone, Professor of Chemical Technology, is another who stands out. In chapter four it was shown how he set up the new department in a way that proved highly successful until the introduction of an undergraduate course in chemical engineering. Only then did his model, favouring chemical technology, begin to show its weaknesses.

Nonetheless, a strong chemical technology and chemical engineering tradition at Imperial owes much to him.

While the 'feudal' system of governance, covered in the early chapters, and discussed in more detail in chapter nine, contributed to the success of the college in the first half of the twentieth century, 'baronial' rule did not fit the more democratic mood following the Second World War. It was contested from the 1950s on, and came to an end in the early 1970s. This was in part a consequence of the large number of young staff who joined the college during the expansionary period bringing new political demands, many related to college governance, with them. Many young lecturers had held post-doctoral positions in the United States where they experienced more democratic academic procedures.

The older pattern of headship entailed also a degree of paternalism, something that had its roots in the three constituent colleges. Professors were expected to care for the staff and students in their charge and, when well managed, departments had something of a family spirit. The City and Guilds College saw itself as replacing the traditional guilds' role of apprenticeship training, and cared for its students accordingly. In the Royal School of Mines there were strong ties between professors and their students, many of whom were to leave Britain for work overseas.[4] With Imperial's major emphasis on practical experience, geology and mining students carried out much collective field work and, as in the biological sciences, this helped to build closely knit departments. In physics and chemistry, the long hours spent with others in the laboratory helped in the creation of strong bonds there. Among retired staff members are some who were students in the 'feudal' era and who look back to what they remember as a more caring and freer intellectual environment. It is hard to know how to judge these matters, far easier to simply describe what has happened. With increased numbers at the college, and with more democratic procedures, came also more bureaucracy. Well defined rules relating to appointments, promotions and so on have made much discretionary power a thing of the past. In many ways this is for the better, but new procedures were bought at a cost.

In the early period the mandate to serve industry was interpreted rather differently than later in the century, at least in the pedagogical area. Large industrial-style laboratories, such as the Bessemer and Whiffen, were constructed, and the Tywarnhaile mine was purchased in Cornwall. The idea was to give students something as close as possible to an industrial (or mining) experience while at the college. Later in the century, the engineering departments began following patterns pioneered in the polytechnics, with many students working in industry as part of their training, or carrying out experimental work in actual industrial settings. Imperial College was a leader among major research universities in offering what are known elsewhere as 'sandwich' courses. The chemistry department gave up the Whiffen Laboratory in mid-century.[5] The space was converted for research, something that became more central to Imperial's identity after the Second World War. The Bessemer Laboratory was also converted after the war as new pedagogical ideas entered the metallurgy department. Also important was that the mandate to serve industry extended to the industries of empire. This point is discussed in chapter seven, but a consequence of the mandate was a more generally outward-looking stance, not limited to Empire and Commonwealth. For example, Imperial College was a pioneer in foreign student exchanges, and one of its staff members founded the International Association for the Exchange of Students of Technical Experience shortly after the Second World War.

In a college specializing in science and technology it is not surprising to find that there has always been a strong technocratic spirit, something evident already in the views of H. G. Wells, a student at the Normal School of Science (forerunner to the Royal College of Science). The idea that the country would be better managed were those in power more knowledgeable in science and technology, more prepared to use what is seen as 'scientific method', and to support the work of scientists and engineers, has influenced the ways in which the college has responded to the outside world for much of its history.[6] This was well illustrated in those chapters dealing with the early governance of the college and in those covering the two world wars but the spirit is still detectable today. We have seen also how Imperial College lived up to its name, at least until the end of empire. The connections to empire made by biologists, geologists, mining engineers, and others during the imperial period were often lasting and have been of great importance also in more recent times. This is true both in terms of students coming to the college from the Commonwealth, and in terms of the kind of research work carried out. For example, another strong head of department, James Munro, made applied entomological work one of Imperial's major strengths. He built on foundations laid by Adam Sedgwick before Sedgwick's early death. Both Sedgwick and Munro had imperial interests in mind, but work related to theirs continues to this day, and still attracts many overseas students. This, and similar examples discussed elsewhere in the book, suggest that once an institution makes a name for itself in a particular field, it can, with good management, remain in the forefront of that field for a very long time. In the case of applied entomology, however, Imperial College was also lucky in that Munro insisted that his department was not closed down after the war, and that continuity was maintained.

In carrying out research I was struck by the relative importance of botany and zoology in the early years of Imperial College, despite their attracting fewer students than other fields. This may well have had something to do with the intellectual legacy of T. H. Huxley, as well as with the dynamism of people like Farmer. But more significant was biology's perceived importance to imperial industries such as cotton, rubber, tea and coffee. Some of the largest research grants that came to the college in its early years were for the biological sciences. Similarly, large donations were made in support of the new botany buildings, one of which was later given to zoology. By the end of the Second World War, however, the biological sciences were in some difficulty and the departments came close to being shut down. When I was a student at the college during the early 1960s the emphasis was very much on engineering, mathematics and the physical sciences. Somehow I had assumed that this had always been the case. Today, the life sciences are once again in the ascendent. This has much to do with the revolution in molecular biology and the founding of the medical school. But, much more than earlier, the life and medical sciences are today heavily dependent on the physical sciences, engineering, and information technology for their advancement.

Archival research has made clear the existence of networks influencing who came to Imperial College, at least until the 1960s. One such network, discussed in chapter eight, was related to aeronautics. The network influenced both senior and junior appointments at the college from the 1920s to the 1960s — not just in the aeronautics department. It brought three Rectors to the college: Henry Tizard, Richard Southwell and Roderic Hill, and a number of professors, including Professors Levy, Pippard, Hall, Jackson, Squire, Owen, Lighthill, Blackett and Christopherson. One of the first interdisciplinary units to come to the college, namely the Physiological Flow Unit, was also a consequence of the aeronautical network. In addition to showing how some networks were important to the history of the college, this book has demonstrated some interesting intellectual lineages, notably in chemistry, physics, applied entomology and electrical engineering.

These networks and lineages were predominantly male. Male fraternity is an important theme, implicit throughout the book. Commentary on the role of women at the college has been integrated into the various chapters and not kept apart. I believe this to be methodologically important since separate treatment only furthers the gender divide. Women have always been integrated in the life of the college and it was important to show how this happened. While women students were, until fairly recently, a small minority, there were more women staff (including technical staff) working at the college than I had imagined — though still a relatively small number.[7] It is surprising that, even in the early period, academic women were seen by most of their colleagues as being fully as capable as men of doing good research work in science and engineering. Their intellectual contributions were respected by their close associates. However, as in the wider society, women were seen as being unsuited to the world of affairs. This serious prejudice worked against their being promoted, being consulted on important college affairs, being included in informal social gatherings, and it largely kept them in positions as research assistants. Many professors had female research assistants, and many publicly acknowledged their indebtedness to them. Even the brilliance of some of these women was publicly acknowledged and they were officially rewarded by learned and professional societies. But, until relatively recently, the idea of promoting women, encouraging them to publish independently, to be active in learned or professional societies, or to aim for professorships, rarely, if ever, occurred to their male colleagues. Further, women graduates were not encouraged to enter industrial or engineering firms. Women were suited to the domestic, not worldly, sphere. They had families to care for, or might have families in the future. And, psychologically, they were unsuited to competition, to management, and to leadership. In the safety of the laboratory, however, women with ability could perform at the highest level. Of course there were some exceptions and a few women did advance beyond assistantship positions, though with more difficulty than their male colleagues. For example, Helen Porter was given the opportunity to run her own laboratory. But she was given a readership only after she had been elected a Fellow of the Royal Society. A few years later, in 1959, she became the first woman to be given a chair at the college. In the chapters on the two world wars we have seen how women joined the college maintenance parties, and how they worked alongside men in war-related work. Further, in accordance with the college charter that there be no sexual discrimination in remuneration, women were paid the same as single men when working in the college munitions factories or on other wartime projects — true in both wars.[8] In overriding the discriminatory wartime pay bonuses set out by the government, the college was out of step with other institutions.

The type of broad prejudice mentioned above affected also women students. As a consequence, they rarely received the kind of mentoring that some of their promising male counterparts did. Women were not expected to make the kind of careers where time spent in mentoring them would pay off in the future. The result was a degree of social exclusion. However, in more recent years (and possibly also in the past), women students at Imperial have in one important respect fared better than women in science and engineering faculties elsewhere. Data collected at the college in the 1990s shows that, whereas in most other universities women did less well than men in the class of degree awarded, at Imperial there was no discernible difference. It is difficult to explain this institutional difference, but perhaps the long tradition of acceptance on an intellectual, if not social, level may have played a role.

During the 1970s attitudes changed, largely as a consequence of increasing feminist activity. Women at the college during this decade faced a more hostile environment than earlier or later. Young women who voiced their opinions, who questioned old assumptions, were seen as pushy and were either ignored, or criticized for their outspokenness. Further, with paternalism under attack, women lost the kind of mentorship and protection they had earlier enjoyed — a good thing in the longer term. But the ideological climate was in transition and, during the 1980s, a younger generation of men began to see things differently. Further, there was a perceived need to attract more women students to the college. Women on the staff, who earlier had rarely managed to rise much above a lectureship, were slowly being promoted to more senior positions. Since the 1980s the situation for academic women seeking advancement at the college has improved; though, as was discussed in earlier chapters, they still face some difficulties.

By and large, the engineering and physical sciences departments were consistently strong throughout the twentieth century, despite complaints by some of the incoming professors. For example, Professors Pippard, Hall, Heilbron, and Jackson all decried the state of their departments on arrival at Imperial. This was unfair since their various predecessors were mostly very good. However, complaints by newcomers can be a healthy sign, especially when coincident with sensible plans for renewal and modernization. Engineering, more generally, changed after the Second World War with less emphasis on contract research and more on academic research and engineering science. But old strengths in the physical sciences and engineering, such as those in energy, internal combustion, thermodynamics, organic chemistry, astrophysics, applied optics and engineering structures, continued to serve the various departments well. Notable in this regard is mechanical engineering which, like applied entomology, built on its traditional areas over many years, and with great success. This is not to say that mere conservatism works well. What I am attempting to describe is what might be termed a fecund natural growth, as opposed to continual innovation which can be risky.

Departments in the Royal School of Mines were less fortunate than others. The end of empire led to a decline in the demand for British trained experts in geology and mining — and, to a lesser degree, in metallurgy. The speed of the decline is surprising given that economic imperialism continued long after the political collapse of empire. Many British companies retained their control of resource industries overseas. But the nationalisation of many of these industries in newly independent countries, international growth in higher education, and the desire to hire locally trained staff, meant that the glory days of the Royal School of Mines were over. We have seen how a difficult transition was managed, and how mining expertise remaining at the college became refocused; at first on environmental work such as mine and quarry reclamation, and hydrology and, later, in other areas. One area where Imperial retained its expertise is in oil exploration and petroleum technology. These fields, built up by Professor Illing and others during and after the First World War, went from strength to strength; related work continues to this day not only in earth science and engineering, but in other departments also.[9]

It was not just departments in the Royal School of Mines that turned towards environmental work. This area attracted interest across the college during the 1970s. For example, the Department of Civil Engineering renamed itself Civil and Environmental Engineering in 1998 in recognition of a transition that had begun in the mid-1970s. In addition, the department's traditional strengths in concrete and steel structures, and its postwar interest in soil mechanics, have served it well. Environmental work flourished also in other departments, and at Silwood where Professor Southwood promoted it alongside work in ecology. The impetus for creating the first major interdisciplinary centre at the college, namely the Centre for Environmental Technology (ICCET), came from Southwood, and from some of his staff, notably Gordon Conway, in the department of zoology and applied entomology. This centre was born also of the financial difficulties of the 1970s since the environment was one of the few areas where the Rector, Brian Flowers, could see the possibility of bringing major research funding to the college.

We have seen that after the Second World War engineering and the physical sciences came to the fore, and that during the first expansionary period they fared better than did the biological sciences. This in part reflected a wider perception of their importance at the dawn of the nuclear age, but it also had much to do with the personal vision of the Rector, Patrick Linstead, and the interests of Professors Jackson, Blackett and Saunders, then the most powerful voices in the college. Linstead promoted also biochemistry, but the pendulum began swinging back towards the life sciences only during the rectorship of Brian Flowers. New directions in the chemical, physical and engineering sciences after the war led also to a wealth of new materials, from polymers to ceramics to composites. Relatedly, the metallurgy department turned more to physical metallurgy and increasingly to new materials. Mineral technology overtook mining in importance. Two other aspects of the postwar boom were a rapid increase in the size of Britain's chemical industry, and the start of nuclear power generation. Chemistry, chemical technology, chemical engineering and mechanical engineering expanded as a result.

The boom brought also more general expansion and much growth in the university sector from the 1950s on. The Robbins Report of 1963, discussed in chapter nine, prompted the government to aim at providing higher education for all those who could profit from it. The consequences flowing from that decision provide the backdrop to several of the book's later chapters. One obvious result of Robbins has been far greater competition for the monies available for undergraduate and postgraduate education, as well as for research. This has led to the introduction of new procedures, both nationally and internal to the college. Imperial has managed to situate itself well for the new kinds of competition that have evolved over the past thirty years and has been ranked not only among the best universities in the United Kingdom, but among the best in the world.[10] While the college still carries the name 'Imperial', it is now a world institution having also closer ties to Europe than to the old empire. Britain's membership in the European Union has brought many European students and staff to the college; though anyone walking through the college will be struck by the yet larger international presence. Like so much else, this relates to an earlier history when foreign students were encouraged to come to the college, when ties to countries overseas were built and nurtured through connections to college alumni, and through shared research.

This book, especially chapter eleven, has shown something also of student and corporate life. It is an aspect that should not be overlooked in any evaluation of the whole. We have seen how the college developed its own forms of ceremonial and ritual, how clubs and collegiality among students and staff evolved over the century, and how social and cultural life fed also on the location of the college in South Kensington. While the nature of social life at the college underwent major change over the century, traces of the old can still be seen. But the Imperial College Union, once a social club for both staff and students, has become a bureaucracy overseeing the many clubs and facilities now available. This was inevitable given the enormous growth in the student body which is now far less cohesive than it once was. After the Second World War students became wealthier and, increasingly, had more disposable income.[11] They began to look beyond the college for social and cultural activity, something encouraged by Rectors Hill and Linstead, both of whom dissented from the Oxbridge collegiate ideals brought to Imperial by Richard Southwell. Hill and Linstead were themselves educated in London and believed that students at a London college should make use of what the city had to offer. Students who attended the college during the 1950s and 60s were encouraged in this direction — to visit museums and galleries, to attend concerts and the theatre. While none of these activities was new, college administrators were more active in promoting them than before — or later. Ties to the neighbouring music and art colleges were strengthened and new ones, such as to the Royal Court Theatre, were made. As has been discussed, this cultural activity was in part a reaction to the postwar image of the scientist, creator of the atomic bomb, as being dangerously remote from humanistic culture and values. Leaders at the college wanted to show that they were responsive to such criticism, and to the need to broaden the minds of their students. As was discussed in chapter fourteen, the introduction of humanities and social sciences education into the college curriculum was a further result of these postwar concerns. Humanities education has grown over the second half of the century but, as the college moves forward, it will need to think clearly about what kind of non-technical education it offers to, and demands of, its students. It is perhaps pertinent that this book, coming from the Centre for the History of Science, Technology and Medicine, gives the college a way of seeing itself within a larger context.

The postwar period saw the arrival of new affordable restaurants and coffee bars in Chelsea and South Kensington, drawing students yet further away from the union. Local pubs, however, have been a permanent fixture in the lives of staff and students, and many that were present one hundred years ago are still in existence today. By the 1970s the Southside bar had become a favourite meeting place, sometimes with musical entertainment. Pop music became more central to student's lives than earlier, and popular culture, more generally, came to the fore. In part this was due to increased wealth and access to consumer goods such as radios, televisions, and record players. In part it was a consequence of the larger cultural and political movements of the period, leading to an increased consciousness of 'the other' in terms of class, race, and ethnicity, the desire to show solidarity across social boundaries, and a determination to change the political *status quo*. Students at Imperial shared the concerns of students worldwide — concerns related to civil rights, nuclear weapons, peace activism, racism, and sexism. They also turned their gaze inward and demanded a greater say in college governance and curriculum reform. In this they were joined by many of the younger members of staff and much changed as a result. Chapters twelve and thirteen have outlined some of the events that led to the modern college. Not only were they influenced by the politics of the 1960s and 70s but also, and relatedly, by the economic downturn of the 1970s.

Rectors Penney and Flowers had to rethink old ways and it was during their tenure that the college began to take on its modern form.

Another theme running through the book is the somewhat troubled relationship between Imperial College and the University of London. The older colleges that federated in 1907 were justifiably proud of their records and were not keen on becoming subservient to the university. Those working at the college in its early years wanted independence and wanted the government to allow Imperial to offer its own degrees. But governments of the period, under pressure from the university, were unwilling to pass the necessary legislation. By the 1920s many employers were insisting that prospective employees have university degrees for the kind of jobs that Imperial College graduates had traditionally filled. The college associateship diplomas were put in doubt. Indeed, increasing numbers of students were taking two sets of examinations in order to acquire external degrees as well as their diplomas. Had the university simply accepted Imperial's associateships as degree equivalent, the college would probably have joined it earlier. But the university senate, dominated by people with classical schooling, and with degrees in the arts, humanities, and medicine, was insistent on common, university-wide examinations for each of the subject areas, and on common matriculation requirements. The senate was also suspicious of Imperial's pedagogical bias towards the practical, and viewed some of Imperial's subjects, notably mining, as vocational and inappropriate for the awarding of university degrees. Had Imperial bowed to the senate's demands, it would have had to change the type of student it accepted, the nature of the courses being taught, and its own modes of student evaluation. In the dispute between college and university we see a classic struggle between modernisers and conservatives in higher education. Were new kinds of people to be admitted to the university? Was Latin necessary for matriculation? Were seemingly vocational subjects worthy of degrees?

Questions such as these were central to the early debates. The Rector of the period, Thomas Holland, proved to be a good negotiator and, in the end, won the concessions needed for Imperial to join the university on its own terms, which it did in 1929. While the fight for independent university status was lost, the agreement was an important victory for the modernisers. The college became the first of the large London colleges to set its own examinations, and all of its courses were accepted for degrees. Imperial was proud of its independence, proud to be different. By the 1960s it gained yet more independence when Patrick Linstead argued successfully for Imperial's Rector to join the Committee of Vice Chancellors and Principals and, at the same time, for the college to gain direct access to the University Grants Committee.[12] Once again, Imperial led the way for the other large London colleges. At the end of the twentieth century yet more independence was achieved after a determined effort by Eric Ash. Among other things, the college won the freedom to award (University of London) degrees and make senior appointments without university approval. In the early twenty-first century Imperial began taking steps towards full independence from the university; it expects to become an independent body on its centenary in 2007.

As it turned out, Imperial College's early success soon led to the shedding of what remained of its trade-school image in the outside world. In reality it never was a vocational college, but one in which science, technology and engineering science were taught at an advanced level. This was not always an easy distinction for outsiders to make so that when, in 1957, the Minister of Education, Lord Hailsham, publicly acknowledged that technology and engineering belonged in universities, those at the college breathed a collective sigh of relief.[13] The second half of the century saw a great change in attitude. Universities were encouraged to take increasingly greater numbers of students in science and engineering. Imperial was well placed to do this and, as we have seen, soon doubled, and then tripled, in size.

The way in which academic staff receive recognition has changed over the century. Undergraduate teaching has become less valued, despite rhetoric to the contrary. This is not to imply that people think such teaching no longer important, but rather that no one knows how to reward it in ways that would encourage academic staff to devote more time to it. This is not simply a financial issue, though having greater monetary resources directed towards teaching would help. Rather, it is a broader cultural issue related to how people gain social respect. In the academic world research and publication are central to this — even more so now that individuals and departments are publicly judged during the Research Assessment Exercise.[14] The result is that it has become increasingly difficult to find people willing to devote more than the bare minimum of time to undergraduate teaching or committee work. Even finding willing heads of departments, once positions of great prestige, has become more difficult. Voluntary service, too, receives less social recognition today than in the past. Staff are less willing to give time to student affairs, or to support clubs and other social activities at the college. Some of the older voluntarism that once made for a pleasant social environment has become professionalized. Paid employees now carry out work that earlier was performed freely by members of the governing body, the academic staff, and by many of their wives.

Imperial College staff are still regularly called on by the government for advice on a myriad subjects and, as has always been the custom, this is rewarded with social recognition, and increasingly also financially. The college's history, its foundation by the state, and its proximity to Westminster, are important factors in its continuing connection to those in government.[15] Strong industrial ties, too, continue to draw staff away from the college as consultants. Today, as we have seen, many people have a personal stake in one or more of the increasing number of college spin-out companies. Staff have become more entrepreneurial and, in this, they mirror the larger society. As elsewhere, financial success has become an important measure of social status among today's academics — though not yet trumping academic achievement.

The boundary between college and outside work has always been somewhat porous. Today it is more so than ever. One of the older problems associated with this is that the college has taken in contract work described as research, but really more routine. It was often cheaper for companies to outsource work to universities than to pay their own employees to do it. For example, until the 1970s engineering and mining departments used to carry out much routine structural testing for both government and industrial bodies in exchange for further funding. Up to a point this was no bad thing, since students learned something from the testing. Further, this kind of work fitted the older ideal of giving students industrial experience while at the college. But the financial benefits were sometimes outweighed by the costs. Some research students spent too much time on routine tasks that did little to advance their education or future careers. This is the type of problem that could well recur in a new form as more of the academic staff hold stakes in spin-out companies. The present administration is facing up to problems such as these by insisting on realistic costing for all work coming to the college — something that will help also to pay the bills.[16] Trying to achieve realistic costing is nothing new. Henry Tizard was the first of several Rectors to attempt something similar. The results to date have been mixed, due both to larger market forces and fierce university competition for contracts; and to staff who, in the past, were more willing than now to freely donate college time to outside bodies, with or without personal remuneration. How the new system will fare remains to be seen but the potential problems are many.

Another old problem has come to the fore with the boom in biomedical science and the increasing support coming to the college from the large pharmaceutical companies. Already in the 1920s the Meston Committee commented on the fact that much research at the college was not fully disinterested. At that time, Meston was referring to the many strings attached to research mainly in the engineering departments. The problem, then as now, was how best to work in cooperation with industry without compromising the integrity of research. But yet other serious ethical problems, especially with respect to medical research, synthetic biology and nanotechnology, appear to be looming.[17] These problems are not unique to Imperial, but the need for institutional and self-monitoring is today more important than ever.

The most influential of all post-war technologies has been the computer. It has changed life everywhere, and the intellectual and administrative life of the college is no exception. Administrators are today far more remote from other staff, and from students at the college, than they once were. This is partly a consequence of the increased size of the college, but also has much to do with new technology. Today people navigate the various college websites to find the kind of information that once was passed directly from person to person. E-mail has further replaced face-to-face discussion between academic staff and between teachers and students. The personal written note has been preserved in the form of e-mail, and written communication has increased as a result. Indeed, e-mail and texting have become major forms of social glue, replacing earlier modes of interpersonal communication. All of this has become possible because computers are now very powerful and very cheap. For the same reasons computers have radically changed the face of research. The consequences of various hypotheses can now be modelled in ways previously unthinkable. This represents a huge advance, though serious problems can arise from overconfidence in initial hypotheses.[18] Space once used for experimental research has been taken over by computers. Indeed, over the past thirty years the word 'laboratory' has gained a new meaning; it has come to mean rooms full of computers as much as rooms filled with more traditional apparatus and equipment. Especially in engineering, computer modelling has obviated the need for much heavy testing machinery. One can only hope that enough attention is being paid to the vagaries of the real world. Computers have also changed the nature of experimental science which has become increasingly responsive to computer simulations, and to developments in information technology more generally. For example, the rapidity with which work in genetics has progressed would have been impossible without cheap computation. The new pace of discovery has led to an overabundance of data and the emergence of new fields such as bioinformatics.

Another major theme of the later twentieth century has been the growth of interdisciplinary work. In part this has been driven by funding patterns, and in part by genuine intellectual need. It is not always easy to know which of these has been dominant in a particular case since such work is almost always defended on intellectual grounds by the participants. Throughout its history Imperial College has been well attuned to governmental directives and industrial needs. In recent years, however, it has responded to these by developing a range of interdisciplinary units, some of which have affected the welfare of traditional departments. The college calendar, which in the early twentieth century listed only departments, now lists a myriad of divisions, centres and institutes. Many of these reflect the increasing importance of medical science and technology. Even before medical science came to the college in a formal sense, engineers and scientists were collaborating with medical researchers in a large number of areas. Today they are doing so more than ever. There is little doubt that the major event of recent years has been the founding of the Imperial College School of Medicine (now Faculty of Medicine). The history of medical science will probably be a major focus for anyone writing the history of the college in the twenty-first century.

Another area of growing importance is business studies. Business is more valued in the larger culture than earlier, something that has resulted in the foundation and growth of business schools worldwide, including the Tanaka Business School at Imperial. This change in the cultural climate has affected the college more generally. Until the 1970s a more gentlemanly ethos, never as strong at Imperial as elsewhere, was typical of university life. It reflected the values of an older societal elite. But, as we have seen, this pattern was challenged by a range of demo-cratic forces in the postwar period, and by the rise of new kinds of business-related power brokers within society. As a consequence, students now entering the college are keen to learn business skills alongside the more traditional scientific and technological ones.

Since staff are increasingly carrying out research within new centres, are working with colleagues from other disciplines, and are becoming more entrepreneurial, there has been a natural erosion of old departmental loyalties. In some cases, there has been an ensuing negative effect on undergraduate teaching. As was discussed in chapter fifteen, for teaching to remain a strong part of what Imperial College offers, basic disciplines, albeit differently construed than in the past, will need some kind of solid structural base, and increased college and governmental support.[19] The problem has become more acute because the much larger college of today has called also for a more vertical governance structure, with further loss of departmental autonomy. Richard Sykes, in having introduced a faculty system, has brought the college's governance more in line with that of other academic institutions. But new research groupings are forming more rapidly than ever before and it is therefore difficult to predict what forms of internal governance will emerge in the future. One can only conclude that restructuring still has a way to go. Sykes is the latest in a succession of Rectors discussed in this book. While his term has not been analysed in the same detail as those of his predecessors, it is fair to conclude that, like them, he was well chosen. Imperial College's Governing Body (now Council) has always kept abreast of national and college affairs, and the various Rectors appear to have fitted their times well, a reminder of Imperial's typically good grasp of political realities. In the early twenty-first century, and after a period of major expansion, the college's internal governance needed a major overhaul. Sykes brought considerable managerial experience to this task, but only time will tell how successful the changes now underway will be.

We have seen how the role of the older colleges in Imperial College governance withered later in the century, and how the college became less of a federation and more of a unity. While the old college associateship diplomas are still awarded, the RSM, C&G College and

RCS now play only symbolic roles in college life. Students, staff and alumni still identify with the old college associations but, with the creation of the faculties, they are less likely to do so in the future. In this connection I see three problems. First, and this may be a historian's bias, I believe heritage is important and that there should be room for the traditions of the older colleges in an evolving Imperial. The reason for this is not sentimental, but rather that a proud institutional heritage gives authority to decisions taken in the present. Philosophers such as Francis Bacon and John Stuart Mill may have shown that, from a logical point of view, tradition should play no role in decision making and that everything should be decided on its own merits. But socio-cultural life is not logical and there is little doubt that traditional authority lends weight to, and encourages acceptance of, the new. Indeed, such authority often determines which new ideas are worth accepting. Tradition nearly always trumps logic (they need not clash), and institutions do best when able to exploit their historical authority in the wider culture. This is something the ancient universities have learned to do with great effect and Imperial could learn something from them. This is not to say that the college should remain static, or hang on to a governance structure that no longer meets modern needs; but rather that ways should be found to more effectively incorporate the old within the new. The college's Victorian and Edwardian inheritance has value also in the present and will have increasing value over time.

A second socio-cultural point is that the association of students and staff with their constituent colleges in the early period engendered great loyalty. This was probably true also of students in the old medical colleges. In South Kensington, the loyalty was not just to the older colleges but increasingly also towards Imperial. Identification with one of the constituent colleges helped sustain both morale and high standards within the college as a whole. As Imperial College grew, so did loyalty towards departments which, by mid-century, had developed highly inclusive cultures of their own. Not all members of staff were dynamic researchers; some contributed in other ways by taking heavy teaching or administrative loads. But, as we have seen, ritualistic bonding mechanisms gave both staff and students a sense of belonging, and contributed much to departmental and college well-being. The new faculties are too large, and new research sections and centres too transient, for the old social technologies to work well. Besides there has to be a place for undergraduates, including medical students. We have seen how departments (some now named divisions) have become less cohesive with the growth of centres. Their self-sufficiency may now be further threatened by a new bureaucracy at faculty level. In making this point, I have no wish to make an argument against faculties, nor one for an older *status quo*, but simply to point out the historical role of the old colleges and the departments in maintaining morale, and encouraging work of the highest standard. As we have seen, departmental autonomy led to some wastage of resources. However, this was more than compensated for by its many positive features. Of especial interest is the familial role of departments, one that gives people a sense of belonging, support, and intellectual space. This role needs to be considered, and possibly reinvented, as the college moves forward. Third, while the association of students and alumni with the old colleges may diminish only gradually, in the longer term the association may well become meaningless unless steps are taken to preserve it. If stability at the departmental level is also in question one has to ask in what way will loyalty to the college be engendered in its future graduates. If alumni are to be looked to for financial or other support then loyalty has somehow to be instilled in them as students. Students and alumni need to see themselves as being associated with something of permanence, and with a recognizable culture and historical tradition.

Intellectually, Imperial College has had a very successful first century. We have seen something of the many ideas that have originated at the college, and of the varied work carried out. Unfortunately a book of this kind is not able to give more than a sketch of the wide ranging research activity at the college, much of it deserving independent treatment. The flux of talent at Imperial has been remarkable and is something in need of explanation. Attempts to explain it have been made in some of the above chapters, where attention has been drawn to the culture of independence already mentioned. Because of patterns laid down when the central administration was very weak, Imperial people have had the freedom to follow their own interests with little interference. Further, the college has always been accepting of a wide range of work.[20] Individual freedom has extended also to teaching. The fact that instructors early set their own examinations gave both departments and individuals control over curricula.[21] This has allowed staff some leeway and, when important new ideas arise, they can immediately be introduced to students. The situation was different in the other London colleges where, until late in the century, most students sat examinations set by the university. Something similar can be said of the systems at Oxford and Cambridge, though they, too, have modernized in recent years. College tutors prepared students for examinations set by the university and so had to teach to a set curriculum which, by necessity, was rather conservative. Further, the Oxbridge colleges compete among themselves to get good examination results. Tutors therefore spend much time teaching also examination skills.[22] Had Imperial not stood up to the University of London its instructors would have been in a similar position.

Imperial College has not only provided a climate for independent thinking, it has also been very egalitarian and meritocratic. What people can do has always counted for more than their family or educational backgrounds. The culture of independence and equality is one that has been welcomed both by students, and by a wide range of people at the start of their careers. Another book could be written about those who spent time at the college making their names only to move on to professorships or other important posts elsewhere. That departments have been large and autonomous has made them also good training grounds for future vice-chancellors. Imperial has lost a fair number of department heads in this way. These various factors have led to an occasional leadership deficit at the college, something compensated for by space being opened up to new talent. When Eric Ash was Rector, he encouraged senior staff to help stem the deficit by promoting their younger colleagues within learned societies and other professional bodies. He wanted senior people to work harder at mentoring so that Imperial would have more of its staff elected to the Royal Society or to the Royal Academy of Engineering. This would have been easier had more of the gifted transients remained at the college and not taken off after (or just before) themselves receiving professional recognition.[23] Oxbridge does not have this problem. Indeed, people with positions at the ancient universities, or at the associated colleges, have been very reluctant to leave, even for professorships elsewhere. This generates other kinds of problems but the resulting continuity makes for a social cohesion that helps in the championing of younger colleagues within professional bodies. In mentioning this, and the great flux of staff at Imperial, I do not wish to imply that people cannot reach the top of the professional ladder by remaining at the college, on the contrary. Many of the most successful academics in the world lead research groups at the college. Further, those in senior positions at the college, often younger than elsewhere, are an interesting mix of college insiders and others brought in from outside, including from other countries. One notable feature of Imperial College professors is how many of them have had major industrial experience. The resulting mix has had very positive results. Despite concerns such as those expressed by Ash, Imperial College has had an impressive number of Fellows of the Royal Society, Fellows of the Royal Academy of Engineering and, more recently, Fellows of the Academy of Medical Sciences. Many of its professors have become presidents of major professional bodies and learned societies, and one or two have become President of the Royal Society. The college has also had its share of Nobel Laureates. While most of the college Nobel Prizes date from the 1960s and 70s, their absence in recent years, while worth pondering, is perhaps reflective of more international competition and a relative decline in the number of laureates nationally.

While this book has pointed to some of the many successes of the college no overall explanation has been given for its being among the best scientific, biomedical and technological universities in the world. As mentioned in the introduction, an anthropological study might provide more insight into the more general problem of what makes for a successful institution than the history written here. Nonetheless, having stated that I would speculate a little on the subject of success, I will make a few general comments in the next three paragraphs. In abstract terms the college can be said to have succeeded due to its being a site where human realities have been channelled in productive ways. For a variety of reasons, discussed in earlier chapters, people at the college have responded well to the complexities of the real world — natural, political, economic and cultural. In other words, they have approached their research and teaching, and have sought support in the larger world, in pragmatic and realistic ways; and they have been quick to adjust to change where and when necessary.

The public sphere of science is unlike other public arenas in that no overall synthetic perspective is possible. In politics strong leaders such as Margaret Thatcher can declare what is the proper way of looking at the world, and carry masses of people with them. Even those who disagree can accept that politicians are in the business of seeking unifying visions. This is not true of scientists. The only roughly scenic vision in science is one dating from the seventeenth and eighteenth centuries, namely the Enlightenment view of progress and the rational pursuit of knowledge. But this is not a truly unifying vision in the sense of letting people know what to believe, or what to do. Rather, it is a view of science forever expanding outwards. In some disciplines, such as theoretical physics, there are those who seek unifying principles but, overall, science is a many-centred market place with no controlling hand. This aspect of scientific practice is embodied in the governance structure of Imperial College. The central administration, recognizing that it can never seriously guide what goes on at the college, has deferred instead to market forces. While doing so, it has at the same time encouraged the academic staff not to give in totally to the distractions of the market place. As a consequence, a good balance between deference to, and independence from, market forces has been maintained at the college over the century and is, perhaps, the basis for its overall success.

Further, the cultural climate at the college has encouraged people to discipline themselves to produce rather than to consume. Consumption in an open market place is not limited to what money can buy. It includes also intellectual consumption which, paradoxically, can be a major distraction for academics. To what degree should people educate themselves in order to be maximally productive? As is well known, there are many scholars who are very learned but produce little that is new. Such people, often very good teachers and authors of fine synthetic works, are more likely to be appreciated in academic institutions where the ethos differs from the one dominant at Imperial College. The discipline that has led people at Imperial to focus on knowledge production has attracted both governmental and industrial investment. This, too, is paradoxical in that by being over-productive people can subvert the market. Too many producers and not enough consumers can be bad thing. But industrial leaders recognize that islands of productivity are necessary, and they value institutions such as universities that are not motivated solely by immediate returns. However, while the market needs creativity, the creative process cannot be totally free for it to work well in economic terms. It has to respond constructively to current fashion which, as mentioned, is something Imperial has always understood, and has used to its advantage. From its foundation in 1907, the absence of an endowment cushion, and the culture of independence, have encouraged people to pay attention to the outside world and its needs. Imperial College is no ivory tower.

In the marketplace of science and technology there is little room for the sceptic, despite rhetoric to the contrary. In this, technological disciplines are fundamentally different from those in the humanities where scepticism and iconoclasm can still pay off in academic rewards — though less so than was once the case. Technological disciplines lend themselves better to team effort, something that people at Imperial have understood. Research group culture is well embedded, and most people recognize that there is little market value in articulating alternate visions of what needs to be done, unless they can be sure of persuading others that their vision will work. While teamwork has always been an integral part of the college culture, it has become far less easy in recent years to be an individualist. The lone, or at least close to lone, genius seems to be a thing of the past. Teamwork has been helped also by the size of the college. There has always been great depth of expertise in South Kensington and, when people need to know something, there is usually someone close by to talk to. This is a huge benefit and a time saver in matters of research and in team formation. Further, in accordance with the college charter, the founding principles of the college, and its attachment to the state, college research teams have always paid close attention to the needs of industry and government, including the military. This attention has served the college well and will likely do so in the future.

But teamwork requires that someone have the ideas, the ability to raise funds, and the determination to see things through. With this in mind, I have given some final thought to which individuals stand out at Imperial with regard to leadership. Who best pushed or encouraged their staff to form productive research teams, and who have themselves led good teams? I have chosen a few exemplars from the early to mid-twentieth century only, recognizing both that there are many others worthy of mention — including the Rectors who, as we have seen, created the space for people to reach their potential — and that there are other ways in which people have successfully contributed to the college. Already mentioned are two people notable for their sheer drive, and for building major departments early in the century. Botanist John Farmer and chemical technologist William Bone were both excellent scientists, but their legacy at Imperial has more to do with their personal vision of what was needed in a college founded to serve the industries of empire. Both encouraged the formation of excellent research teams and Bone, himself, led an important research group. James Munro, too, built an excellent zoology and applied entomology department on foundations well laid by earlier professors. Almost single-handedly he brought two field stations to the college, first at Hurworth and then at Silwood. Later Richard Southwood put Silwood on the map by creating an outstanding school of ecology there. He was a team builder. Civil engineering owes much to Sutton Pippard who began to reinvigorate his department just before the Second World War. After the war, Willis Jackson and Owen Saunders, respectively, rebuilt the departments of electrical and mechanical engineering, and created major new research sections. Their legacies are still recognizable today. But the principal legacy in both civil and mechanical engineering is Victorian and owes much to the vision of William Unwin. Something similar can be said of departments in the RSM which inherited traditions from their Victorian founders. Geology, especially, owed much to its Victorian roots.

Perhaps the greatest all-rounder of the century was Patrick Blackett. Not only was he a brilliant scientist who made major contributions to more than one area of physics, he had remarkable foresight and great organizational ability. He also effectively represented science to the larger world, had a major voice in wartime science, and later also in peacetime science policy — both in Britain and India. In some respects he must have been a difficult head but, on coming to the college in the early 1950s, he encouraged team formation, rebuilt the physics department, and made it one of the best in the world. His legacy has been lasting. I have mentioned just a few of Imperial's more creative heads of department, but accomplishment comes in many forms. For example, in the form of outstanding inventive genius as demonstrated by both Dennis Gabor and Dudley Newitt, or as a remarkable ability to connect with students, as shown by James Philip. Willis Jackson and Hugh Ford showed creative leadership not only in research but also in pedagogy. Some professors were brilliant students of science, not simply during their student years but through life; Alfred Fowler, Helen Porter, Janet Watson, Derek Barton and Geoffrey Wilkinson all come to mind. Barton and Wilkinson also built major research schools and put the chemistry department firmly on the map in mid-century.

One could give more examples, but they can be found in earlier chapters. The point I wish to make is simply that people matter, and that individual contributions have counted for much at Imperial. Collectively the many stories told in this book give meaning to the history of the college in the twentieth century. While difficulties have not been over-looked, the main narrative is one of remarkable success. The book has told of how a college with no significant endowment and, at first, few physical resources, was able to respond to society's need for well-trained specialists and for socially relevant research. In telling the story of the college I have attempted to remain dispassionate, but it has been difficult not to engage with the lives of those who worked at the college over its first one hundred years. As a result of their collective effort, Imperial College has become an academic institution respected around the world. It can enter its second century with well-founded optimism.

...

1. This has been illustrated by earlier discussion of the very different cultures within departments. For example, the mechanical engineering department, descendent of a Victorian foundation, carries a trace of its nineteenth-century heritage even today. The computing department, a child of the late 1950s and 1960s, still carries something of the 1960s culture. This is not to imply that computing is more modern than mechanical engineering — both are highly competitive modern departments.

2. The early Rectors should not have devolved this responsibility to the degree they did, and should have thought more about building an endowment.

3. The deanships of the RCS and RSM were suspended from the foundation of Imperial College until the end of the Second World War.

4. As noted in chapter 2, in the early twentieth century about one-third of the RSM graduates spent their working lives overseas, and about three-quarters spent at least some time overseas. This pattern of overseas work was typical until the 1950s.

5. The Whiffen name was retained for the major organic chemistry research laboratory.

6. While there was some truth to technocratic criticism of this kind, it is far from being the whole story since science and technology were not neglected by the twentieth-century state. Indeed, many historical accounts of Britain in this period have gone wrong because of their uncritical acceptance of technocratic rhetoric. See David Edgerton, *Warfare State*: Britain, 1920–1970 (Cambridge, 2006), especially chapter 5.

7. Before the electronic computer age many women calculators were working at the college. This was especially the case in the structures sections of the aeronautics and civil engineering departments, as well as in the mathematics department. Several women worked on the bubble chamber work in physics both as inspectors of film and as calculators. But women also worked in laboratories, notably in the biological sciences, chemistry, and metallurgy; and in the field, notably in geology and botany.

8. War bonuses for married men were higher than those for single men. They were seen as the family breadwinners, and thus deserving of more. The college charter forbids sexual, racial and religious discrimination.

9. This in part is due to the continued and growing presence of British Petroleum throughout the world. Today it is the world's third largest oil/petroleum company with interests also in other energy sectors. Imperial College has had close ties to this company since its early days as the Anglo-Iranian Oil Company.

10. See, for example, university rankings in *Times Higher Education Supplement* (November, 2005). Imperial placed fifth in Europe and thirteenth overall; first in Europe and fifth in the world for technology; third in Europe and tenth in the world for science; fourth in Europe and sixth in the world for biomedicine. In 2006 Imperial was ranked ninth overall.

11. There is something of a paradox in that while, overall, students had more disposable income (in real terms), the real value of student grants declined during the 1970s and 80s. The explanation for greater disposable income lies in greater parental support, and in the fact that many more students found work in this period than earlier. But wealth is also relative and, just as the larger society became wealthier, so, too, did students.

12. This allowed Imperial to argue for increased numbers of chairs and opened the way to the multi-professorial department, a situation exploited first by Patrick Blackett and Harry Jones when they headed the physics and mathematics departments respectively.

13. There were engineering departments in many other universities at this time, but in the 1950s most budding engineers still took the apprenticeship route. Many educators and politicians thought this was how things should be. The situation was not unlike the reluctance to admit medical and legal education into universities earlier, and business education later. Engineering schools and science departments in the more traditional universities had different matriculation requirements from those at Imperial for much of the twentieth century. For example, O-level proficiency in Latin was often required until the 1960s.

14. The RAE rating system reflects both the number of research active staff, and the quantity and quality of research produced. A 5A rating is only achievable with almost all departmental academic staff being submitted as research active. Since the college has decided that a 5A as well as a 5* (for internationally rated research) is desirable, departments have lost people (sometimes by terminating the contracts of those with many years of service) on whom they depended for much of their teaching. This is a serious problem in need of a national solution. In the meantime colleges and universities, including Imperial, are trying to import more teaching staff by the back door; namely by employing them in non-academic positions, and on part-time contracts. This cannot lead to the improvement of undergraduate teaching.

15. The fact that most of the important learned and professional societies have their headquarters also in Westminster is another important factor in the making of connections.

16. As we have seen, this is in response to government pressure on top universities to become close to self-sufficient in terms of research, and to charge the full economic costs to those 'buying' research time. As currently envisaged, the scheme would have all units in the college behave as small businesses. Any research work undertaken, whether supported by industry, charities or government research councils, would have to bring in sufficient funds to cover all costs, including a share of the upkeep and running of the college. This may be possible in some fields but, as someone working in the humanities, I am well aware that it is not possible in all. Further, it may be difficult for a London college with a relatively small endowment to compete for research contracts under too stringent conditions. In a business world, the quality of personnel carrying out the work is only one factor to be considered in assigning a contract, and may not always win out. Further, some things that are worth doing simply don't pay.

17. Technical as well as ethical mistakes are likely to occur as scientists attempt to persuade nature to fashion what we think we need. Even without sharing the dystopian vision of some science fiction writers, such Michael Crichton in his novel, *Prey*, there is room for concern.

18. The American tag 'garbage in, garbage out' is entirely appropriate.

19. An alternative would be for the college to become independent of the state and charge very high fees. This seems an unlikely prospect in the near future.

20. This statement needs to be qualified in that heads of department could decide which broad areas of research their departments would engage in.

21. The college has always had external examiners to ensure standards are consistent with similar institutions elsewhere and, since 1929, with other colleges within the university.

22. This may encourage the development of some 'bullshitting' ability, perhaps not a bad thing for future career advancement.

23. Among the reasons for leaving the college are escaping the high cost of living in London, and no longer having to commute long distances to work. In the case of the many who have moved to chairs at Oxford or Cambridge, increased social status as well as a comfortable working environment are important motivators. Others have moved to become heads of departments elsewhere, to take higher salaried positions in the United States or in industry; or to take senior government posts, including within the research councils. The opportunity of leading a research council research unit can be a major motivator. The reasons are many, but there is little doubt that Imperial provides an excellent environment for young people to advance their careers, and an excellent jumping-off point for senior posts elsewhere.

Appendices

Compiled by Yorgos Koumaridis.
The appendices are based on records in the Imperial College Archives and Registry.

Contents

Table 1

Rectors	
Years	
1908–1910	Henry Taylor Bovey FRS
1910–1922	Sir Alfred Keogh CH
1922–1929	Sir Thomas Henry Holland FRS
1929–1942	Sir Henry Thomas Tizard FRS
1939–1941	James Charles Philip FRS (*Deputy Rector*)
1942–1948	Sir Richard Vynne Southwell FRS
1948–1954	Sir Roderic Maxwell Hill
1955–1966	Sir Reginald Patrick Linstead FRS
1966–1967	Sir Owen Alfred Saunders FRS FREng (*Acting Rector*)
1967–1973	The Lord Penney (William George Penney) FRS OM
1973–1985	The Lord Flowers (Brian Hilton Flowers) FRS
1985–1993	Sir Eric Albert Ash FRS FREng
1993–2000	The Lord Oxburgh (Ernest Ronald Oxburgh) FRS FREng
2001–	Sir Richard Brook Sykes FRS FMedSci

Table 2

Pro-Rectors	
Years	
1952–1955	Herbert Harold Read FRS
1955–1956	Alfred John Sutton Pippard FRS
1956–1961	Dudley Maurice Newitt FRS
1961–1964	The Lord Blackett (Patrick Maynard Stuart Blackett) FRS OM CH
1964–1966	Sir Owen Alfred Saunders FRS FREng
1967–1970	The Lord Jackson (Willis Jackson) FRS
1970–1972	Harry Jones FRS
1972–1974	Bernard George Neal FREng
1974–1978	Marston Greig Fleming FREng
1978–1980	Sir Hugh Ford FRS FREng
1980–1983	John Sutton FRS
1983–1986	Sydney Alan Vasey Swanson FREng
1986–1991	Bryan Randell Coles FRS
1991–1994	John Stuart Archer FREng

Table 3

Deputy Rectors

Years	
1994–1997	John Stuart Archer FREng
1997	Sydney Alan Vasey Swanson FREng
1997–2001	William Arnot Wakeham FREng
2001–2004	Peter William Bearman FREng
2004–	Sir Leszek Borysiewicz FMedSci

Table 4

Chairmen of the Governing Body (Council since 1998)

Years	
1907–1922	Robert Offley Ashburton Crewe-Milnes, 1st Marquess of Crewe
1923–1934	Stanley Owen Buckmaster, 1st Viscount Buckmaster
1934–1936	Victor Alexander John Hope, 2nd Marquess of Linlithgow
1936–1947	Robert John Strutt FRS, 4th Baron Rayleigh
1947–1962	Evelyn Hugh John Boscawen, 8th Viscount Falmouth
1962–1974	Roger Mellor Makins FRS, 1st Baron Sherfield
1975–1988	Sir Henry Fisher
1988–1996	Sir Frank Cooper
1996–2004	The Lord Vincent (Richard Frederick Vincent)
2005–	The Lord Kerr (John Olav Kerr)

Table 5a

Deans of the City and Guilds College

Years	
1906–1931	William Ernest Dalby FRS
1931–1933	Stephen Mitchell Dixon
1933–1936	Cecil Lewis Fortescue
1936–1938	Cecil Howard Lander
1938–1943	Alfred John Sutton Pippard FRS
1943–1945	Cecil Lewis Fortescue
1946–1949	Alfred John Sutton Pippard FRS
1949–1950	Sir Owen Alfred Saunders FRS FREng
1951–1952	The Lord Jackson (Willis Jackson) FRS
1953–1954	Alfred John Sutton Pippard FRS
1955–1964	Sir Owen Alfred Saunders FRS FREng
1964–1967	Bernard George Neal FREng
1967–1970	Stanley Robert Sparkes
1970–1973	Alan Wilfred Bishop
1973–1976	Roger William Herbert Sargent FREng
1976–1979	Sydney Alan Vasey Swanson FREng
1979–1982	Eric Hugh Brown
1982–1984	Henryk Sawistowski FREng
1984–1988	Bruce McArthur Sayers FREng
1988–1991	Patrick Holmes
1991–1993	Bruce McArthur Sayers FREng
1993–1997	Dame Julia Stretton Higgins FRS FREng
1997–2000	Brian James Briscoe
2000–2001	Christopher Leslie Hankin

Table 5b

Deans of the Royal School of Mines	
Years	
1943–1945	Herbert Harold Read FRS
1945–1947	William Richard Jones
1947–1949	John Anthony Sydney Ritson
1950–1951	Cecil William Dannatt
1952–1959	David Williams
1959–1962	James Cecil Mitcheson
1962–1965	John Geoffrey Ball
1965–1968	John Sutton FRS
1968–1971	Marston Greig Fleming FREng
1971–1974	John Geoffrey Ball
1974–1977	John Sutton FRS
1977–1980	Peter Lynn Pratt
1980–1983	John Lawrence Knill
1983–1986	Edwin Thomas Brown FREng
1986–1989	Donald William Pashley FRS
1989–1991	John Stuart Archer FREng
1991–1995	Charles Timothy Shaw
1995–1998	Rees David Rawlings
1998–2000	John Anthony Kilner
2000–2001	Andrew John Monhemius

Table 5c

Deans of the Faculty of Engineering[a]	
Years	
2002–2003	Andrew John Monhemius
2002–2003	Christopher Leslie Hankin
2003–2006	Richard Ian Kitney FREng

[a]In 2001 the academic and administrative structure of the college changed. Four faculties (Medicine, Engineering, Physical Sciences and Life Sciences) replaced the old structure based on the constituent colleges. Even though the post of Dean remained, titles changed to Dean of the Faculty of Engineering (former Deans of C&G and of RSM) and Dean of the Faculty of Life and Physical Sciences (former Dean of RCS). With the structural changes a new senior academic and administrative post, the Principal of Faculty, was introduced. For the Principals see Table 6, below.

Table 5d

Deans of the Royal College of Science	
Years	
1945–1946	Sir David Brunt FRS
1947–1952	Hyman Levy
1953–1954	Sir Reginald Patrick Linstead FRS
1955–1960	The Lord Blackett (Patrick Maynard Stuart Blackett) FRS OM CH
1960–1963	Harry Jones FRS
1963–1966	Richard Maling Barrer FRS
1966–1969	Clifford Charles Butler FRS
1969–1971	Charles Percival Whittingham
1971–1972	Sir Thomas Richard Edmund Southwood FRS
1972–1975	Paul Taunton Matthews FRS
1975–1978	Ronald Karslake Starr Wood FRS
1978–1981	Walter Kurt Hayman FRS
1981–1984	David Mervyn Blow FRS
1984–1986	Bryan Randell Coles FRS
1986–1989	Wyndham John Albery FRS
1989–1990	James Barber FRS
1990–1993	John Trevor Stuart FRS
1993–1996	Sir John Brian Pendry FRS
1996–1999	David Michael Patrick Mingos FRS
1999–2001	Malcolm Haines

Table 5e

Deans of the Faculties of Life Sciences and Physical Sciences	
Years	
2002	Malcolm Haines (Life and Physical Sciences)
2002–2005	David Phillips (Life and Physical Sciences)
2005–	Robert Edward Sinden FMedSci (Life Sciences)
2005–	Christopher John Isham (Physical Sciences)

Table 6

Years	Faculty and Business School Principals
	Imperial College School of Medicine (Faculty since 2001)
1995–2001	Christopher Richard Watkin Edwards FMedSci
2001–2004	Sir Leszek Borysiewicz FMedSci
2004–	Stephen Kevin Smith FMedSci
	Faculty of Engineering
2001–2004	John Douglas Perkins FREng
2004–2006	Julia Elizabeth King FREng
2006–	Dame Julia Stretton Higgins FRS FREng
	Faculty of Physical Sciences
2001–2002	Sir John Brian Pendry FRS
2002–2004	Frank Leppington
2004–2005	Sir Peter Leonard Knight FRS (*acting*)
2005–	Michael Duff
	Faculty of Life Sciences
2001–2004	Michael Patrick Hassell FRS
2004–2005	Sir Leszek Borysiewicz FMedSci (*acting*)
	Tanaka Business School
2003–	David Knox Houston Begg
	Faculty of Natural Sciences[b]
2005–	Sir Peter Leonard Knight FRS

---

[b]The Faculty of Natural Sciences was created in autumn 2005 after the merger of the Faculties of Physical and Life Sciences.

Table 7

Founding and Re/Naming of Departments

Year	
1907	Botany, RCS
	Chemistry, RCS and C&G
	Geology, RCS
	Mathematics and Mechanics, RCS
	Physics, RCS
	Zoology, RCS
	Civil and Mechanical Engineering, C&G
	Electrical Engineering, C&G
	Mechanics and Mathematics, C&G
	Metallurgy, RSM
	Mining, RSM
1911	Biology: merger of Botany and Zoology
1912	Chemical Technology, RCS
1913	C&G Chemistry closed; all Chemistry moved to RCS
	Oil Technology (ARSM) in Geology
	Civil Engineering and Surveying, C&G
	Mechanical Engineering and Motive Power, C&G
	C&G Mechanics and Mathematics closed; Mathematics moved to RCS
1917	Technical Optics, RCS
1920	Aeronautics, RCS
	Meteorology (section of Aeronautics)
1931	Technical Optics became a section of Physics
1932	Aeronautics moved to C&G
	Meteorology moved from Aeronautics to Physics
1937	Biology split into Biochemistry, Botany, and Zoology and Applied Entomology, (all in RCS)
1939	Chemical Technology moved to C&G
	Meteorology, RCS
1942	Chemical Technology renamed Chemical Engineering and Applied Chemistry, C&G
1945	Biochemistry became a section of Botany
1947	Civil Engineering and Surveying renamed Civil Engineering, C&G
	Mechanical Engineering and Motive Power renamed Mechanical Engineering, C&G
1951	Mathematics and Mechanics renamed Mathematics, RCS
1954	Chemical Engineering and Applied Chemistry renamed Chemical Engineering, C&G
1955	Botany renamed Botany and Plant Technology, RCS

Table 7 (cont'd)

Founding and Re/Naming of Departments

Year	
1959	Chemical Engineering renamed Chemical Engineering and Chemical Technology, C&G
1961	Biochemistry, RCS Mining renamed Mining and Mineral Technology, RSM
1963	History of Science and Technology, RCS Meteorology moved from Aeronautics to Physics
1966	Geology moved to RSM Centre for Computing and Automation, C&G
1969	Industrial Sociology Unit, C&G
1970	Computing and Control, C&G
1971	Management Science, C&G
1973	Associated Studies (Humanities, Social Sciences and Languages) Metallurgy renamed Metallurgy and Materials Science, RSM
1974	Meteorology became a section of Physics Division of Life Sciences including Biochemistry, Botany and Plant Technology, and Zoology and Applied Entomology
1976	Mining and Mineral Technology renamed Mineral Resources Engineering, RSM
1978	Social and Economic Studies, C&G: merger of Industrial Sociology Unit and Economics
1980	Humanities: merger of Associated Studies and History of Science and Technology
1981	Pure and Applied Biology, RCS: merger of Botany and Plant Technology and Zoology and Applied Entomology Computing, C&G
1986	Metallurgy and Materials Science renamed Materials, RSM
1987	Management School, C&G: merger of Social and Economic Studies and Management Science. Humanities Department became Humanities Programme
1988	St. Mary's Hospital Medical School became the fourth constituent college
1989	Pure and Applied Biology renamed Biology, RCS
1991	Electrical Engineering renamed Electrical and Electronic Engineering, C&G Science and Technology Studies Unit within Humanities

Table 7 (cont'd)

Founding and Re/Naming of Departments

Year		Year	
1994	Centre for the History of Science, Technology and Medicine	2001	Creation of four faculties: Faculty of Engineering, Faculty of Life Sciences, Faculty of Medicine, Faculty of Physical Sciences
1995	Mineral Resources Engineering renamed Earth Resources Engineering, RSM National Heart and Lung Institute		T.H. Huxley School closed Agricultural Sciences Earth Science and Engineering Environmental Science and Technology Biological and Medical Systems renamed Bioengineering Biological Sciences: merger of Biology, Biochemistry and Biophysics
1997	Charing Cross and Westminster Medical School Royal Postgraduate Medical School Imperial College School of Medicine Environmental Technology, C&G Biological and Medical Systems, C&G	2003	Management School renamed Business School (renamed Tanaka Business School in 2004)
1998	Civil Engineering renamed Civil and Environmental Engineering, C&G T.H. Huxley School of Environment, Earth Science and Engineering, RSM (Departments of Geology, Earth Resources Engineering, and Environmental Technology)	2005	Restructuring of the Faculty of Life Sciences: creation of the divisions of Biology, Biomedical Sciences (previously in the Faculty of Medicine), Cell and Molecular Biology, Molecular Biosciences; and the Centre for Environmental Policy (created from the former departments of Environmental Science and Technology and Agricultural Sciences)
2000	Kennedy Institute of Rheumatology Wye College (T.H. Huxley School and Department of Biology)		Faculty of Natural Sciences: merger of Faculty of Physical Sciences and Faculty of Life Sciences

Table 8a

Estate History, South Kensington Campus

1871	Normal School of Science (formalised in 1881) and Royal School of Mines began moving into the Naval Architecture School (later Science Schools) building
1885	Waterhouse Building, City and Guilds College
1890	Normal School of Science renamed Royal College of Science
1893	Imperial Institute
1903	New Royal School of Needlework Building
1906	Royal College of Science Building (Physics and Chemistry)
1911	Imperial College Union
1913	Royal School of Mines Building
	Bessemer Laboratory
1914	Botany Building (later named Farmer Building)
	Bone Building (Chemical Technology)
1915	Goldsmiths' Extension of City and Guilds Building (occupied by military personnel)
1921	Bone Building extension
1923	Plant Technology building
1926	Hostel
	Goldsmiths' Extension Building refitted for use by City and Guilds
1931	Hostel extension
1932	RCS and RSM (Science Schools) building renamed for T. H. Huxley (Mathematics, Aeronautics, Zoology and Entomology)
	Botany and Hostel quadrangle named for Otto Beit
1934	Imperial College acquired the Royal School of Needlework Building
1949	Royal School of Needlework Building renamed for W. C. Unwin
1950	Ayrton Hall Refectory in Unwin building
1955	Floor added to the Goldsmiths' Extension
1956	Hostel extension
1957	Roderic Hill Building (Aeronautics and Chemical Engineering)
	Floor added to Imperial College Union
1959	Weeks Hall, north side of Prince's Gardens Mechanical Engineering (phase one)

Table 8a (Cont'd)

Estate History, South Kensington Campus

Year	Event	Year	Event
1960	Blackett Laboratory (Physics)	1975	Huxley Building (Mathematics, Physics and Computing)
1962	Start of demolition of the Imperial Institute (with the exception of the Queen's Tower) Demolition of Waterhouse Building, W. C. Unwin Building and the western end of Royal College of Science Building Mechanical Engineering (phase two) Electrical Engineering Imperial College Road closed (except to college traffic)		Old Huxley building given to V&A museum (Henry Cole Wing)
		1980	Linstead Hall extension
		1988	William Penney Laboratory (Computing)
		1993	Chemistry Two
1963	Keogh, Falmouth, Selkirk and Tizard Halls, (Southside Halls of Residence), Prince's Gardens Civil Engineering	1997	Health Centre, Prince's Gardens
1964	Bessemer Laboratory extension	1998	Sir Alexander Fleming Building (School of Medicine, and Biosciences)
1965	Mechanical Engineering (phase three) Biochemistry	2000	Pilot Plant remodelled and renamed Flowers Building (Biological Sciences, School of Medicine, Chemistry) Blyth Music and Arts Centre Library extension
1966	Aeronautics and Chemical Engineering (ACE) extension		
1968	Linstead Hall, Sports Centre, Prince's Gardens	2004	Faculty Building Tanaka Business School
1969	Centre Block (renamed Sherfield Building in 1975) Library	2005	Demolition of Southside Halls of Residence
1970	Chemistry One	2006	Ethos Sports Centre, Prince's Gardens
1974	Demolition of central part of Royal College of Science Building		

Table 8b

Estate History, Silwood Park

1947	Purchase of the Silwood Estate (Manor House and approximately 80 acres)
1948	Purchase of Ashurst Lodge
1949	Silwood Park Field Station
1953	Purchase of Silwood Park Farm
1961	Shell Parasitology building at Ashurst
1965	University of London Nuclear Reactor/Small Reactor Hall
1967	New Biological Laboratories
1970	Reactor Hall extension
1971	Nuclear Reactor Laboratories
1984	Commonwealth Agricultural Bureaux International (CABI) Institute for Biological Control (IBC) moved its headquarters to Silwood
1987	Science Park
1988	Sale of Ashurst Lodge Technology Transfer Centre CABI IBC new building Garden Wood Laboratory
1990	Centre for Population Biology
1998	Manor House refurbished

Figure 1a

Full-time students. Total, Men, Women, 1908–1954[c]

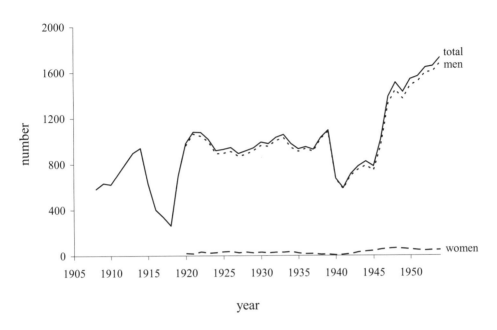

Figure 1b

Full-time students. Total, Men, Women, 1955–2005

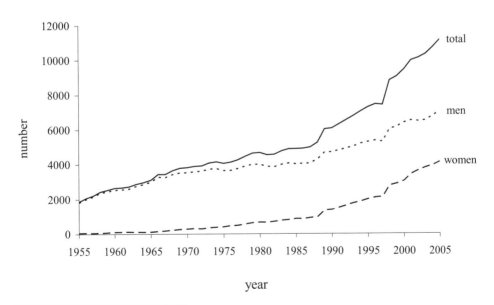

[c]No reliable data on the numbers of men and women students for the period up to 1920. Data missing for the years 1911 and 1912.

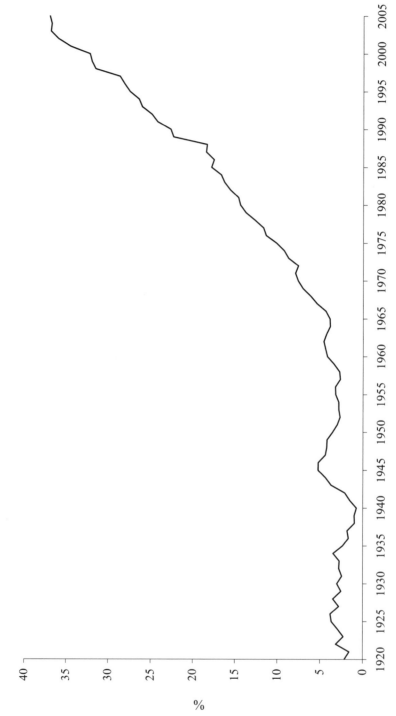

Figure 2
Women full-time students. % of total, 1920–2005

Figure 3a

Full-time students. Undergraduates and Postgraduates, 1913–1954[d]

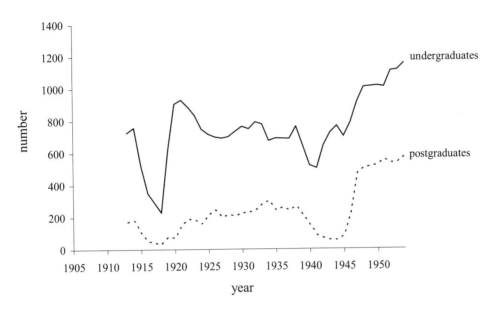

Figure 3b

Full-time students. Undergraduates and Postgraduates, 1955–2005

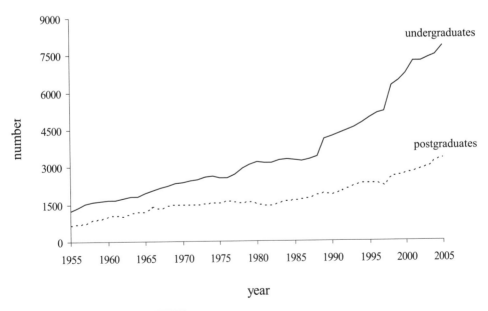

[d]No reliable data on undergraduate and postgraduate student numbers for the period up to 1913. Data missing for the years 1939 and 1949.

Table 9

Full-time students per constituent college, 1910–2000[e]

Year	RCS men	RCS women	RSM men	RSM women	C&G men	C&G women	School of Medicine men	School of Medicine women	Interdepartmental[f] men	Interdepartmental[f] women	Total men	Total women	total[g]
1910	211[h]	—	49	—	361	—	—	—	—	—	—	—	621
1915	206	—	111	—	311	—	—	—	—	—	—	—	628
1920	314	19	186	0	459	2	—	—	—	—	959	21	980
1925	341	33	131	1	423	0	—	—	—	—	895	34	929
1930	396	29	116	0	449	1	—	—	—	—	961	30	991
1935	421	20	106	0	382	2	—	—	—	—	909	22	931
1940	242	4	93	0	332	1	—	—	—	—	667	5	672
1945	258	34	71	0	411	7	—	—	—	—	740	41	781
1950	622	52	182	0	681	3	—	—	—	—	1485	55	1540
1955	686	55	169	2	955	3	—	—	—	—	1810	60	1870
1960	999	99	291	2	1255	10	—	—	—	—	2545	111	2656
1965	1231	102	382	6	1380	13	—	—	—	—	2993	121	3114
1970	1356	222	554	18	1626	51	—	—	—	—	3536	291	3827
1975	1322	275	561	51	1740	82	—	—	25	4	3648	412	4060
1980	1455	385	592	82	1935	191	—	—	20	16	4002	674	4676
1985	1346	477	586	92	2004	271	—	—	79	33	4015	873	4888
1990	1518	557	500	112	2232	379	309	265	138	63	4697	1376	6073
1995	1677	867	495	148	2792	591	317	390	—	—	5281	1996	7277
2000	1948	1036	542	280	2881	614	1028	1106	—	—	6399	3036	9435

[e]No data for college affiliation for the years 1985 and 1990 (the figures here are based on departmental classification).
[f]No data for college affiliation.
[g]Plus 4 sabbatical officers for the years 1985, 1990, 1995 and 6 for 2000.
[h]No reliable data on the number of men and women students for the years 1910 and 1915 (the figures here are for the total number of students per college).

Figure 4

Home residence of full-time students, 1925–1995[i]

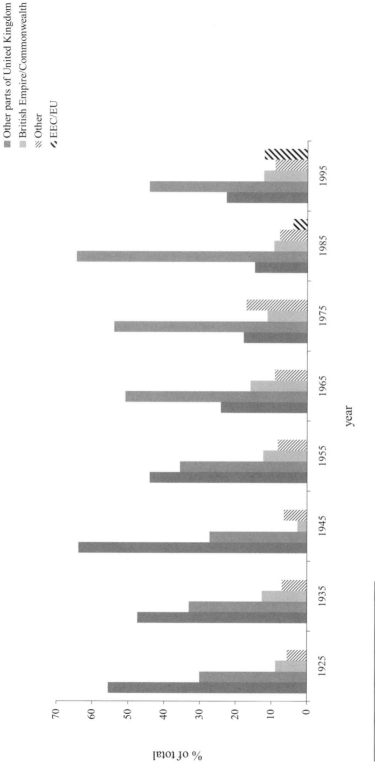

[i]Since 1981 students ordinarily resident in the EEC/EU have been counted in official statistics as "Home" students. In the figure, EEC/EU residents are included in the "Other" category until 1985.

Figure 5a

Full-time students. Geographical Origin, 1940–2005[j]

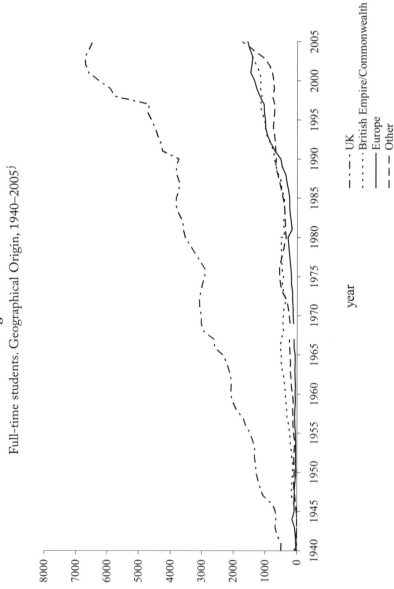

[j]Data for Figures 5a and 5b and for Table 10 are based on nationality figures with the exception of the years 1958–1972 which are based on home residence figures (nationality figures not found for that period). Data missing for the years 1949, 1977–1979. Data for 1968 do not include 'Europe' and 'Other' students. St Mary's Hospital Medical School students are included from 1991.

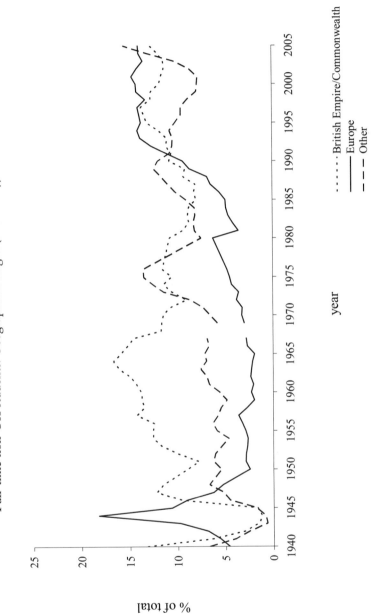

Figure 5b

Full-time non-UK students. Geographical origin (% of total), 1940–2005

Table 10

Full-time students. Largest non-UK nationality/home residence groups per period.

Year	1940–1959[k]									
	India	Poland	Egypt	Canada	Pakistan	Australia	South Africa	Germany	Iraq	Thailand
1940	50	4	14	2	—	3	6	14	4	9
1945	9	41	3	1	—	0	1	12	0	0
1950	52	17	45	6	8	17	18	3	2	3
1955	87	13	23	33	27	13	21	3	10	1
Average/ year[l]	60	24	19	19	14	11	11	8	6	6

Year	1960–1980[m]									
	India	Pakistan	Iran	Canada	Malaysia[n]	Greece	Nigeria	Iraq	Australia	Ceylon/Sri Lanka[o]
1960	107	45	16	65	9	6	17	24	16	9
1965	112	87	26	58	24	8	44	34	28	15
1970	72	63	44	49	39	25	17	28	22	23
1975	76	28	140	24	101	78	47	27	29	25
1980	49	18	86	19	127	78	61	19	29	37
Average/ year	87	53	52	50	44	33	32	28	24	22

Year	1981–2005									
	Greece	Malaysia	Singapore	China	Germany	France	India	Italy	Iran	Sri Lanka
1985	120	86	29	25	24	11	44	14	87	21
1990	166	119	64	88	72	52	68	27	71	35
1995	316	257	178	63	157	119	62	79	75	113
2000	396	302	251	92	182	207	75	124	38	52
2005	399	371	246	874	194	251	123	125	75	75
Average/ year	251	209	133	132	114	108	69	65	61	54

[k]World War II to the Jubilee Expansion of Imperial College.

[l]The average figures for each period are based on annual intakes.

[m]Jubilee Expansion to the introduction of full-cost fees for overseas students (in 1980).

[n]Federation of Malaysia created in 1963 after merger of Malaya, Singapore, Sarawak and Sabah. Singapore seceded in 1965. Numbers for 1964 and 1965 do not include students from Singapore. Numbers for 1960–1963 are for students from Malaya.

[o]Renamed in 1972.

Table 11a

Income by source (total in £), 1908–2005[P]

Year	Government/ UGC/UFC/ HEFCE	Fees	Research Grants and Contracts[q]	Endowment	Grants from non-governmental bodies, Donations and Subscriptions[r]	London County Council[s]	Others	Total
1908	22,978	9,069	—	—	—	—	—	32,047
1910	20,000	11,594	—	—	—	8,000	59	39,653
1915	30,000	18,183	1,339	—	6,425	12,513	12,807	81,267
1920	49,752	35,107	3,002	14,413	5,889	11,881	6,037	126,081
1925	80,500	47,823	5,724	14,317	5,000	22,020	18,703	194,087
1930	90,500	51,061	6,218	14,851	5,000	26,265	16,718	210,613
1935	126,415	46,826	9,356	10,777	6,500	—	21,143	221,017
1940	138,732	32,334	9,307	11,633	8,250	—	32,105	232,361
1945	156,449	43,252	8,594	11,114	8,575	—	29,415	257,399
1950	463,032	70,907	15,942	10,331	15,652	—	25,033	600,897
1955	852,730	93,053	199,196	11,666	28,903	—	30,114	1,215,662
1960	1,796,998	148,116	335,350	24,299	23,050	—	38,988	2,366,801
1965	3,527,540	285,688	733,555	26,523	30,951	—	50,695	4,654,952
1970	5,898,966	503,828	1,826,690	29,227	41,987	—	107,052	8,407,750
1975	12,103,085	568,646	3,339,614	69,596	56,958	—	495,949	16,633,848
1980	22,474,692	3,698,660	9,469,305	124,168	48,211	—	630,609	36,445,645
1985	31,765,000	7,376,000	17,571,000	163,000	494,000	—	1,710,000	59,079,000
1988	36,078,000	10,199,000	21,362,000	178,000	685,000	—	2,486,000	70,988,000
1996	62,063,000	27,012,000	76,376,000	3,802,000	4,352,000	—	31,161,000	204,766,000
2000	103,140,000	35,825,000	121,992,000	2,668,000	2,604,000	—	72,743,000[t]	338,972,000
2005	137,498,000	63,104,000	176,705,000	4,197,000	4,138,000	—	72,880,000[u]	458,522,000

[P] In 2003 pounds, £1 equals: 1907, £76.08; 1920, £28.27; 1940, £35.40; 1960, £14.57; 1980, £2.71. Based on Jim O'Donoghue, Louise Goulding, Graham Allen, 'Consumer Price Inflation since 1750', *Economic Trends*, no. 604, March 2004, pp. 38–46. The figures are approximate.

[q] For an analysis of the income from Research Contracts and Grants see Table 12a.

[r] For an analysis of the income from Grants from non-governmental bodies, Donations and Subscriptions see Table 12b.

[s] After 1930 the LCC grant was included in the UGC grant.

[t] Including £20,229,000 from Health and Hospital authorities and £12,951,000 from Residences, Catering and Conferences.

[u] Including £24,413,000 from Health and Hospital authorities and £17,762,000 from Residences, Catering and Conferences.

Table 11b

Income by source (% of total), 1908–2005

Year	Government/ UGC/ UFC/HEFCE	Fees	Research Grants and Contracts	Endowment	Grants from non-governmental bodies, Donations and Subscriptions	London County Council	Others	Total
1908	71.7	28.3	—	—	—	—	—	100
1910	50.4	29.2	—	—	—	20.2	0.1	100
1915	36.9	22.4	1.6	—	7.9	15.4	15.8	100
1920	39.5	27.8	2.4	11.4	4.7	9.4	4.8	100
1925	41.5	24.6	2.9	7.4	2.6	11.3	9.6	100
1930	43.0	24.2	3.0	7.1	2.4	12.5	7.9	100
1935	57.2	21.2	4.2	4.9	2.9	—	9.6	100
1940	59.7	13.9	4.0	5.0	3.6	—	13.8	100
1945	60.8	16.8	3.3	4.3	3.3	—	11.4	100
1950	77.1	11.8	2.7	1.7	2.6	—	4.2	100
1955	70.1	7.7	16.4	1.0	2.4	—	2.5	100
1960	75.9	6.3	14.2	1.0	1.0	—	1.6	100
1965	75.8	6.1	15.8	0.6	0.7	—	1.1	100
1970	70.2	6.0	21.7	0.3	0.5	—	1.3	100
1975	72.8	3.4	20.1	0.4	0.3	—	3.0	100
1980	61.7	10.1	26.0	0.3	0.1	—	1.7	100
1985	53.8	12.5	29.7	0.3	0.8	—	2.9	100
1988	50.8	14.4	30.1	0.3	1.0	—	3.5	100
1996	30.3	13.2	37.3	1.9	2.1	—	15.2	100
2000	30.4	10.6	36.0	0.8	0.8	—	21.5	100
2005	30.0	13.8	38.5	0.9	0.9	—	15.9	100

Table 12a

Income (total in £) from Research Grants and Contracts, 1915–2005

Year	Government-related/ Research Councils	Industry & Commerce	Charity	Others[v]	Total
1915	1,339	—	—	—	1,339
1920	3,002	—	—	—	3,002
1925	5,724	—	—	—	5,724
1930	6,218	—	—	—	6,218
1935	5,756	—	3,600	—	9,356
1940	5,707	—	3,600	—	9,307
1945	7,783	—	500	311	8,594
1950	14,874	—	1,068	—	15,942
1955	128,182	—	—	71,014	199,196
1960	193,328	—	—	142,022	335,350
1965	483,444	—	—	250,111	733,555
1970	1,237,045	—	—	589,645	1,826,690
1975	2,638,949	—	—	700,665	3,339,614
1980	6,803,783	—	—	2,661,522	9,465,305
1985	11,856,000	2,332,000	1,217,000	2,166,000[x]	17,571,000
1988	13,475,000	2,888,000	1,129,000	3,870,000[y]	21,362,000
1996	38,939,000	8,314,000	16,096,000	13,027,000[z]	76,376,000
2000	45,097,000	16,987,000	33,903,000	26,005,000[aa]	121,992,000
2005	59,457,000	20,428,000	56,614,000	40,206,000[bb]	176,705,000

Table 12b

Income (total in £) from Grants from non-governmental bodies, Donations and Subscriptions, 1915–2005

Industry & Commerce	Charity	University of London	Total[w]
—	5,000	1,425	6,425
—	4,583	1,306	5,889
—	5,000	—	5,000
—	5,000	—	5,000
—	6,500	—	6,500
1,000	7,250	—	8,250
1,000	7,575	—	8,575
5,537	10,115	—	15,652
13,043	15,860	—	28,903
13,500	9,550	—	23,050
22,500	8,451	—	30,951
22,960	19,027	—	41,987
43,416	13,542	—	56,958
—	—	—	48,211
—	—	—	494,000
—	—	—	685,000
—	—	—	4,352,000
—	—	—	2,604,000
—	—	—	4,138,000

[v] No further data on the "Other" sources between 1955 and 1980.

[w] No further data on the Donations' sources between 1980 and 2005.

[x] Of which £1,674,000 from overseas sources.

[y] Of which £3,455,000 from overseas sources.

[z] Of which £8,062,000 from EU Commission and £4,285,000 from other EU and overseas sources.

[aa] Of which £7,925,000 from EU Commission and £13,827,000 from other EU and overseas sources.

[bb] Of which £26,134,000 from EU Commission and £8,140,000 from other EU and overseas sources.

Index